T0205563

Communications in Computer and Information Science 663

Commenced Publication in 2007
Founding and Former Series Editors:
Alfredo Cuzzocrea, Dominik Ślęzak, and Xiaokang Yang

Editorial Board

More information about this series at http://www.springer.com/series/7899

Tieniu Tan · Xuelong Li
Xilin Chen · Jie Zhou
Jian Yang · Hong Cheng (Eds.)

Pattern Recognition

7th Chinese Conference, CCPR 2016
Chengdu, China, November 5–7, 2016
Proceedings, Part II

Springer

Editors

Tieniu Tan
Chinese Academy of Sciences
Institute of Automation
Beijing
China

Xuelong Li
Xi'an Institute of Optics and Precision
  Mechanics
Chinese Academy of Sciences
Xi'an
China

Xilin Chen
Chinese Academy of Sciences
Institute of Computing Technology
Beijing
China

Jie Zhou
Tsinghua University
Beijing
China

Jian Yang
Nanjing University of Science
  and Technology
Nanjing
China

Hong Cheng
University of Electronic Science
  and Technology
Chengdu, Sichuan
China

ISSN 1865-0929   ISSN 1865-0937   (electronic)
Communications in Computer and Information Science
ISBN 978-981-10-3004-8   ISBN 978-981-10-3005-5   (eBook)
DOI 10.1007/978-981-10-3005-5

Library of Congress Control Number: 2016950420

Printed on acid-free paper

This Springer imprint is published by Springer Nature
The registered company is Springer Nature Singapore Pte Ltd.
The registered company address is: 152 Beach Road, #22-06/08 Gateway East, Singapore 189721, Singapore

Preface

Welcome to the proceedings of the 7th Chinese Conference on Pattern Recognition (CCPR 2016), which was held in Chengdu! Over the past decade, CCPR had been hosted by Beijing (2007, 2008, and 2012), Nanjing (2009), Chongqing (2010), and Changsha (2014) with great success. After the emergence of the Asian Conference on Pattern Recognition (ACPR) in 2011, CCPR and ACPR started being held on alternating years since 2012, making CCPR a biennial conference ever since.

Today, pattern recognition is applied in an increasing number of research domains such as autonomic understanding of vision, speech, language, and text. The recent innovative developments in big data, robotics, and multimodal interface have forced the area of pattern recognition to face both new opportunities and unprecedented challenges. Furthermore, like other flourishing new techniques, including deep learning and brain-like computation, we believe that pattern recognition will certainly exhibit greater-than-ever advances in the future. With the aim of promoting the research and technical innovation in relevant fields domestically and internationally, the fundamental objective of CCPR is defined as providing a premier forum for researchers and practitioners from academia, industry, and government to share their ideas, research results, and experiences. The selected papers included in the proceedings not only address challenging issues in various aspects of pattern recognition but also synthesize contributions from related disciplines that illuminate the state of the art.

This year, CCPR received 199 submissions, all of which are written in English. After a thorough reviewing process, 121 papers were selected for presentation as full papers, resulting in an acceptance rate of 60.53 %. An additional nine papers concerning emotion recognition were included. We are grateful to Prof. Mu-ming Poo from the Chinese Academy of Sciences, Prof. Mark S. Nixon from the University of Southampton, and Prof. Matthew Turk from the University of California for giving keynote speeches at CCPR 2016.

The high-quality program would not have been possible without the authors who chose CCPR 2016 as a venue for their publications. We are also very grateful to the members of Program Committee and Organizing Committee, who made a tremendous effort in soliciting and selecting research papers with a balance of high quality, new ideas, and new applications. We appreciate Springer for publishing these proceedings; and we are particularly thankful to Celine (Lanlan) Chang, Leonie Kunz, and Jane Li from Springer for their effort and patience in collecting and editing these proceedings.

We sincerely hope that you enjoy reading and benefit from the proceedings of CCPR 2016.

November 2016

Tieniu Tan
Xuelong Li
Xilin Chen
Jie Zhou
Jian Yang
Hong Cheng

Organization

Steering Committee

Tieniu Tan	Institute of Automation, Chinese Academy of Sciences, China
Chenglin Liu	Institute of Automation, Chinese Academy of Sciences, China
Jian Yang	Nanjing University of Science and Technology, China
Hongbin Zha	Peking University, China
Nanning Zheng	Xi'an Jiaotong University, China
Jie Zhou	Tsinghua University, China

General Chairs

Tieniu Tan	Institute of Automation, Chinese Academy of Sciences, China
Xuelong Li	Xi'an Institute of Optics and Precision Mechanics, Chinese Academy of Sciences, China

Program Chairs

Xilin Chen	Institute of Computing Technology, Chinese Academy of Science, China
Jie Zhou	Tsinghua University, China
Jian Yang	Nanjing University of Science and Technology, China
Hong Cheng	University of Electronic Science and Technology of China, China

Local Arrangements Chairs

Lu Yang	University of Electronic Science and Technology of China, China
Kai Liu	Sichuan University, China
Lei Ma	Southwest Jiaotong University, China

Workshop Chairs

Ce Zhu	University of Electronic Science and Technology of China, China
Lei Zhang	Sichuan University, China

Publicity Chairs

Hua Huang	Beijing Institute of Technology, China
Xiao Wu	Southwest Jiaotong University, China

Program Committee

Haizhou Ai	Tsinghua University, China
Xiaochun Cao	Institute of Information Engineering, Chinese Academy of Sciences, China
Hong Chang	Institute of Computing Technology, Chinese Academy of Sciences, China
Xiaotang Chen	Institute of Automation, Chinese Academy of Science, China
Yingke Chen	Sichuan University, China
Badong Chen	Xi'an Jiaotong University, China
Cunjian Chen	West Virginia University, USA
Xuewen Chen	Wayne State University, USA
Fei Chen	Italian Institute of Technology, Italy
Long Cheng	Institute of Automation, Chinese Academy of Sciences, China
Jian Cheng	Institute of Automation, Chinese Academy of Sciences, China
Rongxin Cui	Northwestern Polytechnical University, China
Daoqing Dai	Sun Yat-Sen University, China
Cheng Deng	Xidian University, China
Weihong Deng	Beijing University of Posts and Telecommunications, China
Yongsheng Dong	Chinese Academy of Sciences, China
Jing Dong	National Laboratory of Pattern Recognition, China
Junyu Dong	Ocean University of China, China
Leyuan Fang	Hunan University, China
Jun Fang	University of Electronic Science and Technology of China, China
Yachuang Feng	Xi'an Institute of Optics and Precision Mechanics, Chinese Academy of Sciences, China
Jufu Feng	Peking University, China
Jianjiang Feng	Tsinghua University, China
Shenghua Gao	ShanghaiTech University, China
Xinbo Gao	Xidian University, China
Yue Gao	Tsinghua University, China
Quanxue Gao	Xidian University, China
Yongxin Ge	Chongqing University, China
Xin Geng	Southeast University, China
Zhenhua Guo	Tsinghua University, China
Junwei Han	Northwestern Polytechnical University, China
Hongsheng He	The University of Tennessee, USA
Chenping Hou	National University of Defense Technology, China
Jiangping Hu	University of Electronic Science and Technology of China, China
Dewen Hu	National University of Defense Technology, China
Kaizhu Huang	Xi'an Jiaotong-Liverpool University, China
Qinghua Huang	South China University of Technology, China
Kaiqi Huang	Institute of Automation of the Chinese Academy of Sciences, China

Rongrong Ji	Xiamen University, China
Wei Jia	Chinese Academy of Sciences, China
Jia Jia	Tsinghua University, China
Dongmei Jiang	Northwestern Polytechnical University, China
Lianwen Jin	South China University of Technology, China
Qin Jin	Renmin University of China, China
Xin Jin	Tsinghua University, China
Wenxiong Kang	South China University of Technology, China
Jianhuang Lai	Sun Yat-sen University, China
Xirong Li	Renmin University of China, China
Ya Li	Institute of Automation, Chinese Academy of Sciences, China
Ming Li	SYSU-CMU Joint Institute of Engineering, Sun Yat-sen University, China
Xi Li	Zhejiang University, China
Wujun Li	Nanjing University, China
Chun-Guang Li	Beijing University of Posts and Telecommunications, China
Yanan Li	Imperial College London, UK
Jia Li	Beihang University, China
Yufeng Li	Nanjing University, China
Zhouchen Lin	Peking University, China
Liang Lin	Sun Yat-Sen University, China
Huaping Liu	Tsinghua University, China
Xinwang Liu	National University of Defense Technology, China
Qingshan Liu	Nanjing University of Information Science and Technology, China
Heng Liu	Anhui University of Technology, China
Kang Liu	Xi'an Institute of Optics and Precision Mechanics, Chinese Academy of Sciences, China
Min Liu	Hunan University, China
Chenglin Liu	Institute of Automation, Chinese Academy of Sciences, China
Dong Liu	University of Science and Technology of China, China
Wenju Liu	Institute of Automation, China Academy of Sciences, China
Jiwen Lu	Tsinghua University, China
Guangming Lu	Harbin Institute of Technology, China
Yue Lu	East China Normal University, China
Huchuan Lu	Dalian University of Technology, China
Xiaoqiang Lu	Xi'an Institute of Optics and Precision Mechanics, Chinese Academy of Sciences, China
Bin Luo	Anhui University, China
Siwei Lyu	State University of New York, University at Albany, USA
Zhanyu Ma	Beijing University of Posts and Telecommunications, China
Yajie Miao	Carnegie Mellon University, USA
Zhenjiang Miao	Beijing Jiaotong University, China
Jing Na	University of Bristol, UK
Feiping Nie	University of Texas, USA

Yanwei Pang   Tianjin University, China
Yu Qiao   Shenzhen Institutes of Advanced Technology, Chinese
                       Academy of Sciences, China
Hongliang Ren   National University of Singapore, Singapore
Nong Sang   Huazhong University of Science and Technology, China
Björn Schuller   Imperial College London, UK
Shiguang Shan   Institute of Computing Technology, Chinese Academy of
                       Sciences, China
Feng Shao   Ninbo University, China
Linlin Shen   Shenzhen University, China
Li Su   University of Chinese Academy of Sciences, China
Ning Sun   Nankai University, China
Zhenan Sun   Institute of Automation, Chinese Academy of Sciences, China
Jun Sun   Fujitsu R&D Center Company Ltd., China
Jian Sun   Xi'an Jiaotong University, China
Fuchun Sun   Tsinghua University, China
Xiaoyang Tan   Nanjing University of Aeronautics and Astronautics, China
Sheng Tang   Institute of Computing Technology, Chinese Academy of
                       Sciences, China
Huajin Tang   Sichuan University, China
Masayuki Tanimoto   Nagoya Industrial Science Research Institute, Japan
Jianhua Tao   Institute of Automation, Chinese Academy of Sciences, China
Wenbing Tao   Institute for Pattern Recognition and Artificial Intelligence,
                       Huazhong University of Science and Technology, China
Qi Wang   Northwestern Polytechnical University, China
Hongxing Wang   Nanyang Technological University, Singapore
Qiao Wang   Southeast University, China
Meng Wang   Hefei University of Technology, China
Yunhong Wang   Beihang University, China
Jinjun Wang   Xi'an Jiaotong University, China
Zengfu Wang   University of Science and Technology of China, China
Liwei Wang   Peking University, China
Hanzi Wang   Xiamen University, China
Ruiping Wang   Institute of Computing Technology, Chinese Academy of
                       Sciences, China
Liang Wang   National Laboratory of Pattern Recognition, Institute of
                       Automation of the Chinese Academy of Sciences, China
Xiangqian Wu   Harbin Institute of Technology, China
Yihong Wu   Institute of Automation, Chinese Academy of Sciences, China
Jianxin Wu   Nanjing University, China
Ying Wu   Northwestern University, China
Shiming Xiang   National Laboratory of Pattern Recognition, Institute of
                       Automation, Chinese Academy of Sciences, China
Mingxing Xu   Tsinghua University, China
Zenglin Xu   University of Electronic Science and Technology, China

Yong Xu	Harbin Institute of Technology, China
Long Xu	Chinese Academy of Sciences, China
Qianqian Xu	Institute of Information Engineering of Chinese Academy of Sciences, China
Hui Xue	Southeast University, China
Haibin Yan	Beijing University of Posts and Telecommunications, China
Jinfeng Yang	Tianjin Key Lab for Advanced Signal Processing, Civil Aviation University of China, China
Chenguang Yang	South China University of Technology, China
Meng Yang	Shenzhen University, China
Wankou Yang	Southeast University, China
Gongping Yang	Shandong University, China
Yujiu Yang	Tsinhua University, China
Jucheng Yang	Tianjin University of Science and Technology, China
Wenming Yang	Tsinghua University, China
Mao Ye	University of Electronic Science and Technology of China, China
Ming Yin	Guangdong University of Technology, China
Xucheng Yin	University of Science and Technology Beijing, China
Zhou Yong	China University of Mining and Technology, China
Shiqi Yu	Shenzhen Institute of Advanced Technology, Chinese Academy of Sciences, China
Yuan Yuan	Xi'an Institute of Optics and Precision Mechanics, Chinese Academy of Sciences, China
Dechuan Zhan	Nanjing University, China
Shishuai Zhang	Guangxi Normal University, China
Zhaoxiang Zhang	Institute of Automation, Chinese Academy of Sciences, China
Daoqiang Zhang	Nanjing University of Aeronautics and Astronautics, China
Tianzhu Zhang	Institute of Automation, Chinese Academy of Sciences, China
Changshui Zhang	Tsinghua University, China
Minling Zhang	Southeast University, China
Hongzhi Zhang	Harbin Institute of Technology, China
Lijun Zhang	Nanjing University, China
Lin Zhang	Tongji University, China
Weishi Zheng	Sun Yat-sen University, China
Wenming Zheng	Southeast University, China
Ping Zhong	National University of Defense Technology, China
Guoqiang Zhong	Ocean University of China, China
Xiuzhuang Zhou	Capital Normal University, China
Jie Zhou	Tsinghua University, China
Jun Zhu	Tsinghua University, China
Liansheng Zhuang	University of Science and Technology of China, China
Yuexian Zou	Peking University, China
Wangmeng Zuo	Harbin Institute of Technology, China

Contents – Part II

Image and Video Processing

Speech and Language

Emotion Recognition

Contents – Part I

Robotics

Computer Vision

Basic Theory of Pattern Recognition

Image and Video Processing

Saliency Region Detection via Graph Model and Statistical Learning

Ling Huang[1(⊠)], Songguang Tang[2], Jiani Hu[1], and Weihong Deng[1]

[1] Beijing University of Posts and Telecommunications, Beijing, China
huangling0805@gmail.com, {jnhu,whdeng}@bupt.edu.cn
[2] Wuhan Research Institute of Posts and Telecommunications, Wuhan, China
751037267@qq.com

Abstract. Saliency region detection plays an important role in computer vision aiming at discovering the salient objects in an image. This paper proposes a novel saliency detection algorithm (named as GMSL) via combining graph model and statistical learning. Firstly, the algorithm generates an initial saliency map by manifold ranking and optimizes it with absorbing Markov chain, both of which are based on graph model. Then, Bayes estimation with color statistical models is utilized as statistical learning to assign the saliency values to each pixel and further purify the map. Extensive experiments comparing with several state-of-the-art saliency detection works tested on different datasets demonstrate the superiority of the proposed algorithm.

Keywords: Saliency detection · Graph model · Statistical learning

1 Introduction

Human eyes have the capability to recognize visual stimuli and localize the most interested regions in a scene. Recently, visual attention has become an important part in computer vision serving as a pre-processing procedure for many visual tasks, such as object recognition, image retargeting, and etc.

Saliency detection can be computed in bottom-up approaches using low-level features [10,11,15,20] and top-down approaches which are task-driven and obtain saliency map with supervised learning [8,19]. In other aspect, Hou et al. turn to frequency domain to find the saliency distribution [6]. This paper focus on bottom-up methods to generate saliency map.

Over past years, contrast-based method guides the main direction of saliency region detection [7,11]. However, these methods usually miss small local region and lack the reliable prior knowledge. To overcome these problems, some researchers pay their attention to graph model which are based on boundary prior [3,4,16].

This paper puts forward to a novel algorithm by combining both graph model and statistical learning approach, and the contribution is as following:

© Springer Nature Singapore Pte Ltd. 2016
T. Tan et al. (Eds.): CCPR 2016, Part II, CCIS 663, pp. 3–13, 2016.
DOI: 10.1007/978-981-10-3005-5_1

(1) Graph model is introduced because of the powerful prior knowledge and the strong relationship between every two elements in an image. Manifold ranking, which focus on ranking elements' similarity degree with candidate seeds, is introduced to generate an initial saliency map, while absorbing Markov chain is applied to enhance foreground and suppress background.

(2) Based on statistical learning, Bayes estimation is utilized in this work, for it has high learning efficiency and good prediction consequence. This paper uses Bayes estimation to assign saliency values in pixel-wise which can rectify the inaccuracy caused by segmenting and further purify the saliency detecting results.

Details of the proposed algorithm are elaborated in Sect. 2, and the performance is testified in Sect. 3 with extensive experiments.

2   Algorithm

The algorithm contains three major steps: (1) generating an initial saliency map by manifold ranking; (2) optimizing the map with absorbing Markov chain; (3) assigning saliency values to all pixels and further purifying the detecting result using Bayes estimation.

2.1   Initial Saliency Map Generation

At the beginning of the algorithm, the input image is over-segmented into N superpixels with simple linear iterative clustering (SLIC) algorithm [9] to decrease computational complexity, and $N = 400$. A graph $G = (V, E)$ is then constructed. In this paper, each node in V represent a superpixel, and the matrix $W = w_{ij}$ is then established to weight each edge connecting node i and j in E as following:

$$w_{ij} = \exp(-\frac{||c_i - c_j||^2}{\sigma^2}) \tag{1}$$

where c_i and c_j are the mean CIELAB colors of node i and j, and parameter $\sigma = 0.4$. The degree matrix $D_1 = diag\{d_1, d_2, ..., d_N\}$ is then established where d_i is calculated as:

$$d_i = \sum_{j=1}^{N} w_{ij} \tag{2}$$

The initial saliency map is generated using manifold ranking proposed in [16], which defines the ranking function as

$$f^* = (D_1 - \alpha W)^{-1} y \tag{3}$$

where $\alpha = 0.99$ is a balance coefficient. $y = [y_1, ..., y_N]^T$ is a binary vector where $y_i = 1$ if node i is the boundary superpixel while $y_i = 0$ otherwise, for the boundary sites are always the background.

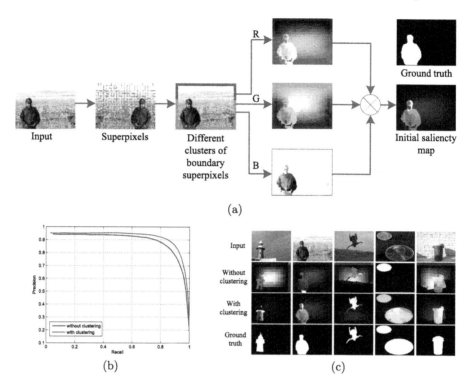

(a)

(b)   (c)

Fig. 1. Detail of generating the initial saliency map with some comparison, where (a) shows the generation flow chart, where the boundary superpixels are divided into in three clusters labeled as red (R), green (G) and blue (B), while (b) and (c) shows the different detection performance between whether the map is generated directly with all superpixles or by integrating the boundary superpixels in different clusters using Eq. (5). (Color figure online)

However, it's worth noting that boundaries which are regarded as background seeds may contain some object elements, and difference always exists among those background seeds, as shown in Fig. 1(c). In order to be more accurate, this paper divides the boundary elements into K cluters by k-means algorithm and then K ranking results $\{f_l^*\}_K, l = 1, ..., K$ are acquired based on Eq. 3. By empirically, K is set to be 3 in this paper. Therefore, each saliency value is obtained by

$$S_l(i) = 1 - f_l^*(i), \ i = 1, ..., N \tag{4}$$

The initial saliency map is finally put into element-by-element multiplication as

$$S^b(i) = \prod_{l=1}^{K} S_l(i), \ l = 1, ..., K \tag{5}$$

Figure 1 tells the implement detail of generating the initial saliency map, and shows the advantage by clustering boundary superpixels, which facilitates saliency accuracy by strengthening the contrast in local regions.

2.2   Foreground Enhancement via Absorbing Markov Chain

It can be seen from the Fig. 1(c) that the background is not suppressed deeply and the foreground is not highlighted enough, because it only considers the boundary cues. In this paper, absorbing Markov chain, focusing more on foreground, is introduced to optimize the initial map by polarizing the opposite sides with absorbing probability, which can enhance the difference between foreground and background.

Because the value of each pixel in saliency map denotes the possibility of this pixel belonging to foreground, and the higher value always means the higher possibility, there are two thresholds t_h and t_l defining based on the initial map as follow:

$$t_h = \frac{mean(S^b) + \max(S^b)}{2} \tag{6}$$

$$t_l = mean(S^b) \tag{7}$$

Superpixels whose values are larger than t_h are selected as the foreground seeds to make the division more convincible, while those whose values are lower than t_l are considered as the background ones.

Besides W used in Sect. 2.1, another affinity matrix $A_1 = (a1_{ij})$ is needed to show the relationship between transient state and absorbing state [12]:

$$a1_{ij} = \exp(-\frac{||c_i' - c_j'||^2}{\sigma^2}) \tag{8}$$

where c_i' and c_j' are the mean colors of transient node i and absorbing node j respectively. The corresponding degree matrix $D_1' = diag\{d_1', ..., d_N'\}, d_i' = \sum_{j=1}^{N} a1_{ij}$ is obtained. At this moment, the foreground seeds are regarded as the absorbing nodes while others are transient nodes, and the corresponding Markov transition matrix P_1 can be decomposed to

$$P_1 = \begin{pmatrix} Q_1 & R_1 \\ O_1 & I_1 \end{pmatrix} \tag{9}$$

where $Q_1 = D^{-1}W$, $R_1 = D^{-1}A_1$, and D is the sum of D_1 and D_1'. O_1 and I_1 are the zeros matrix and the identity matrix respectively.

Also, the background seeds are considered as the absorbing nodes [17], and the corresponding affinity matrix $A_2 = (a2_{ij})$ is:

$$a2_{ij} = \exp(-\frac{||c_i'' - c_j''||^2}{\sigma^2}) \tag{10}$$

where c_i'' and c_j'' are the mean colors of the new transient node i and new absorbing node j, and degree matrix $D_2' = diag\{d_1'', ..., d_N''\}$, $d_i'' = \sum_{j=1}^{N} a2_{ij}$ is acquired. Another Markov transition matrix P_2 is created as following:

$$P_2 = \begin{pmatrix} Q_2 & R_2 \\ O_2 & I_2 \end{pmatrix} \tag{11}$$

where $Q_2 = D'^{-1}W$, $R_1 = D'^{-1}A_2$, where $D' = D_1 + D_2'$. O_2 and I_2 are respectively the zeros matrix and the identity matrix.

As for any absorbing chain, there is a corresponding fundamental matrix, and N_2 is the fundamental matrix of P_2 defined as

$$N_2 = (I_2 - Q_2)^{-1} = I_2 + Q_2 + Q_2^2 + ... \tag{12}$$

By now the absorbing probability matrix $B = N_2 R_1$ is obtained and the values b_{ij} of it should be ranked to all the nodes j to get the new matrix B'. Follow by the above works, the saliency of each node i in the image is defined as

$$S^f(i) = \sum_{j=1}^{d} b_{ij}' \tag{13}$$

where b_{ij}' is the element in B', and d is the number of the top elements of each row.

Figure 2(b) and (c) shows the examples where the background is better suppressed and the foreground is highlighted after absorbing Markov chain.

2.3 Saliency Map Construction by Bayes Estimation

Former saliency detection completely depends on the over-segmentation, which may lead to undesirable result because of the imprecise in segmentation itself. Furthermore, graph-based model may not well grasp the global feature distribution characteristics of an image. To overcome these disadvantages, Bayes estimation is utilized to learn and construct a more accurate saliency map.

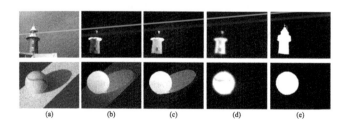

Fig. 2. Some examples of the contrast in each step. From left to right: (a) Input, (b) Initial saliency map via manifold ranking, (c) Optimized map via absorbing Markov chain, (d) Further purified map via Bayes estimation, (e) Ground truth.

As a generative learning model, Bayes formula is calculated with prior probability and learning probability model to obtain posterior probability as the final result. In this paper, the saliency prior probability of each pixel $p(Sal)$ is its saliency value generated in former steps, and the background prior probability is calculated as

$$p(Bg) = 1 - p(Sal) \tag{14}$$

Two color statistic models built in CIELAB color space are used as the likelihood probability:

$$p(x|Sal) = \frac{n_{Sal}[Lab(x)]}{n_{Sal}} \tag{15}$$

$$p(x|Bg) = \frac{n_{Bg}[Lab(x)]}{n_{Bg}} \tag{16}$$

where n_{Sal} and n_{Bg} are the whole number of pixels belonging to saliency part and background part respectively, and $n_{Sal}[Lab(x)]$ as well as $n_{Bg}[Lab(x)]$ are the number of pixels in the same color bin correspondingly. In addition, the pixels whose saliency values are larger than $t_s = 1.5 * mean(S^f)$ are chosen as saliency part Sal, while the others are regarded as background Bg.

Following the Bayes estimation formula, the saliency result for each pixel can be seen as the posterior probability:

$$p(Sal|x) = \frac{p(Sal)p(x|Sal)}{p(Sal)p(x|Sal) + p(Bg)p(x|Bg)} \tag{17}$$

All the values got from Eq. 17 are instigated to a map as S_0^{final}. In order to remove the noise as well as smooth the map, Gaussian filter is applied and the final saliency detection result S^{final} is obtained, showing as the examples in Fig. 2(d).

3   Experiments and Results

3.1   Datasets and Metrics

The proposed algorithm is tested on several public datasets. The first one is ASD dataset [10], the subset of MSRA dataset [13], which is most widely used and contains 1000 images. Another dataset contains 5000 images randomly chosen from MSRA dataset named as MSRA-5000 datasets. The next one is SED2 dataset [15], which only contains 100 images and each in it has two objects. The last is iCoSeg dataset [14] containing 643 images, each of which has more complex scenes than the above and has one or more objects.

Precision, recall and F-Measure are taken into evaluation as following:

$$precision = \frac{|SF \cap GF|}{|SF|} \tag{18}$$

$$recall = \frac{|SF \cap GF|}{|GF|} \tag{19}$$

$$F\text{-}Measure = \frac{1 + \beta^2 \times precision \times recall}{\beta^2 \times precision + recall} \tag{20}$$

where SF denotes the set of segmented foreground pixels after a binary segmentation with a threshold varying from 0 to 255, and GF denotes the set of the ground truth foreground pixels. Since both *precision* and *recall* are expected to be larger, *F-Measure* is used to balance the two values. $\beta^2 = 0.3$ as suggested in [10] to grant more importance to the precision.

In supplement, ROC curves are also introduced to measure performance of each algorithm in another aspect, which show the relationship between the true positive rate and the false positive rate.

MAE (Mean Absolute Error) value is also adopted to measure the similarity between the saliency result and the ground truth calculated as:

$$MAE = \frac{1}{M} \sum_{m=1}^{M} |S^{final}(m) - G(m)| \tag{21}$$

where M is the count of all pixels and G is the ground truth.

3.2   Experiment and Results

Examination of Design Option. The proposed algorithm is firstly examined on MSRA-5000 dataset with precision-recall curves to show each step's effectiveness. The three curves which are shown in Fig. 3 respectively denote the initial results, optimized results, and the final results. It is easily to find from the comparison that absorbing Markov chain can optimize the initial background-based results to some extent, and the Bayes estimation can obviously further refine the former ones.

Table 1. The MAE values of different algorithms in different datasets. Each col is the comparison of different datasets in same method while each row is the comparison of different methods in same dataset. The best result is shown in **Bold**.

	SR	HC	FT	UFO	AMC	RRWR	MR	GMSL
ASD	.254	.182	.204	.109	.093	.073	.075	**.058**
MSRA-5000	.256	.216	.242	.183	.141	.117	.122	**.100**
iCoSeg	.236	.173	.177	.191	.141	.127	.129	**.104**
SED2	.241	.182	.203	.177	.184	.163	.166	**.143**

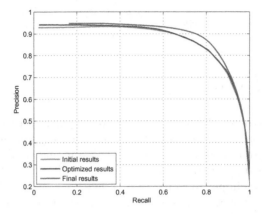

Fig. 3. The precision-recall curves contrast of each step based on MSRA-5000 dataset for the proposed algorithm.

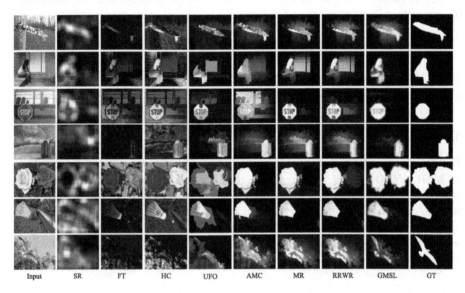

Fig. 4. Comparison of saliency maps in different methods.

Qualitative Comparison with Other Algorithms. The proposed algorithm is then evaluated against some state-of-the-art algorithms, including SR [6], FT [10], HC [11], UFO [7], AMC [12], MR [16], and RRWR [1]. These results are got by running open codes provided by their authors and Fig. 4 shows their qualitative comparison. Results generated by the proposed GMSL highlight the salient object well with less noise, because this algorithm combines both the graph-based model and statistical learning, where the former one is effective picking up the boundary prior information and the latter one is effective finding out the distribution regularities of foreground and background.

Quantitative Comparison with Other Algorithms. Some results of saliency maps generated by the above methods for quantitative comparison are presented in Fig. 5. It demonstrates that the proposed algorithm gets the highest precision value in almost the whole recall interval [0,1] and the F-measure curve is almost the topmost one in each dataset. Also, the ROC curves shows that

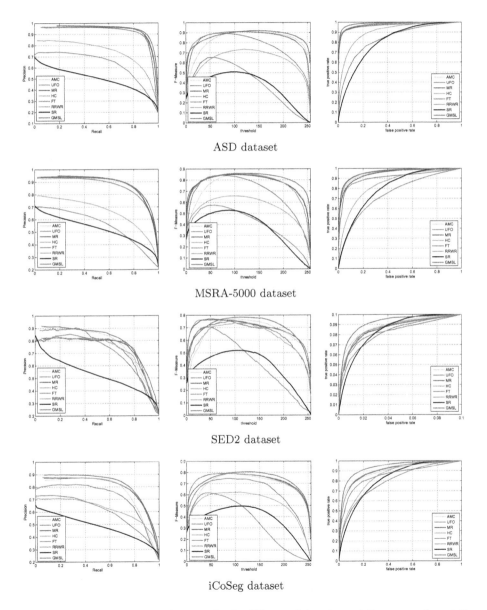

ASD dataset

MSRA-5000 dataset

SED2 dataset

iCoSeg dataset

Fig. 5. Quantitative comparison results in different datasets, and the curves from left to right are: Precision-Recall curves, F-measure curves and ROC curves.

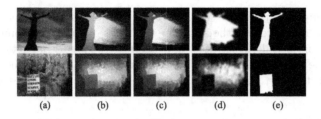

Fig. 6. Failure cases of the proposed algorithm. Left to right: (a) input, (b) initial map, (c) optimized map, (d) final map and (e) ground truth.

the proposed GMSL achieves the best true negative rate while in the same false negative rate with other methods. Furthermore, it's obvious form Table 1 that the proposed algorithm achieves best MAE values.

Failure Cases. Some failure cases of the proposed algorithm are shown in Fig. 6, as the initial graph-based do not perform well so that the statistical part works wrong.

4   Conclusion

This paper proposes a novel and strong model which combines graph model including manifold ranking as well as absorbing Markov chain and statistic learning approach, Bayes estimation. It uses only the simple color feature but has good performance which is powerfully proved by experiments. This algorithm can be applied to other saliency models for significant improvement and more potential applications.

Acknowledgments. This work was partially sponsored by Project 61375031, 61471048, and 61573068 supported by National Natural Science Foundation of China. This work was also supported by the Beijing Higher Education Young Elite Teacher Program, Beijing Nova Program, CCF-Tencent Open Research Fund, and the Program for New Century Excellent Talents in University.

References

1. Li, C., Yuan, Y., Cai, W., Xia, Y., Feng, D.D.: Robust saliency detection via regularized random walks ranking. In: Proceedings of the IEEE Conference on Computer Vision and Pattern Recognition (2015)
2. Lin, Y., Kong, S., Wang, D., Zhuang, Y.: Saliency detection within a deep convolutional architecture. In: Workshops at the Twenty-Eighth AAAI Conference on Artificial Intelligence (2014)
3. Li, X., Lu, H., Zhang, L., Ruan, X., Yang, M.: Saliency detection via dense and sparse reconstruction. In: IEEE International Conference on Computer Vision (2013)

4. Qin, Y., Lu, H., Xu, Y., Wang, H.: Saliency detection via cellular automata. In: Proceedings of the IEEE Conference on Computer Vision and Pattern Recognition (2015)
5. Farabet, C., Couprie, C., Najman, L., LeCun, Y.: Learning hierarchical features for scene labeling. IEEE Trans. Pattern Anal. Mach. Intell. **38**, 1915–1929 (2013)
6. Hou, X., Zhang, L.: Saliency detection: a spectral residual approach. In: IEEE Conference on Computer Vision and Pattern Recognition (2007)
7. Jiang, P., Ling, H., Yu, J., Peng, J.: Salient region detection by UFO: uniqueness, focusness and objectness. In: IEEE Conference on Computer Vision (2013)
8. Marchesotti, L., Cifarelli, C., Csurka, G.: A framework for visual saliency detection with applications to image thumbnailing. In: IEEE Conference on Computer Vision (2009)
9. Achanta, R., Shaji, A., Smith, K., Lucchi, A., Fua, P., Susstrunk, S.: SLIC superpixels compared to state-of-the-art superpixel methods. IEEE Trans. Pattern Anal. Mach. Intell. **34**, 2274–2282 (2012)
10. Achanta, R., Hemami, S., Estrada, F., Susstrunk, S.: Frequency-tuned salient region detection. In: IEEE Conference on Computer Vision and Pattern Recognition (2009)
11. Cheng, M., Mitra, N.J., Huang, X., Torr, P.H., Hu, S.: Global contrast based salient region detection. In: IEEE Conference on Computer Vision and Pattern Recognition (2010)
12. Jiang, B., Zhang, L., Lu, H., Yang, C., Yang, M.: Saliency detection via absorbing Markov chain. In: IEEE Conference on Computer Vision (2013)
13. Jiang, H., Wang, J., Yuan, Z., Wu, Y., Zheng, N., Li, S.: Salient object detection: A discriminative regional feature integration approach. In: IEEE Conference on Computer Vision and Pattern Recognition (2013)
14. Batra, D., Kowdle, A., Parikh, D., Luo, J., Chen, T.: iCoseg: interactive co-segmentation with intelligent scribble guidance. In: IEEE Conference on Computer Vision and Pattern Recognition (2010)
15. Alpert, S., Galun, M., Brandt, A., Basri, R.: Image segmentation by probabilistic bottom-up aggregation and cue integration. IEEE Trans. Pattern Anal. Mach. Intell. **34**, 315–327 (2012)
16. Yang, C., Zhang, L., Lu, H., Ruan, X., Yang, M.: Saliency detection via graph-based manifold ranking. In: IEEE Conference on Computer Vision and Pattern Recognition (2013)
17. Sun, J., Lu, H., Liu, X.: Saliency region detection based on Markov absorption probabilities. IEEE Trans. Image Process. **24**, 1639–1649 (2015)
18. Kim, J., Han, D., Tai, Y., Kim, J.: Salient region detection via high-dimensional color transform. In: IEEE Conference on Computer Vision and Pattern Recognition (2014)
19. Yang, J., Yang, M.: Top-down visual saliency via joint CRF and dictionary learning. In: IEEE Conference on Computer Vision and Pattern Recognition (2012)
20. Cheng, M., Warrell, J., Lin, W., et al.: Efficient salient region detection with soft image abstraction. In: Proceedings of the IEEE International Conference on Computer Vision, pp. 1529–1536. IEEE (2013)

An Efficient Gabor Feature-Based Multi-task Joint Support Vector Machines Framework for Hyperspectral Image Classification

Sen Jia[✉] and Bin Deng[✉]

College of Computer Science and Software Engineering,
Shenzhen University, Shenzhen, China
senjia@szu.edu.cn, szubing@qq.com

Abstract. In this paper, a novel multi-task learning (MTL) framework for a series of Gabor features via joint probabilistic outputs of support vector machines (SVM), abbreviated as GF-MTJSVM, has been proposed for Hyperspectral image (HSI) classification. Specifically, we firstly use a series of Gabor wavelet filters with different scales and frequencies to extract spectral-spatial-combined features from the HSI data. Then, we apply these Gabor features into the multi-task learning framework via joint probabilistic outputs of SVM. Experimental results on two widely used real HSI data indicate that the proposed GF-MTJSVM approach outperforms several well-known classification methods.

Keywords: Hyperspectral image classification · Multi-task support vector machines · Gabor features

1 Introduction

Hyperspectral sensors collect information as a set of images. Each image represents a narrow wavelength range of the electromagnetic spectrum, also known as a spectral band. Each pixel in a hyperspectral image (HSI) is a high-dimensional vector whose entries are the spectral responses of various spectral bands. The very informative spectral information of the HSI pixels can be utilized to distinguish objects in the image scene. As a major application of hyperspectral data analysis, pixel-oriented classification has been widely adopted [6,22,26]. In the hyperspectral supervised classification case, the class label of each pixel, denoted by a vector whose entries correspond to the narrow spectral band responses, is determined by a given training set from each class [5,7,13,17]. To tackle the problem, many classifiers have been employed, including K-nearest neighbors (KNN)

This work was jointly supported by grants from National Natural Science Foundation of China (61671307 and 61271022), Guangdong Foundation of Outstanding Young Teachers in Higher Education Institutions (Yq2013143), Shenzhen Scientific Research and Development Funding Program (JCYJ20140418095735628, JCYJ20160422093647889 and SGLH20150206152559032).

© Springer Nature Singapore Pte Ltd. 2016
T. Tan et al. (Eds.): CCPR 2016, Part II, CCIS 663, pp. 14–25, 2016.
DOI: 10.1007/978-981-10-3005-5_2

[9], support vector machines (SVM) [8], sparse representation-based classification (SRC) [33]. However, due to the high dimensionality in spectral domain of HSI, finding optimal parameters for the supervised classifiers is time-consuming. Meanwhile, due to the extremely high spectral dimensionality of the data, the small sample size scenario (it is very difficult and time consuming to collect sufficient training samples in practice) is one crucial problem that limits the performance of many existing classification methods [18, 27].

Recently, many spectral-spatial classification methods in HSI have been proposed to improve the accuracy further. The spectral-spatial classification methods can be roughly divided into two categories. The first one is that the spectral and spatial contextual information is exploited separately, that is, the spatial dependencies are extracted through various spatial filters (such as morphological [3,11,14,20], range [24], entropy [32], low-rank representation [36]) in advance, and then combined with the spectral features (or dimension reduced ones) to perform pixel-wise classification, or the spatial information is used to refine the classification results through a regularization process in the postprocessing stage [31], including Markov random field [30] and graph cuts [19]. However, a large number of training samples is generally required to adequately characterize the large variability of the objects, which is difficult to meet in most circumstances. The second one is that the spatial information is directly fused with the spectral features to produce joint features [21]. For example, a set of filters (such as wavelets [28], Gabor wavelets [2,29]) can be applied on the hyperspectral data to extract spectral-spatial-combined features. Meanwhile, in recent years, multiple features combines to design a classifier has been a growing trend for the hyperspectral classification. It is obvious that multiple features can provide diverse information in characterizing object from different viewpoints. So it is always can achieve a better classification accuracy through combination of a set of modalities of features [35]. Several methods, such as Multiple Kernel Learning (MKL, in which the similarity functions between images are linearly combined) [15,16], SVM ensemble (inspired by linearly programming Boosting) [12], have been developed for multi-class object classification. Zheng et al. [37] applied the multi-task joint sparse representation classification (MTJSRC) proposed in [35] with spatial filtering postprocessing into large-scale satellite image annotation and obtained excellent results. These methods all can obtained excellent results, but require either a large amount of computation time, or very complex model to construct a multi-task learning framework.

In this paper, a novel and simple multi-task learning (MTL) framework for a series of Gabor features via joint probabilistic outputs of SVM, called as GF-MTJSVM, was proposed for HSI classification. Firstly, a series of Gabor features were extracted by applying a set of predefined Gabor filters (with different scales and orientations) on the original hyperspectral data, which contains the wealth of information about signal changes in the local area, and provide an important classification identification information. Then, the multi-task learning via joint probabilistic outputs of SVM was applied for material identification. In the GF-MTJSVM framework, we simply combined category probabilistic outputs for the

SVM of each Gabor cube features as the final category probabilistic outputs. Experimental results on two real hyperspectral data with different spectral and spatial resolutions demonstrated the effectiveness of the proposed Gabor feature-based multi-task joint support vector machines Framework for hyperspectral image classification.

The rest of the paper is organized as follows. In Sect. 2, we introduce the basic Gabor filters and the classical support vector classification is described later on. Section 3 presents the proposed GF-MTJSVM approach, in detail. Experiment was run on two real hyperspectral data set and the results are shown in Sect. 4. Section 5 concludes the paper with a summary of the proposed work.

2   Related Work

2.1   Gabor Functions and Wavelets

Texture analysis has a long history and texture analysis algorithms range from using random field models to multiresolution filtering techniques such as the wavelet transform [23]. A two dimensional Gabor function $\Psi = (x, y)$ can be written as:

$$\Psi(x, y) = \left(\frac{1}{2\pi\sigma_x\sigma_y} \right) exp \left[-\frac{1}{2} \left(\frac{x^2}{\sigma_x^2} + \frac{y^2}{\sigma_y^2} \right) + 2\pi j\omega x \right] \tag{1}$$

Gabor function form a complete but nonorthogonal basis set. Expanding a signal using this basis provides a localized frequency description. A class of self-similar functions, referred to as Gabor wavelets in the following discussion, is now considered. Let $\Psi(x, y)$ be the mother Gabor wavelet, then this self-similar filter dictionary can be obtained by appropriate dilations and rotations of $\Psi(x, y)$ through the generating function:

$$\Psi_{mn}(x, y) = a^{-m}\Psi(x', y'), \quad a > 1, \quad m, n = integer$$
$$x' = a^{-m}(x \cos\theta_n + y \sin\theta_n), \quad and \quad y' = a^{-m}(-x \sin\theta_n + y \cos\theta_n). \tag{2}$$

Where $\theta_n = n\pi/K$ and K is the total number of orientations. The scale factor $S_m = a^{-m}$ in (2) is meant to ensure that the energy is independent of m. Through change the scaling size and the direction of rotation, we can get a group of Gabor wavelets.

2.2   Classical Support Vector Classification (C-SVC)

Given training vectors $\mathbf{x}_i \in \mathbb{R}^n, i = 1, ..., l$, in two classes, and an indicator vector $\mathbf{y} \in \mathbb{R}^l$ such that $y_i \in \{1, -1\}$, C-SVC [4,10] solves the following primal optimization problem:

$$\min_{\mathbf{w}, b, \xi} \frac{1}{2}\mathbf{w}^T\mathbf{w} + C\sum_{i=1}^{l} \xi_i$$
$$subject \ to \quad y_i\left(\mathbf{w}^T\phi(\mathbf{x}_i) + b\right) \geq 1 - \xi_i, \tag{3}$$
$$\xi_i \geq 0, i = 1, ..., l.$$

Where $\phi(\mathbf{x}_i)$ maps $\mathbf{x}_i$ into a higher-dimensional space and $C > 0$ is the regularization parameter. Due to the possible high dimensionality of the vector variable $\mathbf{w}$, usually we solve the following dual problem:

$$\min_{\alpha} \quad \tfrac{1}{2}\alpha^T \mathbf{Q}\alpha - \mathbf{e}^T\alpha$$
$$subject\ to \quad \mathbf{y}^T\alpha = 0, \quad 0 \le \alpha_i \le C, i = 1, ..., l. \tag{4}$$

Where $\mathbf{e} = [1, ..., 1]^T$ is the vector of all ones, $\mathbf{Q}$ is an l by l positive semi-definite matrix, $\mathbf{Q}_{ij} \equiv y_i y_j \mathbf{K}(\mathbf{x}_i, \mathbf{x}_j) \equiv \phi(\mathbf{x}_i)^T \phi(\mathbf{x}_j)$ is the kernel function.

After problem 4 is solved, using the primal-dual relationship, the optimal $\mathbf{w}$ satisfies:

$$\mathbf{w} = \sum_{i=1}^{l} y_i \alpha_i \phi(\mathbf{x}_i) \tag{5}$$

Then we can get the threshold output of a vector $\mathbf{x}$:

$$f(\mathbf{x}) = \sum_{i=1}^{l} y_i \alpha_i \mathbf{K}(\mathbf{x}_i, \mathbf{x}) + b. \tag{6}$$

And the decision function is:

$$h(\mathbf{x}) = sgn(f(\mathbf{x})). \tag{7}$$

3 Gabor Feature-Based Multi-task Joint Support Vector Machines for Hyperspectral Classification

After introducing the basic Gabor function and classical support vector classification in the previous section, a novel multi-task framework for hyperspectral classification, which is named the GF-MTJSVM, is proposed in this section. The framework GF-MTJSVM consists of two main steps, Gabor features extraction and combine the probabilistic outputs of SVM for each Gabor features. Figure 1 illustrates the schematic diagram of the proposed strategy.

3.1 Gabor Features Extraction for Hyperspectral Image

The nonorthogonality of Gabor wavelets implies that there is redundant information in the filtered images. In order to reduce this redundancy, B.S. Manjunathi et al. [23] proposed an effective strategy to design the Gabor filter dictionary. From the design of the Gabor filter dictionary, we can get a series of Gabor wavelet filters with different scales and frequencies. Denoted a set of 2-D Gabor filters as $\{\Psi_i, i = 1, 2, ..., I\}$ (I is the number of Gabor filters) and the original hyperspectral data as $\mathbf{R} \in \mathbb{R}^{X \times Y \times B}$. For each Ψ_i at each spectral band λ, the magnitude $\mathbf{M}_i(x, y, \lambda) = |(\mathbf{R} * \Psi_i)(x, y)|$ contains rich signal change information around location (x, y), where $*$ is the convolution operation, and $\mathbf{M}_i(x, y) = [\mathbf{M}_i(x, y, 1), \mathbf{M}_i(x, y, 2), ..., \mathbf{M}_i(x, y, B)]$ is the responses of the i-th

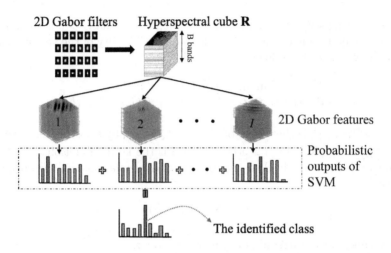

2D Gabor filters   Hyperspectral cube **R**

2D Gabor features

Probabilistic
outputs of
SVM

The identified class

Fig. 1. Framework of GF-MTJSVM

Gabor filter at all bands. Through applying Ψ_i on all pixels of the hyperspectral image data, a Gabor cube $\mathbf{M}_i \in \mathbb{R}^{X \times Y \times B}$ can be obtained, which has the same size as the original hyperspectral data $\mathbf{R}$. Further, after each Gabor filter $\Psi_i, i = 1, 2, ..., I$ has been convolved with the hyperspectral image data, a total of I Gabor cubes $\mathbf{M}_i, i = 1, 2, ..., I$ were extracted.

3.2   Combine Probabilistic Outputs of SVM for Gabor Features

Given k classes of hyperspectral data $\mathbf{R} \in \mathbb{R}^{X \times Y \times B}$, and its corresponding i-th Gabor cube $\mathbf{M}_i \in \mathbb{R}^{X \times Y \times B}$ from Sect. 3.1. For each pixel $\mathbf{x} \in \mathbb{R}^B$ need to predict class label in the hyperspectral data $\mathbf{R}$, and the $\mathbf{x}^{(i)} \in \mathbb{R}^B$ is it corresponding i-th Gabor feature from Gabor cube $\mathbf{M}_i$, our first goal is to estimate

$$p_m^{(i)} = P\left(y = m | \mathbf{x}^{(i)}\right), \quad m = 1, ..., k. \tag{8}$$

Following the setting of the one-against-one approach for multiclass classification, we first estimate pairwise class probabilities

$$r_{mn}^{(i)} \approx P\left(y = m | y = m \ or \ n, \mathbf{x}^{(i)}\right) \tag{9}$$

If f_i is the decision value at $\mathbf{x}^{(i)}$ from the formula 6, then we assume

$$r_{mn}^{(i)} = \frac{1}{1 + e^{af_i + b}} \tag{10}$$

where a and b are estimated by minimizing the negative log likelihood of training data (using their labels and decision values) [25].

After collecting all $r_{mn}^{(i)}$ values, Wu et al. [34] propose two approaches to obtain $p_m^{(i)}, \forall m$. we consider their second approach and solve the following optimization problem.

$$\min_{\mathbf{p}^{(i)}} \frac{1}{2} \sum_{m=1}^{k} \sum_{n:n\neq m} \left(r_{nm}^{(i)} p_m^{(i)} - r_{mn}^{(i)} p_n^{(i)} \right)^2$$

$$subject\ to \quad p_m^{(i)} \geq 0, \forall m, \sum_{m=1}^{k} p_m^{(i)} = 1 \tag{11}$$

Finally, the class label of $\mathbf{x}$ is predicted to the class with the biggest total probability over all the I tasks, i.e.,

$$\text{Class}(\mathbf{x}) = \arg\max_{m} \sum_{i=1}^{I} p_m^{(i)} \tag{12}$$

Where I is the total number of the Gabor cubes.

4 Experimental Results

In this section, the performance of the proposed GF-MTJSVM method is tested in classification of two real hyperspectral imagery. The classification results are compared with state-of-the-art methods, i.e., support vector machines (SVM), sparse representation-based classification (SRC), and extended morphological and attribute profiles (EMAP) [11]. In the experiments, the Homotopy method is used to recover the sparse signals [1], the parameter of sparsity factor in SRC was set to be 0.01, while rbf-kernel and one-against-all scheme in SVM was used for the remaining methods. Besides, the C parameter of SVM is estimated by ten-fold cross validation. After the training set has been randomly partitioned into ten groups, nine groups are used to train a set of models that are evaluated on the remaining group. This procedure is then repeated for all ten possible choices for the held-out group, and the performance scores from the ten runs are then averaged. The performance of the compared techniques is evaluated with different samples (i.e., 5, 6, 7, ..., 15) are selected from each class to form the training set. And the remaining samples are then used as the test set for evaluation. Each experiment is repeated ten times with different training sets to reduce the influence of random effects, and both the mean and standard deviation are reported. In the experimental results, overall accuracy (OA) and kappa coefficient (κ) are used as measures of accuracy.

4.1 Indian Pines Data Set

The first real-world data set to be used is the commonly used Indian Pines data set acquired by the AVIRIS instrument over the agricultural area of Northwestern Indiana in 1992, which has a spatial dimension of 145×145 and 224 spectral

Table 1. Land cover classes with number of samples for the Indiana Pines data

Class	Land cover type	No. of samples
$C1$	Alfalfa	54
$C2$	Buildings-grass-trees-drives	380
$C3$	Corn	234
$C4$	Corn-min till	834
$C5$	Corn-no till	1434
$C6$	Grass-pasture-mowed	26
$C7$	Grass/pasture	497
$C8$	Grass/trees	747
$C9$	Hay-windrowed	489
$C10$	Oats	20
$C11$	Soybean-clean till	614
$C12$	Soybean-min till	2468
$C13$	Soybean-no till	968
$C14$	Soybean-steel-towers	95
$C15$	Wheat	212
$C16$	Woods	1294
	Total	10366

bands. The spatial resolution of the data is 20 m per pixel. After discarding 4 zero bands and the 35 lower SNR bands affected by atmospheric absorption, 185 channels are preserved. The data set contains 10366 labeled pixels and 16 ground-truth classes, most of which are different types of crops (see Table 1). Figure 2 shows the ground-truth map containing 16 mutually exclusive land-cover class.

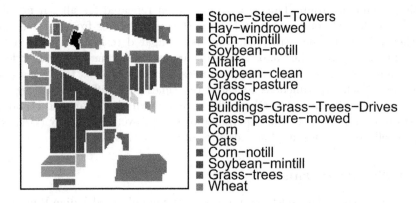

Fig. 2. Ground-truth map of the Indian Pines data set (sixteen land cover classes).

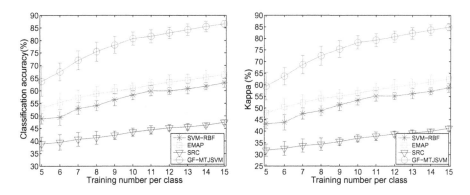

Fig. 3. Overall accuracy and kappa coefficient of the Indian Pines data set.

Figure 3 displays the overall accuracy (OA) and kappa coefficient measure as functions of the number of training samples per class (5, 6, 7, ..., 15). From the figure we can see, the performance of the GF-MTJSVM method is significantly higher than those methods (SVM, SRC, EMAP), demonstrating the superiority of the GF-MTJSVM framework.

Table 2. Land cover classes with number of samples for the Kennedy Space Center data

Class	Land cover type	No. of samples
$C1$	Scrub	761
$C2$	Willow swamp	243
$C3$	Cabbage palm hammock	256
$C4$	Cabbage palm/oak hammock	252
$C5$	Slash pine	161
$C6$	Oak/broadleaf hammock	229
$C7$	Hardwood swamp	105
$C8$	Graminoid marsh	431
$C9$	Spartina marsh	520
$C10$	Cattail marsh	404
$C11$	Salt marsh	419
$C12$	Mud flats	503
$C13$	Water	927
	Total	5211

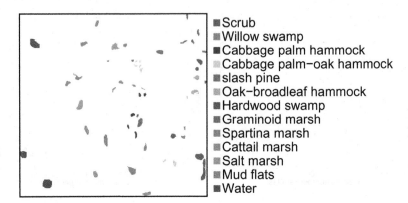

Fig. 4. Ground-truth map of the Kennedy Space Center data set (thirteen land cover classes).

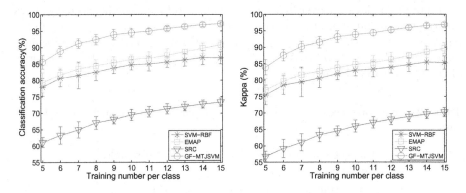

Fig. 5. Overall accuracy and kappa coefficient of the Kennedy Space Center data set.

4.2   KSC Data Set

The second data set that we used in our experiments was acquired by the AVIRIS sensor over the Kennedy Center (KSC), Merritt Island, FL, USA, on March 23, 1996. Figure 4 depicts the ground-truth map of the land covers. In the original 224 bands, 48 bands are identified as water absorption and low-SNR bands (numbered 1–4, 102–116, 151–172, and 218–224), which are discarded, and only 176 bands remain. The spatial resolution of the data is 18 m per pixel. For classification purposes, 13 classes representing the various land-cover types that occur in this environment were defined for the site (see Table 2).

Figure 5 displays the overall accuracy (OA) and kappa coefficient measure as functions of the number of training samples per class (5, 6, 7, ..., 15). Same as the first experiment, the results obtained by our GF-MTJSVM method are more accurate than those of the other three ones. For these two experimental both demonstrating the superiority of the GF-MTJSVM framework.

5 Conclusion

In this paper, we propose the GF-MTJSVM framework for hyperspectral image classification. We observe that the multi-task joint support vector machines is a simple yet effective way to fuse multiple complementary Gabor features to improve the classification accuracy for the small training sample classification task. Experiments on two real-world hyperspectral data show that our method performs quite competitive to several representative state-of-the-art methods. In summary, we can conclude with observations that Gabor feature-based multi-task joint support vector machines is an effective method for hyperspectral image classification with small training sample.

References

1. Asif, M.S.: Dynamic compressive sensing: sparse recovery algorithms for streaming signals and video. Ph.D. thesis, Georgia Institute of Technology (2013)
2. Bau, T.C., Sarkar, S., Healey, G.: Hyperspectral region classification using a three-dimensional Gabor filterbank. IEEE Trans. Geosci. Remote Sens. **48**(9), 3457–3464 (2010)
3. Benediktsson, J.A., Palmason, J.A., Sveinsson, J.R.: Classification of hyperspectral data from urban areas based on extended morphological profiles. IEEE Trans. Geosci. Remote Sens. **43**(3), 480–491 (2005)
4. Boser, B.E., Guyon, I.M., Vapnik, V.N.: A training algorithm for optimal margin classifiers. In: Proceedings of the Fifth Annual Workshop on Computational Learning Theory, pp. 144–152. ACM (1992)
5. Camps-Valls, G., Bruzzone, L.: Kernel-based methods for hyperspectral image classification. IEEE Trans. Geosci. Remote Sens. **43**(6), 1351–1362 (2005)
6. Chang, C.I.: Hyperspectral Imaging: Techniques for Spectral Detection and Classification, vol. 1. Springer, New York (2003)
7. Chang, C.I.: Hyperspectral Data Exploitation: Theory and Applications. Wiley, Chichester (2007)
8. Chang, C.C., Lin, C.J.: LIBSVM: a library for support vector machines. ACM Trans. Intell. Syst. Technol. (TIST) **2**(3), 27 (2011)
9. Chen, Y., Lin, Z., Zhao, X.: Riemannian manifold learning based k-nearest-neighbor for hyperspectral image classification. In: IEEE International Geoscience and Remote Sensing Symposium (IGARSS), pp. 1975–1978 (2013)
10. Cortes, C., Vapnik, V.: Support-vector networks. Machine Learn. **20**(3), 273–297 (1995)
11. Dalla Mura, M., Atli Benediktsson, J., Waske, B., Bruzzone, L.: Extended profiles with morphological attribute filters for the analysis of hyperspectral data. Int. J. Remote Sens. **31**(22), 5975–5991 (2010)
12. Gehler, P., Nowozin, S.: On feature combination for multiclass object classification. In: 2009 IEEE 12th International Conference on Computer Vision, pp. 221–228. IEEE (2009)
13. Genc, L., Inalpulat, M., Kizil, U., Mirik, M., Smith, S.E., Mendes, M.: Determination of water stress with spectral reflectance on sweet corn (Zea mays L.) using classification tree (CT) analysis. Zemdirbyste Agric. **100**(1), 81–90 (2013)

14. Ghamisi, P., Dalla-Mura, M., Benediktsson, J.: A survey on spectral-spatial classification techniques based on attribute profiles. IEEE Trans. Geosci. Remote Sens. **53**(5), 2335–2353 (2015)
15. Gu, Y., Gao, G., Zuo, D., You, D.: Model selection and classification with multiple kernel learning for hyperspectral images via sparsity. IEEE J. Sel. Top. Appl. Earth Observ. Remote Sensing, **7**(6), 2119–2130 (2014)
16. Gu, Y., Wang, C., You, D., Zhang, Y., Wang, S., Zhang, Y.: Representative multiple kernel learning for classification in hyperspectral imagery. IEEE Trans. Geosci. Remote Sens. **50**(7), 2852–2865 (2012)
17. Harsanyi, J.C., Chang, C.I.: Hyperspectral image classification and dimensionality reduction: an orthogonal subspace projection approach. IEEE Trans. Geosci. Remote Sens. **32**(4), 779–785 (1994)
18. Lee, M.A., Prasad, S., Bruce, L.M., West, T.R., Reynolds, D., Irby, T., Kalluri, H.: Sensitivity of hyperspectral classification algorithms to training sample size. In: First Workshop on Hyperspectral Image and Signal Processing: Evolution in Remote Sensing, WHISPERS 2009, pp. 1–4. IEEE (2009)
19. Li, J., Bioucas-Dias, J.M., Plaza, A.: Semisupervised hyperspectral image segmentation using multinomial logistic regression with active learning. IEEE Trans. Geosci. Remote Sens. **48**(11), 4085–4098 (2010)
20. Li, J., Huang, X., Gamba, P., Bioucas-Dias, J., Zhang, L., Atli-Benediktsson, J., Plaza, A.: Multiple feature learning for hyperspectral image classification. IEEE Trans. Geosci. Remote Sens. **53**(3), 1592–1606 (2015)
21. Li, J., Marpu, P.R., Plaza, A., Bioucas-Dias, J.M., Benediktsson, J.A.: Generalized composite kernel framework for hyperspectral image classification. IEEE Trans. Geosci. Remote Sens. **51**(9), 4816–4829 (2013)
22. Lu, D., Weng, Q.: A survey of image classification methods and techniques for improving classification performance. Int. J. Remote Sens. **28**(5), 823–870 (2007)
23. Manjunath, B.S., Ma, W.Y.: Texture features for browsing and retrieval of image data. IEEE Trans. Pattern Anal. Mach. Intell. **18**(8), 837–842 (1996)
24. Pacifici, F., Chini, M., Emery, W.J.: A neural network approach using multi-scale textural metrics from very high-resolution panchromatic imagery for urban land-use classification. Remote Sens. Environ. **113**(6), 1276–1292 (2009)
25. Platt, J., et al.: Probabilistic outputs for support vector machines and comparisons to regularized likelihood methods. Advances Large Margin Classif. **10**(3), 61–74 (1999)
26. Plaza, A., Benediktsson, J.A., Boardman, J.W., Brazile, J., Bruzzone, L., Camps-Valls, G., Chanussot, J., Fauvel, M., Gamba, P., Gualtieri, A., et al.: Recent advances in techniques for hyperspectral image processing. Remote Sens. Environ. **113**, S110–S122 (2009)
27. Prasad, S., Bruce, L.M.: Overcoming the small sample size problem in hyperspectral classification and detection tasks. In: IEEE International Geoscience and Remote Sensing Symposium, IGARSS 2008. vol. 5, pp. V-381. IEEE (2008)
28. Qian, Y., Ye, M., Zhou, J.: Hyperspectral image classification based on structured sparse logistic regression and three-dimensional wavelet texture features. IEEE Trans. Geosci. Remote Sens. **51**(4), 2276–2291 (2013)
29. Shen, L., Jia, S.: Three-dimensional gabor wavelets for pixel-based hyperspectral imagery classification. IEEE Trans. Geosci. Remote Sens. **49**(12), 5039–5046 (2011)
30. Tarabalka, Y., Fauvel, M., Chanussot, J., Benediktsson, J.A.: SVM- and MRF-based method for accurate classification of hyperspectral images. IEEE Trans. Geosci. Remote Sens. **7**(4), 736–740 (2010)

31. Tarabalka, Y., Chanussot, J., Benediktsson, J.A.: Segmentation and classification of hyperspectral images using watershed transformation. Pattern Recogn. **43**(7), 2367–2379 (2010)
32. Tuia, D., Volpi, M., Dalla Mura, M., Rakotomamonjy, A., Flamary, R.: Automatic feature learning for spatio-spectral image classification with sparse. SVM **52**(10), 6062–6074 (2014)
33. Wright, J., Yang, A.Y., Ganesh, A., Sastry, S.S., Ma, Y.: Robust face recognition via sparse representation. IEEE Trans. Pattern Anal. Mach. Intell. **31**(2), 210–227 (2009)
34. Wu, T.F., Lin, C.J., Weng, R.C.: Probability estimates for multi-class classification by pairwise coupling. J. Mach. Learn. Res. **5**, 975–1005 (2004)
35. Yuan, X.T., Liu, X., Yan, S.: Visual classification with multitask joint sparse representation. IEEE Trans. Image Process. **21**(10), 4349–4360 (2012)
36. Zhao, Y.Q., Yang, J.: Hyperspectral image denoising via sparse representation and low-rank constraint. IEEE Trans. Geosci. Remote Sens. **53**(1), 296–308 (2015)
37. Zheng, X., Sun, X., Fu, K., Wang, H.: Automatic annotation of satellite images via multifeature joint sparse coding with spatial relation constraint. IEEE Geosci. Remote Sens. Lett. **10**(4), 652–656 (2013)

Single Image Haze Removal Based on Priori Image Geometry and Edge-Preserving Filtering

Fan Guo[1(✉)] and Changjun Zhu[2]

[1] School of Information Science and Engineering, Central South University,
Changsha 410083, Hunan, China
guofancsu@163.com
[2] Powder Metallurgy Research Institute of Central South University,
Changsha 410083, Hunan, China

Abstract. Images captured in foggy weather conditions often suffer from bad visibility. Therefore, in this paper, we propose an efficient and automatic method to remove hazes from a single input image. Our method benefits much from an exploration on the priori image geometry on the transmission function. This prior, combined with an edge-preserving texture-smoothing filtering, is applied to estimate the unknown transmission map. Next, the restored image is obtained by taking the estimated transmission map and the atmospheric light into the image degradation model to recover the scene radiance. The proposed method is controlled by just a few parameters and can restore a high-quality haze-free image with fine image details. The experimental results demonstrate that our method shows competitive performance against other methods.

Keywords: Single image · Haze removal · Priori image geometry · Edge-preserving filtering · Image degradation model

1 Introduction

The images captured in hazy or foggy weather conditions often suffer from poor visibility. The distant objects in the fog lose the contrasts and get blurred with their surroundings, as illustrated in Fig. 1(a). This is because the reflected light from these objects, before it reaches the camera, is attenuated in the air and further blended with the atmospheric light scattered by some aerosols. Also for this reason, the colors of these objects get faded and become much similar to the fog, the similarity of which depending on the distances of them to the camera. As can be seen in Fig. 1(a), the image contrast is reduced and the surface colors become faint. There are many circumstances that effective dehazing algorithms are needed. In computer vision, most automatic systems for surveillance, intelligent vehicles, object recognition, etc., assume that the input images have clear visibility. However, this is not always true in bad weather.

Previous methods for haze removal mainly rely on additional depth information or multiple observations of the same scene. Representative works include [1–4]. For example, Schechner et al. [1] develop a method to reduce hazes by using two images taken through a polarizer at different angles. Narasimhan et al. propose a physics-based

© Springer Nature Singapore Pte Ltd. 2016
T. Tan et al. (Eds.): CCPR 2016, Part II, CCIS 663, pp. 26–41, 2016.
DOI: 10.1007/978-981-10-3005-5_3

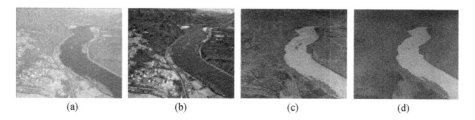

| (a) | (b) | (c) | (d) |

Fig. 1. Image dehazing result by our method. (a) The foggy image. (b) The dehazing result by our method. (c) The initial transmission map. (d) The refined transmission map. (Color figure online)

scattering model [2, 3]. Using this model, the scene structure can be recovered from two or more images with different weather. Kopf et al. [5] propose to remove haze from an image by using the scene depth information directly accessible in the geo-referenced digital terrain or city models. However, the multiple images or the user-interaction that needed in these methods limit their applications.

Recently, some significant progresses have also been achieved [6–11] for a more challenging single image dehazing problem. These progresses benefit much from the insightful explorations on new image models and priors. For example, Fattal's work [15] was based on the assumption that the transmission and the surface shading are locally uncorrelated. Under this assumption, Fattal estimated the albedo of the scene and then inferred the medium transmission. This approach is physically sound and can produce impressive results. However, the method fails when handling heavy haze images. Tan's work [13] was based on the observations that the clear-day images have higher contrast compared with the input haze image and he removed the haze by maximizing the local contrast of the restored image. The visual results are visually compelling but may not physical valid. He et al. [8] proposed the dark channel prior to solve the single image dehazing problem. The prior is based on the observation that most local patches in haze-free outdoor images contain some pixels which have very low intensities in at least one color channel. Using this prior, estimated transmission map and the value of atmospheric light can be obtained. For a better purpose, soft matting is used for the estimated trans-mission map. Combined with the haze image model, a good haze-free image can be recovered by this approach. Tarel et al. [9] used a fast median filter to infer the atmos-pheric veil, and further estimated the transmission map. However, the method is unable to remove the fog between small objects, and the color of the scene objects is unnatural for some situations. Kratz et al. [10] proposed a probabilistic model that fully leverages natural statistics of both the albedo and depth of the scene. The key idea of the method is to model the image with a factorial Markov random field in which the scene albedo and depth are two statistically independent latent layers. Kristofor et al. [11] made an investigation of the dehazing effects on image and video coding for surveillance systems. They first proposed a method for single image, and then consider the dehazing effects in compression.

From the above dehazing algorithms, we can deduce that single image dehazing is essentially an under-constrained problem, and the general principle of solving such problems is therefore to explore additional priors. Following this idea, we begin our study in this paper by deriving inherent priori image geometry for estimating the initial scene transmission map, as show in Fig. 1(c). This constraint, combined with an edge-preserving texture-smoothing filtering, is applied to recover the final unknown transmission, as shown in Fig. 1(d). Experimental results show that our method can restore a haze-free image of high quality with fine edge details. Figure 1(b) illustrates an example of our dehazing result.

Our method mainly uses two techniques to remove image haze. The first is a priori image on the scene transmission. This image prior, which has a clear geometric interpretation, is proved to be effective to estimate initial transmission map. The second technique is a new edge-preserving filtering that enables us to obtain the final transmission map. The filter helps in attenuating the image noises and enhancing some useful image structures, such as jump edges and corners. Experimental results show that the proposed algorithm can remove haze more thoroughly without producing any halo artifacts, and the color of the restored images appears natural in most cases for a variety of real captured haze images.

2 Background

2.1 Atmospheric Scatting Model

The haze image model (also called image degradation model) consists of a direct attenuation model and an air light model. The direct attenuation model describes the scene radiance and its decay in the medium, while the air light results from previously scattered light. The formation of a haze image model is as follows [8, 12]:

$$\mathbf{I}(x) = t(x)\mathbf{J}(x) + (1 - t(x))\mathbf{A} \tag{1}$$

where $\mathbf{I}(x)$ is the input foggy image, $\mathbf{J}(x)$ is the recovered scene radiance, $\mathbf{A}$ is the global atmospheric light, and $t(x)$ is the transmission map. The transmission function $t(x)$ $(0 \leq t(x) \leq 1)$ is correlated with the scene depth. Let assume that the haze is homogenous, the $t(x)$ can this be written as:

$$t(x) = e^{-\beta d(x)} \tag{2}$$

In (2), $d(x)$ is the scene depth, and β is the medium extinction coefficient. Therefore, the goal of image dehazing is to recover the scene radiance $\mathbf{J}(x)$ from the input image $\mathbf{I}(x)$ according to Eq. (1). This requires us to estimate the transmission function $t(x)$ and the global atmospheric light $\mathbf{A}$. As can be seen from Eq. (1), this problem is an ill-posed problem since the number of unknowns is much greater than the number of available equations. Thus, additional assumptions or priors need to be introduced to solve it. In this paper, the priori image geometry and edge-preserving filtering are applied to obtain our final transmission map $t(x)$.

2.2 Priori Image Geometry and Edge-Preserving Filtering

For priori image geometry, a pixel $\mathbf{I}(x)$ contaminated by fog will be "pushed" towards the global atmospheric light $\mathbf{A}$ according to Eq. (1). Thus, we can reverse this process and recover the clean pixel $\mathbf{J}(x)$ by a linear extrapolation from $\mathbf{A}$ to $\mathbf{I}(x)$. The appropriate amount of extrapolation is given by [13]:

$$\frac{1}{t(x)} = \frac{\|\mathbf{J}(x) - \mathbf{A}\|}{\|\mathbf{I}(x) - \mathbf{A}\|} \tag{3}$$

Consider that the scene radiance of a given image is always bounded, that is,

$$\mathbf{C}_0 \leq \mathbf{J}(x) \leq \mathbf{C}_1, \quad \forall x \in \Omega \tag{4}$$

where $\mathbf{C}_0$ and $\mathbf{C}_1$ are two constant vectors that are relevant to the given image. For any x, a natural requirement is that the extrapolation of $\mathbf{J}(x)$ must be located in the radiance cube bounded by $\mathbf{C}_0$ and $\mathbf{C}_1$. The above requirement on $\mathbf{J}(x)$ imposes a boundary constraint on $t(x)$. Suppose that the global atmospheric light $\mathbf{A}$ is given. Thus, for each x, we can compute the corresponding boundary constraint point $\mathbf{J}_b(x)$. Then, a lower bound of $t(x)$ can be determined by using Eqs. (3) and (4), leading to the following boundary constraint on $t(x)$:

$$0 \leq t_b(x) \leq t(x) \leq 1 \tag{5}$$

where $t_b(x)$ is the lower bound of $t(x)$, given by

$$t_b(x) = \min\left\{ \max_{c \in \{r,g,b\}} \left(\frac{A^c - I^c(x)}{A^c - C_0^c}, \frac{A^c - I^c(x)}{A^c - C_1^c} \right), 1 \right\} \tag{6}$$

where I^c, A^c, C_0^c and C_1^c are the color channel of $\mathbf{I}$, $\mathbf{A}$, $\mathbf{C}_0$ and $\mathbf{C}_1$, respectively. In our experiment, the boundary constraint map that used for obtaining out initial transmission map is computed from Eq. (6) by setting the radiance bounds $\mathbf{C}_0 = (20, 20, 20)^T$ and $\mathbf{C}_1 = (300, 300, 300)^T$.

In the proposed method, the above priori image geometry is used for estimating the initial transmission map and an edge-preserving filtering method is performed over the estimated initial transmission map to obtain a refined one. The reason why the filtering process needs preserving edge is that the wrong edge of scene objects in transmission map will cause halo artifacts in final restoration results. For the edge-preserving filtering, the typical filters that can be used for removing the redundant details of the transmission map are Gaussian filter, median filter and bilateral filter [14]. The Gaussian smoothing operator is a 2-D convolution operator that is used to 'blur' images and remove detail and noise. It uses a different kernel that represents the shape of a Gaussian ('bell-shaped') hump to achieve this purpose. The main idea of median filtering is to run through all pixels, replacing each pixel value with the median of neighboring pixel values. Bilateral filtering [15] is done by replacing the intensity value of a pixel by the average of the values of other pixels weighted by their spatial distance and intensity similarity to the original pixel. Specifically, the bilateral filter is defined a normalized convolution in which the weighting for each pixel q is determined by the spatial distance from the center pixel p, as well as it relative difference in intensity. Let I_p be the intensity at pixel p and I_p^F be the filter value, the filter value can thus be written as [14]:

$$I_p^F = \frac{\sum_{q \in \Omega} G_{\sigma_{s'}}(\|p - q\|) G_{\sigma_R}\left(\left|I_p - I_q\right|\right) I_q}{\sum_{q \in \Omega} G_{\sigma_{s'}}(\|p - q\|) G_{\sigma_R}\left(\left|I_p - I_q\right|\right)} \tag{7}$$

where the spatial and range weighting functions $G_{\sigma_{s'}}$ and G_R are often Gaussian, and σ_R and $\sigma_{s'}$ are the range and spatial variances, respectively. Median filter has strong smooth ability but can blur edges, while bilateral filter with small range variance cannot achieve enough smoothing on textured regions in the haze removal algorithm and that large range variance will also blur edges. Therefore, in this paper, a edge-preserving texture-smoothing filtering is adopted which can generate better results in transmission map refined than existing filters. For the new edge-preserving filtering, the image intensities are normalized such that $I_p \in [0, 1]$ and σ_R is normally chosen between $[0, 1]$. Let the image width and height be w and h, we choose $\sigma_{s'} = \sigma_s \cdot \min(h, w)$, such that σ_s is also normally chosen between $[0, 1]$. Besides bilateral filter, we also use $\sigma_{s'}$ to represent the spatial variance of Gaussian filter and the radius of median filter in this paper.

3   The Proposed Algorithm

3.1   The Algorithm Flowchart

The flowchart of the proposed method is depicted in Fig. 2. One can clearly see that the proposed algorithm employs three steps in removing haze from a single image. The first one involves computing the atmospheric light according to the three distinctive features of the sky region. The second step involves the computing of the initial transmission map with the priori image geometry and the edge-preserving filtering. The goal of this

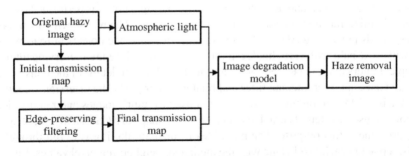

Fig. 2. Flowchart of the algorithm.

step is to obtain the initial transmission map using the boundary constraint from radiance cube and to smooth textured regions while preserving large scene depth discontinuities using the edge-preserving filtering. Finally, with the estimated atmospheric light and the refined transmission map, the scene radiance can be recovered according to the image degradation model.

3.2 Atmospheric Light Estimation

Estimating atmospheric light A should be the first step to restore the hazy image. To estimate A, the three distinctive features of sky region are considered here, which is more robust than the "brightest pixel" method. The distinctive features of sky region are: (i) bright minimal dark channel, (ii) flat intensity, and (iii) upper position. For the first feature, the pixels that belong to the sky region should satisfy $I_{min}(x) > T_v$, where $I_{min}(x)$ is the dark channel and T_v is the 95 % of the maximum value of $I_{min}(x)$. For the second feature, the pixels should satisfy the constraint $N_{edge}(x) < T_p$ where $N_{edge}(x)$ is the edge ratio map and T_p is the flatness threshold. Due to the third feature, the sky region can be determined by searching for the first connected component from top to bottom. Thus, the atmospheric light A is estimated as the maximum value of the corresponding region in the foggy image $I(x)$.

3.3 Initial Transmission Map Estimation

The boundary constraint of $t(x)$ provides a new geometric perspective to the famous dark channel prior [8]. Let $C_0 = 0$ and suppose the global atmospheric light A is brighter than any pixel in the haze image. One can directly compute $t_b(x)$ from Eq. (1) by assuming the pixel-wise dark channel of $J(x)$ to be zero. Similarly, assuming that the transmission in a local image patch is constant, one can quickly derive the patch-wise transmission $\tilde{t}(x)$ in He $et\ al.$'s method [8] by applying a maximum filtering on $t_b(x)$, i.e.,

$$\tilde{t}(x) = \max_{y \in \omega_x} t_b(y) \tag{8}$$

where ω_x is a local patch centered at x. It is worth noting that the boundary constraint is more fundamental. In most cases, the optimal global atmospheric light is a little darker than the brightest pixels in the image. Those brighter pixels often come from some light sources in the scene, e.g., the bright sky or the headlights of cars. In these cases, the dark channel prior will fail to those pixels, while the proposed boundary constraint still holds. Note that the commonly used constant assumption on the transmission within a local image patch is somewhat demanding. For this reason, the patch-wise transmission $\tilde{t}(x)$ based on this assumption in [8] is often underestimated. Here, a more accurate patch-wise transmission is adopted in the proposed method, which relaxes the above assumption and allows the transmissions in a local patch to be slightly different. The new patch-wise transmission is given as below:

$$\hat{t}(x) = \min_{y \in \omega_x} \max_{z \in \omega_y} t_b(z) \tag{9}$$

The above patch-wise transmission $\hat{t}(x)$ can be conveniently computed by directly applying a morphological closing on $t_b(x)$. Figure 3 illustrates a comparison of the dehazing results by directly using the transmissions map derived from dark channel prior and the boundary constraint map, respectively. One can observe that the estimated transmission map from dark channel prior works not well since the dehazing result contains some halo artifacts, as shown in Fig. 3(b). In comparison, the new patch-wise transmission map derived from the boundary constraint map can produces fewer halo artifacts, as shown in Fig. 3(c).

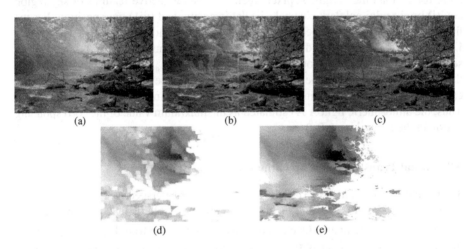

(a)   (b)   (c)

(d)   (e)

Fig. 3. Image dehazing by directly using the patch-wise transmissions from dark channel prior and boundary constraint map, respectively. (a) Foggy image. (b) Dehazing result by dark channel prior. (c) Dehazing result by boundary constraint. (d) Estimated transmission map from dark channel. (e) Estimated transmission map from boundary constraint map.

3.4 Refined Transmission Map Estimation

After obtaining the initial transmission map, an edge-preserving filtering is used here to refine the initial transmission map.

Specifically, given a gray-scale image I (if the input is multi-channel image, we process each channel independently), we first detect the range of the image intensity values, say $[I_{min}, I_{max}]$. Next, we sweep a family of planes at different image intensity levels $I_k \in [I_{min}, I_{max}]$, $k = \{0, 1, ..., N - 1\}$ across the image intensity range space. The distance between neighboring planes is set to $(I_{max} - I_{min})/(N - 1)$, where N is a constant. Smaller N results in larger quantization error while larger N increases the running time. When $N \to +\infty$ or $N = 256$ for 8-bit grayscale images, the quantization error is zero. Also, stronger smoothing usually requires smaller N. We define the distance function of a voxel $[I(x, y), x, y]$ on a plane with intensity level I_k as the truncated Euclidean distance between the voxel and the plane. The process can be written as:

$$D(I_k, x, y) = \min(|I_k - I(x,y)|, \eta \cdot |I_{\max} - I_{\min}|) \tag{10}$$

where $\eta \in (0, 1]$ is a constant used to reject outliers and when $\eta = 1$, the distance function is equal to their Euclidean distance. At each intensity plane, we obtain a 2D distance map $D(I_k)$. A low pass filter F is then used to smoothen this distance map to suppress noise:

$$D^F(I_k) = F(D(I_k)) \tag{11}$$

Then, the plane with minimum distance value at each pixel location $[x, y]$ is located by calculating

$$K(x,y) = \underset{k=\{0,1,\ldots,N-1\}}{\arg\min} \; D^F(I_k, x, y) \tag{12}$$

Let intensity levels $i_0 = I_{K(x,y)}$, $i_+ = I_{K(x,y)+1}$, $i_- = I_{K(x,y)-1}$. Assume that the smoothed distance function $D^F(I_k, x, y)$ at each pixel location $[x, y]$ is quadratic polynomial with respect to the intensity value I_k, the intensity value corresponding to the minimum of the distance function can be approximated using quadratic polynomial interpolation (curve fitting):

$$I^F(x,y) = i_0 - \frac{D^F(i_+, x, y) - D^F(i_-, x, y)}{2(D^F(i_+, x, y) + D^F(i_-, x, y) - 2D^F(i_0, x, y))} \tag{13}$$

In (13), $I^F(x, y)$ is the final filtered result of our framework at each pixel location $[x, y]$. Apparently, any filter F can be integrated into this framework by using it to smoothen the distance map $D(I_k)$ as shown in Eq. (11). Figure 4 presents the experimental filtered results of our framework for a synthetic noisy image. In this experiment, the

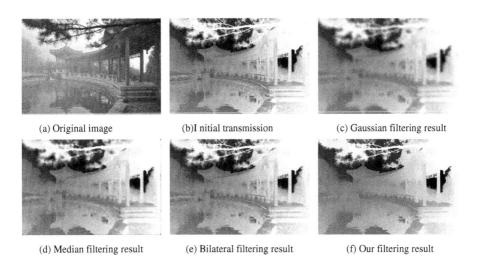

(a) Original image   (b) Initial transmission   (c) Gaussian filtering result

(d) Median filtering result   (e) Bilateral filtering result   (f) Our filtering result

Fig. 4. Filtered images obtained using different filtering methods.

number of intensity planes N is set to 16 and $\eta = 0.1$. The parameter settings for low-pass filters are $\sigma_S = 0.05$ and as for bilateral filter, $\sigma_R = 0.2$. As can be seen in Fig. 4(f), compared with other filtering methods, our refined transmission map maintains the sharp intensity edges and greatly suppresses the noise inside each region.

3.5 Scene Radiance Recovery

Since, now, we already know the input haze image $\mathbf{I}(x)$, the final refined transmission map $t(x)$ and the global atmospheric light $\mathbf{A}$, we can obtain the final haze removal image $\mathbf{J}(x)$ according to the image degradation model, as shown in Eq. (1). The final dehazing result $\mathbf{J}(x)$ is recovered by:

$$\mathbf{J}(x) = \frac{\mathbf{I}(x) - \mathbf{A}}{\max(t(x), t_0)} + \mathbf{A} \tag{14}$$

where t_0 is application-based and is used to adjust the haze remaining at only the farthest reaches of the image. If the value of t_0 is too large, the result has only a slight haze removal effect, and if the value is too small, the color of the haze removal result seems over saturated. Experiments show that when t_0 is set between 0.1 and 0.5, we can get visually pleasing results in most cases. An illustrative example is shown in Fig. 5. In the figure, Fig. 5(a) shows the input hazy images, Fig. 5(b) shows the refined transmission map estimated by using the proposed method, and Fig. 5(c) is our final haze removal result obtained by using the refined transmission map.

(a) Original image   (b) Estimated refined transmission   (c) Final haze removal result

Fig. 5. Image haze removal example.

4 Experimental Results

In order to verify the effectiveness and validity of the proposed image dehazing method, two criteria have been considered: (i) qualitative comparison, and (ii) quantitative evaluation. In the experiments, all the dehazing results are obtained by executing MATLAB R2008a on a PC with a 3.10-GHz Intel® CoreTM i5-2400 CPU.

4.1 Qualitative Comparison

Figure 6 illustrates some examples of our dehazing results and the recovered scene transmission functions. As can be seen from the results, our method can recover rich

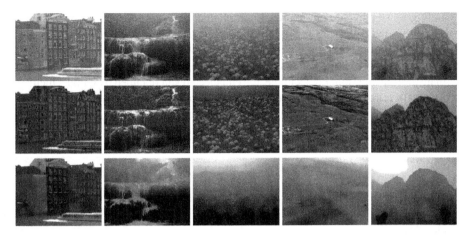

Fig. 6. Image dehazing results by our method. Top: input haze images. Middle: the dehazing results. Bottom: the recovered transmission functions. The recovered transmission gives an estimation of the density map of hazes in the input image.

Fig. 7. Comparison with Tan's work. Left: input image. Middle: Tan's result. Right: our result. (Color figure online)

details of images with vivid color information in the haze regions. One can clearly see that the image contrast and detail are greatly improved compared with original foggy images, especially in the distant region with dense fog. Besides, the transmission function also reflects the density of the hazes in the captured scene. From the Fig. 6, we can see the estimated transmissions by our method are quite consistent with our intuitions.

We also compare our method with several state-of-the-art existing methods. Figure 7 illustrates the comparisons of our method with Tan's work [7]. Tan's method can augment the image details and greatly enhance the image visual effect. However, the colors in the recovered images are often over saturated, since the method is not a physically based approach and the transmission may thus be underestimated. For example, one can clearly see that the color of traffic sign in Fig. 7 has changed to orange after dehazing, while our results have no such problem. Moreover, some significant halo artifacts usually appear around the recovered sharp edges (e.g., trees). The proposed method can improve the visual effect of image structures in very dense haze regions while restoring the natural colors, and the halo artifacts in our results are also smaller, as shown in Fig. 7.

In Fig. 8, we compare our approach with Fattal's method [6]. Fattal's method can produce a visually pleasing result. However, the method is based on statistics and requires sufficient color information and variance. When the haze is dense, the color information that needed in Fattal's method is not enough for the method to reliably estimate the transmission. For example, the enhanced result obtained by Fattal's method in Fig. 8 still remains some fog in the region far away. In comparison, our results are much visually pleasing, and the haze at the farthest reaches of the image can be largely removed.

Fig. 8. Comparison with Fattal's work. Left: input image. Middle: Fattal's result. Right: our result. (Color figure online)

Figure 9 shows the comparisons of our approach with Tarel et al.'s method [9]. Tarel et al.'s method is a filtering based approach. They estimate the atmospheric veil by applying a fast median filter to the minimum components of the observed image. The main advantages of the method is its speed, while the weakness is the haze removal results always contain some halo artifacts and the color seems not very natural. As shown in Fig. 9, the color of the sky and maintain seem too dark. Compared with Tarel's method, the color of our method seems much more natural.

Fig. 9. Comparison with Tarel's work. Left: input image. Middle: Tarel's result. Right: our result. (Color figure online)

We also compare our method with He et al.'s work [8] in Fig. 10. As can be seen in the figure, both methods produce comparable results in regions with heavy hazes (e.g., the distant buildings). However, in regions with many depth jumps (e.g., trees and grasses at close range), our method performs better. The color in our haze removal result also seems more vivid and colorful. Moreover, our method tends to generate a clearer result of image details, as illustrated in Fig. 10. These benefits from the incorporation of a filter bank into image dehazing. These filters can help to exploit and augment the interesting image structures, e.g., jump edges and corners.

Fig. 10. Comparison with He's work. Left: input image. Middle: He's result. Right: our result. (Color figure online)

Figures 11 and 12 allow the comparison of our results with four state of the art visibility restoration algorithms: Tan [7], Fattal [6], Tarel [9] and He [8]. One can clearly see that all the dehazing methods can effectively remove the haze in both near and far regions for the testing images. However, the results obtained with our algorithm seems visually close to the result obtained by He, with better color fidelity and less halo artifacts compared with Tan. However, we find, depending on the image, each algorithm is a trade-off between color fidelity and contrast enhancement. Results on a variety of haze or fog images show that the proposed method can achieve a better enhancement effect in terms of both image color and the profile of the scene objects.

Fig. 11. Experimental results of various dehazing methods. First row: the input image and the results obtained by Tan and Fattal, respectively. Second row: the results obtained by Tarel, He and our method, respectively. (Color figure online)

Fig. 12. Experimental results of various dehazing methods. First row: the input image and the results obtained by Tan and Fattal, respectively. Second row: the results obtained by Tarel, He and our method, respectively. (Color figure online)

4.2 Quantitative Evaluation

An assessment method dedicated to visibility restoration proposed in [16] is used here to measure the dehazing effect. We first transform the color level image to the gray level image, and use three indicators to compare the two gray level images: the input image and the haze removal image. The visible edges in the image before and after restoration are selected by a 5 % contrast threshold according to the meteorological visibility distance proposed by the international commission of illumination (CIE). To implement this definition of contrast between two adjacent regions, the method of visible edges segmentation proposed in [17] has been used.

Once the map of visible edges is obtained, we can compute the rate e of edges newly visible after restoration. Then, the mean $\bar{r}$ over these edges of the ratio of the gradient norms after and before restoration is computed. This indicator $\bar{r}$ estimates the average visibility enhancement obtained by the restoration algorithm. At last, the percentage of pixels σ which becomes completely black or completely white after restoration is computed. Since the assessment method is based on the definition of visibility distance, the evaluation conclusion which complies with human vision characteristic can be drawn.

These indicators e, $\bar{r}$ and σ are evaluated for Tan [7], Fattal [6], Tarel [9], He [8] and our method on six images (see Table 1). For each method, the aim is to increase the contrast without losing some visual information. Hence, good results are described by high values of e and $\bar{r}$ and low value of σ. From Table 1, we deduce that depending on the image, Tan's algorithm generally has more visible edges than our method, Tarel's, Fattal's and He's algorithms. Besides, generally we can order the five algorithms in a decreasing order with respect to average increase of contrast on visible edges: Tan, our method, Fattal, Tarel and He algorithms. This confirms our observations on Figs. 7, 8, 9, 10, 11 and 12. Table 1 also gives the percentages of pixels becoming completely black or white after restoration. Compared to others, the proposed algorithm generally gives the smallest percentage.

Table 1. Comparison with the state of art haze removal algorithms using the three indicators

Indicator	e	$\bar{r}$	$\sigma\,(\%)$	e	$\bar{r}$	$\sigma\,(\%)$
Method	Fig. 7 (600 × 400)			Fig. 8 (512 × 384)		
Tan	0.8907	3.8688	1.8352	0.4917	2.8549	0.8638
Fattal	0.2198	2.2524	1.3604	0.3636	2.4407	0.1609
Tarel	0.6261	2.1354	0	0.3959	1.9833	0.0131
He	0.5924	1.5202	1.4542	0.2857	1.3001	0.6646
Proposed	0.8228	3.6069	1.2701	0.3940	2.5256	0.1355
Method	Fig. 9 (512 × 384)			Fig. 10 (441 × 450)		
Tan	0.1848	2.0759	4.1573	0.2994	1.8604	4.3982
Fattal	0.0347	1.8669	4.5339	0.1168	1.6844	4.2095
Tarel	0.1588	1.2941	3.4831	0.1821	1.5841	3.8532
He	0.0191	1.1236	3.7001	0.1670	1.1219	3.6338
Proposed	0.1732	2.0296	2.8948	0.2863	1.9610	2.6222
Method	Fig. 11 (576 × 768)			Fig. 12 (1024 × 768)		
Tan	0.0967	1.8269	0.4039	0.3035	2.5443	0.3222
Fattal	0.0735	1.3397	0.5951	−0.1088	1.8641	0.1692
Tarel	0.0385	1.2273	0.1398	−0.3539	1.7411	0.0786
He	0.0457	1.1854	0.2068	−0.0173	1.6517	0.2358
Proposed	0.0896	1.7556	0.0325	0.2713	2.3574	0.0884

Computation time is also a very important criterion to evaluate algorithm performance. For an image of $s_x \times s_y$, the fastest algorithm is Tarel's method. The complexity of Tarel's algorithm is $O(s_x s_y)$, which implies that the complexity is a linear function of the number of input image pixels. Thus, only 2 s are needed to process an image of size 600 × 400. For He's method, its time complexity is relatively high since the matting Laplacian matrix L in the method is so huge that for an image of size $s_x \times s_y$, the size of L is $s_x s_y \times s_x s_y$, so 20 s is needed to process a 600-by-400 pixels image. The computational times of Fattal's and Tan's methods are even longer than He's method. They take about 40 s and 5 to 7 min to process an image which is of size 600 × 400, respectively. Our proposed method has a relatively faster speed, 17 s is needed to obtain a haze removal image of the same size by using our method. This can be further improved by using a GPU-based parallel algorithm.

5   Conclusion and Future Work

In this paper, we have proposed an efficient method to remove hazes from a single image. Our method benefits much from an exploration on the priori image geometry on the transmission function. This prior, together with an edge-preserving filtering, is applied to recover the unknown transmission. Experimental results show that in comparison with the state-of-the-arts, our method can generate quite visually pleasing results with finer image details and structures.

However, single image dehazing is not an easy task since it often suffers from the problem of ambiguity between image color and depth. A clean pixel may have the same color with a fog-contaminated pixel due to the effects of hazes. Therefore, without sufficient priors, these pixels are difficult to be reliably recognized as fog-contaminated or not fog-contaminated. This ambiguity revealing the unconstraint nature of single image dehazing often leads to excessive or inadequate enhancements on the scene objects. Therefore, in this paper a priori image geometry is adopted to estimate the transmission map from the radiance cube of an image. Although the constraint imposes a much weak constraint on the dehazing process, it proves to be effective when combined with the edge-preserving filtering to remove fog from most natural images. Another way to address the ambiguity problem is to adopt more sound constraints or develop new image priors. Therefore, in the future we intend to use the scene geometry or directly incorporate the available depth information into the estimation of scene transmission.

Acknowledgements. This work was supported by the National Natural Science Foundation of China (61502537), and the Postdoctoral Science Foundation of Central South University (No. 126648).

References

1. Schechner, Y.Y., Narasimhan, S.G., Nayar, S.K.: Instant dehazing of images using polarization. In: IEEE Computer Society Conference on Computer Vision and Pattern Recognition (CVPR), pp. 325–332 (2001)
2. Narasimhan, S.G., Nayar, S.K.: Vision and the atmosphere. Int. J. Comput. Vision **48**(3), 233–254 (2002)
3. Narasimhan, S.G., Nayar, S.K.: Contrast restoration of weather degraded images. IEEE Trans. Pattern Anal. Mach. Intell. **25**(6), 713–724 (2003)
4. Shwartz, S., Namer, E., Schechner, Y.Y.: Blind haze separation. In: IEEE Computer Society Conference on Computer Vision and Pattern Recognition (CVPR), pp. 1984–1991 (2006)
5. Kopf, J., Neubert, B., Chen, B., Cohen, M., Cohen-Or, D., Deussen, O., Uyttendaele, M., Lischinski, D.: Deep photo: model-based photograph enhancement and viewing. In: ACM SIGGRAPH Asia, pp. 116:1–116:10 (2008)
6. Fattal, R.: Single image dehazing. In: ACM SIGGRAPH, pp. 72:1–72:9 (2008)
7. Tan,R.T.: Visibility in bad weather from a single image. In: IEEE Computer Society Conference on Computer Vision and Pattern Recognition (CVPR), pp. 1–8 (2008)
8. He, K., Sun, J., Tang, X.: Single image haze removal using dark channel prior. In: IEEE Computer Society Conference on Computer Vision and Pattern Recognition (CVPR), pp. 1956–1963 (2009)
9. Tarel, J.P., Hautiere, N.: Fast visibility restoration from a single color or gray level image. In: IEEE Computer Society Conference on Computer Vision and Pattern Recognition (CVPR), pp. 2201–2208 (2009)
10. Kratz, L., Nishino, K.: Factorizing scene albedo and depth from a single foggy image. In: IEEE International Conference on Computer Vision (ICCV), pp. 1701–1708 (2009)
11. Kristofor, B.G., Dung, T.V., Truong, Q.N.: An investigation of dehazing effects on image and video coding. IEEE Trans. Image Process. **21**(2), 662–673 (2012)

12. Middleton, W.: Vision through the atmosphere. University of Toronto Press, Toronto (1952)
13. Meng, G.F., Wang, Y., Duan, J.Y., Xiang, S.M., Pan, G.H.: Efficient image dehazing with boundary constraint and contextual regularization. In: IEEE International Conference on Computer Vision (ICCV), pp. 617–624 (2013)
14. Bao, L.C., Song, Y.B., Yang, Q.X., Ahuja, N.: An edge-preserving filtering framework for visibility restoration. In: 21st International Conference on Pattern Recognition (ICPR), pp. 384–387 (2012)
15. Tomasi, C., Manduchi, R.: Bilateral filtering for gray and color images. In: IEEE International Conference on Computer Vision (ICCV), pp. 839–846 (1998)
16. Hautiere, N., Tarel, J.P., Aubert, D.D., Dumont, E.: Blind contrast enhancement assessment by gradient ratioing at visible edges. Image Anal. Stereol. J. **27**(2), 87–95 (2008)
17. Kohler, R.: A segmentation system based on thresholding. Comput. Graph. Image Process. **15**(3), 319–338 (1981)

Semantic Segmentation with Modified Deep Residual Networks

Xinze Chen, Guangliang Cheng, Yinghao Cai, Dayong Wen, and Heping Li[(✉)]

NLPR, Institute of Automation, Chinese Academy of Sciences, Beijing 100190, China
{xinze.chen,guangliang.cheng}@nlpr.ia.ac.cn,
{yinghao.cai,dayong.wen,heping.li}@ia.ac.cn

Abstract. A novel semantic segmentation method is proposed, which consists of the following three parts: (I) First, a simple yet effective data augmentation method is introduced without any extra GPU memory cost during training. (II) Second, a deeper residual network is constructed through three effective techniques: dilated convolution, LSTM network and multi-scale prediction. (III) Third, an online hard pixels mining is adopted to improve the segmentation performance. We combine these three parts to train an end-to-end network and achieve a new state-of-the-art segmentation accuracy of 79.3 % on PASCAL VOC 2012 test set at the time of submission.

Keywords: Semantic segmentation · Data augmentation · Residual networks · LSTM · Multi-scale prediction

1 Introduction

Semantic segmentation is one of the most challenging problems in computer vision. The task of semantic segmentation is to make a label prediction for each pixel in the image. In recent years, deep convolutional neural networks (CNNs) have been demonstrated to achieve great progress in semantic pixel-wise segmentation. There have been attempts to apply networks designed for structured prediction to segmentation. Specifically, Long et al. [17] showed that a fully convolutional network (FCN) trained end-to-end could be successfully applied to semantic segmentation. Chen et al. [2] simplified the architecture of [17] by making further efforts to dilation and reducing the number of channels at the fully connected layers. Badrinarayanan et al. [22] presented a deep convolutional neural network with an encoder network and a corresponding decoder network. This network is efficient in terms of both memory and computation during inference. All the methods above use VGG16 network [21]. Although VGG16 network has achieved remarkable success on semantic segmentation, its learning ability may be restricted to its relatively shallow layers compared to the deep residual network [10].

In this paper, we put forward a deeper network named SegResNet, a modification of the deep residual network [10], which has been demonstrated to

© Springer Nature Singapore Pte Ltd. 2016
T. Tan et al. (Eds.): CCPR 2016, Part II, CCIS 663, pp. 42–54, 2016.
DOI: 10.1007/978-981-10-3005-5_4

greatly improve the classification accuracy. SegResNet further adopts three effective techniques: dilated convolution [23] (also called hole algorithm [2,18]), long short-term memory (LSTM) network [11] and multi-scale prediction [17].

Dilated convolution can be used to expand the receptive field of the network. It has been demonstrated that it can improve the segmentation performance in VGG16 [2]. At the first glance, considering the high deepness of SegResNet, consequently its large receptive field, it seems that there is no need to use dilated convolution. Nevertheless, we find that using dilated convolution still could drastically improve the segmentation performance of SegResNet and further speed up the convergence.

LSTM networks have been demonstrated their strong capabilities for modeling long-range dependencies in speech recognition and language understanding [6,7]. Inspired by [1], we design an LSTM network, which includes two four-directional (left, right, top, bottom) Recurrent Neural Networks (RNNs). Different from [1], each four-directional RNN in this work is interdependent and shares weights. The underlying rationale is that each four-directional RNN could share memory and reduce GPU memory cost.

Multi-scale prediction is mainly used for reducing the loss of information, such as region boundaries, by combining the features from intermediate layers during up-sampling. Following the promising results of [17], we employ a multi-scale prediction method to increase the boundary localization accuracy.

Data augmentation, such as horizontal flip, random crop and multi-scale, has been proved a useful step to improve the segmentation accuracy. Specifically for multi-scale images, the input image is resized to several scales, each is passed through a shared deep network and then the results are merged. However, this technique will bring extra GPU memory cost during training. For example, this technique needs a concat layer [14] or even an attention model [3]. Therefore, we propose a novel data augmentation method called SAR-based method, which is a combination of random scaling (S), aspect ratio setting (A) and rotation (R). Our method does not bring any extra GPU memory cost during training.

Inspired by [20], we further propose an online hard pixels mining. Shrivastava et al. [20] proposed to select hard region-of-interests (RoIs) by sorting the input RoIs by loss and taking a small number of RoIs for updating the model. For semantic segmentation, we regard the RoIs as pixels. During training, we select hard pixels by sorting the pixels by loss and taking a subset of pixels for updating the model. In our experiments, we observe this method can substantially improve the segmentation performance.

The main contributions of our approach are highlighted as follows:

1. We propose a deeper network named SegResNet, which combines three effective techniques: dilated convolution, LSTM network and multi-scale prediction. It shows stronger learning ability compared with VGG16.
2. We further propose a simple yet effective data augmentation method, which does not bring any extra GPU memory cost during training.
3. We also propose an online hard pixels mining, which further improves the segmentation performance.

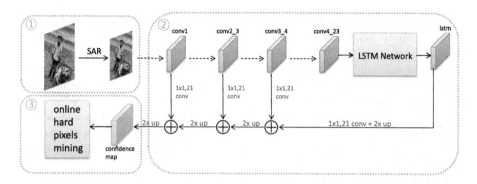

Fig. 1. Overview of our method. It consists of three parts: a SAR-based data augmentation method, a deeper residual network and an online hard pixels mining.

4. We evaluate our method on PASCAL VOC 2012 dataset and achieve a new state-of-the-art segmentation accuracy of 79.3 %[1].

The remainder of this paper is organized as follows. We introduce our method in Sect. 2. Section 3 shows experimental results and analysis. A summary of our work is described in Sect. 4.

2   Our Method

Figure 1 shows our method, which consists of three parts: a SAR-based data augmentation method, a deeper residual network and an online hard pixels mining. We combine these three parts to train an end-to-end network. In the following, these three parts are elaborated separately.

2.1   Network Architecture

We use a modified deep residual 101-layer network [10] as our baseline network architecture and further consider three effective techniques: dilated convolution, LSTM network and multi-scale prediction. In our experiments, we find these three techniques can greatly improve the learning ability of the residual network.

Baseline Network. Table 1 shows our baseline network architecture, *SegRes-Net*. Following the experiments of [10], we do not use all of the 101 layers. We remove all layers of conv5_x and up to increase the output resolution, since some of the information is lost after sub-sampling, *e.g.* ResNet-101 reduces the input size by 32 times, from 512×512 to 16×16. As a result, the smallest size of feature map in SegResNet is 32×32, keeping much more information compared with ResNet-101. The number of parameters of SegResNet is only 27.5M.

---

[1] http://host.robots.ox.ac.uk:8080/anonymous/GHOLEA.html.

Table 1. The baseline network architecture. Input image size: $512 \times 512 \times 3$.

layer name	conv1	pool1	conv2_x	conv3_x	conv4_x		
91-layer	7 × 7, 64, stride 2	3 × 3, max pool, stride 2	$\begin{bmatrix} 1 \times 1, 64 \\ 3 \times 3, 64 \\ 1 \times 1, 256 \end{bmatrix} \times 3$	$\begin{bmatrix} 1 \times 1, 128 \\ 3 \times 3, 128 \\ 1 \times 1, 512 \end{bmatrix} \times 4$	$\begin{bmatrix} 1 \times 1, 256 \\ 3 \times 3, 256 \\ 1 \times 1, 1024 \end{bmatrix} \times (23 - \alpha - \beta)$	$\begin{bmatrix} 1 \times 1, 256 \\ 3 \times 3, 256, d2 \\ 1 \times 1, 1024 \end{bmatrix} \times \alpha$	$\begin{bmatrix} 1 \times 1, 256 \\ 3 \times 3, 256, d12 \\ 1 \times 1, 1024 \end{bmatrix} \times \beta$
output size	256	128		64	32		
parameters	9.4K	213K		1.2M	26M		

Dilated Convolution. The main reason of using dilated convolution in VGG16 is to enlarge the receptive field. However, SegResNet uses shortcut connections. Since the information can spread through shortcut, the definite receptive field of SegResNet is difficult to calculate. If the information does not spread through the shortcut, the biggest receptive field could be 835, which seems large enough to make the dilated convolution unnecessary. However, if all of the information spreads through shortcut, the smallest receptive field is only 23. Therefore, it is still essential to explore the influence of dilated convolution in SegResNet.

In Table 1, two situations are considered: dilation is 2 and dilation is 12, following the experiments of [2]. Because parameter dilation does not work on 1×1 convolutional layer, we use it on 3×3 convolutional layer. By changing the value of parameter α and β, we can obtain different receptive fields. The details are discussed in Subsect. 3.3.

LSTM Network. Inspired by [1], we design an LSTM network. As shown in Fig. 2. We adopt LSTM units proposed by [11], in which an LSTM unit consists of a memory cell c, an input gate i, a forget gate f, an output gate o and a hidden state h. Bell *et al.* [1] uses four directional (left, right, top, bottom) RNNs which are considered independent. Here we consider that these four directional RNNs are interdependent, the output of the last RNN's hidden state h and memory cell c are used as the input of the next RNN. Meanwhile the weights of four directional RNNs are shared. The underlying rationale is that four directional RNNs can share memory to better capture long-term dependencies, meanwhile as weight sharing can further reduce GPU memory cost.

After the first four-directional RNN, we concatenate the output of four RNNs. To prevent overfitting, we add a dropout layer. We add a 1×1 convolution layer further to make channels return to the original size of 1024. Then we add a second four-directional RNN. The input and output direction of hidden state h and memory cell c are opposite to the first one. The purpose is to make the second one able to store different memory.

Multi-scale Prediction. Since the smallest size of the feature map is 32×32, up-sampling directly to the input resolution size will result in too much lossy information, such as region boundaries. Following [17], we employ the features from intermediate layers. As shown in Fig. 1, we use the features from conv3_4, conv2_3 and conv1 layer. As discussed in Subsect. 3.4, introducing these features

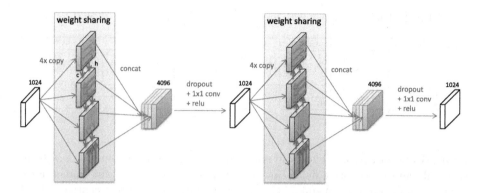

Fig. 2. The architecture of LSTM network. We use two four-directional RNNs, they share weights and are considered interdependent.

from fine-resolution layers can significantly improve the boundary localization performance.

2.2 SAR-Based Data Augmentation Method

Our SAR-based data augmentation method consists of random scaling, aspect ratio setting and rotation. In our experiments, we find that these three techniques can improve the learning ability of network.

For random scaling, the new width and height are:

$$\forall s : \quad w_s = w * s, \; h_s = h * s. \tag{1}$$

For random aspect ratio setting, the new width and height are:

$$\forall a : \quad w_a = w/a, \; h_a = h * a. \tag{2}$$

For rotation, although overlarge rotated degrees will destroy the spatial relationship between objects, *e.g.* people sit on the chair generally, the chair will appear upon people when rotating 180°, we find that small rotation can still improve the learning ability of network efficiently. The new width and height can be defined as:

$$\forall \theta : \quad w_\theta = h * \sin\theta + w * \cos\theta, \; h_\theta = w * \sin\theta + h * \cos\theta. \tag{3}$$

During the training, at each iteration we choose only one technique in the above randomly, and then choose a random value (s or a or θ) within a preset range to transform the image. Compared with multi-scale images, which are resized from the input image to several scales at each iteration, our method does not bring any extra GPU memory cost during training. In other words, multi-scale images are limited by GPU memory and only a limited number of scales can be considered in practice. Our method can use any number of scales without any extra memory cost.

2.3   Online Hard Pixels Mining

It is a common phenomenon that some pixels in image can be distinguished easily while others are hard to distinguish. Training those easy pixels can cause the network overfitting them and ignoring the hard pixels, which severely dampens the learning ability of network. Inspired by [20], we select hard pixels by sorting the pixels by confidence scores and only taking a subset of pixels with the lowest scores for updating the model during training. We call this method as online hard pixels mining.

Let x denote the image values, y the ground truth segmentation map and p the confidence map. In particular, $y_m \in \{0, \dots, L\}$ is the pixel label at position $m \in \{1, \dots, M\}$, p_{my_m} is the predicted score at position m belonging to label y_m, assuming that we have the background as well as L foreground labels and M is the number of pixels. First, we sort all pixels in the increasing order of p_{my_m} and obtain newly ordered pixel position $\phi(m) \in \{1, \dots, M\}$. Then, the loss function is defined as:

$$l = -\frac{1}{\lambda M} \sum_{m}^{\lambda M} \log p_{\phi(m)y_{\phi(m)}}. \tag{4}$$

where λ controls how many pixels are selected to train in an image. As discussed in Subsect. 3.6, we choose λ through cross validation.

3   Experiments

3.1   Dataset

We evaluate our proposed method on the PASCAL VOC 2012 dataset [5], which consists of 20 object categories and one background category. This dataset contains 1464 images for training, 1449 images for validation and 1456 images for testing. Following the common practice, we also use the extra annotations provided by Hariharan et al. [8], resulting in 10,582 training images. We have also experimented with the Microsoft COCO 2014 dataset [15]. Following [24], we select images from the training and validation set with at least one object from the VOC12 dataset, and the ground truth segmentation has at least 200 pixels. Annotated objects from other categories are treated as background. With this selection, we use extra 98,451 images from the COCO dataset.

3.2   Implementation Details

The parameters from layer conv1 to conv4_23 are initialized from ResNet-101 pre-trained model[2] and the parameters of LSTM network are initialized by sampling from a uniform distribution over $[-0.1, 0.1]$. Parameters in other convolutional layers are initialized using the technique described in [9]. Following [10],

---

[2] We use the released ResNet-101 model, which is public available at https://github.com/KaimingHe/deep-residual-networks.

we fix the BN layers while fine-tuning for reducing memory consumption and train the models with a min-batch size of 8 images. The learning rate starts from 0.001 and is divided by 10 when the error plateaus. The SegResNet is trained using stochastic gradient descent (SGD) with a momentum of 0.9 and a weight decay of 0.0001. When using COCO dataset, we first train the models with a combination of the training set from VOC12 and the train-val set from MS-COCO, evaluating the result on VOC12 validation set. We then fine-tune on VOC12 training and validation set to evaluate the result on VOC12 test set. Our implementation is based on the Caffe tool [12].

3.3   Dilated Convolution

In Table 2, we compare different values of parameter α and β to allow the network to obtain different receptive fields. The baseline SegResNet achieves a mean IoU 64.2 % ($\alpha = 0$, $\beta = 0$). Following [2], we first consider the situation $\beta = 1$. We find that when $\alpha = 5$, the mean IoU is improved largely by 6.7 % (70.9 % vs. 64.2 %). However, when we continue to increase the value of α, the network with a larger receptive field does not continue to improve the mean IoU. This may be due to the fact that the receptive field of the network at this point is larger than the input image size. Furthermore, we find that it would affect the result instead when we set $\beta = 2$.

Interestingly, we observe that enlarging the receptive field of network can speed up convergence. As shown in Fig. 3, this phenomenon can be seen at the iteration within the range of 5k and 20k. We train the baseline SegResNet for three times repeatedly and the similar phenomena are observed.

For simplicity, when $\alpha = 5$, $\beta = 1$, we call the network as *Seg-ResNet_LargeFOV*. In all the following experiments, *SegResNet_LargeFOV* is regraded as the baseline.

Table 2. Results of using dilated convolution on VOC12 val set.

Method	α	β	mIoU(%)
SegResNet	0	0	64.2
	3	1	70.4
	5	1	70.9
	7	1	70.9
	9	1	70.8
	4	2	70.3
	5	2	70.6

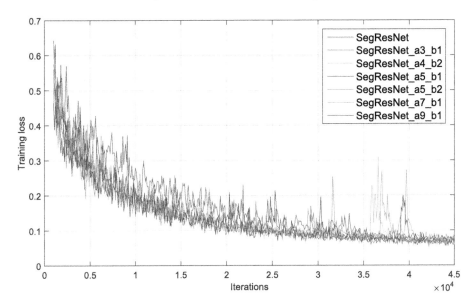

Fig. 3. Training curves. "a3" means "$\alpha = 3$", "b1" means "$\beta = 1$".

3.4   LSTM and Multi-scale Prediction

In Table 3, we observe that including LSTM network can slightly improve the mean IoU, from 70.9 % to 71.8 %. We further exploit features from intermediate layers. When using features from conv3_4 layer, the performance can be further improved by 1.1 %. If we further use features from conv3_4, conv2_3 and conv1 layer, the mean IoU is further improved by 1.6 %.

Table 3. Results of using LSTM and Multi-scale prediction on VOC12 val set.

Method	lstm	conv3_4	conv2_3	conv1	mIoU(%)
SegResNet_LargeFOV	√				71.8
	√	√			72.0
	√	√	√		72.4
	√	√	√	√	72.5

3.5   SAR-Based Data Augmentation Method

In Table 4, we evaluate the impact of of choosing different random scales, aspect ratios and rotated degrees. At first, we consider the effect of single technique on the result. We find that choosing random scales can improve the mean IoU mostly, followed by aspect ratios. Choosing smaller rotated degrees can also

Table 4. Results of using SAR-based method on VOC12 val set.

Method	Scales (s)	Aspect ratios (a)	Rotated degrees (θ)	mIoU (%)
SegResNet_LargeFOV	1	1	0	70.9
	[0.8, 1.0, 1.2]	1	0	71.9
	[0.6, 0.8, 1.0, 1.2, 1.4]	1	0	72.9
	1	[0.9, 1.0, 1.1]	0	71.5
	1	[0.7, 1.0, 1.3]	0	72.3
	1	1	[−10, 0, 10]	72.1
	1	1	[−15, 0, 15]	72.0
	1	1	[−20, 0, 20]	70.8
	[0.6, 0.8, 1.0, 1.2, 1.4]	[0.7, 1.0, 1.3]	0	73.3
	[0.6, 0.8, 1.0, 1.2, 1.4]	1	[−10, 0, 10]	73.5
	[0.6, 0.8, 1.0, 1.2, 1.4]	[0.7, 1.0, 1.3]	[−10, 0, 10]	73.8

improve the mean IoU. But setting larger degrees, such as 20°, would begin to adversely affect the result. Then we consider the results of combinations and find that combining random scales with aspect ratios or rotation can improve the mean IoU. When combining three techniques, the mean IoU is further improved by 2.9 % (73.8 % vs. 70.9 %).

In our experiments, we also assess the impact of random noise. We find that adding random noise does not improve the mean IoU. This may be because the VOC12 validation set does not contain other noise apparently. On the contrary, VOC12 validation set may contain some other scales or degrees, hence the above three techniques improve the results significantly.

3.6   Online Hard Pixels Mining

In Table 5, we show the results of choosing different values of parameter λ in formula (4). We find that selecting hard pixels can improve the mean IoU significantly. In particular, the best result is to select half of the top hardest pixels, which improves the mean IoU from 70.9 % to 72.1 %. However, by continuing to reduce the number of hard pixels not only degrades results, but also slowdowns the convergence.

Table 5. Results of using online hard pixels mining on VOC12 val set.

Method	λ	mIoU(%)
SegResNet_LargeFOV	0.3	71.5
	0.4	71.9
	0.5	72.1
	0.6	71.5
	0.7	71.3

3.7   Results of PASCAL VOC 2012 Dataset

In Table 6, we evaluate our method on PASCAL VOC 2012 test set. We combine the above techniques to train an end-to-end network and apply a common post-processing method [13] to refine the prediction. When only using PASCAL VOC12 data to train our network, our method improves the previous state-of-the-art method Context [14] by 1.6 % (76.9 % vs. 75.3 %). When extra data from MS COCO are used, our method yields a new state-of-the-art segmentation accuracy of 79.3 % and achieves the highest IoUs for 11 out of the 20 objects. Some segmentation results of our method are presented in Fig. 4.

Table 6. Results on PASCAL VOC 2012 test set.

	Only using PASCAL VOC12 data						Using VOC12 + COCO data				
	[2]	[24]	[19]	[16]	[14]	Ours	[24]	[4]	[16]	[14]	Ours
Aeroplane	84.4	87.5	89.9	87.7	**90.6**	90.3	90.4	89.8	89.0	**94.1**	93.8
Bicycle	**54.5**	39.0	39.3	59.4	37.6	41.2	55.3	38.0	**61.6**	40.4	42.2
Bird	81.5	79.7	79.7	78.4	80.0	**88.2**	88.7	89.2	87.7	83.6	**93.1**
Boat	63.6	64.2	63.9	64.9	**67.8**	65.5	68.4	**68.9**	66.8	67.3	68.6
Bottle	65.9	68.3	68.2	70.3	74.4	**77.4**	69.8	68.0	74.7	**75.6**	75.3
Bus	85.1	87.6	87.4	89.3	92.0	**93.3**	88.3	89.6	91.2	93.4	**95.3**
Car	79.1	80.8	81.2	83.5	85.2	**86.8**	82.4	83.0	84.3	84.4	**88.8**
Cat	83.4	84.4	86.1	86.1	86.2	**92.4**	85.1	87.7	87.6	88.7	**92.5**
Chair	30.7	30.4	28.5	31.7	**39.1**	34.3	32.6	34.4	36.5	**41.6**	36.5
Cow	74.1	78.2	77.0	79.9	**81.2**	77.7	78.5	83.6	86.3	**86.4**	84.3
Table	59.8	60.4	62.0	**62.6**	58.9	62.5	64.4	**67.1**	66.1	63.3	64.2
Dog	79.0	80.5	79.0	81.9	83.8	**85.0**	79.6	81.5	84.4	85.5	**86.8**
Horse	76.1	77.8	80.3	80.0	**83.9**	82.8	81.9	83.7	87.8	**89.3**	87.8
Motorbike	83.2	83.1	83.6	83.5	84.3	**85.6**	86.4	85.2	85.6	85.6	**87.5**
Person	80.8	80.6	80.2	82.3	84.8	**87.6**	81.8	83.5	85.4	86.0	**88.5**
Plant	59.7	59.5	58.8	60.5	62.1	**69.0**	58.6	58.6	63.6	67.4	**69.2**
Sheep	82.2	82.8	83.4	83.2	83.2	**86.9**	82.4	84.9	87.3	**90.1**	89.7
Soft	50.4	47.8	54.3	53.4	**58.2**	57.7	53.5	55.8	61.3	62.6	**64.1**
Train	73.1	78.3	80.7	77.9	80.8	**83.2**	77.4	81.2	79.4	80.9	**86.8**
Tv	63.7	67.1	65.0	65.0	72.3	**73.1**	70.1	70.7	66.4	72.5	**74.6**
mIoU(%)	71.6	72.0	72.5	74.1	75.3	**76.9**[a]	74.7	75.2	77.5	77.8	**79.3**[b]

[a] http://host.robots.ox.ac.uk:8080/anonymous/1CJHQ2.html
[b] http://host.robots.ox.ac.uk:8080/anonymous/GHOLEA.html

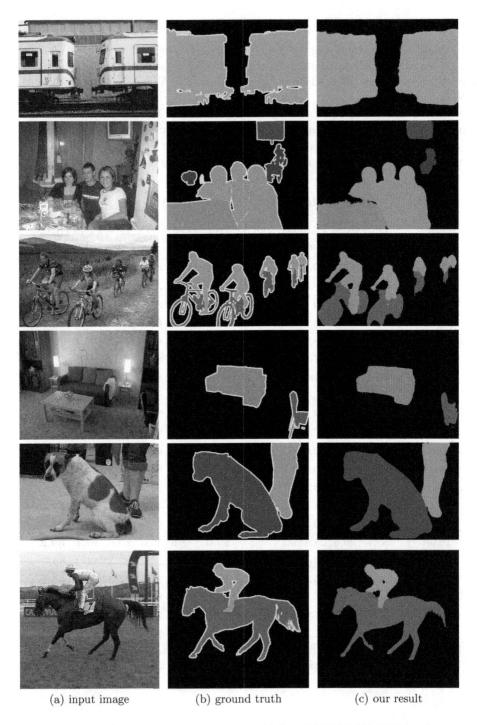

(a) input image   (b) ground truth   (c) our result

Fig. 4. Some segmentation results of our method on PASCAL VOC 2012 val set.

4   Conclusions

In this paper, we propose a novel deeper residual network by incorporating three techniques: dilated convolution, LSTM network and multi-scale prediction. Extensive experiments demonstrate that these three techniques are crucial for the residual network to improve its learning ability. We further propose a simple yet effective SAR-based data augmentation method. The method can improve the segmentation performance significantly without extra GPU memory cost during training. More importantly, we demonstrate that network is easy to overfit easy pixels and ignore hard pixels, which dampens the learning ability of network. To overcome this shortcoming, an online hard pixel mining strategy is proposed. Finally, we show that by combining the above techniques, we train an end-to-end network and achieve a new state-of-the-art segmentation accuracy on PASCAL VOC 2012 test set.

Acknowledgements. This work is supported by the National Natural Science Foundation of China (NSFC) (Grant No.: 61305048, Grant No.: 61503381).

References

1. Bell, S., Zitnick, C.L., Bala, K., Girshick, R.B.: Inside-outside net: detecting objects in context with skip pooling and recurrent neural networks. arXiv preprint arXiv:1512.04143 (2015)
2. Chen, L.C., Papandreou, G., Kokkinos, I., Murphy, K., Yuille, A.L.: Semantic image segmentation with deep convolutional nets and fully connected CRFs. In: ICLR (2015)
3. Chen, L.C., Yang, Y., Wang, J., Xu, W., Yuille, A.L.: Attention to scale: scale-aware semantic image segmentation. In: CVPR (2016)
4. Dai, J., He, K., Sun, J.: BoxSup: exploiting bounding boxes to supervise convolutional networks for semantic segmentation. In: ICCV, pp. 1635–1643 (2015)
5. Everingham, M., Eslami, S.M.A., Gool, L.J.V., Williams, C.K.I., Winn, J.M., Zisserman, A.: The PASCAL visual object classes challenge: a retrospective. Int. J. Comput. Vis. **111**(1), 98–136 (2015)
6. Graves, A.: Generating sequences with recurrent neural networks. arXiv preprint arXiv:1308.0850 (2013)
7. Graves, A., Jaitly, N.: Towards end-to-end speech recognition with recurrent neural networks. In: ICML, pp. 1764–1772 (2014)
8. Hariharan, B., Arbelaez, P., Bourdev, L.D., Maji, S., Malik, J.: Semantic contours from inverse detectors. In: ICCV, pp. 991–998 (2011)
9. He, K., Zhang, X., Ren, S., Sun, J.: Delving deep into rectifiers: surpassing human-level performance on imagenet classification. In: ICCV, pp. 1026–1034 (2015)
10. He, K., Zhang, X., Ren, S., Sun, J.: Deep residual learning for image recognition. In: CVPR (2016)
11. Hochreiter, S., Schmidhuber, J.: Long short-term memory. Neural Comput. **9**(8), 1735–1780 (1997)
12. Jia, Y., Shelhamer, E., Donahue, J., Karayev, S., Long, J., Girshick, R.B., Guadarrama, S., Darrell, T.: Caffe: convolutional architecture for fast feature embedding. In: Proceedings of the ACM International Conference on Multimedia, pp. 675–678 (2014)

13. Krähenbühl, P., Koltun, V.: Efficient inference in fully connected CRFs with Gaussian edge potentials. In: NIPS, pp. 109–117 (2011)
14. Lin, G., Shen, C., van den Hengel, A., Reid, I.D.: Exploring context with deep structured models for semantic segmentation. In: CVPR (2016)
15. Lin, T.-Y., Maire, M., Belongie, S., Hays, J., Perona, P., Ramanan, D., Dollár, P., Zitnick, C.L.: Microsoft COCO: common objects in context. In: Fleet, D., Pajdla, T., Schiele, B., Tuytelaars, T. (eds.) ECCV 2014. LNCS, vol. 8693, pp. 740–755. Springer, Heidelberg (2014). doi:10.1007/978-3-319-10602-1_48
16. Liu, Z., Li, X., Luo, P., Loy, C.C., Tang, X.: Semantic image segmentation via deep parsing network. In: ICCV, pp. 1377–1385 (2015)
17. Long, J., Shelhamer, E., Darrell, T.: Fully convolutional networks for semantic segmentation. In: CVPR, pp. 3431–3440 (2015)
18. Mallat, S.: A Wavelet Tour of Signal Processing, 2nd edn. Academic Press, San Diego (1999)
19. Noh, H., Hong, S., Han, B.: Learning deconvolution network for semantic segmentation. In: ICCV, pp. 1520–1528 (2015)
20. Shrivastava, A., Gupta, A., Girshick, R.B.: Training region-based object detectors with online hard example mining. In: CVPR (2016)
21. Simonyan, K., Zisserman, A.: Very deep convolutional networks for large-scale image recognition. In: ICLR (2015)
22. Badrinarayanan, V., Alex Kendall, R.C.: SegNet: a deep convolutional encoder-decoder architecture for image segmentation. arXiv preprint arXiv:1511.00561 (2015)
23. Yu, F., Koltun, V.: Multi-scale context aggregation by dilated convolutions. In: ICLR (2016)
24. Zheng, S., Jayasumana, S., Romera-Paredes, B., Vineet, V., Su, Z., Du, D., Huang, C., Torr, P.H.S.: Conditional random fields as recurrent neural networks. In: ICCV, pp. 1529–1537 (2015)

A Quantum-Inspired Fuzzy Clustering for Solid Oxide Fuel Cell Anode Optical Microscope Images Segmentation

Yuhan Xiang[1,2], Xiaowei Fu[1,2(✉)], Li Chen[1,2], Xin Xu[1,2], and Xi Li[3]

[1] College of Computer Science and Technology,
Wuhan University of Science and Technology, Wuhan 430065, China
fxw_wh0409@wust.edu.cn
[2] Hubei Province Key Laboratory of Intelligent Information Processing
and Real-Time Industrial System, Wuhan 430065, China
[3] College of Automation, Huazhong University of Science and Technology,
Wuhan 430074, China

Abstract. For better three-phase identification of Solid Oxide Fuel Cell (SOFC) microstructure, this paper presents a novel quantum-inspired clustering method for YSZ/Ni anode Optical Microscopic (OM) images. Motivated by Quantum Signal Processing (QSP), a quantum-inspired adaptive fuzziness factor is introduced to adaptively estimate the parameters of the spatial constraint term in the fuzzy clustering based on Markov Random Filed (MRF). Experimental results show that the proposed method is effective to identify the three phases. The combination of image processing and micro-investigation provides an innovative way to analyze the performance of SOFC.

Keywords: Solid Oxide Fuel Cell (SOFC) · Microstructure · Fuzzy clustering · Image segmentation · Quantum Signal Processing (QSP)

1 Introduction

Global stability, economic prosperity and national security are closely linked to the sufficient supply of clean energy. Solid Oxide Fuel Cell (SOFC) is a kind of electrochemical device with high efficiency and low pollution [1]. Porous Ni-YSZ cermet is the indispensable material for anode supported planar cells with thin film electrolyte, which provides proper electrochemically active sites and gives appropriate fuel-supplying path for electrode reaction. Three phases including metallic Ni, YSZ ceramic and pores are the significant characteristics of porous Ni-YSZ electrodes [2, 3]. The composition, ratio and spatial distributions of three phases, mean grain sizes as well as the amounts of Triple Phase Boundary points (TPBs) are directly related to the electrochemical property such as conductivity and gas permeability. Accurate quantification of these microstructure parameters is a key discriminator for optimizing the design of high performance and robust fuel cells [4].

As SOFC microstructure investigations mainly focused on improvement of materials and porous structure, most of them just utilize artificial observation, simple image

© Springer Nature Singapore Pte Ltd. 2016
T. Tan et al. (Eds.): CCPR 2016, Part II, CCIS 663, pp. 55–64, 2016.
DOI: 10.1007/978-981-10-3005-5_5

processing methods or mature image analysis software to separate three-phase from complicated structure of YSZ/Ni anode, resulting in obtaining rough three-phase distribution [5, 6]. By means of microscopic analysis and the cross section images of cell, applying the specialized image segmentation method to SOFC micro-investigation provide a brilliant way for non-destructively quantifying the microstructure parameters and characterizing the correlation between the Ni-YSZ anode microstructure and its physicochemical properties [7].

Because of complicated structural features of three phases in anode substrate, clustering algorithms are feasible to differentiate all of the phases, which can reflect different attribute of images. However, conventional clustering methods are extremely sensitive to noise and illumination [8]. A trade-off weighted fuzzy factor without any parameters setting was exploited in fuzzy C-means algorithm to improve the robustness of the algorithm to noise [9]. Nevertheless, it is time-consuming in the process of iteration. Markov Random Field (MRF) theory has been widely applied for image segmentation to take into account the mutual influences of neighboring pixels. Nguyen [10] took into account the mutual influences of neighboring pixels by combing a new fuzzy logic model with MRF function for dealing with MRI segmentation. And then he imposed the spatial relationships in expectation-maximization (EM) to optimize the parameters of Gaussian mixture model based on MRF [11]. Although the segmentation results of Nguyen's methods are quite good, both of them need trial-and-error method to select best appropriate parameters. Quantum Signal Processing (QSP) frame is a novel derivative mechanism of quantum mechanics in information processing. Li [12] transformed the segmentation problem based on partition clustering into a combinatorial optimization problem for SAR image segmentation.

Hence, a novel fuzzy clustering is proposed in this paper to solve the problem of three-phase differentiation of SOFC microstructure. Based on the superposition principle of quantum bits [13, 14], a quantum-inspired adaptive weight is defined to replace the manually parameters in the punishment term of spatial constraint information based on MRF, which can atomically correct the inaccurate membership degree caused by noises and edge points. The proposed method is a non-destructive inspection to accurately divide three phases of SOFC electrode.

2   Proposed Approach

In this paper, we introduce a quantum-inspired adaptive factor in the punishment term of MRF to adaptively constrain spatial information into fuzzy clustering.

Suppose that an image I with N pixels has to be segmented into K classes. Fuzzy clustering method provides a fuzzy membership set to model the uncertainty. The probability of pixel x_i in the kth cluster is strongly related to that of its neighbor pixels. Quantum systems can exist in a superposition of many different states at once, which means the quantum mechanical property of a particle can occupy all of its possible quantum states simultaneously [14]. Due to the superposition of quantum state, a qubit for the probabilistic representation that can represent a linear superposition of states.

The assignment state of pixel at spatial index (a, b). can be modeled as two quantum states: $|0\rangle$ and $|1\rangle$ in QSP.

$$|P(a,b)\rangle = p_0|0\rangle + p_1|1\rangle. \tag{1}$$

where $|0\rangle$ represents the pixel belongs to the current cluster and $|1\rangle$ represents that the pixel doesn't belong to the current cluster. p_0 and p_1 are probability amplitudes of $|0\rangle$ and $|1\rangle$. $|p_0|^2$, $|p_1|^2$ are the measurement probability of two state $|0\rangle$ and $|1\rangle$, which satisfies the constraints $|p_0|^2 + |p_1|^2 = 1$. Quantum rotation gate $\begin{pmatrix} \cos\theta & -\sin\theta \\ \sin\theta & \cos\theta \end{pmatrix}$ is used to express a pixel belonging to the cluster or not. The probability that pixel belongs to the cluster is defined as follows:

$$|p_0|^2 = \cos\left(\underset{x_i \in \Omega}{med}(z_{ik})\right)^2. \tag{2}$$

where Ω is the 5×5 median filter of quantum-inspired factor. z_{ik} is the fuzzy membership of pixel x_i in the kth cluster. Larger $|p_0|^2$ means the more likely pixel at (a,b) belongs to the current partition k. The quantum-inspired adaptive fuzziness factor controls the robustness to noise and quality of image segmentation results.

After applying the quantum-inspired adaptive fuzziness factor into fuzzy clustering based on MRF, λ_{ik} is the quantum-inspired adaptive evolution formula of MRF function.

$$\lambda_{ik} = \exp\left(\cos\left(\underset{x_i \in \Omega}{med}(z_{ik})\right)^2 \left(\frac{1}{n}\sum_{m \in N_i} z_{mk}\right)\right)^2. \tag{3}$$

where N_i is the 5×5 local window of the center pixel x_i. n is the number of neighbors of falling into the window around pixel x_i.

Y is the objective function to assign labels to each pixel:

$$Y = \sum_{i=1}^{N}\sum_{k=1}^{K} z_{ik}^{\alpha} \lambda_{ik} \|x_i + \hat{x} - c_k\|^2. \tag{4}$$

where z_{ik} is the fuzzy membership of pixel x_i in the kth cluster. z_{ik} must satisfy the constraints $0 < z_{ik} < 1$ and $\sum_{i=1}^{K} z_{ik} = 1$. The parameter α controls the weight of fuzziness for the partition which is usually set to 2. $\hat{x}$ is image pixel after adopting median filter with a 3×3 window. c_k is the kth cluster center.

The objective function Y can get minimum by means of taking the derivative of Y with respect to the parameter z_{ik} and c_k. The membership z_{ik} and cluster center c_k are updated as follows:

$$z_{ik} = \frac{1}{\sum_{j=1}^{K}\left(\frac{\lambda_{ik}\|x_i + \hat{x} - c_k\|^2}{\lambda_{ij}\|x_i + \hat{x} - c_j\|^2}\right)^{\frac{1}{\alpha-1}}}. \tag{5}$$

$$c_k = \frac{\sum\limits_{i=1}^{N} z_{ik}^{\alpha} \lambda_{ik} (x_i + \hat{x})}{\sum\limits_{i=1}^{N} z_{ik}^{\alpha} \lambda_{ik}}. \tag{6}$$

When algorithm has converged, maximum membership procedure method [15, 16] is used to convert the fuzzy partition of each pixel to exact partition. Every pixel x_i belongs to the class with the largest value z_{ik}.

$$C_k = \arg \max(z_{ik}), \qquad (k = 1, 2, \ldots, K). \tag{7}$$

The procedure of the proposed method can be summarized as follows:

(1) Set the number of cluster K. Use k-means to initialize z_{ik} and c_k.
(2) Evaluate quantum-inspired adaptive evolution formula λ_{ik} using (3).
(3) Update the new parameters z_{ik} and c_k using (5) and (6).
(4) Maximum membership defuzzification method is used to get the crisp segmented image using (7).

3 Experimental Results

In this section, as the three-phase features of anode OM images, three-class synthetic images and OM images of Ni-YSZ anode are used to verify the feasibility of the proposed method. The proposed approach is compared with algorithms including FFCM [10] where parameter β is manually set a value of 0.5, FRGMM [11] where parameter β is manually set a value of 12, KWFLICM [9]. All the experiments are carried on under the MATLAB programming environment on a 2.60 GHz AMD Athlon64 X2 5000 + N680 PC.

3.1 Experiments on Synthetic Images

The segmentation performance is evaluated using three criterions: Misclassification ratio (MCR) [17], Signal to Noise Ratio (PSNR) [13], EPM [14]. MCR reflects the incorrectly classified pixels between result of segmentation and reference segmented image. The lower the value of MCR is, the better the quality of the segmentation is. PSNR reflects the intensity difference between original picture and segmented picture. The higher the value of PSNR is, the better the quality of the segmentation is. EPM reflects the edge difference between original picture and segmented picture. The value of EPM is more close to 1, the ability of keeping edge details is better.

Figure 1 compares the visual quality of the experiments on a synthetic image (128 × 128 image resolution) with luminance values [0, 0.5, 1] contaminated by Gaussian noise (0 Mean, 0.08 Var). The segmentation of FFCM and KWFLICM exist a lot of isolated points. FRGMM improperly identifies the black region into gray region on the right side of image. Figures 2, 3 and 4 present the objective performance evaluation with different levels of Gaussian noise (0 mean, Var). Although the EPM of

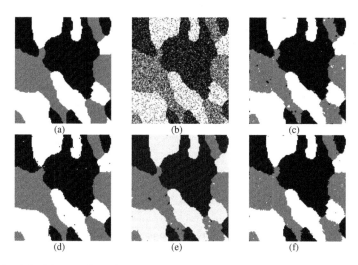

Fig. 1. (a) original image; (b) noisy image (0 Mean, 0.09 Var Gaussian noise); (c) ~ (f) results by FFCM [10], FRGMM [11], KWFLICM [9], the proposed method, respectively.

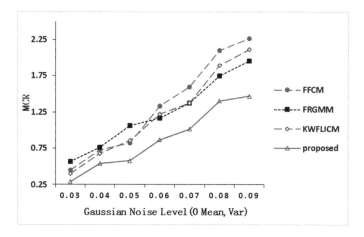

Fig. 2. Results of MCR performance between the proposed method and compared methods

the proposed method is a little smaller than that of FRGMM method in lower noise level, MCR and PSNR performance of ours is stable and much better than the others with the increase of noise density, which means the least misclassification rate, best noise elimination capability and best edge preservation.

As seen in these figures and Table 1, the proposed method not only achieves better visual quality, but also obtains better objective performance evaluation in a relatively fast speed.

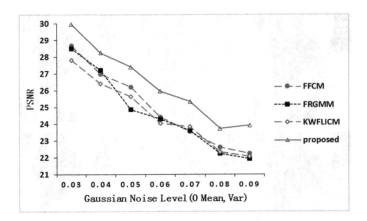

Fig. 3. Results of PSNR performance between the proposed method and compared methods

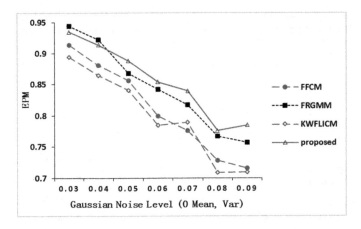

Fig. 4. Results of EPM performance between the proposed method and compared methods

Table 1. Run-time comparison for synthetic images

	FFCM	FRGMM	KWFLICM	Proposed
Average run-time (s)	1.19	1.35	43.1	3.1

3.2 Results on SOFC Porous Electrodes Images

Anode supported planar cells with thin film electrolyte are fabricated through screen-printing technology [18]. Samples are prepared by the traditional preparation of metallographic specimen for cross-sectional views. And then the cross-section micrographs of SOFC anode are obtained by Optical Microscopy (LW200-3JT).

There are three parts shown on OM images. As shown in Fig. 5(a), the relatively bright ones are Ni, intermediate brightness ones are YSZ and the relatively dark ones

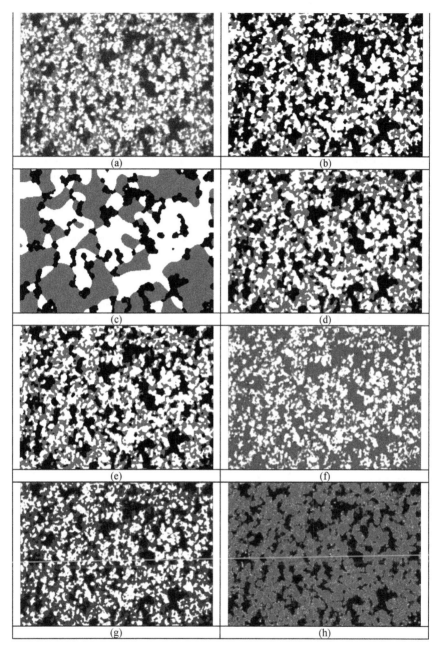

Fig. 5. Results of compare methods (a) anode image in 400 magnification by scanning electron microscope; (b)~(e) results by FFCM [10], FRGMM [11], KWFLICM [9], the proposed method, respectively. (f)~(h) results by the overlapping of original image and the segmented pore, YSZ and Ni phase of the proposed method, respectively.

are pores. Due to very complicated structural features with three different phases, these OM images have no ground truth [6]. The segmentation performance is evaluated using entropy-based evaluation function E [9, 19]. E measures not only the uniformity in the regions of the segmentation, but also the simplicity of the segmentation itself. The smaller the values of E, the better the quality of the segmentation.

In order to verify the robustness and effectiveness of the proposed method, we sample different parts of anode to obtain 25 different size of 400 magnifications OM images. Figure 5 compares the visual quality of the experiments on a 400 magnifications OM image (336×248 image resolution). Obviously, FGRMM could not identify the three-phase successfully because of ignoring too much details. There is low contrast between Ni phase and YSZ phase, especially in corners of the image, resulting that the FFCM and KWFLICM have misclassification of two phases' edge.

The segmentation results of proposed method are overlapped with original image to precisely compare the details of segmentation, as shown in Fig. 5(f)-(h). The proposed method can better classify three phases, keep consistency of three phases and also remain more details.

Figure 6 lists the E performance of these 25 images, the smallest value is obtained by the proposed method. Although there is litter difference between the results of KWFLICM and the proposed method, KWFLICM takes the most amount of time for segmentation in Table 2. As seen in these figures and tables, the proposed method is free of any parameters determination and has the best image quality in rather short computational time.

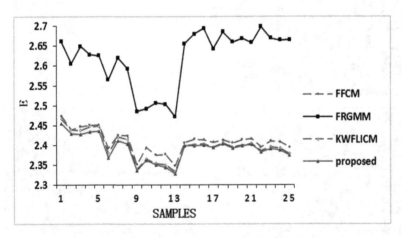

Fig. 6. Results of E performance on $400 \times$ OM images

Table 2. Run-time comparison for SOFC anode OM images

	FFCM	FRGMM	KWFLICM	Proposed
Average run-time (s)	21.99	6.51	2530.20	76.78

4 Conclusions

A novel quantum-inspired clustering method has been proposed to accurately differ-entiate three-phase of Ni/YSZ anode OM images. Quantum properties with probability estimation is applied to fuzzy clustering. The quantum-inspired adaptive fuzziness factor is defined to automatically correct the inaccurate probability distribution caused by noise and edge points. Experiments demonstrate that the proposed approach effectively improves the accuracy and effectiveness of three-phase detection. The segmentation results of proposed method can be used to characterize microstructure parameters for microstructure evolution during SOFC operation.

Acknowledgments. This work is supported by the National Natural Science Foundation of China (No. 61201423, 61440016, 61573162), the China Scholarship Council (201408420205), the open fund project of Hubei Province Key Laboratory of Intelligent Information Processing and Real-time Industrial System (No.2016znss02A) and Hubei Province Key Technology R&D Program (2015BCE059).

References

1. Lichtner, A.Z., Jauffrès, D., Roussel, D., Charlot, F., Martin, C.L., Bordia, R.K.: Dispersion, connectivity and tortuosity of hierarchical porosity composite SOFC cathodes prepared by freeze-casting. J. Eur. Ceram. Soc. **35**, 585–595 (2015)
2. Lee, Y.H., Muroyama, H., Matsui, T., Eguchi, K.: Degradation of nickel–yttria-stabilized zirconia anode in solid oxide fuel cells under changing temperature and humidity conditions. J. Power Sources **262**, 451–456 (2014)
3. Li, W.Y., Shi, Y.X., Luo, Y., Cai, N.S.: Theoretical modeling of air electrode operating in SOFC mode and SOEC mode: the effects of microstructure and thickness. Int. J. Hydrogen Energy **39**, 13738–13750 (2014)
4. Delette, G., Laurencin, J., Usseglio-Viretta, F., Villanova, J., Bleuet, P., Lay-Grindler, E., Le Bihan, T.: Thermo-elastic properties of SOFC/SOEC electrode materials determined from three-dimensional microstructural reconstructions. Int. J. Hydrogen Energy **38**, 12379–12391 (2013)
5. Wilson, J.R., Duong, A.T., Gameiro, M., Chen, H.Y., Thornton, K., Mumm, D.R., Barnett, S.A.: Quantitative three-dimensional microstructure of a solid oxide fuel cell cathode. Electrochem. Commun. **11**, 1052–1056 (2009)
6. Lanzini, A., Leone, P., Asinari, P.: Microstructural characterization of solid oxide fuel cell electrodes by image analysis technique. J. Power Sources **194**, 408–422 (2009)
7. Shimura, T., Jiao, Z.J., Hara, S., Shikazono, N.: Quantitative analysis of solid oxide fuel cell anode microstructure change during redox cycles. J. Power Sources **267**, 58–68 (2014)
8. Qiu, X., Qiu, Y.Y., Feng, G.C., Li, P.X.: A sparse fuzzy C-means algorithm based on sparse clustering framework. Neurocomputing **157**, 290–295 (2015)
9. Gong, M.G., Liang, Y., Shi, J., Ma, W.P., Ma, J.J.: Fuzzy c-means clustering with local information and kernel metric for image segmentation. IEEE Trans. Image Process. **22**, 573–584 (2012)
10. Nguyen, T.M., Wu, Q.M.J.: A fuzzy logic model based Markov random field for medical image segmentation. Evolving Syst. **4**, 171–181 (2013)

11. Nguyen, T.M., Wu, Q.M.J.: Fast and robust spatially constrained gaussian mixture model for image segmentation. IEEE Trans. Circuits Syst. Video Technol. **23**, 621–635 (2013)
12. Li, Y.Y., Feng, S.X., Zhang, X.G., Jiao, L.C.: SAR image segmentation based on quantum-inspired multiobjective evolutionary clustering algorithm. Inf. Process. Lett. **114**, 287–293 (2014)
13. Fu, X.W., Wang, Y., Chen, L., Tian, J.: An image despeckling approach using quantum-inspired statistics in dual-tree complex wavelet domain. Biomed. Signal Process. Control **18**, 30–35 (2015)
14. Fu, X.W., Wang, Y., Chen, L., Dai, Y.: Quantum-inspired hybrid medical ultrasound images despeckling method. Electron. Lett. **51**, 321–323 (2015)
15. Broekhoven, E.V., Baets, B.D.: Fast and accurate center of gravity defuzzification of fuzzy system outputs defined on trapezoidal fuzzy partitions. Fuzzy Sets Syst. **157**, 904–918 (2006)
16. Rouhparvar, H., Panahi, A.: A new definition for defuzzification of generalized fuzzy numbers and its application. Appl. Soft Comput. **30**, 577–584 (2015)
17. Nikou, C., Galatsanos, N.P., Likas, C.L.: A class-adaptive spatially variant mixture model for image segmentation. IEEE Trans. Image Process. **16**(4), 1121–1130 (2007)
18. Luo, J., Yan, D., Fang, D.W., Liang, F.L., Pu, J., Chi, B., Zhu, Z.H., Li, J.: Electrochemical performance and thermal cyclicability of industrial-sized anode supported planar solid oxide fuel cells. J. Power Sources **224**, 37–41 (2013)
19. Zhang, H., Fritts, J.E., Goldman, S.A.: An entropy-based objective evaluation method for image segmentation. Proc. SPIE Storage Retrieval Methods Appl. Multimedia **5307**, 38–49 (2004)

Document Image Super-Resolution Reconstruction Based on Clustering Learning and Kernel Regression

Li Li[1,2], Haibin Liao[1,3(✉)], and Youbin Chen[1,2]

[1] Guangdong Micropattern Software, Co., Ltd., Guangzhou, Guangdong, China
{li.li,haibin.liao,youbin.chen}@micropattern.cn
[2] School of Automation, Huazhong University of Science and Technology,
Wuhan, China
[3] School of Computer Science and Technology,
Hubei University of Science and Technology, Xianning, China

Abstract. There are lots of blank areas and similar or redundant characters in document image. To make use of these characteristics, we propose a weighted kernel regression super-resolution reconstruction model based on steering kernel regression and clustering learning methods in this paper. By this model, we can learn the local structure of characters and achieve document image super-resolution reconstruction. In our method, a large number of unrelated samples are used for local structure clustering, which make the reconstruction process can not only use structure information of local neighborhood, but also make use of lots of non-local neighborhood structure information learning from the cluster sub-sample sets. This proposed approach ensures robustness of reconstruction. Document image super-resolution experiments with subjective evaluation and objective indicators have proved the effectiveness of our method.

Keywords: Document image processing · Super-resolution reconstruction · Clustering learning · Weighted kernel regression

1 Introduction

Small size characters, low resolution images, and fuzziness phenomenon make it hard to recognize and index document images, especially the fax document images. An effective way to recognize low quality character images is to improve their resolution and quality by Super-Resolution (SR) Reconstruction [1,2]. Compared with the traditional interpolation method, SR has the ability of image restoration, and can make full use of the hidden information to reconstruct image details [3,4].

In terms of the prior information, there are two kind of SR reconstructions: SR reconstruction based on single image, and SR reconstruction based on multi images [5,6]. SR reconstruction based on single image is always a challenge for

© Springer Nature Singapore Pte Ltd. 2016
T. Tan et al. (Eds.): CCPR 2016, Part II, CCIS 663, pp. 65–77, 2016.
DOI: 10.1007/978-981-10-3005-5_6

little information to be used. However, SR reconstruction for document image belongs to this kind of reconstruction [7].

In terms of the object to be reconstructed, there are also two kind of reconstructions: reconstruction of normal object, and reconstruction of special object, like face image and document image [8,9]. We can use reconstruction method of normal object for special object. But it should be better to take the particularity of special object and other prior information into consideration.

Image process methods based on nonlocal similarity have been popular recently. Bilateral Filter method and Non-Local Means (NLM) [10] have been successful in image filtering. In this method, neighborhood of the pixel i is expanded to a search window of certain range and centered at i, named nonlocal neighbor window.

Zhang et al. [11] proposed the Non-Local Kernel Regression (NL-KR) method for reconstruction of normal object, by combining NLM and kernel regression. Their method make full use of not only local neighborhood structure information, but also nonlocal structure information (neighborhood structure information of similar pixel in nonlocal region).

While NL-KR has achieved good result for reconstruction of normal object, it can not reconstruct document image well: Firstly, many empty region containing no texture exists between characters in document image. These region can not provide useful nonlocal structure information, and is a waste of computation resource. Secondly, both English and Chinese characters consist of similar basic stroke. So the characters has the similarity of local structure. Taken these similarity into account, the method will have a better performance.

Methods based on sparse learning have been hot spot in research recently. Method that construct over-complete dictionary from many character images [12, 13] can avoid disadvantages above, and offer enough redundancy and samples. However, solving by sparse representation also means great computation cost. In terms of sample learning, any similar pixel or structure that have similar feature can be treated as potential learning sample for each other. By clustering the patches [14], we can provide different local structure with lots of effective sample with similar structure for learning later. These samples may come from any unrelated character image. By steering kernel regression, we can learn the similar structure information from patches of the same class. Figure 1 illustrates several class image after clustering segments from any character images. Figure 1 proves that, many sample image segmentation can be retrieved from any unrelated character images by segmentation clustering.

Based on analysis above, we propose a new super-resolution reconstruction algorithm based on clustering learning and kernel regression, for reconstruction task of low quality document images. Our method adapt advantages of kernel regression and clustering learning, to construction of text image. Our method can take advantage of not only structure information in neighborhood, but also lots of nonlocal structure information from similar segmentations in clusters. And steering kernel regression is used to realize learning and merging of structure

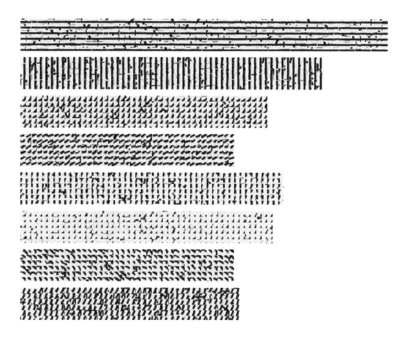

Fig. 1. Several class images after clustering

information. At the same time, our method can improve accuracy of estimation, and avoid high computation cost of complicated sparse representation.

2   Super-Resolution Reconstruction Model

Observation model for super-resolution reconstruction of single frame can be simplified as:

$$y = DHX + N \qquad (1)$$

where y is the known low resolution image, X is the corresponding high resolution image, D represents down-sampling operation, H represents blurring operation, N represents gaussian noise.

Super-resolution reconstruction aims to estimate X from y. Reconstruction for X can be divided into 2 procedures for simplified solution. Set $Z = HX$, so we focus on estimating Z from y by SR reconstruction, and can estimate X by deblurring Z later. The division simplify algorithm implementation, and keep the uncertain blur coefficient from affecting reconstruction.

$Z(x)$ represents the image patch centered at $x = [x_1, x_2]$. We mark the corresponding patch in y as $y(x)$. Neighborhood pixels of $y(x)$ can be represented by vector $\acute{y}_x = V_x y$, where V_x represents operation to convert neighborhood pixels to vector.

According to the theory of steering kernel regression [15], we can get the optimization equation as follow:

$$\hat{\beta}^x = \arg\min_{\beta^x} \|\acute{y}_x - A_x \beta^x\|_{W_x}^2 \tag{2}$$

where, $W_x = diag\left[K_H(X_1 - x), K_H(X_2 - x), \ldots, K_H(X_p - x)\right]$ is steering kernel weight matrix calculated by initial high resolution image patch $y^H(x)$ (after interpolation [16] about $y(x)$), and $K_H(x_i - x)$ is the kernel function, $(x_i - x)$ is the space distance between center point and neighborhood; A_x and β^x is the matrix and coefficient matrix constructed by N-order Taylor polynomial basis. The basis matrix is as follow:

$$A_x = \begin{bmatrix} 1 & (x_1 - x)^T & vech^T\left\{(x_1 - x)(x_1 - x)^T\right\} & \cdots \\ 1 & (x_2 - x)^T & vech^T\left\{(x_2 - x)(x_2 - x)^T\right\} & \cdots \\ \vdots & \vdots & \vdots & \vdots \\ 1 & (x_p - x)^T & vech^T\left\{(x_p - x)(x_p - x)^T\right\} & \cdots \end{bmatrix} \tag{3}$$

where, $vech(\cdot)$ is the half angle vector operator. The N-order Taylor polynomial coefficient matrix is:

$$\beta^x = \left[\beta_0^{xT}, \beta_1^{xT}, \ldots, \beta_N^{xT}\right] \tag{4}$$

where, $\beta_0^x = Z(x)$ is the pixel value to be estimated; coefficient β_i^x is the i-th partial derivative matrix.

According to observation model for super-resolution reconstruction of single frame, Eq. 2 can be represented as:

$$\hat{\beta}^x = \arg\min_{\beta^x} \|V_x y - D_p A_x \beta^x\|_{W_x^D}^2 \tag{5}$$

where, D_p represents down-sampling operation, which ensures the same grid center in high resolution and low resolution, and $W_x^D = D_p W_x D_p^T$.

Least squares estimation of $Z(x)$ can be deduced from Eq. 5 to:

$$\hat{z}(x) = \hat{\beta}_0^x = e_1^T \left(A_x^T D_p^T W_x^D D_p A_x\right)]^{-1} A_x^T W_x^D V_x y \tag{6}$$

where e_1 is a column vector, whose first element is 1, and other element is 0.

Clustering image patches from several high resolution image which can contain any characters, into K subclasses. $Z(x)$ is the pixel to be estimated, and the k-th subclass is the subclass which is the closest to the image patch center at $Z(X)$. Define center of this subclass as $\phi^{(k)}$. Define $\{Z^m(x)\}_{m=1}^M$ as the collection of M image patches which are the most similar to $Z(X)$, where the similarity between $Z^m(X)$ and $Z(X)$ is represented by coefficient w_m, and calculated by following equation:

$$w_m = \exp\left(-\frac{\|Z^m(x) - Z(x)\|_2^2}{2\sigma^2}\right) \tag{7}$$

where, σ is the empirical parameter.

$\{Z^m(x)\}_{m=1}^M$ should be quite similar to $Z(x)$, so they can also be viewed as samples of $Z(x)$ to estimate center pixel $Z(x)$, i.e. learning structure information by kernel regression according to the non-local neighborhood information in $Z^m(x)$. Neighbor pixels in the corresponding low-resolution grid construct vector $\acute{y}_x^m$. So the kernel regression which combine information of similar patches in subclass $\phi^{(k)}$ can be represented as:

$$\hat{\beta}^x = \arg\min_{\beta^x} \sum_{m=1}^X \tilde{w}_m \|\acute{y}_x^m - D_p A_x \beta^x\|_{w_{x,m}^D}^2 \tag{8}$$

where $\tilde{w}_m$ is the normalized similarity weight coefficient.

Combine Eq. 5 with Eq. 8, and use the local neighborhood information of $Z(x)$ and non-local neighborhood information in subclass $\phi^{(k)}$ to estimate value of $Z(x)$. By view patch $Z(x)$ as the 0-th similar patch in subclass $\phi^{(k)}$, the similarity coefficient $w_0 = 1$, and $\acute{y}_x = \acute{y}_x^{m=0}$, we can get the non-local kernel regression model based on clustering learning:

$$\hat{\beta}^x = \arg\min_{\beta^x} \sum_{m=0}^M \tilde{w}_m \|\acute{y}_x^m - D_p A_x \beta^x\|_{W_{x,m}^D}^2 \tag{9}$$

Solve Eq. 8 by $\tilde{W}_{x,m}^D = \tilde{w}_1 W_{x,m}^D$, and we can get:

$$\hat{\beta}^x = \left[A_x^T \left(\sum_{m=0}^M D_p^T \tilde{W}_{x,m}^D D_p \right) A_x \right]^{-1} A_x^T \sum_{m=0}^M D_p^T \tilde{W}_{x,m}^D \acute{y}_x^m \tag{10}$$

After $\hat{\beta}^x$ is resolved, $\hat{Z}(x)$ can be estimated by $\hat{Z}(x) = \hat{\beta}_0^x = e_1^T \hat{\beta}^x$, and the whole high resolution image Z can also be estimated. Finally, X can be estimated after deblurring Z with so many present deblurring algorithm, and we can get a clear high resolution image in the end.

3   Super-Resolution Reconstruction Implementation

Our method can be separated into 2 procedures: training and super-resolution reconstruction.

3.1   Training

Construct Dataset D. Several high resolution images which can contain any characters are collected. These images are segmented into patches of the same size, which may also overlap adjacent patch. All these patches construct our dataset $D = [D_1, D_2, \ldots, D_n]$, where D_i represents any single patch.

Image patch which clipped totally from foreground region nor background region has little effect on Dictionary Learning. So we filter out these patches, by setting a threshold of local variance. It also improve the efficiency of clustering.

Clustering. Dataset D is partitioned into K subclasses by K-Means algorithm, and every subclass represent an unique local structure. The high-frequency component of image patch can reflect structure information, so we process all patches in D with a high-pass filter before clustering, and get a new dataset $D^h = [D_1^h, D_2^h, \ldots, D_n^h]$, where $D_i^h = D_i * G_h$, G_h is the kernel of high-pass filter, and "$*$" represent convolution operation.

K-Means is a popular unsupervised clustering algorithm based on minimum error criterion, and can partition data into K clusters, where K is preset. After K-Means clustering, D^h is partitioned into K subclasses $\{C_1, C_2, \ldots, C_K\}$, and the k-th $(1 \leq k \leq K)$ subclass can be represented as follows:

$$C_k = \left\{ D_i^h | D_i^h \in D^h; \|D_i^h - \mu_k\|_2 \leq \|D_i^h - \mu_l\|_2, l = 1, \ldots, K, l \neq k \right\} \quad (11)$$

where, μ_k is the center of C_k, whose value is the average vector of the subclass:

$$\mu_k = \frac{1}{|C_k|} \sum_{D_i^h \in C_k} D_i^h \quad (12)$$

One patch in D corresponds to a certain patch in D^h. So, dataset D is also partitioned into K subclass, and the k-th $(1 \leq k \leq K)$ subclass can be represented as $D_k = \left\{ D_i | D_i^h \in C_k, i = 1, 2, \ldots, n_k \right\}$.

3.2   Super-Resolution Reconstruction

Construct Sub-dictionary. Initial estimation y^H of high resolution image X is computed by interpolating the input low resolution image y. First of all, we choose the subclass whose distance to the patch centered at pixel $Z(x)$ is minimal among other subclass. Then, M patches are chosen from this subclass according to higher similarity measurement calculated by Eq. 7. These M patches combined with $Z(X)$ construct sub-dictionary $\phi^{(k)}$ which maintain uniformity about local structure feature. The sub-dictionary will be used for learning reconstruction later.

Reconstruct High Resolution Image. High resolution image will be estimated by Eq. 9, and solved by iterative algorithm. Previous high resolution estimation will be replaced by new estimation to estimate local gradient and kernel weight of image. This update procedure will be conducted once after several iteration to save computation cost. High resolution construction implemented as follows:

Initialization. The bi-cubic interpolation result of low resolution image y is the initial estimation result, marked as $Z^{(0)}$. Initialize iterations to T.

Iterations. Solve Eq. 8 as follow, until the iteration limits reach.

1. Skip to next step if iteration count does not reach. Choose the subclass with minimum distance to each estimation patch based on the new high resolution estimation image again. Construct sub-dictionary $\phi^{(k)}$ and calculate the similarity coefficient w_m.
2. Calculate local gradient by $Z^{(k-1)}$, and update kernel weight matrix W_x.
3. Estimate every pixel value in high resolution image by:

$$\hat{Z}^k(x) = e_1^T \left[A_x^T \left(\sum_{m=0}^{M} D_p^T \tilde{W}_{x,m}^D D_p \right) A_x \right]^{-1} A_x^T \sum_{m=0}^{M} D_p^T \tilde{W}_{x,m}^D \acute{y}_x^m \qquad (13)$$

Post-Process. High resolution reconstruction image X will be get by deblurring Z.

4 Experiment Result and Analysis

In this section, we prove the effectiveness of our methods by 2 super-resolution reconstruction experiments: one for simulated text image, another one for actual text image. In our experiments, training samples come from high resolution document images that do not have any relationship with the image to be reconstructed. These document images are 5 images of 355×300 size, which come from a Chinese word document consisting of 925 any Chinese characters. These images are segmented to image patches of 7×7 size, which overlap with each other. These patches are clustered into 200 subclasses following the procedure mentioned in last section. All reconstruction experiments in our paper use these subclasses.

4.1 Reconstruction Experiment for Simulated Text Image

In this subsection, we conduct two experiments for simulated Chinese document image and English document image respectively. We evaluate our method and compare our method with other methods by subjective evaluation and objective indicators. Peak Signal to Noise ratio (PSNR) and Structural Similarity (SSIM) are two main indicators we use.

PSNR is defined via the mean squared error (MSE). Given a noise-free $m \times n$ monochrome image I and its noisy approximation K, MSE is defined as:

$$MSE = \frac{1}{mn} \sum_{i=0}^{m-1} \sum_{j=0}^{n-1} [I(i,j) - K(i,j)]^2$$

The PSNR (in dB) is defined as:

$$PSNR = 10 \cdot \log_{10} \frac{MAX_I^2}{MSE}$$

Table 1. PSNR(dB) and SSIM of experiment results for simulated text image

Test methods	Bi-cubic interpolation		NL-KR		Our method	
Evaluation indicator	PSNR	SSIM	PSNR	SSIM	PSNR	SSIM
Chinese text image	30.94	0.572	31.97	0.872	33.13	0.827
English text image	33.65	0.737	35.66	0.886	36.41	0.895

MAX_I is the maximum possible pixel value of image I. When the pixels are represented using 8 bits per sample, it equals to 255.

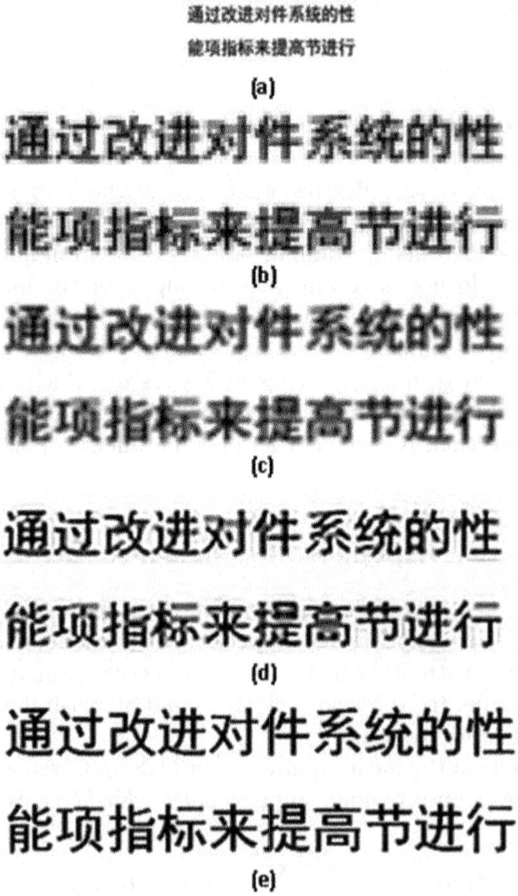

Fig. 2. Reconstruction result of simulated Chinese text image: (a) is low resolution image, (b)–(d) are reconstruction results of BC, NL-KR, and our method, (e) is original high resolution image

The SSIM is calculated on various windows of an image. The measure between two windows x and y of common size NxN is:

$$SSIM(x,y) = \frac{(2\mu_x\mu_y + c_1)(2\delta_{xy} + c_2)}{(\mu_x^2 + \mu_y^2 + c_1)(\delta_x^2 + \delta_y^2 + c_2)}$$

with μ_x the average of x, μ_y the average of y, δ_x^2 the variance of x, δ_y^2 the variance of y, δ_{xy} the covariance of x and y, c_1, c_2 two variables to stabilize the division with weak denominator, $c_1 = 6.50, c_2 = 58.52$ by default.

In the simulation, we get the low quality image to be reconstructed by procedures as follow:

1. Conduct gaussian blur whose size is 7×7, and standard variance is 1.6.
2. Conduct 3 times down-sampling horizontally and vertically respectively.
3. Add gaussian noise whose standard deviation is 0.6.

Fig. 3. Reconstruction result of simulated English text image: (a) is low resolution image, (b)–(d) are reconstruction results of BC, NL-KR, and our method, (e) is original high resolution image

Reconstruct this low quality image by Bi-cubic Interpolation (BC), NL-KR proposed in [11], and our method. In our method, the iterations set to 400, and high quality image segmented into patch of 7×7 size.

Figures 2 and 3 are reconstruction results of different methods for Chinese and English text image respectively. (a) is low resolution image, (b)–(d) are reconstruction results of BC, NL-KR, and our method, (e) is original high resolution image. Reconstruction result of our method is better than BC and NL-KR, and is almost as perfect as the original high resolution image.

In Table 1, we list values of PSNR and SSIM for different methods. And our method has better performance in PSNR and SSIM than other method in both experiments.

4.2 Reconstruction Experiment for Actual Text Image

In this subsection, we conduct two experiments for actual Chinese document image and English document image respectively. After preprocess like denoising to fax document image which contains characters of small size, we conduct BC, NL-KR and our method for 3 times up-sampling reconstruction. In our method,

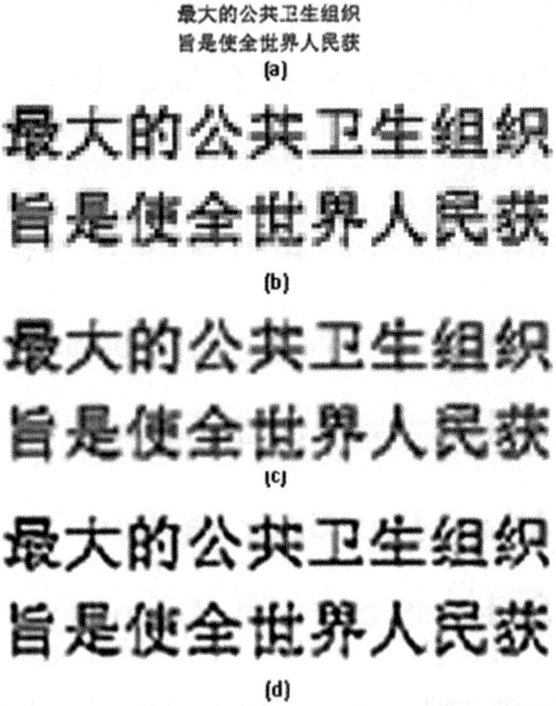

Fig. 4. Reconstruction result of actual Chinese text image: (a) is low resolution image, (b)–(d) are reconstruction results of BC, NL-KR, and our method

the iterations set to 280, and high quality image segmented into patch of 7×7 size.

Randomly select the fax document image, content of which is Chinese characters of small size. Size of fax document image is 120×43. Reconstruction results of different methods are illustrated in Fig. 4. (a) is low resolution image, (b)–(d) are reconstruction results of BC, NL-KR, and our method. Reconstruction results of different methods for actual English character images are illustrated in Fig. 5. So, we can draw the conclusion that, our method has a obvious visual effect improvement, and achieve better result compared with BC and NL-KR.

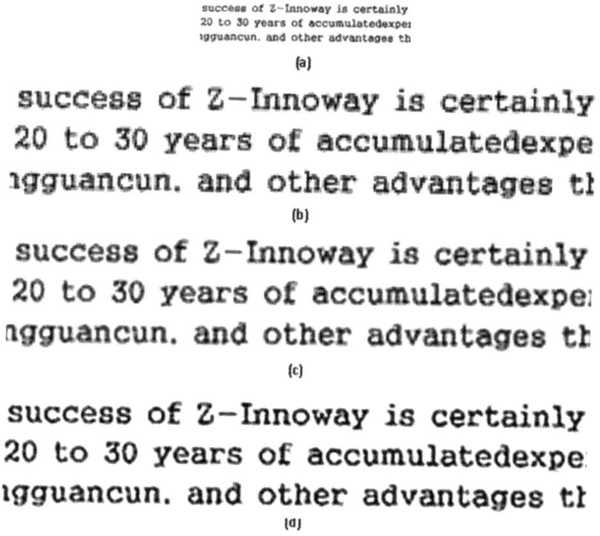

success of Z–Innoway is certainly
20 to 30 years of accumulatedexpe
ıgguancun. and other advantages th

(a)

success of Z–Innoway is certainly
20 to 30 years of accumulatedexpe
ıgguancun. and other advantages tl

(b)

success of Z–Innoway is certainly
20 to 30 years of accumulatedexpe
ngguancun. and other advantages tl

(c)

success of Z–Innoway is certainly
20 to 30 years of accumulatedexpe
ıgguancun. and other advantages tl

(d)

Fig. 5. Reconstruction result of actual English text image: (a) is low resolution image, (b)–(d) are reconstruction results of BC, NL-KR, and our method

Experiments above prove the effectiveness of our method. Our method can be used for reconstruction task of actual fax document image, can improve the image quality of character of small size, and can be used in the preprocess procedure for character recognition and indexing later.

5 Conclusion

SR reconstruction is used to solve the difficulty of recognition and indexing about low quality document images and character of small size, while traditional method can not achieve ideal performance for this problem. In this paper, We propose a new reconstruction model based on clustering learning and weighted kernel regression, by combining steering kernel regression and clustering learning methods. Our method reconstruct character image by learning local structure information from character image. Different from other kernel regression method, our method can learn local structural information not only from neighborhood, but also from many unrelated samples by clustering. Effectiveness and robustness of our method has been proved by the SR reconstruction experiments.

Acknowledgments. We want to thank the help from the researchers and engineers of MicroPattern Corporation. This work is supported partially by China Postdoctoral Science Foundation (No: 2015M582355), the Doctor Scientific Research Start project from Hubei University of Science and Technology (No: BK1418) and the Team Plans Program of the Outstanding Young Science and Technology Innovation of Colleges and Universities in Hubei Province (T201513).

References

1. Walha, R., Drira, F., Lebourgeois, F., Garcia, C., Alimi, A.M.: Multiple learned dictionaries based clustered sparse coding for the super-resolution of single text image. In: Proceedings of the 2013 12th International Conference on Document Analysis and Recognition, ICDAR 2013, Washington, DC, USA, pp. 484–488. IEEE Computer Society (2013)
2. Capel, D., Zisserman, A.: Super-resolution enhancement of text image sequences. In: International Conference on Pattern Recognition, vol. 1, p. 1600 (2000)
3. Shao, W.: Super-resolution reconstruction of multiple image regularization based on image modeling theory. Ph.D. thesis. Nanjing University of Science and Technology (2008)
4. Nasrollahi, K., Moeslund, T.B.: Super-resolution: a comprehensive survey. Mach. Vis. Appl. **25**(6), 1423–1468 (2014)
5. Protter, M., Elad, M.: Super resolution with probabilistic motion estimation. IEEE Trans. Image Process. **18**(8), 1899–1904 (2009)
6. Danielyan, A., Foi, A., Katkovnik, V., Egiazatian, K.: Image and video super-resolution via spatially adaptive block-matching filtering. In: Proceeding of International Workshop on Local and Non-local Approximation in Image Processing (LNLA) 01/2008 (2008)
7. Gao, G., Yang, J.: Context-aware single image super-resolution using sparse representation and cross-scale similarity. Sig. Process. Image Commun. **32**, 40–53 (2015)
8. Gao, G., Yang, J.: Novel sparse representation based framework for face image super-resolution. Neurocomputing **134**, 92–99 (2014). Special issue on the 2011 Sino-foreign-interchange Workshop on Intelligence Science and Intelligent Data Engineering (IScIDE 2011) Learning Algorithms and Applications Selected papers from the 19th International Conference on Neural Information Processing (ICONIP2012)
9. Walha, R., Drira, F., Lebourgeois, F., Garcia, C., Alimi, A.M.: Sparse coding with a coupled dictionary learning approach for textual image super-resolution. In: 2014 22nd International Conference on Pattern Recognition (ICPR), pp. 4459–4464, August 2014
10. Buades, A., Coll, B., Morel, J.M.: A review of image denoising algorithms, with a new one. Multiscale Model. Simul. **4**(2), 490–530 (2005)
11. Zhang, H., Yang, J., Zhang, Y., Huang, T.S.: Image and video restorations via nonlocal kernel regression. IEEE Trans. Cybern. **43**(3), 1035–1046 (2013)

12. Kang, L.W., Hsu, C.C., Zhuang, B., Lin, C.W., Yeh, C.H.: Learning-based joint super-resolution and deblocking for a highly compressed image. IEEE Trans. Multimedia **17**(7), 921–934 (2015)
13. Liao, H., Dai, W., Zhou, Q., Liu, B.: Non-local similarity dictionary learning based face super-resolution. In: 2014 12th International Conference on Signal Processing (ICSP), pp. 88–93, October 2014
14. Dong, W., Zhang, L., Shi, G., Xiaolin, W.: Image deblurring and super-resolution by adaptive sparse domain selection and adaptive regularization. Trans. Imgage Process. **20**(7), 1838–1857 (2011)
15. Takeda, H., Farsiu, S., Milanfar, P.: Kernel regression for image processing and reconstruction. IEEE Trans. Image Process. **16**(2), 349–366 (2007)
16. Yan, Q., Xu, Y., Yang, X., Nguyen, T.Q.: Single image super resolution based on gradient profile sharpness. IEEE Trans. Image Process. **24**(10), 3187–3202 (2015)

Image Fusion and Super-Resolution with Convolutional Neural Network

Jinying Zhong, Bin Yang[✉], Yuehua Li, Fei Zhong, and Zhongze Chen

School of Electrical Engineering, University of South China,
Hengyang, 421001, China
yangbin01420@163.com

Abstract. Image fusion aims to integrate multiple images of the same scene into an artificial image which contains more useful information than any individual one. Due to the constraints of imaging sensors and signal transmission broadband, the resolution of most source images is limited. In traditional processing framework, super-resolution is conducted to improve the resolution of the source images before the fusion operations. However, those super-resolution methods do not make full use of the multi-resolution characteristics of images. In this paper, a novel jointed image fusion and super-resolution algorithm is proposed. Source images are decomposed into undecimated wavelet (UWT) coefficients, the resolution of which is enhanced with convolutional neural network. Then, the coefficients are further integrated with certain fusion rule. Finally, the fused image is constructed from the combined coefficients. The proposed method is tested on multi-focus images, medical images and visible light and near infrared ray images respectively. The experimental results demonstrate the superior performances of the proposed method.

Keywords: Image fusion · Super-resolution · Convolutional neural network

1 Introduction

Image fusion is an essential image preprocessing technology which aims to combine multiples images into an artificial one that contains more useful information than any individual image. In the last two decades, many image fusion approaches have been proposed by the researchers all over the world. Most of these methods can be roughly divided into two categories: the spatial domain methods and the transform domain methods. The spatial domain methods select weights of the image pixels and fuse directly. The guide filtering fusion (GFF) [1], morphological filtering [2], and pulse-coupled neural network (PCNN) [3] are usually used to achieve weight consistency verification. Alternatively, the transform domain methods fuse the decomposed coefficients in the transform domain. The stationary wavelet transform (SWT) [4], dualtree complex wavelet transform (DTCWT) [5], or non-subsampling contourlet transform (NSCT) [6] is usually employed to achieve the fusion task.

Due to the constraints of imaging sensors and signal transmission broadband, the resolution of most source images is limited. The fused results from such low-resolution

© Springer Nature Singapore Pte Ltd. 2016
T. Tan et al. (Eds.): CCPR 2016, Part II, CCIS 663, pp. 78–88, 2016.
DOI: 10.1007/978-981-10-3005-5_7

source images would still be low resolution, which would hinder many further image processing tasks. In general, image super-resolution (SR) is pre-performed to improve the resolution of source images. Then, various fusion methods are conducted to get a high-resolution fusion image. However, most SR methods do not make full use of the multi-resolution characteristics of an image. The multi-resolution transform coefficients of the source images, which are used to guide the fusion operation, also provide important cues for SR [7–10]. Therefore, a novel jointed image fusion and super-resolution algorithm based on the trained convolutional neural network (CNN) is proposed in this paper. Firstly, the source images are decomposed into undecimated wavelet coefficients. Then, the resolutions of the high frequency coefficients are enhanced with a trained CNN-based SR model and the approximation coefficients are unchanged. Next, both approximation coefficients and high frequency coefficients are integrated with certain fusion rule. Finally, the fused image is constructed with inverse wavelet transform. The multi-focus images, medical images, and visible light (VL) and near infrared ray (NIR) images are used to test the validation of the proposed method. To perform the comparison, the source images in comparison methods are also enhanced with the state-of-the-art super-resolution method in [11], which is denotes as SRCNN. Then various fusion methods are used to get the fused results. The experimental results show that the proposed method demonstrates superior results in terms of both visual and quantitative evaluations.

The remainder of this paper is organized as follows. Section 2 gives the proposed algorithm for SR in frequency domain and image fusion. Section 3 presents wavelet coefficients SR with the trained CNN. Section 4 shows comparison experiment results and discussions on performance of the proposed method. Finally, the conclusion of this paper is given.

2   The Basic Algorithm

Suppose that there are two registered source images. The proposed fusion and SR framework is demonstrated in Fig. 1. Notice that the framework can be directly extended to handle the case with more than two source images. In details, the proposed method takes the following fusion steps:

(1.) The source images A and B are decomposed into one low frequency sub-image and 9 high frequency sub-images with three levels UWT. For convenient description, let LFC and HFC represent the decomposed low/high frequency coefficients respectively.

(2.) The LFC sub-image represents the approximation intensity of source image, and HFC sub-image is responding to the detail components of the source images. More reasonable, they should be fused separately. In the proposed method, the SR method with trained CNN is applied to improve the resolution of the HFC sub-image as

$$HRFC^X = \mathbf{F}(HFC^X) \tag{1}$$

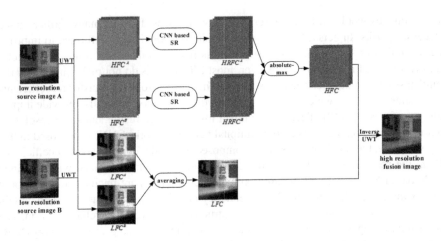

Fig. 1. The fusion and SR framework of the proposed method

where the superscript $X \in \{A, B\}$, $F(X)$ is the SR mapping function which would improve the resolution of the high frequency sub-image X. The trained CNN is used to achieve the SR task, which is described in the following section in details. In order to avoid the intensity information missing in the fused image caused by SR operation, the trained CNN-based SR is only conducted on high frequency coefficient sub-images. The resolution enhanced high frequency sub-image is denoted as HRFC for convenient.

(3) The HRFC sub-images are fused with absolute-max rule as

$$HRFC(i,j) = \begin{cases} HRFC^A(i,j), |HRFC^A(i,j)| \geq |HRFC^B(i,j)| \\ HRFC^B(i,j), |HRFC^B(i,j)| > |HRFC^A(i,j)| \end{cases} \tag{2}$$

where $HRFC^A$ and $HRFC^B$ denote the high frequency sub-images with high-resolution. (i, j) represent the coordinates of a coefficient in sub-images.

(4) The LFC sub-images are fused by

$$LFC(i,j) = \frac{1}{2}(LFC^A(i,j) + LFC^B(i,j)) \tag{3}$$

where LFC^A and LFC^B are the low frequency coefficients of source image A and B respectively.

(5) The final fused image is constructed from the fused coefficients sub-images HRFC and LFC by inverse UWT.

3   Wavelet Coefficients SR with Trained CNN

The deep convolutional neural network theory has extracted many attentions in recently years. An image super-resolution method with CNN denoted as SRCNN was proposed

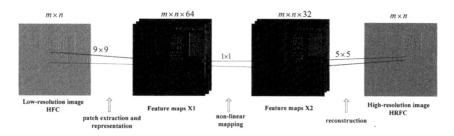

$m \times n$ $m \times n \times 64$ $m \times n \times 32$ $m \times n$

9×9 1×1 5×5

Low-resolution image   Feature maps X1   Feature maps X2   High-resolution image
HFC   HRFC

patch extraction and   non-linear   reconstruction
representation   mapping

Fig. 2. An overview of the CNN model for frequency domain image SR

in [11], which achieves end-to-end mapping between the low/high-resolution images. SRCNN outperforms the traditional SR methods such as interpolation, sparse representation [12]. In addition, the shepard convolutional neural networks (ShCNN) [13] realizes end-to-end trainable translation variant interpolation and performs superior in both image SR and inpainting applications. Since CNN-based SR methods are free from solving any optimization problem on usage, they are more efficient than most of the example-based methods [14–18]. However, the existing CNN-based SR methods are only trained aiming at image pixels in spatial. For convenient of image fusion, a CNN-based model for frequency domain SR is proposed.

The trained CNN in this paper is an end-to-end SR mapping from the low-resolution HFC sub-images to the high-resolution HRFC sub-images. Three layers CNN model used in this paper is shown in Fig. 2. The first convolutional layer contains 64 filters each of which is set as 9×9. The output of first layer contains 64 feature maps denoted as X_1. The second convolutional layer contains 32 filters with size of $64 \times 1 \times 1$. This layer achieves non-linear mapping from 64-dimensional feature to a 32-dimensional vector which denotes as X_2. Thus, each feature in X_2 is a linear combination of the 64-dimensional features in X_1. Therefore, X_2 represents more complex features which are used to construction the high-resolution image in the proposed image. The reconstruction operation aggregates X_2 with a set of $32 \times 5 \times 5 \times 1$ liner filters acting like an averaging filter to produce the final high-resolution image. Training the end-to-end mapping F can be converted to update the CNN parameters (weights and basis) of filters with low/high resolution training data. In the proposed method, the parameters are updated by stochastic gradient descent with the standard back propagation [19]. Due to the CNN parameters are trained with UWT coefficients, the trained CNN certainly makes full use the multi-resolution characteristics of the source images. In addition, the reconstructed high resolution UWT coefficients can directly be used for further image fusion operation. The coefficients with higher absolute value indicate more dominant features of the source image, so the absolute-max fusion rule is used in the proposed method for high frequency coefficients fusion. Due to the CNN is full feed-forward and no optimization problem involved, the time consume of the SR operation is also acceptable.

4   Experiment and Analysis

All the experiments are implemented on an AMD 2.70 GHz PC with the simulation software Matlab 2010a. 90 images are selected from the ILSVRC 2013 ImageNet which are further decomposed with three levels UWT and 810 HFC training images are obtained. Those HFC images are cropped into 143,172 image patches with size 33×33 which regard as the train sets of CNN. Notice that if more train data is used, the performance of network may be further improved. The 'db1' wavelet basis is used for UWT. To better test the effectiveness of the proposed method, the state-of-the-art fusion methods such as NSCT [6], SWT [4], DTCWT [5], and GFF [1] are used to perform the comparison. For a fair comparison, the SRCNN [11] is used to enhance the resolution of the source images before the fusion of various compared fusion methods. For the NSCT-based fusion, 'maxflat' and 'dmaxflat7' are applied as pyramid filter and directional filter respectively. Four decomposition levels with 2, 4, 8, 16 directions from coarser scale to finer scale are exploited. The comparison approach SWT-based fusion method parameters settings are the same as UWT used in this paper. The DTCWT is also decomposed into three levels and filters are 'near_sym_b' and 'qshift_b'. The low frequency coefficients are fused using averaging rule, and the high frequency coefficients are fused with maximum selection rule. The code of GFF method is provided by authors in [1], and the default parameters are set.

4.1   Multi-focus Image Fusion

In this section, the proposed method is tested on multi-focus source images named as "Pepsi", "lab", "clock", and "desk" respectively. Those images are all down sampled with bicubic interpolation. The aim of the proposed method is to integrate the low-resolution source images and get a fused image which has the same resolution as the ground truth images. The fused results of the "Pepsi" images are illustrated in Fig. 3. In each the fused images, the objects are all clearly preserved, which demonstrate the effectiveness of all the methods. To show the advantages of the proposed method, the selected regions labeled in red boxes are magnified as presented in the corresponding green boxes. By visually comparing the fused images with the ground truth images, we can see that the distribution of black and white stripes in the fused image of the proposed method is more uniform and closer to the focus areas in Fig. 3(b). In addition, it's easy to see that the stripes in Fig. 3(i) are brighter than them in Fig. 3(e)–(h) owing to the proposed method doesn't change the approximation components of source images. The Fig. 4 briefly shows the GFF and the proposed method fused results of other three image pairs, since GFF subjectively presents slightly better results than those of the NSCT, SWT and DTCWT methods. Careful inspections of the third and last column in Fig. 4 indicate that the proposed method transfers more information from source images to fused image while retaining more edge features.

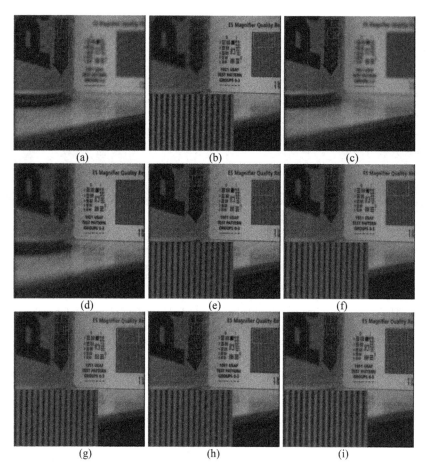

Fig. 3. The "Pepsi" ground truth images (a and b), source images (c and d), the fused image by NSCT (e), SWT (f), DTCWT (g), GFF (h), and the proposed method (i) (Color figure online)

Furthermore, in order to assess the performances of different methods objectively, two metrics are utilized to implement quantitative evaluation. The metric Q_w [20] measures the distortion of the fused image by a combination of correlation loss, luminance distortion, and contrast distortion. For the ideal fused image, the Q_w value should be close to 1. The metric $Q^{AB/F}$ [21] denotes the ability of fused image retaining amount of the edge information from the two source images. Larger $Q^{AB/F}$ value indicates better fused result. All the fused images are assessed with the ground truth source images. In order to avoid the adverse effect resulting from the sequence between fusion and SR in comparison methods, two cases are considered for comprehensive assessment. For the first case, the SRCNN is used to achieve SR operation followed by the fusion operation. The fused results in Figs. 3 and 4 are responding to the first case. For the second case, the low-resolution source images are fused firstly. Then the fused images are enhanced with SR operation. The objective assessments of different fusion methods in two cases

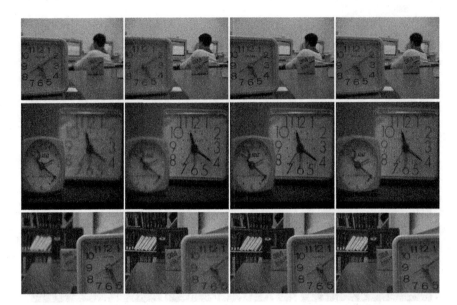

Fig. 4. Three sets source images ("lab", "clock" and "desk") and fused images: for each row, the first column is near focus source image, the second column is far focus image, the third column is the fused image of GFF, and the last column is the fused image of the proposed method

and the proposed methods are reported in Table 1. The corresponding largest value is labeled by bold.

Table 1. The fused results of multi-focus images using the five methods.

Image		Lab		Clock		Desk		Pepsi	
Criteria		Q_w	$Q^{AB/F}$	Q_w	$Q^{AB/F}$	Q_w	$Q^{AB/F}$	Q_w	$Q^{AB/F}$
SR followed by fusion	NSCT	0.7563	0.5097	0.7227	0.5956	0.7295	0.5170	0.5677	0.4103
	SWT	0.7289	0.4997	0.6920	0.5936	0.6965	0.5052	0.5546	0.4018
	DTCWT	0.7482	0.5034	0.6966	0.5802	0.7156	0.5046	0.5605	0.4034
	GFF	0.7582	0.5168	0.7354	0.6046	0.7301	0.5220	0.5636	0.4005
Fusion followed by SR	NSCT	0.7525	0.5056	0.7008	0.5883	0.7309	0.5148	0.5739	0.4108
	SWT	0.7217	0.4890	0.6699	0.5765	0.7030	0.4995	0.5598	0.4019
	DTCWT	0.7463	0.4995	0.6796	0.5742	0.7171	0.5034	0.5621	0.4037
	GFF	0.7554	0.5125	0.7163	0.5968	0.7316	0.5188	0.5631	0.4012
Proposed		**0.8021**	**0.5440**	**0.7582**	**0.6173**	**0.7853**	**0.5533**	**0.6450**	**0.4385**

As shown in Table 1, the proposed method outperforms the NSCT, SWT, DTCWT and GFF based methods in terms of the Q_w and $Q^{AB/F}$ in both cases, which indicates the fused images obtained by the proposed method are the closest to the ground truth images. Due to the CNN model trained with low resolution HFC sub-images and its corresponding high resolution HFC sub-images used as label to adjust network filters weights, the proposed method can easily be applied such model to map the low-resolution source HFC sub-image into a high-resolution one. In addition, the proposed method makes full use of the multi-resolution characteristics of images. Therefore, the fused results extract

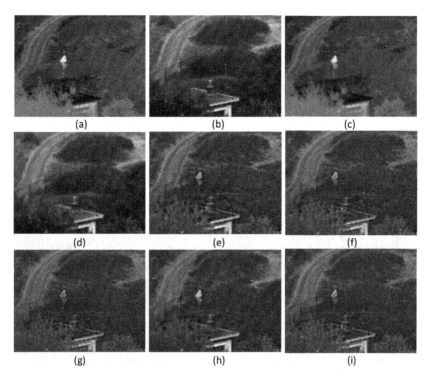

Fig. 5. The ground truth images ((a) is NIR image and (b) is LV image), source images (c and d), and fused image by NSCT (e), SWT (f), DTCWT (g), GFF (h), the proposed method (i)

more edges and details, and light information is better preserved. The second case displays the proposed method still performs the best.

4.2   Medical and VL and NIR Image Fusion

In order to show the potential of proposed method can extend in other image fusion occasions, the VL and NIR image fusion, and medical images fusion based on the proposed method are performed in this part. The objects, like the persons, in the NIR image stand out well against cool backgrounds. While in VL image, the details of the background, like the fence, is more visible. Figure 5 reveals the fused image of NIR and VL image of the first case using different methods and the proposed method. Due to the NSCT, SWT and DTCWT based and proposed methods apply averaging fusion rule for low frequency fusion, thus the light of the person in Fig. 4(h) is slight brighter. However, it can be easy found that the proposed method can generate more satisfactory fused results on the whole. The details such as roof edges and fences in Fig. 4(i) are clearer. What's more, Fig. 4(i) owns more clearly hierarchy than Fig. 4(e)–(h). The objective evaluation of NIR and VL image fusion in two cases and the proposed method is shown in the left of Table 2, which agrees with the subjective results.

Table 2. The fused results of NIR and VL and medical images using the five methods

Source image types		NIR and VL		CT and MRI	
Criteria		Q_w	$Q^{AB/F}$	Q_w	$Q^{AB/F}$
SR followed by fusion	NSCT	0.5058	0.3289	0.6603	0.6398
	SWT	0.4935	0.3323	0.6438	0.6354
	DTCWT	0.4450	0.2972	0.5949	0.5761
	GFF	0.4673	0.3377	0.6791	0.6608
Fusion followed by SR	NSCT	0.5013	0.3261	0.6345	0.6295
	SWT	0.4888	0.3191	0.6318	0.6163
	DTCWT	0.4443	0.2972	0.5883	0.5742
	GFF	0.4705	0.3424	0.7005	0.6794
Proposed		**0.5436**	**0.3522**	**0.7051**	**0.6949**

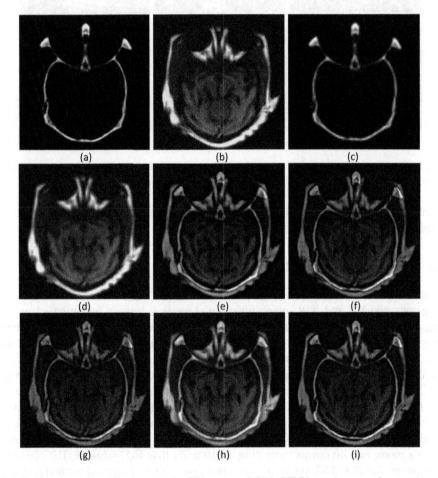

Fig. 6. The ground truth images ((a) is CT image and (b) is MRI image), source images (c and d), and fused image by NSCT (e), SWT (f), DTCWT (g), GFF (h), the proposed method (i)

For medical images, computed tomography (CT) is usually applied for visualizing dense structures and not suitable for soft tissues. While magnetic resonance imaging (MRI), commonly used for detection of tumor and other tissue abnormalities, can provide better visualization of soft tissues. In order to better clinical diagnosis and treatment, integrating such two kind images is highly necessary. The proposed method can also perform well in CT and MRI image fusion and extract desirable and sufficient information for further diagnosis. The subjective evaluation of the first case and the proposed method is shown in Fig. 6, and the right of Table 2 summarizes the objective evaluation with different methods for CT and MRI fusion. The CT image shallows the light in Fig. 6(e)–(g) and Fig. 6(i), which seem worse than Fig. 6(h) compared to Fig. 6(b) because of the low frequency fusion approach. However, fused image of the proposed method owns better edge and texture details than other four comparison methods. Overall, the proposed method performs the CT and MRI fusion.

According to the both visual comparison and objective evaluation fusion of multi-focus images, medical images, and the NIR and VL images, the proposed method surpasses the previous four fusion methods. No matter the fusion first or SR first, the proposed method always works competitive.

5   Conclusions

A novel image fusion algorithm based on CNN is proposed in this paper. We train a CNN model for super-solution process in frequency domain. Source images are decomposed into UWT coefficients. The resolutions of the high frequency coefficients are enhanced with the trained CNN-based SR and the approximation components are unchanged. The image fusion and SR processes can be conducted simultaneously. Compared with several spatial domain and transform domain based methods, the experimental results show the proposed method performs desirable in terms of both visual and quantitative evaluations. The CNN-based fusion algorithm in this paper is only a premature attempt. A more reasonable fusion strategy for LFC sub-images will provide better fused result. It is worthy to further investigate the effectiveness of the proposed method for other related tasks of image fusion.

Acknowledgments. This paper is supported by the National Natural Science Foundation of China (Nos. 61102108), Scientific Research Fund of Hunan Provincial Education Department (Nos. YB2013B039), Young talents program of the University of South China, and the construct program of key disciplines in USC (No. NHXK04).

References

1. Li, S.T., Kang, X.D., Hu, J.W.: Image fusion with guided filtering. J. IEEE Trans. Image Process. **22**, 2864–2875 (2013)
2. Li, H., Li, L., Zhang, J.X.: Multi-focus image fusion based on sparse feature matrix decomposition and morphological filtering. J. Opt. Commun. **342**, 1–11 (2015)

3. Chai, Y., Li, H.F., Guo, M.Y.: Multifocus image fusion scheme based on features of multiscale products and PCNN in lifting stationary wavelet domain. J. Opt. Commun. **284**, 1146–1158 (2011)
4. Mirajkar, P.P., Sachin, D.R.: Image fusion based stationary wavelet transform. J. Int. J. Adv. Eng. Res. Stud. **2**, 99–101 (2013)
5. Lewis, J.J., Callaghan, R.J., Nikolov, S.G.: Pixel- and region-based image fusion with complex wavelets. J. Inf. Fusion **8**, 119–130 (2007)
6. Sravya, K., Govardhan, P., Goud, M.N.: Image fusion on multi focused images using NSCT. J. IJCSIT **5**, 5393–5396 (2014)
7. Lui, S.-F., Wu, J.-Y., Mao, H.-S., Lien, J.-J.J.: Learning-based super-resolution system using single facial image and multi-resolution wavelet synthesis. In: Yagi, Y., Kang, S.B., Kweon, I.S., Zha, H. (eds.) ACCV 2007, Part II. LNCS, vol. 4844, pp. 96–105. Springer, Heidelberg (2007)
8. Liu, W.R., Li, S.T.: Image super-resolution based on hybrid multiresolution analysis. In: The First Asian Conference on Pattern Recognition, Beijing, China, pp. 307–310. IEEE press (2011)
9. Zhao, G., Zhang, K., Shao, W.: A study on NSCT based super-resolution reconstruction for infrared image. In: TENCON 2013 – 2013 IEEE Region 10 Conference, Xi'an, China, pp. 1–5. IEEE press (2013)
10. Lu, X., Li, X.: Multiresolution imaging. J. IEEE Trans. Cybern. **44**, 149–160 (2014)
11. Dong, C., Loy, C.C., He, K., Tang, X.: Learning a deep convolutional network for image super-resolution. In: Fleet, D., Pajdla, T., Schiele, B., Tuytelaars, T. (eds.) ECCV 2014, Part IV. LNCS, vol. 8692, pp. 184–199. Springer, Heidelberg (2014)
12. Yang, J., Wright, J., Huang, T.S.: Image super-resolution via sparse representation. J. IEEE Trans. Image Process. **19**, 2861–2873 (2010)
13. Ren, J.S., Xu, L., Yan, Q.: Shepard convolutional neural networks. In: Proceeding of 28th Annual Conference on Neural Information Processing Systems, Montreal, Canada. MIT press (2015)
14. Bevilacqua, M., Roumy, A., Guillemot, C.: Low-complexity single-image super-resolutionbased on nonnegative neighbor embedding. In: Proceedings of the 23rd British Machine Vision Conference (BMVC), pp. 135.1–135.10. BMVC press, Guidford (2012)
15. Chang, H., Yeung, D.Y., Xiong, Y.: Super-resolution through neighbor embedding. In: IEEE Conference on Computer Vision and Pattern Recognition, California, USA, pp. 275–282. IEEE Computer Society (2010)
16. Jia, K., Wang, X., Tang, X.: Image transformation based on learning dictionaries across image spaces. J. IEEE Trans. Pattern Anal. Mach. Intell. **35**, 367–380 (2013)
17. Timofte, R., De, V., Van Gool, L.: Anchored neighborhood regression for fast example-based super-resolution. In: IEEE International Conference on Computer Vision, Sydney, Australia, pp. 1920–1927. IEEE press (2013)
18. Zeyde, R., Elad, M., Protter, M.: On single image scale-up using sparse-representations. In: Boissonnat, J.-D., Chenin, P., Cohen, A., Gout, C., Lyche, T., Mazure, M.-L., Schumaker, L. (eds.) Curves and Surfaces 2011. LNCS, vol. 6920, pp. 711–730. Springer, Heidelberg (2012)
19. Rumelhart, D.E., Hinton, G.E., Williams, R.J.: Learning representations by back-propagating errors. J. Nat. **323**, 533–536 (1986)
20. Piella, G., Heijmans, H.: A new quality metric for image fusion. In: Proceedings of the IEEE International Conference on Image Processing, Barcelona, Spain, pp. 173–176. IEEE press (2003)
21. Xydeas, C.S., Petrovic, V.: Objective image fusion performance measure. J. Electron. Lett. **36**, 308–309 (2000)

Robust Segmentation for Video Captions
with Complex Backgrounds

Zong-Heng Xing, Fang Zhou[✉], Shu Tian, and Xu-Cheng Yin[✉]

Department of Computer Science and Technology,
School of Computer and Communication Engineering,
University of Science and Technology Beijing, Beijing, China
zonghengxing@xs.ustb.edu.cn, zhoufang@ies.ustb.edu.cn,
tshu23@gmail.com, xuchengyin@ustb.edu.cn

Abstract. Caption text contains rich information that can be used for video indexing and summarization. In this paper, we propose an effective caption text segmentation approach to improve OCR accuracy. Here, an AlexNet CNN is first trained with path signature for text tracking. Then we utilize an improved adaptive thresholding method to segment caption text in individual frames. Finally, the multi-frame integration is conducted with gamma correction and region growing. In contrast to conventional methods which extract video text in individual frames independently, we exploit the specific temporal characteristics of videos to perform segmentation. Moreover, the proposed method can effectively remove the complex backgrounds with similar intensity to text. Experimental results on different videos and comparisons with other methods show the efficiency of our approach.

Keywords: Caption text segmentation · Convolutional neural networks · Path signature · Multi-frame integration

1 Introduction

As the technology of digital multimedia develops rapidly, video has become one of the most popular media forms delivered via TV broadcasting and Internet. Text in video, especially caption text, contains rich information as a significant high-level semantic feature, which directly describes the video content. For instance, scores in sports program can help you quickly know the game situation, headlines in news videos summarize the content of news and captions in films conduce to better understanding of the storyline. Thus, caption text extraction and recognition play an important role in video indexing [1] and summarization. However, the state-of-the-art Optical Character Recognition (OCR) does not work well for text in video, although it has achieved excellent performance on printed documents. In contrast to binary document images, video text presents much more difficulties due to complex backgrounds, variety of text fonts and low contrast caused by the lossy compression.

Before text recognition is implemented, there are generally three phases involved: text detection, text tracking and text segmentation. Text detection aims for locating the

© Springer Nature Singapore Pte Ltd. 2016
T. Tan et al. (Eds.): CCPR 2016, Part II, CCIS 663, pp. 89–100, 2016.
DOI: 10.1007/978-981-10-3005-5_8

text region within a single video frame, while text tracking maintains the consistence of the text between consecutive frames and text segmentation intends to separate text from background (i.e., binarization). After segmentation, we can obtain a binary image, i.e., a mask, which is more adaptable for OCR engines. Although many methods have been proposed for video text segmentation in the past decades, most of them extract text only in individual frames independently. Even if several methods had integrated the information of multiple frames, they could not produce efficient masks in the case that the background has similar intensity with text pixels.

In this paper, we propose an effective method for caption text segmentation from complex backgrounds. Considering that captions are commonly stationary, we do not take the scrolling captions into account. Focusing on text tracking and text segmentation processes, we regard manually annotated caption text regions in video frames as detection results. We first train a Convolutional Neural Networks (CNNs) classifier with path signature for text tracking. Then, text segmentation is carried out in a single frame by an adaptive thresholding based on Kim's method [2]. We improve the transition point labeling strategy presented by Kim and employ the canny edge detector to get better performance. Finally, we use multi-frame information to refine the segmentation. Gamma correction and region growing are applied for effectively removing the background which has similar intensity with text. The rest of the paper is organized as follows. Section 2 reviews related work. Our proposed method is described in details in Sect. 3. Then, experimental results are presented and discussed in Sect. 4. The last section concludes this paper.

2   Related Work

Video text segmentation can be divided into two parts: single-frame text segmentation and multi-frame integration. Single-frame text segmentation extracts text in individual frames independently. Multi-frame integration combines information from multiple frames to depress video backgrounds and enforce segmentation results.

Thresholding, probability models and clustering have been commonly used for text segmentation in a single frame. Liu and Srihari [3] proposed a method applying optimal thresholding based on texture features to extract text in document images. Cheriet et al. [4] presented a recursive thresholding approach extending from Otsu's method [5]. These algorithms are simple in computation and efficient on document images. However, the quality of segmentation results obtained by global thresholding techniques is easily effected by noise and distortion. To solve this problem, adaptive thresholding could be applied, which labels text pixels using local information. Ohya et al. [6] divided the original image into blocks of specific size, in each of which the threshold was determined locally, and subsequently they used a linearly interpolation for each pixel's threshold determination. In [2], Kim et al. proposed an overlay text extraction method extending Lyu et al.'s approach [7]. They utilized a sliding window to conduct adaptive thresholding, which is followed by inward filling procedure.

Since compression can make the text pixels transform in color, and text boundaries often blend with the background, it is difficult to select a credible threshold value for segmentation. In these situations, Gaussian Mixture Models (GMMs) are employed to build color models of text pixels. In [8], Ye et al. trained a GMM of intensity and hue channels in HIS color space, which is further used for text segmentation with spatial connectivity information. Wang et al. [9] developed a two-step text segmentation approach: coarse segmentation carried out by adopting a Gaussian function and noise elimination based on color distribution homogeneity. Mishra et al. [10] formulated the text segmentation problem in a Gaussian Mixture MRF framework and minimized an energy function to label text pixels.

Besides the above-mentioned approaches, clustering is another effective means of text segmentation. Wu et al. [11] and Wakahara and Kita [12] applied K-means algorithm to identify textual pixels. In [13], Mancas-Thillou and Gosselin presented a color segmentation with spatial information using clustering distances to merge text pixels. The performance of clustering is mostly associated with the similarity measurement it used. However, it is arduous to define an optimal evaluation of similarity in different cases.

When extracting text in video, the multi-frame integration strategy is generally used for improving text segmentation results. Lienhart and Wernicke [14] exploited temporal redundancy to refine the candidate text lines identified by a trained feed-forward network and improve the segmentation performance. In the work of Liu and Wang [15], a temporal "and" operation was applied to filter out the background pixels with similar color to text, then a post-processing step was carried out to refine the segmentation results by a priori knowledge of character structures. Phan et al. [16] integrated the binarized results within the same text's frame span at first, a refining process was later implemented by using local information of the averaged intensity image and the newly obtained coarse segmentation.

3   Proposed Method

In this paper, we propose a text segmentation method on the assumption that text pixels have high intensities compared to background pixels, whose outputs are binary images in which white color represents text pixels while black represents background pixels. Our method mainly consists of three steps: text tracking, single-frame text segmentation and multi-frame integration.

3.1   Text Tracking

In this step, we train an AlexNet CNN which is classic and efficient for image classification to identify the consistency of captions in successive video frames. In our network, the input layer is enhanced by the path iterate-integral signature of successive frames.

The theory of path signature was pioneered by Chen [17] to solve any linear differential equation and uniquely depict a path of finite length. It was first introduced by Graham [18] to Machine Learning field to address recognition of handwritten characters [19]. Since video frames are displayed by temporal order, we can consider a pixel's color variation in the same location of the two successive frames as a kind of path. In our approach, we utilize intensity and saturation components in HSI color space as pixels' feature vector, i.e. (I, S). Considering that the hue value of a pixel is greatly effected by surrounding pixels due to the color bleeding, we do not use it for feature extraction. Intensity and saturation are calculated as follows:

$$\begin{cases} I = \dfrac{R+G+B}{3} \\ S = 1 - \dfrac{3 \min(R,G,B)}{R+G+B} \end{cases} \cdot \tag{1}$$

Let $[s, t] \subset R$ denotes a time interval between the successive frames and let $P_{s,t}^k$ denotes the k-th iterated integral of the path P during the interval $[s, t]$. The signature can be computed as follows:

$$P_{s,t}^0 = 1, P_{s,t}^1 = \triangle_{s,t}, P_{s,t}^2 = \frac{\triangle_{s,t} \otimes \triangle_{s,t}}{2!}, P_{s,t}^3 = \frac{\triangle_{s,t} \otimes \triangle_{s,t} \otimes \triangle_{s,t}}{3!}, \cdots, \tag{2}$$

where $\triangle_{s,t} = P_t - P_s = (I_t - I_s, S_t - S_s)$. Considering $\triangle_{s,t}$ as a row vector, the tensor product corresponds to the Kronecker matrix product. As k increases, the dimension of iterated integral grows rapidly, which increases computational complexity while carrying less useful information [18]. Hence, we take a truncated signature defined as $S(P)_{s,t} = (P_{s,t}^1, P_{s,t}^2)$, which is a collection of the first and second iterated integrals. Conventionally, the first iterated integral represents the path displacement, the second iterated integral is related to the curvature of the path, while the zeroth iterated integral is a constant which is useless for tracking. Therefore, each pair of pixels in the same location of two successive frames can generate a six-dimensional feature vector. In another words, two arbitrary images with the same size would produce a path signature map with six channels as the input of our CNN.

The AlexNet CNN that we refer to contains five convolutional layers, each of which is followed by a max-pooling layer except the third and fourth convolutional layers, and two fully connected layers. Label layer is the final layer of the network, where we reset the number of class to two corresponding to the same and different captions. The entire architecture of our CNN is shown in Fig. 1.

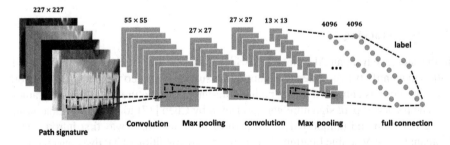

Fig. 1. The architecture of AlexNet CNN with path signature

3.2 Single-Frame Text Segmentation

We propose a text segmentation based on Kim's approach [2] composed of adaptive thresholding, dam point labeling and inward filling. Kim's approach has a good performance in most images. However, in the case that the background connected to text has the similar intensity with text, part of the text may be lost during the filling procedure. To solve this problem, we modify transition point labeling strategy in [2] and employ the canny edge detector. The proposed text segmentation in a single frame can be divided into five steps: adaptive thresholding, transition map generation, edge map generation, dam point labeling, and inward filling.

Adaptive Thresholding. Firstly, each border of the text region is extended by two pixels to cover more background pixels. We define the region inside the origin four borders as OR and the extended outer region as ER. The region including OR and ER is defined as EO, i.e., $EO = OR \cup ER$. Then, a sliding window is applied on EO horizontally and vertically with different sizes, the region in each window is binarized by Otsu method [5]. As shown in Fig. 2, a $(EO\_height/2) \times (EO\_height)$ window is moved from left to right with stepping $EO\_height/4$, and then a $(EO\_width) \times (EO\_height/4)$ window is moved from top to bottom with stepping $EO\_height/8$. Let $G(x, y)$ and $B(x, y)$ denote the grayscale image EO and the resulting binary image, respectively, where $0 \leq x < EO\_width$, and $0 \leq y < EO\_height$. $B(x, y)$ is initialized as all "white". In each window, if the value of $G(x, y)$ is less than the local threshold calculated by Otsu method [5], the corresponding $B(x, y)$ is set to be "black". Finally, we can obtain an elementary segmentation of EO.

(a)   (b)

Fig. 2. Adaptive thresholding with sliding window (a) Horizontal adaptive thresholding (b) Vertical adaptive thresholding

Transition Map Generation. In [2], Kim et al. noted that there are transitional colors between inserted text and its adjacent background in video and the intensities at the boundary of the text have the logarithmical change. The transition maps are generated for further labeling dam points. To better enhance the variation of the text boundary pixels' intensities in the low contrast image, we modify the constraint of transition point by changing the weight factor as follows:

$$W(x, y) = \begin{cases} 2 \times \tilde{I}(x,y) & \tilde{I}(x,y) > 0.5 \\ 2 \times (1 - \tilde{I}(x,y)) & otherwise \end{cases}, \tag{3}$$

$\tilde{I}(x, y)$ denotes the normalized intensity. The transition points in horizontal direction can be defined as follows:

$$D_L(x, y) = \left(1 + \left|\frac{S(x-1, y)}{W(x-1, y)} - \frac{S(x, y)}{W(x, y)}\right|\right) \times |I(x-1, y) - I(x, y)|, \tag{4}$$

$$D_H(x, y) = \left(1 + \left|\frac{S(x, y)}{W(x, y)} - \frac{S(x+1, y)}{W(x+1, y)}\right|\right) \times |I(x, y) - I(x+1, y)|, \tag{5}$$

$S(x, y)$ denotes the saturation value. If a pixel at (x, y) satisfies:

$$\left|D_L(x, y) - D_H(x, y)\right| > T, \tag{6}$$

where T is a threshold, three horizontal consecutive pixels centered by the pixel are labeled as transition points. Compared to Kim's approach [2], our method also labels transition points in vertical direction in the same way to identify the top and bottom text boundaries. Then, a transition map is generated:

$$T(x, y) = \begin{cases} 1 & \text{if } (x, y) \text{ is an transition point} \\ 0 & \text{otherwise} \end{cases}. \tag{7}$$

Edge Map Generation. Considering the canny edge detector has a good performance on text boundary detection in the complex background image, we introduce the edge feature complementary to prevent text loss. However, on the assumption that text pixels have high intensities compared to background pixels, the detected edge by canny operator is outside the caption text while we expect to obtain the text edge which consists of text pixels. Since there is a higher contrast between text and its boundary than others, which is essential to highlight the text in video, we can identify the text edge by calculating the grayscale difference values between each detected edge pixel and the four pixels connected to it. The pixel corresponding to the maximum difference is labeled as a text edge point. Then, an edge map is generated:

$$E(x, y) = \begin{cases} 1 & \text{if} (x, y) \text{ is an text edge point} \\ 0 & \text{otherwise} \end{cases}. \tag{8}$$

Dam Point Labeling. In consideration of the background continuity, Kim et al. filled the background inward from ER and employed dam point definition to prevent filling from flooding into text. With the transition map and the edge map, the modified dam point definition is as follows:

$$\begin{aligned} Dam\, points = \{(x, y) | B(x, y) = \text{"White"} \\ \bigwedge MIN\_W \leq min(H\_len(x, y), V\_len(x, y)) \leq MAX\_W \\ \bigwedge max(H\_len(x, y), V\_len(x, y)) \leq EO\_height - 4 \\ \bigwedge \left(T(x, y) = 1 \bigvee E(x, y) = 1\right)\} \end{aligned} \tag{9}$$

where $MIN\_W = 1$ and $MAX\_W = EO\_height/10$. $H\_len(x, y)$ and $V\_len(x, y)$ denote the length of horizontal and vertical white connectivity passing (x, y), respectively. For example, the *H_len* of the bottom left pixel in Fig. 3(a) is 3 and its *V_len* is 2. The labeled dam points are set to be "Gray" in B(x, y), as shown in Fig. 3(b).

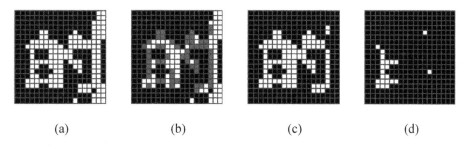

(a)   (b)   (c)   (d)

Fig. 3. Text segmentation procedure (a) Adaptive thresholding (b) Modified dam point labeling (c) Inward filling (d) Result of segmentation using Kim's method

Inward Filling. Obtaining dam point labeled image, we use flood fill algorithm addressed in [2] on each pixel in *ER* of *B(x, y)*. If the pixel is "White", all the "White" pixels connected to it, including itself, are set to be "Black". After the filling is completed, all non-"Black" pixels are set to be "White". The inward filling result of Fig. 3(b) is shown in Fig. 3(c). Figure 3(d) shows the result of segmentation using Kim's method [2], which illustrates that our approach can effectively prevent character stroke loss during segmentation.

3.3   Multi-frame Integration

After the text segmentation process in individual frames, there are still several background areas in the resulting masks, which are regarded as text due to the similarity in intensity and shape. Since caption in video is commonly stationary over time while background may change drastically, we integrate the information of multiple frames to remove the noises. By using our proposed tracking technique, we can find the first and the last occurrences of a caption, which are defined as t_{start} and t_{end}, respectively. We calculate the average mask of the masks between t_{start} and t_{end} as follows:

$$mask_{avr} = \frac{1}{n} \sum_{t=t_{start}}^{t_{end}} mask_t, \tag{10}$$

$n = t_{end} - t_{start} + 1$ denotes the number of frames between t_{start} and t_{end}. The $mask_{avr}$ is subsequently binarized by a fixed threshold:

$$mask_{avr}(x, y) = \begin{cases} 255 & if\ mask_{avr}(x, y) \geq T_m \\ 0 & otherwise \end{cases}, \tag{11}$$

where T_m is set to be 255×0.8, indicating that if a pixel is regarded as text pixel over 80 percent in the text masks, we identify it as the true text pixel. As a result, parts of the background which are not stable passing frames are removed and some missed strokes in a single frame can be retrieved using information from other frames. Then, we calculate the averaged grayscale image $G_{avr}(x, y)$ between t_{start} and t_{end} as follows:

$$G_{avr}(x, y) = \frac{1}{n} \sum_{t=t_{start}}^{t_{end}} G^t(x, y). \tag{12}$$

We use $mask_{avr}$ to filter $G_{avr}(x, y)$:

$$G_{avr}(x, y) = \begin{cases} G_{avr}(x, y) & if\ mask_{avr}(x, y) = 255 \\ 0 & otherwise \end{cases}. \tag{13}$$

In order to further eliminate the erroneous background with similar intensity to text, we normalize the value of $G_{avr}(x, y)$ to $[0, 1]$ and apply gamma correction on $G_{avr}(x, y)$ to stretch the range of high intensity and compress the range of low intensity. The correction is defined as follows:

$$G_{avr}(x, y) = \begin{cases} G_{avr}(x, y)^3 & if\ G_{avr}(x, y) \neq 0 \\ 0 & otherwise \end{cases}. \tag{14}$$

Then, $G_{avr}(x, y)$ is segmented as follows:

$$\hat{G}_{avr}(x, y) = \begin{cases} 255 & if\ G_{avr}(x, y) \geq 0.9 \times M \\ 0 & otherwise \end{cases}, \tag{15}$$

where M denotes the mean of non-zero pixel values in $G_{avr}(x, y)$. Due to the real text pixels blended with background color may be dropped in this procedure, region growing is performed to retrieve these text pixels. All the text pixels in $mask_{avr}$ are set to be "Gray" first. Then we scan each pixel in $\hat{G}_{avr}(x, y)$, if $\hat{G}_{avr}(x, y) = 255$, the corresponding pixel in $mask_{avr}$ is regarded as a seed point. All the "Gray" pixels connected to the seed point in $mask_{avr}$ and the seed itself will be set to be "White". After this process, all non-"White" pixels in $mask_{avr}$ are set to be "Black". Finally, the masks between t_{start} and t_{end} are replaced by the resulting $mask_{avr}$.

4   Experimental Results

To our best knowledge, there is no public dataset for video text recognition, we annotated ten Chinese videos from Internet as our dataset. These videos, including news programs, cartoons, movies, series, and variety shows, are different on the aspects of resolution and contrast. There are 6559 frame images totally in the dataset, including 452 different captions. The captions have different sizes and are all embedded in complex backgrounds. In this section, we first validate the efficiency of our tracking technology. Then, the text segmentation method in individual frames and multi-frame integration strategy are evaluated, respectively.

4.1   Text Tracking

The dataset is divided into five parts based on the 5-folder cross-validation. We first extract path signatures in each part, maintaining the balance of positive and negative samples. To be adapted to AlexNet CNN, the signature maps are then resized to 227×227. Our CNN model is trained on the Caffe framework [20] using SGD with initial learning rate 0.001, 0.9 momentum, 0.0005 weight decay, and batch size 100. Finally, the recognition accuracy reaches 96 %, which demonstrates the good performance of our tracking strategy with path signatures.

4.2   Single-Frame Text Segmentation

To assess the performance of text segmentation, we analyze its contribution to improving the Character Recognition Rate (CRR). Tesseract-OCR in MATLAB 2014 is assembled as the OCR engine.

$$\mathrm{CRR} = \frac{|correct\ characters|}{|ground\ truth\ characters|}. \tag{16}$$

Our proposed text segmentation approach in individual frames is compared with Otsu method [5] and Kim's method [2]. The Otsu method is a simple and classic algorithm cited by many researchers, while the Kim's method which is presented recently works well on text segmentation with complex backgrounds. Since we assume that text pixels have high intensities compared to background pixels, the color polarity computing procedure in Kim's method is omitted. Figure 4 shows the results of segmentation.

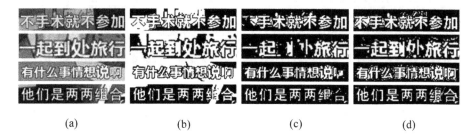

(a)   (b)   (c)   (d)

Fig. 4. Results of single-frame text segmentation (a) Original images (b) Otsu method (c) Kim method (d) Our single-frame segmentation method

As shown in Fig. 4(b), it is difficult for the Otsu method to extract clear text from complex backgrounds, although it can be applied for clean documentary images. Since the Otsu method segments the entire image by a single global threshold, it fails to remove the background whose intensity is similar to text. Figure 4(c) displays the results obtained by Kim's method [2]. We can see that most text are well separated from complex backgrounds. However, a few character strokes are missed because of the blur between text and background in the low contrast and resolution image. And only considering transition points in one direction also partly affects dam point labeling procedure which is used for preventing filling from flooding into text. Therefore, we take account of four direction transition points and add a constraint of edge feature to retrieve the missing strokes. The results of our method are shown in Fig. 4(d). The extra noises introduced by our approach can be further eliminated using multi-frame integration. Table 1 demonstrates the efficiency of our approach.

Table 1. Recognition rates of single-frame text segmentation methods

Method	CRR (%)
Otsu [5]	36.93
Kim [2]	46.14
Our single-frame segmentation	48.81

4.3 Multi-frame Integration

In this experiment, the integration strategies that proposed by Liu [15] and Phan [16] are implemented for comparison on the masks obtain by our single-frame text segmentation. We choose them because the logic "and" operation in former is simple and effective to remove the erroneous background, while the latter is a newly proposed method based on possibility map and shape refining. Figure 5 shows the integrated masks.

| (a) | (b) | (c) | (d) |

Fig. 5. Results of multi-frame integration (a) Single frame masks (b) Liu method (c) Phan method (d) Our integration method

We can observe that all the multi-frame integration strategies mentioned above have improved the performance of single-frame text segmentation to varying degrees. Since caption generally stays in the same position through sequential frames, while background may change drastically, erroneous background can be removed by using the information of multiple frames. However, as shown in Fig. 5(b), logic "and" operation may unexpectedly remove text pixels due to the real text pixel may be erroneously labeled as background in several frames. In Fig. 5(b) (c), we can see that although the methods that proposed by Liu [15] and Phan [16] can eliminate part of noises, they still cannot effectively remove the background with similar intensity to text. Thus, we employ gamma correction in our integration strategy to stretch the range of high intensity and compress the range of low intensity. Then a clear caption text can be extracted. Table 2 shows the CRRs of these methods. The effectiveness of our method is confirmed by achieving the best CRR among them. The average running time to process a video frame of the entire text segmentation is approximately 1.38 s.

Table 2. Recognition rates of multi-frame integration methods

Method	CRR (%)
Our single-frame segmentation	48.81
Liu [15]	51.22
Phan [16]	52.21
Our integration method	57.16

5   Conclusion

We have proposed an effective caption segmentation approach for video. A CNNs classifier is first trained with path signature for tracking. Then, a coarse text segmentation is conducted in individual frames, based on Kim's extraction method [2]. We modify the definition of transition point and introduce edge feature to optimize the segmentation performance. Multi-frame integration is finally performed to refine the segmentation through gamma correction followed by region growing. The experimental results have validated the effectiveness of the proposed approach. Furthermore, in addition to handling Chinese caption, the proposed approach is credible to address captions in other languages because of its independence of characters. And it is also robust to different fonts and character sizes. In the future work, we will add color polarity judgement, optimize the running time to process each frame and expand the approach in scrolling caption text.

References

1. Wang, X., Huang, L., Liu, C.: A video text location method based on background classification. Int. J. Doc. Anal. Recogn. (IJDAR) **13**(3), 173–186 (2010)
2. Kim, W., Kim, C.: A new approach for overlay text detection and extraction from complex video scene. IEEE Trans. Image Process **18**(2), 401–411 (2009)
3. Liu, Y., Srihari, S.N.: Document image binarization based on texture features. IEEE Trans. Pattern Anal. Mach. Intell. **19**(5), 540–544 (1997)
4. Cheriet, M., Said, J.N., Suen, C.Y.: A recursive thresholding technique for image segmentation. IEEE Trans. Image Process. **7**(6), 918–921 (1998)
5. Otsu, N.: A threshold selection method from gray-level histograms. Automatica **11**(285–296), 23–27 (1975)
6. Ohya, J., Shio, A., Akamatsu, S.: Recognizing characters in scene images. IEEE Trans. Pattern Anal. Mach. Intell. **16**(2), 214–220 (1994)
7. Lyu, M.R., Song, J., Cai, M.: A comprehensive method for multilingual video text detection, localization, and extraction. IEEE Trans. Circ. Syst. Video Technol. **15**(2), 243–255 (2005)

8. Ye, Q., Gao, W., Huang, Q.: Automatic text segmentation from complex background. In: 2004 International Conference on Image Processing, ICIP 2004, vol. 5, pp. 2905–2908. IEEE (2004)
9. Wang, X., Huang, L., Liu, C.: A novel method for embedded text segmentation based on stroke and color. In: 2011 International Conference on Document Analysis and Recognition (ICDAR), pp. 151–155. IEEE (2011)
10. Mishra, A., Alahari, K., Jawahar, C.V.: An MRF model for binarization of natural scene text. In: 2011 International Conference on Document Analysis and Recognition (ICDAR), pp. 11–16. IEEE (2011)
11. Wu, V., Manmatha, R., Riseman, E.M.: Textfinder: an automatic system to detect and recognize text in images. IEEE Trans. Pattern Anal. Mach. Intell. **11**, 1224–1229 (1999)
12. Wakahara, T., Kita, K.: Binarization of color character strings in scene images using K-means clustering and support vector machines. In: 2011 International Conference on Document Analysis and Recognition (ICDAR), pp. 274–278. IEEE (2011)
13. Mancas-Thillou, C., Gosselin, B.: Spatial and color spaces combination for natural scene text extraction. In: 2006 IEEE International Conference on Image Processing, pp. 985–988. IEEE (2006)
14. Lienhart, R., Wernicke, A.: Localizing and segmenting text in images and videos. IEEE Trans. Circ. Syst. Video Technol. **12**(4), 256–268 (2002)
15. Liu, X., Wang, W.: Robustly extracting captions in videos based on stroke-like edges and spatio-temporal analysis. IEEE Trans. Multimedia **14**(2), 482–489 (2012)
16. Phan, T.Q., Shivakumara, P., Lu, T. et al.: Recognition of video text through temporal integration. In: 2013 12th International Conference on Document Analysis and Recognition (ICDAR), pp. 589–593. IEEE (2013)
17. Chen, K.T.: Integration of paths–a faithful representation of paths by noncommutative formal power series. Trans. Am. Math. Soc. **89**(2), 395–407 (1958)
18. Graham, B.: Sparse arrays of signatures for online character recognition. arXiv preprint arXiv: 1308.0371 (2013)
19. Yang, W., Jin, L., Xie, Z. et al.: Improved deep convolutional neural network for online handwritten Chinese character recognition using domain-specific knowledge. In: 2015 13th International Conference on Document Analysis and Recognition (ICDAR), pp. 551–555. IEEE (2015)
20. Jia, Y., Shelhamer, E., Donahue, J. et al.: Caffe: convolutional architecture for fast feature embedding. In: Proceedings of the 22nd ACM international conference on Multimedia, pp. 675–678. ACM (2014)

Single Low-Light Image Enhancement Using Luminance Map

Juan Song[✉], Liang Zhang, Peiyi Shen, Xilu Peng, and Guangming Zhu

School of Software, Xidian University, Xi'an 710071, China
songjuan@mail.xidian.edu.cn

Abstract. In this paper, a fast enhancement method based on de-hazing is proposed for single low-light images. Instead of dark channel prior (DCP) used in the de-hazing related literature, the luminance map is used to estimate the global atmospheric light and the transmittance according to the observed similarity between the luminance map and DCP. Through this substitution, on the one hand the computation complexity is greatly reduced; on the other hand the block artifacts is also avoided brought by discontinuous transmittance estimated from DCP. Experimental results indicate that the proposed method has a significant improvement in both enhancement effects and processing speed compared with state-of-art enhancement algorithms.

Keywords: Image enhancement · Low-light conditions · De-hazing · Luminance map

1 Introduction

Images obtained under low-light conditions tend to have characteristics of low dynamic range and indistinguishable details. Image degradation caused by low-light conditions not only affects the recognition of images by human eyes but also influences the performance of the computer vision system, such as object detection, object recognition, and so on. The traditional image enhancement algorithms, histogram equalization [1], enhanced the low-light images by adjusting the gray level distribution of the original images. Although it is simple to be implemented, it may result in super saturation and structure information loss. The Retinex-based enhancement algorithm [2] transformed the original image data into two parts, luminance map and the reflectance map, and then enhanced the image by reducing the influence of the luminance map on the reflectance map, but Retinex based methods often lead to gray-out results. In addition, Malm et al. proposed an adaptive enhancement method for low light-level video according to the structure tensor [3], Fu et al. proposed an enhancement algorithm in terms of the color estimation model (CEM) [4], and Fotiadou et al. proposed an enhancement algorithm in the light of sparse representations (SR) [5]. These algorithms achieved good enhancement performance, but the computation complexity is extremely high, which hinders their applications to real time fields.

© Springer Nature Singapore Pte Ltd. 2016
T. Tan et al. (Eds.): CCPR 2016, Part II, CCIS 663, pp. 101–110, 2016.
DOI: 10.1007/978-981-10-3005-5_9

Dong et al. proposed an enhancement algorithm for low-light video [6] using the de-hazing technique based on the dark channel prior (DCP) [7] through analyzing the similarities between inverted low-light intensity and dense fog images. This algorithm can significantly improve the image brightness and enhance the visual details, but the obvious drawback is that it will introduce block artifacts because the estimated transmittance is not continuous.

In this paper, a fast low-light enhancement algorithm is proposed based on de-hazing technique. In our proposed algorithm, luminance map of the inverted low-light image is used to estimate global atmospheric light and transmittance instead of DCP used in [6], based on the analysis of the similarity of the luminance map and the DCP. By doing so, not only the computation complexity can be reduced, but also the block artifacts could be avoided.

2   The Proposed Algorithm

The low-light images can be enhanced through de-hazing the inverted low-light images since the inverted low-light images have high similarity with the hazy ones [6] as is shown in Fig. 1.

The inverted low-light image is computed in Eq. (1)

$$I^c_{inv}(x) = 255 - I^c(x), \tag{1}$$

where $I_{inv}(x)$ is the inverted low-light image, c denotes one of the three color channels (RGB).

According to the hazy image degradation model, the physical model of inverted low-light image can be expressed as follows:

$$I^c_{inv}(x) = J^c_{inv}(x)t(x) + A^c(1 - t(x)), \tag{2}$$

Fig. 1. Top: low lighting images. Middle: inverted low-light images Bottom: hazy images

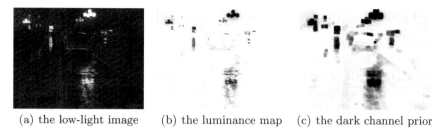

(a) the low-light image   (b) the luminance map   (c) the dark channel prior

Fig. 2. Similarities between the luminance map and the dark channel prior of the inverted low-light images

where $J_{inv}(x)$ is the inverted underlying scene radiance, $t(x)$ is the transmittance, and A is the global atmospheric light. Therefore, in order to recover the inverted scene radiance $J_{inv}(x)$ from $I_{inv}^c(x)$, the global atmospheric light A and transmittance $t(x)$ should be estimated.

In [6], Dong estimated A and $t(x)$ using the DCP $D^{dark}(x)$ computed in (3)

$$D^{dark}(x) = \min_{c \in \{R,G,B\}} (\min_{y \in Omega(x)} (I^c(y)))$$ (3)

On one hand, the minimum filtering is needed to compute the DCP, which is quite time-consuming; on the other hand, the transmittance in a local patch is assumed to be constant and it is not refined by soft matting because of computation complexity. Thus severe block artifacts would be introduced in the places where transmittance is not continuous.

This paper aims to overcome the above drawbacks. Our proposed method is inspired by the observation that luminance maps of the inverted low-light images have high similarity with the DCP as shown in Fig. 2. Therefore we will use the luminance map to estimate the global atmospheric light and transmittance instead of the DCP.

2.1   Computing the Luminance Map

Three color channels of the inverted-low light are weighted summed to computed the luminance map $Y(x)$ in Eq. (4). Since different color channels have different sensitivity degrees to the noise, different weights are assigned to each color channel.

$$Y(x) = 0.299 \times I_{inv}^R(x) + 0.587 \times I_{inv}^G(x) + 0.114 \times I_{inv}^B(x)$$ (4)

As shown in Fig. 2, the computed luminance map of the inverted low-light image is quite similar with the DCP. In fact, the luminance map has even finer and more continuous structure information compared with the DCP. Then we will use the luminance map instead to estimate the global atmospheric light and transmittance.

2.2   Estimating the Global Atmospheric Light

In many previous works, the pixels that have the highest intensity values are considered to be the most haze-opaque, and these pixels are often used as the global atmospheric light A [8]. Although this method is simple and fast, it may lead to inaccurate estimation when the brightest pixel is brighter than the global atmospheric light in the haze image (such as a white building). In the dark channel prior algorithm, He et al. in [7] chose the 0.1 % pixels which have the maximum intensity in the DCP, and then selected the pixel with the highest intensity in the original image from the 0.1 % pixels as the global atmospheric light. This method is quite reasonable; however, the process of computing DCP is so time-consuming that it is not suitable to be used in real-time application. To simplify the estimation process, Dong directly estimated the global atmospheric light from the inverted low-light image. Computation complexity is decreased at the cost of estimation accuracy without the assistance of DCP [6].

Considering the similarity between the luminance map and the DCP, we substitute the luminance map for the DCP to estimate the global atmospheric light. The pixel with the highest intensity in the inverted low-light image is selected as the global atmospheric light from the 0.1 % pixels with the highest intensity in the luminance map.

On the one hand, the luminance map is much easier to be computed compared with the DCP; on the other hand, the luminance map can also offer a good guidance to estimate the global atmospheric light.

2.3   Estimating the Transmittance

It is noted that in real hazy images the transmittance is attenuated exponentially with the scene depth. The farther the scene is, the smaller the magnitude of the transmittance and the denser the haze, vice versa. However, the inverted low-light image is not a real haze one. Instead of attenuating with the scene depth, its transmittance is closely connected to luminance. The darker the scene is, the denser the corresponding haze of the inverted low-light image. Therefore it is more reasonable to estimate the transmittance using the luminance map.

The initial transmittance map $\hat{t}(x)$ is estimated using the luminance map $Y(x)$ as:

$$\hat{t}(x) = 1 - \omega Y(x)/255 \qquad (5)$$

where ω is a parameter. The smoother transmittance may allow the underlying scene radiance map to contain more details [9,10], so the mean filter is used to obtain the final transmittance $t(x)$:

$$t(x) = meanfilter(\hat{t}(x)) \qquad (6)$$

The transmittance $t(x)$ tends to be very low, thus the lower bound of the transmittance is limited to 0.01 in order to avoid being zero as the denominator in the step of recovering the scene radiance.

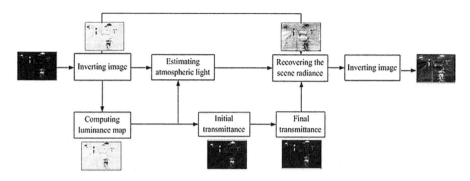

Fig. 3. The flowchart of the proposed algorithm

2.4 Recovering the Inverted Scene Radiance

After obtaining the estimation values of the global atmospheric light and transmittance, the inverted underlying scene radiance $J_{inv}(x)$ can be recovered according to Eq. (7)

$$J_{inv}^c = \frac{I_{inv}^c - A^c}{t(x)} + A^c \tag{7}$$

Finally the enhancement result can be obtained by inverting the recovered scene radiance again.

To summarize the algorithm, the flowchart of the proposed algorithm is shown in Fig. 3. In addition the processing result of each module is also visualized in the figure.

3   Experiment Results

In this section, we compared the proposed algorithm with several other common low-light image enhancement algorithms in terms of subjective quality, objective quality and running speed. The benchmark enhancement algorithms include the histogram equalization (HE) [1], the color estimation model (CEM) [4], the enhancement algorithm based on sparse representations (SR) [5], the de-hazing-based low-light video frame enhancement algorithm proposed by Dong et al. (Dong) [6]. All the test images are downloaded from the high dynamic range image dataset provided by Sen et al. [11]. In our proposed algorithm, the parameter ω in Eq. (5) is set to be 0.98 and the patch size of the mean filter is set to be 15 experimentally for the tradeoffs between the complexity and performance.

3.1   Subjective Quality Comparison

Figures 4, 5 and 6 separately show the enhancement results of the aforementioned benchmark enhancement algorithms and the proposed algorithm. As shown in Figs. 4, 5 and 6, HE causes loss of details (such as the music score region in

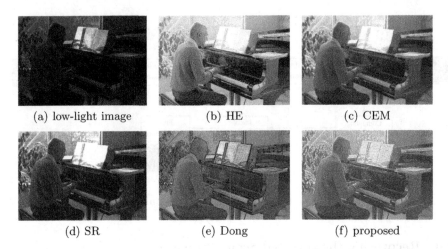

(a) low-light image   (b) HE   (c) CEM

(d) SR   (e) Dong   (f) proposed

Fig. 4. Piano man dataset: the visual comparison with common low-light image enhancement algorithms

Fig. 4(b), the window region in Fig. 6(b)) and supersaturation (such as the grass and sky region in Fig. 5(b)). The brightness and the contrast degree are still not boosted enough by the SR algorithm and the CEM. Dong's results introduce severe block artifacts (such as the music score region in Fig. 4(d), the outline of the baby in Fig. 5(d) and the window in Fig. 6(d)). Our proposed algorithm can enhance the image contrast effectively without introducing saturation effects and block artifacts. In addition, it will not cause the loss of details, only our proposed algorithm can recover the details of the skylight in the memorial dataset as shown in Fig. 6.

3.2 Objective Quality Comparison

Table 1 compares the objective quality of the common low-light image enhancement algorithms and the proposed algorithm using structural similarity image quality index (SSIM) [12]. The well-exposed images in the same scene are chosen as the reference images. As shown in Table 1, the SSIM indices of the proposed algorithm are the highest in all the test datasets.

In general, the proposed method outperforms other state-of-art enhancement algorithms in terms of objective quality and visual effects.

3.3 Running Speed Comparison

Firstly, the runtime of the proposed algorithm is compared with HE and Dong algorithms in Table 2 using a single core of a 1.60 GHz Intel Core i5-4200U CPU in Matlab since the open source codes of the CEM and SR algorithms are unavailable. As shown in Table 2, the proposed algorithm can significantly shorten the processing time for images with different sizes. Among the three algorithms, our

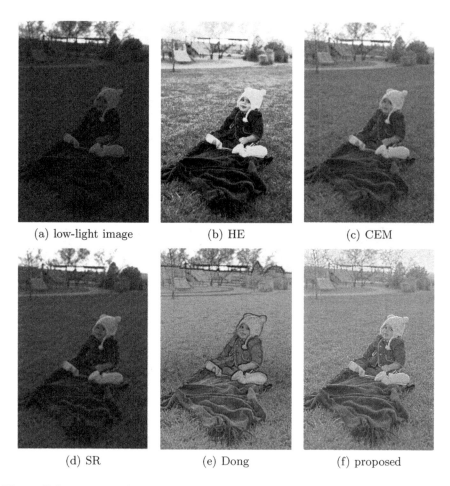

(a) low-light image   (b) HE   (c) CEM

(d) SR   (e) Dong   (f) proposed

Fig. 5. Baby on grass dataset: the visual comparison with common low-light image enhancement algorithms

Table 1. Quality assessment of common enhancement algorithms

Dataset	HE [1]	CEM [4]	SR [5]	Dong [6]	Proposed
PianoMan	0.612	0.791	0.852	0.840	0.861
BabyOnGrass	0.407	0.590	0.650	0.573	0.703
Memorial	0.679	0.792	0.677	0.659	0.810
FeedingTime	0.457	0.683	0.706	0.633	0.792
Church	0.533	0.742	0.659	0.681	0.839

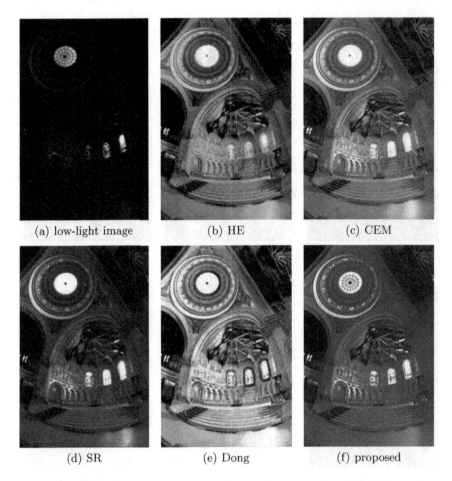

Fig. 6. Memorial dataset: the visual comparison with common low-light image enhancement algorithms

proposed algorithm runs the fastest. Specifically, the speedup can reach more than 20 times compared with Dong algorithm.

Secondly, the proposed algorithm is further accelerated with GPU in parallel since the proposed algorithm is operated in the pixel level. For a 1024×1024

Table 2. Runtime comparison between Dong and the proposed algorithm for images with different sizes

Image size	HE(s) MATLAB	Dong(s) MATLAB	Proposed(s) MATLAB	Proposed(s) GPU
256×256	0.463	0.609	0.071	0.004
512×512	0.492	1.760	0.090	0.013
1024×1024	0.561	6.456	0.351	0.060

high-resolution image, the processing time can be further shortened to 0.06 s on NVIDIA GeForce GTX970 GPU with 2 GB global memory (excluding the reading and writing of the image), which can meet the real-time requirements in practical applications.

4   Conclusion

A fast enhancement algorithm based on de-hazing technique is proposed for a single low-light image in this paper. In our proposed algorithm, the luminance map of the inverted low-light image is used to estimate the global atmospheric light and transmittance instead of the DCP according to the high similarity between the luminance map and the DCP. Experimental results demonstrate that our proposed algorithm achieves excellent enhancement results in terms of subjective and objective quality. In addition, our proposed algorithm can meet the real-time requirements in practical applications since the computation complexity is greatly reduced without computing DCP using minimum filtering.

Acknowledgment. This project is supported by NSFC Grant (No. 61401324, No. 61305109, No. 61072105), by 863 Program (2013AA014601), and by Shaanxi Scientific research plan (2014K07-11, 2013K06-09).

References

1. Pizer, S.M., Amburn, E.P., Austin, J.D., Cromartie, R., Geselowitz, A., Greer, T., Romeny, B.H., Zimmerman, J.B., Zuiderveld, K.: Adaptive histogram equalization and its variations. Comput. Vision Graph. Image Process. **39**(3), 355–368 (1987)
2. Rahman, Z., Jobson, D.J., Woodell, G.: Multi-scale retinex for color image enhancement. In: Proceedings of the Third IEEE International Conference on Image Processing, Lausanne, Switzerland, vol. 3, pp. 1003–1006, 16–19 September 1996
3. Malm, H., Oskarsson, M., Warrant, E., Clarberg, P., Hasselgren, J., Lejdfors, C.: Adaptive enhancement and noise reduction in very low light-level video. In: Proceedings of the 11st IEEE International Conference on Computer Vision, Rio de Janeiro, Brazil, pp. 1–8, 14–21 October 2007
4. Fu, H., Ma, H., Wu, S.: Night removal by color estimation and sparse representation. In: Proceedings of the 21st IEEE International Conference on Pattern Recognition, Tsukuba, Japan, pp. 3656–3659, 11–15 November 2012
5. Fotiadou, K., Tsagkatakis, G., Tsakalides, P.: Low light image enhancement via sparse representations. In: Campilho, A., Kamel, M. (eds.) ICIAR 2014. LNCS, vol. 8814, pp. 84–93. Springer, Heidelberg (2014). doi:10.1007/978-3-319-11758-4_10
6. Dong, X., Wang, G., Pang, Y., Li, W., Wen, J., Meng, W., Lu, Y.: Fast efficient algorithm for enhancement of low lighting video. In: Proceedings of the IEEE International Conference on Multimedia and Expo, Barcelona, Spain, pp. 1–6, 11–15 July 2011
7. He, K., Sun, J., Tang, X.: Single image haze removal using dark channel prior. IEEE Trans. Pattern Anal. Mach. Intell. **33**(12), 2341–2353 (2011)
8. Tan, R.: Visibility in bad weather from a single image. In: Proceedings of IEEE Conference on Computer Vision and Pattern Recognition, pp. 1–8, June 2008

9. Guo, F., Cai, Z., Xie, B., Tang, J.: Automatic image haze removal based on luminance component. In: Proceedings of the Sixth IEEE International Conference on Wireless Communications Networking and Mobile Computing, Chengdu, China, pp. 1–4, 23–25 September 2010

10. Zhang, X., Shen, P., Luo, L., Zhang, L., Song, J.: Enhancement and noise reduction of very low light level images. In: Proceedings of the 21st IEEE International Conference on Pattern Recognition, Tsukuba, Japan, pp. 2034–2037, 11–15 November 2012

11. Sen, P., Kalantari, N.K., Yaesoubi, M.: Robust patch-based HDR reconstruction of dynamic scenes. ACM Trans. Graph. **31**(6), 439–445 (2012)

12. Wang, Z., Bovik, A.C., Sheikh, H.R., Simoncelli, E.P.: Image quality assessment: from error visibility to structural similarity. IEEE Trans. Image Process. **13**(4), 600–612 (2004)

Image Copy Detection Based on Convolutional Neural Networks

Jing Zhang[1], Wenting Zhu[2], Bing Li[2(✉)], Weiming Hu[2], and Jinfeng Yang[1]

[1] College of Electronic Information and Automation,
Civil Aviation University of China, Tianjin 300300, China
zhangjingfighting@gmail.com
[2] National Laboratory of Pattern Recognition,
CAS Center for Excellence in Brain Science and Intelligence Technology,
Institute of Automation, Chinese Academy of Sciences,
No. 95, Zhongguancun East Road, Beijing 100190, China
bli@nipr.ia.ac.cn

Abstract. In this paper, we present a model that automatically differentiates copied versions of original images. Unlike traditional image copy detection schemes, our system is a Convolutional Neural Networks (CNN) based model which means that it does not need any manually-designed features. In addition, a convolutional network is more applicable to image copy detection whose architecture is designed for robustness to geometric distortions. Our model uses fully connected layers to compute a similarity between CNN features, which are extracted from image pairs by a deep convolutional network. This method is very efficient and scalable to large databases. In order to see the comparison visually, a variety of models are explored. Experimental results demonstrate that our model presents surprising performance on various data sets.

Keywords: Image copy detection · Feature extraction · CNN

1 Introduction

Due to the rapid development of technologies like communication, computer network and multimedia, the online multimedia resources are growing exponentially. These advanced technologies are like a double-edged sword, bringing a variety of convenience as well as much challenges to the information security. The convenience of digital image acquisition makes more and more unauthorized copies on the internet. In order to effectively protect the copyright of legitimate users, copyright piracy needs to be monitored. This involves the primary technology, detecting the copies of images.

There are usually two kinds of methods of copyright protection, digital watermarking and copy detection. The digital watermarking technology is to embed watermark information into an image before the image published. Images with no embedded watermark information can not be detected, but the image copy detection only works on the image itself. Because of legal and other reasons,

© Springer Nature Singapore Pte Ltd. 2016
T. Tan et al. (Eds.): CCPR 2016, Part II, CCIS 663, pp. 111–121, 2016.
DOI: 10.1007/978-981-10-3005-5_10

it can not modify image content. Image copy detection becomes a reasonable complement of digital watermarking technology.

In recent years, a technique of mimicking human brain learning—deep learning has made a big leap. It evolved from the initial neural network and achieved great success in a multitude of pattern recognition problems. Take CNN for example, incredible strides had been made on image representation [14]. Because its invariance for translation, scaling, tilting, and other forms of distortion, we build our method surrounding this powerful tool to automatically learn copy image detection.

The contributions of this paper are as follows:

1. The convolutional neural networks are used to directly learn the copy detection task, which does not need any manually-designed features.
2. Different neural network models are proposed and formed a clear contrast, which pave the way for further studies.
3. We apply our models to several data sets, showing tremendous successes in accuracy. It proves that using CNN architecture is perfectly competent for image copy detection.

2   Related Work

2.1   Traditional Image Copy Detection

A traditional image copy detection, includes three main technology parts: feature extraction, feature based index construction and feature comparison. In this section, we briefly review previous works on feature extraction and index construction methods.

The extraction of feature information has two forms, global feature extraction and local feature extraction. The global information can be obtained (like texture) on the whole image. Just like Li in [16], he proposed Gabor texture descriptors using the adjustability of Gabor on the direction and scale. The descriptor is invariant to rotation and scaling invariance. But there is a high false alarm rate for large angle rotation. Copy detection algorithm based on global features is simple in calculation and has high efficiency. But the poor resistance to geometric attacks, especially cropping and rotation, makes scholars prefer detecting image on local features. Berrani [2] computed local differential descriptors for each image which corresponded to the local regions of interest in the image. As [18] puts it, the best performance among all the local descriptors is the Scale Invariant Feature Transform (SIFT) descriptor. This inspires many scholars to do more in-depth research in this field. For example, a method of VLAD (vector of locally aggregated descriptors) based on a compact representation of SIFT was proposed in [11]. Similarly, Cao et al. [3] introduced a new tilt parameters in the affine SIFT transform, using geometric consistency constraints in similarity detection. Comparing with the global features, local features have a good image recognition ability, but its computational complexity is high.

In the second part, Bags-of-visual-Words (BOW) and hash algorithms are the most representative methods in the feature based index construction. The BOW [20] extracts image features as visual words. Some others improved it by changing the grouped visual words into visual phrases [28] and visual sentences [23]. H. Ling [17] abandoned previous methods and proposed a local binary fingerprint as visual word. This method is available on large data sets and more efficient. But the visual vocabulary discards spatial information of local feature, which will influence the matching work. For indexing algorithm using hashing function, Locality-Sensitive Hashing (LSH) [9] is a popular algorithm in multimedia applications. Recently, many new kinds of hashing schemes have been proposed such as Spectral hashing [15] and Self-taught hashing [27]. These schemes appear a significant improvement over other methods. However, it brings a high computational cost based on local embedding techniques. A robust image hashing based on Discrete Cosine Transform (DCT) and Discrete Wavelet Transform (DWT) for copy detection was presented by Tang [22]. The algorithm for conventional digital operation has good robustness and uniqueness. Subsequently, he put forward a hash based on fan-beam transform algorithm [21] which could resist rotation attacks at any angle. Excellent results in image copy detection have been achieved using traditional methods, but the artificial intervention technique makes the experimental process more complicated.

2.2   Research Progress of CNN

In contrast to the above approaches, we learn features end-to-end directly from input pixels. So a simple but efficient deep learning method is proposed for image copy detection, and it achieves more favorable results on publicly available data sets. Deep learning is a successful model for learning useful representation in research field of images, owing to its strong computing power especially in a large dataset [5, 14]. In the task of copy detection, we conduct experiment on two images, and let the model make a decision whether the detected image is a copy of original image. Therefore, in so many models of deep learning, the siamese CNN model is more suitable for our study.

More recently, researchers in the siamese network have presented different approaches. In [26], researchers compared image patches via a similar architecture. The results showed an good performance with a high computational cost. Because they focused on the center of images too much. In older to learn a high precision arithmetic, Han et al. [7] used a multi-layer network followed by a fully connected network. In [25], the best results were obtained on the KITTI benchmark using CNN to compute the stereo matching cost. The success of these methods means that, for the study of two inputs, just like our work, siamese CNN model has a broad research space. Inspired by these methods, our network structure is similar to them. But there is a notable difference that we show how to build small but powerful model using discriminative strategies. Contrast with [25] we use pooling layers to increase the robustness for the different variations of copy images. And softmax-with-loss is used to make a simple classification for input images compared to [26]. Our models still have other differences in architecture, like an additional dropout layer and Local Response Normalization (LRN) layer. The proposed method will be described in next section.

3   The Proposed Method

3.1   Overview

Our goal is to learn an image copy detection model. When two images C (copy image) and O (original image) are input to the model, the model give the corresponding output 0 or 1 in which the 1 represents a copy relationship between C and O, and the 0 opposite.

The early Siamese networks [6] used contrastive loss functions to train a similarity metric from real data, which is able to place similar images nearby and keep dissimilar images separated. But in our work we want to identify a copy relationship between the pair of images instead of a metric. Here we put our work as a binary classification task. Input images are divided into two categories, copy pair (1) or not (0).

Our proposed model in this paper include multiple convolutional and spatial pooling layers to obtain feature vectors, followed by fully connected layer to compare features of input images. This has some similarities with [7, 26]. But in order to adapt to our tasks, on the top of models, softmax-with-loss instead of cross-entropy loss or hinge-based loss leads to the objective function learning. After the fundamental models, we offered an additional method which also achieved a good result.

3.2   Network Architectures

Fundamental Models. As shown in Fig. 1, depending on whether or not the weights of the two branches are shared, the two branches models have two forms, siamese and pseudo-siamese [26]. Following Simonyans [19] advice, 3×3 kernels are adopted to make decision function more discriminative. Rectified Linear Units (ReLU) [14] as a non-linearity for the convolution layers can be used. At the same time, we applied dropout [8] with probability 0.5 between the fully connected layers to avoid overfitting. For the spatial pooling layers, we use max-pooling layers to deal with scale changes. In addition, the proposed 2-channel thought in [26] is also used, which is deemed that two pictures as two channels of a picture is directly fed to a single branch network. The parameters setting in this model is similar to CaffeNet model [12] which is a slightly modified of [14]. But we used fewer layers to construct our model and replaced the final fully connected layer with a two neuron layer.

Hybrid 2-Channel Siamese Model. According to the results of experiment about 2-channel and two branches models (siamese and pseudo-siamese), jointing two images then extracting features has better performance than extracting features separately before concatenating. So a hybrid 2-channel siamese model

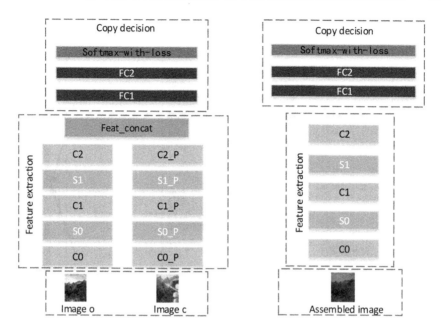

Fig. 1. The fundamental models: On the right side is two branches model. It represents siamese (parameter sharing) network and pseudo-siamese (without parameter sharing) network. On the left side is 2-channel model.

(see Fig. 2) is constructed. On the one hand, we convert color images to gray scale images just like data preprocessing of 2-channel method (see Fig. 3). On the other hand, the two color images are spliced together from top to bottom. So we build a deep CNN to learn feature representations from color and gray scale images separately, which are then connected and fed into the final fully connected layer. As illustrated in Fig. 2, the convolution property of CNN is to remain convolution order of each layer unchanged. This ensures that the features extracted on up and down spliced image are still the features of two original images. So even though we only use a pseudo-siamese model to extract the characteristics of the input, it can also be regarded as a hybrid of 2-channel and siamese model. The system not only use gray scale images to prevent the impact of brightness but also discover the most discriminative features using RGB color space.

4   Learning

Training. A strongly supervised manner is applied to all models for training. The output $y_i \in (0,1)$ is the corresponding label (denoting a non-copy and a copy pair, respectively). We use softmax to compute the loss function and to initialize the backpropagation. Stochastic gradient descent with momentum 0.9 and weight decay $\lambda = 0.0005$ are applied to all architectures.

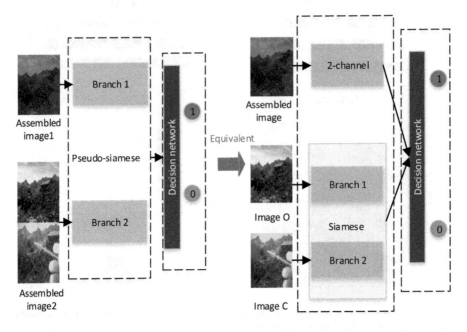

Fig. 2. Hybrid 2-channel siamese model: A simple image preprocessing and a pseudo-siamese model (left) achieves a combination of 2-channel and siamese network (right), but the model parameters have not increased with the change.

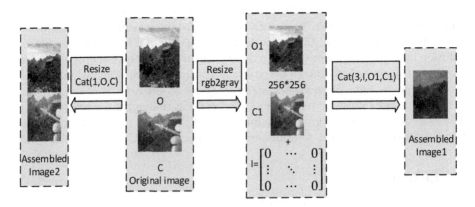

Fig. 3. Image preprocessing: For image pair, O and C, when they are processed into two channels of a picture, they will be used in the 2-channel model. Besides above process, in the hybrid 2-channel siamese model, jointing the two picture up and down is also needed.

Different learning rates have been tried on the proposed models. We find that setting the initial learning rate 0.00001 then decreasing it every 100,000 iterations produces better results than using larger learning rates. Depending on the network architecture, it takes about 2 to 3 days to train the full network. The weights are initialized from a zero-mean Gaussian distribution with standard deviation 0.01. Finally, all models are trained from scratch.

Data Preprocessing. For 2-channel model and hybrid 2-channel siamese model, a process of image preprocessing is needed. It is showed in Fig. 3.

5 Experiments

5.1 Dataset

For deep learning networks, the training data is critical for final results. To collect large data sets, four different data sets were put together to train models, which ultimately constituted more than 52,000 image pairs. We set aside seventy percent of the total data for training. The data sets are widely applied in image copy detection, and can be downloaded conveniently. They are INRIA Copydays dataset [10], CoMoFoD database [24], Image Manipulation Dataset [4], and MICC-F220, MICC-F2000 dataset from [1]. A brief summary is showed in Table 1. In addition, the INRIA Copydays dataset on the copy image has great transformation called strong copy picture. In order to expand data set and increase the difficulty of experiment, a copy relationship is also thought to exist between the strong copy pictures and any other copy pictures. We chose several combinations and finally generated approximately 10500 image pairs. We also had 25800 image pairs as negative samples, which were crawled from "Web Queries" dataset [13].

Table 1. Data set summary and operation. Table shows corresponding numbers. Each of the original picture and its copy form a pair of input images, then go to test set or train set. Copy image contains a variety of transformation, such as, rotation, scaling, distortion, noise adding and image blurring.

	Dataset name	Original image	Copy transformation	Generate pair	Train	Test
1	INRIA Copydays	157	>19	10490	7746	3744
2	CoMoFoD	200	50	10000	10000	0
3	Image manipulation	48	83	3984	0	3984
4	MICC-F220	121	10	110	0	110
5	MICC-F2000	50	14	700	700	0

5.2   Result

In our experiment, we have first trained our proposed models in a small database about 4000 image pairs, which only chose INRIA Copydays dataset as positive samples. It turns out that 2-channel architecture exhibit best performance in all networks. In the two branches networks, pseudo-siamese model is better than siamese model. This result is in line with [26] inference. However, as we expanded our data set to a new training, a little change happened. It is presented in Table 2. The expansion of the data set lead to an significant advancement of overall performance, and result of siamese network also catch up with and surpass the other models.

Table 2. Test accuracies (%) of models on small and large dataset

Networks	Siamese	Pseudo-siamese	2-channel	Hybrid 2-channel siamese
Small dataset	84	86.5	94.5	92
Large dataset	99.07	99.04	98.6	97.93

The results of experiment show that our models have desired robustness against JPEG, added noise and filtering attacks. We find that each model has advantages and disadvantages of its own, although the results of models are approximate. The judgment errors of two branches model (siamese and pseudo-siamese) is closely related to the colors of original images. If the color of original picture is too simple, the network will not be able to correctly identify the copies of it (image 2 in Fig. 4, it is a black and white picture). But these models are robust to image cropping. Conversely, for the 2-channel, its false results mainly come from cropping. But for a picture similar to the image 2 in Fig. 4, there is a good ability to identify, which is related to gray scale transformation in preprocessing.

The hybrid 2-channel siamese model is more like a combination of two branches and 2-channel method. Although its results are not outstanding, but it can effectively make up for the defects of two branches and 2-channel method. It can recognize a simple color picture and improve the capability against cropping attack to some extent. We analyzed the results of this model and discovered that most of the errors are due to the strong copy pictures (image 4 in Fig. 4). And such pictures accounts for large percentage in test set, which may be the reason for the low accuracy rate. In addition, we choose uniform parameters for each model in the experiment, but in fact if the best outcomes need to be presented, the parameters of each model are different.

In experiment, we chose a unified batch-size 100 for training, 60 for testing, and 300 test-iteration. We also tried to increase feature dimension, or add fully connected layer in two branches respectively, but the effect was not significant.

In order to compare with the traditional methods, we simply designed and extracted SIFT feature in the large dataset, and then utilized the feature matching to obtain matching point set. Lastly we got the final accuracy rate of 96.99 %

Image O	Image C		siamese	pseudo-siamese	2-channel	hybrid 2-channel siamese
 1		0	0.00019	0.000077	0	0
		1	0.99981	0.999923	1	1
		0	0	0	0.0004	0.009533
		1	1	1	0.9996	0.990467
 2		0	0.956165	0.833159	0.0806	0.152655
		1	0.043835	0.166841	0.9194	0.847345
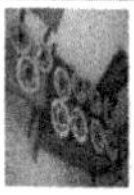 3	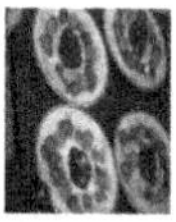	0	0.0256	0.024573	0.9982	0.401183
		1	0.9744	0.975427	0.0018	0.598817
 4		0	0	0	0.0002	0.563827
		1	1	1	0.9998	0.436173

Fig. 4. Performance comparison of siamese, pseudo-siamese, 2-channel and hybrid 2-channel siamese: Results for the classification are presented. Image 1 is a positive example and three of the other images are negative examples. The red line box shows the error result of classification. (Color figure online)

through setting a proper threshold. SIFT algorithm has good robustness against the changes of the object shape, translation and rotation. But for the fuzzy image and edge smooth image, the number of detected feature is too few to do the following matching work. If we depend on traditional method, it is time-consuming to get ideal result while being strenuous.

6   Conclusions

In this paper we present an effective and efficient image copy detection method based on CNN, which learns a comparing function directly from raw image pixels. Several architectures are studied and each of them displays extremely good performance. These results indicate that CNN based methods are specifically suited to copy detection task. Meanwhile we compare these models, and summarize their advantages and disadvantages. In the following work, the advantages are to be inherited, the disadvantages help us target our improvement efforts.

Finally, we have to say, such a good result, has a great relationship with the obviously discriminated database. In the next step, we will increase the complexity of the database, by utilizing a larger training set and a deeper network to see if our scheme can still perform well. (since our training set in the present experiments is considered smaller than todays standards).

Acknowledgement. This work is partly supported by the 973 basic research program of China (Grant No. 2014CB349303), the Natural Science Foundation of China (Grant No. 61472421), the National Nature Science Foundation of China (No. 61370038) and the Strategic Priority Research Program of the CAS (Grant No. XDB02070003).

References

1. Amerini, I., Ballan, L., Caldelli, R., Del Bimbo, A., Serra, G.: A sift-based forensic method for copycmove attack detection and transformation recovery. IEEE Trans. Inf. Forensics Secur. **6**(3), 1099–1110 (2011)
2. Berrani, S.A., Amsaleg, L., Gros, P.: Robust content-based image searches for copyright protection. In: ACM International Workshop on Multimedia Databases, Acm-Mmdb 2003, New Orleans, Louisiana, USA, November, pp. 70–77 (2003)
3. Cao, Y., Zhang, H., Gao, Y., Guo, J.: An efficient duplicate image detection method based on affine-sift feature. In: 2010 3rd IEEE International Conference on Broadband Network and Multimedia Technology (IC-BNMT), pp. 794–797, October 2010
4. Christlein, V., Riess, C., Jordan, J., Riess, C., Angelopoulou, E.: An evaluation of popular copy-move forgery detection approaches. IEEE Trans. Inf. Forensics Secur. **7**(6), 1841–1854 (2012)
5. Girshick, R., Donahue, J., Darrell, T., Malik, J.: Rich feature hierarchies for accurate object detection and semantic segmentation. In: 2014 IEEE Conference on Computer Vision and Pattern Recognition, pp. 580–587, June 2014
6. Hadsell, R., Chopra, S., LeCun, Y.: Dimensionality reduction by learning an invariant mapping. In: 2006 IEEE Computer Society Conference on Computer Vision and Pattern Recognition (CVPR 2006), vol. 2, pp. 1735–1742 (2006)
7. Han, X., Leung, T., Jia, Y., Sukthankar, R., Berg, A.C.: MatchNet: unifying feature and metric learning for patch-based matching. In: 2015 IEEE Conference on Computer Vision and Pattern Recognition (CVPR), pp. 3279–3286, June 2015
8. Hinton, G.E., Srivastava, N., Krizhevsky, A., Sutskever, I., Salakhutdinov, R.R.: Improving neural networks by preventing co-adaptation of feature detectors. Comput. Sci. **3**(4), 212–223 (2012)
9. Indyk, P., Motwani, R.: Approximate nearest neighbors: towards removing the curse of dimensionality. In: Proceedings of the Thirtieth Annual ACM Symposium on Theory of Computing, STOC 1998, NY, USA (1998). http://doi.acm.org/10.1145/276698.276876
10. Jégou, H., Douze, M., Schmid, C.: Hamming embedding and weak geometric consistency for large scale image search. In: Proceedings of the 10th European Conference on Computer Vision, p. 1.1, October 2008
11. Jégou, H., Douze, M., Schmid, C., Pérez, P.: Aggregating local descriptors into a compact image representation. In: 2010 IEEE Conference on Computer Vision and Pattern Recognition (CVPR), pp. 3304–3311, June 2010
12. Jia, Y., Shelhamer, E., Donahue, J., Karayev, S., Long, J., Girshick, R., Guadarrama, S., Darrell, T.: Caffe: Convolutional architecture for fast feature embedding. Eprint Arxiv, pp. 675–678 (2014)

13. Krapac, J., Allan, M., Verbeek, J., Juried, F.: Improving web image search results using query-relative classifiers. In: 2010 IEEE Conference on Computer Vision and Pattern Recognition (CVPR), pp. 1094–1101, June 2010
14. Krizhevsky, A., Sutskever, I., Hinton, G.E.: Imagenet classification with deep convolutional neural networks. Adv. Neural Inf. Process. Syst. **25**(2), 1097–1105 (2012)
15. Li, P., Wang, M., Cheng, J., Xu, C., Lu, H.: Spectral hashing with semantically consistent graph for image indexing. IEEE Trans. Multimedia **15**(1), 141–152 (2013)
16. Li, Z., Liu, G., Jiang, H., Qian, X.: Image copy detection using a robust gabor texture descriptor. In: Proceedings of the First ACM Workshop on Large-Scale Multimedia Retrieval and Mining, LS-MMRM 2009, NY, USA, pp. 65–72 (2009). http://doi.acm.org/10.1145/1631058.1631072
17. Ling, H., Yan, L., Zou, F., Liu, C., Feng, H.: Fast image copy detection approach based on local fingerprint defined visual words. Signal Process. **93**(8), 2328–2338 (2013)
18. Mikolajczyk, K., Schmid, C.: A performance evaluation of local descriptors. IEEE Trans. Pattern Anal. Mach. Intell. **27**(10), 1615–1630 (2005)
19. Simonyan, K., Zisserman, A.: Very deep convolutional networks for large-scale image recognition. Eprint Arxiv (2014)
20. Sivic, J., Zisserman, A.: Video google: a text retrieval approach to object matching in videos. In: Ninth IEEE International Conference on Computer Vision, 2003. Proceedings, vol. 2, pp. 1470–1477, October 2003
21. Tang, Z., Huang, L., Yang, F., Zhang, X.: Robust image hashing based on fan-beam transform. ICIC Express Lett. **8**(8), 2365–2372 (2014)
22. Tang, Z., Yang, F., Huang, L., Wei, M.: DCT and DWT based image hashing for copy detection. ICIC Express Lett. **7**(11), 2961–2967 (2013)
23. Tirilly, P., Claveau, V., Gros, P.: Language modeling for bag-of-visual words image categorization. In: Proceedings of the 2008 International Conference on Content-based Image and Video Retrieval, CIVR 2008, NY, USA, pp. 249–258 (2008). http://doi.acm.org/10.1145/1386352.1386388
24. Tralic, D., Zupancic, I., Grgic, S., Grgic, M.: CoMoFoD - new database for copy-move forgery detection. In: 55th International Symposium ELMAR-2013, pp. 49–54, September 2013
25. Žbontar, J., LeCun, Y.: Computing the stereo matching cost with a convolutional neural network. In: 2015 IEEE Conference on Computer Vision and Pattern Recognition (CVPR), pp. 1592–1599, June 2015
26. Zagoruyko, S., Komodakis, N.: Learning to compare image patches via convolutional neural networks. In: 2015 IEEE Conference on Computer Vision and Pattern Recognition (CVPR), pp. 4353–4361, June 2015
27. Zhang, D., Wang, J., Cai, D., Lu, J.: Self-taught hashing for fast similarity search. In: Proceedings of the 33rd International ACM SIGIR Conference on Research and Development in Information Retrieval, SIGIR 2010, NY, USA, pp. 18–25 (2010). http://doi.acm.org/10.1145/1835449.1835455
28. Zheng, Q.F., Wang, W.Q., Gao, W.: Effective and efficient object-based image retrieval using visual phrases. In: Proceedings of the 14th ACM International Conference on Multimedia, MM 2006, NY, USA, pp. 77–80 (2006). http://doi.acm.org/10.1145/1180639.1180664

Perceptual Loss with Fully Convolutional for Image Residual Denoising

Tao Pan[1,2(✉)], Fu Zhongliang[1,2], Wang Lili[1,2], and Zhu Kai[1,2]

[1] Chengdu Institute of Computer Application, Chinese Academy of Sciences,
Chengdu 610041, China
taopanpan@gmail.com, Fzliang@netease.com
[2] University of Chinese Academy of Sciences, Beijing 100049, China
{wanglili8773,hustzhukai}@qq.com

Abstract. In this paper we propose a fully convolutional encoder-decoder framework for image residual transformation tasks. Instead of only using per-pixel loss function, the proposed framework learn end-to-end mapping combined with perceptual loss function that depend on low-level features from a pre-trained network. Pointing out the mapping function in order to handle noise-free image by introduce identity mapping. And through an analysis of the interplay between the neural networks and the underlying noisy distribution which they seeking to learn. We also show how to construct a uniform transform, which is then used to make a single deep neural network work well across different levels of noise. Comparing with previous approaches, ours achieves better performance. The experimental results indicate the efficiency of the proposed algorithm to cope with image denoising tasks.

Keywords: Residual denoising · Encoder-decoder · Perceptual loss

1 Introduction

Image denoising aims at recovering a clean image from a noisy image, which is a classical problem in low-level vision task. This problem is inherently ill-posed since the clean image usually is unknown. In other words, it is an underdetermined mapping problem, of which image transformation is not unique. Generally speaking, a residual image F can be represented as $F = y - f(x)$, where x is the noisy image and f is the mapping function, which receives input image and transforms it into an output image. And y is the ideal clean image. By accommodating different types of mapping functions, the same mathematical model applies to most other low-level imaging problems such as image deblurring, demosaicking and super-resolution.

Recently, deep neural networks have shown their superior performance in computer vision, ranging from high-level to low-level tasks. It is well known that neural networks are capable of approximating any measurable function to desired degree of accuracy [10]. In image denoising setting, neural networks in

© Springer Nature Singapore Pte Ltd. 2016
T. Tan et al. (Eds.): CCPR 2016, Part II, CCIS 663, pp. 122–132, 2016.
DOI: 10.1007/978-981-10-3005-5_11

a regression framework seek to approximate the latent conditional expectation under some input noise distribution. When train a feed-forward neural network with a supervised manner, one key factor is to chose a loss function to measure the difference between output and ground-truth images. The most widely used is per-pixel loss, which computed by intensity differences of distorted and reference image pixels, along with the related quantity of peak signal-to-noise ratio (PSNR) [27]. But the per-pixel loss don't capture perceptual differences and is well known to correlate poorly with the perceived image quality [28,29]. This is because a number of assumptions implicitly made when using per-pixel loss are not satisfied. Arguably, the most important one is that treats noise independently of the local characteristics of the image; on the contrary, the sensitivity of the Human Visual System (HVS) to noise depends on the local luminance, contrast and structure [27].

Based pure learning strategy, a set of deep neural networks designed for the image denoising has been shown to outperform others widely accepted methods as the state-of-the-art [2]. But all these work have a problem: if the input is noise-free, the learned model also degrade the clean image quality. So their only work at the given noise levels which they were trained for. A standard general-purpose algorithm for noise removal ought to be able to handle different levels of noise, this limitation seems to require a series of such networks, one for each noise level. This is impractical and even unrealistic. Because we don't know the noise level and the type of real image.

Our main contributions are briefly outlined as follows:

1. We propose a very deep fully convolutional architecture for image residual denoising. Aiming at model the residual mapping function by the image transformation, directly learning the noise distribution.
2. Combine the benefits of per-pixel and perceptual loss function, train a transformation networks with low and high level information, produce high-quality denoised image.
3. To make a single neural network work for all noise levels, we investigate the statistical regularity of the network: make the input adding random sample from the different level of noise, the input also can be clean image, because the learned maping function must be identity for the clean image, and the trained network can auto handle different levels.
4. We experiment on some common benchmark images. The results demonstrate the advantages of our network and proposed novel loss layer overcome other recent state-of-the-art methods on image denoising.

2   Related Work

Numerous approaches have been proposed for image denoising. Some selectively smooth parts of a noisy image with the aim of "smoothing out" the noise while preserving image details. Some methods transfer the image signal to an alternative domain where noise can be easily separated from the signal. More recent approaches exploit the "non-local" statistics of images: Different patches in the same image are often similar in appearance. The block-matching and 3D filtering(BM3D) algorithm [3] that groups non-local similar patches by collaborative filtering in a transformed domain. The BM3D has become a benchmark in image denoising.

While BM3D is a well-engineered algorithm, learning-based methods have found widespread using in image denoising. The most significant difference between neural network methods and the others is that they typically automatically learn image transformation directly from pairs of clean and noisy images rather than relying on human priors. Recently, due to the fast development of deep neural network, many new types of neural networks have been applied to the image denoising problem, such as stacked sparse auto-encoder [1,15,21,22,26], multi-layer perceptron [2,24], convolutional networks [7,12,18,23,25,29], which all have shown good performance.

Stacked denoising auto-encoder [22] establish the value of using a denoising criterion as an unsupervised objective to guide the learning of useful higher level representations. Denoising performance can easily be measured and directly optimized. But this approach's objective is classification. Xie et al. [26] combined sparse coding and deep networks pre-trained with denoising auto-encoder(DA), propose an alternative supervised training scheme Stacked Sparse Denoising Auto-encoders (SSDA) that successfully adapts DA, originally designed for unsupervised feature learning, to the tasks of image denoising and blind inpainting.

Burger et al. [2] presented a patch-based algorithm learned on a large dataset with a plain multi-layer perceptron, which achieve better performance than BM3D. However their method tailored to a single level of noise and does not generalize well to other noise levels. Jain et al. [12] proposed deep convolutional neural networks and an unsupervised learning procedure that synthesizes training samples from specific noise models. They found that convolutional networks provide comparable and in some cases superior performance to wavelet and Markov Random Field (MRF) methods.

So far, all the neural networks for image denoising tasks have been solved by training deep neural networks with per-pixel loss functions. But this can be particularly limiting in the context of image processing, since per-pixel loss correlates poorly with perceived image quality [29]. Some recent papers have used optimization to generate images where the objective is perceptual, depending on high-level features extracted from a convolutional network [6]. The work of [13,18] is particularly relevant to ours, as they train a feed-forward neural network to image transformation, they use a loss network pretrained for image classification to define perceptual loss functions that measure perceptual differences in output and ground-truth. However, their focus on style transfer and image super-resolution. And later work propose convolution and deconvolution layers for encoder and decoder image details, and with symmetric skip connections accelerate training. But the noise of image can capture only by convolution and restore image details by deconvolution. Ours network can be viewed as whole transformation function with symmetric skip connections.

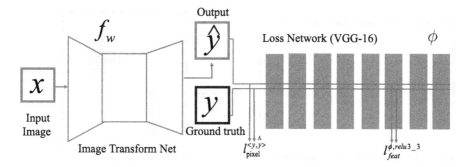

Fig. 1. The overall architecture of our proposed network. The image transformation network contains layers of convolution (encoder) and deconvolution (decoder). We use a loss network pretrained for image classification to define perceptual loss functions that measure perceptual differences in output and ground truth label. The loss network remains fixed during the training process.

Zhao et al. [29] study the performance of several losses, including perceptually-motivated losses, and propose a novel of differentiable error function. From perceptually-motivated metrics design some new loss layer, its still stay focus on low level pixel structure. The other work of Wang et al. [24] is also particularly relevant to ours, investigating the distribution invariance of the natural image patches with respect to linear transforms, they show how to make a single existing deep neural network work well across all levels of Gaussian noise. However, different from the above methods, in this paper we have combined the benefits of image transformation tasks and optimization-based methods for image transformation by training feed-forward transformation networks with perceptual loss functions. Meanwhile make a single deep neural network work well across different levels of additive white Gaussian noise, by explicit training different levels of noise and make original image as input too.

3   Method

The proposed framework mainly contains a chain of convolution layers and deconvolution layers, as shown in Fig. 1. There consists of two components: an *image transformation network* f_W and a *loss network* ϕ that is used to define several *loss functions* $\ell_1, \ldots, \ell_k$. Aim at learning the deep residual convolutional neural network parameterized by weights W; it transforms input images x into output images f_W via the mapping $\hat{y} = f_W(x)$. Each loss function computes a scalar value $\ell_i(\hat{y}, y_i)$ measuring the difference between the output image $\hat{y}$ and a *target image* y_i. Learning objective is trained using stochastic gradient descent to minimize a weighted combination of loss functions:

$$W^* = \arg\min_W \mathbf{E}_{x,\{y_i\}} \left[\sum_{i=1} \lambda_i \ell_i(f_W(x), y_i) \right] \tag{1}$$

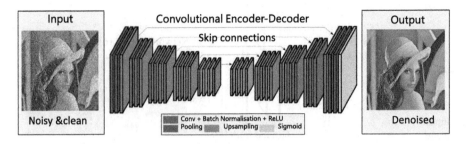

Fig. 2. The architecture of RED-NET and our image transformation network DeNET can be viewed as the RED-NET inserted some residual block.

From this formulation, we can see that the task here is to find a mapping function f_W that best approximates the image transformation. Meanwhile we also want the $f_W(y_i)$ approxiamates the image y_i, so we now treat the image denoising problems in a unified framework by choosing appropriate weights W with different situations. The loss network ϕ is used to define a *feature space loss* ℓ^ϕ_{feat} and a *per-pixel loss* ℓ_{mse} that measure differences in feature and image space. For image denoinsing, the input image x is a noisy input, the ground-truth image y is also input image and target image.

3.1   Encoder-Decoder Architectures

The framework is fully Encoder-decoder convolution model. Combining layers of encoder and decoder [5,9,16,18,19] have been proposed for unsupervised and supervised deep learning.

Our image transformation networks roughly follow the architectural guidelines set forth by [18]. They propose RED-NET for image denoising, the network as shown in Fig. 2. Base on the architecture, The batch normalization [11] and ReLU nonlinearities layers are added after convolution. And insert some residual blocks [8] into the network. The output layer, which uses a sigmoid function to ensure that the output image has pixels in the range [0, 1]. But we do not use any pooling layers, instead using strided and fractionally strided convolutions for downsampling and upsampling. Other than the first and last layers which use 9×9 kernels, all convolution layers use 3×3 kernels. Since the image transformation networks are fully-convolutional, at test-time they can be applied to images of any size.

The difference between RED-NET and ours DeNET is that our network insert some residual block and introduce residual connection. The noise is eliminated step by step after each layer. During this process, the details of the image content can be compensate by the perceptual loss function. The specific configurations of the two networks are described in Table 1.

Two learning strategy is applied to inner blocks of the encoding-decoding network to make training more effective. Skip connections are passed every two convolutional layers to their mirrored deconvolutional layers. He et al. [8] use residual connections to train very deep networks for image classification. They argue that residual connections make the network to learn the identify function easily; this is an appealing property for image transformation networks, since in most cases the output image should share structure with the input image. The body of our network thus consists of several residual blocks, each of which contains two 3×3 convolution layers.

Table 1. Configurations of the DeNET-R and RED-NET networks. "conv3" and "deconv3" stand for convolution and deconvolution kernels of size 3×3. 32,128 and 512 is the number of feature maps after each convolution and deconvolution. "c" is the number of channels of input and output image. i.e., $c = 3$.

DeNET-R	RED-NET
(conv9−32) × 6	(conv3−128) × 6
(conv3−64) × 6	(conv3−256) × 6
(conv3−128) × 3	(conv3−512) × 3
Residual block×5	
(deconv3−64) × 2	(deconv3−512) × 2
(deconv3−32) × 6	(deconv3−512) × 6
(deconv9−3) × 6	(deconv3−512) × 6
(deconv3−c)	(deconv3−c)

3.2 Per-pixel Loss Functions

The *pixel loss* is the (normalized) Euclidean distance between the output image $\hat{y}$ and the target y. If both have shape $C \times H \times W$, then the pixel Euclidean loss is defined as Mean Squared Error(MSE):

$$\ell_2(\hat{y}, y) = \frac{1}{CHW}\|\hat{y} - y\|_2^2 \tag{2}$$

This loss function can introduce splotchy artifacts, So we also examine the ℓ_1-norm loss. The two losses weigh errors differently:ℓ_1 does not over-penalize larger errors and consequently, they may have different convergence properties. Computing the ℓ_1 loss is straightforward:

$$\ell_1(\hat{y}, y) = \frac{1}{CHW}|\hat{y} - y| \tag{3}$$

The derivatives for the back-propagation are also simple, for each pixel p in the whole image,

$$\partial\ell_1/\partial p = sign\left(\hat{y}(p) - y(p)\right) \tag{4}$$

Note that, although L_{ℓ_1} is computed on the whole image, the derivatives are back-propagated for each pixel in the image. The network trained with ℓ_1 provides a significant improvement for several of the issues discussed above.

3.3   Perceptual Loss Functions

We define *perceptual loss functions* that measure perceptual and semantic differences between images, other than the hand design SSIM loss in [29]. They make use of a *loss network* ϕ pretrained for image classification. In all our experiments ϕ is the 16-layer VGG network [20] pretrained on the ImageNet dataset [4]. Rather than encouraging the pixels of the output image $\hat{y} = f_W(x)$ to exactly match the pixels of the target image y, we instead encourage them to have similar feature representations as computed by the loss network ϕ. Let $\phi_j(x)$ be the activations of the jth layer of the network ϕ when processing the image x; if j is a convolutional layer then $\phi_j(x)$ will be a feature map of shape $C_j \times H_j \times W_j$. The *feature feat loss* is the (squared, normalized) Euclidean distance between feature representations:

$$\ell_{feat}^{\phi,j}(\hat{y}, y) = \frac{1}{C_j H_j W_j} \|\phi_j(\hat{y}) - \phi_j(y)\|_2^2 \tag{5}$$

The Euclidean distance also can be alternated by ℓ_1-norm distance. As demonstrated in [17], finding an image $\hat{y}$ that minimizes the feature feat loss for early layers tends to produce images that are visually indistinguishable from y. Using a feature feat loss for training our image transformation networks encourages the output image $\hat{y}$ to be perceptually similar to the target image y, but does not force them to match exactly. To encourage spatial smoothness in the output image $\hat{y}$, we follow prior work on feature inversion [17] and make use of *total variation regularizer* $\ell_{TV}(\hat{y})$.

4   Experiments and Results

In this section, We provide an analysis on our experiments setting with fully encoder-decoder convolutional neural network. Then evaluate denoising performance of our models under some different loss function setting. At the end we explore how to make one neural network handle different levels of noise.

4.1   Analysis on Model Details

We train models to perform single and multi level of standard deviation σ by minimizing some loss function: ℓ_2-Mean Squared Error(MSE) loss,ℓ_1-norm loss and feature feat loss at layer `relu1_1` from the VGG-16 network ϕ. We train image size with 256×256 from the training set, and prepare noisy inputs by add a Gaussian kernel of width σ. We train with a batch size of 10 using Adam [14] with a learning rate of 1×10^{-3} without weight decay or dropout.

Denoising experiments are performed on the standard 14 common benchmark images Set14. As a common experimental setting in the literature, additive Gaussian noises with zero mean and standard deviation σ are added to the test image to test the performance of denoising methods. We report PSNR and SSIM [27], computing both on the three channel color image, following [18,29].

Table 2. Quantitative single-level image denoising results on the Set14; we report average PSNR and SSIM on each dataset. Each σ value we train identical networks, one with a per-pixel loss ℓ_1, ℓ_2 and another with a feature feat loss ℓ_{feat}. ℓ_{mix} is combine with ℓ_1 and ℓ_{feat}. Best results are shown in bold.

Sigma	Noisy	RED-NET [18]	Ours (ℓ_2)	Ours (ℓ_1)	Ours (ℓ_{feat})	Ours (ℓ_{mix})
	PSNR/SSIM	PSNR/SSIM	PSNR/SSIM	PSNR/SSIM	PSNR/SSIM	PSNR/SSIM
$\sigma = 10$	28.16/0.7041	**34.81/0.9402**	34.35/0.8912	33.40/0.8930	31.05/0.7680	33.16/0.7680
$\sigma = 30$	18.88/0.3389	29.17/0.8423	28.73/0.8205	29.76/0.8591	26.70/0.6845	**30.15/0.8681**
$\sigma = 50$	14.79/0.2038	26.81/0.7733	26.40/0.8205	26.79/**0.8325**	25.69/0.6411	**27.09**/0.8312
$\sigma = 70$	12.43/0.1391	25.31/0.7206	25.39/0.7105	26.13/**0.7250**	17.89/0.6650	**26.20**/0.7180
$\sigma = 100$	10.26/0.0901	-	18.40/0.4215	20.19/0.4680	17.31/0.3640	**19.16/0.4695**

As a baseline model we use RED-NET [18] for its state-of-the-art performance. Its is a fully convolutional network with convolutonal and deconvolutonal layers trained to minimize per-pixel loss. To account for differences between RED-NET and our model in data, training, and architecture, we train image transformation networks for the same standard deviation σ using ℓ_2; these networks use identical data, architecture, and training as the networks trained to minimize other loss function. We train denoising networks with the per-pixel loss typically used [18, 29], also with a feature feat loss (see Sect. 3) to allow transfer of semantic knowledge from the pretrained loss network to the denoising network as supervised signal guided denoising.

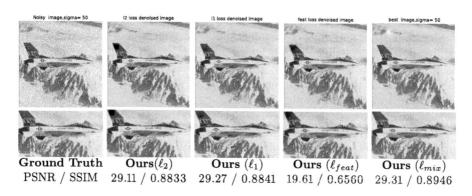

Ground Truth	Ours(ℓ_2)	Ours (ℓ_1)	Ours (ℓ_{feat})	Ours (ℓ_{mix})
PSNR / SSIM	29.11 / 0.8833	29.27 / 0.8841	19.61 / 0.6560	29.31 / 0.8946

Fig. 3. Denoising results with different loss type on an image from the Set14 dataset. We report PSNR / SSIM for the F16-plane image as a example.

First of all, Compared to the per-pixel loss ℓ_1 and ℓ_2 result, ℓ_1 does a better good job at denoising performance and meanwhile restore sharp edges and fine details. As show in Fig. 3 the wing in the ℓ_2 image and the red color block elements of the body in the ℓ_2 image. This is because ℓ_2 penalizes larger errors, but is more tolerant to small errors, regardless of the underlying structure in the image; The conclusion is consistent with the literature [29].

Fig. 4. Comparison denoising performance of four other noise types, which transformation networks trained only with Gaussian noise. **Up:** The four types noise image:Speckle noise, Poisson distributed noise, Salt noise, Pepper noise. **Down:** Denoising suboutputs from corresponding noise type.

Moreover, Results for ℓ_{feat} are Show in Fig. 3 when only with the feature feat loss gives rise to a slight cross-hatch pattern visible under magnification, which harms its PSNR and SSIM compared to baseline methods. Again we see that our ℓ_{feat} model does a good job at edges and fine details compared to other models, such as the wing. The ℓ_{feat} model does not sharpen edges indiscriminately; compared to the ℓ_{pixel} model, the ℓ_{feat} model sharpens the boundary edges of the wing and rider but the background mountain remain diffuse, suggesting that the ℓ_{feat} model may be more aware of image semantics.

Since our ℓ_{pixel} and our ℓ_{feat} models share the same architecture, data, and training procedure, all differences between them are due to the difference between the ℓ_{pixel} and ℓ_{feat} losses. The ℓ_{pixel} loss gives fewer visual artifacts and higher PSNR values but the ℓ_{feat} loss does a better job at reconstructing fine details, leading to pleasing visual results.

Last, we can observe that a single model can work across all levels of Gaussian noise, thereby allowing to reduce significantly the training time for a general-purpose neural network powered denoising algorithm. The reason for this may be that the learned image transformation function can model the Gaussian-like distribute from any levels and other types. Interesting is that shown in Fig. 4, even for other type of noise, such as speckle noise, poission-distributed noise, salt noise or pepper noise, which the model trained for Gaussian noise have ability for image denoise.

5   Conclusion and Outlook

In this paper we have combined the benefits of fully convolutions image transformation tasks and learning-based methods for image residual denoising. We train image transformation networks with per-pixel loss layer and perceptual loss layer. We have applied this method to single and multi-levels of Gaussian noise where we achieve comparable performance compared to existing methods, and we show that train with a mix loss layers allows the model to better restore fine details and edges. In future work we hope to explore the use of proposed method for other low level vision tasks, such as colorization and deblurring. We also plan to investigate the use of deep learning for ultrasound blind denoising.

References

1. Agostinelli, F., Anderson, M.R., Lee, H.: Adaptive multi-column deep neural networks with application to robust image denoising. In: Advances in Neural Information Processing Systems, pp. 1493–1501 (2013)
2. Burger, H.C., Schuler, C.J., Harmeling, S.: Image denoising: can plain neural networks compete with BM3D? In: 2012 IEEE Conference on Computer Vision and Pattern Recognition, pp. 2392–2399. IEEE, June 2012
3. DabovK, F., KatkovnikV, K.O.E.: Image denoising by sparse 3-D transform-domain collaborative filtering. IEEE Trans. Image Process. **16**(8), 2080–2095 (2007)
4. Deng, J., Dong, W., Socher, R., Li, L.J., Li, K., Li, F.F.: Imagenet: a large-scale hierarchical image database, pp. 248–255 (2009)
5. Dong, C., Loy, C.C., He, K., Tang, X.: Image super-resolution using deep convolutional networks. IEEE Trans. Pattern Anal. Mach. Intell. **38**(2), 295–307 (2016)
6. Dosovitskiy, A., Brox, T.: Generating Images with Perceptual Similarity Metrics based on Deep Networks (2016)
7. Eigen, D., Krishnan, D., Fergus, R.: Restoring an image taken through a window covered with dirt or rain. In: 2013 IEEE International Conference on Computer Vision, pp. 633–640. IEEE, December 2013
8. He, K., Zhang, X., Ren, S., Sun, J.: Deep residual learning for image recognition. In: Proceedings of IEEE Conference on Computer Vision Pattern Recognition, vol. abs/1512.03385 (2016)
9. Hong, S., Noh, H., Han, B.: Decoupled deep neural network for semi-supervised semantic segmentation. In: Proceedings of Advances in Neural Information Processing System (2015)
10. Hornik, K., Stinchcombe, M., White, H.: Multilayer feedforward networks are universal approximators. Neural Netw. **2**(5), 359–366 (1989)
11. Ioffe, S., Szegedy, C.: Batch normalization: accelerating deep network training by reducing internal covariate shift, Computer Science (2015)
12. Jain, V., Seung, S.: Natural Image Denoising with Convolutional Networks. In: Bottou, D.K., Schuurmans, D., Bengio, Y., L. (eds.) Advances in Neural Information Processing Systems, vol. 21, pp. 769–776. Curran Associates, Inc. (2009)
13. Johnson, J., Alahi, A., Fei-Fei, L.: Perceptual Losses for Real-Time Style Transfer and Super-Resolution. arXiv Preprint, March 2016
14. Kingma, D., Ba, J.: Adam: a method for stochastic optimization. Eprint Arxiv (2014)
15. Li, H.M.: Deep learning for image denoising. Int. J. Signal Process. Image Process. Pattern Recogn. **7**(3), 171–180 (2014)
16. Long, J., Shelhamer, E., Darrell, T.: Fully convolutional networks for semantic segmentation. In: Proceedings of IEEE Conference Computer Vision Pattern Recognition, pp. 3431–3440 (2015)

17. Mahendran, A., Vedaldi, A.: Understanding deep image representations by inverting them. In: Proceedings of the IEEE Conference on Computer Vision and Pattern Recognition, pp. 1–9 (2015)
18. Mao, X.J., Shen, C., Yang, Y.B.: Image Denoising Using Very Deep Fully Convolutional Encoder-Decoder Networks with Symmetric Skip Connections. arXiv preprint, March 2016
19. Noh, H., Hong, S., Han, B.: Learning deconvolution network for semantic segmentation. In: Proceedings of IEEE International Conference Computer Vision, pp. 1520–1528 (2015)
20. Simonyan, K., Zisserman, A.: Very deep convolutional networks for large-scale image recognition. CoRR abs/1409.1556 (2014)
21. Skribtsov, P.V., Surikov, S.O.: Regularization method for solving denoising and inpainting task using stacked sparse denoising autoencoders. Am. J. Appl. Sci. **13**(1), 64–72 (2016)
22. Vincent, P., Larochelle, H., Bengio, Y., Manzagol, P.A.: Extracting and composing robust features with denoising autoencoders. In: Proceedings of International Conference Machine Learning, pp. 1096–1103 (2008)
23. Wang, X., Tao, Q., Wang, L., Li, D., Zhang, M.: Deep convolutional architecture for natural image denoising. In: 2015 International Conference on Wireless Communications & Signal Processing (WCSP), pp. 1–4. IEEE, October 2015
24. Wang, Y.Q., Morel, J.M.: Can a single image denoising neural network handle all levels of gaussian noise? IEEE Signal Process. Lett. **21**(9), 1150–1153 (2014)
25. Wu, Y., Zhao, H., Zhang, L.: Image denoising with rectified linear units. In: Loo, C.K., Yap, K.S., Wong, K.W., Beng Jin, A.T., Huang, K. (eds.) ICONIP 2014. LNCS, vol. 8836, pp. 142–149. Springer, Heidelberg (2014). doi:10.1007/978-3-319-12643-2_18
26. Xie, J., Xu, L., Chen, E.: Image denoising and inpainting with deep neural networks. In: Proceedings of Advances in Neural Information Processing System, pp. 350–358 (2012)
27. Wang, Z., Bovik, A.C., Sheikh, H.R., Simoncelli, E.P.: Wavelets for Image Image quality assessment: from error visibility to structural similarity. IEEE Trans. Image Process. **13**(4), 600–612 (2004)
28. Zhang, L., Zhang, L., Mou, X., Zhang, D.: A comprehensive evaluation of full reference image quality assessment algorithms. In: 2012 19th IEEE International Conference on Image Processing, pp. 1477–1480. IEEE, September 2012
29. Zhao, H., Gallo, O., Frosio, I., Kautz, J.: Is L2 a Good Loss Function for Neural Networks for Image Processing? arXiv preprint, November 2015

Single Image Super Resolution Through Multi Extreme Learning Machine Regressor Fusion

Xiang Wang, Zongliang Gan$^{(\boxtimes)}$, Lina Qi, Changhong Chen, and Feng Liu

Jiangsu Provincial Key Lab of Image Processing and Image Communication,
Nanjing University of Posts and Telecommunications, Nanjing 210003, China
wangxiang2713@163.com, {ganzl,qiln,chenchh,liuf}@njupt.edu.cn

Abstract. Single image super resolution (SISR) aims to generate a high resolution (HR) image based on a given low resolution (LR) input. The edge priori based SISR methods tend to estimate HR image by edge-preserving constraint. In this paper, a novel learning based SISR method is proposed to reconstruct HR image by using joint HR gradient field and high frequency constraint. In the training phase, interpolated training LR patches with similar structure are partitioned into the same cluster by K-means clustering, and the Extreme Learning Machine (ELM) are used to get gradient and high frequency regressors in each cluster by training LR/HR patch pairs. In the prediction phase, multi-ELM regressor fusion strategy is used to estimate more accurate gradient and high frequency data, in which the fusion weights are based on the distance of the cluster centers with patch isotropic characteristics. Then, the estimated HR image gradient and high frequency are regarded as a joint constraints priori to reconstruct HR image. Experimental results demonstrate that the proposed method achieves better estimating accuracy of gradient and high frequency and have competitive SR quality compared with the other state-of-the-art SISR methods.

Keywords: Super resolution · Extreme learning machine regression · Multi-data fusion

1 Introduction

Single image super resolution (SISR) aims to recover a high resolution (HR) image from a low resolution (LR) one [1], which is still a classic and challenging problem in the imaging processing area. SISR is an inherently ill-posed problem because numerous pixel intensities need to be predicted from limited input data.

Generally, there are three kinds of approaches to deal with the SISR problem, including the reconstruction based [2,3], learning based [4–8] and edge priori

Z. Gan—This research was supported in part by the National Nature Science Foundation, P. R. China. (No. 61071166, 61172118, 61071091, 61471201), Jiangsu Province Universities Natural Science Research Key Grant Project (No. 13KJA510004), and the "1311" Talent Plan of NUPT.

© Springer Nature Singapore Pte Ltd. 2016
T. Tan et al. (Eds.): CCPR 2016, Part II, CCIS 663, pp. 133–146, 2016.
DOI: 10.1007/978-981-10-3005-5_12

based [9–17] methods. Reconstruction based methods [2,3] characterize the LR image by smoothing and downsampling of its HR version and enforce some prior knowledge on the upsampled image (e.g., smooth edges). Learning based methods [4–8] learn an inherent relationship between LR and HR image patch pairs and apply it to the given LR image to predict its HR version. Human vision is very sensitive to image boundary, thus a number of methods focus on reducing edge artifact, which are refered to as edge priori based approaches [9–14]. In this kind of methods, the estimating accuracy of HR gradient is crucial to final reconstruction result, thus some improved methods have been proposed, such as cross-resolution learning for high-resolution gradient estimation [15], defining GPS (Gradient Profile Sharpness) as an edge sharpness metric [16] and applying ramp profiles for learning edge transformations across resolutions [17].

The recently developed local learning based SR methods provide an approach in which partitioning the feature space into a number of clusters and learning a local model for each cluster [18–20]. Inspired by this idea, we propose a novel SISR method using joint constraints by learning based gradient and high frequency estimation in this paper. Our method is composed of training phase and prediction phase. In the training phase, interpolated training LR patches with similar structure are partitioned into the same cluster by K-means clustering. Relationships between LR and HR image patch pairs are learned in each cluster, including gradient and high frequency relationship, Extreme Learning Machine (ELM) [21] is used to learn regression models due to it's a very effective non-linear regression tool. In the prediction phase, HR gradient and high frequency for LR patches are estimated through multi-ELM regressor fusion. First of all, eight copies of each local LR patch are got after flipping and rotating them. Secondly, multiple estimation results of each copy are calculated by regressors corresponding to several cluster centers. Next, the estimation results for each LR patch are produced by fusing all the estimation results of the corresponding eight copies, the fusion weights are based on the distance of the cluster centers. Finally, the estimated HR image gradient and high frequency are regarded as a joint constraints priori to reconstruct HR image. Results show that our proposed method provides state-of-the-art quantitative performance and visual quality in comparison with other SISR methods.

2   Related Work

2.1   Single Image Super Resolution Using Gradient Profile Prior

Given a HR image I_h, LR image I_l is modeled in conventional SISR problem [12,18] as its Gaussian blurred and down sampled version by

$$I_l = (I_h \otimes G) \downarrow_s \tag{1}$$

where $\otimes$ is a convolution operator, G is a spatial filter commonly approximated as a Gaussian function, $\downarrow$ is the down-sampling operation and s is the scaling factor.

Gradient profile prior (GPP) [9] provides an effective constraint for reconstructing LR images, numerous SISR methods using GPP [10–17] have been proposed as enforced edge knowledge is able to produce sharp edges with minimal jaggy or ringing artifacts. Usually these methods aim to reconstruct I_h by enforcing the constraints in both image domain and gradient domain. Given the LR image I_l and the estimated HR gradient field $\widehat{\nabla I_h}$, I_h can be reconstructed as the minimization of the following energy function:

$$I_h^* = arg\min_{I_h} E(I_h|I_l, \widehat{\nabla I_h}) = arg\min_{I_h} E_i(I_h|I_l) + \beta E_g(\nabla I_h|\widehat{\nabla I_h}) \qquad (2)$$

where $E_i(I_h|I_l)$ is the reconstruction constraint in image domain and $E_g(\nabla I_h|\widehat{\nabla I_h})$ is the gradient constraint in gradient domain. The reconstruction constraint measures the difference between the LR image I_l and the smoothed and down sampled version of HR image I_h, i.e.

$$E_i(I_h|I_l) = |(I_h \otimes G) \downarrow_s - I_l|^2 \qquad (3)$$

Increasing the proportion of the reconstruction constraint results in better image color and contrast, yet with ringing or jaggy artifacts along edges [14]. The gradient constraint requires that the gradient ∇I_h of the recovered HR image should be close to the estimated gradient $\widehat{\nabla I_h}$:

$$E_g(\nabla I_h|\widehat{\nabla I_h}) = |\nabla I_h - \widehat{\nabla I_h}|^2 \qquad (4)$$

Increasing the proportion of the gradient constraint conversely contributes to producing sharp edges with little artifacts.

2.2 Extreme Learning Machine Regression

Extreme learning machine (ELM) [21,22] was proposed by Huang et al. as a fast and efficient learning algorithm for single hidden-layer feedforward neural networks (SLFNs) and is widely used for regression and multi-class classification. As shown in Fig. 1, given N distinct training samples (x_i, t_i), where the input $x_i = [x_{i1}, x_{i2}, \cdots, x_{in}]^T \in R^n$ and the target $t_i = [t_{i1}, t_{i2}, \cdots, t_{im}]^T \in R^m$, standard SLFNs with L hidden neurons approximate these N samples as follows

$$\sum_{i=1}^{L} \beta_i g(w_i \cdot x_j + b_i) = t_j, j = 1, \cdots, N \qquad (5)$$

where $w_i = [w_{i1}, w_{i2}, \cdots, w_{in}]^T$ is the input weight vector of the ith hidden neuron, $\beta_i = [\beta_{i1}, \beta_{i2}, \cdots, \beta_{im}]^T$ is the output weight vector of the ith hidden neuron, b_i is the bias of the ith hidden neuron and g is the activation function shared by all the hidden neurons. $w_i \cdot x_j$ denotes the inner product of w_i and x_j.

ELM randomly assigns w_i and b_i, and attempts to solve the following linear system corresponding to the above N equations as

$$H\beta = T \qquad (6)$$

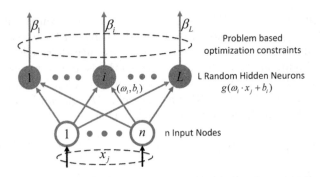

Fig. 1. Extreme learning machine regression

where the ith column of H, $[g(w_i \cdot x_1 + b_i), \cdots, g(w_i \cdot x_N + b_i)]^T$, is the ith hidden neuron output with respect to inputs $x_1, x_2, \cdots, x_N$, $\beta = [\beta_1, \cdots, \beta_L]^T$ and $T = [t_1, \cdots, t_N]^T$.

The minimal norm least-squares solution of the above linear system is

$$\hat{\beta} = H^\dagger T \tag{7}$$

where $H^\dagger$ is the Moore-Penrose generalized inverse of matrix H. Numerous SISR methods based on Extreme Learning Machine (ELM) has been proposed in recent years [23,24].

3   Proposed Super Resolution Method

3.1   Overview of Proposed Method

Previous works indicate that the estimating accuracy of HR gradient field has both theoretical and practical importance on the edge priori based SISR problem. One motivation of our work is estimating HR gradient and high frequency through sample clustering and training, estimation results calculated by ELM regression are used as multi-domain joint constraints to reconstruct HR image. Another motivation is multi Extreme Learning Machine regressor fusion strategy, namely acquiring more accurate estimation result through fusion of multiple estimation results calculated by different ELM regressors according to properties of image patches and K-means clustering algorithm.

The main flow of our method is shown in Fig. 2, which includes phase of training and phase of prediction. In the training phase, LR/HR patch pairs are partitioned into a number of clusters by K-means clustering algorithm, then ELM regressors are trained to estimate HR gradient and high frequency for LR patches in each cluster. In the prediction phase, HR gradient and high frequency of patches dense sampled in interpolated LR image are estimated through multi Extreme Learning Machine regressor fusion, which is based on the trained cluster centers as well as corresponding ELM regressors. The estimated HR horizontal and vertical gradient $\widehat{\nabla_1 I_h}$, orientational gradient $\widehat{\nabla_2 I_h}$ and high frequency $\widehat{I_{hf}}$ of input LR image can be finally obtained by averaging the overlapped pixels. Estimation of HR gradient and high frequency with multi-ELM regressor fusion strategy will be described in Subsect. 3.2 in detail.

Algorithm 1. The proposed SISR method

Input: LR image I_l, LR/HR patch pairs $\{P_l/P_h\}$ for training
Output: Reconstruction image I_h^{*2}
(1) Training:
    (a) Interpolate each patch in $\{P_l\}$ by the scaling factor to get $\{P_{bic}\}$.
    (b) Partition $\{P_{bic}/P_h\}$ into K clusteres by K-means, $\{C_k\}$ denotes cluster centers.
    (c) Train ELM regressors $\{R_k\}$ to estimate HR horizontal and vertical gradient, orientational gradient and high frequency for interpolated patches in each cluster.
(2) Prediction:
    (a) Interpolate I_l to the desired size as I_{bic}.
    (b) Estimate HR horizontal and vertical gradient, orientational gradient and high frequency of patches dense sampled in I_{bic} through multi Extreme Learning Machine regressor fusion, which is based on $\{C_k\}$ and $\{R_k\}$.
    (c) Obtain estimated HR horizontal and vertical gradient $\widehat{\nabla_1 I_h}$, orientational gradient $\widehat{\nabla_2 I_h}$ and high frequency $\widehat{I_{hf}}$ of I_{bic} by (10)-(11).
    (d) Obtain reconstruction image I_h^{*1} with constraints of $\widehat{\nabla_1 I_h}$ and $\widehat{\nabla_2 I_h}$ by (8).
    (e) Obtain final reconstruction image I_h^{*2} with constraints of I_h^{*1} and $\widehat{I_{hf}}$ by (9).

Edge priori based SISR model regards $\widehat{\nabla_1 I_h}$ as the gradient constraint in Eq. 2, while our method regards $\widehat{\nabla_2 I_h}$ and $\widehat{I_{hf}}$ as two other constraints. Reconstructing by using estimation results contains two phase: reconstructing by using gradient constraint and reconstructing by using high frequency constraint. In the first phase, estimated horizontal and vertical gradient $\widehat{\nabla_1 I_h}$ as well as estimated orientational gradient $\widehat{\nabla_2 I_h}$ are used as two constraints:

$$I_h^{*1} = arg \min_{I_h} ||(I_h \otimes G) \downarrow_s -I_l||_2^2 + \lambda_1 ||\nabla_1 I_h - \widehat{\nabla_1 I_h}||_2^2 + \lambda_2 ||\nabla_2 I_h - \widehat{\nabla_2 I_h}||_2^2 \quad (8)$$

I_h^{*1} is the reconstruction result by using constraints of estimated HR image gradient. After I_h^{*1} being solved, estimated HR image high frequency $\widehat{I_{hf}}$ is used as another constraint as it may contain HR image information estimated gradient loses, so as to further improve the reconstruction result:

$$I_h^{*2} = arg \min_{I_h} ||(I_h \otimes G) \downarrow_s -I_l||_2^2 + \lambda_3 ||I_h - I_{bic} - \widehat{I_{hf}}||_2^2 + \lambda_4 ||I_h - I_h^{*1}||_2^2 \quad (9)$$

I_{bic} stands for bicubic interpolated LR image and HR high frequency I_{hf} is calculated as I_h minus I_{bic}, which will be described in next subsection in detail. In each iteration, high frequency of the reconstruction result is further filtered by NLM [25] to reduce jaggy artifacts along edge. Both the objective energy functions are quadratic function with respect to I_h. Therefore they are convex and the global minimum can be obtained by the standard gradient descent by solving the gradient flow equations and λ_1, λ_2, λ_3 and λ_4 are the step sizes.

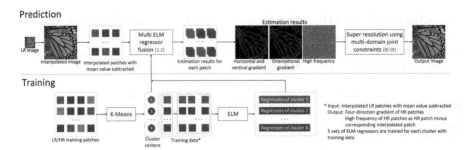

Fig. 2. The processing pipeline of our algorithm

3.2 Estimation of HR Gradient and High Frequency

Feature Training. Given a LR image I_l, it is first interpolated to the desired size as I_{bic} by bicubic interpolation. To reduce computational complexity, we perform estimation method for patches whose standard variance is larger than a threshold TH in I_{bic}. For the other patches, gradient and high frequency of I_{bic} are used instead. Features of each LR/HR patch pair are extracted as the luminance values of interpolated LR patch subtracting the mean value. Through random sampling from numerous training images, a large set of training data are collected to learn K cluster centers by K-means clustering algorithm. For data in each cluster, ELM regression is used to solve the estimation problems for its little training time and stable performance.

Our method trains four sets of ELM regressors for estimating gradient and one set of ELM regressors for estimating high frequency. I_{hf} stands for high frequency of HR image to estimate as one constraint, which is calculated as I_h minus I_{bic}. Four-direction gradient to estimate can be calculated by convoluting I_h with

$$k_1 = [-1/2, 0, 1/2], \ k_2 = k_1^T, \ k_3 = \begin{bmatrix} 0 & 0 & -1/2 \\ 0 & 0 & 0 \\ 1/2 & 0 & 0 \end{bmatrix} \text{ and } k_4 = \begin{bmatrix} -1/2 & 0 & 0 \\ 0 & 0 & 0 \\ 0 & 0 & 1/2 \end{bmatrix},$$

namely $\nabla_x I_h$, $\nabla_y I_h$, $\nabla_{45} I_h$ and $\nabla_{135} I_h$. Horizontal and vertical gradient $\nabla_1 I_h$ as well as orientational gradient $\nabla_2 I_h$ to be used as gradient constraints are calculated as:

$$\nabla_1 I_h = \sqrt{(\nabla_x I_h)^2 + (\nabla_y I_h)^2} \tag{10}$$

$$\nabla_2 I_h = \sqrt{(\nabla_{45} I_h)^2 + (\nabla_{135} I_h)^2} \tag{11}$$

Multi Extreme Learning Machine Regressor Fusion. Given feature $\hat{x}$ of LR image patch x, namely the luminance values of interpolated LR patch subtracting the mean value in our method, previous methods tend to find its cluster and choose corresponding regressor to implement estimation [18, 19]. Notice that natural images are isotropic, that is, flipping or rotating an image still produces a natural image, so do image patches. Motivated by this property of image patches, we get eight different patch features $\hat{x}_i, i = 1, \cdots, 8$, including $\hat{x}$ itself, after flipping and rotating $\hat{x}$. Each one of $\hat{x}_i$ can be seen as a copy of $\hat{x}$ while they belong to different clusters and different sets of regressors take effect. More specific in Fig. 3(a), after opposite operation of flipping and rotating the estimation results of $\hat{x}_i, i = 1, \cdots, 8$, eight estimation results of $\hat{x}$ can be calculated. Estimation results are calculated independently for each of $\hat{x}_i$ next.

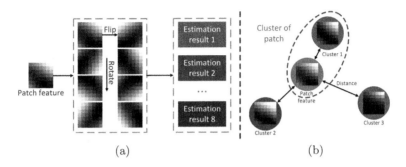

(a)   (b)

Fig. 3. Multi-ELM regressor fusion. (a) Through flipping and rotating, each patch is handled eight times, thus eight sets of estimation results are obtained. (b) Several clusters will be taken into account according to the distance from patch feature to cluster centers.

Original K-means clustering algorithm partitions a sample into one cluster whose center is closest to its feature. Just as shown in Fig. 3(b), the patch is partitioned into cluster 1 according to distance, previous methods then handle the patch using regressor corresponding to cluster 1. Nevertheless, partitioning the patch into cluster 2 or cluster 3 is also reasonable, as the distance between the patch feature and the center of cluster 2 or cluster 3 is also small enough. This means estimation results calculated by regressors corresponding to cluster 2 and cluster 3 are also accurate and useful. Motivated by this, we find n clusters that $\hat{x}_i$ belongs to by the standard of minimum Euclidean distance. Let $k_1^*, \cdots, k_{n-1}^*$ be the labels of the $n - 1$ clusters already found, the label of the nth cluster and the corresponding distance d_n are calculated as follows:

$$k_n^* = arg \min_k ||\hat{x}_i - C_k||_2^2, k = 1, \cdots, K, k \neq k_1^*, \cdots, k_{n-1}^* \tag{12}$$

$$d_n = ||\hat{x}_i - C_{k_n^*}||_2^2 \tag{13}$$

where C_k is cluster center of the cluster labeled by k. Then we can find n regressors to implement estimation of gradient in one direction or high frequency for $\hat{x}_i$, thus $8 * n$ sets of estimation results of gradient and high frequency can be calculated for patch x, as well as $8 * n$ corresponding distances. After that, the $8 * n$ distances are sorted in ascending order, the estimation results corresponding to the m smallest distances are finally averaged to produce the estimation results of gradient and high frequency for x.

Unlike JOR [19], which searches for the most suitable regressor for each patch, our method makes use of several different sets of regressors for more accurate estimation results according to properties of image patches and K-means clustering algorithm.

4   Experiments

4.1   Settings

We test our method on a variety of natural images with rich edges. For color images, we only apply the proposed method on luminance channel (Y) as human vision is more sensitive to luminance information, and chrominance channels (UV) are up-sampled by bicubic interpolation. 12 test examples are presented in Fig. 4.

Parameters: The standard variance of Gaussian blur kernel is set as 1.4 for a scaling factor 3. Threshold TH is set as 10, LR patches of size 7*7 are partitioned by K-means clustering algorithm while their central $5*5$ regions are used for training and estimation, K is set as 128. For ELM, 80 neurons and sigmoid activation function is suitable in our experiment. In the prediction phase, n and m is respectively set as 4 and 6, namely 6 out of 32 sets of estimation results are used for multi-ELM regressor fusion. In the two phase of reconstructing by using estimation results, λ_1, λ_2, λ_3 and λ_4 are all set as 0.1, iterative steps are 0.8 and 1.6 with number of iterations both set as 15 in terms of the objective indicator and visual effect.

4.2   Results

Estimation and reconstruction results of our method during processing are shown in Fig. 5, HR images are reconstructed by using multi-domain joint constraints by learning based gradient and high frequency estimation. In order to measure

Fig. 4. 12 test example images

the SISR results, we compared our method with SRCNN [26], Simple Function [18], and JOR [19]. The PSNR and SSIM [27] results calculated on Y channel are separately listed in Tables 1 and 2. Figures 6, 7 and 8 present visual comparisons of our method with these methods. As shown in these figures, images are blurred and jagged along edges by bicubic interpolation. SRCNN produces less blurred image, but it fails to reconstruct sharp image edges compared to other SISR methods. Edges of the reconstructed image by Simple Function is sharper yet unnatural and artificial. As shown, the results of JOR are sharp with rare ringing and blurring, while our method recovers details better, especially on salient edges. By multi-domain joint constraints, our method preforms best both in quantitative comparison and visual comparison.

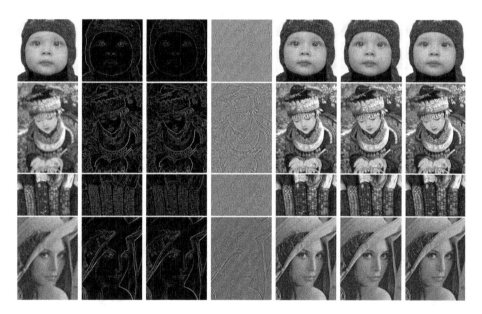

Fig. 5. Estimation and reconstruction results of our method. From left to right: 1. Bicubic upsample. 2. Estimated horizontal and vertical gradient. 3. Estimated orientational gradient. 4. Estimated high frequency. 5. Our reconstruction result by using HR gradient constraint. 6. Our final reconstruction result by using joint the HR gradient field and high frequency constraint. 7. Ground truth.

4.3   Complexity

The total computational complexity of our method is linearly dependent on the number of high frequency patches in the bicubic interpolated image. More specifically, finding several cluster centers for each high frequency patch plays a dominant role in computational time. Both our method and JOR spend time on searching, while our method first flips and rotates each test patch to get eight

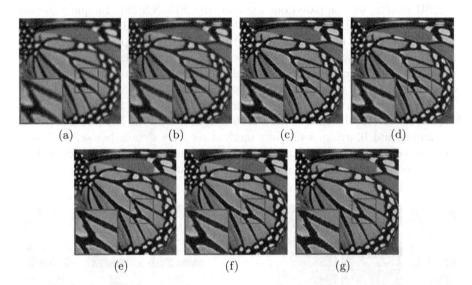

Fig. 6. Super resolution comparison (3×) with other methods. (a) Bicubic upsample (22.85 dB/0.787). (b) SRCNN (25.53 dB/0.875) [26]. (c) Simple Function (26.14 dB/0.875) [18]. (d) JOR (26.90 dB/0.903) [19]. (e) Our reconstruction result by using HR gradient constraint (27.55 dB/0.910). (f) Our final reconstruction result by using joint the HR gradient field and high frequency constraint (27.70 dB/0.914). (g) Ground truth.

Fig. 7. Super resolution comparison (3×) with other methods. (a) Bicubic upsample (35.69 dB/0.967). (b) SRCNN (38.09 dB/0.976) [26]. (c) Simple Function (38.59 dB/0.975) [18]. (d) JOR (39.46 dB/0.980) [19]. (e) Our reconstruction result by using HR gradient constraint (39.67 dB/0.980). (f) Our final reconstruction result by using joint the HR gradient field and high frequency constraint (39.77 dB/0.981). (g) Ground truth.

Table 1. PSNR measurement on 12 examples (Bold: best, underline: second best).

Test images(3×)	SRCNN [26]	Simple Function [18]	JOR [19]	Our Method	
				Gradient Constraint	Final Result
Baby	34.17	34.87	<u>35.09</u>	35.01	**35.15**
Barbara	26.33	**26.68**	26.54	26.64	<u>26.64</u>
Bird	30.26	30.84	<u>31.32</u>	31.61	**31.72**
Building	30.76	31.75	<u>32.16</u>	32.11	**32.28**
Butterfly	25.53	26.14	<u>26.90</u>	27.55	**27.70**
Comic	23.71	24.01	<u>24.24</u>	24.14	**24.27**
Corn	26.52	27.10	<u>27.25</u>	27.17	**27.40**
Flower	30.40	31.18	<u>31.23</u>	31.25	**31.59**
Lena	32.35	33.06	**33.35**	33.10	<u>33.30</u>
Plane	38.09	38.59	<u>39.46</u>	39.67	**39.77**
Woman	31.10	32.06	<u>32.46</u>	32.51	**32.69**
Zebra	27.62	28.61	<u>28.76</u>	28.79	**29.00**
Mean	29.74	30.41	<u>30.73</u>	30.79	**30.96**

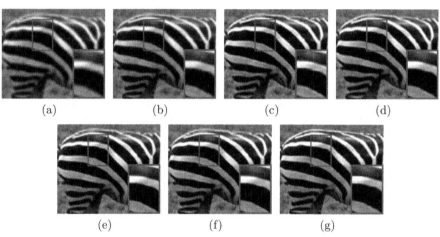

Fig. 8. Super resolution comparison (3×) with other methods. (a) Bicubic upsample (25.12 dB/0.743). (b) SRCNN (27.62 dB/0.815) [26]. (c) Simple Function (28.61 dB/0.845) [18]. (d) JOR (28.76 dB/0.845) [19]. (e) Our reconstruction result by using HR gradient constraint (28.79 dB/0.842). (f) Our final reconstruction result by using joint the HR gradient field and high frequency constraint (29.00 dB/0.847). (g) Ground truth.

patch copies and handles them simultaneously, this renders our method slower than JOR. Computational complexity of our method can be greatly reduced by parallel computing in implementation.

Table 2. SSIM measurement on 12 examples (Bold: best, underline: second best).

Test images(3×)	SRCNN [26]	Simple Function [18]	JOR [19]	Our Method	
				Gradient Constraint	Final Result
Baby	0.908	0.919	<u>0.922</u>	0.920	**0.922**
Barbara	0.759	**0.781**	0.780	0.777	<u>0.780</u>
Bird	0.933	0.937	<u>0.945</u>	0.945	**0.947**
Building	0.924	0.930	**0.939**	0.936	<u>0.938</u>
Butterfly	0.875	0.875	<u>0.903</u>	0.910	**0.914**
Comic	0.732	0.762	<u>0.772</u>	0.764	**0.773**
Corn	0.803	0.828	<u>0.835</u>	0.829	**0.837**
Flower	0.881	0.897	<u>0.904</u>	0.899	**0.906**
Lena	0.868	0.877	**0.884**	0.878	<u>0.882</u>
Plane	0.976	0.975	<u>0.980</u>	0.980	**0.981**
Woman	0.894	0.905	**0.913**	0.909	<u>0.912</u>
Zebra	0.815	<u>0.845</u>	0.845	0.842	**0.847**
Mean	0.864	0.878	<u>0.885</u>	0.882	**0.887**

5   Conclusion

In this paper, a novel learning based SISR method by using joint the HR gradient field and high frequency constraint has been presented based on ELM regression. In proposed method, interpolated training LR patches are partitioned into a number of clusters to get gradient and high frequency regressors by ELM with corresponding HR patches. To reconstruct a HR image, HR gradient and high frequency of patches in interpolated LR image are estimated by multi-ELM regressor fusion, in which the fusion weights are based on the distance of the cluster centers with patch isotropic characteristics. Then, the estimated HR image gradient and high frequency are regarded as a joint constraints priori to reconstruct HR image. Experimental results demonstrate that our proposed method recovers image with sharper edges and fewer artifacts compared with the other state-of-the-art SISR methods.

References

1. Nasrollahi, K., Moeslund, T.B.: Super-resolution: a comprehensive survey. Mach. Vis. Appl. **25**(6), 1423–1468 (2014)
2. Kim, K.I., Kwon, Y.: Single-image super-resolution using sparse regression and natural image prior. IEEE Trans. Pattern Anal. Mach. Intell. **32**(6), 1127–1133 (2010)
3. Sun, J., Zhu, J., Tappen, M.F.: Context-constrained hallucination for image super-resolution. In: 2010 IEEE Conference on Computer Vision and Pattern Recognition (CVPR), pp. 231–238. IEEE (2010)
4. Zhang, K., Gao, X., Tao, D., Li, X.: Multi-scale dictionary for single image super-resolution. In: 2012 IEEE Conference on Computer Vision and Pattern Recognition (CVPR), pp. 1114–1121. IEEE (2012)
5. HaCohen, Y., Fattal, R., Lischinski, D.: Image upsampling via texture hallucination. In: 2010 IEEE International Conference on Computational Photography (ICCP), pp. 1–8. IEEE (2010)

6. Freedman, G., Fattal, R.: Image and video upscaling from local self-examples. ACM Trans. Graph. (TOG) **30**(2), 12 (2011)
7. Yang, J., Wright, J., Huang, T.S., Ma, Y.: Image super-resolution via sparse representation. IEEE Trans. Image Process. **19**(11), 2861–2873 (2010)
8. Wang, S., Zhang, L., Liang, Y., Pan, Q.: Semi-coupled dictionary learning with applications to image super-resolution and photo-sketch synthesis. In: 2012 IEEE Conference on Computer Vision and Pattern Recognition (CVPR), pp. 2216–2223. IEEE (2012)
9. Sun, J., Sun, J., Xu, Z., Shum, H.Y.: Image super-resolution using gradient profile prior. In: 2008 IEEE Conference on Computer Vision and Pattern Recognition, CVPR 2008, pp. 1–8. IEEE (2008)
10. Dai, S., Han, M., Xu, W., Wu, Y., Gong, Y.: Soft edge smoothness prior for alpha channel super resolution. In: 2007 IEEE Conference on Computer Vision and Pattern Recognition, CVPR07, pp. 1–8. IEEE (2007)
11. Sun, J., Xu, Z., Shum, H.Y.: Gradient profile prior and its applications in image super-resolution and enhancement. IEEE Trans. Image Process. **20**(6), 1529–1542 (2011)
12. Zhang, H., Zhang, Y., Li, H., Huang, T.S.: Generative Bayesian image super resolution with natural image prior. IEEE Trans. Image Process. **21**(9), 4054–4067 (2012)
13. Dai, S., Han, M., Xu, W., Wu, Y., Gong, Y., Katsaggelos, A.K.: Softcuts: a soft edge smoothness prior for color image super-resolution. IEEE Trans. Image Proces. **18**(5), 969–981 (2009)
14. Wang, L., Xiang, S., Meng, G., Wu, H., Pan, C.: Edge-directed single-image super-resolution via adaptive gradient magnitude self-interpolation. IEEE Trans. Circuits Syst. Video Technol. **23**(8), 1289–1299 (2013)
15. Han, W., Chu, J., Wang, L., Pan, C.: Edge-directed single image super-resolution via cross-resolution sharpening function learning. In: Zha, H., Chen, X., Wang, L., Miao, Q. (eds.) Computer Vision. Communications in Computer and Information Science, vol. 546, pp. 210–219. Springer, Heidelberg (2015)
16. Yan, Q., Xu, Y., Yang, X., Nguyen, T.Q.: Single image superresolution based on gradient profile sharpness. IEEE Trans. Image Process. **24**(10), 3187–3202 (2015)
17. Singh, A., Ahuja, N.: Learning ramp transformation for single image super-resolution. Comput. Vis. Image Underst. **135**, 109–125 (2015)
18. Yang, C.Y., Yang, M.H.: Fast direct super-resolution by simple functions. In: Proceedings of the IEEE International Conference on Computer Vision, pp. 561–568 (2013)
19. Dai, D., Timofte, R., Van Gool, L.: Jointly optimized regressors for image super-resolution. In: Computer Graphics Forum, vol. 34, pp. 95–104. Wiley Online Library (2015)
20. Si, D., Hu, Y., Gan, Z., Cui, Z., Liu, F.: Edge directed single image super resolution through the learning based gradient regression estimation. In: Zhang, Y.-J. (ed.) ICIG 2015. LNCS, vol. 9218, pp. 226–239. Springer, Heidelberg (2015). doi:10.1007/978-3-319-21963-9_21
21. Huang, G.B., Zhu, Q.Y., Siew, C.K.: Extreme learning machine: theory and applications. Neurocomputing **70**(1), 489–501 (2006)
22. Huang, G.B., Zhou, H., Ding, X., Zhang, R.: Extreme learning machine for regression and multiclass classification. IEEE Trans. Syst. Man Cybern. Part B Cybern. **42**(2), 513–529 (2012)

23. An, L., Bhanu, B.: Image super-resolution by extreme learning machine. In: 2012 19th IEEE International Conference on Image processing (ICIP), pp. 2209–2212. IEEE (2012)
24. Zhu, Q., Li, X., Mao, W.: Image super-resolution representation via image patches based on extreme learning machine. In: 2013 International Conference on Software Engineering and Computer Science. Atlantis Press, Gijon (2013)
25. Buades, A., Coll, B., Morel, J.M.: A review of image denoising algorithms, with a new one. Multiscale Model. Simul. $4(2)$, 490–530 (2005)
26. Dong, C., Loy, C.C., He, K., Tang, X.: Image super-resolution using deep convolutional networks (2015)
27. Wang, Z., Bovik, A.C., Sheikh, H.R., Simoncelli, E.P.: Image quality assessment: from error visibility to structural similarity. IEEE Trans. Image Process. $13(4)$, 600–612 (2004)

Learning-Based Weighted Total Variation for Structure Preserving Texture Removal

Shoufeng Zheng, Chunwei Song, Hongzhi Zhang,
Zifei Yan, and Wangmeng Zuo[✉]

School of Computer Science and Technology,
Computational Perception and Cognition Center,
Harbin Institute of Technology, Harbin 150001, China
cswmzuo@gmail.com

Abstract. An image is generally formed as the composition of salient structures and complex textures. While structures are important for human perception and image analysis, structure extraction from textures remains a challenging issue to be investigated. Even though several methods have been proposed to do this job, they commonly have to balance between texture removing and structure preservation. One problem is that few methods take structural contours into consideration. In this paper, we propose a new learning-based weighted total variation (LTV)model, where the weights are learned from different kinds of texture images to well discriminate pixels belonging to structural contours from pixels belonging to textures. The Chambolles projection method is utilized to solve the optimization problem. Experimental results show that compared with the competing methods, the proposed algorithm performs better in preserving sharp structures while removing textures.

Keywords: Structure extraction · Contour detection · Learning-based total variation · Chambolles projection

1 Introduction

Structure is a fundamental notion for characterizing objects and describing relationships between objects. Together with various textures, it makes up the beautiful world with different views, such as leaves with veins, wood with stripes. Also, artists may look forward to picking up the appropriate combinations between structures and textures to construct aesthetic works, such as paintings on canvas or graffito on brick walls. Some examples can be found in Fig. 1.

Leaving the aesthetics of views unconsidered, it is particularly interesting that human visual system is capable of understanding objects underlying the rich textures effortlessly. Studies have found that structure features are more preferred for human perception to details [1], and salient structures are vital for high-level visual tasks, such as object recognition and classification. Other than that, separations of structures are also important for image processing tasks,

© Springer Nature Singapore Pte Ltd. 2016
T. Tan et al. (Eds.): CCPR 2016, Part II, CCIS 663, pp. 147–160, 2016.
DOI: 10.1007/978-981-10-3005-5_13

such as image segmentation [14], detail enhancement [11], and even producing pencil scratch effects [19]. Therefore, structure extraction has attracted much attention of researchers.

Fig. 1. Examples of texture images. From left to right, artificial lawn, landscape, graffiti on brick wall, marble road.

In contrast to the high efficiency of human visual perception, it is even harder for computer to automatically complete the task of structure extraction. Recently, some novel algorithms are proposed to address this problem. Some representative methods are like bilateral filtering [8,21], anisotropic diffusion [15], guided filtering [12] and the newly proposed rolling guidance filter [22]. These methods essentially implement weighted average to remove small details by taking into consideration the affinity between different neighboring pixel pairs. The weights are dynamically computed to smooth the low-contrast area while preserve the high-contrast structure. To preserve structure information, some other methods, such as total variation (TV) [16], weighted least squares (WLS) [9], L_0 minimization method [19] and L_p-norm minimization method [7], also introduce optimization algorithms into the structure extraction, and usually iterative optimization is necessary. These methods are realized by optimizing a global objective function, and can obtain better results. Besides, some other methods like [17,18,23] also use median/mode filter to remove small details. This kind of methods is suitable for the removal of high-contrast details, but sometimes results in oscillating effects.

Even though many algorithms have been proposed, they usually perform poorly in distinguishing structures from coarse textures. This result in that they cannot effectively preserve structures while removing detailed textures. It seems that these two objectives are contradictory to some extent, but actually this task can be easily conducted by human. In this work, we adopt a data-driven scheme to learn a classifier from a labeled structure-texture dataset. The framework of our learning-based weighted total variation (LTV) algorithm is illustrated in Fig. 2.

The reminder of this paper is organized as follows. Section 2 briefly reviews some prior work that is related to our method. Section 3 presents our LTV algorithm and its solving process. In Sect. 4, some comparative experiments are conducted to prove the effectiveness of our algorithm. And the Sect. 5 concludes our paper.

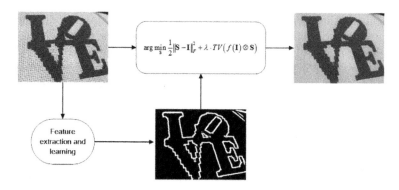

Fig. 2. The flowchart of our algorithm.

2 Related Work

Our proposed LTV method falls into the category of edge-aware filter. This kind of methods aims to removing low-contrast details, while preserving high-contrast edges, and weighted average filter and optimization based filter are commonly utilized.

Among them, bilateral filter [8,21] and its relatives [3,10,13] are widely used, because of their simplicity and effectiveness. This method defines the weights by simultaneously considering spatial and structure relationship. But it usually results in an image with smoother edges, since it cannot effectively describe the local structure.

Guided filter [12] is another kind of edge-aware filter, which instead borrows the structural information from the guidance image. A special case is to set the guidance image to be processing image itself. In that situation, a roughly similar bilateral filter can be achieved. But the difference between them is that, as described in [12], guided filter can avoid the gradient reversal artifacts that may appear in bilateral filtering results.

Combining the above two methods, a new rolling guidance filter [22] is proposed, which is just like the bilateral filter, the only difference is that the structural information is obtained from last time filtering results. An interesting result is that, this method naturally leads to a multi-scale smoothing method, where the method starts from a large-scale guidance image (a constant image), and with the algorithm proceeding, a smooth but edge enhanced result is obtained.

The above mentioned methods are essentially local-filtering methods, and originally designed to remove noise-like structures, therefore not highly suitable for structure extraction. In contrast, some optimization based edge-aware filter can achieve more satisfactory results.

Weighted least squares [9], L_0 minimization method [19] and L_p-norm minimization method [7] can obtain the structure image by optimizing an objective function. They both use the combination of data term and regularization term. The only difference lies in the selection of regularization. In [9], L_2-norm gradient regularization is utilized, and in [22], L_0 regularization is implemented on image gradient, while in [7] the Hyper-Laplacian prior is considered which is statistically more suitable to describe gradient distribution. It is worth noting that the weights in WLS varies w.r.t. the processing image structures, and can dynamically impose different regularization on different areas. This varying weighting scheme is much related to our method, but the difference lies in that in our method the weights are learned from training dataset.

Another important optimization based method is total variation (TV) [16], which has been successfully implemented to image denoising. This method can preserve high-contrast edges by producing a piecewise smooth image. A commonly used model of TV is TV-L_2, which is believed to be more favorable as studied in [2]. To be specific, the TV-L_2 model can be expressed as follows

$$\arg\min_{S} \frac{1}{2}\|\mathbf{I} - \mathbf{S}\|_F^2 + \lambda TV(\mathbf{S}) \tag{1}$$

where $\mathbf{I} \in [0,1]^{m \times n}$ is the image to be processed, $\mathbf{S} \in [0,1]^{m \times n}$ is the algorithm output. The first term $\frac{1}{2}\|\mathbf{I} - \mathbf{S}\|_F^2$ in Eq. 1 is called the data term, which is to measure the similarity between the input image and the output result, while the second term $TV(\mathbf{S})$ is the regularization term, which is introduced to impose proper constraint on the output result, and λ is the weight for tradeoff between the data term and regularization term. The TV regularization has following form

$$TV(\mathbf{S}) = \|D\mathbf{S}\|_i \tag{2}$$

where D is the discrete gradient operation, $(D\mathbf{S})_i$ denoted the gradient for the ith pixel, and the operator $\| \bullet \|_i$ is the L_p-norm, which can be set to be 1 or 2, corresponding to anisotropic TV and isotropic TV, respectively.

It is readily to find that the TV-L_2 model in Eq. 1 can only preserve those high-contrast edges while ignoring whether they are from structures or textures. To improve TV for structure extraction, Xu et al. [20] proposed the so called relative TV (RTV) model. In that paper, the anisotropic TV is utilized, and the weights are defined as the windowed inherent variation (WIV) as follows

$$\psi_x(p) = \left| \sum_{q \in R(p)} g_{p,q} \cdot (\nabla_x \mathbf{S})_q \right|$$
$$\psi_y(p) = \left| \sum_{q \in R(p)} g_{p,q} \cdot (\nabla_y \mathbf{S})_q \right| \tag{3}$$

where $\psi_x(p)$ and $\psi_y(p)$ are the weights of x- and y-direction at pixel p, ∇_x and ∇_y are the x- and y-directional difference operator. q is the neighboring pixel of p, and $g_{p,q}$ is the weighting function defining the spatial affinity just like that in bilateral filter.

$$g_{p,q} \propto \left(\frac{(x_p - x_q)^2 + (y_p - y_q)^2}{2\sigma^2} \right) \tag{4}$$

where x and y are the coordinates, and σ controls the spatial scale. And the authors stated that the proposed RTV and WIV features are more powerful than TV feature to separate structures from textures.

The RTV model extends the TV model to take discriminative nature of structures and textures into consideration to better preserve structures while removing textures. However, in RTV, the weights should be dynamically computed along with the iterations, and may not work well in distinguishing small-scale structures from coarse textures. Therefore, we suggest a learning-based method to have a more powerful discrimination ability. Furthermore, with the learned model, we can recognize the structural part directly from the input image, the weights can be pre-computed and keep invariant along with the iterations.

3   Model and Algorithm

In this section, our model will be presented and the computation method will be described. To combine the discrimination nature with structure extraction, a weighted TV is considered. As we all know, separation of structure from texture is a difficult task, and hand-crafted feature may take much effort to construct. Therefore, in our model, we propose to learn the discriminative features from some labeled datasets, and the learning details will be described later.

3.1   Learning-Based TV Model for Structure Extraction

The structure extraction model can be expressed as follows

$$\arg \min_S \frac{1}{2}\|\mathbf{S} - \mathbf{I}\|_F^2 + \lambda TV(f(\mathbf{I}) \otimes \mathbf{S}) \tag{5}$$

where λ is a constant just like that in Eq. 1, $f(\mathbf{I})$ is a pixel-wise weighting function, which will be set to different values according to the nature of pixels in $\mathbf{I}$, and $\otimes$ is the Hadamard product. In this paper, a simple piecewise function is used to define $f(\mathbf{I})$

$$f(\mathbf{I}) \begin{cases} 1, & if\ \mathbf{I}_p \in structure \\ w, & if\ \mathbf{I}_p \in texture \end{cases} \tag{6}$$

where $\mathbf{I}_p$ represents the p-th pixel of $\mathbf{I}$, and w is set to be a large number, much greater than 1. In order not to interrupt the whole algorithm description, the discrimination process will be left later to present.

3.2   Model Computation

The model in Eq. 5 is a weighted total variation, and we will apply the Chambolles projection method to deal with it. Chambolles projection was first proposed in [6], and as said in [2], this method has some advantages over Euler-Lagrange equations. But the original Chambolles projection was applied to deal with ROF model, and some modifications should be made to apply it to our problem.

In [5], the weighted TV model has been applied to image segmentation, and some results about the computation are given. In this paper, we may borrow some ideas from [5], and present the computational process as the following proposition. **Proposition 1** in [5]: The solution of Eq. 5 is give by:

$$S = I - \lambda div \mathbf{p} \tag{7}$$

where $\mathbf{p} = (\mathbf{p}_1, \mathbf{p}_2)$ is given by

$$f(\mathbf{I})\nabla(\lambda div \mathbf{p} - \mathbf{I}) - |\nabla div \mathbf{p} - \mathbf{I}|\mathbf{p} = 0 \tag{8}$$

Equation 8 can be solved by a fixed point method: $\mathbf{p}^0 = \mathbf{0}$ and

$$\mathbf{p}^{k+1} = \frac{\mathbf{p}^k + \delta_t \nabla(div \mathbf{p}^k - \mathbf{I}/\lambda)}{1 + \frac{\delta_t}{f(\mathbf{I})}|\nabla(div \mathbf{p}^k - \mathbf{I}/\lambda)|} \tag{9}$$

The discrete divergence operator div is given by:

$$(div \mathbf{p})_{i,j} = d\mathbf{p}^1_{i,j} + d\mathbf{p}^2_{i,j} \tag{10}$$

where

$$d\mathbf{p}^1_{i,j} = \begin{cases} \mathbf{p}^1_{i,j} - \mathbf{p}^1_{i-1,j}, & if\ 1 < i < m \\ \mathbf{p}^1_{i,j}, & if\ i = 1 \\ -\mathbf{p}^1_{i-1,j}, & if\ i = m \end{cases} \tag{11}$$

$$d\mathbf{p}^2_{i,j} = \begin{cases} \mathbf{p}^2_{i,j} - \mathbf{p}^2_{i,j-1}, & if\ 1 < j < n \\ \mathbf{p}^2_{i,j}, & if\ j = 1 \\ -\mathbf{p}^2_{i,j-1}, & if\ j = n \end{cases} \tag{12}$$

And the discrete gradient operator is as follows

$$(\nabla \mathbf{u})_{i,j} = ((\nabla \mathbf{u})^1_{i,j}, (\nabla \mathbf{u})^2_{i,j}) \tag{13}$$

where

$$(\nabla \mathbf{u})^1_{i,j} = \begin{cases} \mathbf{u}_{i+1,j} - \mathbf{u}_{i,j}, & if\ i < m \\ 0, & if\ i = m \end{cases}$$

$$(\nabla \mathbf{u})^2_{i,j} = \begin{cases} \mathbf{u}_{i,j+1} - \mathbf{u}_{i,j}, & if\ j < n \\ 0, & if\ j = n \end{cases} \tag{14}$$

We have to notice that the involved multiplication, division and plus operators are all component-wise, therefore the computation is quite highly efficient. And we set two criteria to stop the iterations, the maximum number of iteration and the minimum iteration error limitation. The two consecutive iteration error can be counted by computing the relative error, i.e., $e = \|\mathbf{p}^{k+1} - \mathbf{p}^k\|_F / \|\mathbf{p}^k\|_F$.

The whole process of our algorithm is summarized in **Algorithm 1**. In the following, we will describe the learning and projection process which is Step 3 in **Algorithm 1**.

Algorithm 1. Learning based TV (LTV) Structure Extraction

Input : $image$ **I**$, \lambda, \delta_t, N, Tol$

Output : $image$ **S**

1: **Initialize** $\mathbf{p}^0 = 0, k = 1, e = inf$

2: **Compute**

$$f(\mathbf{I}) = F(\mathbf{I})$$

3: **while** $k < N \ and \ e > Tol$ **do**

4:     $\mathbf{p}^{k+1} = \frac{\mathbf{p}^k + \delta_t \nabla(div\mathbf{p}^k - \mathbf{I}/\lambda)}{1 + \frac{\delta_t}{f(\mathbf{I})}|\nabla(div\mathbf{p}^k - \mathbf{I}/\lambda)|}$

5:     $e = \|\mathbf{p}^{k+1} - \mathbf{p}^k\|_F / \|\mathbf{p}^k\|_F$

6:     $k = k + 1$

7: **end while**

8: $\mathbf{S} = \mathbf{I} - \lambda div\mathbf{p}$

3.3   Structural Contour Learning

Structural contour is the main edge that can depict the structures. They are not supposed to be totally high-contrast; therefore they are difficult to extract. In RTV method [20], the authors constructed the so-called WIV features which can discriminate structures from textures to some extent. But the statistical distributions corresponding to WIV of structures and textures are not totally separable from each other, and may overlap for a large percent. The good news is that we could still produce a more discriminative result based on WIV feature using learning method.

The main point is to construct some features based on WIV and to train a classifier to separate the two kinds of features. If we revisit WIV expression in Eq. 3, we can find that there is a tunable parameter σ which controls the window scale, and is not easy to set since it has to be adapted to local structure. In order to alleviate this problem, in [20], an initial larger σ is selected and its value will decrease with the algorithm iteration. In fact, it is not a problem in our method. Since the stable and invariant features that can adapt to different structure scale are preferred, we could construct the multi-scale features, like the SIFT features for object detection.

Specifically, we set σ to be from 1 to 5.5, spaced by 0.5. Therefore, for each differential direction, we could get a 10-dimensional feature to represent each pixel; thereafter we concatenate the x- and y-directional features to construct the final 20-dimensional feature. After we have obtained the features, we could feed it into some classifiers to get the final classification results.

The contour learning problem is essentially a binary classification problem. And some well-structured algorithms are available. In this paper, we will compare

Table 1. Classification accuracy on 50 testing image by using different classifier.

Classifier	Accuracy(%)
KNN	88.79
Naive Bayes	89.14
Ensemble learning	89.92
Logistic regression	90.59
Discriminative learning	90.70
Random forests	91.18
SVM	91.24
Neural networks	93.02

the KNN method, the naive Bayes method, the ensemble learning method, logistic regression method, discriminative method, random forests method, SVM and neural network. The naive Bayes method and logistic regression method are realized using LibORF[1], a machine learning toolkit in MATLAB. KNN method, the ensemble learning method, discriminative method and random forests method are based on Statistic and Machine Learning Toolbox[2]. The well-known LIB-SVM[3] is used for SVM method, with Gaussian Radial Basis Function. Neural Network method is trained using Deep learning Toolbox[4]. Most of the tunable parameters in these comparative methods are set to be default, except that the K-parameter in KNN is set to be 1, the Tree-number in random forest is set to be 500, the number of layers in neural network is set to be 3 with 40 neurons in hidden layer.

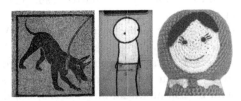

Fig. 3. Three structural contour extraction testing images, from left to right, mosaic dog, graffiti on the door and cross-stitch doll.

A labeled dataset of 200 textured images constructed by Xu et al.[5] is adopted for classifier training and testing. We randomly divide the images into two sets,

---

[1] https://github.com/orhanf/libORF.
[2] http://cn.mathworks.com/help/stats/index.html.
[3] http://www.csie.ntu.edu.tw/~cjlin/libsvm/index.html.
[4] https://github.com/rasmusbergpalm/DeepLearnToolbox.
[5] http://www.cse.cuhk.edu.hk/~leojia/projects/texturesep/database.html.

one set with 150 images for training, and the other with 50 images for testing. Three representative images are shown in Fig. 3. The classification results are exhibited in Table 1. From Table 1 it is readily to find that the neural network method has better performance, therefore it becomes the option of our algorithm.

Using the proposed feature extraction and neural network classification method, we could get the structural contour of these three images in Fig. 3, as shown in Fig. 4. It is easy to find that most of the main structural contour is labeled by our method, while some parts of the contours are not complete and some textures are mislabeled as the structures. But we also have to notice that, the learning process only involves 20-dimensional hand-crafted features and a simple three-layer neural network. With the booming of deep structure machine, we believe that combining the deep feature learning and classification, we may get a better labeling result, and that will be our future research topic.

Fig. 4. The structural contour extraction result corresponding to the images in Fig. 3. The first row consists of the manually labeled contour images which serve as the benchmark, and the lower row contains the result using our learning method.

4   Experiment

4.1   Parameter Setting

In this section we will conduct some comparative experiments to evaluate the performance of our algorithm. Before we proceed to the experimental details, we first give the parameter setting in our algorithm. In **Algorithm 1**, four parameters have to be designated, λ, δ_t, **N** and **Tol**. The parameter λ is not fully invariant, but it usually falls between 0.05 and 0.5, and the temporal step δ_t is set to be 0.24, not 1/8 as suggested in Fast Global Minimization of the Active Contour/Snake Model [5], while **N** is set to be 200, and **Tol** is set to be 0.001, according to our experimental experience. For the learning part, mini-batch size during training is set to be 200, and we just leave the other parameters to be default. Besides, the weight value when pixels belong to texture, i.e., w in Eq. 6, is set to be 100.

4.2 Structure Extraction Comparison

We will compare our method with seven state-of-the-art edge-aware filtering methods, including the ROF total variation (TV) method, weighted least squares (WLS) [9][6], the gradient L_0 minimization method (L_0) [19][7], the Hyper-Laplacian prior method (L_p) [7], the L_1 transform edge-preserving smoothing (L_1) [4][8], the relative total variation method (RTV) [20][9] and the Rolling Guidance Filter [22][10]. All the experiments are conducted using the codes provided by the authors. The parameters involved in all the comparative methods are well

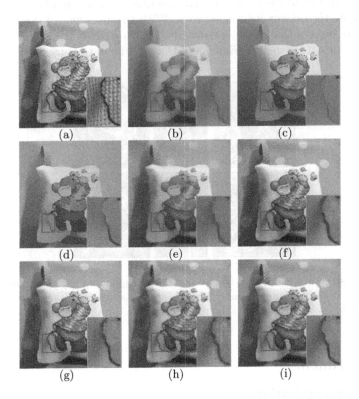

(a)   (b)   (c)

(d)   (e)   (f)

(g)   (h)   (i)

Fig. 5. Experiments on the cross-stitch pillow. (a) The original textured image; (b) Result by WLS ($\lambda = 1.6, \alpha = 1.7$); (c) Result by L_1 transform ($\lambda = 10000, \lambda = 0.2$); (d) Result by L_0 smoothing ($\lambda = 0.07$); (e) Result by L_p smoothing ($p = 0.7, \mu = 0.05$); (f) Result by Rolling guidance filter ($\sigma_s = 40, \sigma_r = 0.1$); (g) Result by RTV ($\lambda = 0.01, \sigma = 4$); (h) Result by TV ($\lambda = 2$); (i) Result by our method with learned contour ($\lambda = 0.05$).

---

[6] http://www.cs.huji.ac.il/~danix/epd/.

[7] http://www.cse.cuhk.edu.hk/~leojia/projects/L0smoothing/index.html.

[8] https://github.com/soundsilence/L1Flattening.

[9] http://www.cse.cuhk.edu.hk/~leojia/projects/texturesep/index.html.

[10] http://www.cse.cuhk.edu.hk/~leojia/projects/rollguidance/.

tuned to produce the best results. Due to space limitation, some representative experiments are presented in Figs. 5, 6 and 7.

In Fig. 5, the experiments on a cross-stitch pillow are displayed. To better observe the differences, zoom-in of the area at the position surrounded by the red rectangle is also given at the bottom-right corner of each result image. That is also the case for other experiments. From Fig. 5 it can be discerned that, TV-L_2 model can remove textures, but it also smooth the main edges. And it is also the case for WLS method. The L_0 minimization method can lead to a better result with sharper edges, while it does not discriminate structures and textures, therefore may produce over-smooth results. On the contrary, the Hyper-Laplacian prior based method makes some improvement, but the edges

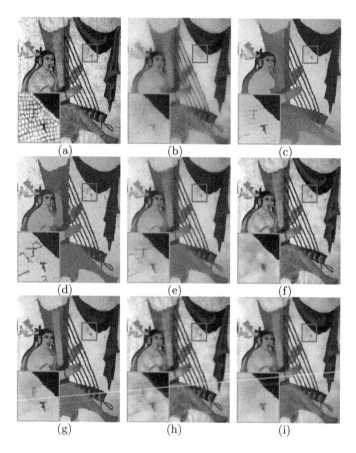

Fig. 6. Experiments on the cross-stitch pillow. (a) The original textured image; (b) Result by WLS ($\lambda = 2.8, \alpha = 2$); (c) Result by L_1 transform ($\lambda = 30000, \lambda = 0.1$); (d) Result by L_0 smoothing ($\lambda = 0.1$); (e) Result by L_p smoothing ($p = 0.7, \mu = 0.05$); (f) Result by Rolling guidance filter ($\sigma_s = 3, \sigma_r = 0.1$); (g) Result by RTV ($\lambda = 0.015, \sigma = 3$); (h) Result by TV ($\lambda = 2$); (i) Result by our method with learned contour ($\lambda = 0.2$).

are still too smoothed. The result of L_1 transform method may generate sharper edges, but are prone to mixing distant pixels and producing some unnatural color. The RTV method is obviously better than the other methods, however still may smooth edges insufficiently. Compare the zoom-in in Fig. 5(g) and (i), it is readily to find that our method can generate images with high contrast, and can preserve the tiny variation along the edges.

Figure 6 is a fresco with a girl playing harp. Different form the image in Figs. 5(a) and 6(a) has more complex textures as shown in the zoom-in area. The results by TV, L_0 and L_p are all smooth to some extent. L_1 transform is good at generating sharp structure, however the color mixture persists. RTV

Fig. 7. Experiments on the cross-stitch pillow. (a) The original textured image; (b) Result by WLS ($\lambda = 3, \alpha = 1.2$); (c) Result by L_1 transform ($\lambda = 5000, \lambda = 0.05$); (d) Result by L_0 smoothing ($\lambda = 0.04$); (e) Result by L_p smoothing ($p = 0.7, \mu = 0.04$); (f) Result by Rolling guidance filter ($\sigma_s = 3.5, \sigma_r = 0.1$); (g) Result by RTV ($\lambda = 0.015, \sigma = 3$); (h) Result by TV ($\lambda = 2$); (i) Result by our method with learned contour ($\lambda = 0.05$).

method is better than aforementioned method, but still left some tiny textures untouched, while our method can generate smoother background meanwhile preserving sharp structure.

The last one is the cross-stitch flowers shown in Fig. 7(a). It is composed of tiny textures to form relatively smooth transit. The difficult lying in this image is that it is easy to produce over-smooth results while removing textures, as shown in zoom-in area. The L_1 transform method has better performance on preserving structures. But the color mixture issue still exists, such as the tiny blue rectangles on the wing. In contrast, our method makes the best balance between texture removal and structure preservation.

5 Conclusion

In this paper, we proposed a new learning-based weighted total variation model (LTV) to deal with structure extraction problem. Different from prior methods, our model explicitly took the discrimination between structure and texture into consideration by learning the TV regularization weights according to the local structure. By constructing features for each pixel, a simple neural network classifier was utilized to assign pixels to any of the two groups, structure or texture. After that, a weighted TV model was proposed to complete structure extraction. Experiments showed that our model was better than some state-of-the-art methods at preserving sharp structures and removing textures.

Acknowledgement. This work is partly support by the National Science Foundation of China (NSFC) project under the contract No. 61271093.

References

1. Arnheim, R.: Art and Visual Perception: A Psychology of the Creative Eye. University of California Press, Berkeley (1954)
2. Aujol, J.F., Gilboa, G., Chan, T., Osher, S.: Structure-texture image decomposition modeling, algorithms, and parameter selection. Int. J. Comput. Vis. **67**(1), 111–136 (2006)
3. Baek, J., Jacobs, D.E.: Accelerating spatially varying gaussian filters. ACM Trans. Graph. (TOG) **29**, 169 (2010). ACM
4. Bi, S., Han, X., Yu, Y.: An l1 image transform for edge-preserving smoothing and scene-level intrinsic decomposition. ACM Trans. Graph. (TOG) **34**(4), 78 (2015)
5. Bresson, X., Esedo?lu, S., Vandergheynst, P., Thiran, J.P., Osher, S.: Fast global minimization of the active contour/snake model. J. Math. Imaging Vis. **28**(2), 151–167 (2007)
6. Chambolle, A.: An algorithm for total variation minimization and applications. J. Math. Imaging Vis. **20**(1–2), 89–97 (2004)
7. Chen, L., Zhang, H., Ren, D., Zhang, D., Zuo, W.: Fast augmented Lagrangian method for image smoothing with hyper-laplacian gradient prior. In: Li, S., Liu, C., Wang, Y. (eds.) CCPR 2014. CCIS, vol. 484, pp. 12–21. Springer, Heidelberg (2014). doi:10.1007/978-3-662-45643-9_2

8. Durand, F., Dorsey, J.: Fast bilateral filtering for the display of high-dynamic-range images. ACM Trans. Graph. (TOG) **21**, 257–266 (2002). ACM

9. Farbman, Z., Fattal, R., Lischinski, D., Szeliski, R.: Edge-preserving decompositions for multi-scale tone and detail manipulation. ACM Trans. Graph. (TOG) **27**, 67 (2008). ACM

10. Fattal, R.: Edge-avoiding wavelets and their applications. ACM Trans. Graph. (TOG) **28**(3), 22 (2009)

11. Fattal, R., Agrawala, M., Rusinkiewicz, S.: Multiscale shape and detail enhancement from multi-light image collections. ACM Trans. Graph. **26**(3), 51 (2007)

12. He, K., Sun, J., Tang, X.: Guided image filtering. In: Daniilidis, K., Maragos, P., Paragios, N. (eds.) ECCV 2010. LNCS, vol. 6311, pp. 1–14. Springer, Heidelberg (2010). doi:10.1007/978-3-642-15549-9_1

13. Kass, M., Solomon, J.: Smoothed local histogram filters. ACM Trans. Graph. (TOG) **29**, 100 (2010). ACM

14. Malik, J., Belongie, S., Leung, T., Shi, J.: Contour and texture analysis for image segmentation. Int. J. Comput. Vis. **43**(1), 7–27 (2001)

15. Perona, P., Malik, J.: Scale-space and edge detection using anisotropic diffusion. IEEE Trans. Pattern Anal. Mach. Intell. **12**(7), 629–639 (1990)

16. Rudin, L.I., Osher, S., Fatemi, E.: Nonlinear total variation based noise removal algorithms. Physica D **60**(1), 259–268 (1992)

17. van de Weijer, J., Van den Boomgaard, R.: Local mode filtering. In: Proceedings of the 2001 IEEE Computer Society Conference on Computer Vision and Pattern Recognition, CVPR 2001, vol. 2, p. 428. IEEE (2001)

18. Weiss, B.: Fast median and bilateral filtering. In: ACM Trans. Graph. (TOG) **25**, 519–526 (2006). ACM

19. Xu, L., Lu, C., Xu, Y., Jia, J.: Image smoothing via l0 gradient minimization. ACM Trans. Graph. (TOG) **30**, 174 (2011). ACM

20. Xu, L., Yan, Q., Xia, Y., Jia, J.: Structure extraction from texture via relative total variation. ACM Trans. Graph. (TOG) **31**(6), 139 (2012)

21. Yang, Q.: Recursive bilateral filtering. In: Fitzgibbon, A., Lazebnik, S., Perona, P., Sato, Y., Schmid, C. (eds.) ECCV 2012. LNCS, vol. 7572, pp. 399–413. Springer, Heidelberg (2012). doi:10.1007/978-3-642-33718-5_29

22. Zhang, Q., Shen, X., Xu, L., Jia, J.: Rolling guidance filter. In: Fleet, D., Pajdla, T., Schiele, B., Tuytelaars, T. (eds.) ECCV 2014. LNCS, vol. 8691, pp. 815–830. Springer, Heidelberg (2014). doi:10.1007/978-3-319-10578-9_53

23. Zhang, Q., Xu, L., Jia, J.: 100+ times faster weighted median filter (WMF). In: Proceedings of the IEEE Conference on Computer Vision and Pattern Recognition, pp. 2830–2837 (2014)

A Novel Texture Extraction Method for the Sedimentary Structures' Classification of Petroleum Imaging Logging

Haoqi Gao[1], Huafeng Wang[1,2(✉)], Zhou Feng[3], Mingxia Fu[1],
Chennan Ma[1], Haixia Pan[1], Binshen Xu[1], and Ning Li[3]

[1] School of Software, Beihang University of Beijing, Beijing 10083, China
gaohaoqi77@163.com
[2] North China University of Technology, Beijing, China
wanghuafeng@buaa.edu.cn
[3] Research Institute of Petroleum Exploration and Development of PetroChina,
Beijing 100083, China

Abstract. The technology for reservoir structure identification has become a challenging problem in the field of imaging logging technology. Because of the huge amount of information and a wide variety, it causes experts with low efficiency on the interpretation of reservoir evaluation and the performance depends highly on the individual experience (including cognitive level, visual decision, etc.). We proposed a new method for texture feature extraction based on macro and micro features. About 3320 imaging logging datasets are fed to support vector machine (SVM) to validate the gains of new method. As a result, the new proposed method achieved an Area Under roc Curve (AUC) value of 0.94.

Keywords: Imaging logging · Texture features · Support Vector Machine (SVM) · Area Under roc Curve (AUC)

1 Introduction

According to statistics reports, 47 % of the world total oil and gas reservation and 60 % of the total production, are located in the reef-bank carbonate reservoir [1, 2]. Especially, the sedimentary structures of the reef-banks is of much value for the petroleum experts to get clearly descriptors of the reservoir underground. Sedimentary structure refers to the space between the various components of sedimentary rock distribution and arrangement. It is sediment deposited on or after deposition, due to the physical, chemical and biological forms. As one of the effective techniques of exploration, the petroleum imaging logging plays a very important role on the evaluation procedure of the reservation of the gas or oil. Though image logging is able to much clearly reflect the sedimentary structure by its characteristic of high resolution, nearly complete borehole coverage and visibility [3], it traditionally needs many experts to do manually drawing for providing a reliable explanation. With the rapid growing of the imaging logging task, a completely automatic or half automatic evaluation system for labelling the sedimentary structure are highly required currently for retrieving the labor force.

© Springer Nature Singapore Pte Ltd. 2016
T. Tan et al. (Eds.): CCPR 2016, Part II, CCIS 663, pp. 161–172, 2016.
DOI: 10.1007/978-981-10-3005-5_14

However, because the strong heterogeneity and various types in reef-banks reservoir, there are much challenges left for building an automatic or half sedimentary structures recognition algorithm. Feature extraction as one of the key steps of the image processing technique, it help us obtain the quantitative and qualitative description of a given image. Comparing to the color and shape features, texture features not only contains the information of local pixels, but also contains the relationship between local and the surrounding pixels [4].

In literature, researchers ever tried several texture analysis techniques such as using the gray histogram, gray difference histogram and gray level co-occurrence matrix methods to extract the texture. Recently, the studies of texture feature are mainly concentrating on hybrid methods to improve texture descriptors. For example, using Fourier transform combined with gray level co-occurrence matrix to extract the texture for the task of identification of volcanic lithology [5]; combining the Gabor filter and the LBP operator to achieve a recognition rate of 95 % in the ORL face database without pretreatment [6]; using LBP, laws, contrast texture factor describing weighted fusion of cracks in a logging image recognition and getting a recognition rate of 92 % [7]; combining the color feature and difference histogram textural features to identify the lithology of cuttings, and achieving a recognition rate at 94.79 % [8]; using the wavelet analysis and hidden Markov combined model to improve the wavelet operator [9]; proposing CLBP improved LBP algorithm [10] and so on.

The remainder of this paper is organized as follows: In Sect. 2, a novel texture feature extraction method is introduced. The results of the improved texture feature extraction are discussed in Sect. 3. In Sect. 4, it summarizes the problems and the future development.

2 Methods

The whole pipeline is shown in Fig. 1. As shown in Fig. 1, at the beginning of the procedure, the preprocess technique is necessary for the aims of removing possible noise and enhancing image contrast; then, hybrid textures (both from LBP and from GLCM) are extracted; finally, support vector machine (SVM) classifier is applied to distinguish the sedimentary structures. As for the evaluation, we exploit the ROC (Receiver operating characteristic) curve to verify the experiment results.

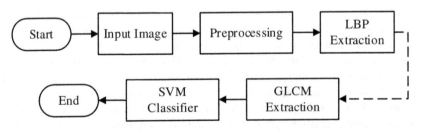

Fig. 1. The flowchart of the proposed approach

2.1   Local Binary Pattern Based Directional Texture Feature Model

Local Binary Pattern (LBP) is a kind of non-parametric operator which is used to describe the local structure of the image. Ojala proposed the LBP operator and proved its high distinguishability in texture classification. The original LBP operator [11] forms labels for the image pixels by thresholding the 3 * 3 neighborhood of each pixel with the center value and considering the result as a binary number. The histogram of these $2^8 = 256$ different labels can then be used as a texture descriptor. This operator used jointly with a simple local contrast measure provided very good performance in unsupervised texture segmentation [12]. The LBP operator was extended to use neighborhoods of different sizes [13]. Using a circular neighborhood and bilinear interpolating values at non-integer pixel coordinates allow any radius and number of pixels in the neighborhood. The gray scale variance of the local neighborhood can be used as the complementary contrast measure. The calculation procedure of LBP operator is shown in Fig. 2.

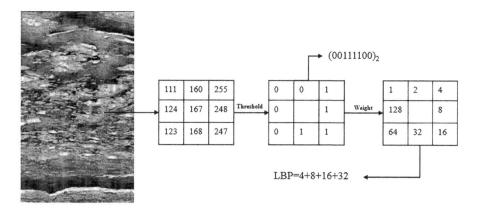

Fig. 2. LBP calculation processing

Where $g_c(x_c, y_c)$ means the center pixel value in the field of preprocessed image, $g_p(x_p, y_p)$ is the pixel value in the field.

$$LBP(x_c, y_c) = \sum_{p=0}^{7} s(g_p - g_c)2^p \tag{1}$$

$$s(x) = \begin{cases} 1 & x \geq 0 \\ 0 & x \prec 0 \end{cases} \tag{2}$$

Since the directional variance described by LBP only focus on the local surrounding change, we need a much large scale of texture descriptor in order to extract both the micro and the macro textual features existing in the images from imaging logging. Luckily, the Gray Level Co-occurrence Matrix (GLCM) performs very well on detailing the directional information. Hence, we use LBP as local feature descriptor and the directional GLCM as global feature descriptor together, which is called "Local Binary Pattern Based Directional Texture Feature model" (LBPBDTFM).

Gray Level Co-occurrence Matrix model was first introduced for analyzing 2D gray-level image. The main idea is to construct the Gray Level Co-occurrence Matrix (GLCM) [14]. GLCM describes the probability gradation occurs between the two points which satisfies a certain distance and a certain direction. As far as our LBPBDTFM concerned, we use four directions: 0° (180°), 45° (225°), 90° (270°) and 135° (315°), and d indicates our choice of value pair from the neighborhood of current center point shown in Fig. 3.

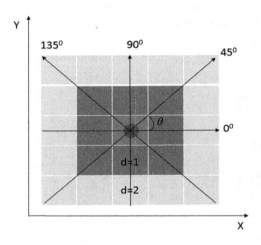

Fig. 3. GLCM extraction model with d = 1 and d = 2 neighbor and four directions.

Suppose that $f(x, y)$ be a two-dimensional digital image of size M × N, $ij = 0, 1 .... L-1$. Thus $p(i, j, \theta)$ can be defined as follows:

$$P(i,j,0^0) = \#\{((k,l),(m,n)) \in M \times N | k - m| = 0, |l - n| = d, f(k,l) = i, f(m,n) = j\} \quad (3)$$

$$P(i,j,45^0) = \#\{((k,l),(m,n)) \in M \times N | k - m| = d, |l - n| = -d, f(k,l) = i, f(m,n) = j\} \quad (4)$$

$$P(i,j,90^0) = \#\{((k,l),(m,n)) \in M \times N | k - m| = d, |l - n| = 0, f(k,l) = i, f(m,n) = j\} \quad (5)$$

$$P(i,j,135^0) = \#\{((k,l),(m,n)) \in M \times N | k - m| = -d, |l - n| = -d, f(k,l) = i, f(m,n) = j\} \quad (6)$$

Where # denotes the number of elements in the set.

LBPBDTFM is just a representation of the image's local texture feature direction, interval and other information, and it cannot directly provide the difference between the two images of the texture. Therefore, we need to calculate some parameters based on the gray level co-occurrence matrix to describe the properties of the texture features.

Based on the literature in [14], fourteen corresponding statistical features such as the entropy, contrast and correlation are exploited to describe the characteristics of various textures of the image. For the sake of statistics, the local LBP values of all the pixels can be normalized into 0-L-1 levels. Our proposed model use seven features (ex: Contrast, Entropy, ASM, Correlation, IDM, Mean, and Variance) as components of feature vectors for further evaluation (as illustrated in Table 1). Multiplied by four directional information, 28 features will be calculated for each given image.

Table 1. Partial characteristic value calculation formula

Name	Formula
Contrast	$f1 = \sum\limits_{n=0}^{L-1} n^2 \left\{ \sum\limits_{|i-j|=n}^{L} \sum\limits^{L} p(i,j) \right\}$
Entropy	$f2 = -\sum\limits_{i=1}^{L} \sum\limits_{j=1}^{L} p(i,j) \log(p(i,j))$
Mean	$f3 = \sum\limits_{i=1}^{L} \sum\limits_{j=1}^{L} p(i,j) * i$
Variance	$f4 = \sum\limits_{i=1}^{L} \sum\limits_{j=1}^{L} (i - \mu)^2 p(i,j)$
Angular Sencond Moment	$f5 = \sum\limits_{i=1}^{L} \sum\limits_{j=1}^{L} p(i,j)^2$
Correlation	$f6 = \dfrac{\sum\limits_{i=1}^{L} \sum\limits_{j=1}^{L} p(i,j) - \mu_x \mu_y}{\sigma_x \sigma_y}$
Inverse Different Moment	$f7 = \sum\limits_{i=1}^{L} \sum\limits_{j=1}^{L} \dfrac{1}{1 + (i-j)^2} p(i,j)$

Binary SVM classifier is employed here to perform the classification task. In implementation, the widely used SVM package LIBSVM [15] with RBF kernel is employed in this study. Following the guideline of LIBSVM, the two parameters (cost: the model slack variable and gamma: the parameter in the RBF kernel) are determined by a grid search process (five-fold cross validation) in the training step, detailed implementation can be found in [16]. The ROC analysis and the AUC merit are used as the measure for evaluation. For each experiment, we randomly split the dataset into training and test set with same size while keeping the original class proportion rate. The SVM model is trained using training set and tested using test set. The whole process is repeated 100 times to avoid time running error and the average results are shown in the following section.

3 Experimental Results and Analysis

3.1 Experimental Evaluation Scale

(1) The accuracy of the test rate is often used to determine the parameters of the classification effect, in which the formula for calculating the accuracy is:

$$p = \frac{true}{count} \tag{7}$$

(2) ROC curve

We use ROC to evaluate the performance of our algorithm in which the vertical coordinates of the curve is defined as TPR (Positive Ratio True), and sometimes we call it sensitivity (Sensitivity). TPR represents the number of positive samples in a class is divided by the number of the total number of positive samples.

$$TPR = \frac{TP}{FN + TP} \tag{8}$$

$$FPR = \frac{FP}{TN + FP} \tag{9}$$

3.2 Experimental Data Description

The input dates are the various categories of regions that picked up by the magic wand after preprocessing. Here are four types of sedimentary structures (muddy bands, muddy thick layer, dissolution pores and induced joints). And these four categories were defined as class"0", "1", "2","3". The first picture in Fig. 4 and the first picture in Fig. 5 are imaging logging images after preprocessing. The second picture in Fig. 4 shows the sedimentary structure (Muddy thick layers and Dissolution pores) of our interesting. And the second picture in Fig. 5 shows the sedimentary structure (Muddy bands and Induced joints).

In this experiment we chose 730 muddy bands 700 muddy thick, 1050 dissolution pores and 840 induced joints to test the effectiveness of experimental methods. 2770 sub-picture images for each type are randomly selected as the training samples, and the rest as the test samples (Table 2).

We analyzed three experiments for the same data resource. The following tables show the four types of sedimentary structural characteristics by using the algorithm of GLCM (Table 3), LBP (Table 4) and the algorithm which proposed in this paper, as shown in the following Table 5:

As shown in Table 3, we listed four directional features extracted by the GLCM and then calculated all 28 features (4 * 7). Here we only chose 20 features to be displayed in Table 3 for illustration, and based on the variation from these listed features the images could be distinguished to some extent. All those extracted 28 features will be fed into a SVM classifier. The results will be summarized in Fig. 6 below.

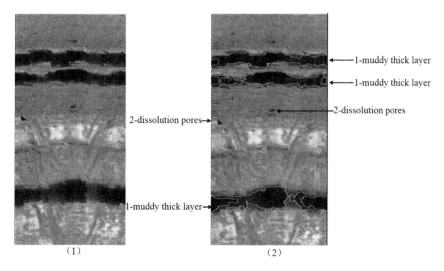

Fig. 4. Muddy thick layers and Dissolution pores

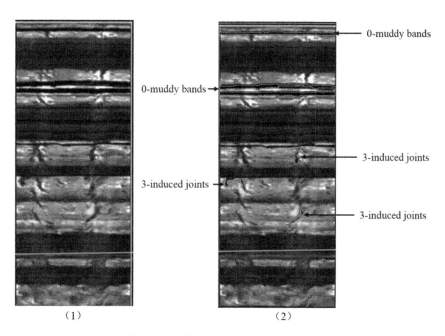

Fig. 5. Muddy bands and Induced joints

The second experiment is the mean and variance of the LBP Operator to extract the texture features of image logging. As shown in Table 4, the listed features seem to be with less dimensional information than those of GLCM.

The third experiment was configured to calculate the features of our new proposed method. Since it is hard to tell the difference of feature extractions between with LBP

Table 2. Experimental data list

Category	Training set	Test set
0	630	100
1	600	100
2	840	210
3	700	140

Table 3. GLCM texture feature

Image	Direction	Contrast	Entropy	Angular second moment	Correlation	Inverse difference moment
(a)	90	498.44	0.286641	1.00053	0.951186	1.03133
	45	1392.46	0.322826	1.00029	0.947896	1.01811
	0	1187.56	0.320273	1.0003	0.9483	1.02019
	135	1387.46	0.323376	1.00028	0.947987	1.00175
(b)	90	172.498	0.067351	1.00007	0.988333	1.00877
	45	730.057	0.0704516	1.00006	0.988008	1.00028
	0	564.4	0.074318	1.00004	0.988011	1.00028
	135	731.022	0.070495	1.00006	0.987998	1.00026
(c)	90	278.842	0.156873	1.00009	0.975755	1.0162
	45	1010.18	0.175647	1.00007	0.97381	1.0039
	0	821.665	0.177818	1.00005	0.974036	1.00674
	135	1004.23	0.175337	1.00007	0.97387	1.00397
(d)	90	418.082	0.603227	1.00133	0.893847	1.07103
	45	1309.17	0.670716	1.00084	0.890297	1.03789
	0	1091.7	0.669524	1.00086	0.890744	1.04407
	135	1308.45	0.670591	1.00085	0.890305	1.03838

Table 4. LBP Texture Features

Image	Mean	Variance
(a)	253.351	196.281
(b)	252.764	304.75
(c)	253.376	189.665
(d)	253.138	218.455

and without LBP, we fed all these features into the same kernel based SVM. Figure 6 illustrates the different performance on the same datasets.

As shown in Fig. 6, we can see that regarding the texture feature extraction methods, the performance the new proposed algorithm is higher than that of the GLCM and LBP. Moreover, the effectiveness of texture feature extraction algorithms is further verified in order to see any gains.

Table 5. LBPBDTFM texture feature

Image	Direction	Contrast	Entropy	Angular second moment	Correlation	Inverse difference moment
(a)	90	608.036	0.453483	1.00013	0.946174	1.02116
	45	1006.86	0.516219	1.00007	0.941124	1.00147
	0	922.724	0.497844	1.00008	0.941878	1.00638
	135	1108.49	0.517865	1.00007	0.940947	1.00178
(b)	90	114.571	0.139481	1.0001	0.977557	1.01671
	45	275.593	0.160536	1.00004	0.97706	1.00016
	0	169.058	0.153977	1.00004	0.977166	1.00521
	135	277.265	0.160746	1.00004	0.97705	1.00015
(c)	90	240.811	0.276289	1.00013	0.963407	1.02021
	45	536.933	0.328362	1.00007	0.959135	1.00096
	0	408.094	0.315619	1.00007	0.95968	1.0062
	135	563.624	0.328127	1.00007	0.959204	1.00104
(d)	90	1326.14	0.893783	1.00018	0.01946	1.0253
	45	2572.16	1.03087	1.00016	0.888966	1.00161
	0	2241.88	0.978038	1.00016	0.892219	1.00715
	135	2731	1.02934	1.00016	0.88903	1.00178

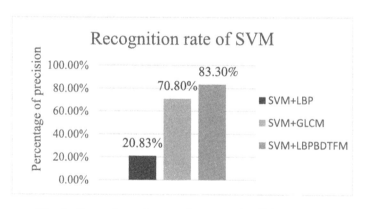

Fig. 6. Comparison of texture features under SVM classifier

Figure 7 illustrates the performance of classification on sedimentary structure such as muddy bands, muddy thick, dissolution pores and induced joints by LBP only or by GLCM only. According to the Fig. 7(a), the AUC of class 2 is about 0.44 by LBP only, while in Fig. 7(b) the AUC of class 2 are nearly 0.92 which shows better performance. However the performance on class 3 illustrates that the LBP only is better than that of GLCM only. Hence it is hard to say that the GLCM is superior to the LBP. In view of this, the new proposed method will make a trade-off based on above two traditional methods. The ROC curve for the new proposed method of feature extraction is shown in Fig. 8.

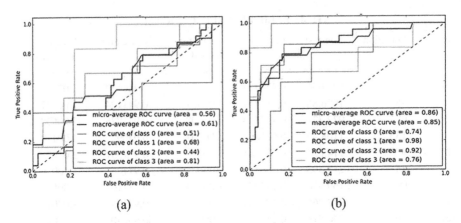

(a)   (b)

Fig. 7. The ROC curve under different operator. Graph (a) shows the ROC curve based only on LBP. Graph (b) shows the ROC curve based only on Directional Texture.

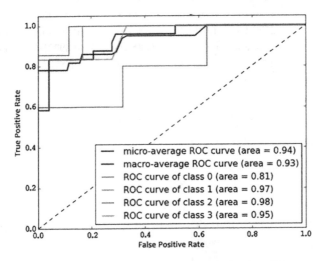

Fig. 8. The ROC curve of the algorithm proposed in this paper

As can be seen from the above figure, the new proposed algorithm achieved the best result. The least AUC of class 0 are nearly 0.81. The classification average in differentiating sedimentary structures' from petroleum imaging logging can reach an AUC value of 0.93-0.94. Though there are slightly difference between the predicated results on class 0 from GLCM and LBPBDTFM, we can still conclude that the novel texture extraction method has significantly improved the recognition accuracy because the overall accuracy is put to a much higher level.

4 Conclusion and Future Work

The experimental results show that LBP operator is more emphasis on the frequency of the local micro structure of the image and ignores the macro structure. And the operator of the gray level co-occurrence matrix can extract the macro structure of the image, but it has a large amount of computation and poor description of the local information. Based on above listed the strengths and weaknesses of each operator, we proposed a method of texture feature extraction combined LBP and GLCM algorithm, which is suitable for well logging image. Finally, we compared their performance with ROC curve. The result shows that by using the method of LBPBDTFM, the classification in distinguishing the different sedimentary structures can reach an average AUC value of 0.94. We can conclude that the improved algorithm has significantly improved the accuracy. In the future, we hope that can improve LBP operator or use 3d GLCM operator to expect achieve better effect.

Acknowledgments. The experimental databases are provided by the China Petroleum Exploration and Development Institute. Here to thank the data support.

References

1. Wei, P.S., Liu, Q.X., Zhang, J.L., et al.: Re-discussion of relationship between reef and giant oil-gas fields. Acts Petrolei Sinica J. Oil. **27**, 38–42 (2006)
2. Chai, H., Li, N., Xiao, C., et al.: Automatic discrimination of sedimentary facies and lithologies in reef-bank reservoirs using borehole image logs. Appl. Geophys. **6**, 17–29 (2009)
3. Russell, S.D., Akbar, M., Vissapragada, B., et al.: Rock types and permeability prediction from dipmeter and image logs: Shuaiba reservoir (Aptian). Abu Dhabi. J. Aapg Bull. **86**, 1709–1732 (2002)
4. Huafeng, W., Yuting, W., Hua, C.: State-of-the-art on texture-based well logging image classification. Comput. Res. Dev. **50**, 1335–1348 (2013)
5. Fei, C., Zhang, P.: Imaging logging image texture feature extraction and lithology identification. Well Logging Technol. **6**, 17–19 (2012)
6. Liao, S., Law, M.W.K., Chung, A.C.S.: Dominant local binary patterns for texture classification. IEEE Trans. Image Process. Publ. IEEE Sig. Process. Soc. **18**, 1107–1118 (2009)
7. Qu Zhong, Z.: Combination of texture features with significant crack extraction algorithm. Comput. Eng. Des. **11**, 3056–3059 (2015)
8. Yao Jin, T., Fu, W.: Cuttings lithology identification based on color and texture features. Sichuan University: Natural Science Edition (2014)
9. Crouse, M.S., Nowak, R.D., Baraniuk, R.G.: Wavelet-based statistical signal processing using hidden Markov models. IEEE Trans. Sig. Process. **46**, 886–902 (1998)
10. Liu, H., Yang, Y., Guo, X., et al.: Improved LBP used for texture feature extraction. Comput. Eng. Appl. **50**, 182–185 (2014)
11. Ojala, T., Pietikäinen, M., Harwood, D.: A comparative study of texture measures with classification based on featured distributions. Pattern Recogn. **29**, 51–59 (1996)

12. Pietikäinen, M., Hadid, A., Zhao, G.: Computer Vision using Local Binary Patterns. Springer Science & Business Media, Heidelberg (2011)
13. Ojala, T., Pietikäinen, M., Mäenpää, T.: Multiresolution gray-scale and rotation invariant texture classification with local binary patterns. IEEE Trans. Pattern Anal. Mach. Intell. **24**, 971–987 (2002)
14. Haralick, R.M., Shanmugam, K.: Textural features for image classification. IEEE Trans. Syst. Man Cybern. **6**, 610–621 (1973)
15. Chang, C., Lin, C.J.: LIBSVM.: a library for support vector machines. ACM Trans. Intell. Syst. Technol. (TIST) **2**, 1–27 (2011)
16. Song, B., Zhang, G., Zhu, W.: ROC operating point selection for classification of imbalanced data with application to computer-aided polyp detection in CT colonography. Int. J. Comput. Assist. Radiol. Surg. **9**, 79–89 (2014)

Rank Beauty

Yanbing Liao[1(✉)], Weihong Deng[1,2], and Can Cui[2]

[1] Beijing University of Posts and Telecommunications, Beijing 100876, China
lyb.bupt@gmail.com, whdeng@bupt.edu.cn
[2] Beijing Jiaotong University, Beijing 100044, China

Abstract. It is useful to automatically select the most attractive face images from large photo collections. Previous works in this area have little attention on facial attractiveness for one subject, but different objects. In this paper, we have a collection of subjects' faces including a range of expression, postures, makeup, lighting and resolutions from Bing Search. Given training data of faces scored based on the majority of subjects' tastes, we train a model to learn how to rank novel faces and show how it can be used to automatically mine attractive photos from personal photo collections. Our system achieves an average accuracy of 73 % on pairwise comparisons of novel faces.

Keywords: Facial aesthetic · Crowdsourcing · Rank

1 Introduction

Beauty is an abstract concept, getting more and more attention in recent years. In general, it's easy for people to select good ones from one subject's portraits but hard to make sure it's right. Which portrait is more attractive? Will other people agree with me? Many factors, such as facial expressions, makeup, lighting or resolutions can contribute to why a face is beautiful. Besides, individuals have different tastes. That means our perception of ourselves is often quite different than that of others [1]. Specially, our method offers users feedback on how their range of portraits are perceived by others. Our model can be used to select the most attractive pictures of people or delete quite bad ones from a photo collection.

At present, many scholars use features to predict facial beauty by machine learning methods [2, 15]. Bottino and Laurentini [3] delivered a study of facial beauty as seen in the pattern analysis literature. These researches suggests more attractiveness levels of one's appearance. Zhu [4] focuses on attractiveness of a given person only based on expression. However, lighting, make up, resolution are also important factors to judge what a portrait is flat. In paper, we consider a more complicated dataset to have a more applicable model.

Ranking and relative ordering have been thoroughly investigated within the machine learning literature especially in applications pertaining to information and document retrieval [5, 6]. We also find image search related applications for image retrieval using similar techniques that were used for document retrieval [7, 8]. Our work differs significantly from those as we learn a facial ranking function [9, 14] for individuals over image pairs. Most comparison works about facial beauty [4] focus on

© Springer Nature Singapore Pte Ltd. 2016
T. Tan et al. (Eds.): CCPR 2016, Part II, CCIS 663, pp. 173–181, 2016.
DOI: 10.1007/978-981-10-3005-5_15

yes or no two answers. However, it's so hard to judge which picture is flatter when compared images are in the same attractiveness level. Differently, we obtain the additional relative attributes what we call "similar" for training. In our comparison works, we find more than 1/3 pairs are similar. Finally, our experimental results show such annotation method can provide more information for our training and get a more accuracy facial ranking model.

This paper is organized as follows. Section 2 introduces our collecting dataset, comparisons and training approach about learning to rank. Section 3 shows our assessment method and experimental results compared with another recent methods. Section 4 shows the limitations and draws the conclusion and feature works.

2   Rank Beauty

In this section we shows our works about collecting portraits dataset, pre-processing and pairwise comparisons (Sect. 2.1). Then we present our main approach we used to build the facial beauty models (Sect. 2.2).

2.1   Data and Crowdsourcing

Our first goal is to collect a set of a subject's portraits including variety of expressions, make up, lighting and resolution. Then rating them along attributes. In this section, we first show how we collect portraits. Then, we pre-process the images to normalize the position and extract the features for our model. Finally, we collect pairwise comparisons of portraits along attractive attributes.

Collecting Data and Pre-processing. We start by collecting a large of photos from Bing Search API that may be appropriate for portraits for each subject. All pictures are from celebrities, in order to make sure the number of portraits being large enough. In total, we collected the data of 108 subjects, 500 to 600 images per subject, including both male and female subjects ranging in age from 20 to 65. We perform several pre-processing steps (Fig. 1) for each image collection to align the facial data, compute facial features and reduce data redundancy.

We first perform recognition and cropping to normalize the face in a common reference frame. We crop faces using the bounding boxes, generated by Vio-Jones Face detector. Then we use a face tracker [10] that accurately estimates 9 facial feature points and localizes different facial parts such as eyes, mouth, nose and facial edge (Fig. 1). We apply a median filter with a window size of 5 frames to smooth the estimated points and suppress tracking temporal jitter. Besides, we filter small face, poor alignment and Non-frontal face. Later, we warp the face into a frontal view using the 3D template model [11] and the 3D-to-2D transformation matrix [4]. We exclude portraits for which little face or the tracker reports tracking failures. Finally, we have 108 subjects and 100 to 200 portraits for every collection. Figure 1 shows several examples of the remaining images. Hani Altwaijry and Serge Belongie [2] find that the most effective feature types for predicting beauty preferences are HOG. So we use a simple and straightforward

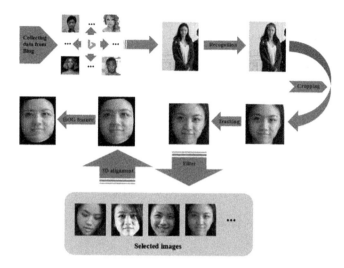

Fig. 1. Key works in pre-processing.

3720-dimensional HOG (Histogram of Oriented Gradients) [12] features that was calculated on five parts of images to capture visual properties of pictures.

Pairwise Comparisons. We next collect human response data that allows us to rank attractive portraits for each subject. We develop our annotation system to collect human response data of every pairwise comparison (e.g., "Is image A more beautiful than B?"). We suggest that volunteers can consider facial expression, postures, makeup, lighting and resolutions to make choose. We sample portraits to form pairwise comparisons in random and provide 3 labels, which are "yes", "no", "similar". Choosing the relative attribute of "similar" not only means photos of the comparison are similar with each other. What's important, our system also allows annotators to chose "similar" when they hesitate between yes or no. There are 20 volunteers to support our work. Each pair is annotated once. A single worker complete about 3 or 4 groups. Finally, we receive 78 subjects' comparisons collections, 12434 pairs in total.

2.2 Learning to Rank

We use ranking functions trained with comparative labels. This method, originally introduced by Parikh and Grauman [9], compare images in terms of how strongly they exhibit a nameable visual property. We use a large-margin approach to model our facial relative attributes. In this work, we require using the "similar" relative attributes we collected before. To learn our ranking function g, it uses a set of portraits $I = \{I_1, I_2, \ldots, I_n\}$ in the dataset, each of which is described by the image features $x_i \in R^d$. The annotated portraits list is given by a subject as a tuple $A = \{A_{12}, A_{23}, \ldots, A_{n-1,n}\}$ and $A_{ij} \in \{0, 0.5, 1\}$, where "$A_{ij} = 1$" denotes that I_i is more attractive than I_j and "$A_{ij} = 0$" means inverse. Moreover, "$A_{ij} = 0.5$" denotes that both I_i and I_j are equivalent in terms of attractiveness. Then, we sort comparisons of all subjects to two sets according to A. The first set

$O = \{(i, j)\}$ consist of ordered pairs of images for which the first image I_i has the attribute more than the second image I_j which means "$A_{ij} = 1$" or "$A_{ji} = 0$". The second set $S = \{(i, j)\}$ consisted of unordered pairs for which both images have the attribute to a similar extend and "$A_{ij} = 0.5$". Our goal is to learn the function:

$$g(x_i) = w^T x_i \tag{1}$$

subject to the constraints:

$$\forall(i,j) \in O \rightarrow g(x_i) > g(x_j) \tag{2}$$

$$\forall(i,j) \in S \rightarrow g(x_i) = g(x_j) \tag{3}$$

While this is an NP hard problem, as described by [9] is modeled as following optimization problem:

$$\text{minimize} \left(\frac{1}{2} \|w\|_2^2 + C \left(\sum \xi_{ij}^2 + \sum \gamma_{ij}^2 \right) \right) \tag{4}$$

$$s.t.\ w^T (x_i - x_j) \geq 1 - \xi_{ij}; \forall(i,j) \in O \tag{5}$$

$$\left| w^T (x_i - x_j) \right| \leq \gamma_{ij}; \forall(i,j) \in S \tag{6}$$

$$\xi_{ij} \geq 0; \gamma_{ij} \geq 0. \tag{7}$$

where ξ_{ij}, γ_{ij} are slack variables, the constant C balances the regularizer and constraints, and controls the satisfaction of strict relative order. Ranksvm is defined without Eq. (6). While we strict with two restrictions imposed by Eqs. (5) and (6) for reasons related to our sorting mechanism which corresponds to the tuple A. Extended ranksvm with Eq. (6) can learn more useful information about similar pairs. Moreover, this setting can enable us to get a more accuracy image ranking.

3   Results

In this section we present our main approach we used to measure accuracy of the ranking model (Sect. 3.1). Finally, we shows our experimental results (Sect. 3.2).

3.1   Measuring Accuracy

To test our method we collect an additional 30 subjects' comparisons collections. To measure the accuracy of our method we turn towards a tool for comparing ranked orders of pairs: Kendall Tau [13].

In our implementation, we first focus on the accuracy of learning "more attractive than" relationships. For one subject, we get the our predicting ranking for pairs defined as $E = \{(i, j)\}$ consisted of ordered pairs for which I_i is more beautiful than I_j.

We measures the number of pairwise accuracy between O and E as follows:
$\tau(E, O) = \sum_{\forall(i,j) \in E} I((i,j) \in O)$, where $I(.)$ is an indicator function.

Based on the Kendall Tau we construct our accuracy measurement to account for correct pairs divided by the total number of pairs to reach a notion of correctness. If N_O is the total number of pairs in set O for one subject, then our accuracy measurement for E matching O is: $\alpha(E, O) = \tau(E, O)/N_O$.

Then we want to make full use of set S to know the ability of our model to rank similar portraits during our testing. We define a set $D = \{D_{ij}\}$ and $(i, j) \in S$, where $D_{ij} = |g(x_i) - g(x_j)|/d$ and d is a normalized parameter. D_{ij} means the normalized margin between mapped image I_i and I_j. In our implementation, we compare our model with the model presented in [3]. Next we set ours as D_s and another as D_{us}. We measure

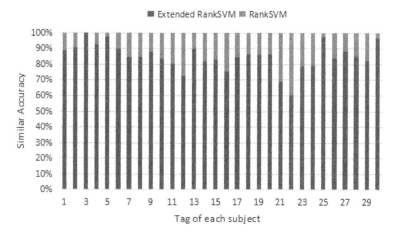

Fig. 2. Accuracy for similar images by RankSVM models

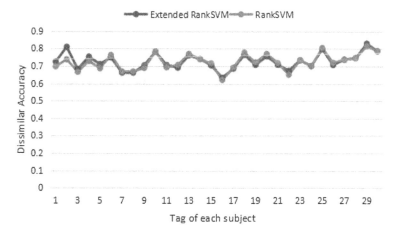

Fig. 3. Accuracy for dissimilar images by RankSVM models

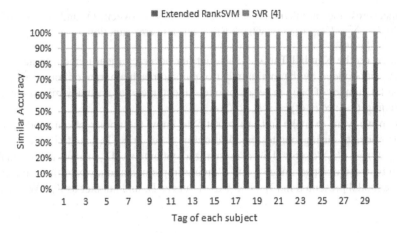

Fig. 4. Accuracy for similar images by RankSVM and SVR [4]

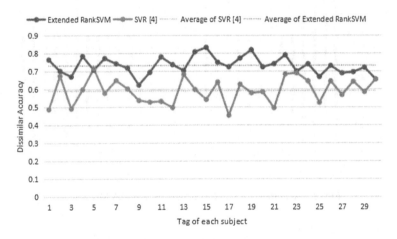

Fig. 5. Accuracy for dissimilar images by RankSVM and SVR [4]

the number of the greater margin pairs between D_s and D_{us} as: $\tau(D_s, D_{us}) = \sum\{D_s(i,j) < D_{us}(i,j)\}$. If N_S is the total number of pairs in set S for one subject, then our accuracy measurement about similar pairs for D_s relative to D_{us} is: $\alpha(D_s, D_{us}) = \tau(D_s, D_{us})/N_S$. Also we can easy know the accuracy of another model about similar pairs is $1 - \alpha(D_s, D_{us})$.

Later, we will show the good performance of our model both in dissimilar and similar image pairs.

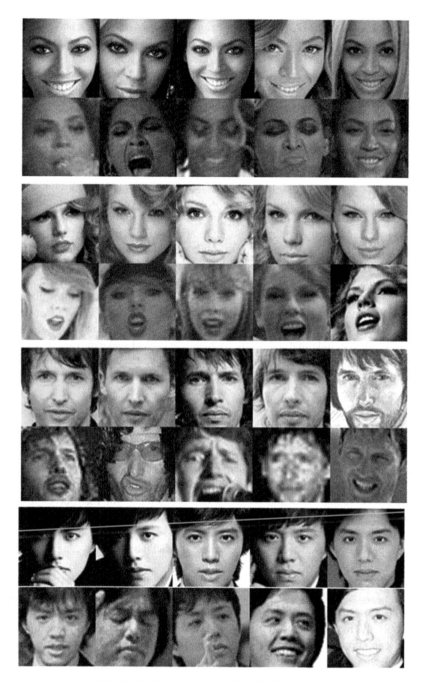

Fig. 6. Ranking results by Extended-RankSVM

3.2 Facial Beauty Ranking

We designed our experiments using the method, which is introduced in Sect. 2.2. We trained our model by the 12434 training examples from 78 subjects and the relative attributes of each subjects from our volunteers. In order to test the accuracy of the rank system, we have 30 testing subjects and collecte pairwise comparisons of each subjects additionally. Running the experiment with four testing subjects showed the ranking results in Fig. 6. In Fig. 6, we show two rows for each of four testing subjects from personal images collections: the five most attractive, and the five least.

For observing how the similar attribute behaved, we decided to train original ranksvm model which is without the restriction imposed by Eq. (6). Noticing that similar attribute has no contribution to this rank model. We ran the experiment with the pairwise comparisons of each subjects and the results as the ranking accuracy plot are showed in Fig. 2. From Fig. 2, two models have exactly the same high rank accuracy for the dissimilar portraits. But original ranksvm model perform bed in Fig. 3 when it comes to the similar images. We believe this is due to the similar attribute. The results shows the similar attribute contributes a more accurate rank model.

Then we contrasted our facial beauty rank system with the existing SVR model proposed in [4]. Figures 4 and 5 shows how the two models perform. In Fig. 5, the average accuracy of 73 % was achieved by our rank system and another model performed the average accuracy of 59 %. Besides, our rank system perform good than another when ranking similar portraits. This can be seen with Fig. 4. Considering Figs. 2, 3, 4 and 5 our facial beauty rank system performs quite accurate.

Finally, we conclude that the portrait which has a good makeup, nice lighting, flat expression and not bad resolution gets higher ranking. Especially, our model loves personal photos with high professional photography level. The result of our facial rank model seems reasonable.

4 Conclusion

Our method has some limitations. Our model is a cross-subject rank model to predict attractiveness automatically. However, different people have different nice poses, expressions, polishing ways and so on, which are suitable for themselves. Besides, although our annotations from 20 workers maybe can stand for the public opinion, people still have many different opinions in details.

In this paper, we described a ranking system for facial beauty that can rank portraits from one subject according to public preferences. More importantly, we made full use of the relative attributes, especially similar attribute. Finally, we achieve an average accuracy of 73 %. Therefore, we can express that our personal facial beauty rank system do not only consider the expressions but also integrate with considering the lighting, clarity, makeup and so on. Similar relative attribute is proved to be important. Next, we can improve our dataset and organized method. Besides, doing some adaptive works is a promising avenue for the future work.

Acknowledgments. This work was partially sponsored by supported by the NSFC (National Natural Science Foundation of China) under Grant No. 61375031, No. 61573068, No. 61471048, and No. 61273217, the Fundamental Research Funds for the Central Universities under Grant No. 2014ZD03-01, This work was also supported by Beijing Nova Program, CCF-Tencent Open Research Fund, and the Program for New Century Excellent Talents in University.

References

1. Springer, I.N., Wiltfang, J., Kowalski, J.T., Russo, R.A.J., Schulze, M., Becker, S., Wolfart, S.: Mirror, mirror on the wall: self-perception of facial beauty versus judgement by others. J. Craniomaxillofac. Surg. **40**(8), 773–776 (2012)
2. Altwaijry, H., Belongie, S.: Relative ranking of facial attractiveness. In: IEEE Workshop on Application of Computer Vision (WACV), pp. 117–124, 15–17 January 2013
3. Bosch, A., Zisserman, A., Muñoz, X.: Scene classification via pLSA. In: Proceedings of ECCV 2006, pp. 517–530 (2006)
4. Zhu, J.-Y., Agarwala, A., Efros, A.A., Shecht-man, E., Wang, J.: Mirror mirror: crowdsourcing better portraits. ACM TOG **33**(6)
5. Cao, Z., Qin, T., Liu, T.-Y., Tsai, M.-F., Li, H.: Learning to rank: from pairwise approach to listwise approach. In: Proceedings of ICML 2007, pp. 129–136. ACM, New York (2007)
6. Li, H.: A short introduction to learning to rank. IEICE Trans. **94-D**(10), 1854–1862 (2011)
7. Gevers, T., Smeulders, A.W.M.: Pictoseek: combining color and shape invariant features for image retrieval. IEEE Trans. Image Process. **9**(1), 102–119 (2000)
8. He, J., Li, M., Zhang, H.-J., Tong, H., Zhang, C.: Manifold-ranking based image retrieval. In: Proceedings of MULTIMEDIA 2004, pp. 9–16. ACM, New York (2004)
9. Parikh, D., Grauman, K.: Relative attributes. In: ICCV (2011)
10. Asthana, A., Zafeiriou, S., Cheng, S., Pantic, M.: Robust discriminative response map fitting with constrained local models. In: Proceedings of 2013 IEEE Conference on Computer Vision and Pattern Recognition (CVPR 2013), Portland, Oregon, USA, June 2013
11. Zhang, L., Snavely, N., Curless, B., Seitz, S.M.: Spacetime faces: high-resolution capture for modeling and animation (2004)
12. Dalal, N., Triggs, B.: Histograms of oriented gradients for human detection. In: IEEE Conference on Computer Vision and Pattern Recognition (2005)
13. Kumar, R., Vassilvitskii, S.: Generalized distances between rankings. In: Proceedings of WWW 2010, pp. 571–580 (2010)
14. Burges, C., Shaked, T., Renshaw, E., Lazier, A., Deeds, M., Hamilton, N., Hullender, G.: Learning to rank using gradient descent. In: ICML (2005)
15. Kalayci, S., Ekenel, H.K., Gunes, H.: Automatic analysis of facial attractiveness from video. In: 2014 IEEE International Conference on Image Processing (ICIP), pp. 4191–4195, October 2014

Robust Optic Disc Detection Based on Multi-features and Two-Stage Decision Strategy

Wang Ying[1(✉)], Zhang Dongbo[1,2,3], Huang Huixian[1,2,3], and Zhang Ying[1,2,3]

[1] The College of Information Engineering, Xiangtan University, Xiangtan 411105, China
kikyo2016@163.com
[2] Institute Control Engineering of Xiangtan University, Xiangtan 411105, China
[3] Robot Visual Perception and Control Technology National Engineering Laboratory,
Xiangtan 411105, Hunan, China

Abstract. This paper proposes a robust method based on multi-features and two-stage decision strategy to locate the OD position. First, we use the global vessel distribution and directional characteristic and local appearance characteristic to find several OD candidates. Then we introduce the HOG to depict local details of OD candidates, and a SVM model is trained to classify OD and non OD candidate regions. Finally the correlation measure is used to remove redundant OD regions. This method has a good detection accuracy in normal images and diseased images. And the detection accuracy reached 97.9 % in four public image databases.

Keywords: Optic disc · Vessel · Detection

1 Introduction

Optic disc (OD) is one of the main physiological structure of the retinal image. In normal retinal fundus image, the mainly physiological structure that can be observed including optic disc (OD) are vasculature and macula. The appearance of OD presents as bright yellow and approximately circular area, moreover the OD is the convergence zone of retinal vasculature. Because OD detection is an important preprocessing for automatic retinal image analysis and diagnosis, therefore automatic OD detection based on computer vision has always gotten attention of researchers.

Vessel and appearance characteristics are the main considerations that were adopted by recent literatures to extract features for optic disc (OD) detection. Usually the shape, intensity, orientation and magnitude of the edge can be used to describe the local appearance characteristics of the OD region. In early approaches, the brightest 1 % pixels are selected as candidate OD in [1, 2]. [3, 4] used the Hough transformation to identify the circular OD region. In recent reported literatures, Wang and Zhang [5] proposed a novel fast OD detection method by multi-scale blob detection technology. Lu designed a line operator [6] or circular transformation [7] to capture the unique circular brightness structure associated with the OD. In a whole, the algorithms related to appearance characteristics usually are simple and can achieve high accuracy in normal retinal fundus

© Springer Nature Singapore Pte Ltd. 2016
T. Tan et al. (Eds.): CCPR 2016, Part II, CCIS 663, pp. 182–190, 2016.
DOI: 10.1007/978-981-10-3005-5_16

images, but will also lead to multitude false positive results caused by lesions or pathologies.

Vasculature is another important characteristic that can be applied to detect OD. Foracchia [8] identified the position of the OD using a geometrical model with two parabolas. Zhang [9] designed a feature combining 3 vessel distribution characteristics to find possible horizontal coordinate of OD. Then, a General Hough Transformation (GHT) is introduced to identify the vertical coordinate of OD. Hoover [10] found the maximum convergence point as OD position through computing the convergence of blood vessels. And in [11], the authors detected the OD by computing the match degree between vessel map and the vessel's direction matched filter.

In this paper, to imitate the human action, we consider a two-stage decision strategy to implement OD detection. Because the vasculature usually expands across the whole retinal image, thus the vessel structure provides enough information to identify the possible OD region. Based on this observation, the vessel directional and distribution characteristics are used to choose OD candidates, and then the local information of the candidates will be adopted to determine the true OD position. Different with previous reported methods that appeal to simple shape, intensity and contrast feature, we introduce local feature descriptor technology to depict local details of OD candidates, which is prevalent for detection and recognition task in recent years and presents powerful distinctiveness to distinguish OD and non OD region. By above analysis, a robust OD detection approach is proposed in this work, which applies multi-features and two-stage decision strategy.

2 Method

To comprehensivelytake advantage of global and local characteristics, a two-stage decision strategy is adopted to achieve OD detection for retinal image. First, some possible OD candidates are found by global vessel and local appearance characteristics, then HOG descriptortechnology and SVM is introduced to distinguish really OD and non OD regions, Finally, the final decision of OD is determined by non maximum suppression with similarity measure.

2.1 Finding Optic Disc Candidates

OD Horizontal Localization by Global Vessel Distribution Characteristic. Vessel structure detection is the premise to find possible OD candidates. A two-dimensional Gabor filter is used for extracting vessel structure in this method. An example retinal image is shown in Fig. 1(a). Figure 1(b) is its corresponding binary manual labeled vessel map. And 3 vertical windows (with the height of the image and twice the width of the main blood vessels) are plotted to help the observation. Usually, there are more than one connected vascular segments in each vertical window. Obviously, there are less vascular segments in the vertical windows including OD center. And they present compact distribution with high vessel density. While in the other vertical windows, there are much of connected vascular segments and scatter in a wider area with relatively even distribution.

Although in position 1 (Fig. 1(e)), it also has less vascular segments, but they are relatively even distribution and show lower local vessel density.

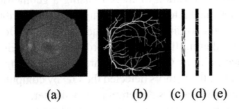

(a)   (b)   (c) (d) (e)

Fig. 1. (a) Example of retinal image, (b) manual labeled vessel map and three marked horizontal positions, and (c)–(e) show the vascular segments appear in these windows

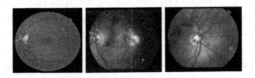

Fig. 2. Examples of ten candidate positions in retinal images that are found by above algorithm

Based on these observations, three aspects of distribution characteristics should be considered to locate the possible horizontal positions of OD center. They are local vessel density, compactness and uniformity. According to these vessel distribution characteristics, A feature is proposed as follow:

$$f_D(x) = std(\mathbf{vc}) \cdot \frac{(-1)\sum_{i=1}^{n_x} p_i \log_2 p_i}{\max_i \{v_i\}}, \quad p_i = \frac{m_i}{M} \tag{1}$$

Where $f_D(x)$ is the vessel distribution feature value at horizontal coordinates x. It is combined with three terms. The first term $std(\mathbf{vc})$ demonstrates the scattering degree of vascular segments in specific vertical window centered on position x. The second term $(-1)\sum_{i=1}^{n_x} p_i \log_2 p_i$ evaluates the uniformity of vascular segments distribution. Where, p_i is the proportion of ith vascular segment in whole vertical window. The third term $\max\{v_i\}$, which can measure local vessel density at a certain extent, is the pixels of the maximum vascular segment in whole vertical window. Generally, the $f_D(x)$ of vertical window centered at OD vicinity is lower than that of vertical windows centered at other horizontal locations. Thus, the lowest k minimum extreme points of the 1–D $f_D(x)$ signal are chosen as possible horizontal locations of OD. And $sort\{f_D(x)\}$ represents a sequence of $f_D(x)$ by ascending order.

$$x_{OD} = \{x_k | sort\{f_D(x)\}\}, k=1,2,3,4,5 \tag{2}$$

OD Vertical Localization with Global Vessel Direction and Local Appearance Characteristics. To further find possible OD vertical ordinate, the global vessel direction and local OD appearance characteristics are used. Because it is rarely presented inimages whose vessel structure and OD appearance are both destroyed, so if we comprehensively apply vessel and OD appearance characteristics, it is believed that the probability of finding really OD region can be greatly improved. By observation, the main arcade of the vasculature can be roughly represented by a parabola model, which describe the global direction characteristics of vasculature. Assume the origin is the up-left corner of the image, then a parabola model with axis parallel to the horizontal image axis is used.

$$(y - y_{OD})^2 = 4p(x - x_{OD}) \tag{3}$$

where p is the focal length, (x_{OD}, y_{OD}) is the vertex of parabola curve, which is also assumed as the coordinate of OD center. Then the ideas of General Hough Transformation (GHT) [3] is adopted to fitting the parabola curve.p and vertical coordinate y_{OD} need to be estimated. To simplify search process, we choose 60 bins for the focal length p (between 9 and 81). And the range of y_{OD} is the integer between 1 and the height of retinal image. To accelerate search process, the step great than 1 is appreciated to identify y_{OD}. In our experiments, this step parameter is set to 2.

More than one parabola will be found by above the GHT technology. The m parabolas fitting most of vessel pixels and according best with the global vessel direction needed to be found from all possible parabolas. And the m vertexes of the parabolas are considered to be optic disc candidate positions.

Then, optic disc appearance characteristics (brightness, edge gradient) are used to find more other n optic disc vertical candidate positions. A rectangular window (length and width are about the diameter of optic disc) slide up and down along the vertical direction at the horizontal coordinate position, as shown in Fig. 3. The feature is calculated by the response graph which is based on local area brightness and Gabor filter in the phase $\theta = 90°$:

$$f(y) = I_N(x, y) * g_N(x, y) \tag{4}$$

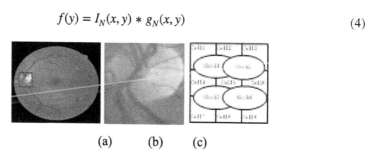

(a)   (b)   (c)

Fig. 3. (a) Ten candidate positions in a retinal image; (b) the detecting neighborhood of the marked candidate point in figure a; (c) the splice of blocks

Where $I_N(x, y)$ is the average luminance value of the rectangular window neighborhood whose center is (x, y).$g_N(x, y)$ is the estimated Gabor filter response value of this field, that is to say, it is the average filter response values of the neighborhood. Since the luminance and Gabor filter responses of the optic disc region should be greater than the value of other regions, therefore, the largest response value has the greater possibility of being the position of the optic disc. When the rectangular windows slides up and down, a one-dimensional curve can be projected to reflect the change in $f(y)$ at the corresponding position, and its maximum point is identified as the vertical coordinate of optic disc.

Through the global vessel direction and local appearance characteristics with above two methods, $m + n$ OD candidates are found. In our experiment, $m = n = 5$, and 10 OD candidates are chosen, and Fig. 2 shown the results for finding OD candidates of some example images as below. The OD centers that are labeled with red "+" are detected by the local appearance characteristics, and others that are labeled with green "+" are based on the global vessel direction.

2.2 Distinguish Really OD Region Based on HOG Feature

By observation, there are some false positive results among the candidates, i.e., some candidates don't locate at really OD region. To remove the non OD candidates, more details of OD candidates neighbourhood should be extracted to depict local region. Histogram of oriented gradients (HOG) feature is a kind of feature descriptor, which has been extensively applied for object detection in computer vision. The process of constructing HOG feature is as follow,

Calculating the gradients of the image pixel (x, y),

$$
\begin{aligned}
G_x(x, y) &= H(x + 1, y) - H(x - 1, y) \\
G_y(x, y) &= H(x, y + 1) - H(x, y - 1)
\end{aligned}
\tag{5}
$$

Where, $G_x(x, y)$ and, $G_y(x, y)$ denote the gradient along horizontal and vertical direction respectively, $H(x, y)$ is the pixel value of the input image pixel (x, y). The gradient magnitude and direction of point (x, y) can be computed by (6)

$$
\begin{aligned}
G(x, y) &= \sqrt{G_x(x, y)^2 + G_y(x, y)^2} \\
\alpha(x, y) &= \tan^{-1}\left(\frac{G_y(x, y)}{G_x(x, y)}\right)
\end{aligned}
\tag{6}
$$

Constructing histogram of gradient direction for each cell unit. Usually a 69×69 candidate OD neighborhood is divided into a plenty of 23×23 pixel unit (cell), and 4 neighbour cells compose a block. The gradient direction $[-\pi, \pi]$ is divided into nine intervals (bin) in even. The gradient magnitudes of all the pixels within each cell are counted in all direction bins, and a 9-dimensional histogram of gradient direction feature vector is obtained. Then a 36-dimensional feature vector of a block can be constructed.

To remove the effect of illumination variation, the feature of each block should be normalized bythe formula (7)

$$V_{norm} = v \left/ \sqrt{\|v\|_2^2 + \epsilon^2} \right. \tag{7}$$

Where v is the block feature before normalization, and ϵ is a small constant.

All of feature vector of the blocks are finally concentrated to achieve HOG feature of candidate OD neighborhood. And a 36 * 4 = 144 dimensional vector is achieved by above method (Fig. 5c).

To recognize really OD and non OD region, many positive and negative samples images are selected to train a SVM model. And all candidate positions of test images can be identified by the trained SVM model.

After classifying the candidate positions, there are may be still exist several candidates that belong to true OD regions. To determine final OD center, correlation coefficient is used to remove redundant OD candidates.

$$P(F_Y, F_M) = \frac{\langle F_Y, F_M \rangle}{\sqrt{\langle F_Y, F_Y \rangle}\sqrt{\langle F_M, F_M \rangle}} \tag{8}$$

In which, F_Y denotes the feature of standard optic disc image in each dataset, F_M is the feature that needs judge, and $\langle , \rangle$ means inner product of two eigenvectors. Then the similarity between the standard HOG feature with the feature of OD candidates is calculated by Eq. (8). According to non-maximum suppress, only the OD candidate with the maximum similarity value is reserved to be final OD center.

3   Experiment Results

3.1   Dataset

In order to assess the performance of the optic disc detection method proposed in this paper, we use four public fundus image database STARE (605 * 700), DRIVE (564 * 584), DIARETDB0 (1500 * 1152), DIARETDB1 (1500 * 1152). In these four fundus image databases, STARE contains a lot of images with a variety of serious lesion, and other datasets contains images that are normal or with fewer lesions.

The training samples are selected from four databases for SVM training. The information of training images and test images are shown in Table 1. There are overlap between the training set and test set, in other words, there are repeated images.

Table 1. Information of training set and test set

Dataset	STARE	DRIVE	DIARETDB0	DIARETDB1
Training images	20	20		40
Test images	81	40	130	89

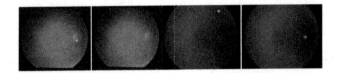

Fig. 4. OD detection by vessel or appearance characteristic

3.2 The Robustness of the Proposed Methods

As many literatures reported, different features show different discrimination. The phenomenon can be observed is the failed images maybe completely different when different features are adopted by OD detection methods. Another fact should be attended is that there are rarely images whose vessel and OD appearance characteristics are both destroyed by lesions or retinopathy. Thus if both vessel and appearance features are used to build the OD model, the success rate of OD detection will be surely improved, and the robustness of the algorithm will be enhanced.

Figure 4 shows OD detection by our approach when only vessel or apperance characteristic are used to find the OD candidates. Figure 4 show two example images' detection results by our method. The detected OD center labeled with white "+" (Fig. 4(a), (c)) is the result of using OD appearance characteristic, and that is labeled with blue "+" (Fig. 4(b), (d)) is the results of using global vessel direction characteristics. The vasculature is not very clear in Fig. 4(a), as result, OD localization fails by only using vessel direction characteristics (Fig. 4(b)). And the appearance is damaged in Fig. 4(c), and OD localization fails by using appearance characteristics (Fig. 4(c)). However, when apperance or vessel characteristic are both used, they will be detected correctly (Fig. 4(a), (d)).

Table 2 shows the accurate rate of the proposed method with other detection methods that have been reported. From the experimental results we can see that this method achieves a high detection in these four fundus image databases. In STARE dataset, detection accuracy is only 92.6 %, and this is because SATRE dataset has many images with serious lesions. The OD structure of lots of the fundus image is damaged, and also only part of the OD can be seen in many of retinal images. At the same time, the destroyed vasculature was also affect the accuracy. But in general, in all of 340 images, this method still reached a accuracy rate of 97.9 %.

Table 2. The comparison with various methods in accuracy, '–' represents data unavailable in the respective publication Ref.

Method	DIARETDB0	DIARETDB1	DRIVE	STARE	Overall
Proposed method	100 %	98.9 %	100 %	92.6 %	97.9 %
Mahfouz [12]	98.5 %	97.8 %	100 %	92.6 %	97.0 %
Sinha [13]	96.9 %	100 %	95.0 %	–	96.3 %
Lu [7]	99.2 %	98.9 %	97.5 %	96.3 %	97.4 %
Qureshi [14]	96.8 %	94.0 %	100 %	–	97.0 %

Figure 5 shows OD detection of 8 pathological images by our method. In each image, the OD center is marked with blue "+.". Among all these images, only Fig. 5(h) is failed.

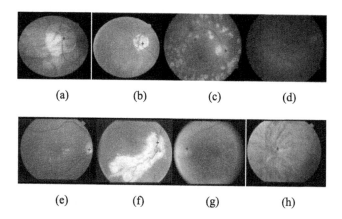

(a)	(b)	(c)	(d)

| (e) | (f) | (g) | (h) |

Fig. 5. OD detection results of example images

4   Conclusion

This paper proposes a robust optic disc detection method based on multi-features and two-stage decision strategy. This algorithm not only use the vessel characteristics and appearance characteristics, but alsoemploy the histograms of oriented gradients (HOG) and support vector machine (SVM). Experimental results demonstrated the viability and robustness of the proposed approach.

References

1. Li, H., Chutatape, O.: Automatic location of optic disc in retinalimages. In: IEEE International Conference on Image Processing, vol. 2, pp. 837–840, 7–10 October 2001
2. Li, H., Chutatape, O.: A model-based approach for automated feature extraction in fundus images. In: 9th IEEE International Conference on Computer Vision (ICCV 2003), vol. 1, pp. 394–399 (2003)
3. ter Haar, F.: Automatic localization of the optic disc in digital colour images of the human retina, M.S. thesis, Utrecht University, Utrecht, The Netherlands (2005)
4. Barrett, S.F., Naess, E., Molvik, T.: Employing the Hough transform to locate the optic disk. Biomed. Sci. Instrum. **37**, 81–86 (2001)
5. Ying, W., Dongbo, Z.: Fast optic disc detection based on multi-scale blob detection. Opt. Tech. (5) (2016)
6. Lu, S.J., Lim, J.H.: Automatic optic disc detection from retinal images by a line operator. IEEE Trans. Image Process. **58**(1), 88–94 (2011). (S1057-7149)
7. Lu, S.: Accurate and efficient optic disc detection and segmentation by a circular transformation. IEEE Trans. Med. Imaging **30**(12), 2126–2133 (2011). (S0278-0062)

8. Foracchia, M., Grisan, E., Ruggeri, A.: Detection of optic disc in retinal images by means of a geometrical model of vessel structure. IEEE Trans. Med. Imaging **23**(10), 1189–1195 (2004). (S0278-0062)

9. Zhang, D.B., Zhao, Y.Y.: Novel accurate and fast optic disc detection in retinal images with vessel distribution and directional characteristics. IEEE J. Biomed. Health Inf. **20**(1), 333–342 (2016)

10. Hoover, G.A., Goldbaum, M.: Locating the optic nerve in a retinal image using the fuzzy convergence of the blood vessels. IEEE Trans. Med. Imaging **22**(8), 951–958 (2003). (S0278-0062)

11. Youssif, A., Ghalwash, A., Ghoneim, A.: Optic disc detection from normalized digital fundus images by means of a vessels' direction matched filter. IEEE Trans. Med. Imaging **27**(1), 11–18 (2008). (S0278-0062)

12. Mahfouz, A.E., Fahmy, A.S.: Fast localization of the optic disc using projection of image features. IEEE Trans. Image Process. **19**(12), 3285–3289 (2010). (S1057-7149)

13. Sinha, N., Babu, R.V.: Optic disk localization using L_1 minimization. In: IEEE International Conference on Image Processing, pp: 2829–2832 (2012)

14. Qureshi, R.J., Kovacs, L., Harangi, B., et al.: Combining algorithms for automatic detection of optic disc and macula in fundus images. Comput. Vis. Image Underst. **116**(1), 138–145 (2012)

15. Zhao, Y., Zhang, D., Wang, Y.: Optic disk segmentation of retinal image with smoothing filtering and CV model. Opt. Tech. **40**(6), 524–530 (2014)

16. Zheng, S., Chen, J., Pan, L., et al.: A novel method of macula fovea and optic disk automatic detection for retinal images. J. Electron. Inf. Technol. **36**(11), 2586–2592 (2014)

Hierarchical Saliency Detection Under Foggy Weather Fusing Spectral Residual and Phase Spectrum

Kun Liu[✉], Jia Tian, Xiu-ping Su, Yu-qiang Zhou, and Jie Wang

School of Control Science and Engineering,
Hebei University of Technology, Tianjin 300130, China
liukun03@mails.thu.edu.cn

Abstract. It brings great difficulty for salient object detection on the road under foggy weather, owing to the low contrast and the low resolution of fog-degraded image. The traditional methods of saliency detection such as spectral residual approach and phase spectrum approach are out of work. According to this problem, a new hierarchical saliency detection approach based on transmission information of foggy images is proposed, which fuses the spectral residual and phase spectrum under transmission information with a new weighted fusing approach. The experiments show that the proposed method can detect salient object such as pedestrians and vehicles on the road more effectively than independent spectral residual and phase spectrum approach, even under heavy fog condition.

Keywords: Saliency detection · Transmission estimation · Spectral residual · Phase spectrum

1 Introduction

The pedestrian and vehicle detection are important tasks in driver assistance system, which can help driver find the interested objects on the road. It aims at detecting the pedestrians and vehicles appearing ahead of the vehicle using a vehicle-mounted camera, so as to assess the danger and take actions in case of danger. To some extent, the system reduces the road accident rates. It has large prospects for application the in the future.

Saliency detection is very helpful for the task of the pedestrian and vehicle detection. Saliency detection is a cognitive process that imitates human visual attention mechanism, which can help humans rapidly select the highly relevant information from a scene. Detection of visually salient image regions is useful for applications like object segmentation, adaptive compression, and object recognition [1–4]. The salient objects can be detected rapidly under the pre-attention and fine adjustment mechanism.

This work is supported by National Natural Science Foundation (NNSF) of China under Grant 61403119 and Natural Science Foundation of Hebei Province under Grant F201402166 and F2015202231.

© Springer Nature Singapore Pte Ltd. 2016
T. Tan et al. (Eds.): CCPR 2016, Part II, CCIS 663, pp. 191–201, 2016.
DOI: 10.1007/978-981-10-3005-5_17

The former Itti model [5] is a famous model using for saliency detection. The bottom features such as color, illumination and orientation are extracted from the original information, and the comparison of center and neighborhood is formed under local range. The saliency map was obtained by using the linear combinations and the salient object is extracted under the winner-take-all competition mechanism. Attention-based information maximization method is similar to Itti model, which utilizes the local features under bottom-to-top framework. The saliency was measured by the self-information which reflects the difference of the information beyond the neighborhoods. However, the above features are undesirable under foggy weather in the image.

The saliency detection algorithm based on frequency-domain obtains the salient objects by transforming the image to the frequency domain. For example, the spectral residual approach [6] presented by Hou utilized the amplitude information in frequency domain and the phase spectrum method presented by Guo [7] utilized the phase information of the Fourier transformation. The spectral residual approach explored the generalized properties of the background of the images, which suppressed the response to frequently occurring features using the log amplitude spectrum residue, while at the same time kept sensitive to features deviate from the normal. The salient object proposed by [7] could be detected only by phase spectrum to suppress the periodic signal among the image. Actually, there are some relations between low frequent part and the fog in the image. High frequent part can be viewed as cues of the objects in the fog. In other words, it's a good choice for the method to work in the frequency domain. But the existing methods didn't work well for the detection under foggy weather since the atmosphere under fog weather is not low frequency information simply.

According to this problem, this paper proposes a new hierarchical saliency detection approach based on the transmission information of foggy images, which can provide atmospheric scattering distribution under foggy weather more accurately. The proposed method fuses the spectral residual and phase spectrum under a new weighting combination method. The experiments show that the proposed method can detect salient object such as pedestrians and vehicles on the road even under heavy fog condition effectively.

2   Transmission Estimation

2.1   Atmospheric Imaging Model

In order to handle the low contrast and the low resolution of fog-degraded images, the atmospheric imaging model needs to be analyzed in detail. The atmosphere scattering model is used to describe the imaging process of scenes under foggy weather, of which the model proposed by McCartney is widely used in computer vision and computer graphics [8]. It includes attenuation model and atmospheric lighting model, the former describes the process that the lighting propagates from the scenes to the camera, and the later describes the influence of atmospheric light after scattering to the observed illumination. The concrete model can be described by (1).

$$I(x) = J(x)t(x) + A(1 - t(x)) \tag{1}$$

Where I is the observed intensity, J is the scene radiance, A is the global atmospheric light, and t is the medium transmission describing the portion of the light that is not scattered and reaches the camera. Generally the goal of haze removal is to recover J, A, and t from I.

In (1), the first term $J(x)t(x)$ on the right-hand side is called direct attenuation, and the second term $A(1 - t(x))$ is called airlight. The direct attenuation describes the scene radiance and its decay in the medium, and the airlight results from previously scattered light and leads to the shift of the scene colors. While the direct attenuation is a multiplicative distortion of the scene radiance, the airlight is an additive one. When the atmosphere is homogenous, the transmission t can be expressed as

$$t(x) = e^{-\beta d(x)} \tag{2}$$

Where β is the scattering coefficient of the atmosphere and d is the scene depth.

2.2   Transmission Estimation

In order to bring down the influence of fog, the intuitive viewpoint is to remove the haze superposed on images. However, the comparison experiments show that the salient object could be detected more effectively under transmission information. Here, we use the dark channel method of [9] to obtain the transmission of foggy images. Assuming that the atmospheric light A is given, the Eq. (1) is firstly normalized by A:

$$\frac{I^c(x)}{A^c} = t(x)\frac{J^c(x)}{A^c} + 1 - t(x) \tag{3}$$

Note that each color channel is normalized independently. The further assumption is that the transmission in a local patch $\Omega(x)$ is constant. The transmission is denoted as $\tilde{t}(x)$. Then, calculate the dark channel on both sides of (3). Equivalently, put the minimum operators on both sides:

$$\min_{y \in \Omega(x)}\left(\min_c \frac{I^c(y)}{A^c}\right) = \tilde{t}(x)\min_{y \in \Omega(x)}\left(\min_c \frac{J^c(y)}{A^c}\right) + 1 - \tilde{t}(x) \tag{4}$$

Since $\tilde{t}(x)$ is a constant in the patch, it can be put on the outside of the min operators. As the scene radiance J is a haze-free image, the dark channel of J is close to zero due to the dark channel prior:

$$J^{dark}(x) = \min_{y \in \Omega(x)}\left(\min_c J^c(y)\right) = 0 \tag{5}$$

As A^c is always positive, this leads to

$$\min_{y\in\Omega(x)}\left(\min_c \frac{J^c(y)}{A^c}\right) = 0 \qquad (6)$$

Putting (6) into (4), we can eliminate the multiplicative term and estimate the transmission $\tilde{t}$ simply by

$$\tilde{t}(x) = 1 - \min_{y\in\Omega(x)}\left(\min_c \frac{I^c(y)}{A^c}\right) \qquad (7)$$

Then, the transmission is evaluated by the above equation. The latest guided filter method proposed by He [10] can be used to refresh the detail of the transmission efficiently.

3   Hierarchical Saliency Detection

3.1   Preliminary Spectral Residual Detection

As for the foggy images, the low contrast and low resolution bring out the difficulties to saliency detection in original spatial domain. Correspondingly, if transform the image to the frequency domain, the high frequency information cut down and low frequency information increase under the influence of fog and haze.

The spectral residual method is one of saliency detection methods in frequency domain, which assumes that natural images are not random and they obey some predictable distributions, under which assumption the coding mechanism is established. From the perspective of information theory, the coding decomposes the image information H into two parts [6]:

$$H \text{ (Image)} = H \text{ (Innovation)} + H \text{ (Prior Knowledge)} \qquad (8)$$

Where H (Innovation) denotes the novelty part, and H (Prior Knowledge) is the redundant information that should be suppressed by a coding system. As for the foggy image, suppressing the response to frequently occurring low-frequency redundant features, while at the same time keeping sensitive to high-frequency features that deviate from the norm, the salient objects can be detected.

Scale invariance is the most famous and widely studied property among the invariant factors of natural image statistics. This property is also known as 1/f law. It states that the amplitude A(f) of the averaged Fourier spectrum of the ensemble of natural images obeys a distribution:

$$E\{A(f)\} \propto 1/f \qquad (9)$$

Given an input image, transform it into frequency space using the Fourier Transform, and the log spectrum $\mathcal{L}(f)$. is computed with the logarithm of the amplitude:

$$A(f) = |F[I(x)]| \tag{10}$$

$$P(f) = \varphi(F[I(x)]) \tag{11}$$

$$L(f) = \log(A(f)) \tag{12}$$

Where F denotes the two-dimension Fourier transform, $|\bullet|$ denotes the amplitude, f denotes the phase and $\mathcal{P}(f)$ denotes the phase spectrum of the image. Since the logarithm curve is similarly linear, the local average filter $h_n(f)$ is used to obtain averaged spectrum that can be approximated by convoluting the input image:

$$V(f) = h_n(f) * L(f) \tag{13}$$

Where $h_n(f)$ is an $n \times n$ matrix defined by

$$h_n(f) = \frac{1}{n^2} \begin{pmatrix} 1 & 1 & \cdots & 1 \\ 1 & 1 & \cdots & 1 \\ \vdots & \vdots & \ddots & \vdots \\ 1 & 1 & \cdots & 1 \end{pmatrix} \tag{14}$$

Therefore the spectral residual $\mathcal{R}(f)$ can be obtained by

$$R(f) = L(f) - V(f) \tag{15}$$

The salient object can be detected utilizing the inverse Fourier transform of spectral residual and phase P(f):

$$S(x) = |F^{-1}[\exp\{R(f) + iP(f)\}]|^2 \tag{16}$$

In the above model, the spectral residual $\mathcal{R}(f)$ contains the innovation part of an image. Using Inverse Fourier Transform, we can in spatial domain construct the output image called the saliency map. The saliency map contains primarily the nontrivial part of the scene. As for the foggy images, the content of the residual spectrum can also be interpreted as the high-frequency parts of the image, which comprises of the objects whose depth information abrupt. For better visual effects, we smoothed the saliency map with a Gaussian filter ($\sigma = 8$). For the transmission image obtained from the foggy image, the log spectrum curve of $\mathcal{L}(f)$, $V(f)$ and $\mathcal{R}(f)$ is shown in Fig. 1.

3.2   Detailed Phase Spectrum Detection

In the Fourier representation of signals, spectral magnitude and phase tend to play different roles and in some situations, the phase contains significant information, especially regarding the edge location. Our purpose is to eliminate the so-called regular patterns while preserving correspondingly "rare" events in the image which can be considered to be saliency. The fact that phase-only reconstruction preserves much of

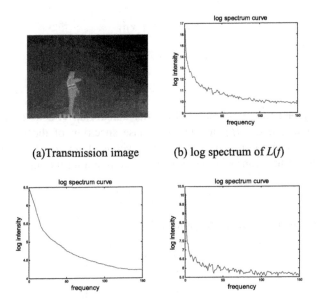

Fig. 1. Spectral residual of transmission information

the correlation between signals would suggest that the location of events tends to be preserved [7]. If we assume that in our application, the background region contains non-localized structures (e.g. regular patterns or homogeneous regions) and that the salient object is well localized, it is reasonable to use the phase to filter all non-localized patterns.

The discrete two-dimensional Fourier transform of an image array F(u, v) is defined in series form as:

$$\mathcal{F}(u,v) = \frac{1}{N}\sum_{j=0}^{N-1}\sum_{k=0}^{N-1} F(j,k)\exp\{\frac{-2\pi i}{N}(uj+vk)\} \tag{17}$$

Where $i = \sqrt{-1}$. The indices (u, v) are called the spatial frequencies of the transformation. The result is a matrix of complex numbers in the frequency domain,

$$\mathcal{F}(u,v) = \mathcal{R}(u,v) + i\mathcal{I}(u,v) \tag{18}$$

or in magnitude and phase-angle form,

$$\mathcal{F}(u,v) = \mathcal{M}(u,v)\exp\{i\Phi(u,v)\} \tag{19}$$

Where

$$\mathcal{M}(u,v) = \sqrt{\mathcal{R}^2(u,v) + \mathcal{I}^2(u,v)} \tag{20}$$

And

$$\Phi(u, v) = \arctan\{\frac{\mathcal{I}(u, v)}{\mathcal{R}(u, v)}\} \tag{21}$$

By normalizing every complex number by dividing both real and image parts by M (u, v) we essentially remove all regular patterns at every scales at once.

$$O(u, v) = \mathcal{F}^{-1}(u, v) \tag{22}$$

Given the transmission information of the input image, the salient object after the inverse Fourier transformation using the phase spectrum is shown in Fig. 2.

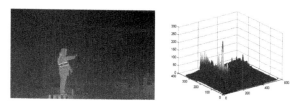

Fig. 2. Saliency detection based on phase spectrum

3.3 Weighted Fusing and Region Localization

Since the spectral residual (SR) method suppresses the response to redundant low-frequency features whose depth information is gradual and denotes the background in the transmission image, at the same time keeping sensitive to novel features that whose depth information is abrupt and denotes the foreground objects standing on the road, it can get the major of the interested objects. The salient object can be selected further through the prior positional distribution that the salient objects are located in the center of the images.

The reconstruction from the phase spectrum (PS) of the image can describe the primary texture information of the original image, which includes the foreground salient objects and some regions from the background scenes. The fusion of the preliminary spectral residual detection and the detailed phase spectrum detection need to consider both the saliency value and spatial position distribution of two results. The weighted fusing approach is proposed to reinforce the foreground objects in the SR map and suppress the disturbance of the background in the PS map.

The fusion can be obtained by the following equation.

$$S(x_i) = w_{SR}(x_i) * S_{SR}(x_i) + w_{PS}(x_i) * S_{PS}(x_i) \tag{23}$$

$$w(x_i) = den(x_i) * pos(x_i) \tag{24}$$

$$den(x_i) = \frac{\#\,num[I(x_i) > thres]}{\sum num} \tag{25}$$

$$pos(x_i) = \frac{1}{\sqrt{(2\pi)^n|C|}} \exp\{-\frac{1}{2}(x-\mu)^T C^{-1}(x-\mu)\} \tag{26}$$

Where x_i is the coordinate position of the image, $S_{SR}(x_i)$ denotes the saliency value of preliminary spectral residual map, $S_{PS}(x_i)$ denotes the saliency value of detailed phase spectrum map, $w(x_i)$ denotes the weighted coefficients on the position x_i of two corresponding saliency detection results. $den(x_i)$ is the spatial density distribution factor around the position x_i, $I(x_i)$ represent the illumination of the coordinate x_i and $num[\bullet]$ count the number of image illumination is larger some threshold in a given window; $pos(x_i)$ is the distance deviation of current position to the maximum value among the indicated region, in which μ and C denote the mean and covariance matrix of image illumination in the given window. Finally, the salient object regions can be localized by thresholding the combined saliency map under Mahalanobis distance.

4 Experiment

To evaluate the performance of the proposed method, a large number of videos under foggy weather on the road are recorded. The images are selected mainly from the videos under heavy foggy weather, not from videos under the thin hazy weather since the transmission information is related to the degree of the atmosphere scattering. Considering that it is important for the driver to find the pedestrians and vehicles on the front under the foggy weather and it is convenient to improve the traffic safety, the salient objects such as the pedestrian, vehicles and obstacles are flagged in the 3000 images as the benchmark. In each image the minimum enclosing rectangles of the salient objects are flagged manually to evaluate the result of the proposed detection algorithm.

To verify the effectiveness of the transmission information, the comparative experiments of salient detection on original foggy image, dehazed image and transmission image are made, in which the spectral residual method is utilized here. Figure 3 gives the corresponding result of salient object detection under the foggy image, dehazed image and transmission image, among which (a) gives the original foggy image and (b) the saliency detection result on foggy image; (c) gives the dehazed image obtained by the method of He [9] and the saliency detection result on it is shown in (d); (e) gives the transmission image and (f) shows the saliency detection result on the transmission image.

From the above experimental results, we can find the salient objects can be detected better by using the transmission information, which can grasp the major of the interested regions. The salient detection under original foggy image gives the ambiguous results because of low resolution and low contrast of the foggy image. Incomplete objects are detected owing to the disturbance of the background information in the

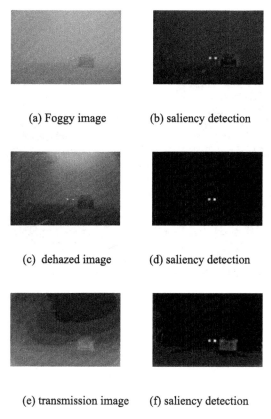

(a) Foggy image   (b) saliency detection

(c) dehazed image   (d) saliency detection

(e) transmission image   (f) saliency detection

Fig. 3. Saliency detection based on foggy image and original image and transmission image using spectral residual method

dehazed image. The transmission image can provide more effective information about the interested object.

Furthermore, the comparison result of our saliency detection with independent spectral residual method and phase spectrum method is given in Fig. 4, among which the first row gives the original foggy image; the second row hows the saliency result of spectral residual method based on transmission independently, in which the profile of the car is incomplete and only part object is detected; the third row gives the saliency result of phase spectrum method based on transmission independently, in which much noise information bring out; the last row shows the result of our method based on hierarchical detection and weighted fusion, which can give the better detection result of the salient pedestrian and car and suppress the background information effectively.

The statistical results are obtained on the flagged images finally. The definition of detection result is given in advance: True positive denotes that the salient objects are detected correctly; false positive explains that the non-salient regions are detected as salient objects; false negative shows that the salient objects are missed and true

Fig. 4. Comparison of our method with spectral residual method and phase spectrum method, the first row gives the original foggy image, the second row gives the result of SR method, the third row shows the result of PS method, the last row is our result.

negative shows that the non-salient objects is classified correctly. Furthermore, the precison and recall ratio are defined as the following equation.

$$precision = \frac{TruePositive}{TruePositive + FalsePositive} \tag{27}$$

$$recall = \frac{TruePositive}{TruePositive + FalseNegative} \tag{28}$$

Further, the F-measure is computed by the following equation based on precision and recall.

$$F_\alpha = \frac{(1 + \alpha) * precision * recall}{\alpha * precision + recall} \tag{29}$$

Here, α is designed to 0.2 in order to increase the weight of precision among F_α. The Table 1 gives the comparison result of our method with independent spectral residual method and phase spectrum method. The statistic result shows that our method can detect the salient objects more effectively than independent spectral residual and phase spectrum approach.

Table 1. Comparison of ours with traditional algorithm

	SR method	PS method	Our method
Precision	0.86	0.83	0.93
Recall	0.81	0.89	0.95
F-measure	0.851	0.839	0.933

5 Conclusion

Aiming at the low contrast and the low resolution of fog-degraded image, we propose a hierarchical method of salient object detection, which is based on the transmission information of foggy images and fuses the spectral residual and phase spectrum according to the transmission information. Furthermore, the high-performance saliency detection method is proposed based on weighted distance which reflects the spatial position and density distribution of the saliency map. The experiments show that the proposed method can detect salient objects such as pedestrians and vehicles on the road even under heavy fog condition effectively.

References

1. Cheng, M., Zhang, G., Mitra, N., Huang, X., Hu, S.: Global contrast based salient region detection. In: IEEE Conference on Computer Vision and Pattern Recognition, pp. 409–416 (2011)
2. Borji, A., Sihite, D.N., Itti, L.: Salient object detection: a benchmark. In: Fitzgibbon, A., Lazebnik, S., Perona, P., Sato, Y., Schmid, C. (eds.) ECCV 2012. LNCS, vol. 7573, pp. 414–429. Springer, Heidelberg (2012)
3. Gao, D., Vasconcelos, N.: Discriminant saliency for visual recognition from cluttered scenes. In: Advances in neural information processing systems, pp. 481–488 (2004)
4. Schwartz, S., Wong, A.: Saliency-guided compressive sensing approach to efficient laser range measurement. J. Vis. Commun. Image Represent. 24(2), 160–170 (2012)
5. Itti, L., Koch, C.: Computational modelling of visual attention. Nat. Rev. Neurosci. 2(3), 194–203 (2001)
6. Hou, X., Zhang, L.: Saliency detection: a spectral residual approach. In: IEEE Conference on Computer Vision and Pattern Recognition, pp. 1–8 (2007)
7. Guo, C., Ma, Q., Zhang, L.: Spatio-temporal saliency detection using phase spectrum of quaternion fourier transform. In: Proceedings of IEEE Conference on Computer Vision and Pattern Recognition, pp. 1–8 (2008)
8. Narsimhan, G., Nayar, S.K.: Vision and the atmosphere. Int. J. Comput. Vis. 48(3), 233–254 (2002)
9. He, K., Sun, J., Tang, X.: Single image haze removal using dark channel prior. IEEE Trans. Pattern Anal. Mach. Intell. 33(12), 2341–2353 (2011)
10. He, K., Sun, J., Tang, X.: Guided image filtering. IEEE Trans. Pattern Anal. Mach. Intell. 35(6), 1397–1409 (2013)

Hierarchical Image Matching Method Based on Free-Form Linear Features

Xiaowei Chen[✉], Haitao Guo, Chuan Zhao, Baoming Zhang, and Yuzhun Lin

Zhengzhou Institute of Surveying and Mapping, Zhengzhou, China
chenxw_2007@aliyun.com

Abstract. In order to better resolve the conflict between the full use and effective description of the Linear Feature information in the study of free form linear feature (FFLF) matching, this paper proposed a remote sensing image matching method using the hierarchical matching strategy. First the edge features of the image were detected and tracked, in order to extract the free-form sub-pixel linear features with better continuity; in the coarse matching process, the closed linear features (CLF), linear feature intersection (LFI) and corner (LFC) were selected as conjugated entity, and then the false match was gradually eliminated based on the area, angle and other geometry information as well as the parameter distribution features of the model determined by the conjugate features combination to be selected, finally the initial value of accurate matching was determined by the conjugate features; in the accurate matching process, based on multi-level two-dimensional iterative closest point (ICP) algorithm, sub-pixel edge points were orderly used with the sampling rate from low to high for matching. Experimental results show that this method has stable performance for the coarse matching; high accuracy and precision of the coarse matching can provide the initial matching parameters of high precision for accurate matching; accurate matching can reach the sub-pixel level precision equal to the least square matching and can better achieve stable and accurate matching for the images with smaller affine deformation.

Keywords: Free-form linear feature (FFLF) · Image matching · Hierarchical matching strategy · Closed linear feature (CLF) · Linear feature corner (LFC) · Linear feature intersection (LFI) · Iterative closest point (ICP) algorithm

1 Introduction

As a basic feature of the image, linear feature contains a wealth of information and is more stable. So, compared with the traditional gray level matching and point feature matching, linear feature matching has considerable advantages in stability and universality, which makes it become a new research hotspot in the field of image matching. Linear and circular features that can be expressed by mathematics analytic formula were first utilized at early stage [1], while the FFLFs such as coastlines and roads are less applied. This is because such features are composed of a series of continuous irregular

© Springer Nature Singapore Pte Ltd. 2016
T. Tan et al. (Eds.): CCPR 2016, Part II, CCIS 663, pp. 202–217, 2016.
DOI: 10.1007/978-981-10-3005-5_18

edge points, which are difficult to be expressed through mathematical analytic model and difficult to be directly applied in the processing [2].

Compared to point matching, little progress has been made in curve matching (including linear matching) in recent years. Only a few methods for curve matching are reported in literature until now. Xiao et al. [3] reported a novel feature called edge-corner that announced corners as the intersection of two or more edges. The initial matching method was done by comparing patches around the edge corners, using Sum of Squared Differences (SSD) and followed up by an optimization process that modeled a more precise affine model for the matching. Zuliani [4] presented a new physically motivated curve/region descriptor based on the solution of Helmholtz's equation which satisfies the six principles set by MPEG-7 and can be generalized in order to take into account also the intensity content of the image region defined by the curve. Wu et al. [5] proposed a feature vector field for images, which is built by the inner products and exterior products of image gradients. The feature vector field can effectively represents image edges and feature points including corners and edge points with big curvature. Experiments show that the descriptors have a good adaptability to small image affine transformation. Kingsbury [6] proposed a matching system that took both appearance and spatial constraints into consideration. SIFT descriptor and a pair-wise relationship descriptor was used to measure the similarity between them. Wu [7] presented an integrated point and edge matching method, which incorporates the edge matching with point matching in the same dynamic matching propagation process. It takes advantage of the edge-constrained Delaunay triangulations with the capability of generating point and edge matches preserving the actual textural features. Zhang et al. [8] proposed a novel affine invariant curve descriptor based on membrane vibration model. They mainly focused on Jardon curve and open curve; the experimental results show the proposed method outperforms the existing Fourier descriptor and Zernike moment descriptor. Saeedi and Mao [9] introduced a novel feature-2EC feature and a unique matching technique, the proposed feature utilizes two straight lines and their intersections. It can establish match correspondences between two oblique aerial images under a large projective transformation.

As a conclusion, these existing FFLF matching methods can be generally divided into two types: one is to extract part of the information of the linear features for matching, such as the shape parameter, center of gravity and rotation invariant moment and other features, which have better stability for the change of imaging conditions but will abandon large amount of the feature information, so as to lack high matching accuracy. Another is to try to use all the coordinate information of the edge points [2] to realize image matching by the chain code edge descriptor matching, although this makes full use of the rich information of the edge, the high dimensionality of character description vector and large amount of calculation make the stability depend on the stability of the edge detection to a large extent, and the edge fracture problems are often hard to overcome; additionally, high-precision initial conditions are needed.

In order to better resolve the conflict between the full use and effective description of the rich linear feature information in the study of FFLF matching, this paper proposed a remote sensing image matching method using the hierarchical matching. First, the FFLFs of the image were extracted, from which a variety of stable features were extracted for coarse matching; and finally, sub-pixel edge points with different sampling rates were orderly used to realize the accurate matching.

2   Research Methods

FFLFs matching method based on hierarchical matching mainly consists of such steps as extraction of FFLFs, extraction of features to be matched, coarse matching and accurate matching. Its basic principle is as follows: first carry out sub-pixel edge detection on the image, and then track and refine the edge points, so as to obtain FFLFs with better continuity; then extract the CLFs, LFIs and LFCs and other stable features for coarse matching; eliminate the false matching by using the geometry information and the parameter distribution features of the model determined by the conjugate features combination and then determine the initial value of accurate matching; finally based on ICP, apply sub-pixel edge points with the sampling rate from low to high for accurate matching. The following is the introduction of main principle of the proposed method.

2.1   Extraction of Free-Form Sub-pixel Linear Features

Extraction of FFLFs consists of two main steps, the edge detection and the edge tracking & refining. Because the edge detection precision directly affects the final precision of the matching method based on edge features, with the application of computer vision technology in every field in recent years, people expect to obtain the accurate location and size of the target based on existing image data. Thus, sub-pixel edge detection has received the extensive attention of the researchers [10, 11]. In order to ensure a higher matching accuracy, the method proposed in 11 of the references, sub-pixel edge detection using extremal gradients, is applied to detect the edge of the image; with good universality, this method can carry out better detection on the edges of various types, especially having higher positioning accuracy for the edge of the corner; the principle is detailed as follows.

The extremal gradients include positive gradient and negative gradient, which respectively represents the maximum increase gradient and minimum decrease in the grey level of each image point; the magnitude is determined by the maximum differential of grey level between current image point and its eight neighborhood image points, and the direction is from current image point towards the neighborhood image point corresponding to the maximum difference. By calculating the extremal gradient of each point, the initial edge consisting of two types of image points which has partially maximum grey increase and maximum grey decrease can be obtained; finally sub-pixel localization models with different types of edges are established according to the features of the initial edge.

The results obtained from all edge detection algorithms are the discrete edge points, which need to be tracked before the application. In general, free-form edges have complex forms and intersect with each other; additionally, the influence of the factors like noise also makes some extracted edges fracture and unsmooth; however, most of the current tracking algorithms have not fully analyzed and discussed these conditions, which causes that the edge features cannot be fully and effectively extracted, restricting the application of FFLFs to a certain extent.

In view of the above problems, reference [12] proposed a sub-pixel edge detection method, which can, based on high positioning accuracy of edge detection, effectively extract the smooth FFLFs with better continuity; thus it is applied to provide tracking & refining of the detected edge. The basic principle of this method is: to propose a tracking strategy of sub-pixel edge points according to the actual distribution of the edge points, so as to record the free-form edges as much as possible; in order to restore the continuity of the edge, trying to avoid the connection of uncorrelated edges and apply an expansion connection method of double-angle and length control to connect the edge; finally using the smoothing algorithm based on curve inner expression to smooth the two-dimensional FFLFs.

2.2   Extraction of the Features for Matching

First of all, this method extracts the CLFs and LFIs and LFCs from FFLFs to realize coarse matching; in order to balance the matching efficiency and precision, extraction of sub-pixel edge points with different sampling rates is applied to gradually realize accurate matching by the edge points with the sampling rate from low to high. So this section mainly introduces the extraction methods and matching entities of the features applied in the two matching processes, as well as the similarity measurement.

(1) Closed Linear Features (CLF)

The CLFs can be determined by identifying the conjugate linear features of the starting point and ending point. This paper selects the CLFs with the area meeting the requirement of threshold for the matching, and 7 invariant moments not changing with rotation, translation and scaling are taken as its matching entities. Let $M_i(p_m)$ and $M_j(q_n)$ $(i, j = 1, 2,\ldots,7; m, n = 1, 2,\ldots)$ represent 7 invariant moments of various CLFs of left and right images; due to large difference of invariant moments, the normalized correlation algorithm shown in Formula (1) is applied to measure the similarity between the CLFs. Without ambiguity, p_m, q_n are used to represent the CLFs and also represent the center of gravity of the surrounding area.

$$R(p_m, q_n) = \sum_{i=1}^{7} \frac{|M_i(p_m) - M_i(q_n)|}{|M_i(p_m)| + |M_i(q_n)|} \tag{1}$$

(2) Linear Feature Intersection (LFI)

The tracking method mentioned in Sect. 2.1 made a complete extraction of the linear features, so the edge points with the number of recording not less than three times in the linear features are the LFIs. Since the extraction of LFI with more than three linear feature branches can be easily affected by the noise and there is small quantity, only the intersections containing three branches and stable direction are selected for matching and the included angles between the branches are taken as its matching entities.

In order to guarantee the uniqueness of the matching entities for LFIs, the following rules should be followed to arrange three included angles: clockwise arranging and the maximum angle are placed at the starting position. Clockwise arrange the direction vector of each branch and calculate the included angle between the adjacent direction vectors; clockwise record the included angles successively from the maximum angle. Through analysis, arbitrary arrangement of three vectors with the corresponding starting point only includes two types, the clockwise and counterclockwise arrangement, and order exchange of the latter two vectors can realize the exchange of two types (Number shown in Fig. 1 represents the serial number of each branch); thus, after the direction vector is determined, the arrangement is identified by comparing the horizontal and vertical components of each vector and then adjusted appropriately to realize the clockwise arrangement.

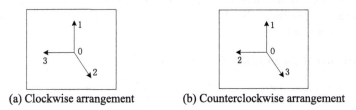

(a) Clockwise arrangement   (b) Counterclockwise arrangement

Fig. 1. Two types of arrangements of three direction vectors

(3) Linear Feature Corner (LFC)

Douglas-Peucker (DP) method is to determine the feature points of the curve from the overall to local, which has invariance to translation and rotation, as well as high computational efficiency. But its threshold is not easy to be determined: a larger threshold is easy to cause excessive compression; a smaller threshold is sensitive to the "thrusting" caused by the noise and is easy to get wrong feature points. The intersection extracting method based on the difference of absolute chain code sum proposed by Li et al. [13] has good information compression ability and anti-interference ability. Therefore, this paper applies this method to test the reliability and accuracy of the corners extracted by DP method, with its basic principle shown in Fig. 2.

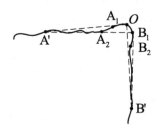

Fig. 2. Basic principle of correction of the corner location

Set O as the corner to be selected (that is the feature point determined by DP method) and consider the O and two edge points on both sides A_i, B_i ($i = 1, 2$): first identify if the difference of absolute chain code sum of O and A_1, B_1 meet the requirements of threshold; if only O can meet the requirement, it should be the corner; if one of point A_1 and B_1 can meet the requirement, its reliability should be compared with that of point O and the point with the greatest reliability is taken as the corner; if all the above three points cannot meet the requirement, then A_2 and B_2 should be further identified. When at least one point meets the requirement, comparing it with the point O, the one with greater reliability should be taken as the corner; otherwise the point O will be eliminated as a false corner.

The reliability of point which is selected as corner may be calculated by the model shown in Formula (2).

$$C = 1 \bigg/ \left(\sum_{i=1}^{2} \sum_{j=1}^{h} d_{ij} \right) \tag{2}$$

In which: d_{1j} and d_{2j} respectively represent the distance from P_j and Q_j to the straight line S_1 and S_2; S_i ($i = 1, 2$) is the connection (i.e. OA', OB') between the current point (point O as shown in Fig. 3) and the edge point (i.e. point A', B') which has h number of points interval with its some linear feature branch; P_j, Q_j ($j = 1, 2, \ldots,$ h) are respectively the edge point of No. j on both sides of the branch.

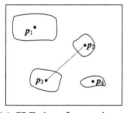 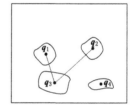

(a) CLFs in reference image   (b) CLFs in image to be matched

Fig. 3. The principle of elimination of false matching by using the geometric information

It should be noted that the method proposed in reference [13] aimed at pixel-level edge curve, while the linear features extracted in Sect. 2.1 and processed by DP method are sub-pixel edge curves; thus each sub-pixel edge point is replaced by the center of the image point (Coordinate as an integer) in the chain code related calculation and the sub-pixel coordinates should be still applied for the reliability calculation.

In order to obtain the corners with uniform distribution and a reasonable number, only the corners which have stable linear feature branch direction and the interval is greater than a certain threshold value should be applied. The angle between the direction vectors of the branches on both sides of the corner is taken as its matching entity. The direction of the angle bisector is taken as the main direction.

(4) Selection of the Features for Accurate Matching
ICP method is considered as three-dimensional object matching algorithm based on pure geometric model, which can directly process the data of 3D object not related to the surface, with no need to assume and divide the object's features. Thus it has been widely used to realize 3D registration. ICP method can obtain higher matching accuracy when the point set has high precision, but it is vulnerable to the interference of noise point, with poor robustness; and there is low efficiency when the point set

quantity is large. Based on the above analysis, in the selection of edge points for accurate matching, the following two strategies are taken to reduce the influence of noise points and improve the efficiency of the algorithm:

(a) smoothing it with the linear features of corner: considering that the ground object especially the artificial ground object has smooth edge curve and in order to keep the corner with higher accuracy and better stability in the edge detection results, in the edge smoothing of Sect. 2.1, the corners extracted in (3) are used and only the edge curves between the corners are smoothed.

(b) using the edge points with different sampling rates in various stages of the iteration: the matching results of two sampling rates show that under the condition of edge points having the same precision, the higher the sampling rate is, more likely the more accurate matching precision will be obtained.

Therefore, in the process of matching, in order to ensure the calculation efficiency, the results of reducing the sampling rate were used for the first few iterations; then all the edge points were used; when there was a certain iteration and the model parameters were stable, the interpolation results were used for the matching in order to improve the matching accuracy. It should be noted that the corners extracted in (3) are still retained while reducing the sampling, which means only reducing the sampling rate to the edge points between the corners.

2.3   Coarse Matching by Using Multiple Linear Features

In remote sensing image processing, two-dimensional or three-dimensional similarity transformation can approximately replace many 2D or 3D transformations; because there is less the transformation parameter, it can independently make the solving and widely used [2]. This paper takes 2D similarity transformation (formula (3)) as the model for coarse matching, in order to determine the conjugate CLFs, intersection and corner and provide the initial value for accurate matching.

$$\begin{pmatrix} x_2 \\ y_2 \end{pmatrix} = s \cdot \begin{pmatrix} \cos \alpha & -\sin \alpha \\ \sin \alpha & \cos \alpha \end{pmatrix} \begin{pmatrix} x_1 \\ y_1 \end{pmatrix} + \begin{pmatrix} t_x \\ t_y \end{pmatrix} \tag{3}$$

In the Formula: (x_1, y_1) and (x_2, y_2) are respectively the image points of the conjugate features on the left and right images, similarly hereinafter; s, α and (t_x, t_y) are respectively the scaling, rotation and translation parameters between the images.

In the coarse matching by a variety of relatively stable features extracted in Sect. 2.2, first identify the conjugate features to be selected of each feature of the left and right images according to their similarity measure and then make the combination of every two to get a number of groups of combinations of the conjugate features to be selected (hereinafter referred to as "combination to be selected"); then gradually eliminate the combination with false matching through the following two methods (that is to say, at least one pair of the conjugate features to be selected is not the correct conjugate feature, hereinafter referred to as "false matching combination").

(1) Elimination of False Matching Combination Using the Geometry Information

Since the 2D similarity transformation is taken as the transformation model, only the scaling, rotation and translation exist between the features through the model transformation, that is to say, the magnitude of various features and the distance between the features show the changes in the same proportion, while the angle between the features has no such changes; based on the above characteristics, the false matching combination can be eliminated by using the area of features to be matched and the distance & angle between the features.

The following is the introduction of the basic principle taking the CLFs as an example: the CLFs of the two images are shown in Fig. 3: set $N_1(p_2, q_1)$, $N_2 (p_2, q_2)$, $N_3 (p_3, q_3)$ and $N_4 (p_4, q_4)$ as four pairs of conjugate features to be selected determined according to the similarity measure, and make the combination of every two to get 6 combinations: N_1N_2, N_1N_3, N_1N_4, N_2N_3, N_2N_4, N_3N_4; then orderly calculate the maximum difference scaling of each combination as shown in Formula (4) and identify whether they meet the ratio threshold T_s; if yes, then keep the combination; if no, it should be considered as the false matching and eliminated.

$$S = \max_{i, j=1, 2, 3;\ i \neq j} \left| S_i - S_j \right| \leq T_s \tag{4}$$

In which: $S_1 = L_{p_2p_3} \big/ L_{q_1q_3}$, $S_2 = \sqrt{A_{p_2} \big/ A_{p_3}}$, $S_3 = \sqrt{A_{q_1} \big/ A_{q_3}}$, L represents the length of connection of center of gravity conjugate to two CLFs, S represents the area of the conjugate CLFs.

Through the above method, the combination containing N_1 - the conjugate features to be selected can be easily eliminated. The principle of eliminating the combination of LFI and LFC is similar and the difference is that the angle between the feature point main directions and the angle between the feature points connection; it will not be detailed here.

(2) Elimination of False Matching Combination Based on Space Distribution of the Transformation Parameters

After preliminary elimination of false matching combination, each pair kept is used to determine a set of transformation parameters; the analysis shows that the parameter determined by the combination composed of two pairs of correct conjugate features will be very close to the real value, i.e. the relatively concentrated distribution; however, there is large difference between the parameters determined by false matching combination, so the transformation parameters determined by each combination are considered as the points of high dimensional space (R^4); and K-D tree and K-means clustering are successively applied to further eliminate the false matching combination.

After the establishment of K-D tree by use of the space points, the closest point of each point (that is the transform parameter with minimum difference) should be searched and the combination of the space points with the distance from the closest point more than the distance threshold T_d should be eliminated; this method can eliminate most of the false matching combinations. And then K-means is applied to divide the remaining points into two types; if the distance of center of two types of points on the various dimensions is less than the threshold T_l, it should be considered small difference of the remaining points and the center of the remaining points is taken as the transform parameter determined by coarse matching, so as to determine the conjugate features; otherwise the one with more points should be kept and K-means method is used to make the classification, until the distance between the center of two types of points meets the threshold requirement or the number of remaining points is less than 3; for the latter case, it should be considered that the conjugate features were not found.

The experiment and analysis show that the CLFs have moderate quantity and stable performance. In order to improve the efficiency, the CLFs should be first matched in the coarse matching; when there are more than three pairs of correct conjugate features realizing the matching, they will be taken as prior reference information to determine the conjugate intersection and corner; otherwise, the above method is further applied to match the intersection and corner extracted, in order to determine the conjugate intersection and corner of two images.

2.4   Accurate Matching Based on Multi-level Two-Dimensional ICP

Based on the principle of the ICP method, this paper proposed an accurate matching method using multi-level two-dimensional ICP. The initial value of the transformation model is calculated according to the conjugate features identified by the coarse matching; the sub-pixel edge points with different sampling rates are used successively for matching. Firstly, the sub-pixel edge points with the lowest sampling rate are used to participate in matching. Each edge points in left image are transformed according to the initial parameters, and the edge points closest to the transformed point are searched in the right image by using the K-D tree. These closest point pairs are regarded as conjugate points to correct the transformation parameters. The above iterative process should be repeated until it meets the convergence condition. Then, edge points with higher sampling rate are used to participate in matching until the edge points with highest sampling rate complete matching to obtain the transformation parameter. The calculation model is introduced briefly as follows.

This paper chooses the projection transformation model shown in Formula (5) for accurate matching to express the transformation relationship between two images; this model is generally applied in the case of complex or unknown transformation model between two images.

$$
\begin{cases}
x_2 = \dfrac{L_1 x_1 + L_2 y_1 + L_3}{L_7 x_1 + L_8 y_1 + 1} \\
y_2 = \dfrac{L_4 x_1 + L_5 y_1 + L_6}{L_7 x_1 + L_8 y_1 + 1}
\end{cases}
\tag{5}
$$

In which: $L = (L_1\ L_2\ L_3\ L_4\ L_5\ L_6\ L_7\ L_8)^{\mathrm{T}}$ is the projection transformation parameter.

Both sides of the equation are multiplied by the denominator at the same time can obtain the linear form as follows:

$$
\begin{cases}
F_{dx} = (L_7 x_1 + L_8 y_1 + 1) \cdot x_2 - L_1 x_1 + L_2 y_1 + L_3 \\
F_{dy} = (L_7 x_1 + L_8 y_1 + 1) \cdot y_2 - L_4 x_1 + L_5 y_1 + L_6
\end{cases}
\tag{6}
$$

Error equation and normal equation are shown in Formula (7) and (8) respectively.

$$V = AL - l\,\mathrm{P} \tag{7}$$

In which: $A = \begin{bmatrix} -x_1 & -y_1 & -1 & 0 & 0 & 0 & x_1x_2 & y_1x_2 \\ 0 & 0 & 0 & -x_1 & -y_1 & -1 & x_1y_2 & y_1y_2 \end{bmatrix}$; $l = \begin{bmatrix} x_2 \\ y_2 \end{bmatrix}$; P is weight matrix, P_i is determined by the distance between the conjugate points.

$$L = (A^T PA)^{-1} \times (A^T Pl) \tag{8}$$

Through the above adjustment model, the transformation parameter is constantly corrected until the change of the parameter meets the demand of threshold.

3   Experimental Results and Analysis

In order to validate the performance of this proposed method, the following three groups of experiments were designed.

Experiment 1: using the simulated data to validate the stability and precision of coarse matching

In order to validate the stability of various features for coarse matching and the reliability and accuracy of coarse matching method, two groups of matching experiments of the images under different rotation angles and scaling factors were designed. Part of an urban area aerial image of 339×353 pixels (Fig. 4(a)) was selected as reference image. Simulated images shown in Fig. 4(b) and (c) were obtained by using the similar transformation parameters in Table 1, they respectively contain a small and large rotation angle.

 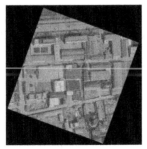 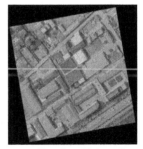

| (a) Reference image | (b) Simulated image 1 | (c) Simulated image 2 |

Fig. 4. Reference image and simulated images generated by the similarity transformation

Two simulated images were provided with the coarse matching to the reference image based on the proposed method; the features extracted and the matching results are shown in Table 2. The first column represents the feature number extracted from the reference image. The three columns below simulated images represent the feature number extracted from simulated image, the true conjugate features number, and the conjugate features number determined by coarse matching and the correct conjugate features which is shown in parentheses. The parameters determined by coarse matching are shown in Table 1.

Table 1. Comparison between the actual transformation parameters and the transformation parameters determined by coarse matching

	Simulated image 1		Simulated image 2	
	Actual parameter	Coarse matching	Actual parameter	Coarse matching
s	1.2000	1.2004	1.1000	1.1010
α (°)	20.0000	19.9978	170.0000	170.0108
t_x (pixel)	180.00	179.92	450.00	450.09
t_y (pixel)	30.00	29.87	400.00	400.27

Table 2. The number of the features which used for coarse matching

	Reference image	Simulated image 1			Simulated image 2		
		Extracted	True	Matched	Extracted	True	Matched
CLF	11	11	4	4 (4)	16	6	6 (6)
LFI	8	14	0	0 (0)	10	1	1 (1)
LFC	64	123	33	33 (33)	92	35	33 (33)

For quantitative evaluation of the matching precision, checkpoints are picked up from the reference image interval of 10 pixels both in row and column direction (a total of 1,155 checkpoints). Then the actual parameters and the parameters determined by coarse matching were used to compute the theoretical coordinates and matching coordinates of the checkpoints in the simulated image. Matching precision is evaluated through the difference between these coordinates. These checkpoints were distributed evenly over the entire image, therefore, they have good performance to evaluate the precision. Quantitative evaluation results (root-mean-square error, RMSE) in column direction (X), row direction(Y) and overall (XY) on these two group experiment data are shown in Table 3.

Table 3. Quantitative evaluation results on the two groups experiment data (pixel)

	RMSE_X	RMSE_Y	RMSE_XY
Simulated image 1	0.70	0.48	0.85
Simulated image 2	0.44	0.57	0.71

As shown in Table 2, the number of CLFs is moderate, CLFs have the largest number. And these two features have good stability for rotation, scaling transformation and noise. The matching results show that all these three kinds of features have good matching performance, high matching precision, low leakage matching rate and without mismatching. Tables 1 and 3 show that the parameters determined by coarse matching are very close to the actual parameters; the error of the checkpoints in row and column direction are both small, the total RMSE is less than 1 pixel.

The experiment results show that the extracted features have stable performance and the coarse matching can effectively eliminate the false matching to ensure high precision and provide more accurate matching initial value for accurate matching.

Experiment 2: using the simulated data generated by affine transformation to validate the accuracy of this method

This experiment utilizes the affine transformation to generate the simulated images; by matching it with the reference image and comparing the matching accuracy between it and SIFT, ASIFT, MSER and least square matching (LSM) based on SIFT. This experiment is designed to validate the matching accuracy of this method. The simulated image obtained from the affine transformation (transformation parameter as shown in Table 4) of the image shown in Fig. 4(a) is shown in Fig. 5(a); the features used for coarse matching are shown in Fig. 5(b) and (c), in which the white area and the asterisk respectively represent the CLF and its center of gravity; blue dot and line respectively represent the LFI and the direction of its branches; red star and red line with arrowhead respectively represent the LFC and its main direction, similarly hereinafter. The results of coarse and accurate matching are respectively shown in Fig. 5(d) and (e). Because there were too many conjugate points determined by the accurate matching (a total of 7,921 pairs), the connection between the conjugate points was not made.

Table 4. Comparison of the matching results between different methods

	a_{11}	a_{12}	a_{13}	a_{21}	a_{22}	a_{23}
Actual parameter	1.0200	−0.1000	50.0000	0.0500	1.0500	10.0000
SIFT	1.0200	−0.1000	49.9060	0.0501	1.0500	10. 0956
ASIFT	1.0201	−0.0999	49.9017	0.0500	1.0501	9.9218
MSER	1.0196	−0.0999	50.0605	0.0498	1.0502	10.0045
SIFT + LSM	1.0200	−0.1000	49.9949	0.0500	1.0500	9.9987
Coarse matching	1.0186	−0.1003	50.3219	0.0481	1.0497	10.1518
Accurate matching	1.0198	−0.1000	50.0417	0.0500	1.0500	9.9844

The parameters determined by various matching methods are shown in Table 4. Quantitatively evaluate the accuracy of each matching method use the same way with Experiment 1, with the results shown in Table 5 below.

The experiment result shows that the extracted features used for coarse matching evenly distributed, based on which the coarse matching has higher accuracy; because a few steps were taken to eliminate the false matching in the process of the coarse matching, there was higher correct rate of matching; in the process of accurate matching, there were a lot of sub-pixel edge points to participate in the matching, which can reach the sub-pixel accuracy equal to that of least square matching. In

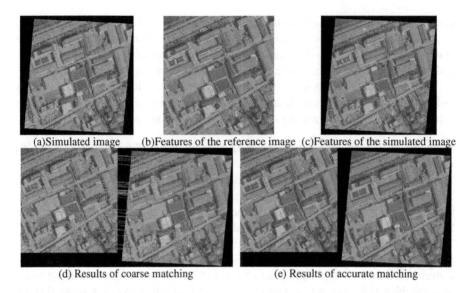

(a)Simulated image   (b)Features of the reference image (c)Features of the simulated image

(d) Results of coarse matching   (e) Results of accurate matching

Fig. 5. Matching between the images with the affine transformation relationship

addition, the experiment result also shows that the proposed method has certain adaptability to the affine transform. When there was not severe affine deformation between the images, it still could realize high accuracy matching.

Table 5. Quantitative evaluation results of the checkpoints on various matching methods (pixel)

	SIFT	ASIFT	MSER	SIFT + LSM	Coarse matching	Accurate matching
RMSE_X	0.09	0.06	0.04	0.01	0.14	0.02
RMSE_Y	0.10	0.06	0.03	0.00	0.30	0.01
RMSE_XY	0.14	0.08	0.05	0.01	0.33	0.02

Experiment 3: Remote sensing stereo images matching

This experiment utilized the IKONOS images of United States of California Santa Barbara (UCSB) campus, which consists of such ground object types as building, sea and lake, with the image size of 1,024 × 1,024 pixels and 1,064 × 1,172 pixels respectively; the experimental images, the features for coarse matching, the results of coarse matching and accurate matching are shown in Fig. 6.

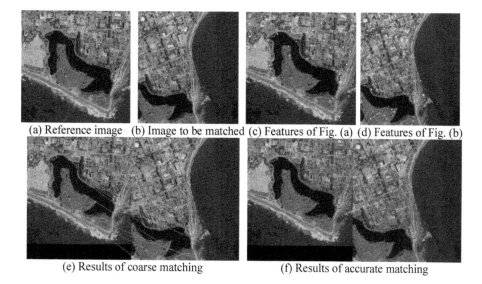

(a) Reference image   (b) Image to be matched (c) Features of Fig. (a) (d) Features of Fig. (b)

(e) Results of coarse matching   (f) Results of accurate matching

Fig. 6. Results of aerial stereo image matching

Theory analysis and the results of the former two experiments indicate that the LSM matching has very high matching precision and can reach 1/10 to 1/100 pixel. Therefore, the parameters determined by SIFT-LSM are regard as actual parameters in this experiment. The matching precision was compared by the quantitative evaluation method of aforementioned groups of experiments. The checkpoints error statistics are shown in Table 6.

Table 6. Quantitative evaluation results of the checkpoints on IKONOS Image (pixel)

	SIFT	ASIFT	MSER	Coarse matching	Accurate matching
RMSE_X	0.35	0.14	0.19	0.12	0.02
RMSE_Y	0.52	0.44	0.34	0.18	0.01
RMSE_XY	0.62	0.46	0.39	0.22	0.02

Figure 6 and Table 6 show that this method can achieve precision matching between high-resolution remote sensing images which only partially overlapping. Coarse matching has comparative accuracy with classical point feature matching methods such as SIFT, ASIFT, MSER, and accurate matching has comparative accuracy with LSM method. Experiment results also fully demonstrate that the proposed method has stable performance of features matching and the matching strategy is feasible to achieve higher matching accuracy and precision.

4 Conclusions

In the study of FFLF matching, full use and effective description of information in the features constitute a conflict, which restricts the development of this kind of method to a certain extent. In addition, because only the linear features are utilized, this method is more sensitive to the error of the linear feature extraction, of which the reliability and accuracy, to a great extent, depend on the reliability and accuracy of the linear feature extraction. In order to better solve the above problems, this paper proposed a remote sensing image matching method using the strategy of hierarchical matching. First the edge features of the image were detected and tracked, in order to extract the free-form sub-pixel linear features with better continuity; then the CLFs, LFI and corner were extracted for coarse matching and the false matching was gradually eliminated based on the area, angle and other geometry information as well as the parameter distribution of the model determined the conjugate features combination to be selected; finally, based on multi-level two-dimensional ICP, sub-pixel edge points were orderly used with the sampling rate from low to high for accurate matching. Experimental results show that this method has stable performance for the coarse matching; high accuracy and precision of the coarse matching can provide the initial matching parameters of high precision for accurate matching; accurate matching can reach the sub-pixel level precision equal to that of the least square matching.

The features extracted for coarse matching in this paper have good stability under the smaller affine condition and can better realize the matching between the images with smaller affine deformation. Further study on FFLF is still needed. The features that can remain the same even under the severe affine deformation may be extracted in future, so as to improve the affine resistance of FFLF matching.

References

1. Zhang, Y.S., Zhu, Q., Wu, B., Zhou, Z.R.: A hierarchical stereo line matching method based on a triangle constraint. Geomatics Inf. Sci. Wuhan Univ. **38**, 522–527 (2013)
2. Vassilaki, D.I., Ioannidis, C.C., Stamos, A.A.: Automatic ICP-based global matching of free-form linear features. Photogram. Rec. **27**, 311–329 (2012)
3. Xiao, J., Shah, M.: Two-frame wide baseline matching. In: 9th IEEE International Conference on Computer Vision (ICCV 2003), pp. 603–609. IEEE Press, New York (2003)
4. Zuliani, M., Bertelli, L., Kenney, C.S., Chandrasekaran, S., Manjunath, B.S.: Drums, curve descriptors and affine invariant region matching. Image Vis. Comput. **26**, 347–360 (2008)
5. Wu, F.C., Wang, Z.H., Wang, X.G.: Feature vector field and feature matching. Pattern Recogn. **43**, 3273–3281 (2010)
6. Ng, E.S., Kingsbury, N.G.: Matching of interest point groups with pairwise spatial constraints. In: 17th IEEE International Conference on Image Processing (ICIP 2010), pp. 2693–2696. IEEE Press, New York (2010)
7. Wu, B., Zhang, Y.S., Zhu, Q.: Integrated point and edge matching on poor textural images constrained by self-adaptive triangulations. ISPRS J. Photogrammetry Remote Sens. **68**, 40–55 (2012)
8. Zhang, G.M., Ma, K., Chu, J.: A new curve matching method based on membrane vibration model. Tien Tzu Hsueh Pao/Acta Electronica Sinica. **41**, 1917–1925 (2013)

9. Saeedi, P., Mao, M.: Two-edge-corner image features for registration of geospatial images with large view variations. Int. J. Geosci. **5**, 1324–1344 (2014)
10. Da, F.P., Zhang, H.: Sub-pixel edge detection based on an improved moment. Image Vis. Comput. **28**, 1645–1658 (2010)
11. Chen, X.W., Xu, Z.H., Guo, H.T., Zhang, B.M.: Universal sub-pixel edge detection algorithm based on extremal gradient. Acta Geodaetica Cartogr. Sin. **43**, 500–507 (2014)
12. Chen, X.W., Zhang, B.M., Guo, H.T., Zhang, H.W., Dang, T.: An edge curve extraction method based on sub-pixel edge. J. Geomatics Sci. Technol. **31**, 624–629 (2014)
13. Li, F.Y., Li, Y.J., Zhang, K.: The use of the chain-code technique in extracting feature point in scene image. J. Image Graph. **13**, 114–118 (2008)

Improved Saliency Optimization Based on Superpixel-Wised Objectness and Boundary Connectivity

Yanzhao Wang$^{(\boxtimes)}$ and Guohua Peng$^{(\boxtimes)}$

School of Natural and Applied Sciences,
Northwestern Polytechnical University, Xi'an 710072, China
wangyz@mail.nwpu.edu.cn, penggh@nwpu.edu.cn

Abstract. Salient object detection is one of the most challenging problems in computer vision and has extensive applications in many fields. Most existing algorithms detect salient object by employing various features. In this work, a bottom-up salient region measurement that integrates superpixel-wised objectness and boundary connectivity is proposed. Furthermore, to improve the result of the salient region measurement, an improved saliency optimization is put forward. Experimental results on three benchmark datasets demonstrate the effectiveness of the proposed method, which can further improve the accuracy of saliency detection than other six state-of-the-art approaches on MSRA10k, DUT-OMRON and SED1.

Keywords: Visual attention · Salient object detection · Objectness · Boundary connectivity

1 Introduction

Visual attention is a basic mechanism in human visual system which filters out meaningless visual information and selects the significant objects efficiently. Recent years, Saliency detection has been an important problem in computer vision. It is motivated by the applications such as image segmentation, image retrieval, object detection and image and video compression. Generally speaking, saliency detection methods can be divided into bottom-up [1,3–6] or top-down strategy [7,8], and they also can be categorized as eye fixation [1] or salient object detection [3–7] as well. In this paper, we focus on bottom-up salient object detection.

Recent years, a lot of models have been proposed for saliency object detection. Inspired by the neuronal structure of visual system, a breakthrough work was made by Itti et al. [1]. In the work, saliency object was defined as the difference of center-surround contrast and the model was proposed based on multi-scale image features like color, brightness, direction, etc. A lot of models had been built based on this work later. Perazzi et al. [9] put forward a high-dimensional Gaussian filtering method by exploiting color and position, and then unifying

© Springer Nature Singapore Pte Ltd. 2016
T. Tan et al. (Eds.): CCPR 2016, Part II, CCIS 663, pp. 218–228, 2016.
DOI: 10.1007/978-981-10-3005-5_19

them to obtain the final saliency map. Based on the prior of contrast, Kim et al. [10] introduced a model which represented salient map as a linear combination of high-dimensional color transform so that salient objects and backgrounds could be separated easily. Tong et al. [12] presented a novel coding-based method by exploring both global and local cues. Wei et al. [11] proposed a novel model called geodesic saliency which combined boundary and connectivity priors measured by geodesic distance. The model was easy to realize, and took full advantage of background prior. Since then, background prior had attracted considerable attentions. Wang et al. [13] collected seeds from foreground and background to construct background-based and foreground-based saliency maps respectively, then integrated them as the final saliency map. Zhu et al. [14] proposed boundary connectivity to characterize spatial layout of image regions, which had intuitive geometrical interpretation and achieved comparable results through principled optimization.

In this paper, we present an improved saliency optimization model based on modified objectness and boundary connectivity. The model first segments the image by SLIC (simple linear iterative clustering) [16] and gets original object-ness proposed in [15]. The modified objectness is obtained by averaging the objectness values of pixels in every segment region. Then, a new saliency mea-surement is achieved by fusing the modified objectness and boundary connectiv-ity proposed in [14]. In order to obtain a better result, improved saliency opti-mization is performed on the saliency map. The method's diagram can be seen in Fig. 1 clearly. To demonstrate the effectiveness, we compare our method with six

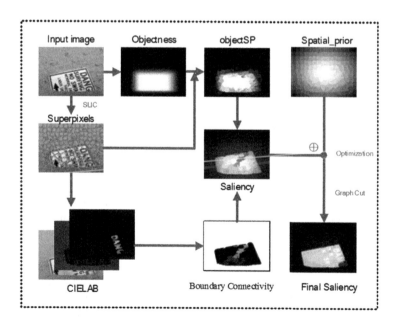

Fig. 1. Diagram of the proposed algorithm

state-of-the-art salient methods on three public benchmark datasets: MSRA10k, DUT-OMRON and SED1. Extensive experiments show that our method can achieve comparable results and outperforms the other six state-of-the-art methods.

2   Backgrounds

In this section, we briefly introduce the related backgrounds that will be used in our model.

2.1   Objectness

Cognitive psychology and neurobiology assume that humans have a strong perceptual ability to perceive targets in their eyes [20]. There is a mechanism called objectness that visual system chooses a rough location before human eyes identifying the targets. Objectness measurement proposed in [15] quantifies how likely it is for a window to contain an object. The higher values indicate the greater probability that the targets appear in the window. Objectness prior can be obtained by sliding windows of different sizes on the image, and accumulating n higher value windows together. For an image I, n windows of highest measured scores can be denoted as $ob(i), i = 1, ..., n$. Therefore, the objectness prior of an image represented by *Object* can be expressed as the frequency that pixel falls into objectness windows:

$$Object = \frac{1}{n} \sum_{i=1}^{n} ob(i) \tag{1}$$

2.2   Boundary Connectivity

In general, salient objects always appear in the center of images, and backgrounds always distribute near image boundaries. Foreground priors usually generated by global and local contrasts or compactness because targets are compact and backgrounds is scattered. However, different from foreground priors, background priors can be achieved in the opposite: the objects can be regarded as the regions having a large difference with image boundaries. Although they have been used in many recent models and obtain better performance than foreground priors, background priors still behave poorly in some situations. For example, good background priors can be acquired when objects and image boundaries are separated, then, when objects connect with image boundaries, the object regions can be easily classified to backgrounds. This problem can be better solved by boundary connectivity proposed in [14], as can be seen below.

For image I, superpixel set $X = \{X_1, ..., X_N\}$ can be obtained through SLIC [16], where N represents the number of superpixels in the image and is usually set as $N = 200$. The method constructs an undirected weighted graph by connecting every adjacent superpixels (p, q), and assign the Euclidean distance $d_{app}(p, q)$ in the CIE-Lab color space as their edge weights, where p, q are regions in X. The geodesic distance $d_{geo}(p,q)$ [11,14] between superpixel p and q can be defined as the accumulated weights along their shortest path on the graph:

$$d_{geo}(p,q) = \min_{p_1=p,...,p_N=q} \sum_{i=1}^{N-1} d_{app}(p_i, p_{i+1}) \tag{2}$$

where $d_{app}(p,p) = 0$. The spanning areas of superpixels can be defined as

$$Area(p) = \sum_{i=1}^{N} \exp\left(-\frac{d_{geo}^2(p,p_i)}{2\sigma^2}\right) = \sum_{i=1}^{N} S(p,p_i) \tag{3}$$

where σ is the scale factor, we set $\sigma = 10$ in our experiments. $S(p,p_i)$ characterizes the contribution of p_i to p. When p_i and p are in the same flat region, $d_{geo}(p,q) = 0$ and $S(p,p_i) = 1$, which indicate that p_i has the largest effect on p and vice versa. The length of path to the boundaries can be defined as

$$Len_{bnd}(p) = \sum_{i=1}^{N} S(p,p_i) \times \delta(p_i \in Bnd) \tag{4}$$

where $\delta(p_i) = 1$ when p_i belongs to image boundaries and $\delta(p_i) = 0$ on the contrary. Finally, the boundary connectivity $BndCon$ can be defined as

$$BndCon(p) = \frac{Len_{bnd}(p)}{\sqrt{Area(p)}} \tag{5}$$

3   Saliency Measurement Based on Modified Objectness and Boundary Connectivity

Through averaging objectness values in superpixel regions, the modified objectness prior $objectSP$ can be obtained in superpixel-wise:

$$objectSP(i) = \frac{1}{N_i} \sum_{k=1}^{N_i} Object(k) \tag{6}$$

where N_i is the number of pixels in superpixel region i. Compared with other objectness, the modified objectness prior proposed in this paper can be more accurate to reflect the spatial distribution and boundary information between foregrounds and backgrounds, as shown in Fig. 2.

Although it can roughly show the position of object, the modified objectness prior cannot distinguish edges between object and background clearly. Oppositely, boundary connectivity $BndCon$ proposed in [14] can recognize the edges effectively. By integrating $objectSP$ and $BndCon$, the saliency measurement $Saliency$ can be obtained as:

$$Saliency(i) = (1 - BndCon(i)) \times objectSP(i) \tag{7}$$

As can be seen in Fig. 2, *Saliency* not only highlights the foreground targets, but also suppress background noises successfully. However, in some images where object and background are similar, the superpixels can be wrongly classified, which will lead to poor saliency object detection in the whole images. Therefore, a global optimization is proposed next.

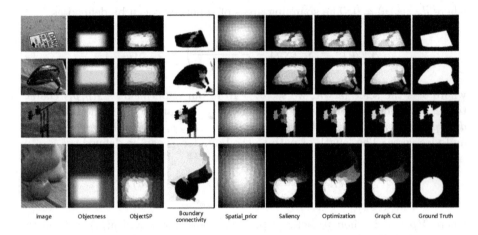

| image | Objectness | ObjectSP | Boundary connectivity | Spatial_prior | Saliency | Optimization | Graph Cut | Ground Truth |

Fig. 2. Results of different steps in the proposed algorithm

4   Improved Saliency Optimization

To enhance the performance of saliency detection and obtain a better result, an improved saliency optimization is proposed for *Saliency*.

Spatial prior based on superpixels can be defined as below. Incorporating with objectness and coordinates of superpixels, spatial prior can be obtained different from other methods. The coordinate (x_i, y_i) of superpixel X_i can be expressed as the mean of pixels coordinates in the same superpixel region. According to superpixels segmentation and their coordinates, spatial prior represented by *spatial_prior* can be defined as

$$spatial\_prior(i) = \exp\left(-\frac{(x_i - px)^2 + (y_i - py)^2}{2\sigma^2}\right) \tag{8}$$

where (px, py) is center coordinate of the image.

The saliency values of superpixels can be represented as $S = \{s_i\}_{i=1}^N$, and the energy function E for optimization can be defined as

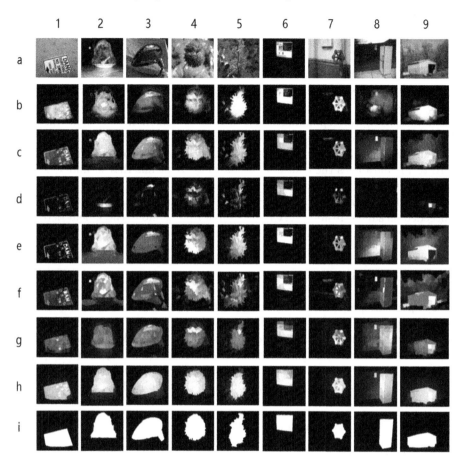

Fig. 3. Comparison of different methods on MSRA10k, DUT-OMRON and SED1. (a) Input image, (b) DSR, (c) RBD, (d) SF, (e) MR, (f) GS, (g) SDGL, (h) Ours, and (i) Ground truth. Such comparisons suggest that our method produces better results closer to ground truth.

$$E = \sum_{i=1}^{N}((1 - spatial\_prior(i))s_i^2 + (1 - saliency(i))s_i^2)$$
$$+ \sum_{i=1}^{N}(spatial\_prior(i)(s_i - 1)^2 + saliency(i)(s_i - 1)^2) \qquad (9)$$
$$+ \lambda \sum_{i,j} w_{ij}(s_i - s_j)^2$$

where
$$w_{ij} = \exp\left(\frac{d_{app}^2(p, p_i)}{2\sigma^2}\right) \qquad (10)$$

σ is the scale factor, and λ is the weight of smooth item controlling the weight of constraint. When $\sigma = 10$ and $\lambda = 4$, we can get the best result. After minimizing the energy function E, the final salient map can be obtained. The first term of E reflects the relation between saliency score and background prior in which background prior is represented by $(1 - spatial\_prior(i))$ and $(1 - saliency(i))$. When background prior of superpixel X_i is large, in order to minimize E, the saliency score s_i corresponding to X_i should be small. In the same way, the second term denotes the relation between saliency score and foreground prior, and the saliency values s_i tend to be one when the foreground prior is large. The third term guarantees the continuity of saliency. It means that s_i and s_j should be similar when X_i and X_j are in the same region.

To further improve the smoothness of salient detection, graph cut [2,12] is proceeded on salient results. As can be seen in Figs. 2 and 8, the salient detection can get further improvement after smoothing.

5   Experiments

We compare the proposed method with stat-of-the-art methods on three benchmark datasets: MSRA10k [4], DUT-OMRON [17] and SED1 [18]. Images contained in these three datasets generally have only one salient object, and all these images have the pixel-wised ground truth correspondingly. In this paper, six stat-of-the-art methods are selected to compare with the proposed method: RBD [14], SF [9], MR [17], GS [11], SDGL [12], and DSR [19].

5.1   Qualitative Evaluation

Fig. 3 shows some results of different methods in three datasets. From the comparison, we can find the proposed method preserves contours more clearly and highlights the whole salient objects more accurately in the image where background is less complicated than others. Besides that, the foreground and background in our results are smoother than others. For images which have complicated backgrounds and different size of objects in column 5, 8 and 9, our method can still get comparable results than others.

5.2   Quantitative Evaluation

For quantitative evaluation, Precision-Recall (PR) curve, Mean Absolute Error (MAE), F-Measure and Area under ROC curve (AUC) are employed to do the comparison.

We show the results in three datasets under different evaluation criterions respectively. In Fig. 4, the proposed method achieves the best performance on MSRA10k under the evaluation of PR curve and F-measure. As can be seen in Fig. 5, PR curve and F-measure show that the proposed method performs slightly better than RBD and outperforms other methods in DUT-OMRON dataset. In Fig. 6, although it performs weak than MR and SDGL in Recall values between

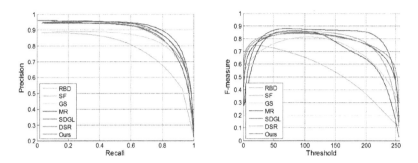

Fig. 4. Evaluation results on the MSRA10k. (left) PR curve and (right) F-measure.Evaluation results on the MSRA10k. (left) PR curve and (right) F-measure.

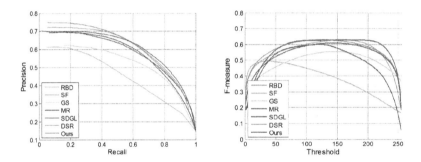

Fig. 5. Evaluation results on the DUT-OMRON. (left) PR curve and (right) F-measure.

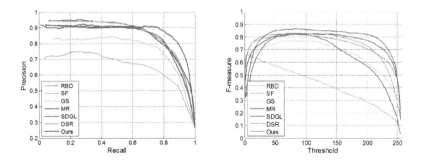

Fig. 6. Evaluation results on the SED1. (left) PR curve and (right) F-measure.

[0, 0.6], our algorithm still can achieve a better result than others. As show in Fig. 7, MAE and AUC verify the effectiveness of the proposed method. Although it is not the best on DUT-OMRON on MAE, the proposed method achieves the best on AUC. Above all, our algorithm obtains the a better performance in MSRA10k and SED1, and reaches the advanced level in DUT-OMRON as well.

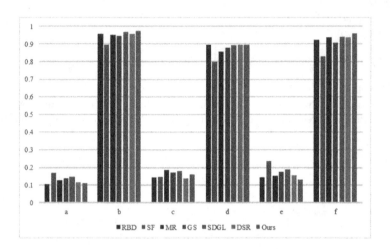

Fig. 7. Evaluation results of MAE and AUC. (a) MAE of MSRA10k, (b) AUC of MSRA10k, (c) MAE of DUT-OMRON, (d) AUC of DUT-OMRON, (e) MAE of SED1, (f) AUC of SED1.

5.3 Model Analysis

The steps of *objectSP*, optimization and smoothing contribute to the performance of our method. Figure 8 shows the results of different steps in SED1 for comparison. Obviously, *objectSP*, optimization and smoothing can respectively improve the performance in different degrees. Although obtaining the worst results among three steps, the method without optimization can still perform better than other methods.

5.4 Run Time

The run time comparison of six state-of-the-arts methods on three datasets is shown in Table 1. As shown in Table 1, our method is time consuming. It is caused by the computation of Objectness. The run time of Objectness accounts for more than 80 % of the whole time.

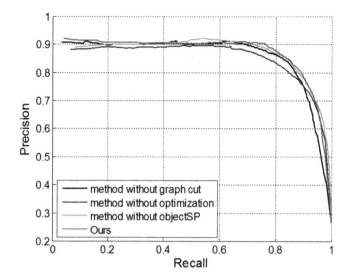

Fig. 8. Evaluation of various steps in our method.

Table 1. Comparison of Run time

Run time (s)	DSR	RBD	SF	MR	GS	SDGL	Ours
MSRA	6.0920	0.2385	0.2228	0.2163	0.2154	5.5932	4.0916
DUT-OMRON	6.1433	0.2418	0.2226	0.2316	0.2176	5.7377	4.0860
SED1	5.1607	0.2009	0.1912	0.1874	0.1905	5.9102	4.3761

6   Conclusion

In this paper, an improved saliency optimization method based on superpixel-wised objectness and boundary connectivity is proposed. The method obtains the salient object measurement by integrating the superpixel-wised objectness measurement and boundary connectivity, and then performs global optimization on the saliency measurement to improve the performance of saliency detection. Experimental results on three benchmark datasets show that our algorithm outperforms the other six state-of-the-arts methods, and is robust to the complicated images. In future, our work will focus on features and theories that will overcome the existing limitation and improve the performance of salient object detection.

References

1. Itti, L., Koch, C., Niebur, E.: A model of saliency-based visual attention for rapid scene analysis. IEEE Trans. Pattern Anal. Mach. Intell. **20**, 1254–1259 (1998)
2. Judd, T., Ehinger, K., Ehinger, F., Durand, F., Torralba, A.: Learning to predict where humans look. In: 12th IEEE International Conference on Computer Vision, Kyoto, pp. 2106–2113 (2009)

3. Chang, K.Y., Liu, T.L., Chen, H.T., Lai, S.H.: Fusing generic objectness and visual saliency for salient object detection. In: 13th IEEE International Conference on Computer Vision, Barcelona, pp. 914–921 (2011)
4. Cheng, M.M., Mitra, N.J., Huang, X., Torr, P.H.S., Hu, S.M.: Global contrast based salient region detection. IEEE Trans. Pattern Anal. Mach. Intell. **37**, 569–582 (2011)
5. Wang, L., Xue, J., Zheng, N., Hua, G.: Automatic salient object extraction with contextual cue. In: 13th IEEE International Conference on Computer Vision, Barcelona, pp. 105–112 (2011)
6. Lu, Y., Zhang, W., Lu, H., Xue, X.: Salient object detection using concavity context. In: 13th IEEE International Conference on Computer Vision, Barcelona, pp. 233–240 (2011)
7. Liu, T., Yuan, Z., Sun, J., Wang, J., Zheng, N., et al.: Learning to detect a salient object. IEEE Trans. Pattern Anal. Mach. Intell. **33**, 353–367 (2011)
8. Yang, J., Yang, M.H.: Top-down visual saliency via joint CRF and dictionary learning. In: 25th IEEE Conference on Computer Vision and Pattern Recognition, Providence, pp. 2296–2303 (2012)
9. Perazzi, F., Krähenbühl, P., Pritch, Y., Hornung, A.: Saliency filters: contrast based filtering for salient region detection. In: 25th IEEE Conference on Computer Vision and Pattern Recognition, Providence, pp. 733–740 (2012)
10. Kim, J., Han, D., Tai, Y.W., Kim, J.: Salient region detection via high-dimensional color transform. In: 27th IEEE Conference on Computer Vision and Pattern Recognition, Columbus, pp. 883–890 (2014)
11. Wei, Y., Wen, F., Zhu, W., Sun, J.: Geodesic saliency using background priors. In: 12h European Conference on Computer Vision, Firenze, pp. 29–42 (2012)
12. Tong, N., Lu, H., Zhang, Y., Ruan, X.: Salient object detection via global and local cues. Pattern Recogn. **48**, 3258–3267 (2015)
13. Wang, J., Lu, H., Li, X., Tong, N., Liu, W.: Saliency detection via background and foreground seed selection. Neurocomputing **152**, 359–368 (2015)
14. Zhu, W., Liang, S., Wei, Y., Sun, J.: Saliency optimization from robust background detection. In: 27th IEEE Conference on Computer Vision and Pattern Recognition, Columbus, pp. 2814–2821 (2014)
15. Alexe, B., Deselaers, T., Ferrari, V.: What is an object? In: 24th IEEE Conference on Computer Vision and Pattern Recognition, San Francisco, pp. 73–80(2010)
16. Achanta, R., Shaji, A., Smith, K., Lucchi, A., et al.: SLIC superpixels compared to state-of-the-art superpixel methods. IEEE Trans. Pattern Anal. Mach. Intell. **34**, 2274–2282 (2012)
17. Yang, C., Zhang, L., Lu, H., Ruan, X., Yang, M.H.: Saliency detection via graph-based manifold ranking. In: 26th IEEE Conference on Computer Vision and Pattern Recognition, Portland, pp. 3166–3173 (2013)
18. Alpert, S., Galun, M., Basri, R., Brandt, A.: Image segmentation by probabilistic bottom-up aggregation and cue integration. IEEE Trans. Pattern Anal. Mach. Intell. **34**, 315–327 (2012)
19. Li, X., Lu, H., Zhang, L., Ruan, X., Yang, M.H.: Saliency detection via dense and sparse reconstruction. In: 14th IEEE International Conference on Computer Vision, Sydney, pp. 2976–2983 (2013)
20. Borji, A., Cheng, M.M., Jiang, H., Li, J.: Salient object detection: a benchmark. IEEE Trans. Image Process. **24**, 5707–5722 (2015)

A Dual-Based Adaptive Gradient Method for TV Image Denoising

Yan Liao[1(✉)], Jialin Hua[2], and Wei Xue[3]

[1] School of Mathematics and Computer Sciences, Gannan Normal University,
Ganzhou 341000, China
nhma0004@126.com
[2] School of Computer and Information Technology, Beijing Jiaotong University,
Beijing 100044, China
[3] School of Computer Science and Engineering,
Nanjing University of Science and Technology, Nanjing 210094, China
http://www.gnnu.edu.cn

Abstract. The total variation model proposed by Rudin, Osher and Fatemi for image denoising is considered to be one of the best denoising models. In this paper, we propose, analyze and test a dual based method to solve the total variation model. This method minimizes a quadratic function with separable constraints, and we make projections onto a feasible set so that it is easy to compute. Under some mild conditions, global convergence of the proposed method is established. The proposed approach could be easily implemented. Preliminary results are reported to demonstrate its performance.

Keywords: Image denoising · Total variation · Nonmonotone line search · Gradient projection

1 Introduction

Image denoising is an important research field in image processing. It forms a significant preliminary step in many machine vision tasks such as object detection and recognition. Image denoising models based on total variation (TV) have become very popular since their introduction by Rudin, Osher and Fatemi (ROF) [17] in 1992. ROF model was designed with the explicit goal of preserving sharp discontinuities in an image while removing noise. This model has proven to be successful in a wide range of applications in image processing.

There exists lots of methods to solve these TV-based models, for instance, primal-dual Newton-based methods, primal-dual active set methods [6,14,27], split Bregman method [13], interior point algorithm [20], second-order cone programming [12], duality-based method [3–5,24], augmented Lagrangian method and alternating directions algorithm [21,22], and references therein.

One method of particular interest to this paper is the dual approach [4,5] introduced by Chambolle etc. The dual formulation of ROF model has a

T. Tan et al. (Eds.): CCPR 2016, Part II, CCIS 663, pp. 229–244, 2016.
DOI: 10.1007/978-981-10-3005-5_20

quadratic objective with separable constraints, making projections onto a feasible set easy to compute. Thanks to this merit, we also focus on the duality based method in this paper. The rest of this paper is organized as follows. In Sect. 2, we give a brief review of the dual formulation of ROF model. In Sect. 3, we state the proposed method along with some remarks, and establish its global convergence under some mild conditions. We report some numerical experiments to show the efficiency of the proposed method in Sect. 4. Finally, we conclude this paper.

2   Dual Formulation

It is well known that the earliest ROF model was designed with the explicit goal of preserving sharp discontinuities in an image while removing noise and other unwanted fine-scale detail. They formulated the following optimization problem

$$\min_u \int_{\mathcal{D}} |\nabla u| \ s.t. \ \|u - f\|_2^2 \leq \sigma^2, \tag{1}$$

where u is the image which we want to restore, $\mathcal{D}$ is the domain of the image definition, which will be assumed to be a bounded domain in R^n with Lipschitz boundary. Frequently, $\mathcal{D}$ is simply a rectangle in R^2. $f: \mathcal{D} \to R$ denotes the observed image, σ^2 is an estimate of the variance of the noise in f. The notation ∇ is the gradient operator. The objective function in (1) is the TV semi-norm of u. Optimization methods often find a solution of (1) by solving the following unconstrained optimization problem

$$\min_u P(u) := \frac{\mu}{2}\|u - f\|_2^2 + \int_{\mathcal{D}} |\nabla u| dx, \tag{2}$$

where $\mu > 0$ is related to the Lagrange multiplier of the constraint in (1), which measures the trade-off between a good fit and a regularized solution. To derive the dual form of ROF model, we first notice that the TV semi-norm has the following equivalent form

$$\int_{\mathcal{D}} |\nabla u| = \max_{w \in C_0^1(\mathcal{D}), |w| \leq 1} \int_{\mathcal{D}} \nabla u \cdot w$$

$$= \max_{|w| \leq 1} \int_{\mathcal{D}} -u \nabla \cdot w, \tag{3}$$

where $w: \mathcal{D} \to R^2$ is the dual variable of u, and $\nabla \cdot$ denotes the divergence operator. Before describing our method, let us fix our main notational conventions. For simplicity, we assume that the domain $\mathcal{D}$ is the unit square $[0,1] \times [0,1]$, and define a discretization via a regular $n \times n$ grid of pixels, indexed as (i,j), for $i = 1, 2, \cdots, n$ and $j = 1, 2, \cdots, n$. We assume that our images are two dimensional matrices of size $n \times n$,

An image u can be regarded as a function

$$u: \{1, 2, \cdots, n\} \times \{1, 2, \cdots, n\} \to R, \tag{4}$$

where $n \geq 2$. To define the discrete total variation, we introduce a discrete gradient operator. The gradient ∇u is a vector given by

$$(\nabla u)_{i,j} = ((\nabla u)_{i,j}^x, (\nabla u)_{i,j}^y), \tag{5}$$

whose two components at each pixel (i,j) are defined as follows

$$(\nabla u)_{i,j}^x = \begin{cases} u_{i+1,j} - u_{i,j} & \text{if } i < n \\ 0 & \text{if } i = n \end{cases} \tag{6}$$

$$(\nabla u)_{i,j}^y = \begin{cases} u_{i,j+1} - u_{i,j} & \text{if } j < n \\ 0 & \text{if } j = n \end{cases} \tag{7}$$

where $u_{i,j}$ denotes the value of the function u at the point (i,j), $(\nabla u)^x$ and $(\nabla u)^y$ denote the two components in the $x-$direction and $y-$direction respectively. Hence, the discrete total variation of u can be represented by the following form

$$TV(u) := \sum_{1 \leq i,j \leq n} |(\nabla u)_{i,j}| \tag{8}$$

with $|x| = \sqrt{x_1^2 + x_2^2}$ for all $x \in R^2$. Therefore, the discretization of (2) can be written as

$$\min_u \frac{\mu}{2} \|u - f\|_2^2 + TV(u), \tag{9}$$

For $(\nabla u)^x$ and $(\nabla u)^y$, the corresponding discrete divergence operator can be defined explicitly as follows

$$(\nabla \cdot w)_{i,j} = (\nabla \cdot w)_{i,j}^x + (\nabla \cdot w)_{i,j}^y, \tag{10}$$

where

$$(\nabla \cdot w)_{i,j}^x = \begin{cases} w_{i,j}^x & \text{if } i = 1, \\ w_{i,j}^x - w_{i_1,j}^x & \text{if } 1 < i < n, \\ -w_{i_1,j}^x & \text{if } i = n, \end{cases}$$

$$(\nabla \cdot w)_{i,j}^y = \begin{cases} w_{i,j}^y & \text{if } j = 1, \\ w_{i,j}^y - w_{i,j-1}^y & \text{if } 1 < k < n, \\ -w_{i,j-1}^y & \text{if } j = n. \end{cases}$$

By (3), the ROF model becomes

$$\min_u \max_{w \in C_0^1(\mathcal{D}), |w| \leq 1} \int_{\mathcal{D}} -u \nabla \cdot w + \frac{\mu}{2} \|\mu - f\|_2^2. \tag{11}$$

The min-max theorem (see [10] for example, Proposition 2.4 in Chapter 6) allows us to interchange the min and max, to obtain

$$\max_{w \in C_0^1(\mathcal{D}), |w| \leq 1} \min_u \int_{\mathcal{D}} -u \nabla \cdot w + \frac{\mu}{2} \|\mu - f\|_2^2. \tag{12}$$

The inner minimization problem can be solved exactly as follows

$$u = f + \frac{1}{\mu}\nabla \cdot w, \tag{13}$$

which leading to the following dual formulation

$$\max_{w \in C_0^1(\mathcal{D}), |w| \leq 1} D(w) := \frac{\mu}{2}(\|f\|_2^2 - \|\frac{1}{\mu}\nabla \cdot w + f\|_2^2), \tag{14}$$

or, equivalently,

$$\min_{w \in C_0^1(\mathcal{D}), |w| \leq 1} \frac{1}{2}(\|\nabla \cdot w + \mu f\|_2^2). \tag{15}$$

3   Motivation and Adaptive Gradient Method

To begin, we give some notation. $\|\cdot\|$ denotes the Euclidean norm. $\langle\cdot\rangle$ denotes the inner product of two vectors. g_k is the abbreviation of $g(w_k)$, a column vector, which denotes the gradient of $\Psi(w)$ at point w_k.

3.1   Motivation

As the analysis in [24], both Chambolle's semi-implicit gradient descent method[1] [4]

$$w_{i,j}^{k+1}(w_k, \tau, g_k)$$
$$= \frac{w_{i,j}^k + \tau[\nabla(\nabla \cdot w_k + \mu f)]_{i,j}}{1 + |\nabla(\nabla \cdot w_k + \mu f)_{i,j}|}, \quad (\forall\ i, j) \tag{16}$$

and Chambolle's projected gradient method[2] [5]

$$w_{i,j}^{k+1}(w_k, \tau, g_k)$$
$$= \frac{w_{i,j}^k + \tau[\nabla(\nabla \cdot w_k + \mu f)]_{i,j}}{\max\{1, |w_{i,j}^k + \tau[\nabla(\nabla \cdot w_k + \mu f)]_{i,j}|\}}, \quad (\forall\ i, j) \tag{17}$$

require the functional value to decrease monotonically at each iteration, which made the iterates of Chambolle's methods slowly approach the minimum in cases when the problem is very ill conditioned. In order to accelerate Chambolle's projected gradient method, we replace the time step τ by a small steplength or a large steplength alternatively at each iteration.

---

[1] Suppose that $0 < \tau < 1/8$, then $\frac{1}{\mu}\nabla \cdot w_k$ converges to $\pi_{\frac{1}{\mu}\mathcal{K}}(-f)$ as $k \to \infty$, where $\pi_{\frac{1}{\mu}\mathcal{K}}(\cdot)$ is the orthogonal projection onto a convex set $\frac{1}{\mu}\mathcal{K}$ with $\mathcal{K} := \{\nabla \cdot w : |w_{i,j}| \leq 1|, \forall i, j = 1, 2, \cdots, n\}$.

[2] Some experiments show that a better convergence can be obtained when $\tau = 0.248$.

Choose a starting point w_0, the gradient method for $\min_{w \in R^n} \Psi(w)$ can be defined by the iteration $w_{k+1} = w_k - \alpha_k g_k$, $k = 0, 1, 2, \cdots$, where α_k is the steplength which is often determined through an appropriate selection rule. In the classical steepest descent method, the steplength α_k is obtained by minimizing $\Psi(w)$ along the ray $\{w_k - \alpha g_k : \alpha > 0\}$. An surprising result was given by Barzilai and Borwein (1988) in [1], where gives formulae for the steplength α_k which lead to superlinear convergence. The main idea of Barzilai and Borwein's approach is to use the information in the previous interation to decide the steplength in the current iteration. The iteration $w_{k+1} = w_k - \alpha_k g_k$ is viewed as

$$w_{k+1} = w_k - Q_k g_k, \tag{18}$$

where $Q_k = \alpha_k I$. In order to force the matrix Q_k having certain quasi-Newton property, it is reasonable to require either

$$\min \|Q_k^{-1} s_{k-1} - y_{k-1}\|, \tag{19}$$

or

$$\min \|s_{k-1} - Q_k y_{k-1}\|, \tag{20}$$

where $s_{k-1} = w_k - w_{k-1}$, $y_{k-1} = g_k - g_{k-1}$. Because in a quasi-Newton method we have that $w_{k+1} = w_k - B_k^{-1} g_k$ and the quasi-Newton matrix B_k satisfies $B_k s_{k-1} = y_{k-1}$. Now, we can obtain two steplengths from $Q_k = \alpha_k I$ and relations (19)–(20)

$$\alpha_k^1 = \frac{\langle s_{k-1}, s_{k-1} \rangle}{\langle y_{k-1}, s_{k-1} \rangle} \text{ and } \alpha_k^2 = \frac{\langle y_{k-1}, s_{k-1} \rangle}{\langle y_{k-1}, y_{k-1} \rangle}. \tag{21}$$

In the strictly convex quadratic case with any number of variables, it has been demonstrated in [16] that the BB method is globally convergent and in [9] that the convergence rate is R-linear. The so-called BB method performs much better than the classical steepest descent method in practice. Inspired by [26], Yu, Qi and Dai [24] proposed the following adaptive gradient method in which they choose a small steplength or a large steplength alternatively at each iteration as follows

$$\alpha_k = \begin{cases} \alpha_k^1 & \text{if } \frac{\alpha_k^2}{\alpha_k^1} > \varrho \text{ or k is odd,} \\ \alpha_k^2 & \text{otherwise,} \end{cases} \tag{22}$$

where $\varrho > 0$ is close to 1.

From now on, we mainly focus on the solution of optimization (15), which we restate here

$$\min_{w \in C_0^1(\mathcal{D}), |w| \leq 1} \frac{1}{2} (\|\nabla \cdot w + \mu f\|_2^2). \tag{23}$$

With reference to the dual problem, we adopt adaptive gradient projection strategy to solve it. Let $g_k := -\nabla(\nabla \cdot w_k + \mu f)$, then $y_{k-1} = -\nabla(\nabla \cdot s_{k-1})$,

$\langle y_{k-1}, s_{k-1} \rangle = \langle -\nabla(\nabla \cdot s_{k-1}), s_{k-1} \rangle$. Hence, (22) becomes

$$\alpha_k = \begin{cases} \dfrac{\|\nabla \cdot s_{k-1}\|^2}{\|s_{k-1}\|^2} & \text{if } \dfrac{\alpha_k^2}{\alpha_k^1} > \varrho \text{ or k is odd,} \\[4mm] \dfrac{\|y_{k-1}\|^2}{\|\nabla \cdot s_{k-1}\|^2} & \text{otherwise.} \end{cases} \qquad (24)$$

Moreover, we impose that the steplength α_k computed through (24) is modified so as to satisfy a condition of the form

$$0 < \alpha_l \le \alpha_k \le \alpha_r, \text{ for all } k, \qquad (25)$$

where α_l and α_r are fixed constants. Replacing the time step τ in (17) by α_k defined by (25), we obtain the following new corresponding Chambolle projection formula

$$w_{i,j}^{k+1}(w_k, \alpha_k, g_k) = \frac{w_{i,j}^k - \alpha_k g_{i,j}^k}{\max\{1, |w_{i,j}^k - \alpha_k g_{i,j}^k|\}}, \quad (\forall \, i, j) \qquad (26)$$

our approaches move from iterate w_k to the next iterate w_{k+1} using the scheme (26). As well known, nonmonotone schemes can improve the likelihood of finding a global optimum; also, they can improve convergence speed in cases where a monotone scheme is forced to creep along the bottom of a narrow curved valley. Encouraging numerical results have been reported in [11,16,19,23,25] when nonmonotone schemes were applied to difficult nonlinear problems. In order to ensure the global convergence, we adopt a nonmonotone line search in our method. Zhang and Hager (2004) proposed a new nonmonotone line search which has the same general form as the scheme of Grippo, Lampariello and Lucidi, except that their "max" is replaced by an average of function values, and this line search technique is also based on an Armijo-type line search on $0 < \delta < 1$ employing an acceptance condition of the form

$$\Psi(w_k + a_k d_k) \le C_k + \delta a_k \langle g_k, d_k \rangle, \qquad (27)$$

where $C_{k+1} = \frac{\eta_k \gamma_k C_k + \Psi(w_{k+1})}{\gamma_{k+1}}$ and $\gamma_{k+1} = \eta_k \gamma_k + 1$, detailed approach of the choice of parameters can consult [25]. In [18], Shi and Shen proposed a new inexact line search, and it is a modified Amijo-type line search

$$\Psi(w_k + a_k d_k) - \Psi(w_k)$$
$$\le \delta a_k (\langle g_k, d_k \rangle + \frac{1}{2} \sigma_2 a_k b L_k \|d_k\|^2), \qquad (28)$$

which can obtain a larger steplength in each iteration, where $\delta \in (0, 1/2]$, $b \in [0, 2)$ and $L_k > 0$, other particular information can refer to [18].

In this paper, we combine the merits of the two above mentioned line search techniques together, and propose the following nonmonotone line search technique

$$\Psi[w_{k+1}(w_k, \alpha_k, g_k)]$$
$$\le C_k + \sigma_1 \alpha_k (\langle g_k, d_k \rangle + \frac{1}{2} \sigma_2 \alpha_k \|d_k\|^2). \qquad (29)$$

Let $L(w_k) = w_{k+1}(w_k, 1, g_k) - w_k$, where $w_{k+1}(w_k, 1, g_k)$ is defined by (26). A full description of the proposed DAGPS algorithm (short for "dual and adaptive gradient projection strategy algorithm") is given as follows.

DAGPS Algorithm

Initialization. Choose a starting point w_0, and parameters $\beta \in (0, 1)$, $\sigma_1 \in (0, 1)$, $\sigma_2 \in [0, 2)$, $0 < \eta_l < \eta_r < 1$. Set $\varrho = 0.9$, $\gamma_0 = 1$, $\eta_0 = 0.1$, $\alpha_0 = 1/\Psi(w_0)$, $C_0 = \Psi(w_0)$. Let $k := 0$.

Termination test. If $\|L(w_k)\|$ sufficiently small, stop.

Direction generation. Compute g_k, $w_{k+1}(w_k, \alpha_k, g_k)$ which defined by (26), then set $d_k = w_{k+1} - w_k$.

Line search update. Find the steplength $\alpha_k = \max\{\bar{\alpha}_k, \beta\bar{\alpha}_k, \beta^2\bar{\alpha}_k, \cdots\}$ satisfying $\Psi[w_{k+1}(w_k, \alpha_k, g_k)] \leq C_k + \sigma_1\alpha_k(\langle g_k, d_k \rangle + \frac{1}{2}\sigma_2\alpha_k\|d_k\|^2)$, where $\bar{\alpha}_k = -(\langle g_k, d_k \rangle)/\|d_k\|^2$.

Update $w_{k+1} = w_{k+1}(w_k, \alpha_k, g_k)$.

Cost update. Choose $\eta_k \in [\eta_l, \eta_r]$, then set $\gamma_{k+1} := \eta_k\gamma_k + 1$, $C_{k+1} := \frac{\eta_k\gamma_k C_k + \Psi(w_{k+1})}{\gamma_{k+1}}$. Let $k := k + 1$ and compute α_k by (25), then return to the termination test.

Remarks

1. In fact, we can control the degree of nonmonotonicity at $(k+1)th$ step by choosing suitable η_k. If $\eta_k = 0$, then $C_{k+1} = \Psi_{k+1}$, which implies that DNTV algorithm is monotone decreasing at $(k+1)th$ step.
2. The line search is the usual nonmonotone line search technique proposed in [25] when $\sigma_2 = 0$.

3.2 Global Convergence

The following lemma comes from [2] directly.

Lemma 1. *Denoting $\mathcal{P}$ as the projection operator onto K and defining the scaled projected gradient $g_\alpha(w) = \mathcal{P}[w_k - \alpha g(w)] - w$, then for all $w \in K$ and $\alpha \in (0, \alpha_r]$, it holds that*

$$\langle g(w), g_\alpha(w) \rangle \leq -\frac{1}{\alpha}\|g_\alpha(w)\|^2 \leq -\frac{1}{\alpha_r}\|g_\alpha(w)\|^2. \tag{30}$$

Corollary 1. Under the same condition of Lemma 1, it is not difficult to derive that

$$\langle g_k, d_k \rangle \leq -\frac{1}{\alpha_r}\|d_k\|^2 < 0, \tag{31}$$

Particularly, we have

$$\|d_k\| \leq \alpha_{max}\|g_k\|. \tag{32}$$

Lemma 2. *Suppose that the sequence $\{w_k\}$ is generated by Algorithm 1, then for all $k > 0$ we have*

$$\Psi(w_{k+1}) \leq C_k, \tag{33}$$

$$\Psi(w_k) \leq C_k, \text{and} \tag{34}$$

$$C_{k+1} \leq C_k - \sigma_1(1 - \frac{1}{2}\sigma_2)(1 - \eta_r)(-\alpha_k \langle g_k, d_k \rangle), \tag{35}$$

i.e., $C_{k+1} \leq C_k$, which implies that the sequence $\{C_k\}$ is monotone nonincreasing.

Proof. (i) By the line search process and (31), we have

$$\begin{aligned}
\Psi[w_{k+1}&(w_k, \alpha_k, g_k)] \\
&\leq C_k + \sigma_1 \alpha_k (\langle g_k, d_k \rangle + \frac{1}{2}\sigma_2 \alpha_k \|d_k\|^2) \\
&\leq C_k + \sigma_1 \alpha_k (\langle g_k, d_k \rangle + \frac{1}{2}\sigma_2 \bar{\alpha}_k \|d_k\|^2) \\
&= C_k + \sigma_1 \alpha_k (1 - \frac{1}{2}\sigma_2) \langle g_k, d_k \rangle.
\end{aligned} \tag{36}$$

Hence, the inequality (33) is valid.
(ii) Combining the definition of C_k with (33), we have

$$\begin{aligned}
C_{k+1} &= \frac{\eta_k \gamma_k C_k + \Psi(w_{k+1})}{\gamma_{k+1}} \\
&\geq \frac{\eta_k \gamma_k \Psi(w_{k+1}) + \Psi(w_{k+1})}{\gamma_{k+1}} \\
&= \Psi(w_{k+1}).
\end{aligned} \tag{37}$$

Consequently, (37) and $C_0 = \Psi(w_0)$ give (34).
(iii) The definition of C_k and (36) yield

$$\begin{aligned}
C_{k+1} &= \frac{\eta_k \gamma_k C_k + \Psi(w_{k+1})}{\gamma_{k+1}} \\
&\leq \frac{\eta_k \gamma_k C_k + C_k + \sigma_1 \alpha_k (1 - \frac{1}{2}\sigma_2) \langle g_k, d_k \rangle}{\gamma_{k+1}} \\
&\leq C_k + \frac{\sigma_1 \alpha_k (1 - \frac{1}{2}\sigma_2) \langle g_k, d_k \rangle}{\gamma_{k+1}} \\
&\leq C_k + \sigma_1 (1 - \frac{1}{2}\sigma_2)(-\frac{1}{\gamma_{k+1}})(-\alpha_k \langle g_k, d_k \rangle).
\end{aligned} \tag{38}$$

Notice that $\eta_{max} < 1$, we can obtain

$$
\begin{aligned}
\gamma_{k+1} &= \eta_k \gamma_k + 1 \\
&= 1 + \eta_k + \eta_k \eta_{k-1} + \cdots + \eta_k \eta_{k-1} \eta_{k-2} \cdots \eta_0 \\
&= 1 + \sum_{j=0}^{k} \prod_{i=0}^{j} \eta_{k-j} \leq 1 + \sum_{j=0}^{k} \eta_r^{j+1} \leq 1 + \sum_{j=0}^{\infty} \eta_r^{j+1} \\
&= \frac{1}{1 - \eta_r}.
\end{aligned}
\tag{39}
$$

It follows from (31), (38) and (39) that

$$
C_{k+1} \leq C_k - \sigma_1 (1 - \frac{1}{2}\sigma_2)(1 - \eta_r)(-\alpha_k \langle g_k, d_k \rangle),
\tag{40}
$$

Therefore, we get (35). $\qquad\qquad\qquad\qquad\qquad\qquad\qquad\qquad\qquad\square$

We now show the global convergence of DAGPS Algorithm. Our global convergence result utilizes the following assumption concerning the search directions.

Assumption 1. *There exist a positive constant ξ such that*

$$
\langle g_k, d_k \rangle \leq -\xi \|g_k\|^2,
\tag{41}
$$

for all sufficiently large k.

Theorem 1. *Suppose that the sequence $\{w_k\}$ is generated by Algorithm 1, then we have*

$$
\liminf_{k \to \infty} \|g_k\| = 0.
\tag{42}
$$

Proof. By contradiction. Suppose that there exists an infinite subsequence $K_1 \subset \{1, 2, 3, \cdots\}$, for all $k \in K_1$ we have

$$
\|g_k\| > \varepsilon
\tag{43}
$$

for some $\varepsilon > 0$. Combining this with (40) gives

$$
\begin{aligned}
C_{k+1} &- C_k \\
&\leq -\sigma_1(1 - \frac{1}{2}\sigma_2)(1 - \eta_r)(-\alpha_k \langle g_k, d_k \rangle) \\
&\leq -\sigma_1(1 - \frac{1}{2}\sigma_2)(1 - \eta_r)(\alpha_k \xi \|g_k\|^2) \\
&\leq -\sigma_1 \xi(1 - \frac{1}{2}\sigma_2)(1 - \eta_r)\alpha_k \|g_k\| \frac{1}{\alpha_r}\|d_k\| \\
&< -\frac{\sigma_1 \xi \varepsilon}{\alpha_r}(1 - \frac{1}{2}\sigma_2)(1 - \eta_r)(\alpha_k \|d_k\|).
\end{aligned}
\tag{44}
$$

Since $\Psi(w)$ is bounded below, C_k is monotone decreasing and bounded below, it follows from (34) and (44) that

$$\lim_{k\in K_1, k\to\infty} \alpha_k\|d_k\| = 0. \tag{45}$$

By the line search process, for all $k \in K_1$ we have

$$\Psi(w_k + \frac{\alpha_k}{\beta}d_k) - C_k$$
$$> \sigma_1\frac{\alpha_k}{\beta}(\langle g_k, d_k\rangle + \frac{1}{2}\sigma_2\frac{\alpha_k}{\beta}\|d_k\|^2). \tag{46}$$

It follows that,

$$\Psi[w_k + (\frac{\alpha_k}{\beta})d_k] - \Psi_k$$
$$> \sigma_1(\frac{\alpha_k}{\beta})[\langle g_k, d_k\rangle + \frac{1}{2}\sigma_2(\frac{\alpha_k}{\beta})\|d_k\|^2]. \tag{47}$$

Furthermore, by the mean-value theorem, we have

$$(\frac{\alpha_k}{\beta})\langle g(\phi_k), d_k\rangle$$
$$> \sigma_1(\frac{\alpha_k}{\beta})[\langle g_k, d_k\rangle + \frac{1}{2}\sigma_2(\frac{\alpha_k}{\beta})\|d_k\|^2], \tag{48}$$

where $\phi_k = w_k + \varphi_k(\frac{\alpha_k}{\beta})d_k$, $\varphi_k \in (0,1)$.
(48) can be wriiten as

$$\langle g(\phi_k), d_k\rangle > \sigma_1[\langle g_k, d_k\rangle + \frac{1}{2}\sigma_2(\frac{\alpha_k}{\beta})\|d_k\|^2]$$
$$> \sigma_1\langle g_k, d_k\rangle. \tag{49}$$

It follows that

$$\|g(\phi_k) - g(w_k)\| = \frac{\|g(\phi_k) - g(w_k)\|\|d_k\|}{\|d_k\|}$$
$$\geq \frac{\langle g(\phi_k) - g(w_k), d_k\rangle}{\|d_k\|}$$
$$> -(1-\sigma_1)\frac{\langle g_k, d_k\rangle}{\|d_k\|}$$
$$\geq \frac{\xi(1-\sigma_1)}{\|d_k\|}\|g_k\|^2$$
$$\geq \frac{\xi(1-\sigma_1)}{\alpha_r}\|g_k\|$$
$$> \frac{\xi(1-\sigma_1)}{\alpha_r}\varepsilon. \tag{50}$$

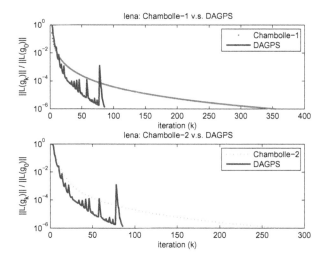

Fig. 1. Relative norm of gradient projection vs. iteration for denoising image "lena" when $\text{TOL} = 10^{-6}$.

The definition of ϕ_k and (45) yield

$$\lim_{k \in K_1, k \to \infty} \|\phi_k - w_k\| = 0. \tag{51}$$

By the Lipschitz continuity of $g(w)$, we have

$$\lim_{k \in K_1, k \to \infty} \|g(\phi_k) - g(w_k)\| = 0. \tag{52}$$

It is not difficult to see that (52) contradicts with (50). Consequently, (42) holds. The proof is complete. □

4   Numerical Experiments

In this section we describe some experiments to illustrate the good performance of the proposed algorithm. All methods are coded in Matlab and run on a PC (Intel Core i3 @ 2.53 GHz, 4 GB) with Windows 7 operating system. Peak signal to noise ratio (PSNR), Signal to noise ratio (SNR) and relative error (ReErr) are used to measure the quality of the restored images.

Table 1. Performance of image denoising via Chambolle-1, Chambolle-2 and DAGPS algorithms.

Image	Algorithm	TOL $= 10^{-4}$			TOL $= 10^{-6}$		
		PSNR	CPU-Time	Iter	PSNR	CPU-Time	Iter
Shape 128 × 128	Chambolle-1	23.858	0.234	83.000	23.794	2.028	710.000
	Chambolle-2	23.862	0.234	84.000	23.794	1.560	566.000
	DAGPS	23.861	0.140	31.000	23.794	0.718	166.000
Lena 256 × 256	Chambolle-1	27.796	1.654	69.000	27.822	6.115	355.000
	Chambolle-2	27.783	1.357	56.000	27.819	4.883	272.000
	DAGPS	27.782	0.998	28.000	27.819	2.465	86.000
Head 512 × 512	Chambolle-1	28.516	10.483	88.000	28.517	74.334	617.000
	Chambolle-2	28.475	9.766	83.000	28.513	58.344	478.000
	DAGPS	28.476	6.334	35.000	28.513	25.756	138.000
Man 1024 × 1024	Chambolle-1	28.328	36.676	74.000	28.303	162.147	427.000
	Chambolle-2	28.303	33.462	67.000	28.300	131.431	336.000
	DAGPS	28.302	22.792	30.000	28.300	70.481	106.000

$$PSNR = 10 \log_{10}\left(\frac{MN \times 255^2}{\|\tilde{u} - u\|^2}\right),$$

$$SNR = 20 \log_{10}\left(\frac{\|u\|}{\|\tilde{u} - u\|}\right),$$

and

$$ReErr = \frac{\|\tilde{u} - u\|^2}{\|u\|^2},$$

where u and $\tilde{u}$ are the original image and the restored image respectively.

In our first experiment, we compare the proposed method (DAGPS) with Chambolle-1 method (16) and Chambolle-2 method (17) for image denoising. The noisy images are generated by adding Gaussian noise to the clean images using the Matlab function *imnoise*, with variance parameter set to 0.01. The fidelity parameter μ is taken to be 0.053. In Chambolle's methods, we take the step length to be 0.248 for near-optimal run time, although global convergence is proved in [4] only for step lengths in the range $(0, 0.125)$. In DAGPS method, we set $\alpha_l = 10^{-10}$ and $\alpha_r = 10^{10}$. The stopping criterion of the three methods are $\|L(g_k)\| \leq TOL\|L(g_0)\|$. Figure 1 shows the nonmonotone descent behavior for DAGPS method which indicate that such a kind of nonmonotone method is a great improvement over the original Chambolle's method.

More detailed data are listed in the Table 1. This table contains the PSNR of the recovered images, CPU time in seconds (CPU-Time) required for the whole denoising process and number of iterations (Iter). From Table 1, we see that the PSNR values attained by these three methods are very similar. However, from Iter and CPU-Time points of view, DAGPS truly saves much computing time and number of iterations. In a word, this simple experiment shows that DAGPS works well, and it can provide an efficient approach to denoise corrupted images.

Fig. 2. The first line is the original image and the noisy image: lena. The second line is the restored images via NESTEROV and DAGPS respectively when $TOL = 10^{-3}$.

Recently, a public-domain software for total variation image reconstruction via Nesterov algorithm [15] was developed by Dahl, Hansen, Jensen and Jensen [7]. Their code is available at the web.[3] In our second experiment, we compare DAGPS method with NESTEROV method. In NESTEROV method, the parameter $\tau = 0.85$. We let $TOL = 10^{-3}$ in this experiment. Additionally, the noisy images are generated by adding white Gaussian noise to the original images with standard deviation $\sigma = 25$. Test results for NESTEROV and DAGPS are

Table 2. NESTEROV vs. DAGPS: performance of image denoising.

Image	NESTEROV				DAGPS			
	PSNR	SNR	CPU-Time	ReErr	PSNR	SNR	CPU-Time	ReErr
Brain 256^2	25.4498	18.6476	1.5197	0.0137	25.8481	19.2501	0.4992	0.0119
Lena 256^2	27.9652	20.7395	1.2812	0.0084	27.8648	20.6934	0.4836	0.0085
Barbara 512^2	25.9535	19.4981	4.5468	0.0112	25.7125	19.3045	2.3400	0.0117
Head 512^2	25.9880	11.3919	9.0205	0.0726	28.4950	14.3273	2.8704	0.0369
Man 1024^2	28.4465	20.8037	21.3964	0.0083	28.2933	20.6849	10.1401	0.0085
Earth 1024^2	26.3697	19.7437	20.1854	0.0106	26.1550	19.5960	9.1261	0.0110

---

[3] http://www2.imm.dtu.dk/~pch/mxTV/index.html.

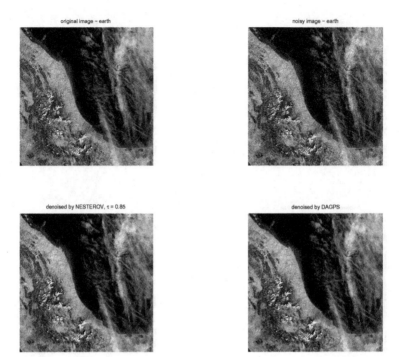

Fig. 3. The first line is the original image and the noisy image: earth. The second line is the restored images via NESTEROV and DAGPS respectively when $\mathrm{TOL} = 10^{-3}$.

listed in Table 2, which contains the PSNR of the recovered images, Signal to noise ratio (SNR), CPU time in seconds (CPU-Time) and relative error (ReErr). From the Table 2, we can see that the DAGPS method is more faster than the NESTEROV method for all of the test images, and saving about fifty percent in CPU-time at least. Moreover, they attain some similar PSNR values, SNR values and ReErr values. Parts of the restoration results by NESTEROV and DAGPS method are listed in Figs. 2 and 3. These results show that the DAGPS method is competitive and can restore corrupted image quite well in an efficient manner.

5   Conclusion

We have proposed an efficient adaptive gradient projection strategy with a non-monotone line search proposed by Zhang and Hager [25] for solving the discretized dual formulation of the total variation image restoration model of Rudin, Osher and Fatemi [17]. The problem has a convex quadratic objective with separable convex constraints, a problem structure that makes gradient projection schemes practical and simple to implement. We compare our method to two popular existing methods proposed by Chambolle [4,5] and Dahl, Hansen, Jensen and Jensen [7]. Our experimental results have shown that the quality of restored images by the proposed method are competitive with those restored by the existing TV restoration methods.

References

1. Barzilai, J., Borwein, J.M.: Two point step size gradient methods. IMA J. Numer. Anal. **8**, 141–148 (1998)
2. Birgin, E.G., Martinez, J.M., Raydan, M.: Nonmonotone spectral projected gradient methods on convex sets. SIAM J. Optim. **10**, 1196–1211 (2000)
3. Carter, J.L.: Dual methods for total variation-based image restoration. Ph.D. thesis, University of California at Los Angeles (Advisor: T.F. Chan) (2001)
4. Chambolle, A.: An algorithm for total variation minimization and applications. J. Math. Imaging Vis. **20**, 89–97 (2004)
5. Chambolle, A.: Total variation minimization and a class of binary mrf models. In: Rangarajan, A., Vemuri, B., Yuille, A.L. (eds.) EMMCVPR 2005. LNCS, vol. 3757, pp. 136–152. Springer, Heidelberg (2005). doi:10.1007/11585978_10
6. Chan, T.F., Golub, G.H., Mulet, P.: A nonlinear primal-dual method for total variation-based image restoration. SIAM J. Sci. Comput. **20**, 1964–1977 (1999). VPL
7. Dahl, J., Hansen, P.C., Jensen, S.H., Jensen, T.L.: Algorithms and software for total variation image reconstruction via first-order methods. Numer. Algorithms **53**, 67–92 (2010)
8. Dai, Y.H., Fletcher, R.: Projected Barzilai-Borwein methods for large-scale box-constrained quadratic programming. Numer. Math. **100**, 21–47 (2005)
9. Dai, Y.H., Liao, L.Z.: R-Linear Convergence of the Barzilai and Borwein gradient method. IMA J. Numer. Anal. **22**, 1–10 (2002)
10. Ekeland, I., Témam, R.: Convex Analysis and Variational Problems: Classics in Applied Mathematics. SIAM, Philadelphia (1999)
11. Grippo, L., Lampariello, F., Lucidi, S.: A truncated Newton method with nonmonotone line search for unconstrained optimization. J. Optim. Theor. Appl. **60**, 401–419 (1989)
12. Goldfarb, D., Yin, W.: Second-order cone programming methods for total variation-based image restoration. SIAM J. Sci. Comput. **27**, 622–645 (2005)
13. Goldstein, T., Osher, S.: The Split Bregman method for L1 regularized problems. SIAM J. Imaging Sci. **2**, 323–343 (2009)
14. Hintermuller, M., Stadler, G.: An infeasible primal-dual algorithm for tv-based infconvolution-type image restoration. SIAM J. Sci. Comput. **28**, 1–23 (2006)
15. Nesterov, Y.: Smooth minimization of non-smooth functions. Math. Program. Ser. A **103**, 127–152 (2005)
16. Raydan, M.: The Barzilai and Borwein gradient method for the large scale unconstrained minimization problem. SIAM J. Optim. **7**, 26–33 (1997)
17. Rudin, L., Osher, S., Fatemi, E.: Nonlinear total variation based noise removal algorithms. Phys. D Nonlinear Phenom. **60**, 259–268 (1992)
18. Shi, Z.J., Shen, J.: New inexact line search method for unconstrained optimization. J. Optim. Theor. Appl. **127**, 425–446 (2005)
19. Toint, P.L.: An assessment of non-monotone line search techniques for unconstrained optimization. SIAM J. Sci. Comput. **17**, 725–739 (1996)
20. Vogel, C.R., Oman, M.E.: Iterative methods for total variation denoising. SIAM J. Sci. Comput. **17**, 227–238 (1996)

21. Wu, C., Zhang, J., Tai, X.C.: Augmented Lagrangian method for total variation restoration with non-quadratic fidelity. UCLA CAM Report 09–82 (2009)
22. Xiao, Y.H., Song, H.N.: An inexact alternating directions algorithm for constrained total variation regularized compressive sensing problems. J. Math. Imaging Vis. **44**, 114–127 (2012)
23. Yu, G.H.: Nonmonotone spectral gradient-type methods for large-scaleunconstrained optimization and nonlinear systems of equations. Pac. J. Optim. **7**, 387–404 (2011)
24. Yu, G.H., Qi, L.Q., Dai, Y.H.: On nonmonotone chambolle gradient projection algorithms for total variation image restoration. J. Math. Imaging Vis. **35**, 143–154 (2009)
25. Zhang, H.C., Hager, W.W.: A nonmonotone line search technique and its application to unconstrained optimization. SIAM J. Optim. **14**, 1043–1056 (2004)
26. Zhou, B., Gao, L., Dai, Y.H.: Gradient methods with adaptive stepsizes. Comput. Optim. Appl. **35**, 69–86 (2006)
27. Zhu, M., Chan, T.F.: An efficient primal-dual hybrid gradient algorithm for total variation image restoration. UCLA CAM Report 08–34 (2008)
28. Zhu, M.Q., Wright, S.J., Chan, T.F.: Duality-based algorithms for total-variation-regularized image restoration. Comput. Optim. Appl. **47**, 377–400 (2010)

Image Inpainting Based on Sparse Representation with Dictionary Pre-clustering

Kai Xu[1], Nannan Wang[2(✉)], and Xinbo Gao[1]

[1] State Key Laboratory of Integrated Services Networks,
School of Electronic Engineering, Xidian University, Xi'an 710071, China
kaixu1993@foxmail.com, xbgao@mail.xidian.edu.cn
[2] State Key Laboratory of Integrated Services Networks,
School of Telecommunications Engineering,
Xidian University, Xi'an 710071, China
nnwang@xidian.edu.cn

Abstract. This paper proposed a new image inpainting algorithm based on sparse representation. In traditional exemplar-based methods, the image patch is inpainted by the best matched patch from the source region. This greedy search will introduce unwanted objects and has huge time consuming. The proposed algorithm directly employs all the known image patches to form an over-complete dictionary. And then, the over-complete dictionary is clustered into several sub-dictionaries. Finally, the unrepaired image patches are repaired over their corresponding closest sub-dictionaries through non-negative orthogonal matching pursuit algorithm. Experimental results show that the proposed method achieves superior performance than state-of-the-art methods. In addition, the time complexity is greatly reduced in comparison with the traditional exemplar-based inpainting algorithm.

Keywords: Image inpainting · Sparse representation · Over-complete dictionary

1 Introduction

Image inpainting aims to restore the missing or corrupted region of the image. In recent years, image inpainting is widely used in image restoration, image target removal, image compression and transmission and so on. Therefore, it attracts growing attentions from both industry and academic.

Existing image inpainting methods could be classified into two categories. The first category is diffusion-based inpainting. These methods are based on the parametric model or partial differential equation (PDE), which diffuses local structure from the external to the internal of the region to be repaired. Diffusion-based image inpainting started from the work of Bertalmio et al. [1]. They employ anisotropic mathematical model to propagate the image Laplasse information from the surrounding to the inside of the region to be repaired. Bertalmio has also carried out an analogy to the transmission of image pixel value along the

© Springer Nature Singapore Pte Ltd. 2016
T. Tan et al. (Eds.): CCPR 2016, Part II, CCIS 663, pp. 245–258, 2016.
DOI: 10.1007/978-981-10-3005-5_21

isophotes and the velocity of the fluid field [2]. The velocity propagation in hydro-dynamic field has been formulated as Navier-Stokes equations, which has a complete theory and good numerical solution. Then the problem of image inpainting can be solved by solving the Navier-Stokes equations. The PDE-based methods need the numerical solution obtained by iteration, thus it is relatively slow for the algorithm. A fast march method [3] estimates the unknown pixel value by the weighted average of all the known pixels in the neighborhood of an unknown pixel. A trace-based PDE model is proposed in [4] to restore multivalued images (with multiple color channels). It uses heat flows constrained on integral curves to better preserve curvatures in image structures. Chan and Shen [5] propose a variation framework based on total variation to restore the missing image information. The diffusion-based approach tends to extend the structure (*i.e.*, isophotes), which is suitable for the image of non-texture image and small area of damage. However, it is not suitable for the texture image and the large area of the repair region. Although the method based on diffusion tries to preserve the structure such as the image edge, image blur occurs when the filling area is large because of the smoothing effect.

The second category of methods is exemplar-based inpainting algorithm. These approaches take the image block as a unit, and spread the image information from the known region of the image to the unknown region. This category of approaches are inspired from the idea of texture synthesis technique [6], which synthesizes textures by searching the best matching image blocks in the known region. However, the natural image is composed of structure and texture, in which the structure constitutes the basic outline of the image (*i.e.*, edges, corners, etc.), while the texture is the flat area with the same pattern or color. The pure texture synthesis technique cannot deal with the condition of the region containing both texture and structure. Bertalmio *et al.* [7] proposed to decompose the image into two layers of structure and texture. They then use the method of diffusion to repair the structural layer, and employ the texture synthesis technology [6] to repair the texture layer. Criminisi *et al.* [8] proposed an exemplar-based inpainting algorithm, which spread the known image patches (*i.e.* examplar) to the image patches that need to be restored gradually. To repair the image structure well, the method defines the image patch priority. The priority makes image patch that contains structural information have priority to be repaired, so as to realize the continuation of the structure information from the known image area into the unknown region. Wong and Orchard [9] proposed a nonlocal average method based on the exemplar. The image patches to be repaired is estimated from a nonlocal mean of the set of candidate image patches rather than a single best matching image patch. Compared with the method based on diffusion, the examplar-based inpainting algorithm has better effect when the inpainting area is large. But the traditional example based method [8] uses a single best matching image patch to infer unknown pixels. It is a kind of greedy idea, so it can cause the unwanted object in the repair area. And Wong [9] uses a weighted average of the fixed number of the best matching image blocks to estimate the unknown area, which causes blurring effect. In addition, because exemplar-based inpainting algorithm need to search the whole image to find the best matching patch when repairing every image patch, has great time complexity.

Recently, sparse representation is introduced to solve the problem of image inpainting [10–12]. Similarly as exemplar-based algorithms [8,9], sparse representation-based methods work on the patch level. The main idea is to represent the image patches by sparse linear combination of the over-complete dictionary, and then deduce the unknown pixel value. In sparse representation, the number of patches and their coefficients are adaptively determined, instead of using only the best matching patch or a fixed number of the best matching patches. This overcomes the artifacts caused by the greedy search strategy used in exemplar-based algorithms. However, the sparse representation-based methods [10–12] also has the problem of high time complexity.

In order to improve the inpainting performance and reduce the time complexity, this paper proposes an improved sparse representation-based algorithm. The main contributions of this paper are twofolds. (1) We directly employs all the known image patches to form an over-complete dictionary. And then, the over-complete dictionary is clustered into several sub-dictionaries. This can avoid the interference of irrelevant atoms effectively. Then we employ orthogonal matching algorithm with non-negative constraints to reconstruct the image patches, which avoids the negative correlation atoms taking part in the linear combination. (2) Experimental results show that the proposed method achieves superior performance than state-of-the-art methods and the time complexity is greatly reduced.

The remainder of this paper is arranged as follows. We introduce the prior knowledge of the sparse representation in Sect. 2. Section 3 gives the proposed algorithm. Experimental results and analysis are presented in Sect. 4. Finally, we concludes this paper in Sect. 5.

2   Sparse Representation

Sparse representation is a kind of signal transformation theory based on over complete dictionary. It is assumed that natural signals can be represented by a linear combination of atoms in sparse dictionary. From the point of view of mathematics, sparse representation is a method of linear decomposition of multidimensional data.

Given an over-complete dictionary $D = [\mathbf{d}_1, \cdots, \mathbf{d}_K] \in \mathbf{R}^{m \times K} (K > m)$, each column is called an atom, represented as $\mathbf{d}_j \in \mathbf{R}^m (j = 1, 2, \cdots, K)$. The target signal $\mathbf{y} \in \mathbf{R}^m$ can be represented as a sparse linear combination of atoms in the dictionary. Specifically, the target signal $\mathbf{y}$ is expressed as $\mathbf{y} \approx D\mathbf{x}(\mathbf{x} \in \mathbf{R}^K)$, $\mathbf{x}$ is a vector containing the sparse representation coefficients. In other words, $\mathbf{x}$ and $\mathbf{y}$ meet $\|\mathbf{y} - D\mathbf{x}\|_2 \leqslant \epsilon$.

If $m < K$ and D is full rank matrix, the sparse representation problem is underdetermined, and there are infinite solutions. A new constraint is introduced and formalized as the following optimization problem.

$$\min_{\mathbf{x}} \|\mathbf{y} - D\mathbf{x}\|_2^2$$

$$s.t. \quad \|\mathbf{x}\|_0 \leqslant T \tag{1}$$

Or the equivalent form

$$\min_{\mathbf{x}} \|\mathbf{x}\|_0$$

$$s.t. \quad \|\mathbf{y} - D\mathbf{x}\|_2^2 \leqslant \epsilon, \tag{2}$$

where $\|\cdot\|_0$ is so-called l^0 norm (actually it is not satisfied with the definition of a norm), which represents the number of non-zero elements in the vector, and T determines the sparsity of the signal.

Solving the sparse representation problem as shown in (1) or (2) is a combinatorial optimization problem. It is proved to be a NP-hard problem [13]. Compressed sensing theory [14] proves that if the vector $\mathbf{x}$ is sparse enough, the l^0 norm can be replaced by the l^1 norm approximately. So the combinatorial optimization problem of the Eq. (1) is reformulated as the following convex optimization problem.

$$\min_{\mathbf{x}} \|\mathbf{y} - D\mathbf{x}\|_2^2$$

$$s.t. \quad \|\mathbf{x}\|_1 \leqslant T \tag{3}$$

For the convex optimization problem (3), there are many effective algorithms to solve this, such as Basis Pursuit [15] and Basis Pursuit Denoising [16]. In addition, through the Lagrange multiplier method the optimization problem (3) can be written in the following form:

$$\min_{\mathbf{x}} \frac{1}{2} \|\mathbf{y} - D\mathbf{x}\|_2^2 + \lambda \|\mathbf{x}\|_1 \tag{4}$$

The problem (4) can be solved by L1 norm penalty least squares method (L1LS) [17].

In addition to using convex optimization problem approximation, many greedy methods are widely studied, such as matching pursuit algorithm (MP) [18], orthogonal matching pursuit (OMP) [19]. In practical applications, the orthogonal matching pursuit method is simple and efficient.

3   Inpainting Algorithm

In this section we will introduce the proposed methods. Given an image I, which contains the missing part U and the known part S, as shown in Fig. 1. The purpose of image restoration is to fill the unknown area U using image information of known part S. The boundary of a missing part is represented as ∂U. The proposed algorithm in this paper gradually fills unknown region in patch level. Ψ_p represents the image patch with the center in pixel p. In this article, if there are no special note, the default image patch size is 9×9.

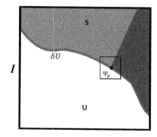

Fig. 1. The inpainting image

3.1   The Creation of Dictionary

The Creation of Original Over-Complete Dictionary. The goal of image inpainting is to restore the distorted image and to make it look natural, which requires consistency between the restoration region and the original known areas. This means not only the texture to be consistent, but also the noise must have the same distribution. So in this paper, we directly use the pixel value of all the whole image patches in the known region to construct the original over-complete dictionary, as shown in Fig. 2. If there are N complete image patches from the known image area, then the original over-complete dictionary has N columns, each column represents an image patch.

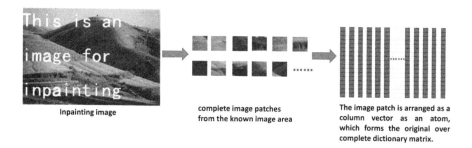

Inpainting image   complete image patches from the known image area   The image patch is arranged as a column vector as an atom, which forms the original over complete dictionary matrix.

Fig. 2. An illustration of the original over-complete dictionary

The Dictionary Pre-clustering. After getting the original complete dictionary, the sparse coefficients are not directly obtained over the original over-complete dictionary. Because the original complete dictionary contains all the known image patches in the image, there will be a large number of unrelated atoms with the image patch to be restored. If the sparse coefficients are solved directly over the original over-complete dictionary, the interference of irrelevant atoms will appear, which will make the inpainting performance worse, and increase the consumption of both time and space.

So in the proposed algorithm, we cluster the atoms in the original over-complete dictionary, which makes the similar atoms be clustered into one class. The class number k can be obtained by the following formula:

$$k = \lceil \frac{N}{c} \rceil \tag{5}$$

where N is the number of atoms in the original over-complete dictionary, and c is a parameter used to control the number of categories. And $c = 1000$ in the whole paper.

For simplicity, we use the prevalent k-means clustering algorithm. In order to make the initial clustering centers contain image texture or color types as much as possible, we choose the initial centers by the method of equal interval, *i.e.* the initial centers evenly distribute in known regions of image restoration.

The Selection of Best Sub-over-complete Dictionary. After clustering the original over-complete dictionary, the k sub-over-complete dictionaries are obtained. In each iteration, after selecting the image patch to be restored, we calculate the distance between the image patch and the centroid of each sub-over complete dictionary. We select the nearest sub-over-complete dictionary as the most similar over-complete dictionary. Then the sparse representation of the image patch to be restored is solved over the most similar over-complete dictionary.

3.2   The Sparse Reconstruction of Image Patches

The image patch is regarded as a signal in the sparse reconstruction of the image patch. The center $\hat{p}$ of the inpainting image patch $\Psi_{\hat{p}}$ is located at the boundary line δU, so the image patch $\Psi_{\hat{p}}$ contains both known part and unknown part. Because the region to be restored is artificially selected, the position of the unknown pixels in the image patch is known. Let P be an operator to extract the unknown pixels in image patch $\Psi_{\hat{p}}$, and $\bar{P}$ is an operator to extract the known pixels. $P\Psi_{\hat{p}}$ represents unknown pixels in $\Psi_{\hat{p}}$ and $\bar{P}\Psi_{\hat{p}}$ represents known pixels in $\Psi_{\hat{p}}$. The illustration of the operator P and $\bar{P}$ is in the Fig. 3. So the sparse reconstruction of image patches can be formulated as follows:

$$\mathbf{x} = \arg \min_{\mathbf{x}} \left\| \bar{P}\Psi_{\hat{p}} - \bar{P}(D\mathbf{x}) \right\|_2^2 \tag{6}$$
$$s.t. \quad \|\mathbf{x}\|_0 \leqslant T$$

where $\bar{P}\Psi_{\hat{p}}$ represents the column vector of the known pixel value in the image patch $\Psi_{\hat{p}}$. $\bar{P}(D\mathbf{x})$ represents the elements of the reconstructed signal $D\mathbf{x}$ at the position corresponding $\bar{P}\Psi_{\hat{p}}$.

After solving sparse reconstruction problem (6), the sparse coefficient $\mathbf{x}$ is obtained. Then the unknown pixels in the $\Psi_{\hat{p}}$ can be inferred by using the following formula:

$$P\Psi_{\hat{p}} = P(D\mathbf{x}) \tag{7}$$

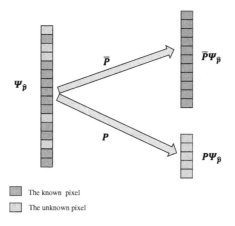

The known pixel

The unknown pixel

Fig. 3. The illustration of the operator P and $\bar{P}$

Algorithm 1. The NNOMP algorithm

1: **Require:** over-complete dictionary D, target signal $\mathbf{y}$, sparsity T;

2: **Ensure:** sparse coefficient vector $\mathbf{x}$;

3: To find out the index λ corresponding to the maximum inner product value between residual $\mathbf{r}$ and the column $\mathbf{d}_j$ in over complete dictionary. That is $\lambda_t = \arg \max\limits_{j=1\cdots N} | < \mathbf{r}_{t-1}, \mathbf{d}_j > |.$

4: Extend index set $\Lambda_t = \Lambda_{t-1} \bigcup \lambda_t$, and extend the set of selected atoms $\Phi_t = [\Phi_{t-1}, \mathbf{d}_{\lambda_t}]$;

5: Solve linear programming problem:

$$\mathbf{x}_t = \arg \min \|\mathbf{y} - \Phi_t \mathbf{x}\|_2^2 \quad s.t. \quad \mathbf{x} > 0$$

6: Update the residual $\mathbf{r}_t = \mathbf{y} - \Phi_t \mathbf{x}_t$, $t = t + 1$;

7: Judge whether to meet $t > T$, if satisfied, then stop; if not satisfied, then continue to perform step 3.

---

where $P\Psi_{\hat{p}}$ represents the column vector of the unknown pixel value in the image patch $\Psi_{\hat{p}}$. $P(D\mathbf{x})$ represents the elements of the reconstructed signal $D\mathbf{x}$ at the position corresponding $P\Psi_{\hat{p}}$.

A number of algorithms for solving sparse coefficients are listed in Sect. 2. In order to avoid the negative correlation atoms with the inpainting image patch taking part in the linear combination, we use the orthogonal matching algorithm with non-negative constraints (NNOMP) [20] to solve the problem of sparse reconstruction (6). The procedure of NNOMP algorithm is in Algorithm 1.

3.3   Inpainting Order

The restoration order of image patches has a significant impact on the final inpainting results. We use the restoration order with structure driven from paper [8], because the structure of the image can be better preserved. The method is

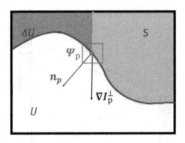

Fig. 4. Priority of image patch

to define a priority for each image patch, and to restore the image patch with the maximum priority in the current iteration step.

As shown in Fig. 4, given an image patch Ψ_p which center point p in the boundary line $(p \in \delta U)$, defined priority $P(p)$ is product of two terms:

$$P(p) = C(p)D(p) \tag{8}$$

$$C(p) = \frac{\sum_{q \in \Psi_p \cap S} C(q)}{|\Psi_p|} \tag{9}$$

$$D(p) = \frac{|\nabla I_p^\perp \cdot n_p|}{\alpha} \tag{10}$$

where C(p) is called the confidence term, D(p) is called data term. $|\Psi_p|$ represents the number of pixels in the Ψ_p; α is a normalization factor (that is, for the general gray image $\alpha = 255$); n_p is a unit vector that is perpendicular to the boundary line δU at the point p; ∇I_p represents the gradient of image I at the point p, and $\perp$ represents vertical operator.

During initialization, $C(p)$ is set to $C(p) = 0 \; \forall p \in U$ and $C(p) = 1 \; \forall p \in S$. Confidence term $C(p)$ is a measure of the number of reliable information around the pixel point p. Data term $D(p)$ is a function of the intensity of the isophotes that intersects with the boundary line. This term makes the image patches that contain isophotes have higher priority, so the linear structure can be spread safely into the interior of the unknown image region.

After image patch $\Psi_{\hat{p}}$ is restored, the confidence of pixels in $\Psi_{\hat{p}}$ is updated according to the following formula:

$$C(p) = C(\hat{p}) \quad \forall p \in \Psi_{\hat{p}} \cap U \tag{11}$$

4   Experimental Results and Comparisons

In this section, we compare the proposed image inpainting algorithm with diffusion-based inpainting, examplar-based inpainting algorithm and the existing method based on sparse representation. The experiments will be carried out

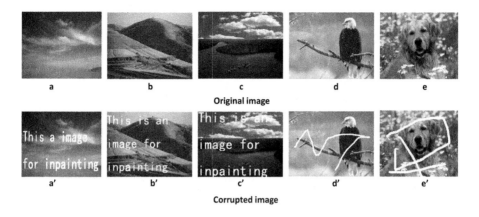

Fig. 5. Examples of removing text and scratch test images

on the images such as text removal, scratch repair and target removal. And then we compare the effects and the running time of different algorithms. For the experimental parameters, the size of the image block defaults to 9 if it does not add a special note. The sparse degree $T = 4$ and $c = 1000$ which determines the dictionary classification number. Experiments are carried out on personal computer with CPU Intel I5 2430M 2.4 GHZ, RAM 8G, and all of the codes are realized by MATLAB. K-means clustering algorithm uses the optimized code [21].

The experiments of text removal and scratch repair are first carried out. In these images, the parts of text coverage and scratches are generally small. For fair, the algorithm in [11] also employs the inpainting order in [8]. The images in the experiment are shown in Fig. 5.

As the results shown in Fig. 6, when the area of text coverage and scratches is small, the proposed method achieves a good inpainting effect. Article [2] is the classical diffusion-based inpainting algorithm. From the restored images, we can see that the regions of texture distribution appear blurs (e.g. Fig. 5e hair parts). This proves the diffusion-based inpainting algorithm is not suitable for repairing the texture images. The algorithms from article [8] and article [11] also have a good repair effect when the inpainting region is small.

Then, the experiment of images with large region to be restored is carried out and it usually contains images with large area of destruction and target removal. Our experimental images are shown in Fig. 7.

Experimental results are shown in Fig. 8. We can see that article [2] diffusion-based inpainting algorithm lead to blurring effect in the repairing of large areas. So the diffusion-based inpainting algorithm does not apply to the large area restoration. Article [8] is the exemplar-based inpainting algorithm. It is based on the image patch so it's suitable for large area of texture image restoration. However, the usage of a single best matching image patch in the repair is a kind of greedy idea, so it will cause the unwanted object in the inpainting region

Fig. 6. Comparison of the results of different repair algorithms in the text and scratch removal. Each column from left to right respectively is the experimental results derived from diffusion-based [2], example-based [8], sparse-based [11] and proposed algorithm.

Fig. 7. Experimental images with large region to be restored. a′ and b′ represent inpainting images with large area of destruction; c′–e′ represent target removal.

Table 1. Quality assessment of image inpainting using PSNR

Image number	Diffusion [2]	Exemplar [8]	Sparse [11]	This paper
Figure 5a′	33.52	32.44	32.68	34.55
Figure 5b′	26.20	24.24	24.28	24.28
Figure 5c′	28.38	26.27	26.28	27.13
Figure 5d′	31.11	30.13	30.77	30.75
Figure 5e′	21.44	20.84	20.57	22.29
Figure 7a′	24.25	27.66	22.39	28.76
Figure 7b′	31.37	32.44	32.41	32.60
Average	28.04	27.72	27.06	28.69

Table 2. Comparisons of time consuming using different restoration algorithms (unit: second)

Image number	Exemplar [8]	Exemplar [9]	Sparse [11]	This paper
Figure 5a′	332.59	327.47	327.35	80.81
Figure 5b′	327.08	317.06	306.69	82.37
Figure 5c′	263.42	240.12	231.72	64.01
Figure 5d′	357.79	336.82	332.03	99.95
Figure 5e′	297.37	274.82	276.36	70.64

(*e.g.* Fig. 7c′ and d′ repair results). The algorithm in [11] and the proposed algorithm in this paper are both based on sparse representation. So the two algorithms can overcome the drawbacks of the exemplar-based inpainting algorithm and effectively repair the damaged areas of the texture images. But from the results of Fig. 7a′, b′ and d′, the proposed method has a better inpainting effect than the algorithm in [11]. So the proposed algorithm achieves the superior performance.

Next, we use the peak signal to noise ratio (PSNR) between original image and the inpainted image as the evaluation index of the image inpainting quality. We compare the effectiveness of different algorithms in PSNR, and the results are shown in Table 1. As can be seen from the Table 1, when the unknown region is small (*e.g.* Fig. 5b′–d′), diffusion-based method can obtain the higher PSNR. But when the unknown area is large (*e.g.* Fig. 7a′ and b′), the PSNR becomes the smallest. It can be seen that the proposed method not only has a good result from the visual evaluation, but also achieves the highest objective evaluation.

Aiming at the five images in Fig. 5, the repair time (*i.e.* from input inpainting image to output the results) of each algorithm is recorded. The time complexity of the algorithms is compared in Table 2. The exemplar-based inpainting algorithm [8,9] traversed the whole image to search for the best matching block when repairing each image block. So the time complex degree is high. Based on sparse inpainting algorithm [11], sparse coefficients are solved by using k nearest neighbors as sparse dictionary. So it also traversed the whole image when repairing each image block. In Table 2, it can be seen that our proposed algorithm is better than the exemplar-based inpainting algorithm and the existed prior sparse-based inpainting algorithm in the time complexity. The time consumption of proposed algorithm reduced to a quarter of the previous three algorithms.

Fig. 8. Comparisons of different restoration algorithms for target removal experiments. Each column from left to right respectively is the experimental results derived from diffusion-based [2], example-based [8], sparse-based [11] and proposed algorithm.

5   Conclusion

In this paper, a new image inpainting algorithm based on sparse representation is proposed. Through experiments on various kinds of images, our proposed algorithm overcomes the shortcomings of the traditional algorithm based on exemplar, and has higher inpainting quality, and greatly reduces the time complexity. In this paper, the original dictionary is clustered to form several more similar sub-complete dictionary. It avoids the interference of the non-related atoms to the sparse reconstruction. So it improves the quality of the inpainting, and greatly reduces the time complexity of the algorithm. Then non-negative orthogonal matching pursuit algorithm is used to solve the sparse coefficients. The non-negative constraint is introduced to avoid the negative correlation atoms with the inpainting image patch involved in the sparse reconstruction, which can further improve the inpainting quality. In the future, we will research on adaptive selection of image patch size, so that it can automatically determine the appropriate size of image blocks according to the size of the image and the size of the texture.

Acknowledgement. This work was supported in part by the National Natural Science Foundation of China under Grant 61432014 and Grant 61501339, in part by the Fundamental Research Funds for the Central Universities under Grant BDZ021403, Grant XJS15049, Grant XJS15068, and Grant JB160104, and in part by the China Post-Doctoral Science Foundation under Grant 2015M580818, and Grant 2016T90893.

References

1. Bertalmio, M., Sapiro, G., Ballester, C., Caselles, V.: Image inpainting. In: Computer Graphics and Interactive Techniques, pp. 417–424 (2000)
2. Bertalmio, M., Bertozzi, A.L., Sapiro, G.: Navier-stokes, fluid dynamics, and image and video inpainting. In: Computer Vision and Pattern Recognition, pp. 355–362 (2001)
3. Telea, A.: An image inpainting technique based on the fast marching method. J. Graph. Tool **9**(1), 23–24 (2004)
4. Tschumperlì, D.: Fast anisotropic smoothing of multi-valued images using curvature-preserving PDE's. Int. J. Comput. Vision **68**(1), 65–82 (2006)
5. Chan, T., Shen, J.: Local inpainting models and tv inpainting. SIAM J. Appl. Math. **62**(3), 1019–1043 (2001)
6. Efros, A., Leung, T.: Texture synthesis by non-parametric sampling. In: The Proceedings of IEEE International Conference on Computer Vision, pp. 1033–1038 (1999)
7. Bertalmio, G.S.M., Vese, L., Osher, S.: Simultaneous structure and texture image inpainting. IEEE Trans. Image Process. **12**(8), 882–889 (2003)
8. Criminisi, P.P., Toyama, K.: Region filling and object removal by exemplar-based image inpainting. IEEE Trans. Image Process. **13**(9), 1200–1212 (2004)
9. Wong, A., Orchard, J.: A nonlocal-means approach to exemplar-based inpainting. In: Proceedings of IEEE International Conference on Image Processing, pp. 2600–2603 (2008)

10. Shen, B., Hu, W., Zhang, Y., Zhang, Y.J.: Image inpainting via sparse representation. In: Proceedings of IEEE Conference on Acoustics Speech and Signal Processing, pp. 697–700 (2009)
11. Xu, Z., Sun, J.: Image inpainting by patch propagation using patch sparsity. IEEE Trans. Image Process. 19(5), 1153–1165 (2010)
12. Ogawa, T., Haseyama, M.: Image inpainting based on sparse representations with a perceptual metric. EURASIP J. Adv. Signal Process. 2013(1), 1–26 (2013)
13. Davis, G., Mallat, S., Avellaneda, M.: Adaptive greedy approximations. Constr. Approximation 13(1), 57–98 (1997)
14. Candés, E.J.: Compressive sampling. In: Proceedings of the International Congress of Mathematicians, pp. 1433–1452 (2006)
15. Chen, S.S., Donoho, D.L., Saunders, M.A.: Atomic decompositionby basis pursuit. SIAM Rev. 43(1), 129–159 (2001)
16. Lu, W., Vaswani, N.: Modified basis pursuit denoising (MODIFIED-BPDN) for noisy compressive sensing with partially known support. In: Proceedings of IEEE Conference on Acoustics Speech and Signal Processing, pp. 3926–3929 (2010)
17. Lee, H., Battle, A., Raina, R., Ng, A.Y.: Efficient sparse coding algorithms. In: Advances in Neural Information Processing Systems, pp. 801–808 (2006)
18. Mallat, S.G., Zhang, Z.: Matching pursuits with time-frequency dictionaries. IEEE Trans. Signal Process. 41(12), 3397–3415 (1993)
19. Chen, L.W., Chen, S., Billings, S.A., Luo, W.: Orthogonal least squares methods and their application to non-linear system identification. Int. J. Control 50(5), 1873–1896 (1989)
20. Yaghoobi, M., Wu, D., Davies, M.E.: Fast non-negative orthogonal matching pursuit. Signal Process. Lett. 22(9), 1229–1233 (2015)
21. Cai, D.: Litekmeans: the fastest matlab implementation of kmeans (2011). http:// www.zjucadcg.cn/dengcai/Data/Clustering.html

Efficient Image Retrieval via Feature Fusion and Adaptive Weighting

Xiangbin Shi[1,2,3(✉)], Zhongqiang Guo[3], and Deyuan Zhang[1,2]

[1] School of Computer, Shenyang Aerospace University, Shenyang 110136, China
sxb@sau.edu.cn
[2] Key Laboratory of Liaoning General Aviation Academy, Shenyang 110136, China
[3] School of Information, Liaoning University, Shenyang 110036, China

Abstract. In the community of content-based image retrieval (CBIR), single feature only describes specific aspect of image content, resulting in false positive matches prevalently returned as candidate retrieval results with low precision and recall. Typically, frequently-used SIFT feature only depicts local gradient distribution within ROIs of gray scale images lacking color information, and tends to produce limited retrieval performance. In order to tackle such problems, we propose a feature fusion method of integrating multiple diverse image features to gain more complementary and helpful image information. Furthermore, to represent the disparate powers of discrimination of image features, a dynamically updating Adaptive Weights Allocation Algorithm (AWAA) which rationally allocates fusion weights proportional to their contributions to matching is proposed in the paper. Extensive experiments on several benchmark datasets demonstrate that feature fusion simultaneously with adaptive weighting based image retrieval yields more accurate and robust retrieval results than conventional retrieval schema.

Keywords: Image retrieval · Feature fusion · Adaptive weighting · Color Names · BoW

1 Introduction

With the rapid growth of digital image data around us, conventional image retrieval methods based on keywords labelling seem incompetent to large scale image retrieval tasks, attributing to two main disadvantages. These disadvantages are heavy overhead on manually labelling, and matching deviation due to different individual subjectivities. In contrast, there is an increasing number of application areas and market demands for content-based image retrieval technology, accompanied by many inevitable challenges in the meantime.

So far, a mass of state-of-the-art CBIR methods are based on Bag-of-Words (BoW) model [5,8,10] which originates from text retrieval field, describing an image by a bag vector consisting of feature descriptor words. First of all, local descriptors (e.g. SIFT [3]) from ROIs are computed with invariant detectors

© Springer Nature Singapore Pte Ltd. 2016
T. Tan et al. (Eds.): CCPR 2016, Part II, CCIS 663, pp. 259–273, 2016.
DOI: 10.1007/978-981-10-3005-5_22

or affine region detectors, outputting a large amount of local high dimensional feature vectors from image patches. After that, constructing the codebook composed of many codewords with unsupervised clustering methods (k-means, approximate k-means, hierarchical k-means, etc.) [6]. Finally, the continuous feature space is subdivided into discrete search space by visual words quantized by the pre-trained codebook. Under BoW model representation, descriptor vectors from images are quantized to the corresponding nearest centroids within the codebook, and then an image is depicted as a frequency vector of those codewords for subsequent matching.

Typically, TF-IDF (Term Frequency - Inverted Document Frequency) [8,10] schema is employed to weight each visual word, similar to the processing in text retrieval. An entropic weight per vocabulary word is calculated, which relies on its probability across the dataset images. Generally utilized in the BoW based image retrieval system, IDF assigns less weight to punish the terms with higher frequency, and more weight to support less frequent terms.

Distance computation and feature matching between a query and target images in datasets rely on visual words, within CBIR system based upon BoW model. Whereas, under this circumstance, weak matching power of single feature tends to result in false matching, as single image feature only excavates one kind of image information and can not reflect the rich image content completely [9]. Apart from low retrieval performance due to the absence of other crucial image characteristics, information loss [8] during feature quantization downgrades the retrieval accuracy as well.

To address the aforementioned problems, as our contributions, an effective image retrieval method based on feature fusion is proposed in this paper, merging RootSIFT [1,2], CN [18] color and Gabor texture features to perform image retrieval, and obtaining more abundant and meaningful image information to increase retrieval accuracy. Multiple features fusion are performed at index level inside the inverted multi-index framework [11], constructed by several codebooks of different image features trained respectively. Furthermore, a dynamically updating Adaptive Weights Allocation Algorithm (AWAA) is presented in this paper, to take full advantage of collected image information and matching powers from diverse features, for more accurate retrieval.

The remainder of this paper is organized as follows. After an overview of related work in Sect. 2, we introduce the feature fusion based image retrieval mode in Sect. 3. Then, AWAA weighting algorithm for image retrieval is described minutely in Sect. 4. In Sect. 5, experimental results are presented and discussed. Finally, we draw the conclusion in Sect. 6.

2   Related Work

There are some works focusing on alleviating quantization error [8]. For instance, soft matching [4] assigns each descriptor to several visual words at the same time, at the cost of query time improvement and high memory overhead. Hamming Embedding (HE) [7] provides binary signatures which can filter out false matches to refine visual words matching. Binary features [12] can be exploited to increase system efficiency and to decrease quantization error.

To improve searching accuracy, spatial constraints are taken in consideration, such as weak geometric consistency (WGC) [7] and RANSAC verification [13]. WGC filters inconsistent matching descriptors in terms of scale and angle. Some research on contextual cues of visual words is done to improve precision in the case of retaining high recall, such as [13,14]. For example, Shen et al. [15] utilize voting-based method to conduct localization and image retrieval synchronously. In [17], Wang and Yang weight visual matching local spatial context similarity. In the meantime, embedding binary features can improve the precision of visual matching to some extent [7], which rebuilds the discriminative power of visual words through mapping feature descriptors into binary features.

Since single feature descriptor only carries partial image content information, it is insufficient to distinguish all kinds of images with limited discriminative power. There are many researchers devoting themselves to fuse local features, achieving complementary information for accurate retrieval [18,19]. Wengert et al. [23] augment the inverted files with local color feature descriptors to supplement hue information. In [21], Zhang et al. combine global and local features on the basis of BoW model, and exploit graph based query specific fusion method to improve retrieval precision. Yue et al. [22] extract color histogram from HSV color space as color feature, and gray level co-occurrence matrix, and then combine these two kinds of image features for better visual perception and retrieval results. Inspired by above-mentioned work, in this paper, we propose to investigate feature fusion and adaptive weighting for accurate and robust image retrieval.

3   Feature Fusion for Image Retrieval

This section gives a formal and detailed description of feature fusion strategy for image retrieval.

3.1   Overview of Image Features

Low-level image features selection and extraction are antecedent bases for content-based image retrieval, which can influence subsequent similarity measurement and the entire retrieval performance. In order to achieve optimal matching power by capturing more diverse valuable and complementary information to the greatest extent among various image features, this paper employs a specific feature fusion method. Besides, the features should keep as independent and heterogeneous as possible, for less redundant information and more helpful information, decreasing unnecessary computational complexity and memory requirement directly.

Following the feature selection principles above, we choose RootSIFT instead of original SIFT feature, color and texture features in our retrieval system.

RootSIFT feature. SIFT [3] is a kind of robust local feature descriptor with scale invariance and rotation invariance introduced by Lowe, and is widely applied vast fields such as image matching, image stitching, image classification and retrieval. The work of Arandjelovic [2] shows the fact that, after a conversion from SIFT to RootSIFT by mapping SIFT space to RootSIFT space through Hellinger kernel computing, the superior performance can be achieved without extra processing and memory overhead. We follow the method in [2], and employ excellent RootSIFT feature to replace conventional SIFT.

Color Names (CN) feature. A potent local color feature, called Color Names (CN, a.k.a., color attributes) [18] is exploited in this paper, for the helpful information supplement to visual content with color cue. Eleven basic color terms in English from linguistic words are chosen and assigned to colors around us to depict the colorful world: black, blue, brown, grey, green, orange, pink, purple, red, white and yellow. Thus, compact and efficient 11-dimensional color descriptor vectors can be extracted from the given image region R, simply shown as in the following formulation:

$$CN = \{p\left(cn_1|R\right), p\left(cn_2|R\right), \cdots, p\left(cn_{11}|R\right)\} \tag{1}$$

where $p\left(cn_i|R\right)$ is the probability value of cn_i, the i-th color name of 11 predefined color names. Towards this end, the mapping from a set of image pixel values within the specified region R to color names probabilities is learnt using Google-collected color names images following the method of Khan and Weijer [18], wherein every 100 images for each color name.

Gabor texture feature. Many works show that Gabor wavelet is an effective and efficient means of texture analysis, and then Gabor texture vectors can be obtained by computing the mean and standard deviation from Gabor filtered images. Generally, Gabor filters are a group of self-similar wavelets generated through rotation and dilation of the mother wavelet by different directions and scales, and each of them captures energy in a particular combination case. Then texture vectors are generated from the group of energy distributions.

3.2   Feature Extraction and Processing

SIFT feature descriptors are detected and extracted from regions of interest within each image using Open SIFT library [24] of Rob Hess. Then, RootSIFT descriptors are obtained via a series of transformations from original SIFT vectors, such as, L_1 normalization, square root operation of each vector element in the SIFT descriptor, L_2 normalization, etc., as introduced in [2]. As for color feature, Color Names descriptors with dimensionality of 11 are computed and acquired from non-overlapping image regions of a specified sampling window size by color names learning model trained on collected color image gallery from Google. Besides, we employ 30 Gabor wavelet filters to describe image texture content, with 5 different scales and 6 kinds of orientations, achieving $30 \times 2 = 60$ dimensional Gabor texture descriptor composed of mean and standard deviation components from such transformation coefficients.

In consideration of speed, different feature files are all stored and accessed in binary forms. We train three independent codebooks for RootSIFT, CN and Gabor features using the corresponding feature descriptors, respectively. For overall retrieval performance of both precision and recall, Multiple Assignment (MA) technique is adopted here for approximate nearest neighbor (ANN) search rather than exact retrieval mechanism which is excessively rigid, inflexible and tends to more generate false matches for cases of false positive matching and false negative matching.

3.3   Multiple Features Fusion and Complementation

Since single feature only focuses generally on partial image content with limited information and restricted discriminative power, multiple feature fusion is exploited for more accurate retrieval, in virtue of complementary image information and increased matching power. We design a specific fusion schema to mine each kind of image feature content to the greatest extent, and then integrate all collected helpful information to improve the discriminative power and enhance the retrieval performance.

Considering that components within a vector may be of disparate physical quantities and diverse magnitudes, consequently intra-normalization is required to guarantee equal significance among components inside vectors when matching. This paper applies Gaussian normalization in the intra-normalization process. Euclidean distance is exploited for distance computing to measure similarities.

Meanwhile, similarity measurement results from diverse image features, i.e., RootSIFT, CN and Gabor, may vary largely with different value basis points and ranges. As a consequence, Gaussian normalization is adopted again for the inter-normalization processing.

RootSIFT, CN and Gabor features are extracted from each image in dataset, respectively, and then save these features in individual files and store them in the features dataset. Features extracted from query image provided by the user, is computed and matched with features in features dataset. Final candidate target images similar to the query image are returned as results. In the matching process, optional weighting mechanism can be leveraged for different contributions of diverse features.

The matching function between two local features x and y is defined as

$$f_m\left(x,y\right) = \delta_{m(x),m(y)} = \begin{cases} 1 & \text{if } m\left(x\right) = m\left(y\right) \\ 0 & \text{otherwise} \end{cases} \tag{2}$$

where $m\left(\cdot\right)$ functions by quantizing local descripotors to the nearest codewords in the corresponding visual vocabulary, and δ is the Kronecker delta function.

In this paper, matching function of two local feature tuples $\boldsymbol{x} = (x^r, x^c, x^g)$ and $\boldsymbol{y} = (y^r, y^c, y^g)$ is defined as

$$f_{m_r,m_c,m_g}\left(\boldsymbol{x},\boldsymbol{y}\right) = w_1 \times \delta_{m_r(x^r),m_r(y^r)} \times \delta_{m_c(x^c),m_c(y^c)} + w_2 \times \delta_{m_g(x^g),m_g(y^g)} \tag{3}$$

$$\text{s.t.} \quad w_1 + w_2 = 1$$

where $m_r (\cdot)$, $m_c (\cdot)$ and $m_g (\cdot)$ are quantization functions for RootSIFT, CN and Gabor features, respectively. In Eq. 3, x^r, x^c and x^g are feature vectors for these three kinds of features, besides w_1 and w_2 are two weights satisfying a linear constraint condition for RootSIFT×CN and Gabor features, respectively.

Furthermore, TF-IDF scheme is applied to weight different visual words in the codebooks based on their frequencies of occurrence in each single image and in the entire image dataset.

Extensive experiments demonstrate that RootSIFT descriptor has the greatest discriminative power among these three features, and CN is in the second place, while Gabor texture feature performs the worst. During feature fusion (Eq. 3), two matched images must be matched successfully both in RootSIFT and CN features simultaneously. Gabor texture is added rather than be multiplied into feature fusion because Gabor texture has the weakest matching power among them, which will restrict the overall recall due to the excessively rigor matching condition if adopting multiplication fusion mode. The addition mode not only improves the precision due to complementary texture information, but also guarantees even increases the recall.

For instance, texture will be the key feature for matching when both RootSIFT and CN lose their efficacy synchronously in particular cases, namely, the matching failure of RootSIFT×CN does not mean the holistic failure thanks to the main force of texture in such extreme circumstances. The feature fusion weights w_1 and w_2 proportional to their contribution, can be assigned by specified weights, or equal values ($w_1 = w_2 = 0.5$), or empirical values based on domain knowledge, or relatively optimal values computed automatically by the computer. An efficient dynamically updating adaptive weights allocation algorithm will be presented in next section.

Binary signatures are calculated from primitive RootSIFT, CN and Gabor descriptors, respectively, to refine visual matching and to lower quantization error, in terms of related approaches [7,8]. In the meantime, L_2 normalization is employed for each image for fair comparison between the query image and every target images in the dataset. Finally, similarity score between query image Q and target image I in dataset are written as follows,

$$sim(Q, I) = \frac{\sum_{\boldsymbol{x} \in Q, \boldsymbol{y} \in I} f_{m_r, m_c, m_g}(\boldsymbol{x}, \boldsymbol{y}) \cdot idf^2}{\|Q\|_2 \|I\|_2} \qquad (4)$$

where $\|Q\|_2$ and $\|I\|_2$ are the query and dataset image after normalization, severally. A higher similarity score means a more relevant relationship between Q and I.

4   Adaptive Weights Allocation Algorithm

Take different contributions of diverse image features into consideration, rational weighting during feature fusion is necessary and helpful. Conventional weights allocation methods are mostly based on human relevance feedback, which suffers two main problems. First, the user is required to rank each returned images by several levels, typically five levels (highly relevant, relevant, no opinion, irrelevant, and highly irrelevant) evaluation, with heavy manual workload. Second, weighting methods based on human relevance feedback are excessively dependent on user knowledge and perception subjectivity. For this reason, an efficient and robust weighting method is proposed in this paper, namely, Adaptive Weights Allocation Algorithm (AWAA).

Suppose the curve of retrieval precision over weight w_1 is plotted, then w_2 is fixed as $1-w_1$ due to the linear constraint with w_1 varying within the range from 0 to 1. Extensive experiments show that the $Precision@w_1$ curve is manifested as a concave function (concave downwards) curve in mathematical form. The proposed AWAA weighting method aims to search automatically and quickly for the approximately optimal weight w_1 with approximately highest precision (Eq. 5). The key steps of AWAA weighting algorithm is described as follows.

$$w_{opt} = arg \max_{w_1} Precision = arg \max_{w_1} f(w_1) \tag{5}$$

Step 1. Initialize the weight w_1 with w_0, where w_0 can be set, with user-specified value (e.g., 0.8), or with empirical value, or random value, or middle value (0.5).

Step 2. Set $w_{mid} = w_1$, besides, two additional weights w_{low} and w_{high} are defined as

$$w_{low} = w_{mid}/2, \ w_{high} = w_{mid} + (1 - w_{mid})/2 \tag{6}$$

Step 3. Given an image dataset with M images, $N_Q = \max(M/100, 30)$ images are chosen randomly as queries, where $M/100$ is rounded off for an integer. Note that these N_Q query images are fixed for precisions computation in subsequent iterations before the convergence. When w_1 is set as w_{low}, w_{mid} and w_{high} orderly, computing the average $Precision^i_{low}$, $Precision^i_{mid}$, $Precision^i_{high}$, as the mean precisions in the i-th iteration.

Step 4. Compare the values of $Precision^i_{low}$, $Precision^i_{mid}$ and $Precision^i_{high}$, simultaneously, the maximum value of these three precisions, i.e., $Precision^i_{max}$ is saved for subsequent comparison. The next process depends on the comparison results of the three precisions, which will be divided into 4 cases and discussed next.

- **Case A.** $Precision^i_{low} \le Precision^i_{mid} \le Precision^i_{high}$, indicating that the precision keeps growing as w_1 varies from w_{low} to w_{mid} and w_{high}. The optimal target weight value w_{opt} is around w_{high}, may higher, or lower than w_{high}, or equal to w_{high}. Next performing weights updating operations as $w_{low} = w_{mid}$, $w_{mid} = w_{high}$, and $w_{high} = 1$, resulting in a new weight solution space denoted as $[w_{low}, w_{mid}, w_{high}]$.
- **Case B.** $Precision^i_{low} \le Precision^i_{mid} \ge Precision^i_{high}$, according to which, the precision increases at first and then decreases as w_1 grows. This time, w_{opt} is around w_{mid}, in the range from w_{low} to w_{high}. In this case, the Weight Search Space (WSS) remains unchanged as before, i.e., $[w_{low}, w_{mid}, w_{high}]$.

- **Case C.** $Precision_{low}^i \geq Precision_{mid}^i \geq Precision_{high}^i$, contrary to Case A, the precision keeps decreasing when w_1 going up. As a consequence, w_{opt} is around w_{low}, and update operations are $w_{high} = w_{mid}$, $w_{mid} = w_{low}$ and $w_{low} = 0$, leading to the new weight searching range $[w_{low}, w_{mid}, w_{high}]$.
- **Case D.** $Precision_{low}^i \geq Precision_{mid}^i \leq Precision_{high}^i$, which means feature fusion based retrieval performance is inferior to that of single feature retrieval, with performance degradation instead of boost. Such extreme case does not exist in terms of multi-features fusion retrieval, which will not be discussed further.

Step 5. After various analyses and adjustments of weight search space under several cases in Step 4, here the obtained weight search space will be compressed further to approach approximately optimal weight value. Firstly, the geometric center point w_{middle} of weight search space is defined as

$$w_{middle} = (w_{low} + w_{high})/2 \tag{7}$$

The precision w.r.t. w_{middle} and w_{mid} are denoted as $f(w_{middle})$ and $f(w_{mid})$ in mathematical form, respectively. Then the updating process of w_{mid}, w_{low} and w_{high} are shown as follows:

$$w_{mid} = \begin{cases} w_{middle} & \text{if } f(w_{mid}) < f(w_{middle}) \\ w_{mid} & \text{otherwise} \end{cases} \tag{8}$$

$$w_{low} = (w_{low} + w_{mid})/2 \tag{9}$$

$$w_{high} = (w_{mid} + w_{high})/2 \tag{10}$$

The compressed weight search space is updated as $[w_{low}, w_{mid}, w_{high}]$.

Step 6. Turn to Step 3, and start the next iteration until meeting the termination conditions. There are three main different iteration termination conditions. The first one is reaching the specified iteration number, and the active termination from the user is the second condition. Besides, the iteration will be terminated when satisfying a certain error constraint condition on precision, i.e.,

$$|Precision_{max}^{i+1} - Precision_{max}^i| \leq \varepsilon \tag{11}$$

where $Precision_{max}^{i+1}$ and $Precision_{max}^i$ are the $(i+1)$-th and i-th round retrieval precisions, $0 < \varepsilon < 1$ and ε is a quite small positive number.

Step 7. The approximately optimal weight value $w_{opt} = w_{mid}$ is achieved, then $w_1 = w_{opt}$ and $w_2 = 1 - w_1$ for fusion weights assignment.

The proposed AWAA weighting method can converge fast to an approximately optimal weight value, based on the combination of images content characteristics and diverse features representations on given image dataset. Once the target retrieval image dataset changes, AWAA algorithm is employed again for weights searching and optimizing on the entire target dataset in advance, and then the approximately optimal fusion weights are fixed for subsequent retrieval task on the same target dataset if there are no dataset variations. On the other hand, the iteration termination condition can be adjusted in terms of requirement, and we can initialize w_1 in lots of ways with great flexibility. Furthermore, our experiments demonstrate that AWAA weighting approach outperforms conventional weighting methods based on human relevance feedback, with higher accuracy and robustness.

5   Experiments

In this section, experiments are conducted on several benchmark datasets, i.e., Corel-1000, Ukbench, Holidays and Oxford building datasets. Brief introductions of datasets are presented firstly, and then experimental parameters settings are discussed. Finally, experimental results and analyses are shown at the end.

5.1   Datasets and Evaluation

Corel-1000 dataset. There are 10 classes of images with diverse objects and scenes, such as Africa, Dinosaurs, Horses, etc., with each class of 100 different images under the same category. Precision and recall are taken to measure retrieval performance.

Ukbench dataset. This dataset consists of 2550 groups of 10200 images, and each group contains 4 similar images with different illumination variations, viewpoints, or scales. The measurement of performance, called N-S score, is the number of relevant images in returned top 4 results, with theoretical maximum value of 4.

INRIA Holidays dataset. An image collection of 1491 personal holiday images, which is composed of 500 image groups (queries) and 991 corresponding relevant images in the dataset. The first image of each group is taken as the query image, and the rest are the true positive images in the group. Here, mAP (mean Average Precision) is adopted for similarity measurement.

Oxford building dataset. A total of 5063 images in this dataset are collected from Flickr labelled with keywords on Oxford buildings. The dataset contains 55 queries, with the ground truth results composed of other images of the same building landmark. Again, mAP serves as the retrieval performance metric.

5.2   Experimental Parameters Settings

In the experiments, 128-D RootSIFT vectors can be achieved after the conversion from original SIFT to RootSIFT, following the method in [2]. As for CN color feature, image patch size is set as 8×8 pixels (width $\times$ height) sampled from entire original image without overlapping. When applying Gabor filters for texture content analysis, we set the numbers of scales and orientations are 5 and 6, respectively, obtaining 60-D Gabor texture vectors. Mask size of each Gabor filter is set as 60×60. Each image is (approximately) equally divided into 16 sub-images in non-overlapping manner, and then one 60-D vector is extracted from each sub-image. Therein, the rightmost and downmost lines of sub-images near the image boundary only have several a little more pixels than other sub-images, due to divisibility problem of image size.

Both Bow model and inverted multi-index are applied in our retrieval system, and codebooks with different sizes are trained offline using k-means clustering algorithm for RootSIFT, CN and Gabor descriptors, respectively. For instance, a codebook including 20K codewords for RootSIFT feature, a codebook containing 200 codewords for CN feature, and a Gabor codebook with size of 500, are all trained severally. Image descriptor vectors are quantized to multiple visual words in the corresponding codebooks according to MA strategy, indexed by inverted files for fast searching. Specially, a large MA value of 100 is set for CN color feature to reduce the detrimental influence from strong illumination variances, while MA values 3 is assigned to both RootSIFT and Gabor features. The 64-bit, 22-bit and 30-bit binary signatures for RootSIFT, CN and Gabor feature, respectively, are exploited in our experiments for visual refining.

5.3   Experimental Results Analysis

A workstation with 3.2 GHz CPU and 16 GB memory is applied to conduct our experiments. When performing experiments on Corel-1000 dataset, 30 images are chosen randomly from 100 images in each class as queries, and computing the means of all of the precision and recall values, respectively. For every retrieval process, N_{RT} images are returned as retrieval results, and N_{RT} is set to 20 in our experiments.

First of all, retrieval results comparisons between retrieval systems based on single feature and multiple features fusion, are demonstrated in Fig. 1. Note that the theoretical maximum of recall in the right plot is limited to 0.2, because only 20 images are returned as results while there are 100 relevant images totally for each query in the class. Firstly, retrieval accuracies are quite different among single features depending on their representation capacities and discriminative powers, and so is the robustness. On the whole, RootSIFT has the best matching power and robustness with consistently highest precision and recall among these three features, showing slight performance floating changes on the 10 image classes in Corel-1000. In contrast, Gabor texture feature works the worst due to barren image information and weak matching power. And CN color feature falls in between the other two features mentioned above. Once in a while, CN outperforms RootSIFT slightly due to abundant color information achievement from images in Buses and Food groups. Similarly, Gabor feaure rarely surpasses CN with rich texture content on Mountains group. Secondly, retrieval method based on feature fusion enhances retrieval performance significantly with higher precision, recall and robustness, compared with single feature retrieval schema, on account of more captured complementary and comprehensive image information.

Besides, a tentative experiment is designed and carried out here, to investigate the rule of retrieval performance changes as the value fusion weight w_1

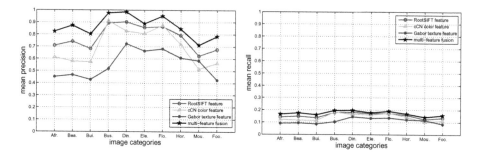

Fig. 1. Performance comparison between single feature and features fusion retrieval

varies range from 0 and 1. Precision is adopted as retrieval performance evaluation protocol here, and multiple assignment (MA) is employed to guarantee recall. The fusion weight interval $[0, 1]$ is equally divided into 10 value subspaces with 11 endpoint values, i.e., 0, 0.1, 0.2, 0.3, $\cdots$, 1.0, with space division for simple test. For each fusion weight w_1 assigned with each endpoint value ($w_2 = 1 - w_1$ accordingly), retrieval precisions is calculated several times for N_Q randomly sampled query images, where N_Q is specified number which is set to 30 in our experiments. During feature fusion with varying fusion w_1 among these 11 weight values, mean precision results with each endpoint weight are computed and plotted in Fig. 2. Experiment results from groups of images on corel-1000 dataset, namely, Africa, Flowers, Horses and Food are shown in Fig. 2, where N_Q images are chosen at random as queries for each group for mean precision computation.

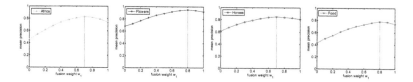

Fig. 2. Retrieval precision changes against different fusion weight w_1 values on Africa, Flowers, Horses and Food groups from Corel-1000 dataset

As can be seen from Fig. 2, different retrieval performances are achieved on different image classes in Core-1000 dataset due to diverse image content characteristics and various information richness properties. For instance, higher accuracies of Flowers and Horses groups than that of Africa and Food groups as demonstrated in Fig. 2, presumably because there exists less effective and helpful image information and more interferential and redundant information junk in Africa and Food images, leading to less discriminative power than that of the former. In the meanwhile, RootSIFT×CN outperforms indeed Gabor texture feature quite significantly, as indicated before. The curves of precisions in Fig. 2

grow slowly as w_1 largens and then reach the peaks when w_1 is around 0.7 or 0.8 roughly, and drop slightly with well power-preserving under high fusion proportion of RootSIFT×CN features. Besides, the fact that retrieval performance based on feature fusion surpasses conventional single feature retrieval schema, is confirmed again. Different fusion weights produce widely different accuracy achievements, and hence proper fusion weights allocation is necessary and significant.

In addition, multi-feature fusion with non-weighting, empirically weighting, relevance feedback weighting and AWAA weighting schemas are tested comparatively as displayed in Fig. 3. Non-weighting schema means that equal weights $w_1 = w_2 = 0.5$ without consideration of the distinction among image features, performs feature fusion retrieval poorly compared with the other three weighting methods. Empirically weighting method outperforms non-weighting mode markedly, but relies directly on the understanding and mastery on specific image dataset content and statistic characteristics. Note that, improper empirical weight values may cause lower accuracies than non-weighting method under the some extreme cases, attributing to more unreasonable allocation of fusion weights. Both relevance feedback weighting and AWAA weighting approaches yield relatively better retrieval performance, owing to iterative fusion weights optimizing and comprehensive visual refining. Differently, rational and approximately optimal fusion weights distribution can be achieved automatically using AWAA weighting method, producing more robust discriminative power and more accurate retrieval results, on the basis of image content characteristics and features properties. Under a few cases, relevance feedback weighting produces slightly worse accuracy than that of empirically weighting method, which is caused by the deviation and unreliability of discrimination from human subjectivity. Experimental results in Fig. 3 demonstrates that AWAA weighting approach is superior to all of the other three weighting methods, without knowledge dependence and subjective interference of the users. In a nutshell, the proposed AWAA weighting method enhances retrieval performance robustly and consistently with high accuracy, and simultaneously improves the interaction between human and system with better user experience.

Retrieval examples comparison between feature fusion with adaptive weighting method introduced in this paper and state-of-the-art method c-MI proposed by Zheng [8], with randomly selected query images from Ukbench, INRIA Holidays and Oxford building datasets are demonstrated in Fig. 4. Insensitive to illumination variations, various scales, viewpoint changes, partial occlusions, etc., the proposed method achieves more accurate and robust searching with more true positive matches and less false positive matches on all of the datasets consistently, compared with c-MI, attributing to beneficial information complementation and rational fusion weights allocating. Additionally, more visually relevant matches ranked closer to the top on the ranking list, more in line with the visual perception of human beings.

From Table 1, we can see that the feature fusion retrieval method introduced in this paper outperforms other retrieval approaches considerably, on the

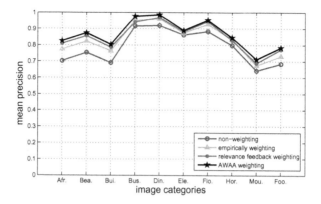

Fig. 3. Average precision against image classes in Corel-1000 dataset

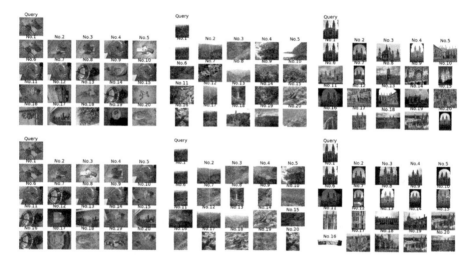

Fig. 4. Retrieval examples via feature fusion and adaptive weighting method in this paper (bottom row) on Ukbench (**left**), Holidays (**middle**), Oxford building (**right**) datasets compared with c-MI method (top row).

aspects of both precision and recall. Compared with conventional state-of-the-art methods, the proposed retrieval method in this paper has considerable precision boost with slight recall gain, thanks to more collected complementary and valuable image information for more powerful image representation, and also to the rational and approximately optimal weights allocation of AWAA weighting.

6   Summary and Outlook

In this paper, an effective and encouraging image retrieval method based on multiple feature fusion and adaptive weighting is proposed, outperforming

Table 1. Retrieval results comparison

Methods	Ours	[8]	[16]	[21]	[20]
Ukbench (N-S score)	**3.86**	3.85	3.75	3.77	3.55
Holidays (mAP, %)	**86.9**	85.8	84.7	84.6	84.8

single feature retrieval mode significantly, and also comparable to conventional state-of-the-art methods favorably, due to comprehensive and complementary information collecting and approximately optimal feature weighting by AWAA. Extensive experiments on several benchmark datasets verify the effectiveness and superiority of the proposed methods in this paper, with considerable accuracy improvement and high robustness.

In the feature, the mapping mechanism from multifarious image features to valuable and efficient image information collection and complementation, as well as the potential information redundancy and information interference among various features, and corresponding dimensionality reduction techniques, will be investigated later.

Acknowledgments. This work was supported by National Natural Science Foundation of China (61170185), Aerospace Science Foundation of China (2013ZC54011), Liaoning Province Doctor Startup Fund (20121034), and Program Funded by Liaoning Province Education Administration (L2014070).

References

1. Zheng, L., Wang, S., Tian, L., et al.: Query-adaptive late fusion for image search and person re-identification. In: Proceedings of the IEEE Conference on Computer Vision and Pattern Recognition, pp. 1741–1750. IEEE (2015)
2. Arandjelovic, R., Zisserman, A.: Three things everyone should know to improve object retrieval. In: Proceedings of the IEEE Conference on Computer Vision and Pattern Recognition, pp. 2911–2918. IEEE (2012)
3. Lowe, D.G.: Distinctive image features from scale-invariant keypoints. Int. J. Comput. Vis. **60**(2), 91–110 (2012)
4. Philbin, J., Chum, O., Isard, M., et al.: Lost in quantization: Improving particular object retrieval in large scale image databases. In: Proceedings of the IEEE Conference on Computer Vision and Pattern Recognition, pp. 1–8. IEEE (2008)
5. Fei-Fei, L., Perona, P.: A Bayesian hierarchical model for learning natural scene categories. In: Proceedings of the IEEE Conference on Computer Vision and Pattern Recognition, pp. 524–531. IEEE (2005)
6. Jegou, H., Douze, M., Schmid, C.: On the burstiness of visual elements. In: Proceedings of the IEEE Conference on Computer Vision and Pattern Recognition, pp. 1169–1176. IEEE (2009)
7. Jegou, H., Douze, M., Schmid, C.: Hamming embedding and weak geometric consistency for large scale image search. In: Forsyth, D., Torr, P., Zisserman, A. (eds.) ECCV 2008. LNCS, vol. 5302, pp. 304–317. Springer, Heidelberg (2008). doi:10.1007/978-3-540-88682-2_24

8. Zheng, L., Wang, S., Liu, Z., et al.: Packing and padding: coupled multi-index for accurate image retrieval. In: Proceedings of the IEEE Conference on Computer Vision and Pattern Recognition, pp. 1939–1946. IEEE (2014)

9. Fu, Y., Cao, L., Guo, G., et al.: Multiple feature fusion by subspace learning. In: Proceedings of the 2008 international conference on Content-based image and video retrieval, pp. 127–134. ACM (2008)

10. Zheng, L., Wang, S., Zhou, W., et al.: Bayes merging of multiple vocabularies for scalable image retrieval. In: Proceedings of the IEEE Conference on Computer Vision and Pattern Recognition, pp. 1955–1962. IEEE (2014)

11. Babenko, A., Lempitsky, V.: The inverted multi-index. In: Proceedings of the IEEE Conference on Computer Vision and Pattern Recognition, pp. 3069–3076. IEEE (2012)

12. Jegou, H., Douze, M., Schmid, C.: Hamming embedding and weak geometric consistency for large scale image search. In: Forsyth, D., Torr, P., Zisserman, A. (eds.) ECCV 2008. LNCS, vol. 5302, pp. 304–317. Springer, Heidelberg (2008). doi:10. 1007/978-3-540-88682-2_24

13. Philbin, J., Chum, O., Isard, M., et al.: Object retrieval with large vocabularies and fast spatial matching. In: Proceedings of the IEEE Conference on Computer Vision and Pattern Recognition, pp. 1–8. IEEE (2007)

14. Zhou, W., Li, H., Lu, Y., et al.: SIFT match verification by geometric coding for large-scale partial-duplicate web image search. ACM Trans. Multimedia Comput. Commun. Appl. (TOMM) 9(1), 4 (2013)

15. Shen, X., Lin, Z., Brandt, J., et al.: Object retrieval and localization with spatially-constrained similarity measure and K-NN re-ranking. In: Proceedings of the IEEE Conference on Computer Vision and Pattern Recognition, pp. 3013–3020. IEEE (2012)

16. Deng, C., Ji, R., Liu, W., et al.: Visual reranking through weakly supervised multi-graph learning. In: Proceedings of the IEEE International Conference on Computer Vision, pp. 2600–2607. IEEE (2013)

17. Wang, X., Yang, M., Cour, T., et al.: Contextual weighting for vocabulary tree based image retrieval. In: Proceedings of the IEEE International Conference on Computer Vision, pp. 209–216. IEEE (2011)

18. Khan, F.S., Anwer, R.M., van de Weijer, J., et al.: Color attributes for object detection. In: Proceedings of the IEEE Conference on Computer Vision and Pattern Recognition, pp. 3306–3313. IEEE (2012)

19. Zheng, Y., Zhang, Y.J., Larochelle, H.: Topic modeling of multimodal data: an autoregressive approach. In: Proceedings of the IEEE Conference on Computer Vision and Pattern Recognition, pp. 1370–1377. IEEE (2014)

20. Herve, J., Douze, M., Schmid, C.: Improving bag-of-features for large scale image search. Int. J. Comput. Vis. 87(3), 316–336 (2010)

21. Zhang, S., Yang, M., Cour, T., et al.: Query specific fusion for image retrieval. In: Zhang, S., Yang, M., Cour, T., Yu, K., Metaxas, D.N. (eds.) ECCV 2012. LNCS, vol. 7573, pp. 660–673. Springer, Heidelberg (2012). doi:10.1007/ 978-3-642-33709-3_47

22. Yue, J., Li, Z., Liu, L., et al.: Content-based image retrieval using color and texture fused features. Math. Comput. Model. 54(3), 1121–1127 (2011)

23. Wengert, C., Douze, M., Jegou, H.: Bag-of-colors for improved image search. In: Proceedings of the 19th ACM international conference on Multimedia, pp. 1437–1440. ACM (2011)

24. Hess, R.: An open-source SIFTLibrary. In: Proceedings of the 18th ACM international conference on Multimedia, pp. 1493–1496. ACM (2010)

The Effect of Quantization Setting for Image Denoising Methods: An Empirical Study

Feng Pan, Zifei Yan[✉], Kai Zhang, Hongzhi Zhang, and Wangmeng Zuo

Harbin Institute of Technology, Harbin 150001, China
cszfyan@gmail.com

Abstract. Image denoising, which aims to recover a clean image from a noisy one, is a classical yet still active topic in low level vision due to its high value in various practical applications. Existing image denoising methods generally assume the noisy image is generated by adding an additive white Gaussian noise (AWGN) to the clean image. Following this assumption, synthetic noisy images with ideal AWGN rather than real noisy images are usually used to test the performance of the denoising methods. Such synthetic noisy images, however, lack the necessary image quantification procedure which implies some pixel intensity values may be even negative or higher than the maximum of the value interval (e.g., 255), leading to a violation of the image coding. Consequently, this naturally raises the question: what is the difference between those two kinds of denoised images with and without quantization setting? In this paper, we first give an empirical study to answer this question. Experimental results demonstrate that the pixel value range of the denoised images with quantization setting tend to be narrower than that without quantization setting, as well as that of ground-truth images. In order to resolve this unwanted effect of quantization, we then propose an empirical trick for state-of-the-art weighted nuclear norm minimization (WNNM) based denoising method such that the pixel value interval of the denoised image with quantization setting accords with that of the corresponding ground-truth image. As a result, our findings can provide a deeper understanding on effect of quantization and its possible solutions.

Keywords: Image denoising · Low level vision · Quantization setting

1 Introduction

The target of image denoising is to estimate the latent clean image $\mathbf{x}$ from its noisy observation $\mathbf{y} = \mathbf{x} + \mathbf{v}$, where $\mathbf{v}$ is commonly assumed to be additive white Gaussian noise of standard deviation (AWGN). Over the past few decades, numerous denoising approaches [2,12,14,18,20,23] have been proposed. In general, those approaches can be divided into two main categories, one is based on maximum a posteriori (MAP) estimation method which is usually equipped with one or several specific natural image priors, the other one is the discriminative learning method which tries to learn a discriminative model from training data.

© Springer Nature Singapore Pte Ltd. 2016
T. Tan et al. (Eds.): CCPR 2016, Part II, CCIS 663, pp. 274–285, 2016.
DOI: 10.1007/978-981-10-3005-5_23

For the MAP-based denoising method, since the image prior plays the role of regularization to alleviate the ill-poseness, thus modeling good image prior is vital important for the success of the denoising method. Typical image priors include nonlocal self-similarity prior adopted in BM3D [7,15], LSSC [17], NCSR [11] and WNNM [13], and natural image patch prior adopted in a generative model named EPLL [26]. For the discriminatively trained model, the representative multi layer perceptron (MLP) approach [4] which learns the relationship between noisy input and clean output has shown competitive performance than BM3D. Note that the recently proposed CSF [21] and TNRD [6] are models which jointly learn the image prior and optimization procedure in a unified framework, and they can be explained from either the MAP or discriminative learning perspectives.

All the above image denoising methods assume the noisy image is generated by adding an additive white Gaussian noise (AWGN) to the clean image. Following this assumption, synthetic noisy images with ideal AWGN rather than real noisy images are usually used to test the performance of the denoising methods. Such synthetic noisy images, however, lack the necessary image quantization procedure which implies some pixel intensity values maybe even negative or higher than the maximum of the value interval (e.g., 255), leading to a violation of the image coding. It has been pointed out that, for realistic evaluation, the noisy input images should be quantized to the interval [0, 255] (i.e., 8-bits quantization) after adding the noise. Consequently, this naturally raises the following two questions: (1) what is the effect of quantization setting for image denoising methods? (2) If there exists an effect which results in inferior performance than that without quantization setting, how can we resolve or sidestep this problem?

In this paper, we address the first question by giving an empirical study on the denoised images of different methods with quantization setting. According to the results, we observed that the pixel value interval of the denoised images with quantization tend to be narrower than that without quantization due to the truncation of the pixel values out of the range of [0, 255]. In addition, different denoising methods have different sensitivity to the quantization setting but they all tend to deliver large difference in denoising results with the increase of noise level. Based on the analysis, we address the second problem by introducing a simple but effective trick to eliminate the negative effect of quantization. To testify the generality of our empirical trick, we apply it to two state-of-the-art denoising algorithms. EPLL is a representative generative model and WNNM [13] is one of low-rank approximation methods [3,5,9,10,13,16,19,22] which have exhibited promising performance on denoising. According to the denoising result of two algorithms with the proposed simple empirical trick, we have obtained very competitive performance with that without quantization setting, demonstrating that the negative effect of quantization setting for image denoising methods can be effectively suppressed by a simple method.

The rest of the paper is organized as follows. Section 2 gives the motivation and provides a detailed analysis of the difference between two kinds of denoised images with and without quantization setting. Section 3 describes an empirical trick suitable for state-of-the-art denoising methods. Section 4 presents the experimental results. Finally, Sect. 5 concludes this paper.

2 Motivation

The noisy observation $\mathbf{y}$ of a clean image $\mathbf{x}$ is usually modeled as

$$\mathbf{y} = \mathbf{x} + \mathbf{v}, \tag{1}$$

where $\mathbf{v}$ is the additive white Gaussian noise (AWGN) with zero mean and standard deviation σ. The goal of image denoising is to estimate the latent clear image $\mathbf{x}$ from $\mathbf{y}$. One common problem of most image denoising methods is that they neglect the quantization setting according to the real imaging process. The real data observation model with quantization is

$$\hat{\mathbf{y}} = \mathbf{x} + \mathbf{v}, \mathbf{y} = \min(255, \max(0, \hat{\mathbf{y}})), \tag{2}$$

where $\mathbf{y}$ is the real noisy observation after quantization of its latent observation $\hat{\mathbf{y}}$. The pixel values in $\hat{\mathbf{y}}$ are set to 255 when they are larger than 255, while set to 0 when they are smaller than 0. The quantization is performed before the denoising. After quantization, most image denoising algorithms can be carried out to complete the process of denoising.

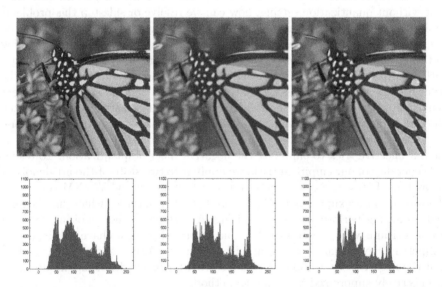

Fig. 1. The images and their respective histograms of original image, unquantized and quantized denoising results by WNNM on image *Monarch* at noise standard deviation $\sigma = 50$.

In order to investigate the changes between $\mathbf{y}$ and $\hat{\mathbf{y}}$, we use the histogram of an image which provides the frequency of the pixel values in the image. The histogram provides a natural bridge between images and a probabilistic description. We conduct an experiment using method WNNM at noise standard deviation $\sigma = 50$ on image *Monarch*. Figure 1 shows the images and their respective histograms of the original image, unquantized and quantized denoising results. From this figure, we have some observations. From the pixel interval perspective, there are few pixel values located in the range of [0–50] and [200–255] after quantization. For unquantized denoising results, there are even some pixels with the value larger than 255. Based on the above, one can draw a conclusion that the pixel interval of the denoised images after quantization tend to be narrower than that without quantization. In view of the pixel distribution, the pixel distribution of the unquantized image is more similar to that of the ground-truth. For the quantized image, however, the narrower interval results in great deviation from the pixel distribution of the ground-truth.

Table 1. Denoising results (PSNR) on image *Monarch* with different noisy standard deviations.

σ	EPLL			WNNM		
	Unquantized	Quantized	Δ_{PSNR}	Unquantized	Quantized	Δ_{PSNR}
15	32.03	32.16	0.13	32.72	32.72	0
25	29.30	29.51	0.21	29.85	29.83	−0.02
35	27.56	27.76	0.20	28.13	27.95	−0.18
50	25.78	25.67	−0.11	26.32	25.63	−0.69
60	24.86	24.37	−0.49	25.45	24.27	−1.18
75	23.72	22.46	−1.26	24.31	22.40	−1.91
90	22.78	20.65	−2.13	23.45	20.66	−2.79
100	22.23	19.52	−2.71	22.95	19.64	−3.31

We then evaluate the impact on the denoising performance. By setting the AWGN standard deviation $\sigma \in \{15, 25, 35, 50, 60, 75, 90, 100\}$, we evaluate two denoising methods, i.e., EPLL [26] and WNNM [13], in terms of the peak signal to noise ratio (PSNR). PSNR measures the physical quality of an image. It is defined as the ratio between the peak power of the image and the power of the noise in the image:

$$PSNR = 10 \log_{10} \left(\frac{\max \mathbf{DR}^2}{MSE} \right) \quad (3)$$

where MSE denotes Mean Square Error and $\mathbf{DR}$ is the dynamic range.

Table 1 shows the denoising results of noisy image *Monarch* with different noise standard deviations. For each denoising method, Δ_{PSNR} denotes the difference of PSNR values with and without quantization setting. Nearly all the difference values are negative, meaning that poor performance will be obtained for the direct use of a denoising method to the noisy image after quantization.

3   Image Denoising Framework

For the denosing of images after quantization, we come up with a simple but efficient trick to eliminate the negative effect of quantization. The key is to estimate the missed pixel values and retrieve information to get it back. All the denoising methods adopt iterative algorithms [1,8,24,25] to remove noise. We input $\mathbf{y}$ into any one of denoising algorithms and get the denoised result of the first iteration denoted by $\mathbf{y}^{(1)}$. We define a matrix $\mathbf{D}$ which has the same size as the original image $\mathbf{x}$ to denote the estimation of missed information. The subscript as follows denotes the location in an image. For some pixel point whose value satisfies the condition that $0 < \mathbf{y}_{ij} < 255$, quantization brings no loss for these points and therefore we set the corresponding $\mathbf{D}_{ij}$ to zero. For other pixel point which has the value of 0 or 255 in the noisy observation $\mathbf{y}_{ij}$, we obtain a new value $\mathbf{y}_{ij}^{(1)}$ after the first iteration and we set $\mathbf{D}_{ij} = 0 - \mathbf{y}_{ij}^{(1)}$ or $\mathbf{D}_{ij} = 255 - \mathbf{y}_{ij}^{(1)}$, respectively. We treat $\mathbf{D}_{ij}$ as the estimation of lost information. Hence, we add this lost information $\mathbf{D}$ to the noisy image $\mathbf{y}$ to simulate the unquantized noisy observation $\mathbf{y} = \mathbf{y} + \mathbf{D}$. At last, we take the simulation of a noisy image $\mathbf{y}$ as the input image and continue the rest of iterations.

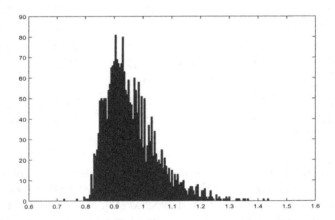

Fig. 2. The histograms of ratios between the estimated noisy image $\mathbf{y}$ and the unquantized noisy image $\hat{\mathbf{y}}$ by WNNM on image *Monarch* at noise standard deviation $\sigma = 50$.

In order to more accurately simulate the original noisy image, we try to find out the difference between the estimate image and the original noisy one. Some statistical experiments are applied to address the above problem. $\hat{\mathbf{y}}$ still denotes the unquantized noisy image, while $\mathbf{y}$ denotes the estimation of $\hat{\mathbf{y}}$. We make statistics for those pixels whose values are located outside the interval of

[0–255] using WNNM at noise standard deviation $\sigma = 50$ on image *Monarch*. We perform analysis of the ratios $\hat{\mathbf{y}}/\mathbf{y}$ between the pixels mentioned above and plot the histogram of the ratios. For pixel values greater than 255, Fig. 2 shows the distribution of the ratios. We can find that most ratios concentrate near the value 0.95 and the mean value of these ratios is 0.95. This indicates that the estimated pixel values are a little bigger than unquantized values and hence we ought to multiply a shrinkage coefficient which we can set 0.95 to simulate the noisy image more accurately. For those pixels whose values are negative, the ratios are unstable because of small denominators. Therefore, we don't need to constrain these points. We also conduct experiments on other images, and nearly all the images follow the rule which can provide guidance for future experiments.

4 Experimental Results

Experiments on denoising with synthetic AWGNs (with the standard deviation $\sigma_n \in \{25, 50, 75, 100\}$) have been carried out on 20 standard benchmark images. These test images have various texture structures, whose scenes are shown in Fig. 3. First of all, we test the effect of quantization in several state-of-the-art algorithms. After that, with the purpose of alleviating quantization effect, we embed our method in two representative image denoising methods: WNNM and EPLL. We test the performance of the modified WNNM and EPLL based image denoising methods, and compare it with other competing algorithms.

Fig. 3. The 20 test images for image denoising.

Firstly, we evaluate the effect of the quantization setting in several state-of-the-art algorithms, including BM3D[1], MLP[2], EPLL[3] and WNNM[4]. The codes of all the competing methods are provided by the authors and we used their default parameters. The experimental results are shown in Tables 2 and 3. The denoising results of MLP at noise standard deviation $\sigma = 100$ are unavailable. When the noise is weak ($\sigma = 25$), the effect of quantization is small. As a result, most of the PSNR values will decrease a little. When the noise is strong ($\sigma = 50, 75, 100$), the

---

[1] http://www.cs.tut.fi/~foi/GCF-BM3D.

[2] http://people.tuebingen.mpg.de/burger/neural_denoising/.

[3] http://www.cs.huji.ac.il/~yweiss/.

[4] http://www4.comp.polyu.edu.hk/~cslzhang/code/WNNM_code.zip.

quantization setting has a severe impact on all the 20 test images. It produces significant decrease in the PSNR on the average of 20 test images with method BM3D (0.054–2.027 dB), MLP (0.035–1.064 dB), EPLL (−0.104–2.521 dB) and WNNM (0.110–3.123 dB).

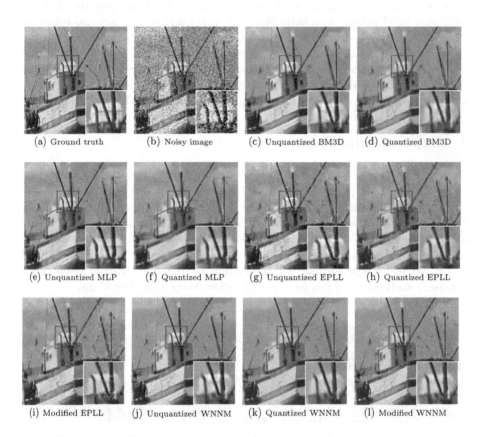

(a) Ground truth   (b) Noisy image   (c) Unquantized BM3D   (d) Quantized BM3D

(e) Unquantized MLP   (f) Quantized MLP   (g) Unquantized EPLL   (h) Quantized EPLL

(i) Modified EPLL   (j) Unquantized WNNM   (k) Quantized WNNM   (l) Modified WNNM

Fig. 4. Denoising performance comparison of different methods on Image *Boat* with noise standard deviation $\sigma = 50$.

We embed the trick into two denoising methods: EPLL and WNNM. The better PSNR result between with and without quantization for each image with each noise standard deviation is highlighted in bold. We have the following observations. The EPLL and WNNM achieve 0.004 dB–1.79 dB and 0.057 dB–2.442 dB improvement after embedding our method. Nearly all the PSNR values outperform their corresponding quantized EPLL and WNNM. The PSNR indices of our methods are almost close to, or lower than, those of their respective unquantized EPLL and WNNM. Especially, when the noise standard deviation $\sigma = 50$, the improvement of our methods is notable and even outperforms unquantized methods.

Table 2. Denoising results (PSNR) by different methods ($\sigma = 25$ and $\sigma = 50$). Unqu. denotes unquantized settings, while Quan. denotes quantization settings.

σ	25									
	BM3D		MLP		EPLL			WNNM		
	Unqu.	Quan.	Unqu.	Quan.	Unqu.	Quan.	Ours	Unqu.	Quan.	Ours
C.Man	29.45	29.22	29.59	29.42	29.24	29.12	**29.24**	29.64	29.37	**29.68**
House	32.86	32.89	32.58	32.61	32.04	**32.30**	32.24	33.23	**33.26**	33.13
Peppers	30.16	30.08	30.39	30.26	30.07	30.14	**30.16**	30.40	30.29	**30.37**
Montage	32.37	32.00	31.97	31.76	31.25	31.14	**31.42**	32.74	32.23	**32.55**
Leaves	28.85	28.59	29.11	29.07	28.18	28.36	**28.49**	29.66	29.04	**29.27**
StarFish	28.56	28.55	28.81	28.84	28.43	28.58	**28.58**	29.03	28.92	**28.99**
Monarch	29.25	29.27	29.60	29.63	29.30	**29.51**	29.44	29.85	29.83	**29.82**
Airplane	28.43	28.45	28.86	28.89	28.56	**28.71**	28.64	28.69	28.71	**28.75**
Paint	29.22	29.23	29.65	29.65	29.25	**29.40**	29.36	29.51	29.51	**29.52**
J.Bean	32.97	33.01	33.25	33.27	32.57	**33.00**	32.85	33.52	**33.55**	33.37
Fence	28.94	28.89	28.58	28.53	28.12	28.19	**28.20**	29.35	29.23	**29.29**
Parrot	28.93	28.74	29.27	29.04	28.91	28.78	**28.92**	29.12	28.78	**29.13**
Lena	29.73	29.74	29.81	29.82	29.41	**29.55**	29.50	29.94	**29.92**	29.84
Barbara	30.19	30.19	29.27	29.28	28.72	**28.89**	28.84	30.65	**30.65**	30.59
Boat	29.06	29.07	29.16	29.15	28.71	**28.82**	28.80	29.30	29.28	**29.29**
Hill	29.20	29.21	29.21	29.21	29.01	**29.05**	29.01	29.35	**29.35**	29.28
F.print	25.21	25.22	25.37	25.39	24.68	**24.80**	24.75	25.44	25.41	**25.47**
Man	28.26	28.27	28.55	28.55	28.24	**28.29**	28.26	28.44	28.43	**28.43**
Couple	29.49	29.43	29.41	29.36	29.24	29.30	**29.32**	29.59	29.49	**29.58**
Straw	25.51	25.52	25.83	25.84	25.42	**25.51**	25.46	25.95	25.95	**25.99**
AVE.	29.333	29.279	29.414	29.379	28.967	29.071	**29.075**	29.67	29.56	**29.617**

σ	50									
	BM3D		MLP		EPLL			WNNM		
	Unqu.	Quan.	Unqu.	Quan.	Unqu.	Quan.	Ours	Unqu.	Quan.	Ours
C.Man	26.13	24.88	26.37	25.74	26.02	25.00	**25.64**	26.42	25.00	**26.25**
House	29.69	29.42	29.62	29.58	28.77	28.90	**29.03**	30.33	29.92	**30.06**
Peppers	26.68	26.11	26.84	26.33	26.63	26.04	**26.51**	26.91	26.11	**26.81**
Montage	27.90	25.84	28.01	27.14	27.17	25.69	**26.69**	28.27	25.87	**27.75**
Leaves	24.68	23.05	25.34	24.64	24.38	23.41	**24.12**	25.47	23.14	**24.29**
Starfish	25.04	24.47	25.34	25.00	25.04	24.41	**24.94**	25.43	24.36	**25.06**
Monarch	25.82	25.52	26.21	26.18	25.78	25.67	**25.94**	26.32	25.63	**26.23**
Airplane	25.10	24.76	25.64	25.51	25.24	25.06	**25.37**	25.42	24.89	**25.41**
Paint	25.67	25.45	26.15	26.02	25.77	25.64	**25.88**	25.97	25.51	**25.93**
J.Bean	29.26	29.00	29.35	29.32	28.75	29.12	**29.22**	29.63	29.15	**29.60**
Fence	25.92	25.54	25.19	24.91	24.58	24.23	**24.59**	26.43	25.73	**26.21**
Parrot	25.90	24.71	26.11	25.49	25.84	24.71	**25.40**	26.09	24.64	**25.81**
Lena	26.49	26.25	26.68	26.45	26.61	26.00	**26.24**	26.68	26.10	**26.63**
Barbara	26.59	26.33	25.35	25.09	24.87	24.60	**24.89**	27.02	26.42	**26.87**
Boat	25.68	25.35	25.99	25.77	25.56	25.32	**25.58**	26.08	25.44	**26.00**
Hill	26.52	26.44	26.59	26.40	26.42	26.38	**26.52**	26.68	26.32	**26.60**
F.print	22.21	22.00	21.83	21.82	21.28	21.09	**21.32**	22.30	21.83	**22.26**
Man	25.45	25.24	25.60	25.35	25.37	25.19	**25.40**	25.60	25.17	**25.59**
Couple	26.04	25.67	26.27	25.93	25.96	25.59	**25.88**	26.20	25.51	**26.10**
Straw	22.04	21.90	22.31	22.20	21.77	21.53	**21.76**	22.51	22.09	**22.54**
AVE.	25.942	25.398	26.040	25.744	25.568	25.179	**25.546**	26.288	25.442	**26.100**

Table 3. Denoising results (PSNR) by different methods ($\sigma = 75$ and $\sigma = 100$). Unqu. denotes unquantized settings, while Quan. denotes quantization settings.

σ	75									
	BM3D		MLP		EPLL			WNNM		
	Unqu.	Quan.	Unqu.	Quan.	Unqu.	Quan.	Ours	Unqu.	Quan.	Ours
C.Man	24.33	22.21	24.58	23.19	24.20	21.91	**23.35**	24.55	21.90	**24.00**
House	27.51	26.39	27.84	27.06	26.68	25.60	**26.67**	28.24	26.48	**27.55**
Peppers	24.73	23.60	24.95	23.79	24.56	22.89	**24.19**	24.92	23.01	**24.10**
Montage	25.52	22.23	25.86	23.93	24.90	22.04	**23.87**	25.73	21.97	**24.70**
Leaves	22.49	19.83	23.18	21.46	22.03	19.67	**21.27**	23.06	19.44	**21.39**
StarFish	23.27	21.91	23.42	22.20	23.16	21.26	**22.71**	23.47	21.17	**22.76**
Monarch	23.91	22.94	24.29	23.57	23.72	22.46	**23.74**	24.31	22.40	**24.12**
Airplane	23.47	22.22	24.01	22.89	23.35	22.20	**23.38**	23.74	21.86	**23.40**
Paint	23.80	22.96	24.23	23.43	23.82	22.52	**23.85**	24.08	22.32	**23.90**
J.Bean	27.22	25.80	27.21	26.32	26.58	25.68	**27.03**	27.42	25.36	**27.17**
Fence	24.22	23.16	23.53	22.53	22.46	21.20	**22.26**	24.68	22.83	**24.18**
Parrot	24.19	21.99	24.52	23.17	24.04	21.59	**23.07**	24.37	21.62	**23.69**
Lena	24.77	23.93	25.04	24.10	24.40	23.26	**24.31**	24.94	23.32	**24.70**
Barbara	24.58	23.72	23.57	22.86	23.03	21.92	**22.95**	24.99	23.15	**24.70**
Boat	23.99	23.15	24.27	23.42	23.78	22.55	**23.59**	24.37	22.72	**24.08**
Hill	25.07	24.35	25.18	24.34	24.95	24.04	**24.96**	25.17	23.85	**25.03**
F.print	20.77	20.18	19.99	18.91	19.16	17.70	**19.15**	20.83	19.44	**20.69**
Man	24.01	23.20	24.12	23.26	23.81	22.79	**23.73**	24.16	22.72	**24.03**
Couple	24.21	23.41	24.57	23.61	24.11	22.77	**23.81**	24.34	22.64	**24.02**
Straw	20.38	19.91	20.43	19.47	19.87	18.88	**19.87**	20.79	19.11	**20.73**
AVE.	24.122	22.855	24.240	23.176	23.630	22.147	**23.387**	24.408	22.366	**23.947**
σ	100									
	BM3D		MLP		EPLL			WNNM		
	Unqu.	Quan.	Unqu.	Quan.	Unqu.	Quan.	Ours	Unqu.	Quan.	Ours
C.Man	23.08	20.25	-	-	22.86	19.45	**21.39**	23.36	19.67	**22.48**
House	25.87	23.80	-	-	25.19	22.61	**24.79**	26.66	23.48	**25.84**
Peppers	23.39	21.51	-	-	23.08	20.26	**22.26**	23.45	20.34	**22.87**
Montage	23.89	19.77	-	-	23.42	19.25	**21.52**	24.16	19.20	**22.73**
Leaves	20.91	17.50	-	-	20.25	16.69	**18.81**	21.57	16.76	**19.45**
Starfish	22.10	19.97	-	-	21.92	18.93	**20.82**	22.23	18.90	**21.22**
Monarch	22.52	20.79	-	-	22.23	19.52	**21.71**	22.95	19.64	**22.47**
Airplane	22.11	19.97	-	-	22.02	19.92	**21.54**	22.55	19.34	**21.92**
Paint	22.51	20.99	-	-	22.50	20.07	**21.96**	22.75	20.04	**22.30**
J.Bean	25.80	22.95	-	-	25.17	22.73	**25.03**	26.03	22.06	**25.29**
Fence	22.92	21.14	-	-	21.11	18.82	**20.60**	23.37	20.47	**22.54**
Parrot	22.96	19.89	-	-	22.71	19.09	**21.10**	23.19	19.19	**22.15**
Lena	23.54	22.05	-	-	23.25	20.99	**22.76**	23.71	21.24	**23.32**
Barbara	23.20	21.63	-	-	21.97	20.14	**21.49**	23.52	20.74	**23.10**
Boat	22.84	21.32	-	-	22.55	20.13	**21.98**	23.23	20.50	**22.70**
Hill	24.05	22.40	-	-	23.96	21.93	**23.59**	24.19	21.71	**23.91**
F.print	19.77	18.67	-	-	17.48	15.82	**17.08**	19.82	17.49	**19.48**
Man	22.93	21.31	-	-	22.73	20.89	**22.31**	23.03	20.73	**22.84**
Couple	23.01	21.70	-	-	22.85	20.67	**22.24**	23.00	20.66	**22.59**
Straw	19.30	18.54	-	-	18.82	17.74	**18.56**	19.52	17.67	**19.47**
AVE.	22.835	20.808	-	-	22.304	19.783	**21.577**	23.115	19.992	**22.434**

In Figs. 4 and 5 we show the denoising results on two typical images with moderate noise corruption and strong noise corruption, respectively. As can be expected from the PSNR values, the results with our methods and other competing methods are visually distinguishable. Figure 5 demonstrates that quantized BM3D, MLP, EPLL and WNNM can over-smooth more image details in wing area of image *Monarch*. Our methods can retrieve some lost information, and reconstruct more textures than quantized methods. On the one hand, EPLL generates more artifacts while quantized EPLL becomes smoother and cannot present some edge structures. EPLL applying our method not only generates much less artifacts but also preserves much better the image edge structures. This also applies to WNNM which shows strong denoising capability among low-rank approximation methods. WNNM embedding our way can achieve the same visual effect as the unquantized WNNM. In a word, the embedding of our trick preserves much better these texture areas, making the denoised image look more natural and visually pleasant.

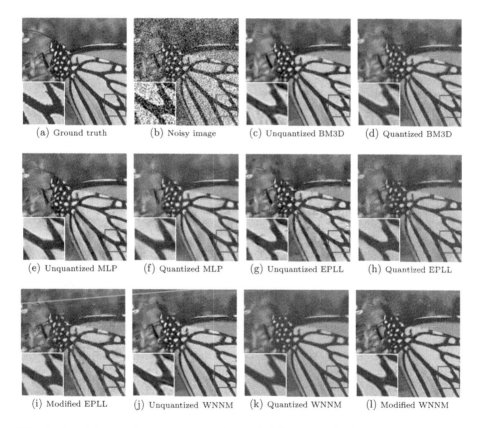

Fig. 5. Denoising performance comparison of different methods on Image *Monarch* with noise standard deviation $\sigma = 75$.

5   Conclusion

In this paper, we emphasis on quantization problems which exist in nearly all image denoising methods and analyze the effect of quantization setting. Quantization problems result in the narrower pixel value interval. Consequently, denoised images with quantization setting has lower PSNR measures and poor visual effects. To resolve the problem, we introduce a simple but effective trick to suppress the negative effect of quantization. We try to estimate unquantized noisy images and make empirical statistics for better simulation. Experimental results verify the effectiveness of our empirical trick applied to two representative denoising methods, EPLL and WNNM, which achieve a significant improvement and more natural and visually pleasant denoising results compared to the original methods.

Acknowledgement. This work is partly support by the National Science Foundation of China (NSFC) project under the contract No. 61271093, 61471146, and 61102037.

References

1. Bioucas-Dias, J.M., Figueiredo, M.A.: A new TwIST: two-step iterative shrinkage/thresholding algorithms for image restoration. IEEE Trans. Image Process. **16**(12), 2992–3004 (2007)
2. Buades, A., Coll, B., Morel, J.M.: A non-local algorithm for image denoising. In: IEEE Conference on Computer Vision and Pattern Recognition (CVPR), vol. 2, pp. 60–65 (2005)
3. Buchanan, A.M., Fitzgibbon, A.W.: Damped Newton algorithms for matrix factorization with missing data. In: IEEE Conference on Computer Vision and Pattern Recognition (CVPR), vol. 2, pp. 316–322 (2005)
4. Burger, H.C., Schuler, C.J., Harmeling, S.: Image denoising: can plain neural networks compete with BM3D? In: IEEE Conference on Computer Vision and Pattern Recognition (CVPR), pp. 2392–2399 (2012)
5. Candes, E.J., Plan, Y.: Matrix completion with noise. Proc. IEEE **98**(6), 925–936 (2010)
6. Chen, Y., Yu, W., Pock, T.: On learning optimized reaction diffusion processes for effective image restoration. In: IEEE Conference on Computer Vision and Pattern Recognition (CVPR), pp. 5261–5269 (2015)
7. Dabov, K., Foi, A., Katkovnik, V., Egiazarian, K.: Image denoising by sparse 3-D transform-domain collaborative filtering. IEEE Trans. Image Process. **16**(8), 2080–2095 (2007)
8. Daubechies, I., Defrise, M., De Mol, C.: An iterative thresholding algorithm for linear inverse problems with a sparsity constraint. Commun. Pure Appl. Math. **57**(11), 1413–1457 (2004)
9. Dong, W., Shi, G., Li, X.: Nonlocal image restoration with bilateral variance estimation: a low-rank approach. IEEE Trans. Image Process. **22**(2), 700–711 (2013)
10. Dong, W., Shi, G., Li, X., Ma, Y., Huang, F.: Compressive sensing via nonlocal low-rank regularization. IEEE Trans. Image Process. **23**(8), 3618–3632 (2014)
11. Dong, W., Zhang, L., Shi, G., Li, X.: Nonlocally centralized sparse representation for image restoration. IEEE Trans. Image Process. **22**(4), 1620–1630 (2013)

12. Elad, M., Aharon, M.: Image denoising via sparse and redundant representations over learned dictionaries. IEEE Trans. Image Process. **15**(12), 3736–3745 (2006)
13. Gu, S., Zhang, L., Zuo, W., Feng, X.: Weighted nuclear norm minimization with application to image denoising. In: IEEE Conference on Computer Vision and Pattern Recognition (CVPR), pp. 2862–2869 (2014)
14. Hel-Or, Y., Shaked, D.: A discriminative approach for wavelet denoising. IEEE Trans. Image Process. **17**(4), 443–457 (2008)
15. Katkovnik, V., Foi, A., Egiazarian, K., Astola, J.: From local kernel to nonlocal multiple-model image denoising. Int. J. Comput. Vis. **86**(1), 1–32 (2010)
16. Ke, Q., Kanade, T.: Robust L1 norm factorization in the presence of outliers and missing data by alternative convex programming. In: IEEE Conference on Computer Vision and Pattern Recognition (CVPR), vol. 1, pp. 739–746 (2005)
17. Mairal, J., Bach, F., Ponce, J., Sapiro, G., Zisserman, A.: Non-local sparse models for image restoration. In: IEEE International Conference on Computer Vision (ICCV), pp. 2272–2279 (2009)
18. Portilla, J., Strela, V., Wainwright, M.J., Simoncelli, E.P.: Image denoising using scale mixtures of Gaussians in the wavelet domain. IEEE Trans. Image Process. **12**(11), 1338–1351 (2003)
19. Rajwade, A., Rangarajan, A., Banerjee, A.: Image denoising using the higher order singular value decomposition. IEEE Trans. Pattern Anal. Mach. Intell. **35**(4), 849–862 (2013)
20. Rudin, L.I., Osher, S., Fatemi, E.: Nonlinear total variation based noise removal algorithms. Phys. D **60**(1), 259–268 (1992)
21. Schmidt, U., Roth, S.: Shrinkage fields for effective image restoration. In: IEEE Conference on Computer Vision and Pattern Recognition (CVPR), pp. 2774–2781 (2014)
22. Srebro, N., Jaakkola, T., et al.: Weighted low-rank approximations. In: ICML, vol. 3, pp. 720–727 (2003)
23. Tomasi, C., Manduchi, R.: Bilateral filtering for gray and color images. In: IEEE International Conference on Computer Vision (ICCV), pp. 839–846 (1998)
24. Zhang, X., Burger, M., Bresson, X., Osher, S.: Bregmanized nonlocal regularization for deconvolution and sparse reconstruction. SIAM J. Imaging Sci. **3**(3), 253–276 (2010)
25. Zibulevsky, M., Elad, M.: L1–L2 optimization in signal and image processing. IEEE Signal Process. Mag. **27**(3), 76–88 (2010)
26. Zoran, D., Weiss, Y.: From learning models of natural image patches to whole image restoration. In: IEEE International Conference on Computer Vision (ICCV), pp. 479–486 (2011)

GPCA-SIFT: A New Local Feature Descriptor for Scene Image Classification

Lei Ju[1,2], Ke Xie[1,2], Hao Zheng[3], Baochang Zhang[4], and Wankou Yang[1,2,3(✉)]

[1] School of Automation, Southeast University, Nanjing 210096, China
[2] Key Laboratory of Measurement and Control of Complex Systems of Engineering, Ministry of Education, Southeast University, Nanjing 210096, China
[3] Key Laboratory of Trusted Cloud Computing and Big Data Analysis, Nanjing Xiaozhuang University, Nanjing 211171, People's Republic of China
wankou.yang@yahoo.com
[4] School of Automation Science and Electrical Engineering, Beihang University, Beijing 100191, China

Abstract. In this paper, a new local feature descriptor called GPCA-SIFT is proposed for scene image classification. Like PCA-SIFT, we get the key points using the detection method in Scale Invariant Feature Transform (SIFT) and extract a 41 * 41 patch for each key point. Then we calculate the horizontal and vertical gradient of each pixel in the patch. However, instead of concatenating two gradient matrices, we directly work with the two-dimensional matrix and apply Generalized Principal Component Analysis (GPCA) to reduce it to a lower-dimensional matrix. Finally, we concatenate the reduced matrix and form a 1D vector. Compared with Principal Component Analysis (PCA), it preserves more spatial locality information. When applied in multi-class scene image classification, our proposed descriptor outperforms other related algorithms in terms of classification accuracy.

Keywords: SIFT · GPCA · GPCA-SIFT · Scene image classification

1 Introduction

Feature extraction is an important aspect of scene image classification. It often determines the merits of classification performance. Local features are invariant to many image transformations. Therefore, they are very suitable to be used as the scene description of images [1].

The Scale Invariant Feature Transform (SIFT) algorithm [2] has been extremely popular in local feature representation of scene images. The extracted features are invariant to image scale and rotation. They are also resistant to change in illumination, viewpoint and noise. However, the computation of feature vector is very complex and the dimensionality of feature vector is relatively high.

Various improvements have been proposed to speed up the calculation process. Herbert Bay et al. [3] firstly proposed SURF, which used an integer

© Springer Nature Singapore Pte Ltd. 2016
T. Tan et al. (Eds.): CCPR 2016, Part II, CCIS 663, pp. 286–295, 2016.
DOI: 10.1007/978-981-10-3005-5_24

approximation to the Hessian matrix and gradients. Edward Rosten et al. [4] introduced an efficient corner detection algorithm, called FAST. It compared pixels in a circle around the selected feature point. Recently, many researchers develop binary descriptors to reduce computational cost. Michael Calonder et al. [5] proposed BRIEF. It directly computed a binary descriptor based on simple intensity difference tests. Ethan Rublee et al. [6] introduced a faster binary descriptor ORB, which computed orientations by intensity centroid moment. Similarly, Stefan Leutenegger et al. [7] proposed BRISK that computed brightness comparisons to build a binary descriptor. Inspired by retina, Alexandre Alahi et al. [8] proposed FREAK. It extends BRISK by comparing intensities over a retinal sampling pattern.

For the high dimensionality of feature vector, many researchers have proposed various dimensionality reduction methods. Ke and Sukthankar [9] proposed PCA-SIFT that firstly applied Principal Component Analysis (PCA) to the normalized gradient patch. Compared with the standard SIFT features, PCA-SIFT are more distinctive and compact. The dimensionality of feature vectors has been reduced to 20. However, this method needs to pre-compute patch eigenspace, and the solving process is very time-consuming. Moreover, PCA only works with vectorized image representations. The spatial structure information may be lost. Jian Yang et al. [10] firstly introduced two-dimensional PCA (2DPCA). 2DPCA is based on image matrices rather than 1D vector. Similarly, Ming Li et al. [11] proposed 2D-LDA, which is based on Fishers linear discriminant analysis. Later, Jian Yang et al. [12] proposed another method: Bi-2DPCA. It implemented 2DPCA twice sequentially. One is in horizontal direction. The other is in vertical direction. The dimension of features has been greatly reduced. Xue-Jun YAN et al. [13] proposed 2DPCA-SIFT that used Bi-2DPCA to reduce dimension of gradient patches. It can speed up the calculation of eigenspace while retaining the spatial information. Jieping Ye [14] proposed Generalized Principal Component Analysis (GPCA). It directly works with two-dimensional matrices. Thus, it can preserve more spatial locality information.

In this paper, inspired by GPCA, we propose a new local feature descriptor, called GPCA-SIFT. Like PCA-SIFT, we extract gradient patches of an image. Then we apply GPCA to directly work with two-dimensional gradient matrices. When compared with SIFT, PCA-SIFT and 2DPCA-SIFT, GPCA-SIFT can achieve higher classification accuracy when applied to scene image classification problems.

The remainder of this paper is organized as follows. Section 2 briefly describes the PCA-SIFT algorithm. Section 3 presents our proposed descriptor GPCA-SIFT in detail. Section 4 conducts some experiments to evaluate our descriptor for scene image classification. Section 5 concludes this paper.

2 PCA-SIFT

Like SIFT, PCA-SIFT detects key points by building a difference-of-Gaussian pyramid. Then through key point localization and orientation assignment, the scale, location and dominant orientation of each key point can be determined. But the computation of descriptors is different. The detailed algorithm can be described in the following steps:

(1) Pre-compute an eigenspace to express the gradient images of local patches.
(2) For each keypoint, extract a patch. Then calculate the horizontal and vertical gradient for each pixel in the patch and concatenate them to form a 3042-dimensional gradient vector.
(3) Normalize the gradient vector and project it using the pre-computed eigenspace to create a more compact feature vector.

When building the eigenspace, PCA is used to calculate the project matrix. Firstly, we extract features from various images and process each feature to derive a 3042-element vector. Then form a covariance matrix of these vectors. Finally, we calculate the eigenvalues and eigenvectors of the covariance matrix. Generally, we apply SVD (Singular value decomposition) to solve this problem. However, the computation of SVD is very time-consuming and which affects the real-time classification performance. Meanwhile, PCA only works with vectorized descriptors. It does not consider the spatial structure information.

3 GPCA-SIFT

3.1 GPCA

GPCA is often used for image compression and retrieval by reducing the dimensionality of image. Unlike PCA, it does not have to pre-compute an eigenspace. GPCA directly works with two-dimensional image matrices, and project them to a vector space that is the tensor product of two lower-dimensional vector spaces.

Assume there are n images in the dataset. Each image can be represented as $A_i(i = 1, 2, ..., n)$, a $r \times c$ matrix. $D_i(n_1 \times n_2)$ is the reduced matrix of A_i. $L(r \times n_1)$ is the left reduction matrix. $R(c \times n_2)$ is the right reduction matrix. The root mean square error (RMSE) is the average reconstruction error. The GPCA algorithm can be summarized in the following steps.

(1) Compute the mean $M = \dfrac{1}{n} \sum_{i=1}^{n} A_i$.

(2) For each image A_i, calculate $\widetilde{A} = A_i - M$.

(3) Set the initial $L = (I_{n_1}, 0)^T$, RMSE=∞.

(4) Form the matrix $M_R = \sum_{i=1}^{n} \widetilde{A_i}^T L L^T \widetilde{A_i}$ and compute the n_2 eigenvectors $\{\phi_i^R\}_{i=1}^{n_2}$ of M_R corresponding to the n_2 largest eigenvalues. Set $R = [\phi_1^R, ..., \phi_{n_2}^R]$.

(5) Form the matrix $M_L = \sum_{i=1}^{n} \widetilde{A_i} R R^T \widetilde{A_i}^T$ and compute the n_1 eigenvectors $\{\phi_i^l\}_{i=1}^{n_1}$ of M_L corresponding to the n_1 largest eigenvalues. Set $L = [\phi_1^L, ..., \phi_{n_1}^L]$.

(6) Calculate $RMSE = \sqrt{\frac{1}{n}\sum_{i=1}^{n}\|\widetilde{A_i} - LL^T\widetilde{A_i}RR^T\|_F^2}$, and $\|\cdot\|$ denotes the Frobenius norm of a matrix.

(7) Repeat (4–6) until RMSE converges to a threshold.

(8) Calculate D_i. $D_i = L^T\widetilde{A_i}R(i-1,...,n)$.

3.2 GPCA-SIFT

Like PCA-SIFT, our proposed new local feature descriptor GPCA-SIFT also extracts the horizontal and vertical gradient patches of each feature in an image. In PCA-SIFT, one needs to concatenate them to form a 3042-element vector. Here, we treat each gradient patch as a two-dimensional matrix. Then we apply GPCA to create a smaller matrix with lower dimensionality. Finally, our descriptor is created by concatenating the two smaller matrices. Figure 1 illustrates the difference between PCA-SIFT and GPCA-SIFT.

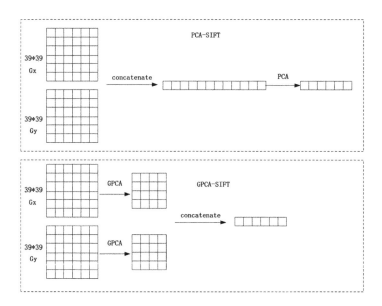

Fig. 1. The difference between PCA-SIFT and GPCA-SIFT

As shown in Fig. 1, GPCA directly works on the original gradient matrices, thus it can preserve more spatial structure information of the pixels. Suppose there are M features in total. For each feature, we extract a 41×41 patch. Then calculate the horizontal and vertical gradient of each pixel in the patch. Let $G_x(39 \times 39)$ denote the horizontal gradient matrix and $G_y(39 \times 39)$ denote the vertical gradient matrix. Next, use GPCA to transform G_x and G_y to a $n_1 \times n_2(n_1 < 39, n_2 < 39)$ matrix. Finally, concatenate the two lower-dimensional matrices. Therefore, the final local feature descriptor is an n-element ($n = 2 \times n_1 \times n_2$) vector.

4   Experiments

In this section, we implemented and compared four different local feature descriptors for scene image classification on two well-known scene image datasets: MIT Scene dataset and Fifteen Scene Categories dataset. The four different methods are: SIFT, PCA-SIFT, 2DPCA-SIFT and GPCA-SIFT. All of our experiments are performed on an i3 3.40 GHz Windows machine with 4 GB memory.

The scene image classification can be divided into two phases: training phase and testing phase. Figure 2 illustrates the flow chart. Firstly, we extract local features by using above four methods. Then, we apply k-means clustering algorithm [15] to organize the features and build the BOW model [16] for each image. Finally, we construct a histogram intersection kernel [17] as the kernel function and implement the SVM classification [18]. For each class in the dataset, we take 50 images for training, 20 images for testing.

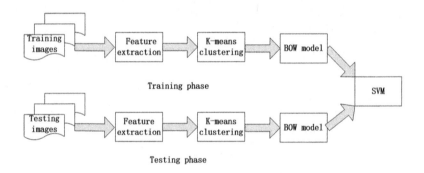

Fig. 2. The flow chart of scene image classification

4.1   Dataset Used

MIT Scene Dataset. The MIT Scene dataset is collected by Oliva and Torralba [19]. The dataset contains eight scene categories: coast (360 images), forest (328 images), highway (260 images), inside city (308 images), mountain (374 images), open country (410 images), street (292 images), and tall building (356 images). Figure 3 gives some samples of eight categories.

Fifteen Scene Categories Dataset. The Fifteen Scene Categories dataset is built by Lazebnik [20]. This dataset contains 4485 images in total. Each image is grayscale image mostly in 300×300 pixels. They are divided into fifteen classes, eight of which come from MIT Scene dataset. Figure 4 gives some samples of newer seven scene classes.

(a) Coast   (b) Forest   (c) Highway   (d) Inside City

(e) Mountain   (f) Open Country   (g) Street   (h) Tall Building

Fig. 3. Samples of MIT Scene dataset

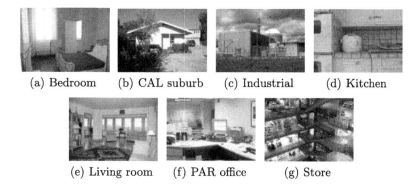

(a) Bedroom   (b) CAL suburb   (c) Industrial   (d) Kitchen

(e) Living room   (f) PAR office   (g) Store

Fig. 4. Samples of Fifteen Scene Categories dataset

4.2   Discussion

As mentioned above, GPCA-SIFT is sensitive to the dimensionality of the feature vector. Let n denotes the dimensionality of a feature vector $(n = 2 \times n_1 \times n_2)$. When we set n a value, n_1 and n_2 can have different combinations. Table 1 shows the relationship between classification accuracy and the dimensionality of feature vector.

As shown in Table 1, for different image datasets, the relationship between classification accuracy and the dimensionality of feature vector is similar. The value of n_1 and n_2 directly determines the final classification performance. When we set $n = 96(n_1 = 8, n_2 = 6)$, GPCA-SIFT in both two datasets can achieve the highest classification accuracy. The corresponding accuracies are 85.625 % and 59 % respectively. Figures 5 and 6 give the confusion matrix on two different datasets.

Table 1. Comparison of classification accuracy

n	n_1	n_2	Accuracy (MIT Scene dataset)	Accuracy (Fifteen Scene Categories dataset)
36	2	9	77.5 %	56.333 %
	3	6	78.125 %	57 %
	6	3	80.625 %	57.667 %
	9	2	78.125 %	56.333 %
72	2	18	80.625 %	52.667 %
	4	9	82.5 %	56.667 %
	6	6	81.875 %	57.667 %
	9	4	81.25 %	57.333 %
	18	2	77.5 %	55 %
96	2	24	78.75 %	52.667 %
	4	12	80.625 %	57.667 %
	6	8	82.5 %	58.333 %
	8	**6**	**85.625%**	**59%**
	12	4	81.25 %	58.333 %
	24	2	79.375 %	54.667 %
72	2	32	78.125 %	56.667 %
	4	16	84.375 %	58.333 %
	8	8	84.375 %	56.667 %
	16	4	77.5 %	56.667 %
	32	2	75.625 %	54.333 %

4.3 Evaluation of GPCA-SIFT

Next, in order to evaluate our proposed algorithm, we have done four experiments using four different local feature descriptors on two well-known scene image datasets. Namely, SIFT, PCA-SIFT, 2DPCA-SIFT and GPCA-SIFT. Table 2 shows the classification results.

Table 2. Comparison of classification accuracy

Algorithm	Accuracy (MIT Scene dataset)	Accuracy (Fifteen Scene Categories dataset)
SIFT	79.375%	53.333%
PCA-SIFT	80.625%	56.667%
2DPCA-SIFT	80.625%	56%
GPCA-SIFT	**85.625%**	**59%**

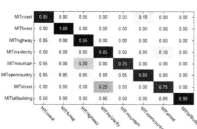

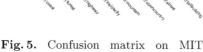

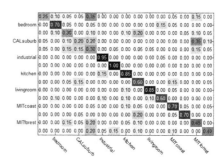

Fig. 5. Confusion matrix on MIT Scene dataset

Fig. 6. The difference between PCA-SIFT and GPCA-SIFT

Table 3. Comparison of running time

Algorithm	Time (seconds) (MIT Scene dataset)	Time (seconds) (Fifteen Scene Categories dataset)
SIFT	1779.99	3350.548
PCA-SIFT	1235.985	2082.735
2DPCA-SIFT	1685.994	2501.353
GPCA-SIFT	**1365.243**	**2126.864**

From Table 2, we can see that our proposed descriptor GPCA-SIFT outperforms other three listed algorithms. The classification accuracy on two datasets has increased by 5 % and 2.333 % respectively compared with PCA-SIFT. Therefore, we can draw the conclusion that GPCA-SIFT can achieve good performance when applied in scene image classification.

Moreover, we examine the efficiency of GPCA-SIFT. The time in Table 3 includes training time and testing time. As shown in Table 3, GPCA-SIFT has less computational time than SIFT and 2DPCA-SIFT. Compared with PCA-SIFT, GPCA-SIFT needs a little more running time. The reason is that GPCA-SIFT performs GPCA twice and PCA-SIFT only calculate PCA once. This can be seen from Fig. 1. However, this extra time can be negligible relative to the improvement of classification accuracy.

5   Conclusion

In this paper, we proposed a new local feature descriptor, named GPCA-SIFT. Like PCA-SIFT, we extract horizontal and vertical gradient patch of features in an image. In PCA-SIFT, we have to concatenate two gradient matrices and transform them to a vector. But in our algorithm, we directly work with the two-dimensional matrix. Then apply GPCA to reduce it to a lower-dimensional matrix. Compared with PCA-SIFT, it preserves more spatial structure information. Experiments show that our method achieves good classification accuracy when applied in multi-class scene image classification.

GPCA-SIFT is based on the keypoint detection method of standard SIFT. As we know, this process is relatively complicated and time-consuming. How to optimize this process or apply our method to other keypoint detection algorithm is our next work.

Acknowledgment. This project is partly supported by NSF of China (61375001, 31200747), the Natural Science Foundation of Jiangsu Province (No.BK20140638, BK2012437), the Fundamental Research Funds for the Central Universities.

References

1. Mikolajczyk, K., Schmid, C.: A performance evaluation of local descriptors. IEEE Trans. Pattern Anal. Mach. Intell. **27**(10), 1615–1630 (2005)
2. Lowe, D.G.: Distinctive image features from scale-invariant keypoints. Int. J. Comput. Vision **60**(2), 91–110 (2004)
3. Bay, H., Tuytelaars, T., Gool, L.: SURF: Speeded Up Robust Features. In: Leonardis, A., Bischof, H., Pinz, A. (eds.) ECCV 2006. LNCS, vol. 3951, pp. 404–417. Springer, Heidelberg (2006). doi:10.1007/11744023_32
4. Rosten, E., Drummond, T.: Machine learning for high-speed corner detection. In: Leonardis, A., Bischof, H., Pinz, A. (eds.) ECCV 2006. LNCS, vol. 3951, pp. 430–443. Springer, Heidelberg (2006). doi:10.1007/11744023_34
5. Calonder, M., Lepetit, V., Strecha, C., Fua, P.: BRIEF: Binary Robust Independent Elementary Features. In: Daniilidis, K., Maragos, P., Paragios, N. (eds.) ECCV 2010. LNCS, vol. 6314, pp. 778–792. Springer, Heidelberg (2010). doi:10.1007/978-3-642-15561-1_56
6. Rublee, E., Rabaud, V., Konolige, K.: ORB: an efficient alternative to SIFT or SURF. In: 2011 IEEE International Conference on Computer Vision (ICCV), pp. 2564–2571. IEEE (2011)
7. Leutenegger, S., Chli, M., Siegwart, R.: Binary robust invariant scalable keypoints. In: 2011 IEEE International Conference on Computer Vision (ICCV), pp. 2548–2555. IEEE (2011)
8. Alahi, A., Ortiz, R., Vandergheynst, P.: Fast retina keypoint. In: 2012 IEEE Conference on Computer Vision and Pattern Recognition (CVPR), pp. 510–517. IEEE (2012)
9. Ke, Y., Sukthankar, R.-S.: A more distinctive representation for local image descriptors. In: Proceedings of the 2004 IEEE Computer Society Conference on Computer Vision and Pattern Recognition, CVPR 2004, vol. 2, pp. II-506–II-513. IEEE (2004)
10. Yang, J., Zhang, D., Frangi, A.F., et al.: Two-dimensional PCA: a new approach to appearance-based face representation and recognition. IEEE Trans. Pattern Anal. Mach. Intell. **26**(1), 131–137 (2004)
11. Li, M., Yuan, B.: 2D-LDA: a statistical linear discriminant analysis for image matrix. Pattern Recogn. Lett. **26**(5), 527–532 (2005)
12. Yang, J., Xu, Y., Yang, J.Y.: Bi-2DPCA: a fast face coding method for recognition. In: Pattern Recognition Recent Advances, pp. 313–340 (2010)

13. Yan, X.-J., Zhao, C.-X., Yuan, X.: 2DPCA-SIFT: an efficient local feature descriptor. Acta Automatica Sinica **40**(4), 675–682 (2014)
14. Ye, J., Janardan, R., Li, Q.: GPCA: an efficient dimension reduction scheme for image compression and retrieval. In: Proceedings of the Tenth ACM SIGKDD International Conference on Knowledge Discovery and Data Mining, pp. 354–363. ACM (2004)
15. Hartigan, J., Wang, M.: A k-means clustering algorithm. Appl. Stat. **28**, 100–108 (1979)
16. Fei-Fei, L., Perona, P.: A Bayesian hierarchical model for learning natural scene categories. In: IEEE Computer Vision, Pattern Recognition, pp. 524–531. IEEE, New York (2005)
17. Barla, A., Odone, F., Verri, A.: Histogram intersection kernel for image classification. Int. Conf. Image Process. **3**(2), III-513–III-516 (2003)
18. Chang, C.-C., Lin, C.-J.: LIBSVM: a library for support vector machines [EB/OL] (2001). http://www.csie.ntu.edu.tw/cjlin/libsvm
19. Oliva, A., Torralba, A.: Modeling the shape of the scene: a holistic representation of the spatial envelope. Int. J. Comput. Vision **42**(3), 145–175 (2001)
20. Lazebnik, S., Schmid, C., Ponce, J.: Beyond bags of features: spatial pyramid classification for recognizing natural scene categories. In: IEEE Computer Society Conference on Computer Vision and Pattern Recognition, New York, pp. 2169–2178 (2006)

Speech and Language

Low-Quality Character Recognition Based on Dictionary Learning and Sparse Representation

Haibin Liao[1,2,3], Li Li[3,4(✉)], Youbin Chen[1,4], and Ruolin Ruan[2]

[1] Guangdong Micropattern Software, Co., Ltd., Guangdong, China
[2] School of Computer Science and Technology,
Hubei University of Science and Technology, Xianning, China
[3] National Laboratory of Pattern Recognition,
Institute of Automation, Chinese Academy of Sciences, Beijing, China
{haibin.liao,li.li}@micropattern.cn
[4] School of Automation, Huazhong University of Science and Technology,
Wuhan, China
youbin.chen@micropattern.cn

Abstract. Dictionary Learning (DL) and Sparse Representation Classification (SRC) have shown great success in face recognition recently. Practice have proven that SRC has strong robustness against noise and occlusion in face images. Our work focused on a new low-quality character recognition method based on DL and Sparse Representation (SR). SRC is introduced to deal with the low quality of character images, such as broken stroke, noise, fuzziness. Simultaneously, we also apply the linear combination of over-complete dictionary to recognize characters with different fonts and sizes. A dictionary learning method based on factor analysis is also proposed to make the dictionary more discriminative. Experiments show our method not only can recognizes characters with different fonts and sizes, but also is robust against broken stroke, noise, and fuzziness. Our method is also efficacious as it does not acquire some complex preprocessing procedures, such as binarization and refinement.

Keywords: Optical character recognition · Dictionary learning · Sparse representation · Factor analysis

1   Introduction

The research of Chinese character recognition started in 1960s, and achieved preliminary result in 1970s [1]. Here lists some Chinese character recognition software: OCR of Thunis, NC-OCR of National Research Center for Intelligent Computing Systems, OCR of Han Wang, and OCR developed by Beijing Information Science and Technology University. These systems can recognize more than 4000 Chinese characters in common use, and their accuracy can reach 95 % up to 99 %. However, most of above are developed for images by electronic scanner. They can reach the high accuracy only when the quality of character image

© Springer Nature Singapore Pte Ltd. 2016
T. Tan et al. (Eds.): CCPR 2016, Part II, CCIS 663, pp. 299–311, 2016.
DOI: 10.1007/978-981-10-3005-5_25

is relatively good. If stain, broken stroke or adhesive stroke exists, those systems cannot achieve ideal result [2].

Broken stroke, noise, fuzziness tend to occur in low-quality fax character images, as Figs. 1 and 2 illustrate. Recognition for well printed characters has achieved ideal result, while it still remains challenging to recognize low-quality characters, whose quality is greatly affected by broken stroke, noise and fuzziness. Simultaneously, there are different fonts and sizes for characters. Character recognition algorithm mainly focused on several font and sizes varying in certain range at present time. It is still a challenge to recognize character with different fonts and sizes.

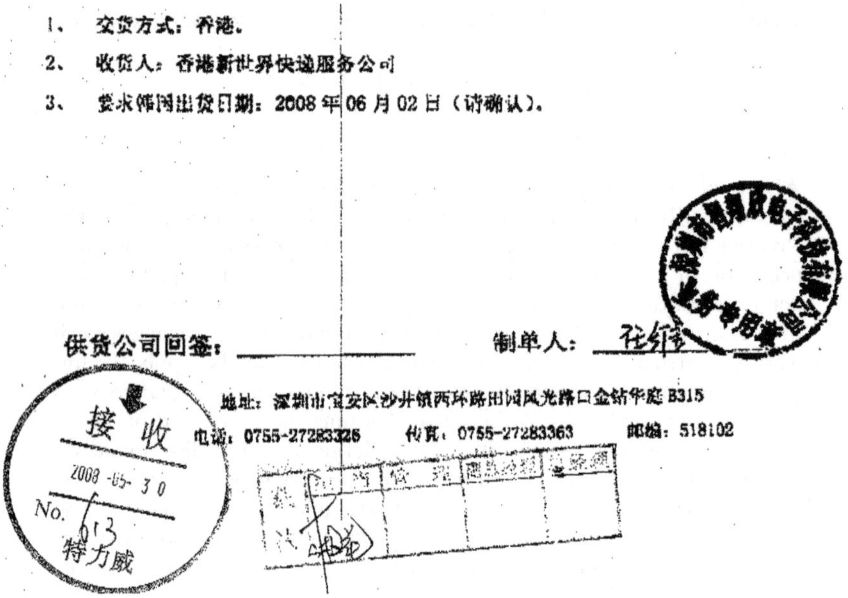

Fig. 1. Broken stroke and noise in low-quality fax character

Combination of Compressed Sensing [3] and Variable Selection [4] to generate better SR has been hot spot in the research of computer vision and machine learning. In terms of the advantages of compressed sensing and variable selection in data processing, Wright and Ma Yi in the University of Illinois at Urbana-Champaign and Microsoft Research Asia introduce feature selection by $L_1\text{-}norm$ constraint for face recognition [5]. There are plenty of features can be extracted from images. How to find effective sparse representation in high dimensional feature space, and semantic understanding of image based on sparse representation, have been popular in machine learning and pattern recognition, and applied in image classification [6], visual character selection [7], image super-resolution [8], and so on.

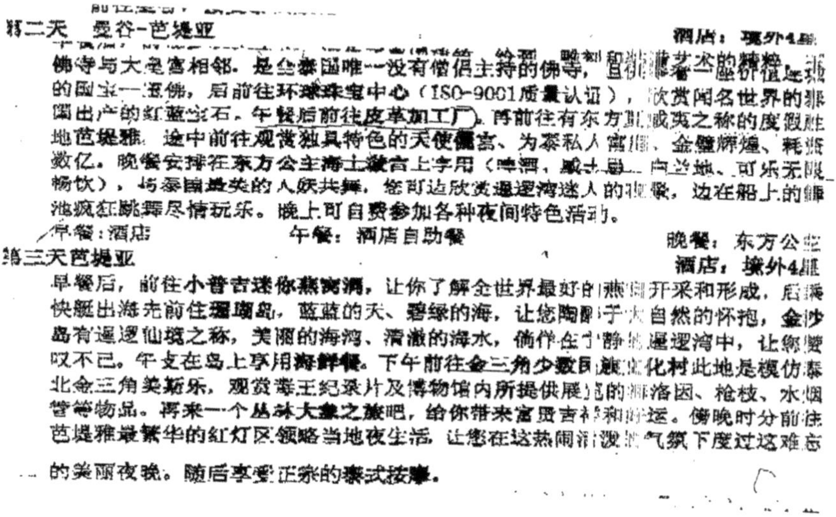

Fig. 2. Broken stroke, noise, partial information missing and fuzziness in low-quality fax character

In sparse representation based classification, there are two phases: coding and classification. First, the query signal/image is collaboratively coded over a dictionary of atoms with some sparsity constraint, and then classification is performed based on the coding coefficients and the dictionary. The dictionary for sparse coding could be predefined. Although this SRC scheme shows interesting face recognition results, the dictionary used in it may not be effective enough to represent the query images due to the uncertain and noisy information in the original training images. The number of atoms of such a dictionary can also be very big, which increases the coding complexity. In addition, using the original training samples as the dictionary could not fully exploit the discriminative information hidden in the training samples.

DL aims to learn from the training samples the space where the given signal could be well represented or coded for processing, which is vital in the application of SRC. Completion should be kept in signal reconstruction, discrimination should be kept in classification, and the computation complexity should also be taken into consideration in DL.

Many discriminating dictionary learning methods have appeared in recent years [9–13]. By structure constraint or discrimination constraint, those methods have achieved remarkable result in face recognition [10], scene classification [11], and object recognition [12].

In this paper, we introduce the theory of dictionary learning and sparse representation into low-quality character recognition, solve the broken stroke, noise and fuzziness of low quality character by SRC, and recognize characters of different fonts and sizes by DL.

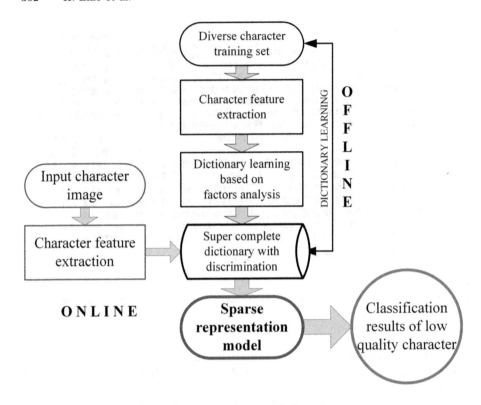

Fig. 3. Research framework flow chart

2   Recognition Framework for Low-Quality Character

Framework is illustrated in Fig. 3. It consist of four procedures, character feature extraction, discriminative dictionary learning, model solving and character classification. Procedures marked as red should be conducted on-line, and others marked as black can be conducted off-line. In this paper, the image Gabor-features are used (5 scales 8 directions) due to the little dependency on feature extraction algorithm with our framework.

2.1   Dictionary Learning

Diverse Character Dictionary Construction. Our training set consist of 3755 simplified Chinese character from GB first-level character database, we select 6 different fonts (Song, Fang Song, Regular Script, Regular GB2312 Script, Boldface, and Official Script), and 3 different sizes (28, 21, and 16). Therefore, there are 3755 classes in total, and each class contains 18 samples. Some examples are illustrated in Fig. 4.

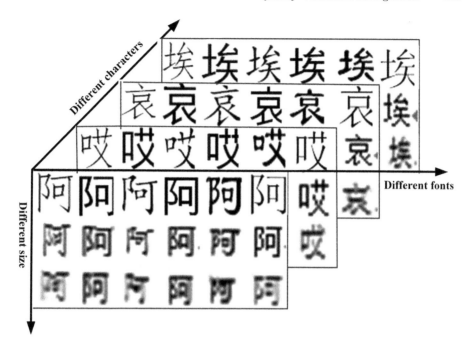

Fig. 4. Examples of characters training set

Discriminative Dictionary Learning. We can directly used above training samples of all classes as the dictionary to code the query character image, and classified the query character image by evaluating which class leads to the minimal reconstruction error. But the dictionary used in it may not be effective enough to represent the query images. The number of atoms of such a dictionary can also be very big, which increases the coding complexity.

Assuming training set $A = [A_1, A_2, ..., A_{3775}]$ is represented as linear combination of dictionary D, and the coding coefficient matrix of sparse representation is X, where $A_i = [a_i^1, a_i^2, ..., a_i^{18}]$ is the subset of the training samples from class i, and a_i^j is feature vector of sample j in class i. We can write X as $X = [X_1, X_2, ..., X_C]$, where $C = 3775$ is the maximum number of classes, and X_i is the sub-matrix containing the coding coefficients of A_i over D.

To ensure D has powerful reconstruction capability and discriminative capability at the same time, we propose the following Factor Analysis based Dictionary Learning (FADL) model:

$$J_{(D,X)} = \operatorname*{arg\,min}_{(D,X)} \left\{ \sum_{C=1}^{C} r\left(A^c, D, X^c\right) + \lambda_1 \left\|X\right\|_1 + \lambda_2 f\left(X\right) + \gamma \sum_{C=1}^{C} L\left(D^C\right) \right\}$$

(1)

where $r\left(A^c, D, X^c\right)$ is reconstruction fidelity term, $\|X\|_1$ is sparse constraint, $f\left(X\right)$ is discriminative factor analysis constraint term, $L\left(D^C\right) = \|D^C\|_*$ is noise processing term with low-rank regularization, and $\|D^C\|_*$ is the nuclear norm of each sub-dictionary D^C, which is the convex envelope of its matrix rank, λ_1, λ_2 is balance factor parameter.

Reconstruction Fidelity Term. We can write X^c, the representation of A^c over D, as $X^c = [X_1^c; ...; X_c^c; ...; X_C^c]$, where X_c^c is the coding coefficient of A^c over the sub-dictionary $D^{c'}$. First of all, the dictionary D should be able to well represent A^c, and there is $A^c \approx DX^c = D^1 X_1^c + \cdots + D^c X_c^c + \cdots + D^C X_C^c$. Second, since D^c is associated with c-th class, it is expected that A^c should be well represented by $D^c, c' \neq c$. This means that X_c^c should occupy most significant coefficients such that $\|A^c - D^c X_c^c\|_F^2$ is small, while $X_c'^c$ should have nearly zero coefficients such that $\|D^{c'} X_c^{c'}\|_F^2$ is small. So define reconstruction fidelity term as follows:

$$r\left(A^c, D, X^c\right) = \left\|A^c - DX^c\right\|_F^2 + \left\|A^c - D^c X_c^c\right\|_F^2 + \left\|D^{c'} X_{c'}^c\right\|_F^2 \qquad (2)$$

Discriminative Factor Analysis Constraint Term. To make dictionary D discriminative for the samples in A, we can make the coding coefficient of A over D, i.e. X, be discriminative. In factors analysis, content and style factors can be seen as two independent elements which influence the object and determine the observation [14]. For example, the major task of character recognition is identifying the content feature ignoring its font, size, and other style information. So it will benefit character recognition if we can filter those negative style factors. The main objective of factor analysis is to find the essential features by minimizing the style factors and maximizing the content factors of the object through the method of projection. We hope the coding coefficients X over D can represent content information of A as much as possible. According factor analysis criterion, discriminative factor analysis constraint term can be defined as following:

$$f(X) = tr(F_s(X)) - tr(F_c(X)) + \eta \left\|X\right\|_F^2 \qquad (3)$$

where F_s is the style factor, F_c is the content factor, η is the constant parameter, $tr(\cdot)$ is the trace of corresponding matrix, $\|X\|_F^2$ is the elastic term to ensure convex optimization and stability of $f(X)$.

Content factor F_c is defined as:

$$F_c = \sum_{i=1}^{C-1} \sum_{j=i+1}^{C} p_i p_j w(i,j)(\mu^i - \mu^j)(\mu^i - \mu^j)^T \qquad (4)$$

where C is class counts, p_i is prior probability for class i, as default. μ^i is average of X^i, $w(i,j) = \frac{1}{(ed)^2}$ is weight of class i and class j, and ed is the Euclidean distance between the two class average. The weight is used to reduce the weight between classes which can be classified well, and add the weight between classed which is similar.

Style factor F_s is defined as:

$$F_s = \sum_{i=1}^{C} p_i \sum_{j=1}^{N_i-1} \sum_{k=j+1}^{N_i} w(j,k)(x_j^{(i)} - x_k^{(i)})(x_j^{(i)} - x_k^{(i)})^T \qquad (5)$$

where $x_j^{(i)}$ is the coding coefficient of the j-h sample in class i, $w(j,k) = \exp -\frac{\|x_j - x_k\|^2}{2t}$ is weight of sample j and k in the same class, and t is experienced constant parameter. Weight here is used to reduce the weight between similar pair-sample which can be classified well, and add the weight between pair-sample which are far apart.

Noise Processing Term. For image classification tasks, training samples belonging to the same class are usually linearly correlated and reside in a low dimensional subspace. The sub-dictionary for representing samples from one class should be reasonably of low rank. Besides, requiring that the sub-dictionaries are of low rank can separate the noisy information and make the dictionary more pure and compact. Of all the possible sub-dictionary D^c that can represent samples from c-th class, we want to find the one with the most compact bases, that is to minimize $\|D^c\|_*$.

Optimization of FADL. To solve the optimization problem in Eq. 1, we divide it into two sub-problems. First, we optimize $X^c(c = 1, 2, \cdots, C)$ when the dictionary D and all $X^{c'}(c' \neq c)$ are fixed. We can get coding coefficient matrix X by putting all the $X^c(c = 1, 2, \cdots, C)$ together. Second, we optimize D^c when X and $D^{c'}(c' \neq c)$ are fixed. We describe the detailed implementations of solving these two sub-problems in this section.

Updating Coding Coefficient Matrix X. Assumed that D is fixed, the original objective function Eq. 1 is then reduced to a sparse coding problem. We update each X^c one by one and make all $X^{c'}(c' \neq c)$ fixed. This can be done by solving the following problem:

$$J_{(X^c)} = \underset{X^c}{\arg\min} \left(\begin{array}{l} \left\|A^c - DX^c\right\|_F^2 + \left\|A^c - D^c X_c^c\right\|_F^2 + \\ \sum\limits_{\substack{c'=1 \\ c' \neq c}}^{C} \left\|D^{c'} X_{c'}^c\right\|_F^2 + \lambda_1 \left\|X^c\right\|_1 + \lambda_2 f_c(X^c) \end{array} \right) \qquad (6)$$

This reduced objective function can be solved by using Iterative Projection Method (IPM) in [10].

Updating Sub-dictionaries D^c. When X is fixed, we can update D^c by fixing all the other $D^{c'}(c' \neq c)$. The objective function Eq. 1 is then reduced to

$$J_{D^c} = \underset{D^c, X_c^c}{\arg\min} \left(\begin{array}{l} \left\|A^c - D^c X_c^c - \sum\limits_{\substack{c'=1, \\ c' \neq c}}^{C} D^{c'} X_{c'}^c\right\|_F^2 \\ + \sum\limits_{\substack{c'=1, \\ c' \neq c}}^{C} \left\|D^{c'} X_{c'}^c\right\|_F^2 + \left\|A^c - D^c X_c^c\right\|_F^2 + \gamma \left\|D^c\right\|_* \end{array} \right) \qquad (7)$$

Denote $R(D^c) = \left\| A^c - D^c X^c_c - \sum\limits_{\substack{c'=1, \\ c' \neq c}}^{C} D^{c'} X^c_{c'} \right\|^2_F + \sum\limits_{\substack{c'=1, \\ c' \neq c}}^{C} \left\| D^{c'} X^c_{c'} \right\|^2_F$, the above objective function can be converted to the following:

$$\min_{D^c, E^c, X^c_c} \left\{ \left\| X^c_c \right\|_1 + \gamma \left\| D^c \right\|_* + \alpha \left\| E^c \right\|_{2,1} + \beta R(D^c) \right\}, \text{ s.t. } A^c = D^c X^c_c + E^c \quad (8)$$

where E^c is the error matrix corresponding to A^c, and $\| \cdot \|_{2,1}$ is the $l_{2,1}$-norm that is usually employed to measure the sample-specific corruption or noise. Then the problem can be solved by using the Inexact Augmented Lagrange Multiplier (ALM) algorithm [15].

2.2   Character Sparse Representation and Classification Model

Coefficient of sparse representation x can be obtained by sparsely represent a query character sample y by the over-complete dictionary D learned in previous procedures. The equation is described as follows:

$$\tilde{x} = \arg\min \{ \left\| y - Dx \right\|^2_2 + \gamma \left\| x \right\|_1 \} \quad (9)$$

where γ is the constant balance factor, and x is the sparse representation coefficients to be resolved. Denoted $\tilde{x} = [\tilde{x}^1, \tilde{x}^2, ..., \tilde{x}^c]$, and $\tilde{x}^c$ is the coefficient vector over sub-dictionary D^c, we can calculate the residual associated with c-th class as

$$e_c = \left\| y - D^c \tilde{x}^c \right\|^2_2 + w \left\| \tilde{x} - \mu^c \right\|^2_2 \quad (10)$$

where μ^c is the learned mean coefficient of class c, and w is a preset weight parameter. The identity of testing sample y is determined according to

$$identify(y) = \arg\min_c e_i \quad (11)$$

2.3   Dealing with Noise and Broken Stroke in Character Image

Noise and broken stroke (called by a joint name - Occlusion) may occur in low quality text images. When a query character image y is occluded, it could be represented as follows,

$$y = y_0 + e_0 = Dx + e_0 = [D \ D_\epsilon] \begin{bmatrix} x \\ x_\epsilon \end{bmatrix} = Bw \quad (12)$$

where $B = [D \ D_\epsilon] \in R^{m \times (n+n_\epsilon)}$, image without occlusion y_0 and error image with occlusion e_0 can be sparsely represented by dictionary D and occlusion dictionary $D_\epsilon \in R^{m \times (n+n_\epsilon)}$ respectively. D_ϵ is usually orthogonal unit matrix in SRC. Dimension of the unit matrix may be high, and unit matrix cannot describe image error and noise directly. We will improve it in this paper.

We collect character images under different noise and broken stroke as training set for occlusion in the first phase. Then, error images are computed by subtracting training samples for occlusion with the average image. Finally, the occlusion dictionary D_ϵ is constructed by the error images in second step. Error images computed this way are more straight-forward. Optimal training by K-SVD will lead to D_ϵ with lower dimensions and better performance.

Therefore, sparse representation model for occlusion character recognition is:

$$\widetilde{w} = \arg\min\{\left\|y - Bw\right\|_2^2 + \gamma\left\|w\right\|_1\} \tag{13}$$

where coding coefficients x without occlusion and x_ϵ with occlusion can be resolved after solving coefficients w. Classification can be done by the minimum residual error calculated by reconstructing character with occlusion y with x and x_ϵ. Therefore, our system can recognize noise and broken stroke phenomenon in low quality fax image successfully, by adding occlusion dictionary D_ϵ.

3   Experiments and Analysis

We conduct two experiments to verify the state-of-art performance of our algorithm. We apply our model to recognize Chinese characters of high quality in the first experiment, for the effectiveness of our algorithm; and to recognize Chinese characters of low quality in the second experiment, for the robustness of our algorithm against broken stroke and noise.

Implement System Environment of Experiment (for both experiments):

1. Hardware: Intel Core i5-4440 3.10 GHz, 4 GB RAM
2. Operating System: Microsoft Windows 7 SP1
3. Software: Matlab R2010b

Tech Specification (for the first experiment):

1. Identify Words: Chinese characters of key level, a total of 3755 (Simplified Chinese).
2. Identify Fonts: Song, Fang Song, Regular Script, Regular GB2312 Script, Boldface, Official Script.
3. Identify Font Sizes: 28, 21, and 16.

3.1   Recognition for Chinese Characters of High Quality

We do our experiments with characters choosing from 6 different fonts, and 3 different sizes, which means $3775 * 6 * 3 = 67950$ samples in total. All character images have been resized to 64×64. Gabor transformation is introduced to extract character features, and PCA is introduced for dimensionality reduction. In training set, there are 18 images for each character. After optimization learning by our algorithm, there will be 10 character images kept for each character, which means 10 sub-dictionaries corresponding to one character.

Choosing optimal parameter: 2 parameters are important in our method, dimensions reduced to by PCA, and equilibrium factor λ in sparse representation model. We choose 500 characters whose font is Song for parameter training. Table 1 shows recognition rate for different parameter values. When λ equals to 0.05, dim equals to 200, our method achieve the best performance.

All experiments following will be conducted with λ set to 0.05 and dim set to 200.

Table 1. The recognition results of different parameter values

λ	30 dim	100 dim	150 dim	200 dim	300 dim	500 dim
0.001	97.6 %	98.0 %	98.4 %	98.8 %	98.8 %	98.0 %
0.005	97.8 %	98.2 %	98.4 %	98.8 %	98.8 %	98.6 %
0.01	98.0 %	98.4 %	98.6 %	99.0 %	99.0 %	99.0 %
0.05	98.4 %	98.8 %	98.8 %	99.2 %	99.0 %	99.0 %
0.1	98.6 %	98.8 %	98.8 %	98.8 %	99.0 %	99.0 %
0.5	97.6 %	98.0 %	98.4 %	98.8 %	98.8 %	98.0 %

Table 2 shows recognition rate and time cost for 6 different fonts. Boldface achieved 100 % recognition rate, and Official achieved 99.8 %, which is very ideal. Even the lowest recognition rate of Regular also achieved 95.6 %. Character recognition based on SRC in this paper has achieved the same performance of other commercial character recognition systems, without preprocess procedure, such as character binarization and stroke refinement, or algorithms based on combination of multiple features and classifiers. Average time cost is 0.13 s, which means our algorithm is capable for real application. So our algorithm has achieved the state-of-the-art performance.

3.2   Recognition for Chinese Characters of Low Quality

To prove the robustness against broken stroke, noise and missing information, we pick up two images for each character of key level from character images of low quality in real as training samples. By subtracting the low quality sample with the average image of corresponding character, we get the error image. Finally, we get the occlusion dictionary by training those error images. And we also choose

Table 2. The recognition results of high-quality printed characters

Performance	Song	Fang Song	Regular GB2312	Regular	Boldface	Official
Rate (%)	99.2	99.2	97.6	95.6	100.0	99.8
Time (s/word)	0.132	0.132	0.132	0.132	0.132	0.132

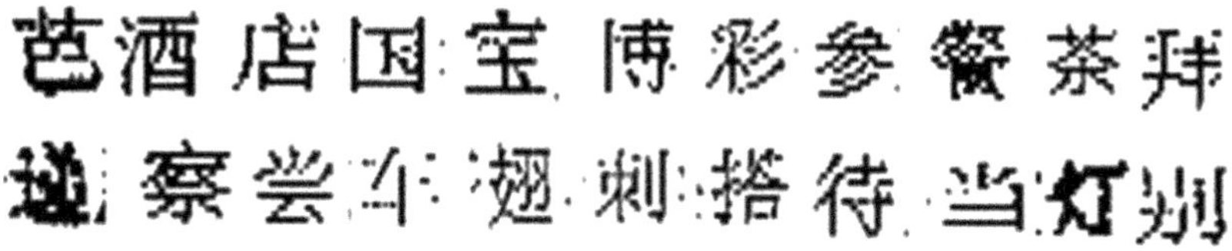

芭酒店国宝 博彩参餐茶拜
�‍察尝车翅刺搭待当灯别

Fig. 5. Examples of the character with broken stroke, noise, or ambiguity

500 character images with broken stroke, noise, or fuzziness as test set (as Fig. 5 illustrated). From Fig. 5, those characters of low quality are hard recognition for both human and computer.

We choose character classification methods based on SVM-linear, SVM-poly, and SVM-rbf as comparison. We also conduct method based on normal SRC (without occlusion dictionary) to prove the effectiveness of adding occlusion dictionary, and method based on traditional K-SVD dictionary learning to prove the effectiveness of dictionary learning based on factor analysis. Experiment results show in Table 3.

Table 3. Recognition accuracy and time cost of the low-quality characters

Methods	Test accuracy (%)	Train time (s)	Test time (s)
Our Methods	75.42	1200	0.318
SRC	54.00	540	0.132
K-SVD+SRC	71.24	1050	0.318
SVM-linear	54.23	217	0.002
SVM-poly	56.84	5714	0.020
SVM-rbf	56.90	7855	0.023
Commercial	51.75		0.006

In view of experiment results, nonlinear SVM achieves better performance than linear SVM, while more computing complexity. K-Fold is used to optimize some hyper-parameters in the training process of character SVM classifier, such as regular coefficient of function based on minimum cross entropy, variance of Gaussian kernel function of nonlinear SVM.

Recognition rate of method based on normal SRC is 54 %, while recognition rate of method based on SRC with occlusion dictionary is 75 %. So Add occlusion dictionary improve the robustness of algorithm for noise and broken stroke. Additionally, due to the factor analysis dictionary learning method, our SRC is better than K-SVD+SRC. Therefore, our method has great advantage for recognition of character of low quality.

4   Conclusion

Methods based on dictionary learning and sparse representation have a remarkable performance in signal processing and image classification. In this paper, we propose a new character recognition method to solve low quality problems, like broken stroke, noise, occlusion. The Chinese character recognition algorithm in this paper has several advantages, as follows:

1. Ability to recognize characters of different fonts and different sizes.
2. Improved rate for characters recognition of low quality (with broken stroke or noise).
3. Done without some preprocess procedure like binarization, stroke refinement.
4. Low complexity compared with other methods which combine multiple features and multiple classifiers.

Acknowledgments. We want to thank the help from the researchers and engineers of MicroPattern Corporation. This work is supported partially by China Postdoctoral Science Foundation (No: 2015M582355) and the Doctor Scientific Research Start project from Hubei University of Science and Technology (No: BK1418).

References

1. Liu, J.: Study and realization on printed Chinese character recognition system. Dalian University of Technology (2011)
2. Nie, J.: Research on feature extraction and matching recognition of printed Chinese character recognition system. Dalian University of Technology (2008)
3. Yang, S.Y., Wang, M., Li, P.: Compressive hyperspectral imaging via sparse tensor and nonlinear compressed sensing. IEEE Trans. Geosci. Remote Sens. **53**(11), 5943–5957 (2015)
4. Thomson, T., Hossain, S., Ghahramani, M.: Application of shrinkage estimation in linear regression models with autoregressive errors. J. Stat. Comput. Simul. **85**(16), 3335–3351 (2015)
5. Wright, J., Yang, A., Ganesh, A., et al.: Robust face recognition via sparse representation. IEEE Trans. Pattern Anal. Mach. Intell. **31**(2), 210–227 (2009)
6. Zhu, X.F., Xie, Q., Zhu, Y.H.: Multi-view multi-sparsity kernel reconstruction for multi-class image classification. Neurocomputing **169**, 43–49 (2015)
7. Dornaika, F., Aldine, I.K.: Decremental sparse modeling representative selection for prototype selection. Pattern Recogn. **48**(11), 3714–3727 (2015)
8. Liao, H., Dai, W., Zhou, Q., et al.: Non-local similarity dictionary learning based on face super-resolution. In: IEEE International Conference on Signal Processing, Hangzhou, pp. 88–93 (2014)
9. Zhang, Q, Li, B.: Discriminative K-SVD for dictionary learning in face recognition. In: IEEE Computer Society Conference on Computer Vision, Pattern Recognition, pp. 2691–2698 (2010)
10. Yang, M., Zhang, L., Feng, X., Zhang, D.: Fisher discrimination dictionary learning for sparse representation. In: IEEE International Conference on Computer Vision (ICCV), pp. 543–550 (2011)

11. Jiang, Z., Lin, Z., Davis, L.: Label consistent K-SVD: learning a discriminative dictionary for recognition. IEEE Trans. Pattern Anal. Mach. Intell. **35**(11), 2651–2664 (2013)
12. Cai, S., Zuo, W., Zhang, L., et al.: Support vector guided dictionary learning. In: 13th European Conference on Computer Vision, pp. 624–639 (2014)
13. Liao, H., Lu, S., Chen, Q.: Face recognition based on non-local similarity dictionary. In: Sun, Z., Shan, S., Sang, H., Zhou, J., Wang, Y., Yuan, W. (eds.) CCBR 2014. LNCS, vol. 8833, pp. 70–77. Springer, Heidelberg (2014). doi:10.1007/978-3-319-12484-1_7
14. Haibin, L., Qinghu, C., Yuchen, Y.: Practical face recognition via factor analysis. J. Electron. Inf. Technol. **12**(7), 1611–1617 (2011)
15. Wu, F., Jing, X.Y., You, X., et al.: Multi-view low-rank dictionary learning for image classification. Pattern Recogn. **50**(C), 143–154 (2015)

A Unified Approach for Spatial and Angular Super-Resolution of Diffusion Tensor MRI

Shi Yin[1], Xinge You[1(✉)], Weiyong Xue[1], Bo Li[1], Yue Zhao[1], Xiao-Yuan Jing[2], Patrick S.P. Wang[3], and Yuanyan Tang[4]

[1] School of Electronic Information and Communications,
Huazhong University of Science and Technology, Wuhan, China
youxg@hust.edu.cn
[2] School of Computer, Wuhan University, Wuhan, China
[3] College of Computer and Information Science,
Northeastern University, Boston, USA
[4] Faculty of Science and Technology, University of Macau, Macau, China

Abstract. Diffusion magnetic resonance imaging (dMRI) can provide quantitative information with which to visualize and study connectivity and continuity of neural pathways in nervous systems. However, the very subtle regions and multiple intra-voxel orientations of water diffusion in brain cannnot accurately be represented in low spatial resolution imaging with tensor model. Yet, the ability to trace and describe such regions is critical for some applications such as neurosurgery and pathologic diagnosis. In this paper, we proposed a new single image acquisition super-resolution method to increase both the spatial and angular resolution of dMRI. The proposed approach called single dMRI super-resolution reconstruction with compressed sensing (SSR-CS), uses a low number of single diffusion MRI in different gradients. This acquisition scheme is effectively in reducing acquisition time while improving the signal-to-noise ratio (SNR). The proposed method combines the two strategies of nonlocal similarity reconstruction and compressed sensing reconstruction in a sparse basis of spherical ridgelets to reconstruct high resolution image in k-space with complex orientations. The split Bregman approach is introduced for solving the SSR-CS problem. The performance of the proposed method is quantitatively evaluated on simulated diffusion MRI, using both spatial and angular reconstruction evaluating indexes. We also compared our method with some other dMRI super resolution methods.

Keywords: Diffusion magnetic resonance imaging (dMRI) · Tensor model · Single dMRI super-resolution · Compressed sensing (CS) · Sparse representation

1 Introduction

Diffusion tensor imaging enables the reconstruction of information revealing the shape, the coherence and the integrity of brain tissue microstructure which can

© Springer Nature Singapore Pte Ltd. 2016
T. Tan et al. (Eds.): CCPR 2016, Part II, CCIS 663, pp. 312–324, 2016.
DOI: 10.1007/978-981-10-3005-5_26

be indirectly analyzed through the assessment of motion of water molecules [1]. This technique has been applied widely in vivo analysis of white matter architecture and has recently been applied to the study of gray matter. Moreover, diffusion tensor imaging has been used to study a large range of neurological disease, or to quantify other causes of tissue degradations such as epilepsy and malformations of cortical development [2]. It has been shown that at a voxel resolution of around 2–3 mm, a simple tensor model used to track the major white matter pathways in the human brain [3]. Despite its interesting properties, diffusion tensor imaging is an inherently low signal-to-noise ratio (SNR) technique and yields to relatively poor spatial resolution [4]. Besides, the tensor model fails to accurately track through regions with more complex fiber arrangements such as crossing, fanning and branching.

It has been shown that the limited spatial and angular resolution introduces a bias in diffusion parameter estimation [4]. Therefore, to improve the sensitivity and robustness of studies based on diffusion tensor imaging, high spatial and angular resolution (HSAR) diffusion MRI with high SNR has to be considered. Such HSAR diffusion MRI with high SNR could provide a better sensitivity for the brain microstructure. However the acquisition of such HSAR diffusion MRI remains a challenging problem in clinical conditions since the improvement in HSAR is obtained at the cost of either lower SNR, longer acquisition time or both [5]. For example, to resolve the crossing fiber direction one can apply more sophisticated local models like the DSI or HARDI, a low spatial resolution of about 3 mm is used to achieve sufficient SNR to enable the acquisition of a high number of diffusion directions and multiple b-values, thereby resolving crossing fiber directions [6]. However, this improvement of high angular resolution with high SNR is obtained at the cost of the longer acquisition time and the spatial resolution.

To enable acquisition of high spatial diffusion MRI without long acquisition times, super-resolution (SR) acquisition techniques have been investigated in the past. Some possible strategies consist in fusing several anisotropic acquisitions with a high in-plane resolution only along one axis [7–10]. In contrast to SR acquisition techniques that require specific acquisitions protocols of multiple LR images, there exists also a category of single image SR methods [4]. Since single image SR techniques are pure post-processing methods and thus are totally independent from the acquisition protocol. The main idea is to use the image content to reconstruct information at higher-resolution. Besides, another SR method for diffusion MRI has been proposed. The main idea is to use an HR image to drive the reconstruction of another modality (e.g. using $\mathbf{B_0}$ image) [4]. However, this type of method is built on assumption that the two modalities have the similar image structures. Another problem in these methods is that few of them have considered the angular resolution. Only increasing the spatial resolution cannot break through the limitations of the tensor model. Besides, the ignorance of angular resolution also means the ignorance of the relationship between the diffusion images of different directions.

In this paper, we investigate the possibility to increase both the spatial and the angular resolution of diffusion imaging using single image SR method and compressed sensing. We proposed a unified approach to reconstruct HSAR image using the single dMRI set which are undersampled in k-space and q-space. Then we acquire the corresponding probabilistic orientation function (ODF) through the HSAR dMRI. To quantitatively assess the performance of the proposed method, we use the simulated dMRI. Some other super spatial dMRI methods are also evaluated for these data for comparing the high spatial reconstruction performance. We should note that, to the best of our knowledge, this is a first instance of using the single diffusion dMRI set to reconstruct HSAR dMRI.

The contributions of this work are twofold.

(1) The proposition of a new SSR-CS method to reconstruct the HSAR dMRI.
(2) The introduction of an efficient optimization algorithm for solving the SSR-CS model.

The remainder of the paper is organized as follows. Section 2 briefly reviews the background of the approach. Section 3 presents the proposed SSR-CS model. The experimental results are demonstrated in Sect. 4 in comparing with the state-of-the-art SR methods. Section 5 concludes the paper.

2   Background

In this section, we provide a brief background on diffusion tensor imaging and spherical ridgelets which will be used subsequently in our proposed SSR-CS algorithm.

2.1   3D Diffusion MRI

In diffusion MRI, measurements are commonly made using the pulsed gradient spin echo (PGSE) method, which samples the Fourier transform of the ensemble average diffusion propagator (EAP) $P(\mathbf{r})$, two magnetic field gradient pulses of duration δ and separation Δ are introduced to the simple spin-echo sequence [11, 12]. Assuming rectangular pulse profiles, the associated diffusion direction is $\mathbf{q} = \gamma \delta \mathbf{g}$, where $\mathbf{g}$ is the component of the gradient in the direction of the fixed field $\mathbf{B}_0$ and γ is the gyromagnetic ratio. The diffusion MRI measurement $S(\mathbf{q}; \mathbf{r})$is defined at each location $\mathbf{r}$ of a finite regular image grid in three-dimensional space and depends on the gradient direction $\mathbf{q}$. For each vector $\mathbf{q}_k$, the diffusion MRI measurement $S(\mathbf{q}_k)$ is a 3D DWI. The normalized $S^*(\mathbf{q}_k)$ is given by

$$S^*(\mathbf{q}_k) = (S(\mathbf{0}))^{-1} S(\mathbf{q}_k) \tag{1}$$

$S(\mathbf{0})$denotes the diffusion signal obtained in the absence of diffusion encoding (i.e.,the so-called "$\mathbf{B}_0$-image"). when the $\Delta^{-1}\delta$ is negligible,

$$S^*(\mathbf{q}_k) = \int_{\mathbf{r} \in \mathbb{R}^3} P(\mathbf{r}) exp(i 2\pi \mathbf{q}_k \cdot \mathbf{r}) d\mathbf{r} \tag{2}$$

2.2 Spherical Ridgelets

Spherical ridgelets are constructed by following the fundamental principles of wavelet theory [13,14]. Specially let $x \in \mathbb{R}_+$ and $\rho \in (0,1)$ be a positive scaling parameter. Further, let $\kappa(x) = exp\{-\rho x(x+1)\}$ be a Gaussian function, which we subject to a range of dyadic scaling which result in

$$\kappa(x) = \kappa(2^{-j}x) = exp\{-\rho\frac{x}{2^j}(\frac{x}{2^j}+1)\} \tag{3}$$

with $j \in \mathbb{N} := \{0,1,2,...\}$.

The semi-discrete frame $\mathbb{U}$ of spherical ridgelets can be defined as

$$\mathbb{U} := \{\psi_{j,v}|v \in \mathbb{S}^2, j = -1,0,1,2,...\} \tag{4}$$

where the spherical ridgelet functions $\psi_{j,v}$ at resolution $j \in \mathbb{N}$ and orientation $v \in \mathbb{S}^2$ is

$$\psi_{j,v} = \frac{1}{2\pi}\sum_{n=0}^{\infty}\frac{2n+1}{4\pi}\lambda_n(\kappa_{j+1}(n) - \kappa_j(n))P_n(\mathbf{u} \cdot \mathbf{v}), \forall \mathbf{u} \in \mathbb{S}^2 \tag{5}$$

where P_n denotes the Legendre polynomial of order n and $\kappa_{-1}(n)=0, \forall n$

$$\lambda_n = \begin{cases} 2\pi(-1)^{\frac{n}{2}} \cdot \frac{1 \cdot 3 \cdots (n-1)}{2 \cdot 4 \cdots n} & \text{if } n \text{ is even} \\ 0 & \text{if } n \text{ is odd} \end{cases} \tag{6}$$

when the $n = 0$, $\lambda_n = 2\pi$.

3 Proposed Model

3.1 3D Non-local Similarity Regularization

To improve the spatial resolution of the diffusion MR imaging, we adopt a 3D NLMs filter in [4,15] to capture the nonlocal similarity in each diffusion MRI. Different from the method in [4], We capture the 3D nonlocal similarity of the normalized diffusion MRI which directly determine the diffusion of water rather than the diffusion MRI. Based upon the philosophy of the NLMs, each target voxel $S(\mathbf{q}_k;\mathbf{r}_i)$ in the reconstructed high spatial resolution diffusion MRI of the kth direction can be represented as the weighted average of the voxels within its similarity neighborhoods, i.e.,

$$\hat{S}^*(\mathbf{q}_k;\mathbf{r}_i) = \sum_{j \in V_i} w_{ij}^k S^*(\mathbf{q}_k;\mathbf{r}_j) \tag{7}$$

where $\hat{S}^*(\mathbf{q}_k;\mathbf{r}_i)$ is the current estimation of $S^*(\mathbf{q}_k;\mathbf{r}_i)$, w_{ij}^k is the NLMs weights. For the kth 3D diffusion MRI, we could rewrite the Eq. (7) in a brief

$$\hat{S}_i^* = \sum_{j \in V_i} w_{ij}S_j^* \tag{8}$$

w_{ij} is defined as

$$w_{ij} = \frac{1}{Z_i} exp\{-\frac{\|N_{3D}(S_i^*) - N_{3D}(S_j^*)\|_2^2}{h^2}\} \tag{9}$$

where the constant Z_i ensures the sum of the weights is equal to 1. The $N_{3D}(S_i^*)$ and $N_{3D}(S_j^*)$ represents the 3D image patches around S_i^* and S_j^*, in practical we use $3 \times 3 \times 3$ voxels 3D windows. The similarity between patches is estimated within a restricted nonlocal volume V_i. With the nonlocal similarity in diffusion MRI, we require the estimation error if the weighted average and the upsampled image X as small as possible. Thus for each 3D diffusion MRI, the corresponding regularization term can be written as:

$$E_{nlms}(S^*) = \sum_{i \in \Omega} \| S_i^* - W_i S^* \|_2^2 \tag{10}$$

where Ω is the image grid of S^*, W_i is a row vector formed by the NLMs weights, which is defined as

$$W_i(j) = \begin{cases} w_{ij}, & j \in V_j \\ 0, & \text{otherwise} \end{cases} \tag{11}$$

We further rewrite (10) into the following concise form:

$$E_{nlms}(S^*) = \| (I - W)S^* \|_2^2 \tag{12}$$

where I is an identity matrix and W is the NLMs similar weight matrix defined by W_i in (11).

3.2   Sparse Representation Regularization

For the fixed $\mathbf{r}_0$, in q-space the $S^*(\mathbf{q}; \mathbf{r}_0)$ is the normalized diffusion signal at a b-shell along the direction $\mathbf{q} \in \mathbb{S}^2$. In practical settings, the $\mathbf{q}$ are discretized and restricted to a discrete set of orientations $\{\mathbf{q}_k\}_{k=1}^K$ which prescribes the acquisition of diffusion data in the form of K diffusion-encoded images $\{S_k^*\}_{k=1}^K$, which each $S_k^* : \mathbb{R}^3 \to \mathbb{R}^+$ corresponding to a given $\mathbf{q}_k$. In this case, for a fixed $\mathbf{r}_0$, the vector $[S_1^*(\mathbf{r}_0), S_2^*(\mathbf{r}_0), ..., S_K^*(\mathbf{r}_0)]^T \in \mathbb{R}^K$ represents a discretization of $S^*(\mathbf{q}|\mathbf{r}_0)$. For this purpose, we let $s(\mathbf{r}_0) \in R^K$ denotes the vector of diffusion signal whose kth entry is equal to $S_k^*(\mathbf{r}_0)$ and let the values of ψ_m for $m = 1, ..., M$ at the K acquisition locations be stored in a $K \times M$ matrix A defined as

$$A = \begin{bmatrix} \psi_1(u_1) & \psi_2(u_1) & \cdots & \psi_M(u_1) \\ \psi_1(u_2) & \psi_2(u_2) & \cdots & \psi_M(u_2) \\ \vdots & \vdots & \cdots & \vdots \\ \psi_1(u_K) & \psi_2(u_K) & \cdots & \psi_M(u_K) \end{bmatrix} \tag{13}$$

Each column of A is normalized through its $\mathbb{L}_2$-norm, to make all the spherical ridgelets have a unit norm. Thus the measurement s can be represented as

$$s(\mathbf{r}_0) = Ac(\mathbf{r}_0) + e(\mathbf{r}_0) \tag{14}$$

where $c(\mathbf{r}_0)$ denotes the vector of representation coefficients and $e(\mathbf{r}_0)$ denotes the measurement noise. Since our main intension is to recover the coefficients c using as few diffusion-encoding gradients as possible (implying $K \ll M$), there is an infinite number of solutions which would fit the constraint $\|A\{c\} - s\|_2 \ll \epsilon$. For each $\mathbf{r} \in \Omega_d$, the vector of representation coefficients $c(\mathbf{r})$ is sparse, we may write a compressed sensing criterion for the estimation of $c(\mathbf{r})$ as follows

$$\min_{c(\mathbf{r})} \|c(\mathbf{r})\|_1 \tag{15}$$

$$s.t. \|A\{c(\mathbf{r})\} - s(\mathbf{r})\|_2 \le \epsilon. \tag{16}$$

independently at each $\mathbf{r} \in \Omega$. This setup has been successfully used in [14] to reconstruct the high angular resolution dMRI signals.

3.3   Our Proposed SSR-CS Approach

To gain the high spatial and angular resolution diffusion MRI, we should not only consider the similarity regularization of single 3D diffusion MRI but also the sparse representation regularization of the different direction 3D diffusion MRI. For reconstructing a collection of different directions image volumes $S = \{S_k^*\}_{k=1}^K$ with high spatial resolution, a unified SSR-CS approach is proposed. The super spatial and angular resolution problem can be deemed as

$$\hat{S} = \arg\min_{V>0,S} \|DHS - L\|_2^2 + \lambda_1 \sum_{i=1}^K \|(I - W)S\|_2^2 + \lambda_2\|V\|_1 \tag{17}$$

$$s.t. \ S = AV$$

where L is a collection of low spatial resolution different directions images $L = \{L_k^*\}_{k=1}^K$; H and D stand for the blurring and down-sampling operations; A is the basis matrix introduced in compressed sensing section and each column vector of V is the representation coefficients at the corresponding voxel; the positive λ_1 and λ_2 determine the relative importance of data fitting terms versus the non-local similarity and sparse representation regularization terms. Note that it is not easy to directly solve the above problem because of the compound nature of the regularization it involves. we could rewrite the (18) into the following constrained optimization problem as

$$\hat{S} = \arg\min_{V>0,S} \|DHS - L\|_2^2 + \lambda_1 \sum_{i=1}^K \|(I - W)S\|_2^2 + \lambda_2\|V\|_1 + \lambda_3\|S - AV - q^t\|_2^2 \tag{18}$$

$$q^{t+1} = q^t + AV^{t+1} - r^{t+1} \tag{19}$$

The split Bregman approach [16] allows one to reduce (18) to a simpler form, the minimization can now be performed by sequentially minimizing with respect to S and V separately. The resulting iteration steps are

$$Step\ 1: S^{t+1} = \arg\min_S \|DHS - L\|_2^2 + \lambda_1\|S - \bar{S}\|_2^2 + \lambda_3\|S - AV^t - q^t\|_2^2 \tag{20}$$

$$(\bar{S}_i)^{t+1} = \sum_{j \in V_i} w(\bar{S}_i^{\,t}, \bar{S}_j^{\,t}) \bar{S}_j^{\,t} \tag{21}$$

$$Step\ 2:\ V^{t+1} = \arg\min_{V>0} \lambda_3 \|AV - (S^t - q^t)\|_2^2 + \lambda_2 \|V\|_1 \tag{22}$$

After we acquired the sparse coefficients, we can further compute the corresponding ODF image [13].

4   Experimental Results and Analysis

4.1   Construction of the Gold Standard

To validate the effectiveness of the proposed method, we conduct experiments on the simulated data sets proposed in [17]. In real clinical cases, diffusion MRI are often contaminated by rician noise. To validate the effectiveness of our model in high level noise condition and low level noise condition respectively, the simulated diffusion-encoded images were contaminated by two different levels of rician noise, giving rise to SNR of 40 db and 24 db. We use the following model in [14] to generate corresponding diffusion-encode images $\{S_k\}_{k=1}^K$ for a range of different values of K

$$S(\mathbf{q};\mathbf{r}) = S(\mathbf{0};\mathbf{r}) \sum_{i=1}^{M(\mathbf{r})} \alpha_i(\mathbf{r}) exp\{-b(\mathbf{q}^T D_i(\mathbf{r})\mathbf{q})\} \tag{23}$$

where $\alpha_i(\mathbf{r}) > 0$ are the positive weights obeying $\sum_{i=1}^{M(\mathbf{r})} \alpha_i(\mathbf{r}) = 1$, b is defined as a function of shape and amplitude of diffusion-encoding gradients, and $D_i(\mathbf{r})_{i=1}^{M(\mathbf{r})}$ are 3×3 diffusion tensors associated with $M(\mathbf{r})$ neural fiber tracts passing through the $\mathbf{r}$ coordinates.

The simulated set had a spatial dimension of $31 \times 31 \times 31$ voxels with 90 gradients directions in a quasi-uniform manner which is straightforward to adapt for sampling of the "northern" hemisphere. The set are consisted of some "fibres" crossing. $b = 2000\,\text{s/mm}^2$ were used for data generation. The diffusion tensors $D_i(\mathbf{r})$ in (24) respectively were: $D_1 = diag([y,x,z])$, $D_2 = diag([x,z,y])$, $D_3 = diag([0,0,1])$, $D_4 = diag([1/\sqrt{3}, 1/\sqrt{3}, 1/\sqrt{3}])$. The orientations of diffusion flows (ODF) are shown in Fig. 1. The ODF are computed by the method in [18,19]. Note that the number of the diffusion components $M(\mathbf{r})$ varied between 1 and 3. One of the diffusion encoding images was shown in Fig. 2.

To again the low resolution diffusion image, we firstly blur and downsample the diffusion encoding image. A downsampling factor of 2 along each axis was used. Then we choose $K = 24$ of the original set of 90 diffusion gradients. Within the subset of the 24 diffusion gradients, their corresponding points in q-space were also in a quasi-uniform coverage of the northern hemisphere. Finally the chosen 24 simulated diffusion-encoded images were contaminated by two level of Rician noise, giving rise to SNR of 24 dB and 40 dB diffusion encoding LR images as shown in Figs. 3 and 4 to reconstruct the high spatial and angular resolution images.

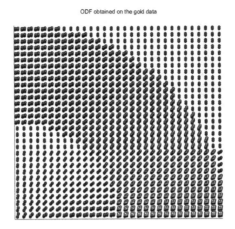

Fig. 1. The orientations of the original high angular diffusion flows

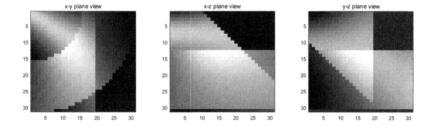

Fig. 2. Simulated 3D diffusion MRI and SNR $= \infty$ dB

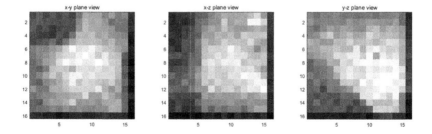

Fig. 3. Simulated downsampling 3D diffusion MRI and SNR $= 24$ dB

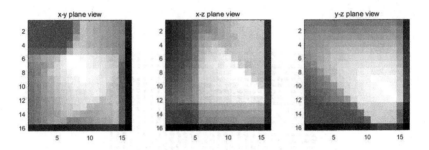

Fig. 4. Simulated downsampling 3D diffusion MRI and SNR = 40 dB

4.2   Experiment Results

To evaluate the quality of image reconstruction, we compare the proposed approach with cubic interpolation, B-spline interpolation. At the same time, the approach will also be compared to previously proposed patch-based superresolution methods (PBSR), the original PBSR method described in [15] used to reconstruct the normalized diffusion MRI instead of MRI was included in the comparison.

To evaluate the quality of reconstruction two different metrics were used, the usual Peak Signal-to-Noise Ratio (PSNR) and the percentage of generalized fractional anisotropy (GFA) were computed. The quality measures were estimated between the reconstructed HR images and the gold standard. The PSNR index was used to measure the spatial resolution quality and the GFA difference was used to measure the angular resolution quality. The experiment results were shown as follows.

Firstly we show the PSNR index in Tables 1 and 2 of our proposed method and the other methods compared when the LR data were in 24 dB and 40 dB. Then we should consider the angular index, i.e., GFA difference, see Tables 3 and 4.

From the results, we could see the proposed method has obvious advantage over the other methods when the LR data was in low SNR. When the LR data

Table 1. PSNR estimated between the gold standard and the reconstructed images from 24 dB data

	Cubic	Spline	PBSR	Proposed
PSNR (dB)	19.57	19.67	23.78	26.00

Table 2. PSNR estimated between the gold standard and the reconstructed images from 40 dB data

	Cubic	Spline	PBSR	Proposed
PSNR (dB)	19.79	20.00	31.79	31.92

Table 3. GFA difference estimated between the gold standard and the reconstructed images from 24 dB data

	PBSR	Proposed
GFA difference	2.46×10^{-2}	2.33×10^{-2}

Table 4. GFA difference estimated between the gold standard and the reconstructed images from 40 dB data

	PBSR	Proposed
GFA difference	3.21×10^{-2}	3.09×10^{-2}

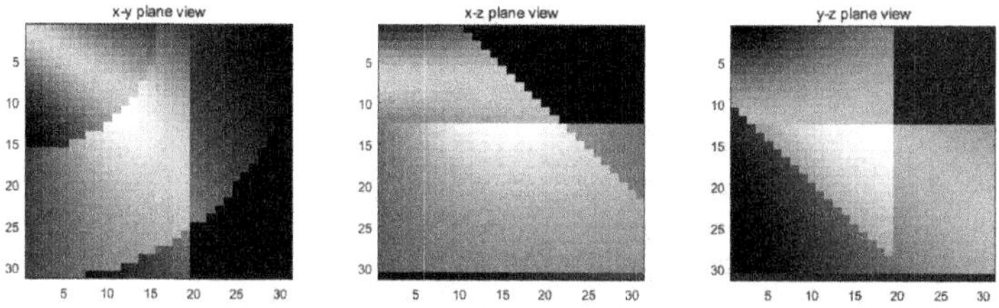

(a) Original high resolution 3D DWI

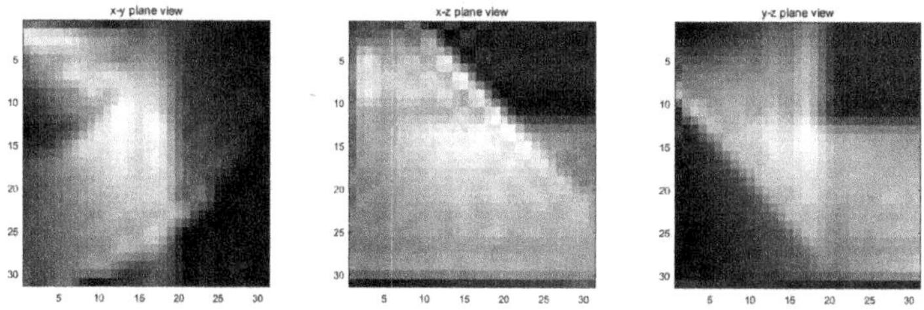

(b) PBSR reconstruction from 3D DWI 40db data

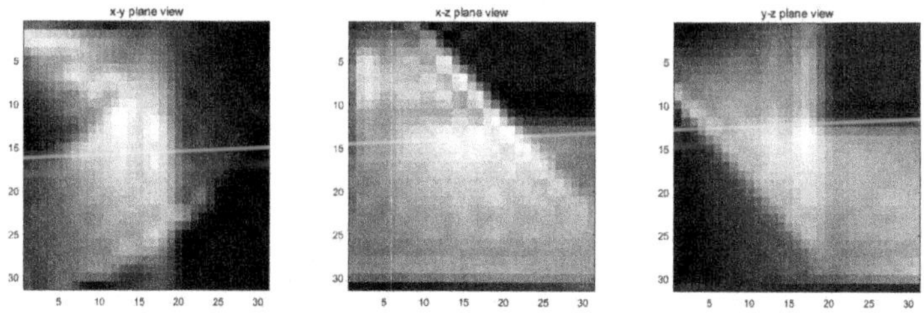

(c) Proposed method reconstruction from 3D DWI 40db data

Fig. 5. 3D DWI reconstruction using compared methods

was in high SNR, the proposed method is still best. Besides, the GFA difference index shows the proposed method is better than PBSR when the LR data was in 24 dB data and 40 dB data.

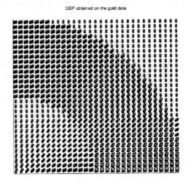

(a) The original ODF

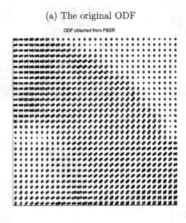

(b) The ODF of PBSR

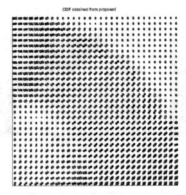

(c) The ODF of our method

Fig. 6. ODF using compared methods

The reconstructed 3D diffusion MRI from the 40 dB data using PBSR and our proposed method are shown in Fig. 5.

The reconstructed ODF from the 40 dB data using PBSR and our proposed method are shown in Fig. 6.

5   Conclusion

In this work, we investigated the possibility to increase diffusion MRI spatial resolution and angular resolution using a new method named SSR-CS. We combined the single image spatial superresolution and CS to propose a practical diffusion acquisition and reconstruction scheme that allows for obtaining HSAR diffusion MRI. The proposed technique is independent of scanner type and can be implemented on any clinically feasible scan time. We found that the proposed method could effectively reconstruct the high spatial and angular resolution diffusion MRI from the compared results.

We also note some limitations of the proposed method. When the LR diffusion MRI was in 40 dB, the proposed algorithm just improve 0.13 dB compared to the PBSR in PSNR. The advantage of the proposed method is not obvious. In the future we will focus on this aspect of the algorithm. At the same time, we should consider the speed of the algorithm because the clinical diffusion MRI were actually 4D data. The high dimension of the data will limit the speed of the super-resolution algorithm.

References

1. Stieltjes, B., Brunner, R.M., Fritzsche, K., Laun, F.: Diffusion Tensor Imaging. Springer, Heidelberg (2013)
2. Eriksson, S.H., Rugg-Gunn, F.J.: Diffusion tensor imaging in patients with epilepsy and malformations of cortical development. Brain **124**, 617–626 (2001)
3. Mori, S., van Zijl, P.: Fiber tracking: principles and strategies-a technical review. NMR Biomed. **15**, 468–480 (2002)
4. Coup, P., Manjn, J.V.: Collaborative patch-based super-resolution for diffusion-weighted images. NeuroImage **83**, 245–261 (2013)
5. Heidemann, R.M., Anwander, A.: k-space and q-space: combining ultra-high spatial and angular resolution in diffusion imaging using ZOOPPA at 7 T. NeuroImage **60**(2), 967–978 (2012)
6. Landman, B.A.: Resolution of crossing fibers with constrained compressed sensing using traditional diffusion tensor MRI. NeuroImage **59**(3), 2175–2186 (2012)
7. Scherrer, B., Gholipour, A., Warfield, S.K.: Super-resolution reconstruction to increase the spatial resolution of diffusion weighted images from orthogonal anisotropic acquisitions. Med. Image Anal. **16**, 1465–1476 (2012)
8. Poot, D.H.J., Jeurissen, B.: Super-resolution for multislice diffusion tensor imaging. Magn. Reson. Med. **69**, 103–113 (2013)
9. Van Steenkiste, G., Jeurissen, B.: Super-resolution reconstruction of diffusion parameters from diffusion-weighted images with different slice orientations. Magn. Reson. Med. **75**, 181–195 (2015)

10. Ning, L., Setsompop, K., Michailovich, O.: A joint compressed-sensing and super-resolution approach for very high-resolution diffusion imaging. NeuroImage **125**, 386–400 (2016)

11. Stejskal, E., Tanner, J.: Spin diffusion measurements: spin echoes in the presence of a time-dependent field gradient. J. Chem. Phys. **42**, 288–292 (1965)

12. Jansons, K.M., Alexander, D.C.: Persistent angular structure: new insights from diffusion MRI data. Dummy version. In: Taylor, C., Noble, J.A. (eds.) IPMI 2003. LNCS, vol. 2732, pp. 672–683. Springer, Heidelberg (2003). doi:10.1007/978-3-540-45087-0_56

13. Michailovich, O., Rathi, Y.: On approximation of orientation distributions by means of spherical ridgelets. IEEE Trans. Image Process. **19**(2), 461–477 (2010)

14. Michailovich, O., Rathi, Y., Dolui, S.: Spatially regularized compressed sensing for high angular resolution diffusion imaging. IEEE Trans. Med. Imaging, **30**(5), 1100–1115 (2011)

15. Manjn, J.V., Coup, P.: Non-local MRI upsampling. Med. Image Anal. **14**(6), 784–792 (2010)

16. Yin, W., Osher, S., Goldfarb, D., Darbon, J.: Bregman iterative algorithms for l1-minimization with applications to compressed sensing. SIAM J. Imaging Sci. **1**, 143–168 (2008)

17. Barmpoutis, A., Jian, B., Vemuri, B.C.: Adaptive kernels for multi-fiber reconstruction. In: Prince, J.L., Pham, D.L., Myers, K.J. (eds.) IPMI 2009. LNCS, vol. 5636, pp. 338–349. Springer, Heidelberg (2009). doi:10.1007/978-3-642-02498-6_28

18. Barmpoutis, A., Hwang, M.S.: Regularized positive-definite fourth order tensor field estimation from DW-MRI. NeuroImage **45**, 153–162 (2009)

19. Barmpoutis, A., Vemuri, B.C.: A unified framework for estimating diffusion tensors of any order with symmetric positive-definite constraints. In: 2010 IEEE International Symposium on Biomedical Imaging: From Nano to Macro, pp. 1385–1388. IEEE (2010)

Dump Truck Recognition Based on SCPSR in Videos

Wenming Yang, Xiaoling Hu$^{(\boxtimes)}$, Riqiang Gao, and Qingmin Liao

Shenzhen Key Laboratory of Information Science and Technology,
Shenzhen Engineering Laboratory of IS and DRM, Department of Electronic
Engineering, Graduate School at Shenzhen, Tsinghua University, Shenzhen, China
hxl14@mails.tsinghua.edu.cn

Abstract. Dump truck recognition plays an important role in the state-owned land surveillance system, which aims at fore-warning illegal construction. However, there is no special algorithm for dump truck recognition. In this paper, we explore a dump truck recognition algorithm combing structure components projection with spatial relationship (SCPSR). Instead of detecting dump truck directly as a whole, we propose a dump truck recognition algorithm based on foreground detection and components detection. An improved three frames difference method is used for foreground detection. Inspired by structure feature of dump truck components, we first locate the wheels by its valley feature on gray-scale image, and then search the candidate cab and hopper zones with the help of spatial relationship. Further, cab and hopper zones are determined by the components projection. Combining foreground detection with components detection method, the system is able to provide real-time and reliable vehicle supervision results. Experiments on real site videos demonstrate promising performance of the proposed algorithm.

Keywords: Dump truck recognition · SCPSR · Valley feature · Structure components projection · Spatial relationship

1 Introduction

Protecting the state-owned land from encroachment has been the main responsibility of the regulatory department. At present, the main regulation methods include remote sensing [1,2], human inspection [3] and vehicle video monitoring [4]. However, there are some shortcomings in these methods, such as poor real-time, high cost and unavailable for the bumpy areas. The intelligent surveillance system will be a replaceable method to realize automation and saving labor. At present, there are few video analysis methods which aim at monitoring construction activities. Especially, because of various views and complex background [5], the existing front-view or rear-view methods mostly used in highway [6,7] are not suitable for the dump track detection. As dump truck can handle multiple activities, dump truck has been widely used in different stages of the project. In

© Springer Nature Singapore Pte Ltd. 2016
T. Tan et al. (Eds.): CCPR 2016, Part II, CCIS 663, pp. 325–333, 2016.
DOI: 10.1007/978-981-10-3005-5_27

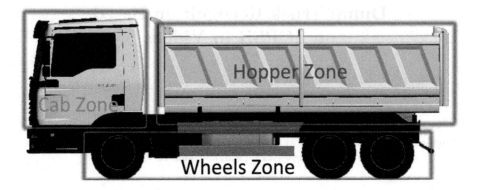

Fig. 1. The structure of the dump truck.

this case, dump truck detection is a key indicator for illegal construction on the state-owned land [8].

There is few literature involving dump truck detection, but the papers about vehicles detection on the road or highway can give us some ideas. Over the past two decades, many methods for vehicle detection in the context of traffic surveillance have been proposed [9–11]. All those methods can be roughly divided into two stages: foreground detection and target recognition [12]. In foreground detection stage, there are three popular methods: frame difference [13], background difference [14] and optical flow [15]. Robust detection algorithms such as Histogram of Oriented Gradients (HOG) [16,17] and Harr-like features [18] have been developed in the last decade. However, these methods failed to detect the target robustly due to the complexity of the background.

Considering the actual circumstances that the majority of dump truck would present side-view profile in the videos, we only consider the situations of side-view cameras in this paper. A typical profile of dump truck is illustrated in Fig. 1. In Fig. 1, it is clear that the dump truck can be divided as cab, hopper and wheels. In this paper, we first extract the valley feature of the wheels areas, and then find the ROI (Region of Interest) areas of cab and hopper respectively according to the structure relationship of dump truck. The final result is obtained by the united-decision of the respective component projection.

2   Method

For a given video sequences, a foreground detection method named improved three frame difference is performed firstly to decide whether there exists a candidate dump truck. Then we find the valley feature in the candidate wheels area and determine the ROI of the cab and hopper according to structure relationship of dump truck. For the cab and hopper detection, we use a method which makes decision based on projection. Detailed handling process is shown in Fig. 2.

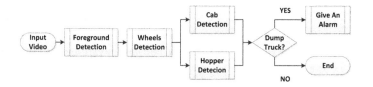

Fig. 2. The flowchart for dump truck detection.

2.1 Foreground Detection

In order to improve the computational efficiency, the improved three frames difference method is used for foreground detection in our method, and the details are shown as Algorithm 1. The previous frame Img_{i-1} and next frame Img_{i+1} are used to compared with Img_i on gray-scale to get Dif_{i-1} and Dif_i, respectively. To enhance the reliable foreground area, the result of frame difference D_i is obtained through Dif_{i-1} plus Dif_i. Inspired by the image segmentation [19], we take background and noise as one class, and the foreground as the other class. The idea of segmentation based on energy is used in this paper to get a binary image Img_{bw}. Open and close operation [20] is used to purify the binary image by removing the annoying noises and pseudo foreground. Take the outline of dump truck into consideration, a contour filter [21] which is composed of restricted width, height, length-width ratio and space occupation ratio is used to obtain the final result.

Algorithm 1. Foreground detection algorithm.

Problem:

   For a given video sequences, we need to extract its foreground and determine where the candidate dump truck area is.

Solution:

   The improved three frames difference algorithm is adopted to extract the foreground areas.

1: **for** each $Frame_i$ **do**
2:    Calculate the absolute value of frame difference of $Frame_i$ & $Frame_{i-1}$ on grayscale to get the Dif_{i-1}.
3:    Calculate the absolute value of frame difference of $Frame_{i+1}$ & $Frame_i$ on grayscale to get the Dif_i.
4:    Calculate $D_i = Dif_{i-1} + Dif_i$.
5:    D_i is binarized by using the idea of energy threshold segmentation to get D_{BW}.
6:    Perform the open and close operation on the binary image to get D_{mor}.
7:    A contour filter is applied on the image to extract the foreground.
8: **end for**

2.2 Wheels Detection

The innovation of this paper is to put forward a wheels detection method on gray-scale image. Then the cab and hopper can be determined according to the structure relationship easily. As we know, the dump truck usually works in the construction sites, and the road surface of the construction site is consisted of loess. So if we count the gray value of the lines which locates in the wheels and loess, the statistical curve would present a valley feature. The valley feature obtained by the scan line on the gray-scale image is shown in Fig. 3.

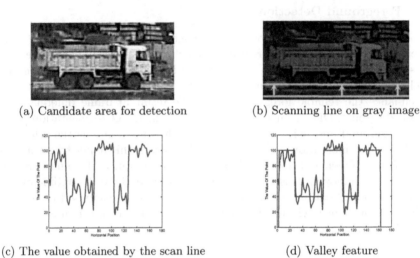

(a) Candidate area for detection   (b) Scanning line on gray image

(c) The value obtained by the scan line   (d) Valley feature

Fig. 3. Detection for the cab of dump truck.

Our method utilizes a scan line to scan the candidate areas to find whether there exists the valley feature or not. Once the valley feature is found, the position and size of wheels are determined by the scan line and the carve of valley feature of wheels area. The structure relationships of wheels, cab and hopper of dump truck are shown in Fig. 4.

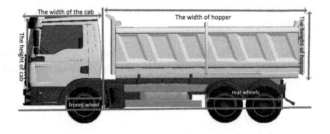

Fig. 4. The structure relationships of dump truck.

2.3   Cab Detection

From the structure relationship of dump truck shown in Fig. 4, we can easily determine the ROI of cab area based on the position and size of wheels. Once the ROI of the cab is determined, the detection algorithm is stated as Algorithm 2.

Algorithm 2. The algorithm for cab detection.

Problem:

    Determine the ROI area of the cab according to the scan line and valley feature of the wheels area and then determine if there exists a cab in the ROI area.

Solution:

1: Determine the position of the front wheel and its size, that is $P_{front-wheel}(x, y)$ and R_{wheel}.
2: Determine the ROI of the cab which is $Rect_{cab}(x, y, w, h)$ based on $P_{front-wheel}(x, y)$ and structure relationship.
3: Randomly select 5 points in specific location on the ROI area according to the size of cab and treat the 5 points as one class, then judge the consistency of the class. If it is inconsistency, get rid of the special points and re-select points until the class is consistency.
4: Clustering the ROI area according to the 5 seed points to get the contour of the cab.
5: Project the contour of the cab horizontally and vertically respectively to get the curve of P_X and P_Y.
6: Judge whether the object region is actually a cab or not according to the carve of P_X and P_Y.

Our purpose is to find the contour of the cab, so projecting it along horizontal and vertical directions respectively is an efficient method. We search for the cab of dump truck according to the projection curve of ROI. In order to extract the contour, we first convert the image from RGB space to HSV space. We randomly select five seeds in space of H according to the proportion and location of ROI area, and then decide the consistency of the five seed points. If the points are inconsistent, we get rid of the special points and re-select some others until they are consistent. Clustering data points in H space according to the five seed points to generate the surface contour. Some image de-noising and morphology operations are used to purify the image. The final step is to project the binary image to the horizontal and vertical direction respectively, and then decide whether the surface contour satisfies the constraint condition according to project curve of the surface contour. The detection results is shown in Fig. 5.

2.4   Hopper Detection

The hopper detection method is similar to the cab of the dump truck, but the particular point of hopper detection is that the location of ROI needs the information from both front-wheels and rear-wheels. It's easy to determine the

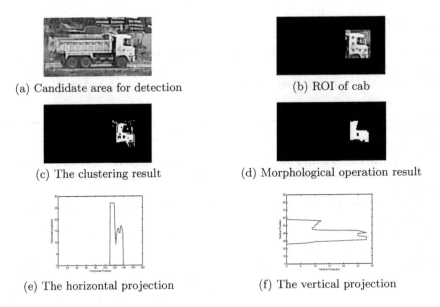

(a) Candidate area for detection

(b) ROI of cab

(c) The clustering result

(d) Morphological operation result

(e) The horizontal projection

(f) The vertical projection

Fig. 5. Detection for the cab of dump truck.

ROI of hopper area based on the position and size of front and rear-wheels. Once the ROI of hopper is determined, the detection algorithm is similar to Algorithm 2. The post processing for the binary image is different from the cab detection process. As the hopper is rectangular strictly, we use a rectangular filter to handle the binary image. To state explicitly, we use a rectangular sliding window to dispose the binary image. If the space occupation ratio is larger than the threshold, the rectangular sliding window is filled with 255, otherwise, the rectangular sliding window is filled with 0.

In order to describe the rectangle of the hopper robustly, we locate the rectangle first and project it to horizontal and vertical direction respectively. The specific method is to locate the ROI of the hopper based on the projection curve of the front an drear wheels, and then the method of clustering and projection is similar to cab detection. The detection results is shown in Fig. 6.

3   Experiments

To prove the effectiveness of the proposed algorithm, two experiments are carried out in this paper. The first experiment is to illustrate the advantage of our algorithm in detection rate, and the second one is to explain the efficiency of the proposed method in video processing.

3.1   Detection Rate

In order to analyze the performance of proposed method in dump truck recognition, we randomly select 482 videos for the Land Planning Supervision of Shen-

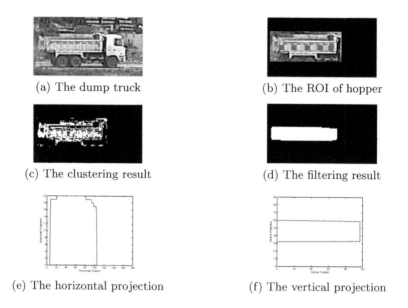

(a) The dump truck

(b) The ROI of hopper

(c) The clustering result

(d) The filtering result

(e) The horizontal projection

(f) The vertical projection

Fig. 6. Detection for the hopper of dump truck.

zhen Municipality for evaluation. The captured videos last about two minutes, and the size of frame sequences is 1280×720 pixels. Among the 482 videos, there are 91 videos containing working dump truck. We first detect the foreground region in videos, and then use the proposed method to detect the dump truck in the ROI. For comparison purpose, we also detect the dump truck in the ROI with the method of HOG+SVM. The detecting results are shown in Table 1. From Table 1, it is clear that the proposed method achieves a higher detection-rate and lower false alarm-rate than HOG+SVM.

3.2   Time Consumption

In order to apply our system into actual application, a good real-time performance is necessary for our algorithm. To test the efficiency of our system, we randomly select 50 videos for testing. The test is performed on a PC with quad Intel Core 2.5G processors, and the Ram is 4G. The detailed processing results are shown in Fig. 7. From the result of the experiments, we can calculate that

Table 1. Detection rates and false-alarm rates on our database by different methods

Methods	HOG+SVM	Our method
Detection rate (%)	81.36	89.01
False-alarm rate (%)	16.36	8.69

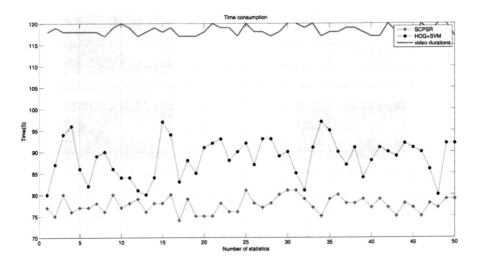

Fig. 7. Time consumption.

the average processing time of our algorithm is 77.54 s for a two minutes video, and the time of HOG+SVM is 88.56 s.

4   Conclusion

SCPSR is introduced to detect dump truck in this paper. This algorithm first detects foreground in video sequence, and then finds valley feature in the candidate areas. The ROI of cab and hopper is determined by the structure relationship of the dump truck. A simple and time efficient method which named component projection is proposed for the cab and hopper detection. This method not only shows promising results in recognizing dump truck in online videos from stationary cameras, but also can be used to detect the other motor vehicles. The general idea of this method is to master an essential feature of one component, and then determine the other component by the structure relationship of the target. A simple and easy method is preferred to judge the rest components. Future work will be focused on spatial-temporal reasoning to enhance the reliability of the judgement and other identification methods.

References

1. Lillesand, T., Kiefer, R.W., Chipman, J.: Remote Sensing and Image Interpretation. Wiley, New York (2014)
2. El Amrani, C., Rochon, G.L., El-Ghazawi, T., Altay, G., Rachidi, T.-E.: Development of a real-time urban remote sensing initiative in the mediterranean region for early warning and mitigation of disasters. In: 2012 IEEE International Geoscience and Remote Sensing Symposium (IGARSS), pp. 2782–2785. IEEE (2012)

3. Jiang, X., Gramopadhye, A.K., Melloy, B.J., Grimes, L.W.: Evaluation of best system performance: human, automated, and hybrid inspection systems. Hum. Factors Ergon. Manuf. Serv. Ind. **13**(2), 137–152 (2003)
4. Boverie, S., Giralt, A., Lequellec, J., Hirl, A.: Intelligent system for video monitoring of vehicle cockpit. Technical report, SAE Technical Paper (1998)
5. Yang, W., Li, D., Sun, D., Liao, Q.: Hydraulic excavators recognition based on inverse "v" feature of mechanical arm. In: Pattern Recognition: 6th Chinese Conference on Pattern Recognition, CCPR, pp. 536–544 (2014)
6. Zheng, Z., Zhou, G., Wang, Y., Liu, Y., Li, X., Wang, X., Jiang, L.: A novel vehicle detection method with high resolution highway aerial image. IEEE J. Sel. Top. Appl. Earth Observ. Remote Sens. **6**(6), 2338–2343 (2013)
7. Deng, H., Guo, Y., Chen, G.: The cavity detection method of highway subgrade, the grouting effects contrast. In: Rock Dynamics: From Research to Engineering: Proceedings of the 2nd International Conference on Rock Dynamics and Applications, p. 423. CRC Press (2016)
8. McCann, M., Cheng, M.-T.: Dump truck-related deaths in construction, 1992–2007. Am. J. Ind. Med. **55**(5), 450–457 (2012)
9. Sun, Z., Bebis, G., Miller, R.: On-road vehicle detection: a review. IEEE Trans. Pattern Anal. Mach. Intell. **28**(5), 694–711 (2006)
10. Razakarivony, S., Jurie, F.: Vehicle detection in aerial imagery: a small target detection benchmark. J. Vis. Commun. Image Represent. **34**, 187–203 (2016)
11. Wang, X., Tang, J., Niu, J., Zhao, X.: Vision-based two-step brake detection method for vehicle collision avoidance. Neurocomputing **173**, 450–461 (2016)
12. Bouwmans, T.: Recent advanced statistical background modeling for foreground detection-a systematic survey. Recent Pat. Comput. Sci. **4**(3), 147–176 (2011)
13. Fan, X., Cheng, Y., Fu, Q.: Moving target detection algorithm based on Susan edge detection and frame difference. In: 2015 2nd International Conference on Information Science and Control Engineering (ICISCE), pp. 323–326. IEEE (2015)
14. Yang, X., Chuang, Z., Shuai, X., Cheng, X.: Moving region detection based on background difference. In: 2014 IEEE Workshop on Electronics, Computer and Applications, pp. 518–521. IEEE (2014)
15. Ayvaci, A., Raptis, M., Soatto, S.: Sparse occlusion detection with optical flow. Int. J. Comput. Vis. **97**(3), 322–338 (2012)
16. Dalal, N., Triggs, B.: Histograms of oriented gradients for human detection. In: 2005 IEEE Computer Society Conference on Computer Vision and Pattern Recognition, CVPR 2005, vol. 1, pp. 886–893. IEEE (2005)
17. Chandrasekhar, V., Takacs, G., Chen, D.M., Tsai, S.S., Reznik, Y., Grzeszczuk, R., Girod, B.: Compressed histogram of gradients: a low-bitrate descriptor. Int. J. Comput. Vis. **96**(3), 384–399 (2012)
18. Zhang, C., Liu, J., Liang, C., Huang, Q., Tian, Q.: Image classification using harr-like transformation of local features with coding residuals. Sig. Process. **93**(8), 2111–2118 (2013)
19. Pal, N.R., Pal, S.K.: A review on image segmentation techniques. Pattern Recogn. **26**(9), 1277–1294 (1993)
20. Gonzalez, R.C., Woods, R.E.: Digital image processing (2007)
21. Catanzaro, B., Su, B.-Y., Sundaram, N., Lee, Y., Murphy, M., Keutzer, K.: Efficient, high-quality image contour detection. In: 2009 IEEE 12th International Conference onComputer Vision, pp. 2381–2388. IEEE (2009)

Parallel Randomized Block Coordinate Descent for Neural Probabilistic Language Model with High-Dimensional Output Targets

Xin Liu[1], Junchi Yan[1,2(✉)], Xiangfeng Wang[1], and Hongyuan Zha[1,3]

[1] East China Normal University, Shanghai, China
xinchrome@gmail.com, {jcyan,xfwang,zha}@sei.ecnu.edu.cn
[2] IBM Research – China, Shanghai, China
[3] Georgia Institute of Technology, Atlanta, USA

Abstract. Training a large probabilistic neural network language model, with typical high-dimensional output is excessively time-consuming, which is one of the main reasons that more simplified models such as n-gram is often more popular despite the inferior performance. In this paper a Chinese neural probabilistic language model is trained using the Fudan Chinese Language Corpus. As hundreds of thousands of distinct words have been tokenized from the raw corpus, the model contains tens of millions of parameters. To address the challenge, popular parallel computing platform MPI (Message Passing Interface) based on cluster is employed to implement the parallel neural network language model. Specifically, we propose a new method termed as Parallel Randomized Block Coordinate Descent (PRBCD) to train this model cost-effectively. Different from traditional coordinate descent method, our new method could be employed in network with multiple layers, allowing scaling up the gradients with respect to hidden units proportionally based on sampled parameters. We empirically show that our PRBCD is stable and is well suited for language models, which contain only a few layers while often have a large amount of parameters and extremely high-dimensional output targets.

Keywords: Language model · Stochastic optimization · Parallel computing

1   Introduction

The purpose of constructing a statistical language model is to learn a joint distribution of words in a word sequence, e.g. sentence, and probability of a word sequence is deduced from that joint distribution. High probability of a sentence implies that this sentence is probably valid, i.e. a sentence that is consistent with language grammar and/or semantics. A well-generalized language model can be applied to various fields, e.g. speech recognition, information retrieval, Chinese Pinyin input system and language translation.

ⓒ Springer Nature Singapore Pte Ltd. 2016
T. Tan et al. (Eds.): CCPR 2016, Part II, CCIS 663, pp. 334–348, 2016.
DOI: 10.1007/978-981-10-3005-5_28

Consider a given word sequence of length T: $S_T = (w_1, w_2, ..., w_t, ..., w_T)$, where w_t is integer tag of a word. The conditional probability of the next word given all the previous ones is represented as $P(w_t|w_1^{t-1})$, where $w_1^{t-1} = (w_1, w_2, ..., w_{t-1})$. One popular and successful way to obtain the well-generalized probabilistic language model is the n-gram model [2], which avoids the *curse of dimensionality* and obtains generalization on the assumption that probability distribution of a word in a sequence is only relevant with the former $n-1$ words in that sequence. According to this n-th order Markov assumption, we have

$$P(w_t|w_1^{t-1}) = P(w_t|w_{t-n+1}^{t-1}),$$

where $w_{t-n+1}^{t-1} = (w_{t-n+1}, ..., w_{t-1})$ and n is the order of language model, which is typically a small integer in practice, e.g. $n = 2$ or $n = 3$. When n is small, one simple idea to estimate $P(w_t|w_{t-n+1}^{t-1})$ is through straightforward statistics: scan the training set, count the frequency of occurrence of tuple $(w_{t-n+1}, ..., w_{t-1})$ and tuple $(w_{t-n+1}, ..., w_{t-1}, w_t)$ respectively. Then the conditional probability of w_t is estimated as $\hat{P}(w_t|w_{t-n+1}^{t-1}) = freq(w_{t-n+1}^t)/freq(w_{t-n+1}^{t-1})$, where $freq(\cdot)$ is the frequency of occurrence of a given tuple. For instance, if $n = 3$, the joint probability of $P(w_t)$ can be represented as a mixture of unigram, bigram and trigram, which is the interpolated or smoothed trigram [9]. Interpolated trigram enables flexible choice of size of context, and has obtained a good generalization.

n-gram model has been compared to the probabilistic neural network language model (NNLM) [1]. Neural models have already been proved to routinely outperform n-gram [13]. Neural models take distributed representations of several given words as input and then output the probability distribution of the next word, such that when input words in two given sequences have similar distributed representation, these words are semantically and/or grammatically similar, and the likelihood of these two sequences are also close to each other.

However, for such a neural network with a large number of training parameters and high-dimensional output targets, the training is rather time-consuming as often measured in weeks [1]. This renders practical burden especially when there is a further need for parameter tuning to choose a best model. This challenge often discourages researchers and engineers, who have to resort to alternative models like n-gram in spite of NNLM's potential superior performance.

Reason for NNLM's slow training speed is that the number of dimensions of output is exactly equal to the size of vocabulary, taking entire vocabulary into consideration will lead to extremely high dimensional output targets. There are several ways to reduce the complexity of computation times, for example, we can use NNLM to only handle the most frequent words, while the rest is handled by n-gram [19]. What is more, if the vocabulary can be organized in a good tree structure, then computation will be greatly accelerated [17]. But it is usually not easy to find out a good tree structure [15]. Another way is low-rank matrix factorization, which is roughly equivalent to a reduction in training time without a significant accuracy loss [18]. There also exists some methods for training NNLM based on noise-contrastive estimation such as [16].

In this paper, we will firstly introduce the architecture of a neural network language model and the key idea of training NNLM under parallel computing environment, and we adopt a state-of-the-art stochastic optimization algorithm [10] for model training. Then, we propose our **P**arallel **R**andomized **B**lock **C**oordinate **D**escent (PRBCD) method for speedup. In particular, we will show that the proposed method is useful when training a typical NNLM with high-dimensional output targets. The difference between PRBCD and ordinary coordinate descent method is that we allow scaling up the gradients with respect to units in hidden layer, which is empirically proved stable in back propagation.

In our experiment, we train a Chinese probabilistic language model using a Chinese Language Corpus. The empirical results suggest that PRBCD is cost-effective, with little compromise in terms of testing accuracy compared to full-dimensional computation. The main contribution of our paper are:

- We propose a new parallel optimization method termed by **P**arallel **R**andomized **B**lock **C**oordinate **D**escent (PRBCD) tailored to the neural network with high-dimensional output targets, such as the neural language model.
- We further combine our method with the state-of-the-art stochastic optimization algorithm Adam. As a result, we lower its complicated mathematics computation of Adam without compromising its other advantages.
- We apply the combined method for training a Chinese language model under parallel computing environment on the Fudan Chinese Language Corpus dataset. The results corroborate the cost-effectiveness of our method.

2   Probabilistic Language Model

For interpreting the problem and technical background clearly, the key notations used in this paper are summarized in Table 1 for cross-reference.

2.1   Basic Concepts and Related Work

Given a vocabulary V, each term in V is associated with a positive integer tag w_t, as well as a distributed representation v_t. Given a word sequence $S_T = \{w_1, w_2, ..., w_t, ..., w_T\}$, where w_t is integer tag of a word from the integers in $\{1, 2, ..., |V|\}$, t is the location of w_t. By adopting a general language model, probability distribution of the next word is relevant with the most recent words. In order to judge the generalization ability of a language model, *perplexity* is introduced which is defined as [1]:

$$perplexity = \sqrt[T]{\prod_t \frac{1}{P(w_t|w_1^{t-1})}},$$

Perplexity is used as an indicator to measure the performance of probabilistic language models. Sentences in training dataset are supposed to be valid, therefore a good language model should assign as high probability as possible to these sequences. The perplexity of given word sequences should be as low as possible.

Table 1. Description of the main notations used in this paper.

$\mathcal{L}$	Log likelihood of model on entire corpus
T	Length of entire corpus
w_t	An interger tag corresponding to one word
J_t	Objective function for an example sequence
d_t	Probability distribution of next word w_t given previous words
h_t	Output vectors of hidden layer
x_t	Concatenation of several word vectors
W_S	Weights matrix in softmax output layer
W_H	Weights matrix in hidden layer
b_S	Biases vector in softmax output layer
b_H	Biases vector in hidden layer
k	Integer identity representing one processor

The probabilistic NNLM comprises several layers which transform distributed representation of the previous words into a probabilistic distribution of next word. To address the curse of dimensionality and data scarcity, the aforementioned n-th order Markov dependency assumption is observed. Moreover, the order of models n can be larger in NNLM compared to others such as n-gram. The perplexity can be still estimated by geometric average of $1/P(w_t|w_{t-n+1}^{t-1})$, and the model is expected to achieve maximum likelihood on training corpus with respect to the model parameter θ, in addition with a prior regularization:

$$\max_{\theta} \mathcal{L} = \frac{1}{T} \sum_t \log P(w_t|w_{t-n+1}^{t-1}; \theta) + R(\theta) \tag{1}$$

where the first term in the right side of the formula is the log-likelihood. Maximizing this term is in general equivalent to minimizing the perplexity. $R(\theta)$ is the regularization term to prevent overfitting. Moreover, it can also encourage the model to automatically learn sparse features.

Specifically, let $d_t = f(w_{t-n+1}, ..., w_{t-1}, \theta)$ be the distribution vector representing probability distribution of next word w_t such that the number of dimensions of d_t is equal to the size of vocabulary V i.e. $d_t \in \mathcal{R}^{|V|}$. Moreover the elements of d_t are supposed to be between $[0, 1]$ and their summation is 1.

The conditional probability of the given sequence $(w_{t-n+1}, ..., w_{t-1}, w_t)$ for Eq. (1) is modeled by:

$$P(w_t|w_{t-n+1}^{t-1}; \theta) = \log d_t^{(w_t)} \tag{2}$$

where $x^{(k)}$ denotes retrieving the k-th element in vector x. Technically, to guarantee d_t is a probability distribution, Softmax is used to normalize the output i.e. $\phi(a) = (e^{a^{(1)}}, \dots, e^{a^{(n)}})/\sum_i e^{a^{(i)}}$ for a is the activation of a layer.

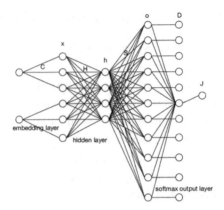

Fig. 1. The feedforward neural network architecture with three layers. Note that soft-max output layer S comprises to parts, the first part converts h_t into a unnormalized output o_t, the second part will normalized o_t into a probability distribution d_t.

Now we further define $x_t = C(w_{t-n+1}, ..., w_{t-1})$ to represent the embedding layer from word tag w_t to word vector, where x_t is the concatenation of $n-1$ word vectors of those words immediately preceding w_t. Let $h_t = H(x_t)$ represent the hidden layer, where h_t is the output vector of hidden layer. Define $o_t = S(h_t)$ to represent the Softmax output layer, and $d_t = (e^{o_t^{(1)}}, ..., e^{o_t^{(n)}})/\sum_i e^{o_t^{(i)}}$ is the normalized form of o_t. Thus, the model f is decomposed into three layers: C, H and S as illustrated in Fig. 1.

2.2   Architecture of Network with Rectifier

Rectified Linear Unit (ReLU) [5] and its variants like PReLU [6], LReLU [12] are widely used in deep learning. It is also utilized in the hidden layer in our feed-forward neural network and its parallel implementation. The formula of ReLU is $\phi(a) = \max\{0, a\}$, where a is the activation of a neuron. Compared to other activation functions such as hyperbolic tangent and sigmoid, the unsaturation of ReLU keeps more information of previous layer, creating sparse representations with true zeros which seem remarkably suitable for sparse data [5].

A logistic regression model comprises *embedding layer* from word tag to word vector C and *Softmax output layer* S, which outputs the probabilistic distrib-ution of next word. Feedforward neural network adds a single *hidden layer* H into logistic regression between embedding layer and Softmax output layer. The general architecture is shown in Fig. 1.

In the feedforward stage, for each sample $e = (w_{t-n+1}, ..., w_{t-1}, w_t)$ we have the following objective function:

$$J_t = \log\left(\frac{\exp(o_t^{(w_t)})}{\sum_{i=0}^{|V|}\exp(o_t^{(i)})}\right) - \frac{1}{2}\lambda||\theta||_2,$$

$$o_t = W_S \cdot h_t + b_S,$$
$$h_t = ReLU(W_H \cdot x_t + b_H),$$
$$x_t = (C^{(w_{t-n+1})}, ..., C^{(w_{t-1})})^T \tag{3}$$

where x_t, h_t and o_t are the output of embedding layer, hidden layer and Softmax layer respectively and $\theta = (W_S, b_S, W_H, b_H, C)$ represents parameters of network. $C^{(k)}$ means the k-th row vector in matrix C, thus x_t is precisely the concatenation of several word vectors. To alleviate overfitting, we further add a weight decay regularization term $-\frac{1}{2}\lambda||\theta||_2$ to the objective function to avoid too large parameters and too complicated model. In back propagation stage, we compute the gradients of J_t with respect to model parameters as follows:

$$\frac{\partial J_t}{\partial x_t^{(j)}} = \sum_i 1_{h_t^{(i)}>0} \cdot W_H^{(i,j)} \cdot \frac{\partial J_t}{\partial h_t^{(i)}} - \lambda x_t^{(j)},$$

$$\frac{\partial J_t}{\partial b_H^{(i)}} = 1_{h_t^{(i)}>0} \cdot \frac{\partial J_t}{\partial h_t^{(i)}} - \lambda b_H^{(i)},$$

$$\frac{\partial J_t}{\partial W_H^{(i,j)}} = 1_{h_t^{(i)}>0} \cdot \frac{\partial J_t}{\partial h_t^{(i)}} \cdot x_t^{(j)} - \lambda W_H^{(i,j)},$$

$$\frac{\partial J_t}{\partial W_S} = \frac{\partial J_t}{\partial o_t} \cdot \left(\frac{\partial J_t}{\partial h_t}\right)^T - \lambda W_S,$$

$$\frac{\partial J_t}{\partial h_t} = (W_S)^T \cdot \frac{\partial J_t}{\partial o_t},$$

$$\frac{\partial J_t}{\partial o_t^{(j)}} = \frac{\partial J_t}{\partial b_S^{(j)}} = 1_{j=w_t} - \frac{\exp(o_t^{(j)})}{\sum_{i=0}^{|V|} \exp(o_t^{(i)})} \tag{4}$$

where $1_{j=w_t}$ is the binary indicator. The update rule is $\theta := \theta + \alpha \frac{\partial J}{\partial \theta}$. The learning rate ϵ is gradually decreased for each training epoch. For embedding layer C, the gradients with respect to parameters is precisely equal to the gradients w.r.t. x_t.

2.3 Parallel Neural Network

The computational bottleneck is mainly in the Softmax layer S, where the output vector has hundreds of thousands of dimensions, implying that the Softmax layer is involved for training tens of millions of parameters. In contrast, embedding layer and hidden layer have much fewer parameters. Specifically, for the Fudan Chinese Language Corpus used in this paper, we have more than 400,000 distinct Chinese words tokenized. For instance, when the number of neurons in hidden layer is roughly set to 100, then the number of parameters in the Softmax layer is more than 40 million, while the hidden layer contains at most tens of thousands of parameters. In backward and updating stage, we need to update all of the parameters in output layer and hidden layer, along with several word vectors (whose number of dimensions is typically less than 100) in the embedding layer. More than 99.9 % of gradients computation are involved in the output layer.

In parallel computing environment, the weights W_S in the Softmax layer will be distributed among many processors for parallel updating. Given K processors $k = 1, 2, 3, ..., K$, denote $W_S(k)$ the parameters that are updated by processor k. In the feedforward stage, each processor performs the identical[1] feedforward procedure on the embedding layer C and the hidden layer H, then each processor will perform 'local' feedforward procedure in the Softmax layer S. For processor k we have $o_t(k) = W_S(k) \cdot h_t + b_S(k)$, in order to normalize $o_t(k)$, global sum $\sum_{i=0}^{|V|} \exp(o_t^{(i)})$ is required, and this sum can be efficiently obtained through *Allreduce*[2] operation on MPI (Message Passing Interface)[3] platform. After computing the normalized output of Softmax $d_t(k)$, if the label w_t exactly locates in current processor, then this processor will send $\mathcal{L} = \log d_t^{(w_t)}$ to the root processor, who will maintain the total log-likelihood of all samples.

In the back propagation phase, the 'local' gradients of log-likelihood with respect to $o_t(k)$ must be firstly computed according to Eq. (4). The second step is to compute the global gradients of the log-likelihood with respect to h_t, which is precisely the sum of local $\frac{\partial J_t}{\partial h_t(k)}$ in each processor k. Therefore it can be also efficiently computed through the *Allreduce* operation. One shall note that when the proposed PRBCD method is not adopted, each processor will perform identical back propagation in both the hidden layer and embedding layer.

3   Proposed Parallel Optimization Algorithm

3.1   The Adam Baseline and Its Computational Limitation

For very sparse gradients like language model, plain stochastic optimization method does not perform well in terms of both training speed and attained optimum. Besides exploiting the sparse rectifier units activation to help our model better represent and transform naturally sparse data, state-of-the-art SGD methods e.g. Adam [10] is also applicable for our problem, which is empirically verified by our preliminary trial on a tiny data set which consists of tens of thousands of popular Chinese words as a subset of the Fudan Chinese Language Corpus. By [10], after computing the gradients $\frac{\partial J}{\partial \theta}$ in the backward stage, instead of updating parameters straightforwardly, the updating rule is given by:

$$m^{(t)} = \beta_1 \cdot m^{(t-1)} + (1 - \beta_1)J_\theta,$$
$$v^{(t)} = \beta_2 \cdot v^{(t-1)} + (1 - \beta_2)J_\theta \odot J_\theta,$$
$$\hat{m}^{(t)} = m^{(t)}/(1 - \beta_1^t),$$
$$\hat{v}^{(t)} = v^{(t)}/(1 - \beta_2^t),$$
$$\theta^{(t)} = \theta^{(t-1)} - \alpha \cdot \hat{m}^{(t)}/(\sqrt{\hat{v}^{(t)}} + \epsilon)$$

---

[1] Here 'identical' means each processor keeps the exactly same parameters and training samples in the embedding layer and hidden layer. While when the proposed PRBCD as will be discussed later in the paper is applied, they are not strictly identical.

[2] http://www.mpich.org/static/docs/latest/www3/MPI_Allreduce.html.

[3] See http://www.mpi-forum.org.

where $\hat{m}^{(t)}$ is the unbiased exponential moving average of gradients, and $\hat{v}^{(t)}$ is the unbiased second raw moment estimate. The biased moving averages $m^{(t)}$ and $v^{(t)}$ are both initialized as vectors of 0's. $J_\theta \odot J_\theta$ indicates the element-wise square of vector J_θ. This method performs well in solving problems that are large in terms of data and/or parameters and problems with noisy and/or sparse gradients. The updating term $\alpha \cdot \hat{m}^{(t)}/(\sqrt{\hat{v}^{(t)}} + \epsilon)$ is relatively stable when performing optimization procedure and is invariant to the scale of gradients. It combines the advantages of AdaGrad [3] and RMSProp [20].

However, one possible disadvantage of the Adam method [10] is the computation cost due to complicated mathematics for every gradient. To address this issue, we propose a method equipped with Adam to train the neural network.

3.2   Proposed Parallel Randomized Block Coordinate Descent

As discussed above, the computational bottleneck is in the Softmax output layer. In a parallel computing environment, parameters of this layer are distributed among a number of processors for simultaneous updating. For the reason that (i) as aforementioned, almost 99.9 % of the gradient computation are in output layer; and (ii) the overhead of communication between processors due to two MPI Allreduce operations in forward and backward stage is negligible in a typical setting e.g. 56G InfiniBand fast network in our setting, therefore the computation efficiency can be almost linearly increased by adding more processors – see Fig. 3 for empirical illustration. Motivated by this fact, Parallel Randomized Block Coordinate Descent (PRBCD) is proposed to further accelerate the parallel optimization. Furthermore, we allow that backward processing in the hidden layer in each processor can be inconsistent to a certain extent.

For such big capacity models encountered in this paper, even the simplest full-dimensional vector operations are very expensive, let alone the Adam optimization method. When performing gradient computation in proposed PRBCD method, as illustrated in Fig. 2, processor k will not compute all the gradients of J with respect to o, i.e. the entire vector J_o. Instead it samples a block of neurons in the output layer. Suppose the selected neurons block is $N = \{n_1, n_2, ..., n_{|N|}\}$, where n_p refers to a neuron, if $j \in N$, then we execute $J_o^{(j)} = 1_{j=w_t} - D^{(j)}$ by Eq. (4). The second step is to compute $\partial J/\partial W_S$, for the reason that only those gradients of neurons in N are computed, we can not and will not compute every gradient in W_S, instead we only update those weights corresponding to selected output neurons. The third step is to compute $\partial J/\partial h$, this step still need MPI Allreduce operation in PRBCD method, for each processor k, sampled neurons set is N_k, we have

$$\frac{\partial J}{\partial h(k, N_k)} = (W_S(k, N_k))^T \cdot \frac{\partial J}{\partial o(k, N_k)} \tag{5}$$

where $\partial J/\partial h(k, N_k)$ means gradients w.r.t. the output of hidden layer is computed only by sampled weights in processor k. There are at least tens of millions of parameters in the Softmax output layer, hence we compute gradients as

$$\frac{\partial J}{\partial h} = \sum_k \frac{dim(o(k))}{N_k} \frac{\partial J}{\partial h(k, N_k)} \tag{6}$$

with far less computation cost, efficiency can be significantly improved. This method is also flexible, when the signal of convergence under sampled updating is observed, one can gradually increase the size of N_k, and eventually full-dimensional gradients can be computed if needed. In fact, we empirically find that using full-dimensional gradients often does not lead to better optimum.

Note that our method is different from traditional coordinate descent method when the gradients propagate to the hidden layer. Traditional coordinate descent algorithm often focuses on the performance and convergence analysis when all the parameters are directly connected to the objective such as [23], while in neural network with multiple layers, objective is a compound function of hidden units, it is not possible to compute gradients with respect to previous layer before calculating all the gradients in succeeding layer, which limits the ability of coordinate descent in multilayer perception. PRBCD provides a solution to this issue, as shown in Eq. (6), we scale up the gradients with respect to hidden units by factor $dim(o(k))/N_k$ based on sampled parameters, thus we can estimate the gradients in hidden layer without calculating every gradient in softmax output layer. Typically we have a huge amount of parameters in output layer, therefore it is believable that this estimation method will not bring too much deviation. Empirical result verifies that this strategy is stable for back propagation.

The next step is updating the hidden layer. Optionally, rather than strictly keep this layer identical for every processor, we could introduce inconsistency in the hope of avoiding overfitting though it can also slightly improve efficiency, after all hidden layer is far smaller than output layer. Specifically, we sample a block of hidden units, and perform sampled updating as we do in the Softmax output layer. Every model on each processor may make 'errors' in different ways, their mixture has the potential to achieve better testing performance. This optional method may have the potential to enable a model to achieve better optimum in regularized empirical risk minimization problem when the dataset is not large enough. This idea seems similar to other counter-overfitting techniques, such as Dropout [8] and stochastic pooling [22]. One notable difference to Dropout is that in feedforward stage we still use the standard deterministic objective function to keep all information from input.

4   Training Set Collection and Model Initialization

4.1   Chinese Language Segmentation

Chinese sentences are written in a different way from English, and there is no white space between Chinese words. Therefore, it is necessary to cut the sentences word by word before training, and the way of segmentation will influence

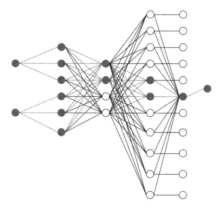

Fig. 2. Illustration for the proposed PRBCD method. Red nodes and lines represent those sampled parameters in one backward step on a single processor. The network structure is the same with the one in Fig. 1. (Color figure online)

the size of universal vocabulary set $|V|$. Segmentation technique employed in this paper is based on popular and open-source project Jieba[4].

Algorithms applied to segment words involve maximum probability path and the Viterbi algorithm [21]. Jieba project has provided a Chinese dictionary, which is represented in data structure Trie Tree [4]. A directed acyclic graph G is generated based on a sentence S, each vertex of G corresponds to one word in S except two special nodes *start node* and *end node*, each path from start node to end node corresponds to one possible way to cut S. The path with highest probability (the product of probability of words in the path) is chosen to be the most correct path. For those words not contained in dictionary, segmentation program resorts to HMM with Viterbi algorithm, each character in a sentence corresponds to four possible hidden states: (1) state 'B' signifies that a character is at the beginning of a word; (2) state 'E' signifies that a character is at the end of a word; (3) state 'M' signifies that a character is at the middle of a word, i.e. neither the beginning nor the end, (4) state 'S' implies that a character itself is a word. Transition probability and emission probability are trained beforehand.

For the purpose of speeding up segmentation to more corpora, the idea of parallelized implementation is straightforward, since segmentation task itself could be 'embarrassingly parallel'[5]. On a computer of multi-cores and shared memory, the master separates the overall documents into several groups and dispatches these groups to workers, every worker constructs a *words dictionary data structure* containing frequency statistics based on one documents group. After all workers finish their jobs, the master receives word sets from each worker and merges them into one, the master is responsible to make sure that every word corresponds to one tag (i.e. a positive integer, which is used to identify this

---

[4] See https://github.com/fxsjy/jieba/.

[5] In parallel computing, an embarrassingly parallel workload or problem is one where little or no effort is needed to separate it into multiple parallel tasks [7].

word). Then the master broadcasts the word dictionary, and each worker converts stream of words into stream of integers based on this 'word-tag' dictionary. We empirically find parallel segmentation on one 4-cores CPU will reduce the overall time of constructing dataset by 54 % compared to single process model.

4.2 Constructed Dataset

Fudan Chinese Language Corpus[6] is collected and organized by the NLP team of School of Computer Science, Fudan University, Shanghai, China. Original corpus contains 19,637 articles in total, these articles are classified into 20 categories, e.g. Agriculture, History, Politics[7].

It is necessary to resolve some minor issues on the original corpus. Firstly, many of these documents are encoded with gbk instead of utf8, while some articles do not use gbk encoding. Secondly, more than four thousand documents are duplicate articles, and therefore need to be eliminated. Thirdly, Chinese sentences in a portion of documents in Law categories have already been segmented, but the effect of segmentation is not satisfying. In this paper, *MD5* algorithm is employed to eliminate duplicate files, encoding was converted utf8, English words and white spaces are all eliminated, numbers are mapped into one symbol, documents in original corpus are reorganized such that 80 % of articles are incorporated in training set, remaining 20 % are incorporated in testing set. After pre-processing and segmentation on original corpus, it becomes a stream of 42,723,332 Chinese words. There are 14,895 different articles remaining in the dataset, the number of different words including punctuation is 441,544, each word corresponds to one non-negative integer tag.

4.3 Initialization of Word Embedding

For training a good distribution representation of word in an NLP task, it is crucial to initialize a good word embedding [11], and random initialization often under-performs. A good embedding should be able to represent the semantic of words, words that are semantically similar to each other are supposed to have a similar distributed representation. Some basic guidelines for initializing a good word embedding include (1) using a larger corpus often yields better results, (2) typically faster models provide sufficient performance in most cases, and more complex models can be used if the training corpus is sufficiently large, (3) the early stopping criterion for iterating should rely on the development set of the desired task rather than the validation loss of training embedding. With the aim to seek a cost-effective solution, we choose Continuous Bag-of-Word Model (CBOW)[8] [14] as the word embedding initializer for NNLM.

---

[6] http://www.datatang.com/data/44139, http://www.datatang.com/data/43543.

[7] Overall categories include Agriculture, Art, Communication, Computer, Economy, Education, Electronics, Energy, Environment, History, Law, Literature, Medical, Military, Mine, Philosophy, Politics, Space, Sports, Transport.

[8] Toolbox: https://code.google.com/p/word2vec/.

Table 2. Performance of word embedding initialized by different models. Field 'min-count' means the minimum frequency of a word. Rare words are mapped to one symbol.

Model	Min-count	Dimension	Iteration	Vocabulary	Validation-error
CBOW1	5	60	15	121,071	0.258370
CBOW2	5	100	15	121,071	0.244370
CBOW3	3	60	15	170,943	0.257235
CBOW4	3	100	15	170,943	0.243386
CBOW5	1	60	15	441,545	0.255240
CBOW6	1	100	15	441,545	0.240525

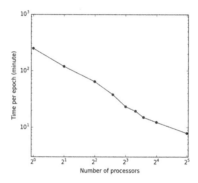

Fig. 3. Average training time per epoch regarding with the number of processors.

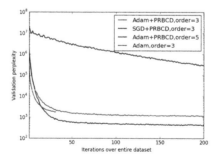

Fig. 4. Perplexity of NNLMs on Fudan Chinese Language Corpus. Adam without PRBCD appears to converge in less epochs (purple curve), whereas in fact its average time cost for each epoch is much more than Adam with PRBCD (red curve). In our test, its time cost for one epoch is less than a third of Adam when sample rate is set 0.1 in the first few epochs (in average 25 min vs. 7 min per epoch). Though the sample rate gradually increases as iteration continues, PRBCD can still accelerate the model training procedure notably. (Color figure online)

Table 3. Time (in minute) of full-dimensional gradients computation with Adam.

Processors	Forward-clock	Forward-time	Backward-clock	Backward-time	Total-clock	Total-time
n = 32	0.65	2.394	1.33	5.4098	1.98	7.804
n = 16	1.15	3.623	2.84	8.5948	3.99	12.217
n = 12	1.45	4.063	3.57	10.858	5.02	14.921
n = 10	1.65	5.087	4.7	14.135	6.35	19.222
n = 8	1.85	5.589	5.76	17.450	7.61	23.039
n = 6	2.60	7.637	9.66	29.937	12.26	37.574
n = 4	3.90	12.734	16.33	51.000	20.23	63.735
n = 2	7.50	23.290	27.73	96.045	35.23	119.330
n = 1	14.60	45.319	61.49	202.810	76.09	248.130

5   Experimental Results

Our experiments are conducted on Linux cluster with MPI parallel computing environment. The cluster contains 8 computational nodes connected through 56G InfiniBand fast network, each node is equipped with two Intel(R) Xeon(R) E5-2680v2 2.8 GHz CPUs and 16 GB of memory and runs Linux CentOS 6.2.

As discussed above, we use CBOW [14] model to efficiently give a better embedding initialization. We have tried a variety of parameters and part of the results are shown in Table 2. Generally, higher dimension of words' distributed representation will lead to lower validation error in CBOW model (more specifically, the word2vec tool) and better semantic properties, whereas empirical results in NNLM (a few trials on tiny data set with tens of thousands popular Chinese words sampled from the Fudan Chinese Language Corpus) show that a dimension of 60 seems already enough for training a probabilistic language model. Higher dimension will lead to slower training efficiency, but will not improve the performance of NNLM on the validation dataset.

To verify that the parallel implementation of neural network can be almost linearly accelerated in parallel computing environment, we have shown the statistics of NNLM's average computation time for each epoch on sub-dataset in Fig. 3. It is clear that even if expanding the system up to 32 CPUs, communication cost in our experiment is not significant. This is due to on one hand the huge amount of parameters and the corresponding excessively long computation time, and the other hand the communication only involves performing a single aggregation over the 32 processors. This fact makes the communication between processors relatively negligible.

Empirical results have suggested that backward time/clock is much more than forward time/clock as shown in Table 3. Such a result is not only due to complicated mathematics in Adam, but also the more comparison and condition judgement in implementation of back propagation. The efficacy of PRBCD lies in accelerating the backward stage, as long as computation cost of backward is much higher that forward, our method can play a significant role. Combining with the Adam method, PRBCD in general can accelerate the total training efficiency several times and even nearly an order of magnitude in the best cases.

In our implementation, the sample rate (i.e. the proportion of gradients that are actually computed) increases from 0.1 to 1 gradually, with a step 0.01 for each epoch. The motivation of this policy is to reduce the uncertainty and to keep accuracy as full-dimensional computation when converging. Implementation of dynamic sample rate leads to the results as shown in Fig. 4, which shows that our PRBCD method is very stable. Compared to Adam without PRBCD (which is certainly much slower), our method does not significantly lose the accuracy, i.e. the perplexity on validation set of corpus, while proposed method will run far more epochs in a certain period of time and will converge faster in practice.

We also show in Fig. 4 that the order of the model (i.e. the length of context words plus 1) is closely relevant with the final perplexity, NNLM with order 5 has better performance. In fact, besides the results shown in the above figures and tables, we have also found some other preliminary phenomenon in experiments by using a small sub-dataset containing tens of thousands of popular Chinese words. For example, NNLM can generally take advantages of longer context than n-gram model, but when the order is larger than 6, perplexity on the validation set tends to converge much slower and models are easier to overfit as well. What is more, number of hidden units also significantly influences the performance according to several trials with the number of hidden units ranging from 50, 100, 150 and 200, generally more units will improve the model. We finally choose NNLMs with 200 hidden units to train over the entire dataset.

6   Conclusion

We have introduced a cost-effective method for training models with a large amount of parameters and high-dimensional output targets such as probabilistic neural network language model. Our method can be combined with a state-of-the-art stochastic optimization algorithm Adam [10]. Empirical results on a Fudan Chinese Language Corpus corroborate that our method is well-suited for training such category of neural network with high-dimensional output targets.

Acknowledgement. This work is partially supported by China Postdoctoral Science Foundation Funded Project (2016M590337), NSFC (11501210), Shanghai YangFan Plan (15YF1403400), and Shanghai Science and Technology Committee Project (15JC1401700).

References

1. Bengio, Y., Schwenk, H., Senécal, J.S., Morin, F., Gauvain, J.L.: Neural probabilistic language models. In: Holmes, D.E., Jain, L.C. (eds.) Innovations in Machine Learning, pp. 137–186. Springer, Heidelberg (2006)
2. Brown, P.F., Desouza, P.V., Mercer, R.L., Pietra, V.J.D., Lai, J.C.: Class-based n-gram models of natural language. Comput. Linguist. **18**(4), 467–479 (1992)
3. Duchi, J., Hazan, E., Singer, Y.: Adaptive subgradient methods for online learning and stochastic optimization. J. Mach. Learn. Res. **12**, 2121–2159 (2011)
4. Fredkin, E.: Trie memory. Commun. ACM **3**(9), 490–499 (1960)

5. Glorot, X., Bordes, A., Bengio, Y.: Deep sparse rectifier neural networks. In: International Conference on Artificial Intelligence and Statistics, pp. 315–323 (2011)
6. He, K., Zhang, X., Ren, S., Sun, J.: Delving deep into rectifiers: surpassing human-level performance on imagenet classification. In: ICCV (2015)
7. Herlihy, M., Shavit, N.: The art of multiprocessor programming. Revised Reprint (2012)
8. Hinton, G.E., Srivastava, N., Krizhevsky, A., Sutskever, I., Salakhutdinov, R.R.: Improving neural networks by preventing co-adaptation of feature detectors. arXiv preprint arXiv:1207.0580 (2012)
9. Jelinek, F.: Interpolated estimation of markov source parameters from sparse data. In: Pattern Recognition in Practice (1980)
10. Kingma, D.P., Adam, J.B.: A method for stochastic optimization. In: International Conference on Learning Representation (2015)
11. Lai, S., Liu, K., Xu, L., Zhao, J.: How to generate a good word embedding? arXiv preprint arXiv:1507.05523 (2015)
12. Maas, A.L., Hannun, A.Y., Ng, A.Y.: Rectifier nonlinearities improve neural network acoustic models. In: Proceedings of ICML, vol. 30, p. 1 (2013)
13. Mikolov, T., Deoras, A., Kombrink, S., Burget, L., Cernocky, J.: Empirical evaluation and combination of advanced language modeling techniques. In: Proceedings of Interspeech, pp. 605–608 (2011)
14. Mikolov, T., Chen, K., Corrado, G., Dean, J.: Efficient estimation of word representations in vector space. arXiv preprint arXiv:1301.3781 (2013)
15. Mnih, A., Hinton, G.: A scalable hierarchical distributed language model (2009)
16. Mnih, A., Teh, Y.W.: A fast and simple algorithm for training neural probabilistic language models (2012)
17. Morin, F., Bengio, Y.: Hierarchical probabilistic neural network language model, pp. 246–252 (2005)
18. Sainath, T., Kingsbury, B., Sindhwani, V., Arisoy, E., Ramabhadran, B.: Low-rank matrix factorization for deep neural network training with high-dimensional output targets, pp. 6655–6659 (2013)
19. Schwenk, H., Gauvain, J.L.: Training neural network language models on very large corpora, pp. 201–208 (2005)
20. Tieleman, T., Hinton, G.: Lecture 6.5-RMSProp: divide the gradient by a running average of its recent magnitude. COURSERA Neural Netw. Mach. Learn. **4**, 2 (2012)
21. Viterbi, A.J.: Error bounds for convolutional codes and an asymptotically optimum decoding algorithm. IEEE Trans. Inf. Theor. **13**(2), 260–269 (1967)
22. Zeiler, M.D., Fergus, R.: Stochastic pooling for regularization of deep convolutional neural networks. In: ICLR (2013)
23. Zhao, T., Yu, M., Wang, Y., Arora, R., Liu, H.: Accelerated mini-batch randomized block coordinate descent method. In: Advances in Neural Information Processing Systems, pp. 3329–3337 (2014)

Multi-label Ranking with LSTM²
for Document Classification

Yan Yan[1], Xu-Cheng Yin[2(✉)], Chun Yang[2],
Bo-Wen Zhang[2], and Hong-Wei Hao[3]

[1] School of Mechanical Electronic and Information Engineering,
China University of Mining and Technology, Beijing, China
yanyanustb@126.com
[2] University of Science and Technology Beijing, Beijing, China
xuchengyin@ustb.edu.cn
[3] Institute of Automation, Chinese Academy of Sciences, Beijing, China

Abstract. Multi-label document classification is an important challenge with many real-world applications. While multi-label ranking is a common approach for multi-label classification. However existing works usually suffer from incomplete and context-free representation, and nonautomatic and part based model implementation. To solve the problem, we propose a LSTM² (Long short term memory) model for document classification in this paper. This model consists of two-steps. The first is repLSTM process which is based on supervised LSTM by introducing the document labels to learn document representation. The second is rankLSTM process. The order of documents labels are rearranged in accordance with a semantics tree, which better exerts the advantages of the LSTM in sequence. Besides by predicting label serially, the model can be trained as a whole. In addition, Connectionist Temporal Classification is used in this process which is a good solution to deal with the error propagation for variable length output (the number of labels in each document). Experiments on three generalization datasets have achieved good results.

Keywords: Document classification · Multi-label ranking · repLSTM · rankLSTM

1 Introduction

Multi-label classification [21] is a generalization of multi-class prediction where the goal is to predict a set of labels that are relevant to a given input. Multi-label classification problems exist in several domains, e.g., document classification [16,30], protein function analysis in bioinformatics [1], semantic annotation of image tasks [28] and music categorisation into emotions [25]. The most widely used approaches divide a multi-label learning task into multiple independent binary labelling tasks [3]. However, most of these approaches suffer from unbalanced data distributions when constructing binary classifiers to distinguish individual classes from the remaining classes and cannot solve the classification in

© Springer Nature Singapore Pte Ltd. 2016
T. Tan et al. (Eds.): CCPR 2016, Part II, CCIS 663, pp. 349–363, 2016.
DOI: 10.1007/978-981-10-3005-5_29

sparse categories. Multi-label ranking [27] is proposed to solve the problem by considering the correlation among classes. In this paper, we address multi-label classification using a multi-label ranking approach.

In the domain, several multi-label ranking methods have been proposed for document classification, which can be broadly considered as a two-step process. The first step is to learn the document representation. Usually in these methods the length of input features need to be fixed when dealing with the documents and the representations are always based on bag-of-words and TF-IDF (term frequencyinverse document frequency). However, these representations are very limited for semantics as they only contain the information of word frequency, and do not consider the sequence of words in the documents. The second step is ranking. This strategy requires a real-valued score predicted by the trained classifiers for each class (label) to conduct label ranking. When training classifiers, most methods are based on a binary relevance approach which independently trains one classifier for each label or each pair of labels [10,18]. Afterwards, the results are sorted, an appropriate threshold is hereafter determined with priori knowledge, such as the professional vocabularies and tools in specific fields, and the output document labels are finally judged. In the ranking based classification procedure, both the scoring process and the threshold selection have a great influence on the classification results.

Recently, deep learning has been successfully applied to solve the above problems. For example, for the challenge about document representation, RNN (Recurrent Neural Network) [9] analyzes text word by word and stores the semantics of all previous text in a hidden layer. The advantage of RNN is the ability to better capture the contextual information. This could be beneficial to capture the semantics of long texts. However, this method faces the problem of gradients vanishing and exploding. One of the solutions is LSTM (Long Short Term Memory) network. LSTM not only act as a type of recurrent neural network architecture which effectively learns features, but also obtains strong results on a variety of sequence modelling tasks recently [24]. Hence, for the challenge about document ranking, in order to address the problem of requested manual or semi-automatic selection, we further use LSTM for ranking with a label sequence. Besides, by predicting the label sequentially (which means we predict the next label by current selected label subset), the scoring process and the threshold selection process can be trained as a whole instead of two separate parts. For multi-label classification, we specifically use the CTC (Connectionist Temporal Classification) [12] method, which can better solve the error updates of labels with variable lengths, and make the model more robust.

Summarily, we propose a $LSTM^2$ framework with label ranking for multi-label document classification. Document representation is first learnt by LSTM (repLSTM), where the document input can have an arbitrary length and the keywords are extracted without limited window-size. Then labels are represented as a semantic tree trained with dependency parsing. This tree can capture correlations between labels. Based on the document representation, another LSTM is afterwards utilized to rank document labels (rankLSTM).

2 Related Work

2.1 Document Classification

The purpose in document classification task is to retrieve the most suitable labels for the testing document. A common solution is to represent the documents in some certain space, then to select labels by the relevance between the labels and the test document. Generally speaking, conventional document classification research can be mainly categorized into two major directions.

One is shallow learning. There are several learning methods for document classification, such as LSI (Latent Semantic Indexing), LDA (Latent Dirichlet Allocation) and pLSI (probabilistic Latent Semantic Indexing) [2]. Nearly all of these methods use SVD (Singular Value Decomposition) to operate on a document vector matrix and remap it to a semantic space (always with a low-dimensional representation), where each dimension represents a "latent topic". However, these methods use linear function computation and are unsupervised. These shallow learning methods need to fix the length of the input features when dealing with the documents and the representations are based on bag-of-words and TF-IDF. However, these representations only contain the word frequency information and ignore the sequence of words in the documents, which are very limited for semantic representation.

Document classification is also investigated with deep learning techniques. The existing deep learning methods, for example CNN (Convolutional Neural Network) [7] and RNN, can be good ways to retain the information of word order in the document. Zeng et al. used CNNs for relation classification [30], and Dos Santos utilized CNNs for semantic analysis of text [8]. Mikolov [19] proposed the Recurrent Neural Network Language Model which improve the performances by providing a contextual real-valued input vector associated with each word. However, these methods have some disadvantages to some different degrees. When using the CNN method, after the convolutions are calculated, we still have to fix the length of representation of each document through k-max or average pooling. RNN model is a biased model, where words in the later sequence are more dominant than those in the earlier sequence.

Lai [16] proposed RCNN (Recurrent Convolutional Neural Networks) model for document classification that can overcome RNN's bias. The RCNN effectively utilizes the advantages of recurrent structure which captures the contextual information, and learns the feature representation of documents by CNN. But this model will also face with problems from gradients vanishing and gradient exploding when updating the error with back-propagation.

2.2 Label Ranking

In recent years, several multi-label ranking methods have been proposed for multi-label classification. Zhang [31] proposed a ML-kNN (Multi-Label K Nearest Neighbours) which extends the popular k Nearest Neighbors (kNN) lazy learning algorithm using a Bayesian approach. For each test instance, its k nearest neighbors in the training set are firstly identified. The maximum posteriori is calculated by them in order to determine the labels of the test sample. Chiang [6] proposed a ranking-based KNN approach for multi-label classification. Compared with Zhang's model, Chiang's model assigns higher weights to those neighbours with high confidence during ranking process. The weights can be calculated by using a generalized pattern search technique. Hüullermeier [13] proposed a RPC (Ranking by Pairwise Comparison) method which can reduce the problem of label ranking to several binary classification problems. The binary classification problems are comparatively simple and efficiently learnable. Tsoumakas [26] proposed a random k-labelsets method for multi-label classification. The main idea in the work is to randomly break a large set of labels into m small-sized labelsets, and train m CC classifiers C_1, C_2, ..., C_m for each of them. Then all classifiers are gathered and combined to predict for the test instance. A threshold is used to select the most popular labels which form the final predicted multi-label set.

2.3 LSTM (Long Short Term Memory)

When using RNN model to learn long-distance correlations in a sequence, it will face the problems such as gradients exploding or vanishing [32]. LSTM replaces the hidden vector of a recurrent neural network with memory blocks which are equipped with gates. It can keep long term memory in principle by training proper gating weights, and it has practically showed to be very useful, achieving the state of the art on a range of problems including speech recognition [11] and bioinformatics [5]. LSTM also has been proven to be a good variant of the RNN model.

Each LSTM (see Fig. 1) unit has a memory cell and the state at time t is represented as c_t. Reading and Modifying are controlled by sigmoidal gates and also impact on the input gate i_t, forget gate f_t and output gate o_t. LSTM is calculated as follows. At the tth moment the model receives inputs from two external sources (h_{t-1} and x_t). The hidden state h_t is calculated by the input vector x_t that the network receives at time t and the previous hidden state h_{t-1}. At the time of calculating hidden layer node status, the input, the output, the forgotten gate and x_t will all affect the state of the node at the same time. Additionally, each gate has an internal source, that is, the cell state c_{t-1} of its cell block. The links between a cell and its own gates are called peephole connections. These updates are summarized for a layer of LSTM units as follows.

$$i_t = \sigma(W_{ix}x_t + W_{reci}h_{t-1} + W_{ic}c_{t-1} + b_i) \tag{1}$$

$$f_t = \sigma(W_{fx}x_t + W_{recf}h_{t-1} + W_{fc}c_{t-1} + b_f) \tag{2}$$

$$c_t = f_t \odot c_{t-1} + i_t \odot g(W_{cx}x_t + W_{recc}h_{t-1} + b_c) \tag{3}$$

$$o_t = \sigma(W_{ox}x_t + W_{reco}h_{t-1} + W_{oc}c_t + b_o) \tag{4}$$

$$h_t = o_t \odot m(c_t) \tag{5}$$

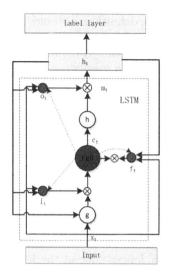

Fig. 1. The flow of LSTM (Long Short Term Memory).

$$y_t = \phi(W_{recy}h_t + b_y) \tag{6}$$

In the above equations, x_t represents the input of document. W_{ic}, W_{fc} and W_{oc} are peephole connections. W_{jx}, W_{recj} and b_j, $j = i, f, o, c$ are input weight matrix connections, recurrent weight matrix connections and bias values of input, forget, output and cell units. $\phi(\cdot)$, $g(\cdot)$ and $m(\cdot)$ are $tanh(\cdot)$ function and $\sigma(\cdot)$ is the sigmoid function. When it reaches the last word, the hidden layer (h_t) of the network provides a semantic representation vector of the entire sentence. The key advantage of using LSTM unit over traditional neuron in RNN is that the cell state in LSTM unit sums activities over time. Since derivatives distribute over sums, the error derivatives don't vanish quickly as they get sent back into time [22].

3   The LSTM^2 Model

To overcome the disadvantages of traditional document classification approaches, we propose a LSTM based Rep-Ranking framework (LSTM^2) for document classification. The LSTM model is used in both representation and ranking processes, but in different ways.

As shown in Fig. 2, the left part is the repLSTM part which consists of three layers: the document input layer, document representation layer and label layer. The document inputs are represented by the word number order (WNO). The word number is the current word's corresponding position number in the vocabulary list. We use LSTM to learn document representation. The keywords of current document are extracted from LSTM's input, forget and output cells.

Keywords in this paper are the definitions obtained from the visualization analysis of Li's paper [17]. For each sentence, the words and the number of the words that are activated are both different. We define these activated words as keywords. These keywords can also be considered as the coarse classification of the documents. In our model the document representation can reflect the impacts of these keywords. Compared with RNN, the advantage of LSTM is to generate keywords with arbitrary length. Although the length of keywords in each document may be quite arbitrary, the dimension of the document representation is fixed. The dimension of the label layer outputs equal the number of the dataset categories.

Then the ranking process, namely rankLSTM (the right part in Fig. 2), is executed with the document representation extracted from the previous process. This process also composed of three layers: the label input layer, LSTM layer and label output layer. The task of ranking process is to predict the document's label according to the document representation from the first process. Taking into consideration the dependencies among the labels, we firstly learn a label tree based on the semantics by dependency parsing. Then for each document, the universal semantic tree is pruned, and a label sequence is produced by a breadth-first traversal of the pruned tree. In addition, we add the Start statue at the beginning of the sequence, and End statue in the end. And ranking process is redefined as follows: with the document representations as a priori knowledge, given the current label when the label state is A (such as A or Start), we predict that the next label statues is B (such as B or End). In this process, we also use the LSTM to implement this functionality. This process is closely related to that of karpathy's paper [15].

In order to implement the model, we train the representation process (repLSTM) and the ranking process (rankLSTM) separately. For the representation process, the optimization goal is the minimum error between the representation and inputs. In addition, we introduce the labels for an supervised learning and use the softmax function to optimize the loss function.

For the ranking process, the model is expected to predict more accurate labels from the coarse document classification results (the keywords of each document). We use Connectionist Temporal Classification (CTC) [12] for the ranking process to calculate the errors. CTC is a good solution to deal with the error propagation from variable length input to the variable length output. And it allows the network to make label predictions at any point in the input sequence, so long as the overall sequence of labels is correct. Moreover, CTC directly outputs the probabilities of the complete label sequences, which means that no external post-processing is required to use the network as a temporal classifier. Experiments on speech and handwriting recognition show that LSTM network with a CTC output layer is an effective sequence labeller [12].

The most ideal way to train the model should be an end-to-end leaning, which means learning the document representation implicitly in the classification process. However, our classLSTM can only update the influence (the weights) of the document representation (inputs) on label prediction (outputs) instead

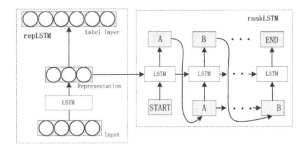

Fig. 2. The LSTM^2 model: repLSTM for document representation and rankLSTM for label ranking.

of updating representation itself. The LSTM calculation consists of two parts. One is the current state x_t and the other is h_{t-1} (we have mentioned in related work). In other words, the ranking process needs to be given an exact document representation (h_0) as prior information. Thus, we extract the representation by an supervised learning before the ranking process.

4   Experiments and Results

4.1   Dataset and Experimental Setup

We select three multi-label document datasets in our experiment. The dataset1[1] is the MEDLINE dataset. Yepes' paper [29] selected the top 10 most frequent MESH headings to avoid the extremely unbalanced distribution dataset. Similar to their motivation, we select the top 150 MeSH headings as only about 150 appear in more than 1 % of the whole medline. To further extend the experiments, in our paper, we introduce some unbalanced samples and select the top 2000 categories. We processed the citations to extract the text from the title and the abstract. The dataset2[2] is the Enron corpus which is a subset of the Enron Email Dataset, as labelled by the UC Berkeley Enron Email Analysis Project. There are many versions of this dataset and we select the Padhye and Pedersen corpus which is a subset of the Bekkerman corpus. This dataset has six top-level categories, each of which contains numerous sub-categories. So this is a tree multi-label dataset. The dataset3[3] is RCV1 which is simply a collection of newswire stories. Details are shown in Table 1. For each dataset, we randomly select 80 % as the training samples and 20 % as the test samples, and make 10-fold cross validation in experiment. We apply the widely-used cuda-convnet package to train our model on a single GPU.

---

[1] http://www.bioasq.org/participate/challenges.

[2] http://www.d.umn.edu/~tpederse/enron.html.

[3] http://www.daviddlewis.com/resources/testcollections/rcv1/.

Table 1. Dataset details.

Dataset	Sample	Classes	Type	Field
Dataset1	100,000	150	Multi-label	Biomedicine
		2000	Multi-label	
Dataset2	3021	6	Tree multi-label	Email
Dataset3	800,000	103	Multi-label	News

4.2   Details

To compare with word-order-based feature representations(WNO), we use BOW method. Besides, we use a restricted vocabulary list to reduce the sparsity of the inputs. To generate the vocabulary, we retain firstly the top 2000 most frequent words for both dataset1 and dataset2, and the top 10,000 for dataset3. Then we conduct all stop words removing and stemming. The frequency-of-occurrence of each word in the vocabulary is used as a feature for training.

LSTM-Rep (**LSTM-R**): only use LSTM model in document representation and use the softmax function to optimize the loss (also have the repLSTM part in LSTM^2). We train the parameters between each pair of layers. After training the model, we can directly predict the testing document labels using only the LSTM model.

RNN-LSTM-parse (**RLP**) is using RNN to replace the LSTM to learn the document representation in the document representation process. RNN can also represent the variable length of the document input, then use the intermediate representation of the documents as the input of the second part. The last variable length label layer of the model still uses the semantic tree based on dependency parsing.

RNN-LSTM (**RL**): different from LSTM^2, this model uses frequency order in the vocabulary about specific categories instead of the parse tree to represent the label output layer.

Besides, we also select Supported Vector Machines (**SVM**)[4] [4], Naive Bayesian (**NB**) [14], **ML-KNN**[5] [31], **CNN**[6] [7] and **RNN**[7] [19] which are now popular deep learning methods as comparison with our method (our code is modified **LSTM**^{2} method (our code is modified based on the code[8] which is for a two-class sentiment classification). We use the Precision (P), Recall (R), Similarity (S) and F$_1$-measure with Micro and Macro as the evaluation criteria.

---

[4] https://www.csie.ntu.edu.tw/~cjlin/libsvm/.
[5] http://lamda.nju.edu.cn/code_MLkNN.ashx.
[6] http://www.vlfeat.org/matconvnet/.
[7] http://www.fit.vutbr.cz/~imikolov/rnnlm/.
[8] http://deeplearning.net/tutorial/lstm.html.

4.3   Experimental Results

In this section, we conducted three sets of comparison experiments. In the first group of comparison, we only observe which feature representation has better performance. We choose the BOW representation to compare with WNO which the representation is based on word sequence. In the second group, we analyse different models' effects with these features representation on all datasets. Through the last group, we analyse the effects of different components on our model. Significant testing is described in each experiment part.

Table 2. Classification results (%) from different models with different features on all datasets (Micro).

Dataset	Method	BOW				WNO			
		MiP	MiR	MiF$_1$	MiS	MiP	MiR	MiF$_1$	MiS
Dataset1 (2,000 C)	SVM	49.28	48.89	49.08	49.55	49.58	47.95	48.75	49.39
	NB	46.23	45.82	46.03	48.09	46.52	45.66	46.08	48.12
	CNN	60.72	58.78	59.73	55.40	65.33	62.72	64.00	58.16
	RNN	60.73	58.19	59.43	55.22	66.64	63.57	65.07	58.90
	ML-KNN	53.90	51.90	52.88	51.49	57.14	54.62	55.86	53.13
	LSTM^2	**63.76**	**61.30**	**62.51**	**57.16**	**69.05**	**67.96**	**68.50**	**61.35**
Dataset2	SVM	56.37	57.10	56.73	53.61	55.69	56.36	56.02	53.20
	NB	50.46	51.42	50.93	50.47	49.50	49.88	49.69	49.85
	CNN	61.86	62.04	61.95	56.79	62.79	63.05	62.92	57.42
	RNN	62.11	61.36	61.73	56.65	62.23	62.42	62.32	57.03
	ML-KNN	54.37	55.00	54.69	52.46	54.03	54.52	54.28	52.23
	LSTM^2	**63.40**	**60.85**	**62.10**	**56.90**	**63.34**	**65.01**	**64.16**	**58.26**
Dataset3	SVM	35.20	34.64	34.92	43.45	33.91	34.05	34.96	43.47
	NB	35.28	34.40	34.83	43.42	35.60	34.54	35.06	43.50
	CNN	45.15	44.74	44.95	47.59	45.24	44.97	45.10	47.66
	RNN	45.66	44.80	45.23	47.72	45.42	45.50	45.46	47.83
	ML-KNN	36.78	38.88	37.80	44.58	38.00	36.55	37.26	44.35
	LSTM^2	**47.07**	**47.91**	**47.49**	**48.78**	**50.28**	**49.72**	**50.00**	**50.00**

Experiment with Different Features. Through the statistics in Tables 2 and 3 we can find that the performances of feature representation based on document order are superior to those based on BOW in all methods. This is mainly because both RNN and LSTM model are based on sliding window to capture information between adjacent words, but the BOW break the adjacent relations. In order to find further advantages of feature representation based on word order, we visualize the representation of the documents. Through comparisons with BOW, the results of WNO get improved not only in amount but also in accuracy of keywords. As a consequence, with these keywords the documents are represented

Table 3. Classification results (%) from different models with different features on all datasets (Macro).

Dataset	Method	BOW				WNO			
		MaP	MaR	MaF$_1$	MaS	MaP	MaR	MaF$_1$	MaS
Dataset1 (2,000 C)	SVM	48.65	47.28	47.96	49.00	47.68	44.82	46.20	48.19
	NB	45.64	44.32	44.97	47.61	44.73	42.67	43.67	47.04
	CNN	59.35	57.51	58.42	54.61	62.09	60.04	61.04	56.22
	RNN	59.65	56.67	58.12	54.44	63.39	60.63	61.98	56.82
	ML-KNN	52.74	50.68	51.69	50.87	54.13	52.85	53.48	51.81
	LSTM^2	**60.87**	**59.71**	**60.29**	**55.74**	**65.69**	**64.09**	**64.88**	**58.75**
Dataset2	SVM	57.93	57.22	57.57	54.10	56.70	57.75	57.22	53.89
	NB	51.48	52.10	51.79	50.91	51.00	52.05	51.52	50.77
	CNN	60.49	60.73	60.61	55.93	61.82	61.99	61.91	56.76
	RNN	60.06	60.72	60.39	55.80	61.21	61.37	61.29	56.37
	ML-KNN	57.12	59.02	58.05	54.39	56.37	57.13	56.75	53.62
	LSTM^2	**64.54**	**62.36**	**63.43**	**57.77**	**66.67**	**65.77**	**66.22**	**59.68**
Dataset3	SVM	34.01	33.57	33.79	43.03	35.08	33.03	34.03	43.12
	NB	34.19	33.37	33.77	43.02	35.20	34.13	34.66	43.35
	CNN	44.11	43.13	43.62	47.00	47.78	46.60	47.18	48.63
	RNN	44.51	43.69	44.10	47.22	48.07	46.97	47.51	48.79
	ML-KNN	37.36	37.01	37.19	44.32	40.93	38.86	39.87	45.41
	LSTM^2	**45.21**	**46.77**	**45.98**	**48.07**	**49.29**	**49.18**	**49.23**	**49.62**

more accurately. The document input based on word sequence contain more information of phrases, but the one based on the BOW representation filters out some keywords, which leads to very little remaining information. Because the WNO input format is based on word order characteristics with variable length, it is rarely used in shallow learning methods, and this is one of the reason why deep learning shows the strengths in the document classification.

Experiment Different Models Under the Same Features. Among all models, CNN, RNN and our model can attain good document representation on the same features. But CNN and RNN models predict the labels directly by classifiers, and are not applicable to unbalanced dataset. On the other hand, compared with our model, the ranking process in ML-KNN is separated into two steps. The first step is to get the candidate list labels from classifiers and the second one is to select a threshold to determine the number of labels. The ranking process of our model is based on the label sequence. This can not only address the unbalance problem on classification especially in a large number of labels, but also deal with the end-to-end learning in various ranking process. The experiments show that our model is more robust than others.

Table 4. Experiment compare with other methods.

Method	Dataset		
	bio (10 C)	Email	News
Yepes's method [29]	0.693	-	-
Padhye's method [20]	-	0.573	-
Srivastava's method [23]	-	-	0.453
LSTM²	0.810	0.681	0.523

Significant testing. Figure 3 shows the F_1 performance comparison of our method (LSTM²) against other approaches. From this figure, we observe the following conclusions: (1) the deep learning methods have better performances than shallow learning methods. (2) the whole ranking process based on the sequence is better than the traditional separate methods.

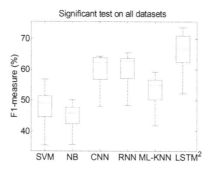

Fig. 3. Significant test with different models.

We also compare with the Yepes's [29], Padhye's [20] and Srivastava's approaches [23] on dataset1, dataset2 and dataset3. Better results are obtained through our proposed models (Table 4). We propose this method based on automatic feature extraction which also obtain similar results. Our main research direction is not for qualitative data sets, but for the generalized dataset to ordinary multi-label document datasets. Furthermore, our idea can be applied to other domains.

Experiment with Different Components in Our Model. Compared with LSTM-R method, the experimental results (Table 5) show that our model (LSTM²) achieves much higher F_1 scores. We check the predicted labels from LSTM-R model, and find that the precision is very high while the recall is relatively low. The more frequent labels are forecast with higher confidence on the whole dataset. This also means that the model prediction on the label is with high accuracy as well as insufficient quantity. Since LSTM-R model is based on independent binary labelling, the training on the label with unbalanced data

Table 5. Classification results (%) with different components in our model on all datasets.

Dataset	Method	BOW						WNO					
		MiP	MiR	MiF$_1$	MaP	MaR	MaF$_1$	MiP	MiR	MiF$_1$	MaP	MaR	MaF$_1$
Dataset1 (2,000 C)	LSTM-R	59.70	56.52	58.07	60.17	57.59	58.85	66.47	64.27	65.35	63.90	60.96	62.40
	RLP	63.22	58.71	60.88	60.27	57.92	59.07	69.02	66.37	67.67	64.29	62.75	63.51
	RL	61.27	56.68	58.89	60.09	57.75	58.89	68.60	65.01	66.75	64.10	62.56	63.32
	LSTM^2	**63.76**	**61.30**	**62.51**	**60.87**	**59.71**	**60.29**	**69.05**	**67.96**	**68.50**	**65.69**	**64.09**	**64.88**
Dataset2	LSTM-R	60.53	58.91	59.71	60.62	61.84	61.22	60.05	59.56	59.80	61.30	61.34	61.32
	RLP	62.54	61.15	61.38	62.48	61.95	62.21	62.09	62.24	62.16	60.68	61.45	61.06
	RL	62.11	59.81	60.94	62.07	61.87	61.97	61.49	61.46	61.48	63.18	63.97	63.57
	LSTM^2	**63.40**	**60.85**	**62.10**	**64.54**	**62.36**	**63.43**	**63.33**	**65.01**	**64.16**	**66.67**	**65.77**	**66.22**
Dataset3	LSTM-R	44.94	44.49	44.71	42.07	43.43	42.74	47.16	46.91	47.03	45.55	44.95	45.25
	RLP	46.32	47.05	46.68	44.33	45.76	45.03	48.53	47.85	48.19	48.25	48.01	48.13
	RL	45.31	46.20	45.75	43.77	44.91	44.33	47.51	47.03	47.27	46.54	46.44	46.49
	LSTM^2	**47.07**	**47.91**	**47.49**	**45.21**	**46.77**	**45.98**	**50.28**	**49.72**	**50.00**	**49.29**	**49.18**	**49.23**

distributions is insufficient. As a result, the model focus on predicting the categories which have more training samples. But the ranking process of our model can adjust the imbalanced distribution samples effectively. The document output has variable length. This is equivalent to directly training model for each category of articles, and when the testing document is similar to training document, it can predict the labels more accurately.

Compared with our model, the RLP model use RNN to extract the document representation, and it also achieves some reasonable results. Just like LSTM, the RNN can also handle the sentences of the document, but it lacks the ability on arbitrarily long sentences. Especially in news dataset, compared with our model, the gap between the RNN model and our model is greater than that on the biomedical dataset. Tracing to the causes, on the news dataset, this progressive or transitions between sentences are more intense, thus the context of the semantic structure between words is even more important, and such a relatively short distance of memory in RNN cannot satisfy the semantic information to be stored.

Compared with RL model, our model adopts a totally different classification process. Through visualizations of semantics based on the parse tree, we find that there are a lot of dependencies among the labels of the child nodes under a same father node. But this dependency is not just a simple synonym relationship.

We also compared the effects of the number of labels (see Fig. 4). Although the samples of 2000 categories in dataset1 had worse results than those of 150 classes, 2000 label tree contains more semantic refinement than 150 label tree according to the analysis of semantic parsing tree of labels. The results are not as good as of 150 classes because when there are 2000 classes, the samples corresponding to the label are extremely unbalanced which impact the results of the classification.

Significant testing. Figure 5 shows the comparisons of the overall F_1 scores among the components of our method (LSTM^2). From this figure, we can summarize as follows: (1) the CTC framework is slightly better than the traditional

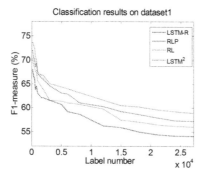

Fig. 4. Classification results vs. label number.

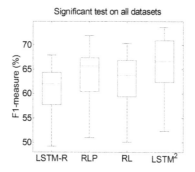

Fig. 5. Significant test with different components.

classification methods. (2) our LSTM² method and RLP get the best and the next-best results which shows that the LSTM with arbitrary length keywords can capture more information of documents than RNN.

5   Conclusion

This paper proposes the LSTM² model based on ranking and the model achieves good results in document classification task. While the LSTM can get information about the sentence with arbitrary length, the selected datasets of our experiment are based on the paragraph context which can be considered as a long sentences, so the performances are still fairly satisfying. But for full-text dataset, the sequence is too long to extract effective representations by LSTM². So how to classify the full text document is our main research in the future.

References

1. Barutcuoglu, Z., Schapire, R.E., Troyanskaya, O.G.: Hierarchical multi-label prediction of gene function. Bioinformatics **22**(7), 830–836 (2006)
2. Blei, D.M., Ng, A.Y., Jordan, M.I.: Latent dirichlet allocation. In: Advances in neural information processing systems, pp. 601–608 (2001)
3. Bucak, S.S., Mallapragada, P.K., Jin, R., Jain, A.K.: Efficient multi-label ranking for multi-class learning: application to object recognition. In: 2009 IEEE 12th International Conference on Computer Vision, pp. 2098–2105. IEEE (2009)
4. Chang, C.C., Lin, C.J.: LIBSVM: a library for support vector machines. ACM Trans. Intell. Syst. Technol. (TIST) **2**(3), 27 (2011)
5. Chen, J., Chaudhari, N.S.: Protein secondary structure prediction with bidirectional LSTM networks. In: International Joint Conference on Neural Networks: Post-Conference Workshop on Computational Intelligence Approaches for the Analysis of Bio-data (CI-BIO), August 2005
6. Chiang, T.H., Lo, H.Y., Lin, S.D.: A ranking-based KNN approach for multi-label classification. ACML **25**, 81–96 (2012)
7. Collobert, R., Weston, J., Bottou, L., Karlen, M., Kavukcuoglu, K., Kuksa, P.: Natural language processing (almost) from scratch. J. Mach. Learn. Res. **12**, 2493–2537 (2011)
8. Dos Santos, C.N., Gatti, M.: Deep convolutional neural networks for sentiment analysis of short texts. In: Proceedings of the 25th International Conference on Computational Linguistics (COLING), Dublin, Ireland (2014)
9. Elman, J.L.: Finding structure in time. Cognit. Sci. **14**(2), 179–211 (1990)
10. Elsas, J.L., Donmez, P., Callan, J., Carbonell, J.G.: Pairwise document classification for relevance feedback. Technical report, DTIC Document (2009)
11. Graves, A., Mohamed, A., Hinton, G.: Speech recognition with deep recurrent neural networks. In: 2013 IEEE International Conference on Acoustics, Speech and Signal Processing (ICASSP), pp. 6645–6649. IEEE (2013)
12. Graves, A., Daojian, L., K., Lai, S., Zhou, G., Zhao, J.: Supervised Sequence Labelling with Recurrent Neural Networks, vol. 385. Springer, Heidelberg (2012)
13. Hüllermeier, E., Fürnkranz, J., Cheng, W., Brinker, K.: Label ranking by learning pairwise preferences. Artif. Intell. **172**(16), 1897–1916 (2008)
14. Jordan, A.: On discriminative vs. generative classifiers: a comparison of logistic regression and naive Bayes. Adv. Neural Inf. Process. Syst. **14**, 841 (2002)
15. Karpathy, A., Fei-Fei, L.: Deep visual-semantic alignments for generating image descriptions. arXiv preprint arXiv:1412.2306 (2014)
16. Lai, S., Xu, L., Liu, K., Zhao, J.: Recurrent convolutional neural networks for text classification. In: Twenty-Ninth AAAI Conference on Artificial Intelligence (2015)
17. Li, J., Chen, X., Hovy, E., Jurafsky, D.: Visualizing and understanding neural models in NLP. arXiv preprint arXiv:1506.01066 (2015)
18. Loza Mencía, E., Fürnkranz, J.: Efficient pairwise multilabel classification for large-scale problems in the legal domain. In: Daelemans, W., Goethals, B., Morik, K. (eds.) ECML PKDD 2008. LNCS, vol. 5212, pp. 50–65. Springer, Heidelberg (2008). doi:10.1007/978-3-540-87481-2_4
19. Mikolov, T., Karafiát, M., Burget, L., Cernocký, J., Khudanpur, S.: Recurrent neural network based language model. In: 11th Annual Conference of the International Speech Communication Association, INTERSPEECH 2010, Makuhari, Chiba, Japan, 26–30 September 2010, pp. 1045–1048 (2010)

20. Padhye, A.: Comparing supervised and unsupervised classification of messages in the enron email corpus. Ph.D. thesis. University of Minnesota (2006)
21. Petterson, J., Caetano, T.S.: Reverse multi-label learning. In: Advances in Neural Information Processing Systems, pp. 1912–1920 (2010)
22. Srivastava, N., Mansimov, E., Salakhutdinov, R.: Unsupervised learning of video representations using LSTMS. arXiv preprint arXiv:1502.04681 (2015)
23. Srivastava, N., Salakhutdinov, R.R., Hinton, G.E.: Modeling documents with deep Boltzmann machines. arXiv preprint arXiv:1309.6865 (2013)
24. Tai, K.S., Socher, R., Manning, C.D.: Improved semantic representations from tree-structured long short-term memory networks. arXiv preprint arXiv:1503.00075 (2015)
25. Trohidis, K., Tsoumakas, G., Kalliris, G., Vlahavas, I.P.: Multi-label classification of music into emotions. In: ISMIR, vol. 8, pp. 325–330 (2008)
26. Tsoumakas, G., Katakis, I., Vlahavas, I.: Random k-labelsets for multilabel classi-fication. IEEE Trans. Knowl. Data Eng. **23**(7), 1079–1089 (2011)
27. Vembu, S., Gärtner, T.: Label ranking algorithms: a survey. In: Fürnkranz, J., Hüllermeier, E. (eds.) Preference Learning, pp. 45–64. Springer, Heidelberg (2011)
28. Xue, X., Zhang, W., Zhang, J., Wu, B., Fan, J., Lu, Y.: Correlative multi-label multi-instance image annotation. In: 2011 IEEE International Conference on Com-puter Vision (ICCV), pp. 651–658. IEEE (2011)
29. Yepes, A.J., MacKinlay, A., Bedo, J., Garnavi, R., Chen, Q.: Deep belief networks and biomedical text categorisation. In: Australasian Language Technology Associ-ation Workshop, p. 123 (2014)
30. Zeng, D., Liu, K., Lai, S., Zhou, G., Zhao, J.: Relation classification via convolu-tional deep neural network. In: Proceedings of COLING, pp. 2335–2344 (2014)
31. Zhang, M.L., Zhou, Z.H.: ML-KNN: a lazy learning approach to multi-label learn-ing. Pattern Recognit. **40**(7), 2038–2048 (2007)
32. Zhu, X., Sobihani, P., Guo, H.: Long short-term memory over recursive structures. In: Proceedings of the 32nd International Conference on Machine Learning (ICML-2015), pp. 1604–1612 (2015)

Attribute Based Approach
for Clothing Recognition

Wang Fan[1]([✉]), Zhao Qiyang[1], Liu Qingjie[2], and Yin Baolin[1]

[1] State Key Laboratory of Software Development Environment,
Beihang University, Beijing, China
`wangfan@nlsde.buaa.edu.cn`
[2] Intelligent Recognition and Image Processing Lab,
Beihang University, Beijing, China

Abstract. Clothing recognition is hot topic for its potential benefits to lots of visual tasks, such as people identification, pose estimation and recommendation system. However, due to the wide variations of clothing appearance and the "semantic gap" between low-level features and high-level category concepts, clothing recognition is very challenging. To narrow this gap, a novel method, which uses intermediate attributes to bridge low-level features and high-level category labels, is proposed. This method first recognizes local attributes from low-level visual features, and then infers clothing category based on these attributes. To this end, DPM models and pixel-level parsing are applied to obtain geometric structure attributes, such as collar shape, and geometric size attributes, such as sleeve length, respectively. Then, Multiple Output Neural Networks are built to predict clothing category based on attributes. Experiments show that the performance of our method is superior to two state-of-the-art approaches on both of attribute and category recognition.

Keywords: Attribute recognition · Clothing recognition · Attribute based

1 Introduction

In recent years, more and more studies have been put forward on clothing analysis, such as pixel-level parsing [1,2], clothing modeling [3] and clothing recognition [4,5], due to its potential values in not only visual tasks, such as people identification [6,7], pose estimation [8,9], recommendation system [10,11] and retrieval [12,13], but also in e-commerce, such as online shopping and adversing. Clothing recognition plays an essential role in these tasks. However, the variation of clothing appearance makes the recognition task very challenging. This paper focuses on recognizing clothing categories as well as multiple kinds of local attributes.

Although more and more works are being presented on clothing recognition, most of them [4,9,14,15] focused on predicting category based on low-level features directly. Chen *et al.* [4] extracted features, including SIFT, color, texture

© Springer Nature Singapore Pte Ltd. 2016
T. Tan et al. (Eds.): CCPR 2016, Part II, CCIS 663, pp. 364–378, 2016.
DOI: 10.1007/978-981-10-3005-5_30

descriptors and skin probabilities from body parts, then employed bag of words model and SVM to predict attributes and clothing categories. Shen *et al.* [9] presented a similar method but processing simultaneously with pose estimation to improve the performance. Yang *et al.* [15] extracted features based on clothing segmentation and use linear SVM for category prediction. The drawbacks of these approaches are obvious: Clothing category is a high-level semantic concept, which is the abstraction of attribute combination instead of low-level features. Predicting clothing category from its low-level features suffers from the curse of "semantic gap". As the appearances of clothes vary dramatically, these methods do not work well in most cases and often have low precision. "Semantic gap" is a common problem in visual tasks. One way to solve this problem is using attributes to bridge the gap between low level features and high level concepts [16,17].

In clothing studies, several existing works have leveraged attributes to improve performance [2,4,18,19]. For example, short style clothes usually have a plain but not skirt bottom; suit is unlikely co-occurrence with skirt. In these works, inter-attribute and inter-object relationships are considered. Our study is distinct from these works in that we focus on the relationships between attributes and clothing categories, i.e. clothing category has compatible attribute configurations, as shown in Fig. 1. Thus, clothing categories can be identified based on their attributes. This is actually straightforward that human recognize category or style of clothes from the combination of its local attributes rather than low-level features. For example, the description "clothes with round collar, short sleeve, long cloth body and skirt bottom..." probably refers to "dress". Furthermore, clothing recognition also suffers from the curse of dimensionality. As a large set of features are needed to represent the variation of cloth appearances, which results in high dimensioned feature vector [4,9,15,18] and increases the difficulty of model parameters optimization [4]. In contrast, attributes are low-dimensioned and more representative for category than low-level features.

Previous studies [4,18] applied an identical feature set and method to recognize attributes of various kinds, which would be unreasonable for a particular attribute. For example, color is useless in recognizing the collar types; geometric structure is crucial for collar type recognition, which could be better dealt with by structure sensitive models, such as Deformable Part Model. In our method, features and models are selected according to the characteristics of the attributes, which obtains significant improvements in accuracy.

Based on the aforementioned observations, in this paper, an attribute based approach is proposed for clothing recognition by exploiting the relationship between clothes and their attributes. Our method describes clothing appearance by local attributes, which are divided into two groups: geometric structure attributes (e.g. collar shape) and geometric size attributes (e.g. sleeve length), see Table 1 for details. The two groups of attributes are recognized by different methods, then fed into Multiple Output Neural Networks (MONN) for clothing category prediction. These attributes work as medium to bridge clothing categories and low-level features. Experiments are conducted on a merged dataset

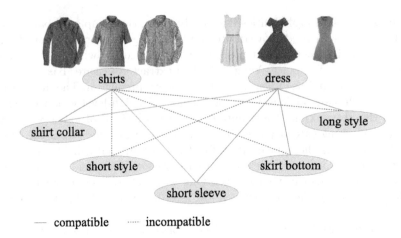

Fig. 1. Relationship between clothing attributes and categories.

including 2172 images, which are annotated with 19 attributes and 10 category labels. Results show that our method not only outperforms two state-of-the-art approaches on both attribute and clothing recognition tasks significantly, but also generalizes well on imbalanced data.

2   Method

The framework of our method is shown in Fig. 2. The method mainly consists of three steps. Firstly, a human pose estimator [20] is trained for attribute localization and a dense clothing parsing approach [21] is applied to obtain the regions of clothing items. Attributes are grouped into two types according to their characteristics: geometric structure and geometric size, as shown in Table 1. Then, DPM detectors are trained to recognize the attributes of geometric structure while geometric size are predicted by logistic regression based on the features extracted from dense parsing results. The confident scores output by all attribute detectors are concatenated, which implies the possibilities of these attributes. At last, two three-layer neural networks with multiple output are built based on the concatenated feature vector for predicting the categories of upper and lower clothing.

2.1   Clothing Attribute Recognition

The accuracy of attribute recognition plays an important role in attribute based category recognition. In order to obtain better accuracy, our method parse attributes by analyzing the characteristic of each rather than applying the same method and features for all attributes, such as in [4,18]. The key point in recognizing collar types, bottom shape and button is modeling their geometric structure while the length of sleeve and cloth body needs measurement on specific

Table 1. Clothing attributes and categories.

Attributes		
Types	Sets	Values
Geometric structure	Collar	Round, v-shape, shirt-collar strapless, one-shoulder
	Bottom	Skirt, plain
	Button	Yes, no
Geometric Size	Sleeve	None, short, half, long
	Ulength	Short, normal, long
	Llength	None, short, long
Categories		
Sets	Values	
Ucategory	Coat, dress, shirt, suit, T-shirt, vest	
Lcategory	Mini-skirt, long-skirt, pants, shorts	

clothing parts. Our method deals the two types of attributes with different methods.

Geometric structure: Attributes of geometric structure discussed in this paper include collar types, bottom types and button. As the name suggests, the distinction of these attributes lies in the composition of element parts and their spatial layout. For instance, two bottom collar-points and side lines compose the shirt-collar. Relative displacement of the element parts makes some variation in the appearance, such as an open shirt-collar. To make use of the information of geometric structure, one possible way is extracting shape features, such as HOG and SIFT, within a grid map as in [4]. However, these features only works in local area and lacks spatial context information, which is important for structure modeling.

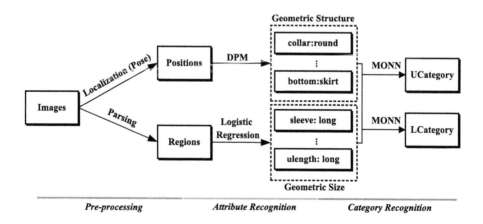

Fig. 2. Framework of our approach.

Fig. 3. Region of interest for each detector.

Our method models such information directly with Deformable Part Model (DPM) [22], which is a well known method for modeling the geometric structure of objects. However, the performance is limited by bad initialization of the model parts [23]. The original DPM model applies a heuristic method for model parts initialization. Such model parts are often meaningless and sometimes out of the target. To eliminate this, the target attributes are decomposed into several element parts. Their spatial layout is predefined but not generated by heuristics. That is the relative position of these element parts. In practice, the root filters for each attribute are trained based on HOG feature firstly. Then, several part filters are manually annotated within it. Figure 4 presents an example for one-shoulder collar. These manually positioned parts work much better than the ones selected by heuristics, which is shown in the experiment. The number of part filters are set according to the complexity of the structure. In practice, 4~5, 6 and 3 part filters are used for collar, bottom and button types. Finally, the full DPM models are trained by Latent SVM for each attribute. In the prediction step, a human pose estimator [20] is firstly trained on the training set, then the Region Of Interest (ROI) and the size of detector window for each attribute is defined with the guideline of human pose, similar to [18]. We do this even for detectors of different collar types because they share the same coarse region but can be more accurately indicated each. All the ROIs are shown in Fig. 3. Then, each attribute detector is performed within the corresponding region of interest by sliding window from top left to bottom right. The maximum score output by each detector implies the possibility of that attribute appears within the corresponding region. As the detectors are trained independently, logistic regression is used to smooth the scores output by different detectors. The final prediction is given by:

$$S(l_i) = g(\theta_{l_i}^{\mathrm{T}} x + b_{l_i}) \tag{1}$$

$$l = \underset{l_i}{\operatorname{argmax}}(S(l_i)), l_i \in L_s \tag{2}$$

where g is the logistic function, θ and b are the regression parameters for each attribute. x is the score vector of attributes within an attribute set L_s.

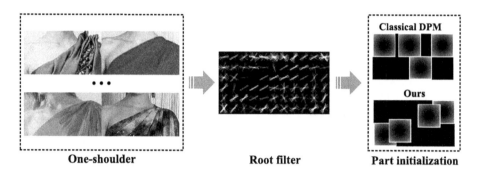

Fig. 4. Part initialization for DPM.

Geometric size: Attributes of geometric size discussed in this paper include the length of sleeve, length of upper and lower clothing body. These attributes are strongly related to the silhouette of clothes. Therefore, we obtain the region of clothes by a dense parsing approach and measure the length of these clothing parts based on the semantic regions directly. This works better than using some auxiliary features, such as skin color probability. Previous work mainly discussed the role of dense clothing parsing in retrieval [12,24]. This paper extend its application scope to attribute recognition. The work [21] is applied to parse the region of *upper-* and *lower-clothing* as well as *skin, hair* and *background* while human pose is used for localization. The key point of the method is to find the endpoint of sleeve, upper and lower clothing bodies. In the method, several patches are set along the corresponding segments of pose skeleton. In practice, 10 patches are set on the segment of shoulder to hand for sleeve; 10 patches on the segment of hip to ankle for lower body; 20 patches on the segment of neck to knee for upper body. One examples is shown in Fig. 5. The percentage of pixels belonging to a certain semantic region is calculated for each path. These percentages for each segment are then concatenated to form feature vectors:

$$\alpha_i = s_i/(s_i + c_i + \epsilon) \tag{3}$$

$$f = <\alpha_1, ..., \alpha_n> \tag{4}$$

where c_i and s_i is the amount of pixels of a certain semantic region in the patch, which is different according to the measurement task: since normally the end of a clothing part is skin or another item, we use the pixel amount of skin and clothing as s_i and c_i respectively for measuring the length of sleeve and lower clothing. As there might be no lower clothing on human body (e.g. a girl in a summer dress), we use the pixel amount of both skin and lower clothing in the path as c_i for measuring the length of upper clothing. Failures in dense parsing sometimes causes a zero denominator. Thus, we add ϵ to it. The α values on each segment are concatenated into a feature vector f, where n is the number of patches. Then, we apply logistic regression to train each label based on the feature and prediction is made by taking the attribute with maximum score within an attribute set.

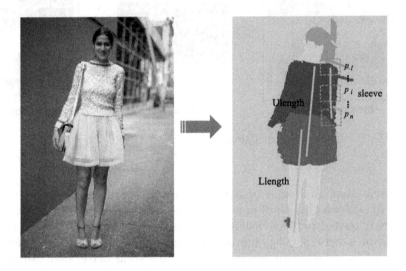

Fig. 5. Patches set for length measurement.

The length of sleeve is predicted on both of the arms. When different predictions have been made, our method chooses the shorter one as the result. This is because different predictions are often caused by parsing failure or wrong pose estimation in one arm, where skin pixels can not be detected. In this case, the detector tends to make predictions of long sleeve. Detection of a shorter sleeve often means that skin area is found, which is much more reliable. We use the same criterion to measure the length of lower clothing.

2.2 Attribute Based Clothing Recognition

The distinction of different clothing categories lies in their attribute configurations, which makes them useful cues for attribute-based recognition. This can also be regarded as the relationship between clothing attributes and categories. Such relationship could be positive, negative or irrelevant. For example, long sleeve is a required attribute for suit (i.e. positive) but not for dress, because its sleeve can be any length (i.e. irrelevant); If short sleeve is detected, the possibility of suit is largely reduced (i.e. negative). Our method uses linear combination of attribute weights to model their configurations. The relationship between attribute and category is implied in the signs and corresponding values of the weights.

One possible way to recognize clothing category is using the predicted attribute labels directly. For example, we can set the recognized attributes to 1 in the feature vector while leaving the others 0 for inference. This is equivalent to do predictions by rigid "rules". It works only when all key attributes are correctly recognized. In the proposed method, the score of predicted attributes as well as the other attribute scores are taken and their similarity to all clothing categories are estimated. This method is much more robust to attribute errors than using attribute labels only, which is shown in our experiments.

Categories have different attribute configurations. Therefore, we use several groups of attribute weights to represent for different configurations. Then, category recognition is formulated as:

$$y = \sum_{i=1}^{N} w_i S(l_i) + \sum_{j=1}^{M} u_j C(l_j) \tag{5}$$

where N and M are the number of geometric structure and geometric size attributes, respectively. The attribute weights in this equation are learned as hidden variables of a three layers neural network with multiple output by back propagation. The number of weight vectors is determined by the amount of hidden units. We train such neural networks for upper and lower clothing, respectively. The predictions are made in the output layer to the category with maximum score.

The resulting category of upper clothing is also influenced by the attributes of lower clothing, and vice versa. For example, if the length of lower clothing is zero (no lower cloth on human body), it is probable that a long style item appears on human body, such as dress. Therefore, the scores of both upper and lower clothing attributes are used in the two networks.

3   Experiments

3.1   Dataset

Fashionista [24] and CCP [25] are two widely used datasets in clothing studies. They consist of full body shots and various types of clothes and attributes. However, the sizes of them are relatively small. Fashionista [24] has only 685 images and CCP [25] consists of 1004 images. Only pixel-level category labels are available in the two datasets. To make the experiment more convincible, we merge them into one dataset and collect more images from the Internet. The

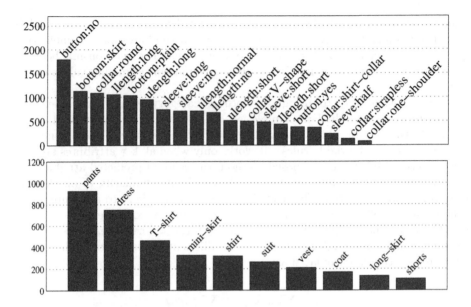

Fig. 6. Label statistics of the dataset.

images with poor visibility or containing clothes with seldom seen attributes are removed. As a result, a new dataset consists of 2172 images are obtained.

We annotate clothes in the new dataset with category and attribute labels, which are listed in Table 1. These attributes are defined according to [12]. Totally, 19 attributes and 10 category labels are obtained and label statistics are shown in Fig. 6. Half of the images are randomly selected as training set and the other half as testing set. The bounding boxes for collar, button and the bottom of clothes are also annotated in the training set, which are used to learn DPMs for geometric structure attributes.

3.2 Results

The proposed method is compared with two state-of-the-art methods: DC [4] and JM [18]. Both of the two method recognize clothing from low-level features directly with an identical model. DC uses SVM based on hand-craft features and JM applies the deep CNN feature extracted using AlexNet(fc7, trained on ImageNet) and logistic regression for classification. We use the comparison of attribute recognition to show how much the combined method outperforms an identical one and the comparison of category recognition to measure how much the attribute based approach improve visual feature based methods. We also compare MONN with radial basis function(RBF) kernel based SVM classifier in attribute based clothing recognition and compare the feature of attribute scores with attribute labels (described in Sect. 2.2). The performance is measured in standard metrics: accuracy, average precise, average recall and average F-1 score.

Table 2. Experimental results and comparison.

Method	Accuracy	Avg. precise	Avg. recall	Avg. F-1
Attributes				
DC [4]	0.7589	0.7734	0.6621	0.6839
JM [18]	0.7133	0.7158	0.6750	0.6795
Ours	**0.8243**	**0.8091**	**0.8124**	**0.8081**
Categories				
DC [4]	0.7003	0.5938	0.5220	0.5457
JM [18]	0.6330	0.5416	0.4652	0.4921
Ours(labels)	0.7463	0.6392	0.5721	0.5787
Ours(SVM)	0.8089	0.7137	0.6889	0.6912
Ours(MONN)	**0.8527**	**0.7799**	**0.7415**	**0.7541**

Table 2 summarizes the performance of these methods. We use different methods to recognizes geometric structure and geometric size attributes, which outperforms approaches with an identical model, such as DC [4] and JM [18]. It shows the importance of selecting appropriate models according to the characteristics of the attributes in recognition. For category recognition, our approach concatenates attribute scores into feature vector and uses neural networks to learn attribute configurations for clothes of different categories. This approach characterizes the appearance of clothing items better than the low-level feature based methods DC [4] and JM [18] and obtains significant improvements. Furthermore, it can be observed from "Ours(labels)" of Table 2 that confident scores of all attributes work much better than labels only. This is because they imply the probability of all attributes which make our method robust to attribute failures. "Ours(SVM)" shows that the MONN method outperforms SVM in attribute based clothing recognition.

The categories of upper and lower clothing are predicted by two MONNs. The hidden units represent for the combination of attributes. Choice on the number of hidden units is important [26]. Therefore, we evaluate the performance of MONN with different hidden units. This experiment is performed by randomly selecting half of the dataset for training and the other half for testing. The evaluation performs 10 times at each number and the average errors are shown in Fig. 7. In practice, we use 20 hidden units in the networks.

F-1 scores for each attribute and category label are shown in Fig. 8. As our method uses a modified version of DPM model, the comparison with classical DPM model on geometric structure attribute recognition is also presented in this figure. The modified model works better than the classical one because it is trained with more accurate parts initialization. As the amount of processed attributes and categories is large, it is hard to create dataset with balanced label distribution. It can be observed from Figs. 6 and 8 that the performance of DC and MJ are limited by number of training samples, such as one-shoulder collar,

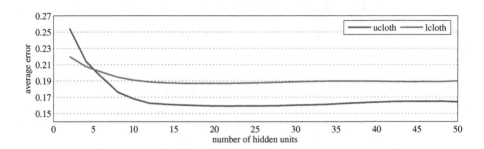

Fig. 7. MONN with different hidden units.

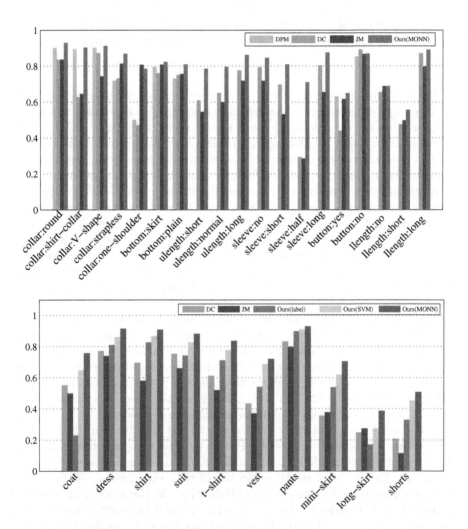

Fig. 8. F-1 scores of each attribute and category label.

DC	JM	Ours
dress	T-shirt	T-shirt
collar:shirt-collar	collar:round	collar:round
sleeve:long	sleeve:long	sleeve:long
ulength:normal	ulength:normal	ulength:short
button:no	button:no	button:no
bottom:plain	bottom:plain	bottom:plain
pants	pants	pants
llength:long	llength:long	llength:long
DC	**JM**	**Ours**
dress	T-shirt	dress
collar:round	collar:round	collar:round
sleeve:no	sleeve:long	sleeve:short
ulength:short	ulength:normal	ulength:long
button:yes	button:no	button:yes
bottom:skirt	bottom:plain	bottom:skirt
llength:no	pants	llength:no
	llength:long	

Fig. 9. Qualitative results of attribute and category recognition.

but our method generalizes much better. More errors occur on lower clothing recognition, such as shorts and long skirt, because of the larger variation on position and limited number of attributes defined for lower clothing. Examples of qualitative results are shown in Fig. 9.

3.3 Discussions on the Networks

The proposed method learn the "rules" for category inference by neural networks. Since the features applied are the confident scores of attributes in semantic domain, it would be interesting to know whether the "rules" learned by networks are the same as the ones used by human beings. The difference is that human usually use rigid rules to recognize clothes because attributes recognized by human are always correct. However, the recognized attributes are noisy, the proposed method needs to learn robust "rules" based on confident scores. We use a method to visualize the "rules" learned by the networks.

The latent units in the networks represent for the configurations of attributes. Therefore, it can be observed from the weight matrix of latent to output layer that which configuration contribute most to a specific clothing category. These latent nodes can be consider as the "rules" for that category. Then, the corresponding attributes weights in the weight matrix of input to latent layer illustrate

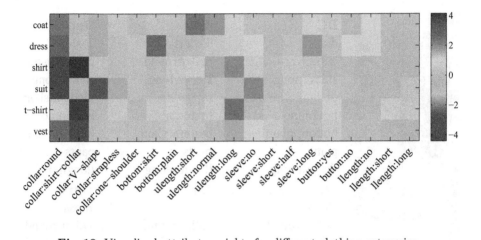

Fig. 10. Visualized attribute weights for different clothing categories.

the details of the "rules". For each category, we visualize the attribute weights of the most significant "rule" in Fig. 10.

It can be observed from Fig. 10 that these learned "rules" are similar to human. For example, the weights of "suit" tells that it usually has long sleeve, normal length body, buttons and V shape collar. The negative correlation between skirt and suit means that suit do not has a skirt bottom. However, there are some small incorrect details, too. For example, there is a strong positive correlation between shirt and *sleeve : no*, although most of shirts have long sleeves. This might be caused by the noisy training data. It can also be inferred from this figure that collar type is the most important attribute in clothing recognition for the drastic changing of the weights. The length of sleeve is less valuable than clothing body because of its multiple possibilities on different clothes.

4 Conclusions and Future Work

An attribute based approach has been discussed to bridge the "semantic gap" between low-level features and high-level category concept in clothing recognition. To obtain better accuracy, different methods have been proposed for geometric structure and geometric size attributes respectively according to the characteristics. The methods works much better than dealing the attributes with an identical model and same features. Clothing categories are predicted based on the attribute scores output by all detectors. Experiments show that our method obtains significant improvements over two related approaches.

Our method recognizes clothes from some predefined local attributes. The number of processed attributes is still limited. Discovering more descriptive attributes would be helpful for clothing recognition, which is one of our future work. The rules learned by neural network is similar to human. However, there are still conflicts. Fusing the learning process with the knowledge of clothes to improve clothing recognition is another future work of us.

References

1. Yamaguchi, K., Kiapour, M.H., Berg, T.L.: Paper doll parsing: retrieving similar styles to parse clothing items. In: ICCV, pp. 3519–3526. IEEE (2013)
2. Liu, S., Feng, J., Domokos, C., Xu, H., Huang, J., Hu, Z., Yan, S.: Fashion parsing with weak color-category labels. IEEE Trans. Multimed. 16(1), 253–265 (2014)
3. Chen, H., Xu, Z., Liu, Z.Q., Zhu, S.C.: Composite templates for cloth modeling and sketching. In: CVPR, vol. 1, pp. 943–950. IEEE (2006)
4. Chen, H., Gallagher, A., Girod, B.: Describing clothing by semantic attributes. In: Fitzgibbon, A., Lazebnik, S., Perona, P., Sato, Y., Schmid, C. (eds.) ECCV 2012. LNCS, vol. 7574, pp. 609–623. Springer, Heidelberg (2012). doi:10.1007/978-3-642-33712-3_44
5. Bossard, L., Dantone, M., Leistner, C., Wengert, C., Quack, T., Gool, L.: Apparel classification with style. In: Lee, K.M., Matsushita, Y., Rehg, J.M., Hu, Z. (eds.) ACCV 2012. LNCS, vol. 7727, pp. 321–335. Springer, Heidelberg (2013). doi:10.1007/978-3-642-37447-0_25
6. Anguelov, D., Lee, K., Gökturk, S.B., Sumengen, B.: Contextual identity recognition in personal photo albums. In: CVPR, pp. 1–7. IEEE (2007)
7. Lin, D., Kapoor, A., Hua, G., Baker, S.: Joint people, event, and location recognition in personal photo collections using cross-domain context. In: Daniilidis, K., Maragos, P., Paragios, N. (eds.) ECCV 2010. LNCS, vol. 6311, pp. 243–256. Springer, Heidelberg (2010). doi:10.1007/978-3-642-15549-9_18
8. Zhang, W., Shen, J., Liu, G., Yu, Y.: A latent clothing attribute approach for human pose estimation. In: Cremers, D., Reid, I., Saito, H., Yang, M.-H. (eds.) ACCV 2014. LNCS, vol. 9003, pp. 146–161. Springer, Heidelberg (2015). doi:10.1007/978-3-319-16865-4_10
9. Shen, J., Liu, G., Chen, J., Fang, Y., Xie, J., Yu, Y., Yan, S.: Unified structured learning for simultaneous human pose estimation and garment attribute classification. IEEE Trans. Image Process. 23(11), 4786–4798 (2014)
10. Liu, S., Feng, J., Song, Z., Zhang, T., Lu, H., Xu, C., Yan, S.: Hi, magic closet, tell me what to wear! In: Proceedings of the 20th ACM International Conference on Multimedia, pp. 619–628. ACM (2012)
11. Kalantidis, Y., Kennedy, L., Li, L.: Getting the look: clothing recognition and segmentation for automatic product suggestions in everyday photos. In: Proceedings of the 3rd ACM Conference on International Conference on Multimedia Retrieval, pp. 105–112. ACM (2013)
12. Liu, S., Song, Z., Liu, G., Xu, C., Lu, H., Yan, S.: Street-to-shop: cross-scenario clothing retrieval via parts alignment and auxiliary set. In: CVPR, pp. 3330–3337. IEEE (2012)
13. Weber, M., Bäuml, M., Stiefelhagen, R.: Part-based clothing segmentation for person retrieval. In: 8th IEEE International Conference on Advanced Video Signal Based Surveillance, AVSS 2011, pp. 361–366. IEEE (2011)
14. Lorenzo-Navarro, J., Castrillón, M., Ramón, E., Freire, D.: Evaluation of LBP and HOG descriptors for clothing attribute description. In: Distante, C., Battiato, S., Cavallaro, A. (eds.) VAAM 2014. LNCS, vol. 8811, pp. 53–65. Springer, Heidelberg (2014). doi:10.1007/978-3-319-12811-5_4
15. Yang, M., Yu, K.: Real-time clothing recognition in surveillance videos. In: ICIP, pp. 2937–2940. IEEE (2011)
16. Li, L.-J., Su, H., Lim, Y., Fei-Fei, L.: Objects as attributes for scene classification. In: Kutulakos, K.N. (ed.) ECCV 2010. LNCS, vol. 6553, pp. 57–69. Springer, Heidelberg (2012). doi:10.1007/978-3-642-35749-7_5

17. Fu, Y., Hospedales, T., Xiang, T., Gong, S.: Learning multimodal latent attributes. IEEE Trans. Pattern Anal. Mach. Intell. **36**(2), 303–316 (2014)
18. Yamaguchi, K., Okatani, T., Sudo, K., Murasaki, K., Taniguchi, Y.: Mix and match: joint model for clothing and attribute recognition. In: BMVC (2015)
19. Dong, J., Chen, Q., Xia, W., Huang, Z., Yan, S.: A deformable mixture parsing model with parselets. In: ICCV, pp. 3408–3415. IEEE (2013)
20. Yang, Y., Ramanan, D.: Articulated pose estimation with flexible mixtures-of-parts. In: CVPR, pp. 1385–1392. IEEE (2011)
21. Fan, W., Qiyang, Z., Baolin, Y., Tao, X.: Parsing fashion image into mid-level semantic parts based on chain-conditional random fields. IET Image Proc. **10**(6), 456–463 (2016)
22. Felzenszwalb, P.F., Girshick, R.B., McAllester, D., Ramanan, D.: Object detection with discriminatively trained part based models. IEEE Trans. Pattern Anal. Mach. Intell. **32**(9), 1627–1645 (2010)
23. Azizpour, H., Laptev, I.: Object detection using strongly-supervised deformable part models. In: Fitzgibbon, A., Lazebnik, S., Perona, P., Sato, Y., Schmid, C. (eds.) ECCV 2012. LNCS, vol. 7572, pp. 836–849. Springer, Heidelberg (2012). doi:10.1007/978-3-642-33718-5_60
24. Yamaguchi, K., Kiapour, M., Ortiz, L., Berg, T.: Parsing clothing in fashion photographs. In: CVPR, pp. 3570–3577. IEEE (2012)
25. Yang, W., Luo, P., Lin, L.: Clothing co-parsing by joint image segmentation and labeling. In: CVPR, pp. 3182–3189. IEEE (2014)
26. Karsoliya, S.: Approximating number of hidden layer neurons in multiple hidden layer bpnn architecture. Int. J. Eng. Trends Technol. **3**(6), 713–717 (2012)

Bloody Image Classification with Global and Local Features

Song-Lu Chen, Chun Yang, Chao Zhu, and Xu-Cheng Yin[✉]

School of Computer and Communication Engineering,
University of Science and Technology Beijing, Beijing, China
chensonglu@xs.ustb.edu.cn, ych.learning@gmail.com,
{chaozhu,xuchengyin}@ustb.edu.cn

Abstract. Object content understanding in images and videos draws more and more attention nowadays. However, only few existing methods have addressed the problem of bloody scene detection in images. Along with the widespread popularity of the Internet, violent contents have affected our daily life. In this paper, we propose region-based techniques to identify a color image being bloody or not. Firstly, we have established a new dataset containing 25431 bloody images and 25431 non-bloody images. These annotated images are derived from the Violent Scenes Dataset, a public shared dataset for violent scenes detection in Hollywood movies and web videos. Secondly, we design a bloody image classification method with global visual features using Support Vector Machines. Thirdly, we also construct a novel bloody region identification approach using Convolutional Neural Networks. Finally, comparative experiments show that bloody image classification with local features is more effective.

Keywords: Bloody image classification · Violent scenes dataset · Support vector machines · Convolutional Neural Networks

1 Introduction

As we enjoy the fast and convenient access to all kinds of information via the Internet, we are also exposed to undesirable contents, such as violent images, gory videos and pornographic information. The flourishing images sharing websites like Instagram, social websites like Facebook and Twitter, search engines like Google and videos sharing websites like Youtube even enable obtaining and sharing harmful information as simple as sliding on the screen or clicking the mouse. For those who are deficient of good sense of judgement and self-control, especially teenagers, long-term exposure to such negative information can seriously affect them both physically and mentally, even may cause aggressive behaviour or crimes. For this, it's desperately needed to limit the propagation of such unhealthy contents. In this paper, we propose a fairly effective method aimed on bloody image identification.

© Springer Nature Singapore Pte Ltd. 2016
T. Tan et al. (Eds.): CCPR 2016, Part II, CCIS 663, pp. 379–391, 2016.
DOI: 10.1007/978-981-10-3005-5_31

Compared with pornographic contents, the investigations of violence detection is still in the initial stage. Recognizing the bloody images can be considered as one category of violence detection. Following are several related works. Wang et al. [1] propose to detect suspected bloodstain in YC_gC_r color space. To accelerate bloodstain pixels detection, the transformed thresholds ($C_{g'}, C_{r'}$) are used. This approach can be used as the basis for bloodstain identification in color images, but it also works for pseudo-bloody images like red apple, red clothes etc. This method by nature is based on the idea of Dios et al. [2,3]. Yoo et al. [4] propose an efficient red-eye removal algorithm which uses an inpainting method and biometric information. Yan et al. [5] extract global features as well as local features from the detected bloody regions in an image. These features are fed into the SVM to classify. Wang et al. [6] adopt the Bag-of-Words model to discriminate violence images and non-violence images. They combine four global image features with the BoW framework for all categories of violence. Li et al. [7] propose a novel horror image recognition based on emotional attention mechanism. Guermazi et al. [8] focus on the color descriptors to recognize violent images. They utilize the MPEG7 color descriptors. Lopes et al. [9] add color information to original SIFT features to detect nude or pornographic images. This method does not rely on any skin or shape models and has considerable improvement than traditional SIFT features. Ulges et al. [10] present a visual recognition system for the detection of child pornographic image material, based on color-enhanced visual word features and SVM classifier. Violence Scenes Dataset (VSD) [11] is a public shared dataset for the detection of violent scenes in Hollywood movies and web videos. The detail of VSD has been described by Demarty et al. [12] and Schedi et al. [13]. The rest [14–19] focus on either video violence detection or audio violence detection or both.

Although there have been plenty of investigations for content-based image classification, very few of them attempt to deal with the issue of violent image detection. A big challenge is how to define a violent image. In our opinion, a violent image should be a picture which can trigger negative emotions such as panic, menace, anxiety and aggressiveness. So the range of violent images should not just include images with gory scenes, but also those embody gunshots, car chases, explosions, fire, fights as well. Anyway, ambiguity certainly exists due to subjectivity in discriminating violent images. For simplicity, in this paper we only concern about bloody image detection which means the image must contain one or more blocks of bloodstain. We define that if the proportion of blood pixels in a color image is greater than 0.5 %[1], it's bloody, otherwise, it is not. This definition can make sure a bloody image objectively instead of intuitive judgement. Thus, the issue will be converted into the following steps: segmenting all suspected bloodstain regions, filtering out all real bloodstain regions, adding up the area of all blood pixels and determining if the image being bloody or not.

Another challenge is that unlike violence detection in videos where various visual, audio and motion features can be combined together to solve the problem,

---

[1] We compared all bloody images and non-bloody images in our dataset and found this number is a quite reasonable threshold to distinguish these two kind of images.

detecting violence in still images can only depend on visual features. In addition, the detection task becomes rather complicated due to the wide range of scenes, backgrounds and luminance, furthermore, pseudo-bloody regions make the task more challenging. To this end, we have established a dataset of a large number of bloody and non-bloody images, which contains the samples of all aforementioned types. Then we propose a region-based method for bloody image classification because we believe that local information is more representative of the content of an image. Finally, we utilize a convolutional neural network (CNN) to classify all red-like regions. With CNN, a large number of features can be extracted automatically rather than collecting complicated color, texture or shape features manually, furthermore, we can constantly enhance the performance of CNN model based on reinforcement learning.

The remainder of this paper is organized as follows: Sect. 2 introduces the new established dataset. Section 3 presents our approach for bloody image identification from both global and local views. Section 4 shows the experimental results as well as corresponding analysis. Section 5 concludes this paper.

2   Dataset

An appropriate dataset is essential to evaluate the effectiveness of our classification method. Fortunately, there is a public shared dataset named Violent Scenes Dataset (VSD) [11] for the detection of violent scenes, which contains many Hollywood movies, web videos and corresponding annotations. These annotations mainly consist of high-level concepts for Hollywood movies. More specifically, six visual concepts and three audio concepts are annotated. They are presence of blood, fights, fire, cold arms, car chases and gory scenes for the visual modality as well as presence of gunshots, explosions and screams for the audio modality.

The VSD provides high-level concepts annotations for seventeen movies, but eventually only five representative Hollywood movies have been picked out as samples for our classification task. Sorted by contribution, they are in turn *Reservoir Dogs*, *Saving Private Ryan*, *Harry Potter 5*, *Armageddon* and *Pirates of the Caribbean 1*. Some downloaded movies over the Internet are not corresponding to the annotations, such as *Leon* and *I am Legend*, and some such as *Billy Elliot* and *Independence Day* lack bloody scenes, or even have no bloody frames, and others are too dark to be used as samples, like *Fight Club*.

Even the movies selected as samples still need further processing. Firstly, these movies all have frame offset ranged from several frames to hundreds of frames, and we can only adjust them according to movies scenes containing blood, car chases and fights etc. Secondly, official released annotations are spot-based, and each spot may contain hundreds of frames. Inevitably, there exist several non-bloody frames in some spots. Moreover, some parts of the annotations are overlapped with other parts, and some are even wrongly annotated. We have modified many frame-wise annotations manually and that's a time-consuming task. Thirdly, we extracted all sample movies to images by frame and removed the black edge of each image. Finally, we conducted many different kinds of classification experiments using features extracted from all sample movies and compared them with official released features to make sure all movies are usable. Eventually, we got five usable movies from VSD by making large adjustments of them.

In practice, the number of non-bloody images is far more than bloody images. For sample balance, we randomly selected equal number of non-bloody images from each movie. According to corresponding annotations, we collected 50862 images including 25431 bloody images and 25431 non-bloody images. For convenience, we named the dataset as PVSD (Part of Violent Scenes Dataset). They are all color images in PNG format with height larger than 500 pixels and width larger than 1000 pixels. As time goes on, we will enrich our dataset constantly with images from different sources, such as digital images, scanned photographs, software edited images, and images with large range of luminance or resolution. Figure 1 demonstrates samples of some bloody images in PVSD.

Fig. 1. Samples of some bloody images in PVSD

In addition, we have also established our training set and test set based on PVSD. Firstly, we randomly pick out 10% images from each sample movie as training set and the other 90% as test set. In order to distinguish them with PVSD, we called the training set TSGF (Training Set for Global Features), and named the test set TID (Test Image Dataset). TSGF contains 5088 images with 2544 bloody images and 2544 non-bloody images. TID contains 45774 images with 22887 bloody images and 22887 non-bloody images. Besides, we have established our training set for local features based on TSGF, which contains 12416 images covered with bloodstain and 11627 images covered with pseudo-blood after manual sortation. In order to test the effect of different magnitude of training set, we have set up five datasets called TSLF-1X (Training Set for Local Features 1X), TSLF-2X, TSLF-3X, TSLF-4X and TSLF-5X respectively. These five sets all contain images centered with bloodstain or pseudo-blood, but the differences lie in that the images in TSLF-1X almost exclusively contain bloodstain or pseudo-blood while images in other four sets contain more surrounding background, expanding 2 to 5 times in height and width respectively. In summary, TSGF is the training set for global features, while TSLF-1X, TSLF-2X, TSLF-3X, TSLF-4X and TSLF-5X are training sets for local features. TID is the test set for both global and local features. Figure 2 shows samples of some images

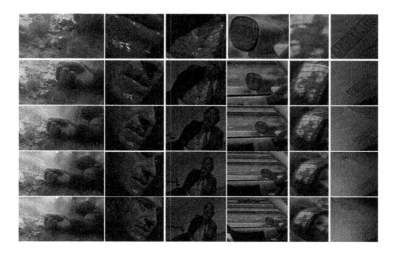

Fig. 2. Samples of some images covered with blood and pseudo-blood: the first three columns of the first row, bloody images in TSLF-1X; and the last three columns of the first row, pseudo-bloody images in TSLF-1X; and the next four rows for TSLF-2X, TSLF-3X, TSLF-4X and TSLF-5X respectively

covered with bloodstain and pseudo-blood in TSLF-1X, TSLF-2X, TSLF-3X, TSLF-4X and TSLF-5X.

3   Bloody Image Classification

In this section, the detection task can be divided into two parts: classification with global visual features as well as bloodstain segmentation and classification with local features. In the first part, three visual features extracted from images as well as their fusion are fed into SVM to classify. In the second part, we first segment all red-like regions in an image and filter out all real bloody ones using Deep Learning method.

3.1   Classification with Global Visual Features

In our view, bloodstain has distinguishing features compared with other objects, such as color, appearance and texture. Normally, bloodstains are red and scattered like irregular clusters or bands. For example, blood droplets always have little or no distortion on their peripheral edges and change smoothly from the center, while splashed blood varies largely in luminance and shape. Based on this, we apply three visual features for bloody image classification. These three features are briefly introduced as follows:

Color moments (CM), a measure to characterise color distribution in an image, contain the first three central moments of an image color distribution: mean, standard deviation and skewness.

The first color moment can be interpreted as the average color in the image, and it can be calculated by the following formula:

$$E_i = \sum_{j=1}^{N} \frac{1}{N} p_{ij} \tag{1}$$

where N is the number of pixels in the image and p_{ij} is the value of the j-th pixel of the image at the i-th color channel.

The second color moment is the standard deviation, which is obtained by taking the square root of the variance of the color distribution.

$$\sigma_i = \sqrt{(\frac{1}{N} \sum_{j=1}^{N} (p_{ij} - E_i)^2)} \tag{2}$$

where E_i is the mean value, or first color moment, for the i-th color channel of the image.

The third color moment is the skewness. It measures how asymmetric the color distribution is, and thus it gives information about the shape of the color distribution. Skewness can be computed with the following formula:

$$s_i = \sqrt[3]{(\frac{1}{N} \sum_{j=1}^{N} (p_{ij} - E_i)^3)} \tag{3}$$

Histogram of Oriented Gradients (HOG) can exploit the local object appearance and shape within an image via the distribution of edge orientations. Local Binary Patterns (LBP) is a type of texture descriptor used for classification. In some detection cases [20], LBP is often used in combination with HOG to improve the detection performance. These three visual features can be used together in our classification task.

According to the introduction document of VSD [13], VSD also extracted these three visual features for all movies besides corresponding annotations. For convenience, we apply official released CM, HOG and LBP features as well as their fusion into our task. The CM and HOG features are all 81-dimensional, and the LBP is 144-dimensional. During the classification stage, SVM with RBF kernel is chosen as the classifier. LIBSVM [21], a library for SVM, makes our experiments much easier. For the TSGF, we randomly select 60 % of the images as training set while the rest as validation set and TID is used as test set. After normalization, different combination of features are separately fed into SVM to classify.

3.2 Detection and Classification with Local Features

In this section, we first determine the color thresholds of bloodstain pixels, then generate TSLF-1X, TSLF-2X, TSLF-3X, TSLF-4X and TSLF-5X using these thresholds based on TSGF, and finally conduct classification with Deep Learning method.

Bloodstain Segmentation. According to Wang et al. [1], their approach can be used as the basis for bloodstain segmentation. Figure 3 shows outline of their algorithm. However, based on our actual conditions, we have made some adjustments, such as, no color histogram equalization for color images, no erosion and dilation to smooth binary images and effective cluster areas being raised to 400 pixels.

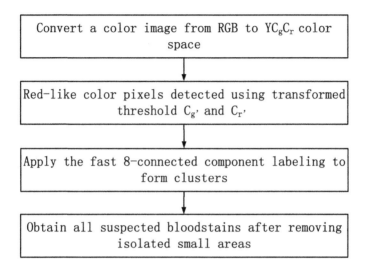

Fig. 3. The flowchart of red-like region segmentation

The YC_gC_r color space can be easily transformed from RGB color space through the following formula.

$$\begin{bmatrix} Y \\ Cg \\ Cr \end{bmatrix} = \begin{bmatrix} 16 \\ 128 \\ 128 \end{bmatrix} + \begin{bmatrix} 65.481 & 128.553 & 24.966 \\ -81.085 & 112 & -30.915 \\ 112 & -93.768 & -18.214 \end{bmatrix} \times \begin{bmatrix} R \\ G \\ B \end{bmatrix} \qquad (4)$$

Four hundred images randomly selected from TSLF-1X are used as training set to simulate the bloodstain color. Figure 4 illustrates the bloodstain color distribution in the $C_g - C_r$ plane, and the bloodstain color region is concentrated in an inclined stripe area.

On the basis of Wang et al. [1], the color stripe should be rotated 36° clockwise to a transformed $C_{g'} - C_{r'}$ plane to accelerate the bloodstain pixels detection. The new color space $YC_{g'}C_{r'}$ can be transformed by the following equations.

$$C_{g'} = C_g \times cos36° + C_r \times sin36° \qquad (5)$$

$$C_{r'} = -C_g \times sin36° + C_r \times cos36° \qquad (6)$$

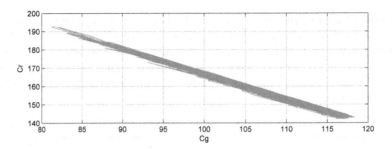

Fig. 4. Bloodstain color distribution in the $C_g - C_r$ plane (Color figure online)

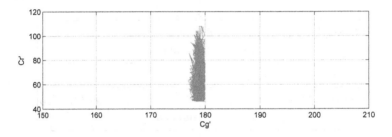

Fig. 5. Bloodstain color distribution in the $C_{g'} - C_{r'}$ plane (Color figure online)

The new bloodstain area is shown in Fig. 5, and it's perpendicular to $C_{g'}$ axis. Undoubtedly, it will make the bloodstain detection more accurate and fast.

The decision thresholds for bloodstain segmentation are determined by the minimum and maximum of $C_{g'}$ and $C_{r'}$. In contrast with Wang et al. [1], our decision thresholds are [176, 180] for $C_{g'}$ and [46, 109] for $C_{r'}$.

Classification with Local Features Using Deep Learning. So far, we have acquired the thresholds of bloodstain pixels in the transformed $YC_{g'}C_{r'}$ color space. According to the method of red-like region segmentation, we get 24043 images covered with bloodstain or pseudo-blood from TSGF and separate them into bloody and non-bloody manually[2]. By this way, we have established TSLF-1X, TSLF-2X, TSLF-3X, TSLF-4X and TSLF-5X, and images in these five datasets are all centered with bloodstain or pseudo-blood but differ in size, as shown in Fig. 2. Convolutional Neural Networks (CNN) have been widely applied to computer vision [22] and shows great effect on image detection and recognition. With CNN, effective features can be extracted automatically rather than being collected manually. Besides, MatConvNet [23], a library for CNN, provides great convenience for us.

---

[2] Normally, an image in TSGF contains mutiple red-like regions. Not all these regions are bloodstain and some are red car, red clothes, red light or human face etc. Thus all extracted red-like regions are mixed together at first and need manual sortation.

Similar to classification with global features, we randomly select 60 % of the images in TSLF-1X, TSLF-2X, TSLF-3X, TSLF-4X and TSLF-5X as training set while the rest as validation set and TID is used as test set. CNN requires all input images to be resized to the same size and for our network, it's 100×180^3, which is based on the mean value of height and width in TSLF-1X. Also, all images are normalized in RGB color space through the following formula:

$$p_{ij} = \frac{p_{ij} - E_i}{\sigma_i} \times 128.0 \tag{7}$$

where p_{ij} is the value of the j-th pixel of the image at the i-th color channel. E_i and σ_i represent the mean and standard deviation value for the i-th color channel respectively.

Figure 6 shows our convolutional neural network structure. At the beginning of the network, all input images should be resized to 100×180 in each color channel of RGB. In the middle part of it, the entire network includes five convolution layers and three pooling (subsampling) layers. Finally the input image would be classified into two types: bloody or non-bloody.

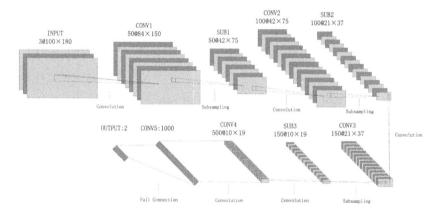

Fig. 6. Convolutional neural network structure (Color figure online)

4   Experiments

So far, we have established datasets for classification with global and local features. For classification with global features, three visual features as well as their fusion are feed into SVM to classify. For classification with local features, firstly, we train the classification model for bloodstain; secondly, all suspected bloodstain regions in an image are segmented out using transformed $YC_{g'}C_{r'}$ color space and all real bloodstain regions are filtered out using this model; finally,

---

[3] Height $\times$ Width.

all blood pixels are added up to determine if the image being bloody or not. As mentioned before, we define that if the proportion of blood pixels in a color image is greater than 0.5 %, it's bloody, otherwise, it is not.

Classification with Global Features. For the TSGF, we randomly select 60 % images in it as training set while the rest as validation set and TID is used as test set. During the classification stage, SVM with RBF kernel is chosen as the classifier. All visual features are normalized to $[-1, 1]$. Finally, different combinations of feature are separately fed into SVM to classify. The classification results with global visual features on TID are presented in Table 1.

Table 1. Classification results with global visual features on TID

Adopted features	Precision	Recall	F_1
CM	78.46 %	83.65 %	80.97 %
HOG	69.56 %	93.57 %	79.80 %
LBP	69.15 %	96.68 %	80.63 %
HOG+LBP	73.80 %	96.10 %	83.45 %
CM+LBP	75.69 %	94.37 %	84.00 %
CM+HOG+LBP	78.62 %	93.00 %	85.21 %

As we can see, CM obtains higher precision rate and lower recall rate while HOG and LBP all have higher recall rate and lower precision rate. When combined together, it shows a relatively better performance than each feature alone based on F_1, indicating that CM with HOG and LBP could provide complementary information.

Classification with Local Features. For the TSLF-1X, TSLF-2X, TSLF-3X, TSLF-4X and TSLF-5X, we also randomly select 60 % images in them as training set while the rest as validation set and TID is used as test set. We choose the network of 100th epoch[4] for five training sets respectively and training errors are all close to 0, but validation errors differ a bit, as shown in Table 2.

From the results in Table 2, we can conclude that surrounding information would enhance the bloodstain classification effect (TSLF-2X and TSLF-3X compared with TSLF-1X). However, too much surrounding information would bring extra noise and cause performance degradation (TSLF-4X and TSLF-5X compared with TSLF-2X and TSLF-3X). To sum up, TSLF-2X should be used as the dataset for local features because of its lower validation error.

Table 3 shows the comparison results of different methods on TID and results show that bloody image classification with local features is more effective based on F_1.

---

[4] That means the neural network passes through training set for 100 times.

Table 2. The validation errors of TSLF-1X, TSLF-2X, TSLF-3X, TSLF-4X and TSLF-5X

Classification model	Validation error
TSLF-1X	10.40 %
TSLF-2X	8.85 %
TSLF-3X	8.87 %
TSLF-4X	9.37 %
TSLF-5X	9.45 %

Table 3. The comparison results of different methods on TID

Method+(Features, Dataset)	Precision	Recall	F_1
SVM+CM+HOG+LBP	78.62 %	93.00 %	85.21 %
CNN+TSLF-2X	93.17 %	82.25 %	87.37 %

5   Conclusion

In this paper, we propose a novel region-based method for bloody image classification using Convolutional Neural Networks. At the beginning, we defined the bloody image by the proportion of blood pixels. Then we established many datasets for our task. Finally, our research is divided into two parts: classification with global visual features as well as classification with local features. In the first part, three visual features as well as their fusion are fed into SVM to classify. In the second part, we first segment all suspected bloodstain regions using transformed $YC_{g'}C_{r'}$ color space, then filter out all real bloodstain regions using Deep Learning method, finally add up the area of all blood pixels to determine if an image is bloody. The local method shows a better performance than the global one.

Future work includes a constant enlargement of our dataset with images from different sources, luminances and resolutions as well as the optimization of the convolutional neural network structure. Another future direction is to combine global and local visual features to enhance the identification performance.

References

1. Ying-Wei, W., et al.: Bloodstain segmentation in color images. In: Proceedings of the 2011 First International Conference on Robot, Vision and Signal Processing (RVSP 2011), pp. 52–55 (2011)
2. de Dios, J.J., et al.: Face detection based on a new color space YCgCr. In: Proceedings of the IEEE International Conference on Image Processing, vol. 3, pp. 902–912, September 2003
3. de Dios, J.J., Garcia, N.: Fast face segmentation in component color space. In: 2004 International Conference on Image Processing (ICIP) (IEEE Cat. No. 04CH37580), vol. 191, pp. 191–194 (2004)

4. Yoo, S., Park, R.H.: Red-Eye detection and correction using inpainting in digital photographs. IEEE Trans. Consum. Electron. **55**(3), 1006–1014 (2009)
5. Yan, G., et al.: Region-based blood color detection and its application to bloody image filtering. In: Proceedings of 2015 International Conference on Wavelet Analysis and Pattern Recognition (ICWAPR), pp. 45–50 (2015)
6. Wang, D., et al.: Baseline results for violence detection in still images. In: 2012 IEEE Ninth International Conference on Advanced Video and Signal-Based Surveillance (AVSS), pp. 54–57 (2012)
7. Li, B., Hu, W., Xiong, W., Wu, O., Li, W.: Horror image recognition based on emotional attention. In: Kimmel, R., Klette, R., Sugimoto, A. (eds.) ACCV 2010. LNCS, vol. 6493, pp. 594–605. Springer, Heidelberg (2011). doi:10.1007/978-3-642-19309-5_46
8. Guermazi, R., et al.: Violent web images classification based on MPEG7 color descriptors. In: 2009 IEEE International Conference on Systems, Man and Cybernetics, pp. 3106–3111 (2009)
9. Lopes, A.P.B., et al.: A bag-of-features approach based on Hue-SIFT descriptor for nude detection. In: 2009 17th European Signal Processing Conference (EUSIPCO 2009), pp. 1552–1556 (2009)
10. Ulges, A., et al.: Automatic detection of child pornography using color visual words. In: 2011 IEEE International Conference on Multimedia and Expo (2011)
11. Violent Scenes Dataset. http://www.technicolor.com/en/innovation/research-innovation/scientific-data-sharing/violent-scenes-dataset. Accessed 25 May 2016
12. Demarty, C.H., et al.: VSD, a public dataset for the detection of violent scenes in movies: design, annotation, analysis and evaluation. Multimedia Tools Appl. **74**(17), 7379–7404 (2015)
13. Schedi, M., et al.: VSD2014: a dataset for violent scenes detection in Hollywood movies and web videos. In: 2015 Proceedings of 13th International Workshop on Content-Based Multimedia Indexing (CBMI), p. 6 (2015)
14. Acar, E., et al.: Detecting violent content in Hollywood movies by mid-level audio representations. In: Czuni, L., Schoffmann, K., Sziranyi, T. (eds.) 2013 11th International Workshop on Content-Based Multimedia Indexing, pp. 73–78 (2013)
15. Gong, Y., Wang, W., Jiang, S., Huang, Q., Gao, W.: Detecting violent scenes in movies by auditory and visual cues. In: Huang, Y.-M.R., Xu, C., Cheng, K.-S., Yang, J.-F.K., Swamy, M.N.S., Li, S., Ding, J.-W. (eds.) PCM 2008. LNCS, vol. 5353, pp. 317–326. Springer, Heidelberg (2008). doi:10.1007/978-3-540-89796-5_33
16. Penet, C., et al.: Multimodal information fusion and temporal integration for violence detection in movies. In: 2012 IEEE International Conference on Acoustics, Speech and Signal Processing (ICASSP), pp. 2393–2396 (2012)
17. Bermejo Nievas, E., Deniz Suarez, O., Bueno García, G., Sukthankar, R.: Violence detection in video using computer vision techniques. In: Real, P., Diaz-Pernil, D., Molina-Abril, H., Berciano, A., Kropatsch, W. (eds.) CAIP 2011. LNCS, vol. 6855, pp. 332–339. Springer, Heidelberg (2011). doi:10.1007/978-3-642-23678-5_39
18. Deniz, O., et al.: Fast violence detection in video. In: Proceedings of 9th International Conference on Computer Vision Theory and Applications (VISAP 2014), pp. 478–485 (2014)
19. Zhijie, Y., et al.: Violence detection based on histogram of optical flow orientation. In: Proceedings of the SPIE - The International Society for Optical Engineering (2013)
20. Wang, X., Han, T.X., Yan, S.: An HOG-LBP human detector with partial occlusion handling. In: Proceedings of IEEE ICCV, September 2009

21. Chang, C.C., Lin, C.J.: LIBSVM: A Library for Support Vector Machines. In: ACM Trans. Intell. Syst. Technol. **2**(3) (2011)

22. Xu-Cheng, Y., et al.: Shallow classification or deep learning: an experimental study. In: 2014 22nd International Conference on Pattern Recognition (ICPR), pp. 1904–1909 (2014)

23. MatConvNet: CNNs for MATLAB. http://www.vlfeat.org/matconvnet/. Accessed 25 May 2016

Local Connectedness Constraint and Contrast Normalization Based Microaneurysm Detection

Mengxue Liu[1], Qi Yu[2], Jie Yang[1(✉)], Yu Qiao[1], and Xun Xu[2]

[1] Institute of Image Processing and Pattern Recognition,
Shanghai Jiao Tong University, Shanghai, China
{liumengxue,jieyang}@sjtu.edu.cn
[2] Shanghai General Hospital, Shanghai Jiao Tong University, Shanghai, China
{yu.qi,yuqiao}@sjtu.edu.cn

Abstract. Diabetic retinopathy is an eye disease that damages the retina and leads to vision loss. As the first sign of diabetic retinopathy, microaneurysm (MA) usually appears as a round small spot, which seems similar to incontinuous vessels and background noise dots. It is hard for the individual algorithms published so far to distinguish them. In this paper, we model the MA detection problem as finding the interest spots from an image. Based on the characteristics of the target, connectedness rules are set up to constrain the MA inside a window. Hemorrhages and noise points are removed that way. Local contrast normalization is also used to better classify MAs from dots within vessels. Through comprehensive experiments on DIARETDB1 dataset and ROC dataset, we show that dots within vessels and noise points in the background can be well removed. Our method outperforms others with high sensitivity and specificity.

Keywords: Microaneurysm detection · Local connectedness constraint · Region growing · Local contrast normalization

1 Introduction

Diabetic Retinopathy (DR) is an eye disease caused by diabetes, which may lead to serious sight loss. As the earliest sign to detect DR, microaneurysms (MAs) are visible in fundus photography. Regular eye screening is an effective way for DR detection and treatment but it brings a large amount of eye screening images for ophthalmologists to diagnose. Therefore, an automatic detection system dealing with digital retinal images can well solve the problem.

MAs are discrete saccular distensions of the weak retinal capillary walls. They are very hard to detect due to their low contrast against the background and similarity to vessels or hemorrhages. Nonetheless, the basic characteristics of MAs, being of a uniform shape and size, have led to the use of basic morphology for identification of MAs. Baudoin et al. [1] firstly employed a top-hat transformation with linear structuring elements at different orientations to distinguish connected, elongated structures (i.e. vessels) from unconnected circular

T. Tan et al. (Eds.): CCPR 2016, Part II, CCIS 663, pp. 392–403, 2016.
DOI: 10.1007/978-981-10-3005-5_32

objects (*i.e.* MAs) in fluorescein angiograms (FAs). Although FAs have a higher contrast between MAs and the background, the harm of fluorescein makes people take more delight in digital color photographs these years. Niemeijer *et al.* [2] combined the top-hat method with a supervised classification in which first, each image is preprocessed and then the candidate red lesions are extracted, and in the final step, true lesions are selected by a classifier. Mizutani *et al.* [3] used the difference of pixels' average gray scale between inner and outer ring areas to estimate the existence of MAs. However, this method ignores features around lesions and cannot handle the false positives caused by noise or dots within vessels. Recent years, methods for the accurate detection of MAs have developed a lot. An ensemble-based framework for the detection of MAs is proposed in [4]. [5] present a hybrid classifier which combines the Gaussian mixture model (GMM), support vector machine (SVM) and an extension of multi-model method based modeling approach to improve the accuracy of classification.

In this paper, we design a special window as the constraint of MA. Pixels exceeding the window, which violate the connectedness rules, are removed. This step can filter noise points and large regions like hemorrhages. Then to discriminate true MAs from dots within vessels, we adopt the local contrast normalization method [6] to enlarge the difference between MAs and other structures. A special local-contrast feature map is generated, where the difference in the characteristics of each structure is magnified.

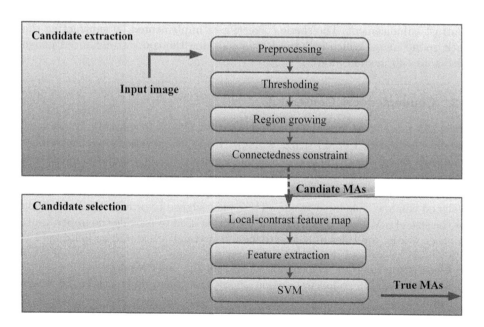

Fig. 1. The flowchart of the overall detection scheme for detection of microaneurysms.

The rest of this paper is organized as follows. Section 2 introduces the candidate extraction. The connectedness constraint and the region growing method used in our approach are described in details. In Sect. 3, the selection strategy is then presented for filtering the MA candidates based on local contrast normalization. Section 4 shows the experimental results and we conclude our approach in Sect. 5. The flowchart of the overall detection scheme is shown in Fig. 1.

2   Candidate Extraction

The proposed method in this section utilizes a pixel-wise sliding window strategy to get a decision map. If the sliding window covers a MA, then the value of its central point on the decision map is set to "1" and "0" for vice versa.

2.1   Preprocessing

The quality of retinal images counts significantly for the performance of the MA detection. Any factors such as noise, low contrast, non-uniform illumination can result in a poor quality image. Since red lesions such as MAs and vessels have the highest contrast against the background in the green plane, a median filtering operation is first applied on the green plane to reduce the noise. Then we apply the contrast limited adaptive histogram equalization for contrast enhancement and a shade correction algorithm to remove the dark region generated after contrast enhancement. The shade correction is implemented by subtracting the background image which is derived by smoothing the contrast-enhanced image with a $40 * 40$ median filter.

2.2   Connectedness Constraint

Since MA presents as an isolated round point, we expect to constrain it inside an elaborately designed double-rings window. Fu *et al.* proposed a connectedness constraint based algorithm in [7] for small target detection in infrared image and promising results were achieved. Based on that, a template filter with connectedness constraint is used to guarantee the MA is totally constrained inside the designed window. The inner window (W^{in}) is designed to provide the lower limit for the size of a MA and outer window (W^{out}) is used to provide the upper limit. In other words, the rules of connectedness constraint are stated as following:

Condition(I): connected region should fill the inner window;
Condition(II): connected region should be constrained inside the outer window.

Let I_i denotes the intensity of the central point p_i in the preprocessed image. Small size region, which fails to fill the inner window will be discarded. Regions with large coverage or connected distension such as hemorrhages and vessels, which are expected to exceed the boundary of the outer window, are

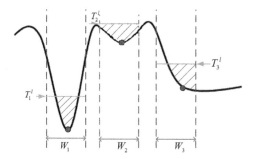

Fig. 2. Local adaptive thresholding in 1-D case. W_1, W_2, W_3 denotes three kinds of windows that cover a MA, a small blob with lower valley, and an edge respectively. The red points represent corresponding central point of each window. The local thresholding $T_i{}^l$ is judged by the maximum between the average value of $I_i{}^b$ and I_i, and the sum value of T^0 and I_i in the window. (Color figure online)

also excluded. Only Target satisfies connectedness constraints can get detected successfully.

The procedure is as following: in each sliding window Ws_i, we derive the binary value of each pixel with a local adaptive threshold. Then we take the central point p_i of Ws_i as a seed and run a region growing algorithm on the binary image. After that, the connectedness rules are applied to judge whether the growed region is totally constrained by window Ws_i. If so, this window is deemed as covering a MA and the location p_i is set to value "1" on the decision map. Otherwise, if the region touch the outer boundary of Ws_i, the window is regarded as covering other structures and the location p_i is set to value "0".

2.3   Region Growing

Microaneurysm is characterized as a red lesion darker than the surrounding retinal regions. Therefore, MAs can be popped out by using an local adaptive threshold whose value is just above the local minimum. The local threshold T_i^l for each sliding window Ws_i should be adapted as:

$$T_i^l = \max\{\tfrac{I_i+I_i{}^b}{2}, I_i + T^0\} \tag{1}$$

where $I_i{}^b$ denotes the intensity of the background of Ws_i. As the outer window size is slightly larger than the actual size of MA, the boundary pixels of the outer window can be treated as a rough estimation of the background. T^0 is a fixed value, which is introduced to avoid noise pixels with lower valleys (illustrated in Fig. 2). The binary image after thresholding can be derived by:

$$B_j = \begin{cases} 1, & if I_j \leq T_i^l \\ 0, & otherwise \end{cases} \text{ s.t. } p_j \in Ws_i \tag{2}$$

where B_j denotes the resulting binary value of each pixel in the sliding window Ws_i after threshold.

After thresholding, we only focus on the positive regions and run a flood fill algorithm inside the sliding window with its p_i as a seed. Then connectedness constraint is applied to the growed image inside the sliding window to get the decision map. Three typical cases are illustrated using 1-D examples in Fig. 2. W_1, W_2, W_3 represent three windows that cover a MA, a small blob with a lower valley, and an edge respectively. The edge is usually a vessel or a hemorrhages reaching outside the outer-window. In W_1, a MA candidate is well segmented after thresholding. But in W_2 and W_3, pixels after thresholding and region growing tend to fill or touch the outer window, which break Condition (II) and are not selected as MA candidates. The window sizes are fixed predefined values which are determined by the actual size of MA.

We run an 8-neighbour connectedness detection on the decision map to get candidate MA regions. Hence, the neighboring pixels of p_i with comparable pixel values are included in one candidate lesion, which effectively reduces the computation complexity in candidate selection step. The center location of each candidate region is collected in L, treated as the candidates location. The pseudo-code for our technique is shown in Algorithm 1:

Algorithm 1. Candidate Extraction

Input: r^{in}; r^{out}; T^0; input retinal image I;
Output: decision map D, location list L
1: **while** for each sliding window-couple W_i^{in} and W_i^{out} **do**
2:      Binarize all pixels in W_i^{out} using T_i^l, resulting in a binary map B;
3:      Run a region growing algorithm on B using the central point p_i as a seed;
4:    **if** Condition(I) and Condition(II) are satisfied **then**
5:        $D_i \leftarrow 1$;
6:    **else**
7:        $D_i \leftarrow 0$;
8:    **end if**
9: **end while**
10:      Run an 8-neighbour connectedness detection on D;
11: **while** for each foreground region R_j **do**
12:      Insert the region center location into L;
13: **end while**

---

3   Candidate Selection

The purpose of candidate selection is to classify each of the candidates as either a MA or a non-MA. The misleading spots mainly concentrated on the dots within vessels caused by the discontinuity of vessels. They appear darker than the surrounding areas, seemingly similar to MAs (indicated by the yellow square in Fig. 4). Obviously, the candidate extraction step only locates the MA candidates. For effective classification, further operations are required to get more distinguishing features of the candidates. Unlike traditional region growing algorithms, our method in this step aims to enlarge the differences between different kinds of targets to get better classification result. We take the candidate MAs as the initial points (denoted by L). For each point $q \in L$, given a threshold value $t \geq 0$, $C(t)$ is defined as the largest 8-connected region which contains q and in which $S(p) \leq S(q) + t$ for all $p \in C(t)$. $t = 0.1, 0.2, ..., t_{max}$, which are used for evaluating the region. t_{max} is the largest value in this sequence for which Area$(C(t)) \leq 3000$ pixels. S denotes the image after preprocessing. With different thresholds, we get a series of region growing results. Now an energy function proposed by Fleming $et\ al.$ [6] is defined as the mean squared gradient magnitude of S around the boundary of $C(t)$:

$$E(t) = mean_{p \in boundary(C(t))} G(p)$$
$$G\ \ \ = grad(S)^2 \tag{3}$$

The gradient magnitude of the image S is calculated by

$$grad(S) = \sqrt{(S * s_x)^2 + (S * s_y)^2} \tag{4}$$

where s_x and s_y are sobel gradient operators in the x and y directions. Values of $E(t)$ attains a peak are likely to be where the boundary separates two regions. To remove peaks generated by noise, $E(t)$ is smoothed with a Gaussian curve ($\sigma = 0.2$) and peaks correspond to the boundary of a MA or of a vessel leaves. $t_1, ... t_{N_{peak}} (t_1 \leq ... \leq t_{N_{peak}})$ are used to denote at which $E(t) * gauss(0.2)$ attains a peak and N_{peak} is the number of peaks.

A typical MA gives rise to a single peak in the curve $E(t) * gauss(0.2)$ at where the boundary of $C(t)$ seperates MA from the background. The dot within a vessel has two peaks, for the gradient attains a peak at the boundary of the dot and again at the boundary between the vessel and the background. The threshold value $t_{N_{peak}}$, as the outermost boundary of the target, can well indicate the boundary between target and the background. A region growing operation is then applied with $t_{N_{peak}}$ as a threshold and q as a seed. Note that $t_{N_{peak}}$ is a local value which varies from different windows. Dots within vessels will grow into strips while MAs will grow into spots (see in Fig. 3(d)). Same operations are performed on all $q \in L$ to get a local-contrast feature map.

A SVM classifier, trained with a total of 28 features (listed in Table 1), is used to discriminate true microaneurysms from the fake ones. The features are partly picked from [2] with the concern on containing enough information and less time consuming. The connected region at each candidate MA's location in the local-contrast feature map is regarded as a sample and all features of each sample are extracted upon this feature map. A z-scale operation is performed on the entire feature sets before training and 5 fold cross-validation is used to estimate the hyper-parameters of the gaussian kernel used for the SVM classifier. For the unbalance problem that the non-MA samples outnumber the MA samples, a bootstrapping approach [8] is used to get a small subset of representative non-MA samples:

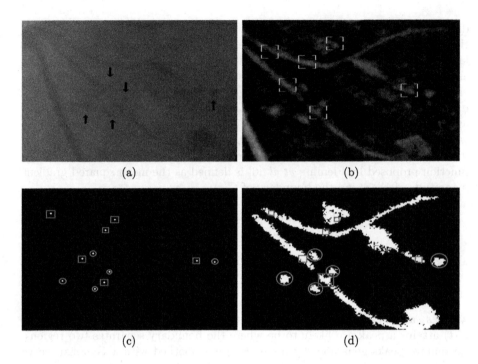

Fig. 3. Comparative results related to the detection of dots within vessels. (a) Original image, (b) Preprocessed image, (c) Candidate extraction result, (d) Local-contrast feature map. Objects encircled in green are true positives, whereas objects inside the squares are FPs. In the local feature map, dots within vessels will grow into strips while MAs will grow into spots. (Color figure online)

(1) Select a fixed size set F, which includes two parts: all the MA samples and a fraction of non-MA samples from the whole training set N;
(2) Train the SVM classifier with F;
(3) Apply the trained SVM on N. Update F by using M incorrectly classified non-MA samples in N to replace M correctly classified non-MA samples in F.
(4) Repeat step (2) and (3) until convergence.

The convergence of this bootstrapping approach is proofed in [9].

4   Results and Discussion

The performance evaluation is conducted on two well established public datasets: DIARETDB1 (standard diabetic retinopathy database [10]) and ROC (retinopathy online challenge [11]). DIARETDB1 (DB1) is an international standard DR database which contains 89 images. They were all captured using the same 50 degree FOV digital camera and compressed in PNG format with the size of

Table 1. Feature sets

A, P, C	The area, perimeter, circularity of the candidate region respectively.
L_{Ma}, L_{Mi}	The major and minor axis length of the candidate region.
R	The ratio of the major axis and minor length of the candidate region.
G_{mean}	The mean of the candidate region in green plane image (I_g).
BG_{std}	The standard deviation of the candidate region in the background image (I_{bg}). I_{bg} is derived by a $68 * 68$ median filter applied to I_g.
GD_{std}	The standard deviation of the candidate region in the deviation image. The deviation image is the deviation of I_g.
H_m, H_{std}	The mean and standard deviation of the candidate region in hue image.
D_b	The intensity difference between p_i and boundary's mean in I_g.
Rrg_m, Rrg_{std}	The median value and standard deviation of the (I_g/red plane image of the candidate region) ratio respectively.
$Prop_{max}$	The most proportion of pixels with same value of the candidate region in I_g.
H_{std}	The standard deviation of probability histogram of the candidate region in I_g.
$DOG_{std,m}$	The mean and standard deviation of 6 Difference of Gaussian (DoG) filter responses. 7 different Gaussian kernels with standard deviations of 0.5, 1, 2, 4, 8, 16, and 32 were used. The total amount of features obtained is $2 * 6 = 12$.

Table 2. Comparison of different FROC results on DB1 at medium range of false positive per image.

Sensitivity of fppi	1	2	4	8	12	16	20
Clutter-reject	0.71	0.74	0.78	0.83	0.85	0.87	0.88
Our method	0.57	0.68	0.79	0.91	0.91	0.91	0.91

Table 3. Comparison of different MA detection methods. Note that, except our method, all the performance values are reported from [13] implementation of each method.

Author	Methodology	SE	FPs/Img
Walter	Diameter Closing	36 %	154.42
Spencer	Top-hat Transform	12 %	20.3
Hough	Circular Hough-transform	28 %	505.85
Zhang	Multiple-Gaussian Masks	33 %	328.3
Lazar	Cross-section Profile	48 %	73.94
Our method		51 %	72.34

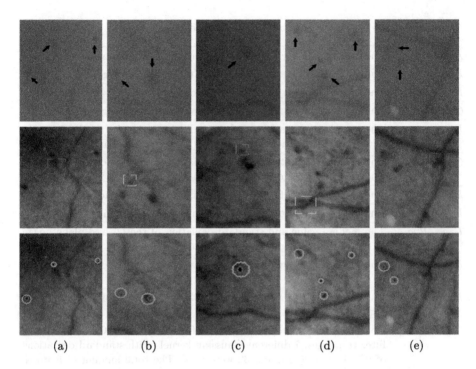

(a)   (b)   (c)   (d)   (e)

Fig. 4. Results of the proposed method. Each column shows a sub-image instance with MA candidates. From top to bottom: origin image, preprocessed image, result image. The red circle marks the ground truth while the green one shows the true MA correctly detected. Different squares denote different kinds of false positives. The candidate in the red square exceeds the boundary of the outer window while the candidate in the blue square fails to fill the inner window. They are all rejected in the candidate extraction step. The yellow squares indicate that dots within vessels will grow into strips in the feature map so that they will be discarded after classification. (Color figure online)

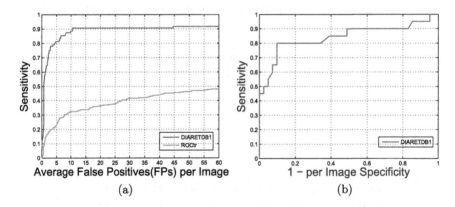

(a)   (b)

Fig. 5. Our performance at lesion-level and image-level. (a) FROC curves: lesion-level performance on two datasets (b) ROC curves: image-level performance on DB1.

1152 * 1500 pixels. Each image in DB1 has the ground truth collected by four experts. A total of 182 MAs are obtained at 75 % consensus level. 28 images in DB1 are used for training while the remaining 61 images are used for testing. The ROC database includes a set of 100 images, spliting into a train set (ROC_{tr}) and a test set (ROC_{te}) with 50 images respectively. Images in ROC are all compressed in JPEG format, but with different resolutions and shaped FOVs. Four retinal experts were asked to annotate the MAs and irrelevant lesions in the dataset. For ROC dataset, we only use the ROC_{tr} as another test set.

Results of the proposed method are shown in Fig. 4. As we can see, true MAs are well recognized without bringing in hemorrhages, noise points and incontinuous vessels. The candidate in the red square covers a large region, exceeding the boundary of the outer window of the filter. While the candidate in the blue square are noise points, failing to fill the inner window. They are all rejected in the candidate extraction step. The yellow square shows the situation that occurs in the candidate selection step. Dots within vessels will grow into strips so that they will be discarded after classification.

The Free-Response Operating Characteristic (FROC) curves of the proposed method on DB1 and ROC are presented in Fig. 5(a). The FROC curve plots sensitivity against the average number of false positives. Sensitivity measures the proportion of positives that are correctly identified and false positives (FPs) are non-MAs detected as MAs. Due to the image quality that DB1 has less inter-image variability than ROC, the highest sensitivity S_{max} achieves on DB1 with 91.67 % at 44.62 false positives per image (fppi). The sensitivity of each curve has an upper limit, which is determined by the candidate extraction accuracy.

A set of fppis on FROC are considered to capture the methods performance. To compare with Clutter-reject method [12], sensitivities obtained at this set of fppis on DB1 are summarized in Table 2. The results show that our method gets better effects at individual stages. And the maximum sensitivity of our method reaches higher (91 %) with a smaller fppi (8 fppi). Another comparative experiment is conducted on ROC_{tr}. Results from our method and referenced published methods [13] are listed in Table 3. This table measures sensitivity obtained at averaging false positives per image on ROC_{tr}. It is important to mention that our method reduces the false detection rate compared to the best performing technique proposed by Lazar *et al.* [14] with higher sensitivity.

Figure 5(b) shows the Receiver Operating Characteristic (ROC) curve on DB1 dataset. An image is declared abnormal if it contains at least one MA. It can be seen that the proposed method can reach a good balance between high sensitivity (80 %) and high specificity (90 %).

5   Conclusion

In this paper, we propose an algorithm based on local connectedness constraint and local contrast normalization to detect MAs in retinal images. It consists of candidate extraction to pick up MA candidates and candidate selection to distinguish true MAs. The approach is evaluated on the public databases DB1 and ROC_{tr}. In conclusion, our method picks up MAs from retinal images more accurately compared with conventional methods. The false positives, such as noise points in the background and dots within vessels, which appear similar to MAs, are efficiently removed. In the future work, we aim at improving the detection of MAs which are sticking together and making the MA detection system more robust.

Acknowledgement. This research is partly supported by NSFC, China (No: 61375048) and Funding for joint research of engineering and medical science, SJTU (15X190020072).

References

1. Baudoin, C.E., Lay, B.J., Klein, J.C.: Automatic detection of microaneurysms in diabetic fluorescein angiography. Revue D pidmiologie Et De Sant Publique **32**, 254–261 (1984)
2. Niemeijer, M., Van Ginneken, B., Staal, J., Suttorp-Schulten, M.S.A., Abramoff, M.D.: Automatic detection of red lesions in digital color fundus photographs. IEEE Trans. Med. Imaging **24**(5), 584–592 (2005)
3. Mizutani, A., Muramatsu, C., Hatanaka, Y., Suemori, S., Hara, T., Fujita, H.: Automated microaneurysm detection method based on double ring filter in retinal fundus images. In: Proceedings of SPIE - The International Society for Optical Engineering, vol. 7260, pp. 72601N-1–72601N-8 (2009)
4. Sabarivani, A.: An ensemble based system for micro aneurysm detection and diabetic retinopathy grading using preprocessing and candidate extractors. Res. J. Pharm. Biol. Chem. Sci. **6**(2), 1887–1903 (2015)
5. Akram, M.U., Khalid, S., Khan, S.A.: Identification and classification of microaneurysms for early detection of diabetic retinopathy. Pattern Recogn. **46**(1), 107–116 (2013)
6. Fleming, A.D., Philip, S., Goatman, K.A., Olson, J.A., Sharp, P.F.: Automated microaneurysm detection using local contrast normalization and local vessel detection. IEEE Trans. Med. Imaging **25**(9), 1223–1232 (2006)
7. Fu, K., Xie, K., Gong, C., Gu, I.Y.H., Yang, J.: Effective small DIM target detection by local connectedness constraint. In: IEEE International Conference on Acoustics, pp. 8110–8114 (2014)
8. Xiaohui, Z., Opas, C.: A SVM approach for detection of hemorrhages in background diabetic retinopathy. In: Proceedings of 2005 IEEE International Joint Conference on Neural Networks, vol. 4, pp. 2435–2440 (2005)
9. El-Naqa, I., Yang, Y., Wernick, M.N., Galatsanos, N.P., Nishikawa, R.M.: A support vector machine approach for detection of microcalcifications. IEEE Trans. Med. Imaging **21**(12), 1552–1563 (2002)
10. Kauppi, T., Kalesnykiene, V., Kamarainen, J.-K., Lensu, L., Sorri, I., Raninen, A.: DIARETDB1 diabetic retinopathy database and evaluation protocol. In: British Machine Vision Conference 2007 (2007)
11. Niemeijer, M., Van Ginneken, B., Cree, M.J., Mizutani, A., Quellec, G., Sanchez, C.I.: Retinopathy online challenge: automatic detection of microaneurysms in digital color fundus photographs. IEEE Trans. Med. Imaging **29**(1), 185–195 (2010)

12. Keerthi, R., Gopal Datt, J., Jayanthi, S.: A successive clutter-rejection-based app-roach for early detection of diabetic retinopathy. IEEE Trans. Bio Med. Eng. **58**(3), 664–673 (2011)
13. Shah, S.A.A., Laude, A., Faye, I., Tang, T.B.: Automated microaneurysm detection in diabetic retinopathy using curvelet transform. J. Biomed. Opt. **21**, 10 (2016)
14. Lazar, I., Hajdu, A.: Microaneurysm detection in retinal images using a rotat-ing cross-section based model. In: IEEE International Symposium on Biomedical Imaging: From Nano To Macro, vol. 7906, pp. 1405–1409 (2011)

Aurora Sequences Classification and Aurora Events Detection Based on Hidden Conditional Random Fields

Baibai Xu, Changhong Chen[(✉)], Zongliang Gan, and Bin Liu

Jiangsu Provincial Key Lab of Image Processing and Image Communication,
Nanjing University of Posts and Telecommunications, Nanjing 210003, China
chenchh@njupt.edu.cn

Abstract. The dynamically evolving process of aurora is closely related to complex and energetic plasma processes of the outer magnetosphere, so aurora image sequences often have complex underlying structures. In this paper, we present a novel aurora sequences classification and aurora events detection method based hidden conditional random fields (HCRF) employing spatial texture features. Firstly, divided uniform local binary patterns (uLBP) are extracted as the spatial texture features; then HCRF model is built for the spatial texture features of aurora sequences; at last, the model is applied in automatic classification and detection for four primary categories of dayside auroral observations. The supervised classification results on labeled data demonstrate the effectiveness of our method. The occurrence distributions of four categories from automatic detection confirm the multiple-wavelength intensity distribution of dayside aurora, and further illustrate the validity of our method.

Keywords: Aurora sequence classification · Aurora events detection · Hidden conditional random fields (HCRF) · Aurora morphology

1 Introduction

Aurora is a photo-excitation phenomenon, which is the consequence of the dynamic plasma processes that mainly take place in the Polar Regions and that are ultimately driven by the solar wind [1]. Different shapes and types of aurora are correlated with cumulative effects of the solar wind-magnetosphere interaction and the physics of the magnetosphere-ionosphere interaction [6, 7]. Therefore, the study of aurora sequences classification and detection is very important in aurora research.

So far, aurora images research has mainly focused on two aspects, i.e., either aurora images are treated independently that ignores temporal information, or aurora images

This research was supported in part by the National Nature Science Foundation, P.R. China. (No. 61571353, 61172118, 61471202), Jiangsu Province Universities Natural Science Research Key Grant Project (No. 13KJA510004), Natural Science Foundation of Jiangsu Province (BK20130867), and a Project Funded by the Priority Academic Program Development of Jiangsu Higher Education Institutions(Information and Communication Engineering).

© Springer Nature Singapore Pte Ltd. 2016
T. Tan et al. (Eds.): CCPR 2016, Part II, CCIS 663, pp. 404–415, 2016.
DOI: 10.1007/978-981-10-3005-5_33

are treated as a sequence including temporal correlation. A mass of methods have been proposed to the former, such as LBP [2], X-GLAM [3], salient coding [4], and Wavelet hierarchical model [5]. For the latter, Han *et al.* [17] proposed ST-PVLBP for classifying arc aurora sequences and non-arc aurora sequences. It is limited to multiclass aurora sequences classification. Yang *et al.* [8] utilized HMM based on affine normalized log-likelihood to represent aurora sequences and gain good results. Wang *et al.* [16] proposed the local vector difference to represent dynamic texture and utilized it to detect the changes of aurora activities successfully.

HCRF is a discriminative model that learns the joint distribution of a class label and a sequence of latent variables conditioned on a given observation sequence, with dependencies among latent variables expressed by an undirected graph [13]. HCRF learns not only the hidden states that discriminate one class label from all the others, but also the structure that is shared among labels. HCRF has been shown strong power for a number of problems, including object recognition [13], gesture recognition [19], speech modeling [14], and multimodal cue modeling for agreement/disagreement recognition [15]. Aurora is essentially a dynamically evolving process and it involves two distinct stochastic processes: one is the resultant observable aurora activities displayed in the sky, and the other is the hidden collision process of charged particles from the earth's magnetosphere and solar wind. HCRF can integrate both aspects well with the former addressed by extracting image features as HCRF observations and the latter reflected via the hidden variables in HCRF. Thus, in this paper, we present an HCRF-based framework to model auroral sequences.

In this paper, we implement automatic aurora sequences classification and detection. The quantitative supervised classification experiments are developed using the aurora observations from 2003–2004 years at the Chinese Yellow River Station (YRS). The good accuracy demonstrates the advantages of the proposed method. Then we detect four categories of aurora images from unlabeled days' sequences during the years 2004–2009 at the YRS. In detection process, we gain a serious of short sequences through segmenting one-day aurora sequence automatically by local vector difference (LVD) algorithm. The occurrence distributions of the four auroral categories consistent with multiple-wavelength intensity distributions of dayside auroras presented in [9] proves the validity of our method.

2 Spatial Texture Representation

The basic LBP was proposed in [10] for texture classification. It labels every pixel in an image with decimal codes transformed from binary codes by thresholding the neighborhood of the pixel with its value of intensity. Those neighbor pixels whose values are smaller than that of the center pixel are marked as '0's, otherwise as '1's, and thus obtain the binary codes. Histograms of decimal codes can be used as a texture descriptor. The process is shown in Fig. 1.

The uniform LBP (uLBP) was proposed in [11], it codes every pixel on the basis of the number of spatial transitions (bitwise 0/1 changes) in LBP binary code. If the numbers of pixels have the value of at least 3, those pixels are labeled with a constant value. If the numbers of pixels have the value of at most 2, those pixels are labeled with

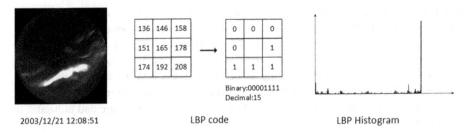

2003/12/21 12:08:51   LBP code   LBP Histogram

Fig. 1. LBP operator. Image name indicates when the image was captured (Universal Time).

the value as same as the basic LBP. The uLBP has shorter feature vector (59 dimensions) and achieves better rotation and translation invariance compared to the basic LBP (256 dimensions).

In our work, the uniform LBP (uLBP) with dividing each image into two rows and two columns is employed as auroral spatial texture representation, as shown in Fig. 2. This is mainly because a stationary auroral sequence usually consists of a dozen of aurora images, long feature vectors should be avoided since they may cause over fitting and thus low reliability of the estimated HCRF parameters. Furthermore, simple partition scheme can obtain both the local texture and global shape information.

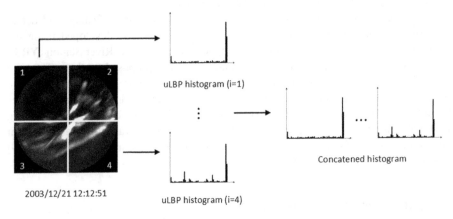

2003/12/21 12:12:51   uLBP histogram (i=4)

Fig. 2. Schematic of building spatial texture representation for aurora image.

3   Proposed Method

3.1   Aurora Classification Method

In this paper, we use HCRF to classify aurora sequences. An HCRF's graph structure E is shown in Fig. 3. It models the conditional probability of a class label given a set of observations by:

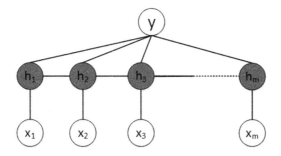

Fig. 3. HCRF model.

$$P(y|x, \theta) = \sum_{h} P(y, h|x, \theta) = \frac{\sum_{h} \exp(-\psi(y, h, x; \theta))}{\sum_{y',h} \exp(-\psi(y', h, x; \theta))} \qquad (1)$$

where x is a vector of m local observations, $x = \{x_1, x_2, \cdots, x_m\}$, and each local observation x_i is represented by a feature vector $\varphi(x_i) \in \mathfrak{R}^d$, that is the divided uLBP in our work. Where y is class labels and $y \in Y$. $Y = \{arc, drapery, radial, hot\text{-}spot\}$, representing the four types of aurora. Where $h = \{h_1, h_2, \cdots, h_m\}$, each $h_i \in H$ captures certain underlying structure of each class and H is the set of hidden states in the model, the size of H is set manually. Where $\psi(y, h, x; \theta)$ is the potential function, which is parameterized by θ, and measures the compatibility among a label, a set of observations and a configuration of the hidden states. As shown in Fig. 3, $\psi(y, h, x; \theta)$ consists of three components:

$$\psi(y, h, x; \theta) = \sum_{i} \varphi(x_i) \cdot \theta_s(h_i) + \sum_{i} \theta_y(y, h_i) + \sum_{i,j \in E} \theta_p(y, h_i, h_j) \qquad (2)$$

where θ_s refers to the parameters that correspond to hidden state $h_i \in H$ and observation x_i, in this paper, that is the connection between collision process of charged particles at the moment and auroral texture and shape. Similarly, θ_y stands for the parameters that correspond to class y and hidden state $h_i \in H$, in this paper, that is the connection between collision processes of charged particles at the moment and the types of aurora. In like manner, θ_p represents the parameters that correspond to class y and the pair of hidden state h_i and h_j, in this paper, that is the connection between collision processes of charged particles at the adjacent time and the types of aurora.

We use the following objective function in training the parameters:

$$L(\theta) = \sum_{i=1}^{n} \log P(y_i|x_i, \theta) - \frac{1}{2\sigma^2} ||\theta||^2 \qquad (3)$$

where n is the total number of training sequences. The first term in Eq. 3 is the log-likelihood of the data; the second term is the log of a Gaussian prior with variance σ^2, i.e.

$$P(\theta) \sim \exp(-\frac{1}{2\sigma^2}||\theta||^2) \qquad (4)$$

We could use gradient ascent [12] to search for the optimal parameter values:

$$\theta^* = \text{argmax}_\theta \, L(\theta) \qquad (5)$$

After figuring out θ^*, we are able to use the model to predict the type of an unknown aurora sequence:

$$y^* = \text{argmax}_y \, P(y|x, \theta) \qquad (6)$$

3.2   Aurora Detection Method

Millions of aurora images are acquired annually from the YRS and it is not practical to manually label all auroral observations. An automatic detection method is very necessary for auroral statistic analysis. Here we present an automatic detection method for aurora images through combining HCRF with LVD.

Our objective is to detect the occurrence distributions of the four auroral categories. It mainly includes four steps. Firstly, a number of aurora sequences are selected and labeled as the labeled dataset. Secondly, the detection model is trained. Then, the abundant short test sequences are obtained from segmenting unlabeled days' sequences automatically. Finally, the test sequences are recognized using the model trained in the first step. The flowchart of aurora events detection is shown in Fig. 4.

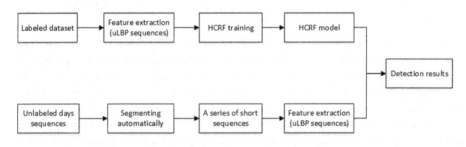

Fig. 4. Flowchart of aurora detection.

The local vector difference (LVD) algorithm [16] was used to represent fluid dynamic texture. For a sequence S:

$$S = \{s(x, t), x \in D, t = 1, 2, \cdots, T\} \qquad (7)$$

where x is the position coordinates of a pixel, D represents the spatial domain, t is the current frame, and T is the length of sequence S. The next step is to calculate the optical flow V [18] for two consecutive frames of sequence S:

$$V = \{v(x,t),\ x \in D,\ t = 1, 2, \cdots, T - 1\} \tag{8}$$

where v is the motion vector of the pixel in x and t. Then LVD denoting the relative motion in the neighborhood is defined as:

$$v_{LVD}^d(x,t) = v(x,t) - v(x^d, t) \tag{9}$$

where $d = 1, 2, 3, 4$, it represents four neighborhood pixels of the center pixel. Then the optical flow is projected into polar coordinate that has been divided into m range areas and n angle areas. Calculating the histogram for each d and concatenating them all as the representation for sequence S, as shown in Fig. 5.

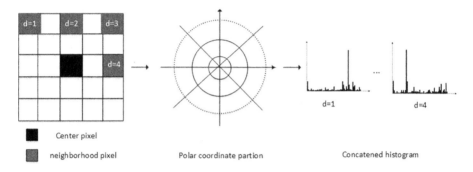

Fig. 5. Schematic diagram of LVD.

Given a one-day aurora sequence, we apply sliding window with the length of $2L$ to segment it automatically. Firstly, we get similarity curve through calculating the similarity of LVDs for former L sequence and later L sequence with moving the sliding window. The process is shown in Fig. 6. Then, a threshold is set to delete mutational points with lower value. Finally, the sequence is segmented into a series of short sequences based on mutational points.

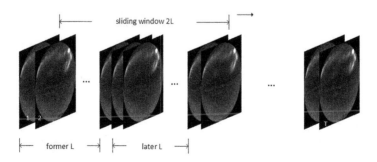

Fig. 6. Sliding window applied to segment aurora sequence.

4 Experiments Results and Analysis

4.1 Dataset and Classification Mechanism

The aurora data used in this paper were obtained from the Chinese Yellow River Station (YRS), Ny-Ålesund, Svalbard [9]. The optical system at YRS contains three identical all-sky imagers, which are developed to measure photoemissions at 427.8, 557.7 and 630.0 nm since December 2003. The optical instruments can make 24-hour surveys of auroral emissions with temporal resolution of 10 s during the winter season from October to March of the following year. In this paper, we concentrate on the dayside aurora [03:00–15:00 Universal Time (UT)/06:00–18:00 Magnetic Local Time (MLT)] and only use observations from November to February to avoid the influence of daylight. Considering the image characteristics, only auroras at 557.7 nm are used.

In this paper, we take the classification scheme proposed in [9], where the dayside aurora is classified into arc, drapery corona, radial corona, and hot-spot categories based on the spectral and morphological characteristics as follows: (1) The arc aurora is single arc or multiple longitudinally extended discrete arcs with intense emission at 557.7 nm, along with east-west bands or stripes on ASI images. The arc may be twisted occasionally into auroral curls, spirals and folds. (2) The drapery corona is an auroral display presenting multiple east-west elongated rayed bands with weak emission at 557.7 nm and pulsant gauzelike diffuse auroras often present in an equatorward direction near the dayside drapery corona. (3) The radial dayside corona presents as a radially rayed structure with weak emission at 557.7 nm, but appears strong at 630.0 nm. Rays in the ASI image are radially directed from the zenith in all directions and rapidly change. (4) The hot-spot aurora is a complex auroral structure, including rayed structures, transient brightening rayed bundles, spots, and irregular patches with intense emission at 427.8, 557.7, 630.0 nm. Some sample images at 557.7 nm of the four categories are shown in Fig. 7.

The supervised classification experiments are carried out on the dataset SD1 [8]. It consists of 609 aurora sequences selected from December 2003 to January 2004, including 284 arc sequences, 119 drapery corona sequences, 138 radial dayside corona sequences and 68 hot-spot sequences. The sequences of SD1 range from 18 frames to 94 frames and their average length is 31.4 frames.

4.2 Classification Results and Analysis

The multiple-times m-fold cross-validation technique is used to estimate the classification accuracy. Our choice for m-fold cross validation is $m = 8$ and the experiments are independently repeated 5 times. Two types of HCRF models are used in our experiments. One is the multi-class HCRF model: we train a single multi-class HCRF model using twelve hidden states. Test sequences are run with this model and the aurora class with the highest probability is selected as the recognized aurora. The other one is the one-vs-all HCRF model: For each aurora class, we train a separate HCRF model to discriminate the aurora class from other classes. Each HCRF is trained using six hidden states. For a given test sequence, we compared the probabilities for each single HCRF, and the highest scoring HCRF model is selected as the recognized aurora type.

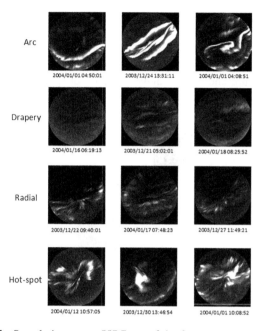

Fig. 7. Sample images at 557.7 nm of the four aurora categories.

Table 1 summarizes the results for supervised classification experiments. $f = 1$ represents uLBP with no partition, while $f = 2$ represents uLBP with diving into two rows and two columns. The best is the multi-class HCRF with $f = 2$, it shows that the combination between dived uLBP and HCRF is able to model aurora sequences effectively. $f = 2$ performs better than $f = 1$, this implies that uLBP designed block partition scheme achieves both shapes and local textures representations well. The improvement in performance is obtained when we used a multi-class HCRF compared to a one-vs-all HCRF, suggesting that it is better to jointly learn the best discriminative structure.

Table 1. Comparisons of classification performance for SD1 dataset.

Methods	Accuracy (%)
Improved HMM in [8]	84.15
ST-PVLBP in [17]	74.01
HCRF($f = 1$, multi-class)	83.47
HCRF($f = 2$, multi-class)	**85.45**
HCRF($f = 2$, one-vs-all)	83.78

In order to examine classification accuracy of each auroral category, we depict the confusion matrix for HCRF ($f = 2$, multi-class) in Fig. 8. The classification accuracy of each category indicates the following: (1) the arc auroras are more easily recognized because of the simple morphological characteristics. (2) Two corona auroras are often confused with each other. This is because they have similar rayed structure and both of them have weak emission at 557.7 nm. (3) The hot-spot auroras are often confused with arc auroras and radial auroras. The primary cause is that the hot-spot auroral display has more complex morphology, such as a distinct brightening spot and irregular patches, rayed structure and rayed bundles. Some hot-spot auroras have local shape similarity with arc auroras or radial corona auroras that result in wrong classification.

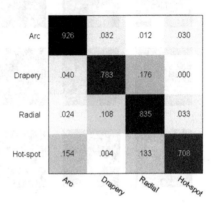

Fig. 8. Confusion matrices for HCRF ($f = 2$, multi-class).

Because aurora classification is still ongoing problem, the number of categories and the characteristics of each category are not clearly defined. This may bring about unknown sequences in SD1 set. In order to confirm whether unknown sequences exist in SD1 set, we normalized the classifier scores from HCRF ($f = 2$, one-vs-all) and set a threshold. If the highest score is below the threshold, we consider the test sequence belong to an unknown aurora class. The result is shown in Table 2. The first row is the threshold and the second row is the ratio of unknown sequences.

Table 2. Ratios of unknown sequences

Threshold	0.5	0.3	0.1	0.05
Ratio	10.3 %	6.22 %	2.23 %	1.13 %

As shown in Table 2, when threshold = 0.05 with a very low value, there still are a few unknown sequences in SD1 set. It indicates that some unknown sequences do exist in SD1 set, but the amount is small.

4.3   Detection Results and Analysis

We select forty days' auroral sequences during the years 2004–2009 at the YRS. The forty days' sequences are selected manually because some days do not contain auroras or are captured under bad weather conditions. In segmentation stage, we set the sliding widow 20 and threshold 0.02 to insure getting a serious of short sequences including only one auroral type as far as possible. In recognition stage, although the classification accuracy of one-vs-all HCRF is a little lower than multi-class HCRF, we still chose it on account of unknown types. We set the HCRF score threshold 0.3 to detect unknown type.

In order to show the temporal distribution of each category, the temporal axis is divided into 432 bins of 10-frame durations. We distribute a sequence label to its frames, and then all frames' labels of the forty days are acquired. Table 3 summarizes detected frames number of each type. The ratio for every auroral category in each bin is calculated and the occurrence distributions of every category are obtained, as shown in Fig. 9.

Table 3. Detected frame number of each type.

Arc	Drapery	Radial	Hot-spot	Unknown
64847	68660	22743	6980	9570

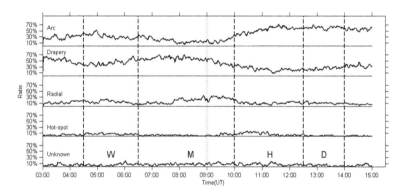

Fig. 9. Temporal occurrence distributions of four auroral categories.

In Fig. 9, W, M, H and D corresponding to four auroral active regions proposed in [9], while M region includes two states partitioned by a thin dashed line. As shown in Fig. 9, four auroral types defined by morphological characteristics dominate different regions of the dayside oval. Their global distributions approximately coincide with the multiple-wavelength intensity distribution of the dayside aurora. Peaks of the four categories fall in different regions respectively. The arc aurora has a distinct double-peak distribution and the post-noon peak is stronger than the pre-noon peak. Both the weak drapery corona and radial corona mostly occur before 10:00 UT, but their peak positions are different. The drapery corona peak predominantly occurs between 06:30 and 09:00 UT, while the radial corona peak dominantly occurs between 09:00 and 10:00 UT. The hot-spot aurora often occurs in region H and there is an evident peak around 10:30 UT. The distribution of unknown images is rather flat and has no distinct peak. Because we select all images from the forty days, unknown images may contain no aurora or be different from the images belong to the four auroral categories. The distribution indicates that the unknown occur randomly.

5   Conclusion

In this paper, we present an effective approach based on HCRF for aurora sequences classification and aurora events detection. With static feature extraction based on divided uLBP, the HCRF integrates auroral two stochastic processes very well. The supervised classification results demonstrate the effectiveness of our method. Because millions of ASI images are acquired annually, it is very significant for automatic aurora events detection. We present an automatic method based on HCRF and LVD. The detection results coincide with the multiple-wavelength intensity distribution of the dayside aurora and confirm our method's validity.

References

1. Syrjäsuo, M.T.: Auroral monitoring network: from all-sky camera system to automated image analysis. Finnish Meteorological Institute (2001)
2. Wang, Q., Liang, J., Hu, Z.J., et al.: Spatial texture based automatic classification of dayside aurora in all-sky images. J. Atmos. Solar-Terr. Phys. **72**(5), 498–508 (2010)
3. Wang, Y., Li, J., Fu, R., et al.: Dayside corona aurora classification based on X-grey level aura matrices and feature selection. Int. J. Comput. Math. **88**(18), 3852–3863 (2011)
4. Han, B., Qiu, W.L.: Aurora images classification via features salient coding. J. Xidian Univ. **40**(6), 180–186 (2013)
5. Yang, X., Li, J., Han, B., et al.: Wavelet hierarchical model for aurora images classification. J. Xidian Univ. **40**(2), 18–24 (2013)
6. Feldstein, Y.I., Elphinstone, R.D.: Aurorae and the large-scale structure of the magnetosphere. J. Geomagn. Geoelectr. **44**(12), 1159–1174 (1992)
7. Kullen, A., Brittnacher, M., Cumnock, J.A., et al.: Solar wind dependence of the occurrence and motion of polar auroral arcs: a statistical study. J. Geophys. Res. Space Phys. **107**(A11), 1362 (2002)
8. Yang, Q., Liang, J., Hu, Z.J., et al.: Auroral sequence representation and classification using hidden markov models. IEEE Trans. Geosci. Remote Sensing **50**(12), 5049–5060 (2012)
9. Hu, Z.J., Yang, H., Huang, D., et al.: Synoptic distribution of dayside aurora: multiple-wavelength all-sky observation at yellow river station in Ny-Ålesund Svalbard. J. Atmos. Solar-Terres. Phys. **71**(8), 794–804 (2009)
10. Ojala, T., Pietikäinen, M., Harwood, D.: A comparative study of texture measures with classification based on featured distributions. Pattern Recogn. **29**(1), 51–59 (1996)
11. Ojala, T., Pietikäinen, M., Mäenpää, T.: Multiresolution gray-scale and rotation invariant texture classification with local binary patterns. IEEE Trans. Pattern Anal. Mach. Intell. **24**(7), 971–987 (2007)
12. Lafferty, J., McCallum, A., Pereira, F.C.N.: Conditional random fields: probabilistic models for segmenting and labeling sequence data (2001)
13. Quattoni, A., Wang, S., Morency, L.P., et al.: Hidden conditional random fields. IEEE Trans. Pattern Anal. Mach. Intell. **10**, 1848–1852 (2007)

14. Gunawardana, A., Mahajan, M., Acero, A., et al.: Hidden conditional random fields for phone classification. In: Interspeech, pp. 1117–1120 (2005)
15. Bousmalis, K., Morency, L.P., Pantic, M.: Modeling hidden dynamics of multimodal cues for spontaneous agreement and disagreement recognition. In: 2011 IEEE International Conference on Automatic Face & Gesture Recognition and Workshops (FG 2011). IEEE, pp. 746–752 (2011)
16. Wang, Q., Liang, J., Hu, Z.J., et al.: A method for detecting the change of auroral activities based on the all-sky image sequence. Scott. J. Geol. **58**(9), 3038–3047 (2015)
17. Han, B., Liao, Q., Gao, X., et al.: Spatial-temporal poleward volume local binary patterns for aurora sequences event detection. J. Softw. **25**(9), 2172–2179 (2014)
18. Corpetti, T., Mémin, É., Pérez, P.: Dense estimation of fluid flows. IEEE Trans. Pattern Anal. Mach. Intell. **24**(3), 365–380 (2002)
19. Wang, S.B., Quattoni, A., Morency, L.P., et al.: Hidden conditional random fields for gesture recognition. In: 2006 IEEE Computer Society Conference on Computer Vision and Pattern Recognition. IEEE, pp. 1521–1527 (2006)

Source Printer Authentication for Printed Documents Based on Factor Analysis

Changjun Jin[1,3], Haibin Liao[2,3(✉)], and Youbin Chen[1,3]

[1] School of Automation, Huazhong University of Science and Technology,
Wuhan, China
[2] School of Computer Science and Technology, Hubei University of Science
and Technology, Xianning, China
haibin.liao@micropattern.cn
[3] Guangdong Micropattern Software, Co., LTD.,
Guangzhou, Guangdong, China
{changjun.jin,youbin.chen}@micropattern.cn

Abstract. Source printer authentication for printed documents is very hard in printed documents identification. When the characters in samples and test material are completely different, the problem becomes extremely challenging. In this paper, a new method primarily based on factor analysis has been proposed to solve the problem. Concretely, we utilize bilinear model to analysis feature matrix to separate the printer factor and text factor entirely. Then, we extract text-independent printer features of the samples approximately. Finally, expectation maximization algorithms are utilized to obtain maximum after posterior probability to classify. Experimental results show that our proposed method was effective to improve the recognition rate.

Keywords: Source printer authentication · Factor analysis · Bilinear model · EM algorithm

1 Introduction

With the development of modern science and technology, people's lives become increasingly digitized and printers have been more common. Meanwhile, the civil disputes and criminal cases related to printed documents become more and more common, for example, forged contracts and documents, dissemination of intimidation, inflammatory and other instruments. Printed files are important evidences or clues to identify whether the problem documents have been tampered and identifying the sources of printers which print the problem documents gives valuable assistance to the detection of cases. In addition, it is of great significance to verify the authenticity of printed documents and bills.

For some particular printed documents, the current identification technology can already use the watermark [1, 2], security fibers, holograms [3] or special ink [4] and other characteristics to identify the authenticity, but this type of security technology often cost greater and need special equipment to implant safety features. For ordinary users, the cost is too expensive. Therefore, using computer image processing and

© Springer Nature Singapore Pte Ltd. 2016
T. Tan et al. (Eds.): CCPR 2016, Part II, CCIS 663, pp. 416–426, 2016.
DOI: 10.1007/978-981-10-3005-5_34

artificial intelligence to automatically identify printed document is of great theoretical signification and application values.

Printed files identification technology researches began in the twenty-first century, and there are few researchers currently. J. Oliver and J. Chen [5] use statistical area features of computer-printable characters and determine whether there is any illegal counterfeit document content by computing the different areas of the same characters, which is simple and fast calculation, but the basis for judgment is not tight and the effect is not ideal. Purdue University Sensors and Printers File Identification laboratory (PSAPF) was established in 2002 and their results include: study the stripe features of light and dark intervals caused by colossal error and the large gear eccentricity error; by computing GLCM of character "e" to obtain texture features, respectively, using a 5-nearest neighbor classifier and SVM method to classify [6–10]. Tsai et al. proposed to use discrete wavelet transform and feature selection methods to identify the color laser printer [11]. To identify the ink-jet printer [12], Akao et al. estimated the number of spur gear by the maximum entropy method. Ning Wang, Guoqiang Han [13] et al. used strokes total area and total perimeter of strokes of printed characters to identify the source of the document type by establishing word repository. Wei Deng, Yankai Tu, Qinghu Chen etc. [14–17] designed and developed the entire image magnification system to collect microscopic details of the image information of printed documents, with using pattern matching algorithms, bipolar Hausdorff distance, penumbral fringes features based on small-scale wavelet domain and varieties of other methods to identify the source machine of printed documents. Overall, it has been achieved some results in computer printed documents identification, but still need to further improve the recognition rate.

When scholars study identification of computer printed documents, training documents and identification papers often take the same character content, such as high frequency of occurrence of the letter "e" or the word "the". In identification of Chinese printed files, they also tend to take a sample of the same characters to match the content classification. While in the Chinese printed files, it is possible to find that there is little of the same characters or even none between the identification documents and training documents, the difficulty has been increased in the meanwhile. This is because the distance of same character within the different machine is significantly less than the distance of different characters within the same machine. To better identify printed files when files contain different characters in training and test stage, this paper will divide the factors impacting the morphological features of the printing characters into two categories from the source, such as printer factor representing differences introduced by different components and devices, namely a style factor; text element representing differences introduced by the character itself, namely a content factor. For the identification of printed files, printer factor is an effective factor while text factor is the interference factor. In this paper, the secondary features of printing characters are obtained by utilizing the factor analysis. By using bilinear method for factorization to extract printer factors of feature matrix and reduce interference of text element for recognition, it can improve the recognition rate of the source machines that print the document.

2   Features Decomposition Based on Factor Analysis Model

Content and style determine the observation, which can be regarded as two independent factors to affect the essence of the thing [18]. For example, in speech signal, content factor represents speech text semantic information while style factor indicates the speaker's tone, talking mood and tone information; In handwriting, content factor indicates that the sample is which character while style factor indicates that the sample is what people write [19]; And in characters printing, content factor represents the character text message while style factor represents the character font information; In face images, regular human face (front, neutral, light normalization) factor is the content, and face pose, illumination, expression and other changes can be regarded as style factor [20]. Similarly, in printed documents, the difference introduced by printers is the style factor, also called the printer factor; the difference introduced by the character of the text is the content factor, also called text factor. The purpose of document object identification is to identity the files printed by which printed machine based on the style of the printer information. If we can isolate the text message which affects the characteristics of printed characters and extract the features unrelated to the contents, it will be beneficial to the printed document identification. Based on the above, we proposed the extraction method of secondary feature based on printable characters on factor analysis model, namely by using bilinear method to separate the printer factor and text factor to extract the text independent print character features approximately.

2.1   Factor Analysis Model Based on Bilinear Method

If the content of printed characters $b_j \in R^J$ has the style of $a_i \in R^I$, the observation of printed characters $y \in R^K$ can be represented by bilinear model:

$$y_k^{sc} = \sum_{i=1}^{I} \sum_{j=1}^{J} w_{ijk} a_i^s b_j^c \tag{1}$$

In which, $k \in [1, K]$ denotes k th dimension of feature vector of the printing characters, and s and c denote the style and content separately, w_{ijk} denotes the interaction relations of the style and content. In order to make factor analysis model more flexible, we assume that the interaction term w_{ijk} changes with content changing. Providing that $b_{ik}^c = \sum_j w_{ijk}^c b_j^c$, the Eq. (1) can be changed into:

$$y_k^{sc} = \sum_i a_i^s b_{ik}^c \tag{2}$$

Assuming that B^c denotes the $K * I$ -dimensional matrix of each element $\{b_{ik}^c\}$, respectively, the Eq. (2) can be written as a more concise analysis of the form factor model:

$$y^{sc} = a^s B^c \qquad (3)$$

For example, the bilinear model is applied to different fonts printed characters set. Therefore, the font information is the style factor, and the character itself is the content factor, the results shown in Fig. 1. Each character can be represented by basic factor matrix content and fonts factor coefficients. If we want to rebuild a particular font specific content of character, just need to combine the coefficients of the basic matrix font weighted linearly together.

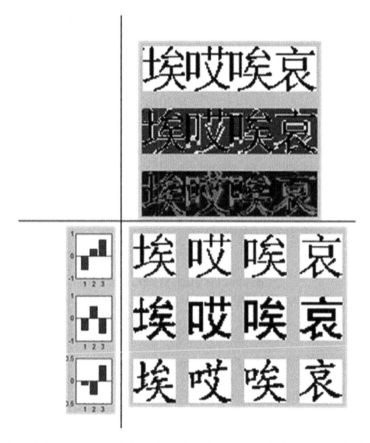

Fig. 1. Exploded bi-linear model view of the three fonts. (Top): factor matrix of basic content: y^{sc}; (Left): Coefficients of character factors: a^s; (Right): Reconstruction of different font character image.

2.2 Factor Analysis Model for Solving Match

The solving matching target of factor analysis model is to minimize total squared error of all samples in training phase. Assuming that the observation value is $y(t)$ in t th training stage, and $t = 1, 2, \ldots T$. Providing that indicator variable is $h^{sc}(t)$, wherein:

$$h^{sc}(t) = \begin{cases} 1, \ y(t) \in (s,c) \\ 0, \ other \end{cases} \tag{4}$$

Therefore, total squared error of all training set of Factor Analysis Model E is:

$$E = \sum_{t=1}^{T} \sum_{s=1}^{S} \sum_{c=1}^{C} h^{st}(t) \|y(t) - a^s B^c\|^2 \tag{5}$$

If the number of observations on a variety of style and content are equal in training samples, the best result of the model factor analysis can be obtained by singular value decomposition (SVD).

In the identification of the printer, providing the printer can be denotes as s, the mean value of the observation of the text c is as follows:

$$\bar{y}^{sc} = \frac{\sum_t h^{sc}(t)y(t)}{\sum_t h^{sc}(t)} \tag{6}$$

Apparently, these observations are 3-dimensional matrix. To take advantage of the standard matrix algorithm, we change the $SC * K$ dimension row vector into a two-dimensional $S * CK$ matrix, and it is as follows:

$$\bar{Y} = \begin{bmatrix} \bar{y}^{11} & \cdots & \bar{y}^{1C} \\ \vdots & \ddots & \\ \bar{y}^{S1} & \cdots & \bar{y}^{SC} \end{bmatrix} \tag{7}$$

In which, $\bar{y}^{sc} = [\bar{y}_1^{sc}, \ldots, \bar{y}_K^{sc}]$ denotes the mean vector of observation of K dimensions. The formula (3) can be expressed as a more compact form of a matrix:

$$\bar{Y} = AB \tag{8}$$

Where $A = [a^1 \cdots a^s]^T$ is the $S * I$ dimensional matrix, indicating the parameter matrix of printer factor; $B = [b^1 \cdots b^c]$ is the $I * (KC)$ dimensional matrix, denoting he parameter matrix of the content factor.

In order to obtain minimum variance estimation of parameters of printer factor and text factor, we utilized the SVD to compute the $\bar{Y} = USV^T$, with the diagonal elements of S taken in descending order. We can get A from the I front columns of the matrix U and get B from the I front rows of the matrix SV^T. The dimensions of the model can be obtained based on a prior knowledge of reference or experimental results.

2.3 Printer Category Based on EM Algorithm

Assuming that testing data is originated from the set of printers S with the same of training data, but the character and content is different from the training data. Supposing that printer factor can be denoted as a^s and the factor of the new text is $B^{\tilde{c}}$;Supposing that a new printer s with observations y of text $\tilde{c}$ is Gaussian distribution with mean bilinear predictive value and variance is σ^2, then

$$p(y|s,\tilde{c}) \propto \exp\left\{-\left\|y - a^s B^{\tilde{c}}\right\|^2 \bigg/ (2\sigma^2)\right\} \qquad (9)$$

The overall probability density distribution of y is:

$$p(y) = \sum_{s,\tilde{c}} p(y|s,\tilde{c})p(s,\tilde{c}) \qquad (10)$$

According to prior knowledge, $p(s,\tilde{c})$ is corresponding to uniform distribution. We then use EM algorithm to obtain the new iteration factor $B^{\tilde{c}}$ and text description of the test data $p(s,\tilde{c}|y)$ in the below.

E-Step: for printer s with text $\tilde{c}$, the observation data y can be used to compute the probability density function:

$$p(s,\tilde{c}|y) = p(y|s,\tilde{c})p(s,\tilde{c})/p(y) \qquad (11)$$

M-Step: Estimating the new text factor $B^{\tilde{c}}$, to make the maximum log-likelihood probability. Let us make it:

$$L = \sum_y \log p(y) \qquad (12)$$

The new solution $B^{\tilde{c}}$ can be made out by the equation of $\partial L/\partial B^{\tilde{c}} = 0$:

$$B^{\tilde{c}} = \left[\sum_s \sum_y p(y|s,\tilde{c}) \cdot a^{p\mathrm{T}} \cdot a^p\right]^{-1}\left[\sum_p \sum_y p(y|s,\tilde{c}) \cdot a^{p\mathrm{T}} \cdot y\right] \qquad (13)$$

The specific steps of EM iterative algorithm are as follows:

(1) Initializing the text factor;
(2) Calculating posterior probability E- step observations by formula (11);
(3) According to formula (13) to update the text value factor;
(4) Repeating steps (2) (3) until the calculated difference is less than twice the threshold or the maximum number of iterations exceed a predetermined number of times.

EM algorithm can converge to the local maximum and test data can be classified based on the posterior probability of the largest such category. For the EM algorithm, the initialization is very important. Because this paper focuses on the recognition performances, it is initialized by using the nearest neighbor method. Namely, for each test vector data, text factor can be taken with the most similar character text factor.

3 Results and Discussion

In order to test the effectiveness of the model based on Factor Analysis, we establish database of printed files including 40 laser printers. These printers include some popular brands and types, see Table 1. 40 printers were sampled and each printer is to print two files, one for training, and one for testing. The contents of file of each printer are 1100 commonly used Chinese characters, with SimSun, small fourth printing, which can be extended to other fonts. We use the image acquisition system to sample 504 characters from each test file for training and sample another 504 characters with completely different content for test. After preprocessing, cutting, automatic identification of character content, it can constitute 40 samples with 504 characters in each sample for training set and another 40 samples for test set.

Table 1. Number and type of printers in experiments

Number	Manufacture	Type
1	BROTHER	HL2240D
2	CANON	LBP 1820
3	CANON	LBP 1810
4	CANON	LBP 910
5	CANON	IR5000
6	CANON	LBP 6750D
7	CANON	LBP 1910
8	CANON	LBP850
...	...	...
38	SHARP	MX-M700 N
39	SHARP	700 N
40	XEROX	Phaser3117

We take two files from one printer, recorded them as A1, A2, and take a file from another printer, and referred it as B2. Then, we extract the 8-dimensional features reflecting the shape of the characters for each character in the above files, and compute distance between A1 and A2, denoted by D1, calculated distance between A1 and B2, denoted by D2. D1 represents the distance between the different characters from the same printers, as shown in red dots in Fig. 2; D2 represents the distance between the same characters from different printers, as shown in blue dot in Fig. 2. As it can be seen, the distance of different characters from the same printers is much greater than the distance of the same characters from different printers. Namely, text factor is significant for the features of files, much larger than printer factor. Therefore, printer factor in the character of the morphological characteristics is a weak signal and is susceptible to interference factors from text strong signal and error factors, which also affect the accuracy of the print file identification.

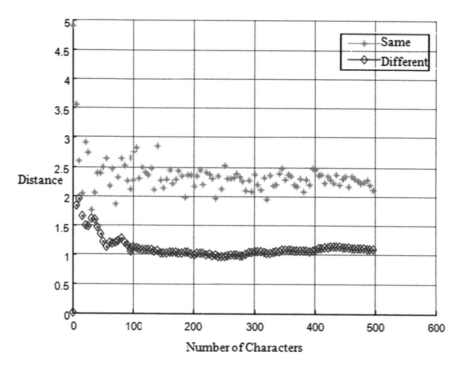

Fig. 2. Comparisons of distances of different characters from same machine and same characters between different machines. (Color figure online)

After the separation of text factor and printer factor by the proposed method, we compute the distance of A1, A2 and B2 by the above described methods, as shown in Fig. 3. As it can be seen, the distance between different characters of the same machine is less than the distance of same characters between different printers. Namely, the significance of text factor to character features is reduced, and the impact of printer factor is more significant.

The following is the identification experiments. We use methods of Moment Features (MF), the direction index histogram (DIH) and Wigner characteristics (WF) to extract features. We utilize the proposed method and Euclidean distance to extract features matrix for identification test. In our method, the posterior probability computing from EM algorithm is compared with a threshold value. However, in Euclidean distance method, the distance of features is compared with a threshold value directly. If it is greater than the threshold value, the files are determined to be printed from the same machine. However, we compute statistics correct identification number, and the experimental results are shown as in Table 2. Further experimental results are also compared with the method used in literature [18] (referred to as method 1), and there are 50 same characters averagely between each of the two documents of training set and test set. In our experiments, there is no same character completely between training set and test set. Experimental data with optimal thresholds of three methods are listed as followed.

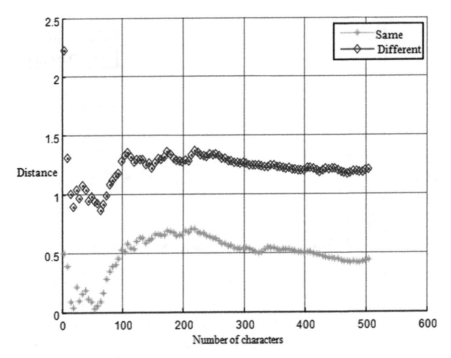

Fig. 3. Comparisons of distance of different characters between same machine and distance of same character between different machines after processing by factor separation

Table 2. Identification results of Euclidean distance, the proposed method and the method 1st

Method	Euclidean distance	Factor Analysis	Method 1st
MF	7.63	80.63	83.35
DIH	5.83	70.38	76.69
WF	8.38	81.03	81.14

As it can be seen, the feature matrix through separating printer factor and text factor by factor analysis model, the impact of the text factor is significantly reduced and we can obtain an approximate text independent characteristics between samples and test and the recognition rate is significantly increased in condition that the samples do not have any same word in case. However, the recognition rate is still poorer that the condition of 50 identical word in samples, which indicates that text factors is not completely mined and the features are still influenced by parts of the text factors after separation, which may be characterized by the fact that matrix with bilinear between the model does not fit entirely. In next step, transformation process is performed firstly in consideration of the characteristic matrix and to make it more in line with the bilinear model, and then transformed. In addition, we will consider the introduction of new models to separate the estimate factors, such as non-linear model.

4 Conclusions

Identification of computer printed files is a new research domain, and it is quite difficult to identify the printed files in conditions that there are a few same characters in the test and training documents. For this problem, we proposed the solution that separated the printer factor and text factor based on factor analysis model. We use the model based on the feature matrix bilinear to finish the separation and extract approximate text-independent features, then use EM algorithm to obtain maximum after posterior probability to classify such samples, which improves the correct rate significantly in the above conditions.

Acknowledgments. We want to thank the help from the researchers and engineers from MicroPattern Corporation. This work is supported partially by China Postdoctoral Science Foundation (No: 2015M582355), the Doctor Scientic Research Start project from Hubei University of Science and Technology (No: BK1418) and the Team Plans Program of the Outstanding Young Science and Technology Innovation of Colleges and Universities in Hubei Province (T201513).

References

1. Chun-tao, C., Lei, P.: The hiden information acquisition of Shile DocuColor laser printer. J. Jiangsu Police Officer Coll. **21**(6), 145–148 (2006)
2. Huang, S., Wu, J.K.: Optical watermarking for printed document authentication. IEEE Trans. Inf. Forensics Secur. **2**(2), 164–173 (2007)
3. Steenblik, R.A., Hurt, M.J., Knotts, M.E.: Advantages of micro-optics over holograms for document authentication. In: Processing of Optical Security and Counterfeit Deterrence Techniques IV. SPIE, San Jose (2002)
4. Gebhardt, J.: Document Authentication using Printing Technique Features. University of Kaiserslautern, Germany (2012)
5. Oliver, J., Chen, J.: Use of signature analysis to discriminate digital printing technologies. In: Processing of International Conference on Digital Printing Technologies. Society for Imaging Science and Technology, San Diego, September 2002
6. Mikkilineni, A.K., Khanna, N., Delp, E.J.: Forensic printer detection using intrinsic signatures. SPIE **7880**, 24–35 (2011)
7. Mikkilineni, A.K., et al.: Printer identification based on graylevel co-occurrence features for security and forensic applications. In: Proceedings of SPIE-IS and T Electronic Imaging - Security, Steganography, and Watermarking of Multimedia Contents VII, pp. 430–440 (2005)
8. Mikkilineni, A.K., Khanna, N., Delp, E.J.: Texture based attacks on intrinsic signature based printer identification. In: Proceedings of the SPIE International Conference on Media Forensic and Security, San Jose, CA, USA (2010)
9. Pei-Ju, C., Allebach, J.P., Chiu, G.T.: Extrinsic signature embedding and detection in electrophotographic halftoned images through exposure modulation. IEEE Trans. Inf. Forensics Secur. **6**(3), 946–959 (2011)

10. Chiang, P.-J., Khanna, N., Mikkilineni, A.K., Segovia, M.V., Allebach, J.P., Chiu, G.T., Delp, E.J.: Printer and scanner forensics: models and methods. In: Sencar, H.T., Velastin, S., Nikolaidis, N., Lian, S. (eds.) Intelligent Multimedia Analysis for Security Applications. SCI, vol. 282, pp. 145–187. Springer, Heidelberg (2010)
11. Tsai, M.J., Liu, J., Wang, C.S., et al.: Source color laser printer identification using discrete wavelet transform and feature selection algorithms. In: Processing of IEEE International Symposium on Circuits and Systems, pp. 2633–2636 (2011)
12. Akao, Y., Yamamoto, A., Higashikawa, Y.: Improvement of inkjet printer spur gear teeth number estimation by fixing the order in maximum entropy spectral analysis. In: Sako, H., Franke, K.Y., Saitoh, S. (eds.) IWCF 2010. LNCS, vol. 6540, pp. 101–113. Springer, Heidelberg (2011)
13. Ning, W., Guoqiang, H., Guosheng, G.: Printer identification based on computer fuzzy recognition of character image. Appl. Res. Comput. 7(3), 953–956 (2008)
14. Wei, D., Luo, X.Q., Yan, Y.C., et al.: Printed character analysis based printed document examination. Appl. Res. Comput. 28(12), 4763–4765 (2011)
15. Yankai, T., Qinghu, C., Wei, D.: Computer laser print document identification and retrieval. J. Electron. Inf. Technol. 33(2), 499–503 (2011)
16. Qinghu, C., Wei, D., Yankai, T.: The high-magnification scanning system of the whole image. utility-model patent, patent number: ZL2009 20084691.2
17. Zhou, Q., Yan, Y., Fang, T., et al.: Text-independent printer identification based on texture synthesis. Multimedia Tools Appl., 1–17 (2015)
18. Tenenbaum, J.B.F.W.: Separating style and content with bilinear models. Neural Comput. 12(6), 1247–1283 (2000)
19. Yuchen, Y., Chen, C.H., Yuan, F., et al.: Writer identification of offline Chinese handwriting documents based on feature fusion. Pattern Recog. Artif. Intell. 32(2), 203–209 (2010)
20. Haibin, L., Qinghu, C., Yuchen, Y.: Practical face recognition via factor analysis. J. Electron. Inf. Technol. 33(7), 45–54 (2011)

Applying Batch Normalization to Hybrid NN-HMM Model For Speech Recognition

Hongjian Zhan[1(✉)], Guilin Chen[2], and Yue Lu[1]

[1] Shanghai Key Laboratory of Multidimensional Information Processing,
Department of Computer Science and Technology,
East China Normal University, Shanghai 200241, China
ecnuhjzhan@gmail.com
[2] Shanghai Youngtone Technology Co., Ltd, Shanghai, China

Abstract. Batch Normalization has showed success in image classification and other image processing areas by reducing internal covariate shift in deep network model's training procedure. In this paper, we propose to apply batch normalization to speech recognition within the hybrid NN-HMM model. We evaluate the performance of this new method in the acoustic model of the hybrid system with a speaker-independent speech recognition task using some Chinese datasets. Compared to the former best model we used in the Chinese datasets, it shows that with batch normalization we can reach lower word error rate (WER) of 8 %–13 % relatively, meanwhile we just need 60 % iterations of original model to finish the training procedure.

1 Introduction

During the past ten years, the methods of deep learning have dramatically improved the state-of-art in speech recognition, visual object recognition, image classification and many other domains such as genomics [1].

Since Hinton [2] showed the power of deep model in handwriting numerals recognition in 2006, the revolution of deep learning first sprung up in image processing areas, but soon extended to almost all artificial intelligence fields. We find that many techniques which were proposed to solve image processing problems are also effective in other domains; for instance, the convolutional neural network (CNN) architecture is first used in digital recognition but also shows strong ability in speech recognition tasks [3].

Compared to image processing, speech recognitions calculation is always larger. In order to accommodate the increasing scale of datasets, more complex and deeper architectures are proposed, which makes the training more difficult and require more efficient ways to train. Although the ability of computing card is increasing rapidly, it is not enough to adapt to the enormous datasets enlarging every minute. Especially in deep learning field, the execution time of training is an important indicator to the whole performance.

Stochastic gradient descent (SGD) is one of the most effective ways of training neural network (NN). In order to train NN more effectively, new improvements

© Springer Nature Singapore Pte Ltd. 2016
T. Tan et al. (Eds.): CCPR 2016, Part II, CCIS 663, pp. 427–435, 2016.
DOI: 10.1007/978-981-10-3005-5_35

have been came up with such as Adagrad [4] and momentum [5]. In SGD, our goal is to minimize the loss function which indicates the distance between the given labels and NNs outputs. While this method is simple, it sensitive to the models hyper-parameters. We need to be more careful with parameter initialization. Because the inputs of each layer are computed with all down layers parameters, a small change in the lower layer can be amplified when it reaches the upper layer. When we train a NN, the distribution of each layers inputs changes during training as the parameters of previous layers change, we refer to this phenomenon as internal covariate shift [6,7].

The ideal goal for automatic speech recognition (ASR) is to efficiently process inputs with various quality such as that with different channels and complicated environmental conditions. However in fact, what we can do now is much more primitive than what we want it to be. In speech recognition system, better acoustic models should be able to model a variety of acoustic variations in speaking and environment. In recent years, there has been a surge of interest in neural networks for acoustic modeling in speech recognition systems and reached the best results. Improved performance of neural network means better ability to recognize. Both deep neural network(DNN) and CNN are proposed to reach the state-of-art, and each of them has its own advantages.

The rest of the paper is organized as follows. In Sect. 2 we introduce the batch normalization method. In Sect. 3 we describe the experiments which add batch normalization to DNN and CNN in NN-HMM system and conclude the paper in Sect. 4.

2   Batch Normalization

Batch normalization is a method of data normalization; it is proposed by Sergy [7] and beat the best published result on ImageNet classification. It is mainly to solve the internal covariate shift phenomenon which indicates the change in the distribution of network activations due to the change in network parameters during training. Suppose we have a NN model with five layers, the inputs of the third layer is generated from the transformation of preceding two layers, through the training procedure, parameters in the first layer and the second layer are changed during each epoch, so even the same data feed to the NN model in different time, when it arrived at the third layer the distribution of parameters is diverse, we refer to it internal covariate shift. Obviously, the unstable input distribution of each layer will slow down the training speed by needing more epoches to reach stable state. Batch normalization is proposed to solve this problem by fixing the input distribution while training.

There is a common data preprocessing step in NN training. It has been long known [8] that the network training converges faster if its inputs are whitened. The goal of whitening is to make the input less redundant; more formally, our desiderata are that our learning algorithms sees a training input where (i) the features are less correlated with each other, and (ii) the features all have the same variance. As each layer observes the inputs produced by the layers below, it would be advantageous to achieve the same whitening of the inputs of each layer. By whitening the inputs to each layer, we would take a step towards achieving the fixed distributions of inputs that would remove the ill effects of the internal covariate shift.

Since the full whitening of each layers inputs is costly and not everywhere differentiable, we make two necessary simplifications. The first is that instead of whitening the features in layer inputs and outputs jointly, we will normalize each scalar feature independently, by making it have the mean of zero and the variance of one. In theory, parameters should be normalized after activation function, because it is the real value sent to next layer, but in fact, we do the normalization before the activation function, i.e., after the affine transformation which can be expressed by Wx+b. On account of the real effect, the initialized value of parameters are sampling from the normal Gaussian distribution, and the number of elements in W is much larger than the dimension of x, so after Wx + b each dimensions mean is closely to 0 and variance closely to 1. With this fact we do batch normalization after affine transformation before activation function can get much stable results.

For a layer with d dimensional input $x = (x_1, x_2, ..., x_d)$, we will normalize each dimension by subtract this dimensions mean and then divide the variance. Because we use SGD to train our NN, this means and variances can be computed only in the current epoch or batch, which is why this method named batch normalization. It should be noted that when the model has a parameter-shared layer like convolution layer in CNN, the means and variances are computed within the whole map.

After normalization describe above, there is not sure about the stability of parameters distribution. They add two parameters, gamma and beta, to refine the value after normalization, re-count like:

$$y = \gamma * x + \beta \qquad (1)$$

noticed that, if we make equal to the mean and equal to the variance we compute before, this transform will restore the parameters. The performance of final affect is sensitive to these two parameters, we can set them as fixed value or iteration solving during training.

For example, we train a NN with sigmoid activation function. Consider a layer $z = g(W * u + b)$ where u is the layer input, the weight matrix W and bias vector b are the layer parameters to be learned, and $g(x) = 1/(1 + \exp(-x))$. As x increases, $g(x)$ tends to zero. This means that for all dimensions of $x = Wu + b$ except those with small absolute values, the gradient flowing down to u will vanish and the model will train slowly. However, since x is affected by W, b and the parameters of all the layers below, changes to those parameters during training will likely move many dimensions of x into the saturated regime of the nonlinearity and slow down the convergence.

When we train BN module with back propagation, we not only calculate the loss function derivative of u in W * u + b, but also compute the final loss function derivative, we generate them by chain rule.

At the last training epoch, we need to count the mean and variance of all training samples in this epoch, so when test samples coming, we should normalize the test samples with the mean and variance which calculate from the training samples standard variance. In this situation, we use the unbiased estimation to generate the mean and variance.

Thanks to the stable distribution, we can apply much larger learning rate. In fact, if the scale in each layer is different, the best or required learning rate is different too. Furthermore, in most case the different dimension in each layer may need its own learning rate, so we should use the smallest learning rate to ensure the loss function decrease effectively. With batch normalization, we keep the scale consistent in every layer and every dimension, then large learning rate is applied to speed up training.

3   Experiments

To verify the performance of batch normalization in speech recognition, we have done a series of experiments on two Chinese speech datasets that we own. We add this new module to the acoustic model DNN and CNN within the Kaldi [9] framework.

3.1   Datasets

There are two datasets used in our experiments. One is 50 h long and the other is 100 h, covering both small and large datasets. We have reasons to believe that our experiments results have representativeness and this method can apply to much larger datasets.

The two Chinese speech datasets are collected in a silence environment. We named them wm50 and wm100, which represent the 50 h dataset and 100 h dataset respectively. The wm50 contains about fifty thousand sentences totally with forty thousand for training and the others for testing. The wm100 contains ninety thousand sentences, the training set and testing set are about eighty thousand and ten thousand respectively. The testing datasets in two dataset are different.

3.2   Architectures

We use both DNNs and CNNs in our experiments. Thanks to many prior knowledge, according to the size of datasets we used, we applied five layers in DNN model, the neuron number of each hidden layer was one third of the output number. In CNNs, as Tara N. Sainath [10] said, six layers with two convolutional layers followed with four full connected layers. By the way, the CNNs we used were just fed with frequency features (Fig. 1).

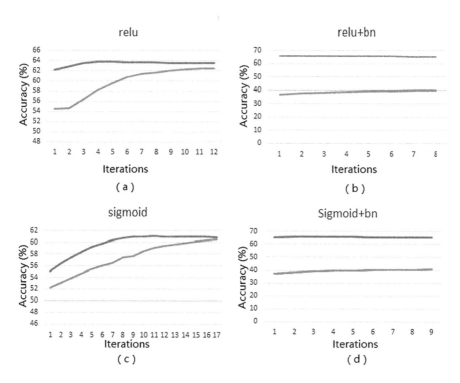

Fig. 1. The relationship between train accuracy and test accuracy with iterations, the blue line represents the train accuracy and the gray line means test accuracy. All these experiments conduct on dataset wm50: (a) is the experiment with ReLU, (b) is the experiment with ReLU and batch normalization, (c) is the experiment with Sigmoid and (d) is the experiment with Sigmoid and batch normalization (Color figure online)

3.3 Batch Normalization in DNNs

DNN plays an important role in modern speech recognition. In this group's experiments, we add batch normalization module to several excellent models, such as the best situations with different activation functions sigmoid and rectified linear unit (ReLU). The models initialization follows to the default setting in Kaldi as the weights sampling from a Gaussian distribution with small mean and variance. The features are Fbanks with delta order equals two and the splice is five. Various values of learning rate are used because as [7] says with batch normalization we can apply much higher learning rate without divergence. We also set different learning rate in the same model to show the influence to batch normalization. We test DNNs on all datasets. The results are shown in Table 1. More details are described in Subsect. 3.5.

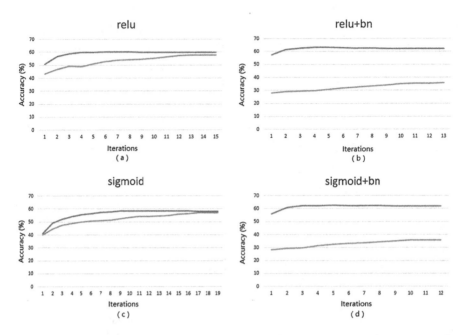

Fig. 2. The relationship between train accuracy and test accuracy with iterations, the blue line represents the train accuracy and the gray line means test accuracy. All these experiments are conducted with dataset wm100: (a) is the experiment with ReLU, (b) is the experiment with ReLU and batch normalization, (c) is the experiment with Sigmoid and (d) is the experiment with Sigmoid and batch normalization (Color figure online)

Table 1. Experiments on DNNs with batch normalization tested on several datasets with different activation functions and learning rate.

Dataset	Activation function	Model	Learning rate	wer (%)
wm50	Sigmoid	DNN+bn	0.08	**38.06**
		DNN	0.008	42.80
	ReLU	DNN+bn	0.08	**38.36**
		DNN	0.008	39.96
wm100	Sigmoid	DNN+bn	0.08	**33.68**
		DNN	0.008	39.36
		DNN+bn	0.008	38.37
	ReLU	DNN+bn	0.08	**34.41**
		DNN	0.008	35.32

3.4   Batch Normalization in CNNs

CNN is another powerful deep learning model, and also get the state-of-art in modern speech recognition, so we also test on this model to verify the performance of batch normalization.

The CNN used in our experiments has eight layers with two convolution layers followed by a pooling layer each, four full connected layers behind them. The convolution layers are different, with different initial parameters, strides and the numbers and sizes of convolution kernels.

We done experiments on wm50 and wm100. The results of CNN are shown in Table 2.

3.5   Analysis

The results of our experiments show that with BN module, both DNN and CNN can improve their performance. All new models with batch normalization beat the original model, and on both wm50 and wm100 we get the best results. Through Tables 1 and 2 we can find that the performance of batch normalization is unstable with different datasets, activation functions and models. Batch normalization can bring higher improvement with sigmoid than ReLU. In dataset wm50, we test the same model with two different learning rate, 0.008 and 0.08. The results show that just adding the new module without enlarge learning rate will get small improvement while after applying larger learning rate the performance increasing fast and get the best result. Dataset wm100 is a large scale dataset and the excellent results show batch normalization also can work well in big datasets. We can apply this new method to the actual industrial application to improve the performance.

In both experiments with DNN and CNN we find a phenomenon that with batch normalization, the train accuracy increases very fast and in several iterations it reaches more than 60 % in dataset wm100 and wm50, but the test accuracy increases slowly and the value is small than the model without batch

Table 2. Results of CNN with batch normalization module on different datasets with different activation functions and learning rate.

Dataset	Activation function	Model	Learning rate	wer (%)
wm50	ReLU	CNN+bn	0.08	**33.40**
		CNN	0.008	35.32
	Sigmoid	CNN+bn	0.08	**35.52**
		CNN	0.008	42.80
wm100	ReLU	CNN+bn	0.08	**32.90**
		CNN	0.008	33.10
	Sigmoid	CNN+bn	0.08	**33.77**
		CNN	0.008	38.50

normalization. The values of test accuracy in the two situations are 30 %, 40 % and more than 60 % respectively. Figure 2 shows the relationship between test accuracy andtrain accuracy in our several experiments. It shows that with batch normalization the test accuracy is small and increasing very slow, but the final WER is larger than models without batch normalization. This shows that the test accuracy is not proportional to the WER, perhaps it due to the complicated system in speech recognition. Compare the two figures in Fig. 2 we find that with batch normalization we just need 30 % less iterations to reach the halt condition.

4   Conclusions

In this paper, we proposed to use the batch normalization principles in automatic speech recognition task within the NN-HMM hybrid framework. The results have shown that batch normalization as a way to train NN is also effective in speech recognition tasks. Both DNN and CNN can benefit from adding batch normalization module to the network architecture. Because of limited fund, our experiments are done on Chinese datasets, and the scale of network is not large enough to fit the forefront research. In the future we will seek much large datasets to refine our method. On the same time, we need set the hyper-parameters of batch normalization carefully to get the best results.

Acknowledgment. This work is jointly supported by the Science and Technology Commission of Shanghai Municipality under research grants 14511105500 and 14DZ2260800.

References

1. LeCun, Y., Bengio, Y., Hinton, G.: Deep learning. Nature **512**, 436–444 (2015)
2. Hinton, G.E., Salakhutdinov, R.R.: Reducing the dimensionality of data with neural networks. Science **313**, 504–507 (2006)
3. Abdel-Hamid, O., Mohamed, A., Jiang, H., et al.: Applying convolutional neural networks concepts to hybrid NN-HMM model for speech recognition. In: 2012 IEEE International Conference on Acoustics, Speech and Signal Processing (ICASSP), pp. 4277–4280. IEEE (2012)
4. Duchi, J., Hazan, E., Singer, Y.: Adaptive subgradient methods for online learning and stochastic optimization, **12**, 2121–2159. JMLR.org (2011)
5. Sutskever, I., Martens, J., Dahl, G., Hinton, G.: On the importance of initialization and momentum in deep learning. In: Proceedings of the 30th International Conference on Machine Learning (ICML-13), pp. 1139–1147 (2013)
6. Shimodaira, H.: Improving predictive inference under covariate shift by weighting the log-likelihood function, **90**, 227–244. Elsevier (2000)
7. Ioffe, S., Szegedy, C.: Batch normalization: accelerating deep networktraining by reducing internal covariate shift (2015). arXiv preprint arXiv:1502.03167
8. Wiesler, S., Ney, H.: A convergence analysis of log-linear training. In: Advances in Neural Information Processing Systems, pp. 657–665 (2011)

9. Povey, D., Ghoshal, A., Boulianne, G., Burget, L., Glembek, O., Goel, N., Hannemann, M., Motlicek, P., Qian, Y., Schwarz, P., et al.: The kaldi speech recognition toolkit. In: IEEE 2011 Workshop on Automatic Speech Recognition and Understanding, no. EPFL-CONF-192584. IEEE Signal Processing Society (2011)
10. Sainath, T.N., Mohamed, A.R., Kingsbury, B., Ramabhadran, B.: Deep convolutional neural networks for LVCSR, pp. 8614–8618 (2013)

Preprocessing Algorithm Research of Touchless Fingerprint Feature Extraction and Matching

Kejun Wang[1], Jinyi Jiang[1], Yi Cao[1], Xianglei Xing[1(✉)], and Rongyi Zhang[2]

[1] College of Automation, Harbin Engineering University, Harbin 150001, China
xingxl@hrbeu.edu.cn
[2] College of Mechanical and Electrical Engineering, Heilongjiang Institute of Technology, Harbin 150050, China

Abstract. Touchless fingerprint recognition with high acceptance, high security, hygiene advantages, is currently a hot research field of biometrics, but because of the different image principle of the non-contact fingerprint image and contact fingerprint image, the difference of the two fingerprint image is large. There are still a small number of fuzzy regions in the non-contact fingerprint image after pretreatment, and the traditional method of extracting the future from the detail points can lead to a serious decline in recognition accuracy because of false points. In this paper, the non-contact pretreatment in our laboratory is used according to the characteristics of the contactless fingerprint image, the LBP operator, LGC operator and their improve algorithms are used for image processing; the nearest neighbor classifier is used for feature matching. The experimental result shows that the contactless fingerprint feature extraction method proposed in this paper can obtain higher division fingerprint feature.

Keywords: Touchless fingerprint · Improved LBP · Improved LGC

1 Introduction

With its high practicability and feasibility, fingerprint identification technology has become the most common and legally binding biometric technology. Even so, since the fingerprint image is collected by a touchable optical sensor or capacitive sensor, there are still many problems in the conventional fingerprint identification like fingerprint deformation, fingerprint residue, sensitive to skin conditions and the spread of germs at the time of collection, etc. However, identifying fingerprint in a touchless way, where the surface of the fingers is not in direct contact with the camera or sensing element when collecting a fingerprint image, can not only eliminate these negative factors, but also has high recognition performance and anti-counterfeiting performance. Therefore, the touchless fingerprint recognition technology has become a promising research direction in the field of fingerprint identification [1].

Considering touchless fingerprint identification relies mainly on shadows which are produced by the ridge line when a light source illuminates the surface of the finger to get the fingerprint image. The background areas of touchless fingerprints are more complex than those of the contact. Fingerprint image will appear rotation and translation

© Springer Nature Singapore Pte Ltd. 2016
T. Tan et al. (Eds.): CCPR 2016, Part II, CCIS 663, pp. 436–450, 2016.
DOI: 10.1007/978-981-10-3005-5_36

phenomenon. What's more, the contrast of the ridge and valley lines is much lower. So, it is very important to find a touchless fingerprint image preprocessing algorithm. Many scholars have studied touchless fingerprint recognition. The shape-from-silhouette method can get a three-dimensional fingerprint image with five cameras from multiple perspectives, and then expand it into an equivalent two-dimensional fingerprint image [2]. R. Donida Labati achieved the fast 3-D fingertip reconstruction using a single two-view structured light acquisition and identified the fingerprint image by expanding fingerprint images [3]. LEE et al. proposed a clustering-based dynamic score selection (CDSS) algorithm for the combination of scores which are generated by different vision touchless fingerprint recognition systems [4]. And found that the performance of CDSS-based multi-vision touchless system can be enhanced efficiently compared to touchless fingerprint recognition and better than those of sum, max, SVM and Fisher linear discrimination algorithms. Choi Hand Kim J proposed a new touchless fingerprint sensing device capturing three different views at one time and a method for mosaicking these view-different images [5]. They can get a high-quality fingerprint template to solve problems caused by a touch-based sensing device such as a view difference problem and a limited usable area due to perspective distortion and rotation. However, the device, including a single camera and two planar mirrors reflecting side views of a finger, is large and expensive and the fingerprint acquisition method is complex, so its application scope is limited.

Many scholars conduct a study of touchless fingerprint identification technology based on simple acquisition devices in order to overcome these difficulties. V. Piuri and F. Scotti get the fingerprint image with a webcam in natural light conditions [6], and then extract the ridge structure characteristics by composite image processing method. Chulhan Lee gets the fingerprint image with mobile cameras and delved into a low contrast between ridges and valleys [7]. Ajay Kumarand Cyril Kwongdeveloped a 3-D fingerprint identification system that uses only single camera and a new representation of 3D finger surface features using Finger Surface Codes which is very effective in matching three dimensional fingerprints [8]. M. Kokou Assogba presented a contactless fingerprint system based on supervised contactless image acquisition with only an ordinary camera and a ridge minutiae extraction method based on orientation computation [9]. This kind of method, using the cheap fingerprint acquisition device and simple fingerprint acquisition method, has broad prospects. However, the image-forming principle of the contactless fingerprint images is different from the contact one, and its fingerprint image quality is relatively poor. The recognition algorithm's effect of touchless fingerprint recognition is so poor now, and the research on touchless fingerprint image recognition algorithm is still not enough.

Fingerprint feature extraction and matching is one of the most important parts of the fingerprint identification, which is directly related to the processing speed and recognition accuracy of fingerprint identification system. The touchless fingerprint identification system in our lab relies mainly on the differences between the finger area formed in blue light, and we can hardly get an ideal effect with the traditional contact fingerprint feature extraction method.

Firstly, in our lab, a fingerprint sampling device, which is made up of common CMOS cameras, is adopted under blue light. The device is so simple that it has a promising prospect. Secondly, a method is put out to preprocess the images reasonably aiming at touchless fingerprint features. The Otsu based on the Cb component of the YCbCr model is adopted to extract the finger area. This method does well in the condition of blue light. Thirdly, when the fingerprint images are enhanced, combining the high frequency emphasis filtering and iterative adaptive histogram equalization technique is adopted firstly and then the simplified Gabor function template is used to enhance them again. Fourthly, due to the phenomena such as fingerprint rotation, translation and edge blurring caused by factors like the touchless acquisition method and low camera depth of field, a new method of extracting the ROI fingerprint area is adopted. Lastly, this paper analyzes the local binary pattern and the local gradient coding algorithm. In the paper, the asymmetric region local binary pattern and asymmetric region local gradient coding algorithm is adopted, and the nearest neighbor classifier based on the chi-square distance is used for feature matching.

2 Feature Extraction of Touchless Fingerprint

The size of fingerprint image collected by the finger fingerprint acquisition system in our lab is 1280×720 pixels, as is in Fig. 1. The processed image, as is in Fig. 2, is used for feature extraction.

Fig. 1. Fingerprint image collected

Fig. 2. Fingerprint image after pretreatment

2.1 Basic Theory of LBP

In a window of 3×3, the threshold is the gray value of the center pixel. Compare the gray scale of its eight neighboring regions. Set the pixel to 1 when its gray value is less than the threshold and set it to 0 when higher than the threshold. After getting an eight bit binary number of each window area, each region is assigned a weighting $2^i (i = 0, 1 \ldots 7)$ according to clockwise direction or counterclockwise direction, and it

by the corresponding binary number. Then get the eigenvalue of LBP by converting the binary number, which is the summation of 8 regions, into a decimal number (Fig. 3).

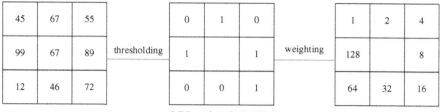

LBP code: 01011001
LBP eigenvalue =2+8+16+128=154

Fig. 3. The encoding process of LBP

LBP operator has advantages like a simple principle, convenient operation and good capacity for texture description of image features. However, there are still many problems in specific application. After analyzing the theory and experiments of LBP operator, the problems existing in the LBP operator are listed as follows:

1. LBP operator has too many binary eigenvalue encoding modes. The LBP eigenvalue of a 3×3 window area is an eight bit binary number. There are $2^8 = 256$ texture modes; however, not all of them can describe the texture feature of the image well. The texture mode, which is useless for describing the image, can be abandoned since too many characteristic patterns will weaken the resolution capability of LBP texture description.
2. LBP operator has the bad robustness to noise and exquisite illumination. Since LBP code is the comparison of two single pixels, LBP codes of adjacent pixels are interrelated. Exquisite illumination and noise will generate different LBP patterns.
3. The sampling of LBP is not stable. Generally, the LBP operator sampling of the local area is sparse. Although the double linear differential method is adopted to calculate the gray value of the neighborhood points which didn't get mapped to a pixel in the image, there is still instability.
4. When using LBP histogram as a feature vector, there are too many LBP operator binary modes. The dimension of the histogram is too high, which increases the amount of calculation. And there are too little feature points of each mode, which has lost its statistical significance and has a bad effect on recognition rate.
5. LBP operator shows the texture pattern between neighborhood regions without considering the relationship between local texture features. To some degree, the LBP has certain limitations in complex texture processing.

Therefore, the study of LBP can start from the above parts, and make sure to improve at least one aspect. Combining with practical application, we also need to improve the LBP operator to get the texture information with distinctive characteristics and solve specific problems.

2.2   Uniform LBP

In practical applications, the histogram has the translation invariance, which means that the similarity between the two histograms shows the similarity between two images. So LBP histogram is widely used as a feature vector in classification recognition. The image is usually separated into several blocks, and then the LBP histogram of the image is made up of histogram of each block in series with each other. The binary pattern of $LBP_{P,R}$ and the amount of calculation increases a lot when the radius is bigger and there are more sample points. However, there are less pixels in the block areas. We get a useless histogram which is too sparse. To solve this problem, Ojala came up with an improved LBP operator, uniform LBP. The LBP encoding binary is head-tail linked into a binary circle. Define U as the times the binary circle change between 0 and 1. The LBP pattern is called uniform LBP when U is less than 3.

$$U(LBP_{P,R}) = |s(g_{P-1} - g_c) - s(g_0 - g_c)| + \sum_{i=1}^{P-1} |s(g_i - g_c) - s(g_{i-1} - g_c)| \quad s(x) = \begin{cases} 1, x \geq 0 \\ 0, x < 0 \end{cases} \quad (1)$$

We can know whether the binary pattern G_P is a uniform pattern with $U(LBP_{P,R})$ in formula (1). Get the number of conversion between 0 and 1 by the sum absolute value of the subtract of G_P and its bitwise binary number. It is a uniform pattern if $U(G_P) \leq 2$, or G_P is a non-uniform pattern. The number of uniform patterns is $P(P-1) + 2$ when the number of sampling points is P, and count other non-uniform patterns as one pattern. As a result, there are $P(P-1) + 3$ patterns for P sampling points. In a window area of 3×3 with 8 sampling points, there are 58 uniform patterns, which is 23 % of that of the original binary mode. The uniform method reduces the dimension of LBP operator effectively. As is shown in Fig. 4, the dimension of original histogram is 256 and the dimension of uniform LBP histogram is 59, which greatly reduced the histogram feature dimension.

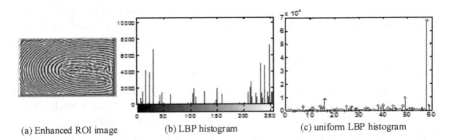

(a) Enhanced ROI image   (b) LBP histogram   (c) uniform LBP histogram

Fig. 4.   The LBP histogram of fingerprint image

2.3   Asymmetric Region Local Binary Pattern

At present, there are a lot of improvements of LBP operator have been carried out. However, these improved LBP operators perform badly in reducing the length of characteristic vector of different images and is weak in identification capability and robustness. For example, the basic LBP is not extendable [9] and the improved ELBP is extendable [10], but the histogram dimension increases with the increase of sampling points. MB-LBP [11] is extendable and has a constant dimensional histogram, however, the resolving ability decreases with the increase of block size. Therefore, we find a new method called Asymmetric Region Local Binary Pattern to extract features of fingerprint images. This operator is extendable and has a constant vector length, what's more, it has good distinguish ability. The AR-LBP operator inherited the advantages of LBP, ELBP and MB-LBP.

AR-LBP operator, which is extendable, can capture the main features in a larger scale. The problem about the length of the feature histogram is solved and the identify ability of AR-LBP is much better than the original one. The length of the histogram of AR-LBP is only related to the number of blocks divided. The AR-LBP operator is composed of sub-blocks of different sizes, which lowers the loss of texture information. Since the pixel average gray scale calculations for different sub-blocks can increase the identification ability of the operator, summed-area tables are adopted to efficiently calculate the average gray scale of AR-LBP sub-blocks.

In Fig. 5, AR-LBP operator consists of 9 regions. The 8 peripheral regions $R_i(i = 1 \ldots 8)$ are marked and the middle one is not. The size of region1, region3, region5 and region7 changes in the horizontal and vertical direction. The size of region2 and region6 changes in vertical direction. The size of region4 and region8 changes in horizontal direction. The middle region doesn't change. The size of AR-LBP operator is $(2m + 1) \times (2n + 1)$, including four $m \times n$ areas, two $1 \times m$ areas, two $n \times 1$ areas and a 1×1 rectangular area. Where m is the height and n is the width of the region. When m = n=1, the AR-LBP operator is nearly equal to the basic LBP operator.

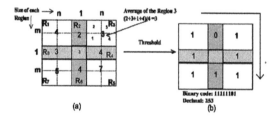

(a)   (b)

Fig. 5. The encoding process of AR-LBP. (a) shows the average gray scale in each area except the center area, and (b) is the AR-LBP encoding process of an image sized 5 × 5. The gray-scale of the center point is 3 and the decimal AR-LBP of this image is 253.

The AR-LBP of point with pixel value (x_c, y_c) in decimal is showed in formula (2):

$$AR - LBP(x_c, y_c) = \sum_{i=1}^{8} s(a_i - a_c) \times 2^i \qquad (2)$$

Where, a_i is the average pixel value of $R_i(i = 1 \ldots 8)$, a_c is the pixel value of the center region.

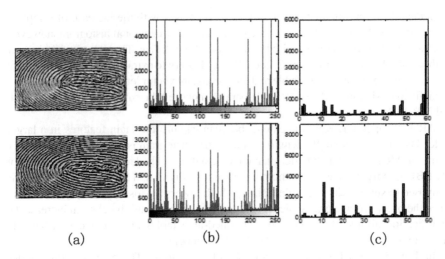

Fig. 6. The diagram and histogram of AR-LBP. (a) The response diagram of region sized 5 × 5 and 7 × 7 of Fig. 4(a). (b), (c) The histogram and uniform histogram of the image.

2.4   Local Gradient Coding

The traditional LBP considered only about the relationship between the center pixel and surrounding pixels, but not the relation of the neighborhood pixels. In Fig. 7, there are two gray-scale images. The LBP operator code shows the same result, but obviously they are not the same. So literature 13 proposed the Local Gradient Codingoperator [12]. In Fig. 8, the neighborhood size of LGC operator is 3 × 3. LGC operator compares the gray-scale between the neighborhood from horizontal, vertical and diagonal gradient. Then the binary code can be converted into a decimal number. The LGC statistical histogram series of each sub-block constitute the whole fingerprint image. Formula (3) shows the algorithm:

$$
\begin{aligned}
LGC(x,y) = {} & s(g_0 - g_2) \times 2^7 + s(g_7 - g_3) \times 2^6 + s(g_6 - g_4) \times 2^5 + s(g_0 - g_6) \times 2^4 \\
& + s(g_1 - g_5) \times 2^3 + s(g_2 - g_4) \times 2^2 + s(g_0 - g_4) \times 2^1 + s(g_2 - g_6) \times 2^0
\end{aligned}
\tag{3}
$$

1	5	5
1	6	5
1	1	5

(a)

1	1	1
1	6	1
1	1	1

(b)

LBP: $(00000000)_2$

LGC: $(00001001)_2$ LGC: $(00000000)_2$

Fig. 7.   Two images with different gray-scale

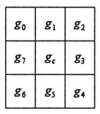

Fig. 8. LGC gray-scale of 3*3

Formula (4) shows S(x) in formula (3):

$$s(g_f - g_c) = \begin{cases} 1, g_f > g_c \\ 0, g_f \le g_c \end{cases} \tag{4}$$

The LBP code of the two images in Fig. 7 is $(00000000)_2$, however, the gray-scale relationship between 8 neighborhood areas in (a) is different from (b). LGC code of image (a) is $(00001001)_2$ and the LGC code of image (b) is $(00000000)_2$, which is different from the first one. Compared with the traditional LBP operator, LGC operator has better identification capability. Figure 9 shows the histogram of Fig. 4(a) after the LGC transformation.

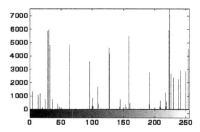

Fig. 9. LGC histogram

Considering the length of LGC operator code is 8, if the image is divided into 8 × 8 sub-blocks, the dimension of the image will increase to 8 × 8 × 256 = 16384, which will greatly affect the speed of recognizing. On the other hand, scholars have deeply theoretical proof about the vertical gradient of a fingerprint, which contains less information [13, 14].

The paper further optimized LGC operator, and get a LGC-HD operator considering only horizontal and diagonal gradient. This operator not only reduced the feature dimension to extract the main texture information, but also improved the recognition rate by reducing useless calculation. Formula (5) shows the algorithm:

$$LGC - HD_p^R = s(g_0 - g_2) \times 2^4 + s(g_7 - g_3) \times 2^3 + s(g_6 - g_4) \times 2^2 + s(g_0 - g_4) \times 2^1 + s(g_2 - g_6) \times 2^0 \tag{5}$$

But the size of the neighborhood area of the LGC-HD operator is fixed, which can hardly extract the texture features well, and is greatly influenced by noise. Figure 10 shows the histogram of Fig. 4(a) after the LGC-HD transformation.

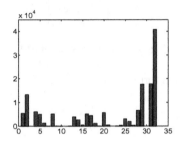

Fig. 10. LGC-HD histogram

2.5   Asymmetric Region Local Gradient Coding

LGC operator takes the gray-scale relationship between the neighborhoods around the center pixel into consideration, however, the size of the LGC operator neighborhood is fixed and cannot be extended. It can hardly extract good texture feature in a large scale. So inspired by the AR-LBP operator, this paper proposes a new descriptor called asymmetric region local gradient coding descriptor. The size of this descriptor is $(2m + 1) \times (2n + 1)$, where $1 \leq m \leq \dfrac{h - 1}{2}$ and $1 \leq n \leq \dfrac{w - 1}{2}$. H is the height and w is the width of the image. Symbol $\lfloor \rfloor$ means round down the number. As is shown in Fig. 11, the region is divided into 9 neighborhoods and each sub-neighborhood is named R_i. The pixel value of each region is $g_{R_i}, i = (1, 2 \ldots 9)$, as is shown in the formula (6).

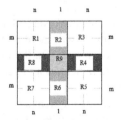

Fig. 11. Regions of AR-LGC

In Fig. 11, R_9, sized 1×1, is the center pixel point. And the size of R_1, R_3, R_5 and R_7 is m × n. The size of R_2 and R_6 is m × 1. The size of R_4 and R_8 is 1 × n.

$$g_{R_i} = \frac{1}{N_i} \sum_{j=1}^{N_i} p_{ij} \tag{6}$$

Formula (6) shows the pixel value of each region, where N_i is the number of pixel points in R_i and P_{ij} is the pixel value of pixel j in R_i.

AR-LGC descriptor compares the relationship between neighborhood from three horizontal gradients, three vertical gradients and two diagonal gradients. Get the pixel value relationship between neighbor regions in turn and generate an eight-bit binary code. The texture description value of the local neighborhood is a decimal number converted from the binary code. The binary sequence of Fig. 11 is shown as below:

$$P_0: s\left(g_{R_1} - g_{R_3}\right) \; P_1: s\left(g_{R_8} - g_{R_4}\right) \; P_2: s\left(g_{R_7} - g_{R_5}\right) \; P_3: s\left(g_{R_1} - g_{R_7}\right)$$
$$P_4: s\left(g_{R_2} - g_{R_6}\right) \; P_5: s\left(g_{R_3} - g_{R_5}\right) \; P_6: s\left(g_{R_1} - g_{R_5}\right) \; P_7: s\left(g_{R_3} - g_{R_7}\right)$$

Where S (x) is shown in the formula (4), formula (7) is the coding of AR-LGC:

$$AR - LGC(x, y) = \sum_{i=0}^{7} P_i \times 2^i \tag{7}$$

The AR-LGC descriptor has extensibility, which can extract rich texture features in large scales with a constant dimension of features.

Figure 12 is the histograms of the image in Fig. 6(a) sized 5×5, and Fig. 13 is the histograms of the image in Fig. 6(a) sized 7×7. As we can see in the diagrams, the dimension of the histogram using AR-LGC-HD operator is much less than the one using AR-LGC operator.

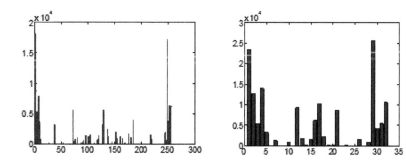

Fig. 12. Histogram of AR-LGC and AR-LGC-HD sized 5×5

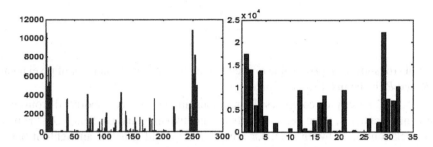

Fig. 13. Histogram of AR-LGC and AR-LGC-HD sized 7×7

2.6 Fingerprint Feature Representation

Figure 14 shows the generation process of fingerprint features based on LBP or LGC operator. The specific steps are as follows:

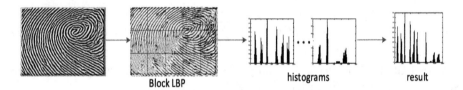

Fig. 14. Generation process of fingerprint features

Step 1. Normalize the preprocessed fingerprint images into 150×250 pixels.

Step 2. Divide the preprocessed image into m $\times$ n blocks and extract the texture features of each block by LBP or LGC algorithm. Cascade the histograms of these blocks to get the features of the whole image. Assume that there are m blocks in each ROI image, and name each block as $r_1, r_2 \ldots r_m$. The corresponding histogram of each block is $H_{r_i}, i = 0 \ldots, m$. As is shown in formula (8), symbol $\Re$ is a feature vector in a series of histograms of each block.

$$\Re = \left(H_{r_0}, \ldots, H_{r_m}\right) \tag{8}$$

Formula (9) is the histogram calculation of an image, of which the range of pixel value is [0, L].

$$t_i = \sum_x f(I(x) = i), i = 0, 1, \ldots, L \qquad f(x) = \begin{cases} 1, I(x) = i \\ 0, I(x) \neq i \end{cases} \tag{9}$$

Where, i is gray level i and t_i is the pixel number of gray level i. I(x) is the gray value of pixel x in the image. Formula (10) defines the histogram descriptor H:

$$H = \left\{t_i | i = 0, 1, \ldots, L\right\} \tag{10}$$

3   Fingerprint Feature Matching

The response of the image using a LBP encoding only changed in the pixel value, because the decimal converted from the binary code extracted from the image is still the gray value of this image. However, LBP response images are not commonly used as a feature vector in recognition. After the histograms of each block are concatenated together to get the LBP feature histogram, we use it as a feature vector to calculate the similarity. Classify the images according to the nearest distance rule. Three commonly used methods to measure the distance between histograms are as follows:

1. Histogram intersection

$$D(S, M) = \sum_i \min(S_i, M_i) \tag{11}$$

2. Log-likelihood Statistic

$$L(S, M) = -\sum_i S_i \log M_i \tag{12}$$

3. Chi Square Statistic

$$\chi^2(S, M) = \sum_i \frac{(S_i - M_i)^2}{S_i + M_i} \tag{13}$$

Where S is the sample image and M is the registered image. S_i and M_i is the χ^2 bin value of the range i in the corresponding histogram of the two images. The commonly used Chi-square statistic calculates the similarity of two histograms. The images are similar when the χ^2 value is smaller, which means the distance between the two histograms is smaller. Define the sample as the category with minimum χ^2 value according to the nearest distance rule after calculating the χ^2 of the sample and all registered images.

4   Analysis of Experimental Results

This paper adopted preprocessed ROI images, including 105 fingers, 5 images of each finger. Extract the features with LBP operator, LGC operator and their improved operators. After matching the features by chi-square distance, compare the performance of basic LBP, uniform LBP, AR-LBP, LGC, LGC-HD, AR-LGC and AR-LGC-HD to find the best algorithm. Analyze the influence of the size of the neighborhood and the number of sub-blocks to fingerprint identification with the AR-LGC operator.

4.1   The Size of AR-LGC Neighborhood

Because the AR-LGC neighborhood size is average, which cannot be too big or too small. A big one cannot show the pixel change of the pixel point while too small neighborhood is too sensitive to noise. The paper analyzed the influence of different size of AR-LGC neighborhood based on the self-built database in our lab, and the experimental results are shown in Table 1, where H is the height and W is the width of sub-neighborhood area.

Table 1. Recognition rate of different neighborhood size

H \ W	3	5	7	9
3	90.58	91.53	89.18	88.21
5	91.96	91.60	92.46	87.76
7	89.64	92.47	93.81	84.01
9	83.04	91.06	88.22	90.11

As is shown in Table 1, the recognition rate increases when the height and width increases, but after reaching a certain size, it declines. The maximum recognition rate is 93.81 % when the size is 7 × 7.

4.2 Recognition Rate of Different Number of Blocks

Table 2 shows the experimental result of the influence of different block number in the AR-LGC algorithm.

Table 2. Recognition rate of different blocks

Number of blocks	1 × 1	3 × 3	5 × 5	7 × 7	9 × 9
Recognition rate	85.72	89.33	93.81	91.51	90.80

As we can know in Table 2, the highest recognition rate is 93.81 when there are 25 blocks.

Experiments of AR-LBP operator and AR-LGC-HD operator shows that the recognition rate is the highest when there are 25 blocks sized 7 × 7.

4.3 LGC-HD Algorithm

The paper compared the performance of 7 LGC operators to show the validity of the LGC-HD algorithm. LGC considers all directions. LGC-HD considers horizontal and diagonal directions. LGC-HV considers horizontal and vertical directions. LGC-VD considers vertical and diagonal directions. LGC-H considers only horizontal direction. LGC-V considers the vertical direction only. LGC-D considers only diagonal direction. The result is listed in Table 3.

Table 3. Performance of LGC in different directions

Algorithm	LGC	LGC-HD	LGC-VD	LGC-HV	LGC-H	LGC-V	LGC-D
Recognition rate	90.58	91.6	86.67	88.33	82.22	85	88.89
Recognition time(ms)	200.19	14.32	14.47	14.50	2.18	2.17	1.30

As Table 3 shows, the LGC-HD operator has better descriptive ability. Also, LGC-HD operator has only 32 patterns while LGC has 256 patterns. It has faster operation speed. So the LGC-HD operator is very suitable for feature extraction for fingerprint identification.

4.4   Performance of Different Algorithm

Tables 4 and 5 Shows the Performance of LBP, LGC, LGC-HD, AR-LBP, AR-LGC and AR-LGC-HD in the Same Experimental Environment

Table 4. Recognition rate of different algorithms

Algorithm		Uniform LBP	LBP	LGC	LGC-HD	AR-LBP	AR-LGC	AR-LGC-HD
Recognition rate (%)	First	90.82	89.64	90.58	91.6	91.53	92.84	93.81
	Second	94.06	93.17	93.39	94.33	94.66	96.67	96.8

Table 5. Performance of different algorithms

Algorithm	Uniform LBP	LBP	LGC	LGC-HD	AR-LBP	AR-LGC	AR-LGC-HD
Feature dimension	1475	6400	6400	800	6400	6400	800
Feature extraction time (ms)	320.9	310.7	340.8	150.5	450.6	460.5	262.5
Recognition time(ms)	42.6	195.4	200.2	14.32	194.6	195.7	14.9

In Tables 4 and 5, the recognition rate based on the database in our lab is high, which, to some extent, satisfied the requirement of real-time. Above all, the AR-LGC-HD algorithm has lower feature extraction time and recognition time and higher recognition rate. The experiment in this paper proved that the AR-LGC-HD algorithm is very appropriate for fingerprint identification.

5   Conclusion

In this paper, the contactless fingerprint feature extraction and matching method are analyzed and compared. The paper includes the analysis and utilization of LBP operator, LGC operator and their improving methods for fingerprint image feature extraction. The nearest neighbor rule of Chi-square is used for feature matching. The analysis of experimental results based on the fingerprint database in our lab shows that the block AR-LGC-HD operator can not only get a higher recognition rate, but also reduce computation time and storage space occupancy, which is the fundament of the following feature-level fusion and multi-modal identification.

Acknowledgments. This work was supported by the Fundamental Research Funds for the Central Universities of China, Natural Science Fund of Heilongjiang Province of China, and Natural Science Foundation of China, under Grand No HEUCF160415, F2015033, and 61573114.

References

1. Jain, A.K., Ross, A., Prabhakar, S.: An introduction to biometric recognition. IEEE Trans. Circ. Syst. Video Technol. **14**(1), 4–20 (2004)
2. Nandini, C., RaviKumar, C.N.: An approach to gait recognition. In: International Symposium on Biometrics and Security Technologies (ISBAST 2008), pp. 1–3 (2008)
3. Deng, P., Liao, H., Ho, C., Tyan, H.: Wavelet-based off-line handwritten signature verification. Comput. Vis. Image Underst. **76**(5), 173–190 (1999)
4. Hosseinzadeh, D., Krishnan, S.: Gaussian mixture modeling of keystroke patterns for biometric applications. IEEE Trans. Syst. Man Cybern. Part C Appl. Rev. **38**(6), 816–826 (2008)
5. Dhruva, N., Rupanagudi, S.R., Sachin, S.K., Sthuthi, B.: Novel segmentation algorithm for hand gesture recognition. In: International Multi-conference on Automation, Computing, Communication, Control and Compressed Sensing (iMac4 s), 383–388 (2013)
6. Li, S., Jain, A.: I NetLibrary. In: Handbook of Face Recognition. Citeseer (2005)
7. Kafai, M.: Reference face graph for face recognition. IEEE Trans. Inf. Forensics Secur. **9**(12), 2132–2143 (2014)
8. Imtiaz, H., Fattah, S.A.: A wavelet-based dominant feature extraction algorithm for palmprint recognition. Digital Sig. Process. Rev. J. **23**(1), 244–258 (2013)
9. Ojala, T., Pietikainen, M., Maenpaa, T.: Multiresolution gray-scale and rotation invariant texture classification with local binary patterns. IEEE Trans. Pattern Anal. Mach. Intell. **24**(7), 971–987 (2002)
10. Ahonen, T., Hadid, A., Pietikainen, M.: Face description with local binary patterns: application to face recognition. IEEE Trans. Pattern Anal. Mach. Intell. **28**(12), 2037–2041 (2006)
11. Liao, S., Zhu, X., Lei, Z., Zhang, L., Li, S.Z.: Learning multi-scale block local binary patterns for face recognition. In: Lee, S.-W., Li, S.Z. (eds.) ICB 2007. LNCS, vol. 4642, pp. 828–837. Springer, Heidelberg (2007)
12. Tong, Y., Chen, R., Cheng, Y.: Facial expression recognition algorithm using LGC based on horizontal and diagonal prior principle. Optik-Int. J. Light Electron Opt. **125**(16), 4186–4189 (2014)
13. Xu, J., Zhang, Y.J.: Expression recognition based on variant sampling method and Gabor features. Comput. Eng. **18**, 67 (2011)
14. Huang, D., Shan, C., Ardabilian, M., et al.: Local binary patterns and its application to facial image analysis: a survey. IEEE Trans. Syst. Man Cybern. Part C Appl. Rev. **41**(6), 765–781 (2011)

Violent Scene Detection Using Convolutional Neural Networks and Deep Audio Features

Guankun Mu, Haibing Cao, and Qin Jin[✉]

Multimedia Computing Lab, School of Information,
Renmin University of China, Beijing, China
{muguankun,coastchb,qjin}@ruc.edu.cn

Abstract. Violent scene detection (VSD) in videos has practical significance in various applications, such as film rating and child protection against violent behavior. Most of previous VSD systems have mainly used visual cues in the video although acoustic or audio cues can also help to detect violent scenes especially when visual cues are not reliable. In this paper, we focus on exploring acoustic information for violent scene detection. Convolutional Neural Networks (CNNs) have achieved the state-of-the-art performance in visual content processing tasks. We therefore investigate using CNNs for violent scene detection based on acoustic information in videos. We apply CNNs in two ways: as a classifier directly or as a deep acoustic feature extractor. Experimental results on the MediaEval 2015 evaluation dataset show that CNNs are effective both as classifiers and as acoustic feature extractors. Furthermore, fusion of acoustic and visual information significantly improves violent scene detection performance.

Keywords: Violent scenes detection · Convolutional neural networks · Video content analysis

1 Introduction

The huge amount of videos on the Internet has attracted great research interests in video content analysis, such as event detection and concept detection [1–3]. Violent scene detection, which is one type of video content analysis, aims to detect the segments with physical violence or violent intentions in videos. There has been a benchmark in the MediaEval Benchmarking Initiative for Multimedia Evaluation (MediaEval) [4], which offers a platform for researchers to evaluate new methods for multimedia retrieval. The VSD evaluation task in MediaEval 2015 defines violent videos as "those one would not let an 8 years old child see because of their physical violence".

Obviously, rich visual information from videos can be used for violent scene detection, as most violent scenes are associated with certain objects and actions, such as blood, guns, and fights and etc. Audio track also contains much information that visual cues cannot represent. For example, screaming, explosions,

© Springer Nature Singapore Pte Ltd. 2016
T. Tan et al. (Eds.): CCPR 2016, Part II, CCIS 663, pp. 451–463, 2016.
DOI: 10.1007/978-981-10-3005-5_37

words of abuse and even thrilling music or sound effects are all violence according to the previous definition. They can be easily recognized from the audio. Some violent elements occur only for a short period of time and in a very sparse locality, for example, screaming or a gun shot only lasts no more than one second and the sound of fighting occurs within the whole video but with its peaks sparsely distributed. So the violent scene detection system based on acoustic information must be effective enough to distinguish the violent audio from the nonviolent audios which are usually several ten times longer.

The rest of the paper is organized as follows: Sect. 2 presents related works on violent scene detection. Section 3 describes the practice of CNN in our audio-based VSD system. Section 4 presents experimental setups and results on the MediaEval 2015 dataset. Section 5 presents some conclusions and future work.

2   Related Work

Most of previous related works have mainly focused on using visual cues for violent scene detection. Dai et al. [5] extracted visual features from CNNs based on AlexNet [6], whose effectiveness has been proved in various visual recognition tasks. They further trained two-stream CNNs in spatial and temporal dimension respectively and utilized improved LSTM units [7] in the nets. Along with some conventional hand-crafted features such as improved dense trajectories (IDT) features [8], they achieved the best result in the MediaEval 2015 violent scene detection task. But in terms of audio features, only standard Mel-Frequency Cepstral Coefficients (MFCC) features are used. Other evaluation teams in MediaEval 2015, such as Yi et al. [9] and Lam et al. [10], also exploit mostly in visual cues.

A convolutional neural network is a kind of feed-forward artificial neural network which is inspired from animals visual cortex [11]. In recent years, CNNs have been demonstrated to be capable of effectively understanding images in a higher level and achieving the state-of-the-art performance in handwriting recognition, image classification, and object detection [6,12]. In video classification, according to Karpathy et al. [13], CNNs are powerful to retrieve robust spatial-temporal features. CNNs are also used in speech recognition and music relevant classification. Sainath et al. [14] and Abdel-Hamid et al. [15] have applied CNNs to large vocabulary speech recognition tasks. They showed CNNs are capable of reducing spectral variations and exploring the locality in frequency space.

In this paper, we build convolutional neural networks to discover violence relevant audio features both in frequency and time domain. The best network architecture and parameter sets for VSD have also been carefully studied.

3   Violent Scene Detection System

Our violent scene detection system is based on audio information only. The architecture of our system is shown as in Fig. 1. The two key components in the system are feature extraction in pre-processing and classification.

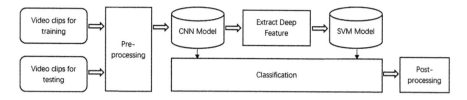

Fig. 1. System architecture illustration of our audio-only VSD System

3.1 Low Level Feature Input

MFCCs are the most popular low-level acoustic features in audio and speech processing tasks. But there are shortcomings of MFCCs for CNN based deep acoustic features. As mentioned in the work by Abdel-Hamid et al. [15], frequency loses locality after the spectral energies are projected into a new basis by cosine transform. So the Mel Filter-Bank (MFB) is used as our low-level feature, which is computed directly from mel-frequency spectrum without DCT. Mohamed et al. [17] have demonstrated that MFB feature maintains stronger local structure than MFCC features. In our VSD system, we use the 40-dimensional MFB features with their delta and delta-delta as the input features to CNN.

We extract MFB with a 25 ms analysis window and 10 ms shift. The videos are segmented into short chunks of 3 s in duration and 1 s shift, resulting in 300 frames of MFB features as one sample fed into the neural network. In order to explore the local feature in the frequency domain, we divide the 120-dimensional MFB features into 3 feature maps (channels), containing the raw MFB feature and its delta, delta-delta. The three 2-D feature maps are formed as shown in Fig. 2.

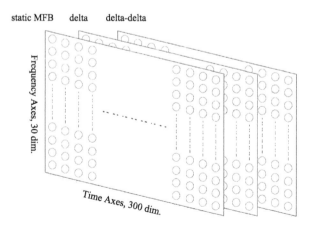

Fig. 2. Input feature map illustration

3.2 Classification

There have been two types of approaches to use CNNs for classification. One is to use CNNs in the end-to-end fashion, taking raw features as input and producing the classification result at its last layer. The other one is to use CNNs for feature representations. The CNN-based features are then used for building SVM classifiers. The violent scene detection is performed on each of the segments that are chunked as mentioned in the above Sect. 3.1. Finally the video-level detection is produced by mean pooling or max pooling over the segment-level detections.

4 Experiments

In this section, we present the detailed experiments conducted on the MediaEval 2015 dataset. Different parameter settings, such as activation function, loss function, number of convolutional-pooling layers, and dropout methods are investigated in this work.

4.1 Dataset Description

We conduct our experiments on the dataset provided by the MediaEval 2015 Affective Impact of Movies task [18]. It is by far the largest dataset for violent scene detection benchmark where all the video clips originated from 100–200 movies, which consist of both professionally made and amateur movies. The dataset is divided into three parts: training set, validation set and testing set. Video clips in the testing set are irrelevant to the training and validation set, while the other two sets have intersecting clips chosen from the same movies. Table 1 shows the number of positive and negative clips and the ratio of positive samples in the datasets. We can see from the table that the positive samples and negative samples are highly unbalanced in all three parts.

4.2 Data Augmentation

As shown in Table 1, the positive and negative samples are highly unbalanced. Generally, input data needs to be relatively balanced in order to make the best use of CNNs. There are multiple ways to solve this problem. We can over-sample

Table 1. Number of positive and negative video clips in the datasets

Dataset	Number of positive	Number of negative	Positive ratio
Training	190	4110	4.42 %
Validation	82	1762	4.45 %
Testing	230	4526	4.84 %

the minority of the data or down-sample the majority of the data, as well as modify the misclassification cost for different classes. Zhou et al. [19] have done detailed experiments on these methods. In our experiments, we conduct data augmentation by over-sampling the positive class. We generate new positive data by adding Gaussian noise to the provided positive data's MFB features. Considering the MFB values range from -1 to 1, we set the average of the Gaussian distribution μ to be the exact data value x, and the standard deviation σ to $x/10$.

4.3   Evaluation Metric

The Average Precision (AP) is used to evaluate the performance of the VSD system. Average Precision is initially used in information retrieval. In the violent scene detection context, it is defined as:

$$AP = \frac{1}{P} \sum_{i=1}^{N} L_i * \frac{P_i}{i} \tag{1}$$

where P is the number of groundtruth violent video clips, N is the total number of testing video clips, $L_i = 1$ when the i_{th} video clip is violent, otherwise $L_i = 0$. The predicted videos are sorted by corresponding scores from highest to lowest. And there are P_i video clips which are correctly predicted being violent among the first i predicted samples.

4.4   Baseline

We built our baseline VSD system based on acoustic and visual information in the MediaEval 2015 evaluation [16]. In the baseline visual-only system, we extracted deep visual features from pre-trained CNNs. In the baseline audio-only system, bag-of-audio-word (BoAW) and fisher vector based on low-level MFCCs features are used. As shown in Table 2, AP of 0.348 and 0.106 on the validation and testing sets are achieved respectively for the audio-only system. AP of 0.308 and 0.120 are achieved on the two sets for the visual-only system. Simple late fusion of the visual-only and audio-only systems improves the performance, increasing AP to 0.500 and 0.211 on the validation and testing sets respectively. The fusion weights are learned on the validation set. We also see

Table 2. Baseline performance

Baseline system setting	Val	Test
Audio-Only (BoAW + FV)	0.348	0.106
Visual-Only (Video-level VGGNet-CNN)	0.308	0.120
Learned Fusion of Audio and Visual	0.500	0.211

that there is a big gap between the performance on the validation set and testing set. This may relate to the big divergence between testing data and training data. The original movies for testing video clips have no overlaps to movies on both training and validation sets, while the training and validation sets have some clips originated from the same movies.

4.5   VSD with CNN as the Classifier

We explore the impact of different parameter settings for CNN on the detection performance, including activation function, dropout, loss function and etc. We experiment with ReLU and Maxout as activation function, multinomial logistic loss which takes the probability contribution computed by softmax function as input (SoftmaxWithLoss) and hinge loss as loss function and exploit the effectiveness of Dropout technique. For these variables, we all use only one pair of conv-pooling layer. After we find the optimal parameters, we investigate the best number of conv-pooling layers with the best parameters.

Activation Function. Activation function produces a non-linear decision boundary with weighted input data in neural networks. There are several popular activation functions used in the practice of deep neural networks, such as sigmoid function, hyperbolic tangent, Rectified Linear Units (ReLUs) and Maxout function.

ReLU function can increase the non-linear properties and efficiency when training networks [6]. Maxout unit is proposed in recent years by Goodfellow et al. [20] whicht solves the problem of the constant 0 in ReLU function that blocks the gradient from flowing through the units. Integrated naturally with dropout, maxout has been reported to outperform the ReLU and showed state-of-the-art performance on multiple computer vision datasets [20].

Table 3 compares the performance based on ReLU and Maxout activation functions, with or without dropout. The experimental results show that the maxout activation function outperforms the ReLU activation function on both validation and testing sets. The two different strategies of generating the video-level detection from the segment/chunk-level detections, mean pooling or max pooling (Average/Max), are also compared. As we can see from the table, it

Table 3. Performance comparison with different activation functions

Models	AP (Average/Max)	
	Val	Test
ReLU, SoftmaxWithLoss, No Dropout	0.193/0.183	0.083/0.143
Maxout, SoftmaxWithLoss, No Dropout	0.222/0.216	0.085/0.160
ReLU, SoftmaxWithLoss, Dropout	0.212/0.204	0.076/0.127
Maxout, SoftmaxWithLoss, Dropout	0.232/0.204	0.083/0.143

Table 4. Performance comparison with different loss functions

Models	AP (Average/Max)	
	Val	Test
Maxout, SoftmaxWithLoss, No Dropout	0.222/0.216	0.085/0.160
Maxout, Hinge loss(L1 Norm), No Dropout	0.220/0.201	0.088/0.164
Maxout, SoftmaxWithLoss, Dropout	0.232/0.204	0.083/0.143
Maxout, Hinge loss(L1 Norm), Dropout	0.240/0.201	0.094/0.161
Maxout, Hinge loss(L2 Norm), Dropout	0.196/0.154	0.090/0.155

seems that the mean pooling strategy works better on the validation set while it is better to use max pooling on the testing set.

Loss Function. In the following experiments, we compare multinomial logistic loss function and hinge loss function. In caffe [21], softmax function and multinomial logistic loss are built in one layer called SoftmaxWithLoss. Generally, hinge loss is wildly used in SVM. Some previous works investigated hinge loss on neural networks, such as Tang [22] which indicated that implementation of hinge loss instead of Softmax can achieve better performance in multiple datasets, especially for squared hinge loss.

Table 4 shows that hinge loss works slightly better than SoftmaxWithLoss on both validation and testing sets in our violent scene detection task. The performance of squared hinge loss (L2 Norm) is also listed in Table 4 which is not as good as L1 norm hinge loss.

Dropout. Due to the highly unbalanced data, we perform dropout to further prevent overfitting. Dropout [23] is a technique to prevent overfitting by approximating exponentially many different neural networks. When training a CNN, each unit is dropout with a certain probability (usually 0.9 for data layers and 0.5 for others) at each iteration. Exponentially many neural networks are generated in this way. Dropout trains each neural network for only one iteration and the weights are shared among all models. At prediction stage, an effective approximation is made to averaging all the models by running the whole network with all the weights divided by 2.

We conduct experiments by applying dropout technique in the first fully connected layer with dropout ratio of 0.5. The results in Table 5 show that dropout has better performance on the validation set but has no further improvement on the testing set. Since better performance is achieved with dropout on the validation set, we use dropout in training in the following experiments. We assume that the divergence on the validation set and testing set is the reason of different performance on the two sets.

Table 5. Performance comparison with and without dropout

Models	AP (Average/Max)	
	Val	Test
ReLU, SoftmaxWithLoss, No Dropout	0.193/0.183	0.083/0.143
ReLU, SoftmaxWithLoss, Dropout	0.212/0.204	0.076/0.127
Maxout, SoftmaxWithLoss, No Dropout	0.222/0.216	0.085/0.160
Maxout, SoftmaxWithLoss, Dropout	0.232/0.204	0.083/0.143
Maxout, Hinge loss(L1 Norm), No Dropout	0.220/0.201	0.088/0.164
Maxout, Hinge loss(L1 Norm), Dropout	0.240/0.201	0.094/0.161

Number of Convolutional-Pooling Layers. In the computer vision field, deep convolutional neural network with several pairs of convolutional and pooling layers can extract high level features which greatly enhanced the capability of convolutional neural network in image classification and detection tasks. AlexNet [6] and GoogLeNet [24] both contains more than 3 pairs of convolutional and pooling layers. In the practice of CNN on audio field, most works on speech recognition [14,15] and music relevant tasks [25,26] use no more than three pairs.

In the violent scene detection task, we conduct experiments with 1, 2, and 3 sets of convolutional-pooling layers. When using Maxout as the activation function, the number of parameters is 82048, 238272 and 275424 respectively. Dropout technique is used in all experiments in Table 6. For SoftmaxWithLoss, more convolutional-pooling layers yields significantly better results on the testing set, while hinge loss tends to work better with simpler network structures.

4.6 VSD Based on Deep Audio Features and SVM as Classifiers

In our VSD system, we use the output of the first fully connected layer (fc1) as our deep audio features. Then a binary SVM classifier is trained based on

Table 6. Performance comparison with different number of Conv-Pooling layers

Models	AP (Average/Max)	
	Val	Test
1conv, Maxout, SoftmaxWithLoss	0.232/0.204	0.083/0.143
2conv, Maxout, SoftmaxWithLoss	0.211/0.202	0.084/0.130
3conv, Maxout, SoftmaxWithLoss	0.205/0.179	0.141/0.192
1conv, Maxout, Hinge loss (L1 Norm)	0.240/0.201	0.094/0.161
2conv, Maxout, Hinge loss (L1 Norm)	0.241/0.191	0.074/0.116
3conv, Maxout, Hinge loss (L1 Norm)	0.253/0.214	0.074/0.087

Table 7. Performance comparison with CNN as classifier or as deep feature extractor

Models	CNN (Average/Max)		SVM (Average/Max)	
	Val	Test	Val	Test
1conv, SoftmaxWithLoss	0.232/0.204	0.083/0.143	0.204/0.203	0.081/0.124
1conv, Hinge loss (L1 Norm)	0.240/0.201	0.094/0.161	0.201/0.175	0.092/0.105
2conv, SoftmaxWithLoss	0.211/0.202	0.084/0.130	0.235/ 0.214	0.094/0.121
2conv, Hinge loss (L1 Norm)	0.241/0.191	0.074/0.116	0.230/0.229	0.092/0.105
3conv, SoftmaxWithLoss	0.205/0.179	0.141/**0.192**	0.238/0.225	0.136/**0.174**
3conv, Hinge loss (L1 Norm)	**0.253**/0.214	0.074/0.087	0.230/**0.261**	0.134/0.167

the deep features using Negative Bootstrap algorithm [27]. We extract deep features from CNN with 1–3 pairs of convolutional-pooling layers, maxout as activation function and dropout technique. Table 7 summarizes results with CNN as a classifier and with SVM as a classifier.

We can observe that when there are only 1–2 pairs of convolutional-pooling layers, results based on SVM are not as good as those based on CNN as a classifier due to the weaker representation capability of deep features. But the performance improved when the neural network goes deeper especially on the validation set.

4.7   Fusion of Visual and Audio

We fuse the visual-only and audio-only systems via late fusion at the score level with learned weights as our baseline fusion system. From Table 8, we can see that fusion of audio and visual cues is beneficial both on the validation and testing sets. Fusion of the end-to-end CNN based audio-only and visual-only systems achieves the best performance on the testing set. It further improves the performance over our previous fusion baseline, which indicates that our new end-to-end CNN based audio-only system performs better violent scene detection than the previous MFCC Bag-of-Audio-Word based audio-only system.

Table 8. Fusion of visual and audio VSD systems

System setting	val	test
Audio-only Baseline [16]	0.348	0.106
End-to-end CNN based on Audio-only	0.205	0.192
Visual-only + Baseline Audio-only	**0.500**	0.211
Visual-only + End-to-end CNN based on Audio-only	0.485	**0.291**

4.8   Experiment Results Analysis

In our experiments, we observe that our VSD system gives very low scores to almost all video clips which are mainly dialogs with no violent content, while the ones which consist of concepts, like shot, scream and heavy metal music, gains high scores for their explicit violent relevance. These are the good cases that we expect a VSD system to output.

But we observe that some videos with non-violent elements such as clapping, cheer, sounds of shutters, and rock music etc. also get high scores. We notice that these elements share the same pattern - bursting in a sudden and with fast tempo. These videos indicate that our VSD system is capable to find some special patterns in the audio track, even though some of them are irrelevant to violence. Figures 3 and 4 present some true positive samples and false positive samples.

The experiment results also show that some video clips which have quiet background sounds are predicted as violent with high scores. It is because some video clips which contain silent or even no background sounds in the training set are labeled as violent. As we mentioned in Sect. 4.4, the groundtruth labels are only based on visual contents. Consequently, some audio-level violent clips are mistaken as non-violent, while some visual-level violent clips have none violent

(a) Fights   (b) Shots

Fig. 3. True Positive Videos. (a) shows a video clip with concept fight, and in (b), a man is shooting with a pistol. These two videos have obvious violence in both visual domain and acoustic domain.

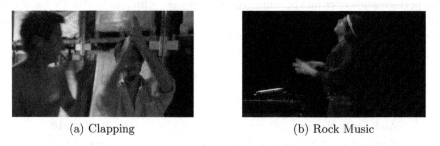

(a) Clapping   (b) Rock Music

Fig. 4. False Positive Videos. (a) shows a clip from the same movie as Fig. 3 (a) which contains concept clapping with no violent element. (b) shows a clip containing strong background music with fast tempo and noisy sounds.

information in audio domain. The system may therefore learn from such samples and make the wrong detection.

5   Conclusions and Future Work

In this paper, we present our audio based violent scene detection system. We apply CNN to extract deep audio features which show promising representative capability. Great progress has been made compared to our previous baseline. Almost all model settings get a better AP on testing set than our baseline audio-only system, which uses Bag-of-Audio-Words and Fisher Vector representations based on MFCCs. The fusion of our end-to-end CNN based audio-only system with visual-only system greatly outperforms the baseline fused system.

In the violent scene detection task, we use CNNs to successfully extract violent scene related audio features. We think CNNs can be generalized in more common concepts or scene detection tasks. Given large amounts of audio database, we can further train CNNs for some specific concepts relevant to violence, such as shot and scream, to help in the VSD tasks. The application of CNNs in more audio classification and scene detection tasks is our focus in the future work.

Acknowledgements. This work was supported by the Beijing Natural Science Foundation (No. 4142029), the Fundamental Research Funds for the Central Universities and the Research Funds of Renmin University of China (No. 14XNLQ01), and the Scientific Research Foundation for the Returned Overseas Chinese Scholars, State Education Ministry.

References

1. Jin, Q., Schulam, P.F., Rawat, S., Burger, S., Ding, D., Metze, F.: Event-based video retrieval using audio. In: Proceedings of INTERSPEECH, p. 2085 (2012)
2. Snoek, C.G., Worring, M.: Concept-based video retrieval. Found. Trends Inf. Retrieval **2**(4), 215–322 (2008)
3. Chang, S.F., Ellis, D., Jiang, W., Lee, K., Yanagawa, A., Loui, A.C., Luo, J.: Large-scale multimodal semantic concept detection for consumer video. In: Proceedings of the International Workshop on Multimedia Information Retrieval, pp. 255–264. ACM (2007)
4. Demarty, C.H., Ionescu, B., Jiang, Y.G., Quang, V.L., Schedl, M., Penet, C.: Benchmarking violent scenes detection in movies. In: 2014 12th International Workshop on Content-Based Multimedia Indexing (CBMI), pp. 1–6. IEEE (2014)
5. Dai, Q., Zhao, R.W., Wu, Z., Wang, X., Gu, Z., Wu, W., Jiang, Y.G.: Fudan-Huawei at MediaEval 2015: Detecting Violent Scenes and Affective Impact in Movies with Deep Learning (2015)
6. Krizhevsky, A., Sutskever, I., Hinton, G.E.: Imagenet classification with deep convolutional neural networks. In: Advances in Neural Information Processing Systems, pp. 1097–1105 (2012)

7. Wu, Z., Wang, X., Jiang, Y.G., Ye, H., Xue, X.: Modeling spatial-temporal clues in a hybrid deep learning framework for video classification. In: Proceedings of the 23rd Annual ACM Conference on Multimedia Conference, pp. 461–470. ACM (2015)

8. Wang, H., Schmid, C.: Action recognition with improved trajectories. In: Proceedings of the IEEE International Conference on Computer Vision, pp. 3551–3558 (2013)

9. Yi, Y., Wang, H., Zhang, B., Yu, J.: MIC-TJU in MediaEval 2015 Affective Impact of Movies Task (2015)

10. Lam, V., Phan, S., Le, D.D., Satoh, S.I., Duong, D.A.: NII-UIT at MediaEval 2015 Affective Impact of Movies Task (2015)

11. Hubel, D.H., Wiesel, T.N.: Receptive fields and functional architecture of monkey striate cortex. J. Physiol. **195**(1), 215–243 (1968)

12. Simard, P.Y., Steinkraus, D., Platt, J.C.: Best practices for convolutional neural networks applied to visual document analysis, p. 958. IEEE (2003)

13. Karpathy, A., Toderici, G., Shetty, S., Leung, T., Sukthankar, R., Fei-Fei, L.: Large-scale video classification with convolutional neural networks. In: Proceedings of the IEEE Conference on Computer Vision and Pattern Recognition, pp. 1725–1732 (2014)

14. Sainath, T.N., Mohamed, A.R., Kingsbury, B., Ramabhadran, B.: Deep convolutional neural networks for LVCSR. In: 2013 IEEE International Conference on Acoustics, Speech and Signal Processing (ICASSP), pp. 8614–8618. IEEE (2013)

15. Abdel-Hamid, O., Mohamed, A.R., Jiang, H., Deng, L., Penn, G., Yu, D.: Convolutional neural networks for speech recognition. IEEE/ACM Trans. Audio Speech Lang. Process. **22**(10), 1533–1545 (2014)

16. Jin, Q., Li, X., Cao, H., Huo, Y., Liao, S., Yang, G., Xu, J.: RUCMM at MediaEval 2015 Affective Impact of Movies Task: Fusion of Audio and Visual Cues (2015)

17. Mohamed, A.R., Hinton, G., Penn, G.: Understanding how deep belief networks perform acoustic modelling. In: 2012 IEEE International Conference on Acoustics, Speech and Signal Processing (ICASSP), pp. 4273–4276. IEEE (2012)

18. Sjberg, M., Baveye, Y., Wang, H., Quang, V.L., Ionescu, B., Dellandra, E., Chen, L.: The mediaeval 2015 affective impact of movies task. In: MediaEval 2015 Workshop (2015)

19. Zhou, Z.H., Liu, X.Y.: Training cost-sensitive neural networks with methods addressing the class imbalance problem. IEEE Trans. Knowl. Data Eng. **18**(1), 63–77 (2006)

20. Goodfellow, I.J., Warde-Farley, D., Mirza, M., Courville, A., Bengio, Y.: Maxout networks. arXiv preprint arXiv:1302.4389 (2013)

21. Jia, Y., Shelhamer, E., Donahue, J., Karayev, S., Long, J., Girshick, R., Darrell, T.: Caffe: convolutional architecture for fast feature embedding. In: Proceedings of the ACM International Conference on Multimedia, pp. 675–678. ACM (2014)

22. Tang, Y.: Deep learning using linear support vector machines. arXiv preprint arXiv:1306.0239 (2013)

23. Srivastava, N., Hinton, G., Krizhevsky, A., Sutskever, I., Salakhutdinov, R.: Dropout: a simple way to prevent neural networks from overfitting. J. Mach. Learn. Res. **15**(1), 1929–1958 (2014)

24. Szegedy, C., Liu, W., Jia, Y., Sermanet, P., Reed, S., Anguelov, D., Rabinovich, A.: Going deeper with convolutions. In: Proceedings of the IEEE Conference on Computer Vision and Pattern Recognition, pp. 1–9 (2015)

25. Li, T.L., Chan, A.B., Chun, A.: Automatic musical pattern feature extraction using convolutional neural network. In: Proceedings of the International Conference on Data Mining and Applications (2010)
26. Ullrich, K., Schlter, J., Grill, T.: Boundary detection in music structure analysis using convolutional neural networks. In: ISMIR, pp. 417–422 (2014)
27. Li, X., Snoek, C.G., Worring, M., Koelma, D., Smeulders, A.W.: Bootstrapping visual categorization with relevant negatives. IEEE Trans. Multimedia **15**(4), 933–945 (2013)

Random Walk Based Global Feature
for Disease Gene Identification

Lezhen Wei, Shuai Wu[✉], Jian Zhang, and Yong Xu

Shen Zhen Graduate School, Harbin Institute of Technology, Shenzhen, China
shuaiwu9@gmail.com

Abstract. Disease gene identification is of great significance for the treatment of genetic disorders. In recent years, the rapid development of high-throughput sequencing technologies has brought great revolution for disease gene identi-fication methods. Network-based methods are now the most efficient component for disease gene identification, while the most of current methods pay only attention to the local topological attributes regardless of the global distribution. In this paper, we proposed to apply the random walk algorithm to extract global features for each gene and finally used binary logistic regression model to identify whether a gene belongs to the given disease. We also integrate the local features and global features into a complex feature vector to improve the identification performance. The experimental results show that the global feature is of great efficiency for disease gene identification. We organize the global feature into different kinds of feature vectors and we can get higher AUC scores than other state-of-the-art methods for all these feature vectors.

Keywords: Gene identification · Logistic regression · Global features · Disease gene

1 Introduction

Most human diseases, e.g. breast cancer, diabetes and lung cancer, are directly related to gene expression disorders. So the fundamental challenge in human health is the identification of disease genes. The conventional biological experiments are very time-consuming and limited by the experimental instruments. Moreover, these exper-imental instruments could handle only a small amount of data each time. In recent years, the accelerating development of high-throughput sequencing technology pro-vides an effective way for analyzing the underlying genetic mechanism of complex diseases [1–14]. All these data can be engaged in constructing and analysis of bio-logical data. The accumulating biological data indeed provide more effective methods for improving disease gene identification performance.

Diseases can be divided into monogenic diseases and polygenic diseases and the corresponding researches on disease-causing genes could be divided into monophyletic network and multi-source network method. RWR [15] is one of the most famous disease gene identification methods. It is based on a computational way that use information come from diverse datas to obtain the ranked list of underlying biological elements (such as protein, gene, disease, etc.). RWR adapted a global perspective to

© Springer Nature Singapore Pte Ltd. 2016
T. Tan et al. (Eds.): CCPR 2016, Part II, CCIS 663, pp. 464–473, 2016.
DOI: 10.1007/978-981-10-3005-5_38

quantify the relationship according to the relevance between the biological elements in the biological molecules network. It could get a list of ranked scores to measure the association between genes and diseases. Similarly, CIPHER [16] proposed to predict the underlying disease-related genes on the basis of the interrelation measured by a linear regression of phenotype and protein-protein network. These studies have successfully tackle with the identification of disease-related genes based on the observation that disease genes associated with the same or similar genetic disease tend to lie close with each other in the biological network. PRINCE [17] proposed to formulate constrains on the prioritization function and to exploit this function for identification.

In recent years, the data integrated method has gained wide attention. One of the ways for integrated analysis is to combine molecular networks. The relationships between varying biological data are too complex. They interact with each other and can be represented by a heterogeneous network. In the biological network, nodes can be expressed as macromolecular (such as genes, proteins, metabolites, viruses) or phenotype (such as disease, normal), and edges represent the relationship of physics functions, or chemistry between two biological entities [18]. In past few decades, the knowledge of graph theory has been applied to biological networks [19]. For example, in order to obtain more useful information from different modules of cells, graph theory could be applied to studies on protein-protein networks (PPI) [20, 21], gene interaction network (GI) [22–24], metabolic interaction networks (MI) [25–27] or gene co-expression networks (Co-Ex) [28–30]. Chen [31] proposed to exploit the local topological property of different biological networks to formulate the feature vector for each gene, and then applied the binary logistic regression method to finally determine whether a gene belong to the disease-causing gene. This method could get a very high AUC score than other state-of-the-art methods. However, it focuses on only the direct neighbor and the second neighbor for a gene and ignores the global distribution of the whole network. In this paper, we proposed to apply the random walk algorithm to extract the global feature for a given gene. Using the random walk algorithm, we could quantify the relationship between a given gene and other genes and then we will choose its top k genes to formulate its feature vector which is called F1. We also combine the local feature with global feature to formulate an integration feature vector which is called F2. In order to integrate multiple data sourses of biological network, we also design F3 which, similar to chen, is the integration feature vector of different features from different biological networks. The experimental results show that our method could get better performance than Chen' method [31] in terms of F1, F2 and F3. The organization of this paper is as follows: Sect. 2 gives the detail of our method and Sect. 3 shows the experimental analysis. Section 4 gives the conclusion.

2   Methods

In this section, we will give the process of our method. Let $D = \{d_1, \ldots, d_m\}$ be the set of human genetic diseases and $G = \{g_1, \ldots, g_N\}$ be the set of all human genes. We can get the association between diseases and different genes from Online Mendelian Inheritance in Man (OMIM) database. Here, we need to point out that not all the genes have explicit association with diseases. Actually, the majority of human genes are not known to associate with any human diseases which are called unknown genes. So we can divide all human genes into two partitions, disease genes which have the explicit association with one or more genetic diseases and unknown genes. Let $\{g_1, g_2, \ldots, g_n\}$ denote the unknown gene set and $\{g_{n+1}, g_{n+2}, \ldots, g_{n+r}\}$ denote the disease gene set, where $N = n + r$.

2.1   Estimation Label for Unknown Genes

In real applications, the number of unknown genes is much larger than the number of disease genes. So for a specific disease d_k, it is not reasonable to set all the labels of unknown genes to 0 which means they all do not have association with d_k. We apply the method in [32] to exploit protein complex to calculate a prior probability for unknown genes. If gene g_i encodes a protein in a complex, then we have

$$p_i = \frac{A}{B}$$

Where A is the number of disease genes of d_k in the complex and B is the number of all disease genes in the complex. If gene g_i does not belong to any protein complex, then A the number of all disease genes of d_k and B is the number of all human genes. Once p_i is available, we will generate a random number following the standard uniform distribution for g_i. If it is larger than p_i, then assign 0 as the prior label for g_i. Otherwise, we assign 1 as the prior label for g_i. Repeating this step, we will get a estimation label for each unknown gene for d_k.

2.2   Feature Construction

For a given disease d_k, the challenge is to find a set of candidate genes which may have associations with d_k. In order to identify whether gene g_i has association with d_k exactly, we need to find reasonable features for g_i. In [31], Chen use $(1, \varphi_{i1}, \varphi_{i2})$ as the feature vector of g_i where φ_{i1} and φ_{i2} represent the number of direct neighbors of g_i with label 1 and 0 for disease d_k respectively. In order to improve the performance, Chen also extends the feature vector as $(1, \varphi_{i1}, \varphi_{i2}, \bar{\varphi}_{i1}, \bar{\varphi}_{i2})$ where $\bar{\varphi}_{i1}$ and $\bar{\varphi}_{i2}$ represent the numbers of the second neighbors of g_i with label 1 and 0 for disease d_k respectively. As we point out previously, such feature vectors focus on only the local topological attributes of the network regardless of its global distribution. We proposed to apply the random walk algorithm to get the global feature for gene g_i (see Fig. 1).

We present the detail of our method below. For gene g_i, we firstly exploit the random walk algorithm to quantify the relationship between gene g_i and other genes. Let $\{r_{g_i}^1, r_{g_i}^2, \ldots, r_{g_i}^N\}$ denote the association value between g_i and other genes in human genome. The sequence has been sorted in descent order. Then we choose the top k genes as the "friends" of gene g_i. Then we use $R_i = \{1, R_{i1}, R_{i2}\}$ as its feature vector, R_i can be calculated in following two different ways: firstly, R_{i1} and R_{i2} could represent the relation value accumulation of friends genes of g_i with label 1 and 0 for disease d_k respectively; next, R_{i1} and R_{i2} could represent the number of friends genes of g_i with

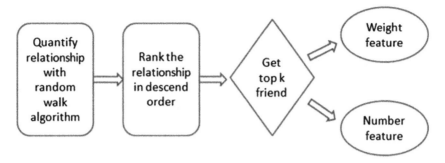

Fig. 1. Feature extraction process

label 1 and 0 for disease d_k respectively. Figure 4 shows the feature extraction process. We call the two kinds of R_i weight-feature and number-feature respectively. Repeating the above step for all the genes, we can get the feature matrix as follows:

$$F_1 = \begin{bmatrix} 1 & R_{i1} & R_{i2} \\ \ldots & \ldots & \ldots \\ 1 & R_{N1} & R_{iN} \end{bmatrix}_{N \times 3} \tag{1}$$

N is the number of all the human genes. As we know, the random walk algorithm adopts a global perspective to capture interactions, which takes into account the entire network structure. So we could consider that R_{i1} and R_{i2} contain more information than φ_{i1} and φ_{i2}. We could also integrate both local and global feature to organize the features vector as $R_i = \{1, \varphi_{i1}, \varphi_{i2}, R_{i1}, R_{i2}\}$. Also, we could get the feature matrix as (2)

$$F_2 = \begin{bmatrix} 1 & \varphi_{11} & \varphi_{12} & R_{11} & R_{12} \\ 1 & \varphi_{21} & \varphi_{22} & R_{21} & R_{22} \\ \ldots & \ldots & \ldots & \ldots & \ldots \\ 1 & \varphi_{N1} & \varphi_{N2} & R_{N1} & R_{N2} \end{bmatrix}_{N \times 5} \tag{2}$$

Nowadays, there are different kinds of data source, e.g. PPI, network, pathway network. How to effectively take full use of different data source is a big challenge. Similar to [31], we integrate the global feature from different data source into a comprehensive feature vector. Suppose there are n biological networks. Let (R_{i1}^j, R_{i2}^j) denotes the global feature from jth biological network, then the final feature vector could be organized as $R_i = (1, R_{i1}^1, R_{i2}^1, \ldots, R_{i1}^n, R_{i2}^n)$. Similarly, the feature vector of all genes could form feature matrix (3)

$$F_3 = \begin{bmatrix} 1 & R_{11}^1 & R_{12}^1 & \cdots & R_{11}^n & R_{12}^n \\ 1 & R_{21}^1 & R_{22}^1 & \cdots & R_{21}^n & R_{22}^n \\ \ldots & \ldots & \ldots & \ldots & \ldots & \ldots \\ 1 & R_{N1}^1 & R_{N2}^1 & \cdots & R_{N1}^n & R_{N2}^n \end{bmatrix}_{N \times (2n+1)} \tag{3}$$

When we get the feature vector for all genes. We ill train a logistic regression model shown in [31] to calculate the posterior probability for a gene, here we do not give the detail. After getting the posterior probability, we use the evaluation criteria as shown in [33], if the posterior probability of a candidate gene larger than most of the unknown gene, we consider it has a good potential to be disease gene.

This section shows the detail of our method. We apply the random walk algorithm to extract the global feature of each gene for a given disease, then use logistic regression to perform the final classification.

3 Experiment

3.1 Experimental Data and Evaluation Methods

In order to do the reasonable comparison, we apply the same experimental data as shown in [31]. Such data comprise different biological networks, e.g. protein-protein networks, pathway networks and gene co-expression networks. Moreover, in order to analyze the biological function of cells in the deep extent, the data also apply 3881 protein complexes from both CORUM [34] and PCDq [35] databases. We finally get 7311 genes in which 6496 genes are unknown genes and 815 genes are disease genes (known genes). Here, we need to point out that in the OMIM database,the number of disease genes for a single disease is relatively small. So the data map the 815 disease genes to 12 disease categories which are manually partitioned by Goh [36].

We choose the leave-one-out cross validation method to evaluate the performance of our method and the receiver operating characteristic (ROC) curve is employed as the most important evaluation criteria. The positive control genes are those known genes associated with disease d_k (d_k stands for a disease category). For the negative control genes, similar to [31], suppose there are s known genes that associate with d_k, then we will randomly select $s/2$ known genes that do not associate with d_k as the negative control genes. The negative control set is also validated by the leave-one-out cross validation method.

3.2 Results

In order to evaluate the performance of our method, we compare it with other state-of-the-art methods including the method in [31], RWR [15] and DIR [33]. Moreover, we test our method on F_1, F_2 and F_3 which are showed in Sect. 2. We also test the performance difference between weight-feature and number-feature.

3.2.1 Results of F_1

In this section, we will show the performance of our methods on F_1. Table 1 gives the AUC score of our method on the PPI network across different k (choose top k genes as the friends genes) and different kinds of features (weight-feature and number-feature). We could see clearly that on the F_1 level, the performance of weight-feature is much better than that of number-feature, and the AUC score of weight-feature is much larger

Table 1. Results of our method on PPI network (F_1)

k	weight-feature	number-feature
90	**0.7540**	0.6908
150	0.7491	0.7010
200	**0.7599**	0.6887
300	**0.7551**	0.6858

than that of number-feature no matter what k is. So we can conclude that on the F_1 level, weight-feature will contain more information than number-feature.

Figure 2(a) shows the comparison on ROC curves between our method and other three different methods on the PPI network. PCF1 represents the method in [31]. We could see that our method owns the highest AUC score which is 0.86 %, 2.75 % better than PCF1 and RWR respectively.

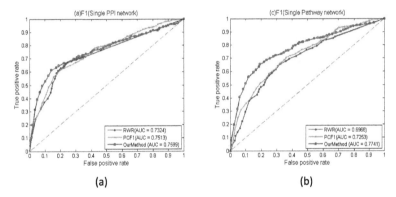

Fig. 2. ROC curve of different methods (F_1): (a) PPI network (b) Pathway network

Similar to Table 1 and Fig. 2(a), Table 2 and Fig. 2(b) shows the corresponding performance on the pathway network. From Table 2, we could get the same conclusion as the PPI network, the performance of weight-feature is much better than number-feature. In terms of pathway network, we could see clearly from Fig. 2 that our method outperforms PCF1 and RWR by 7.8 % and 4.9 %.

3.2.2 Results of F_2

As we point out previously, we integrate the local feature and global feature to formulate F_2, which is more reasonable than the feature in [31]. Chen introduce the second neighbors of a gene to extend its feature vector. No matter using direct neighbors or second neighbors, they only stands for the local information of the network and ignore the global information. On the F_2 level, Table 3 shows the performance of our method on PPI network. Different with F_1, the performance of number-feature is relatively higher than that of weight-feature. We can see the AUC

Table 2. Results of our method on the gene pathway network (F_1)

k	weight-feature	number-feature
100	**0.7547**	0.5838
300	**0.7552**	0.6894
500	**0.7741**	0.6833
700	**0.7698**	0.6784

Table 3. Results of our method on the PPI network (F_2)

k	weight-feature	number-feature
30	0.7443	**0.7847**
100	0.7507	0.7557
150	0.7516	0.7575
300	0.7491	0.7384

score peak on k = 30 for number-feature. This is because the number-feature is more consistent with local feature. After all, the local feature stands for the number of direct neighbors of a gene with label 1 and 0 for a given disease.

Figure 3(a) shows the ROC curves of different methods on the F_2 level. We could see that our method has higher AUC score than PCF2 In [31]. Table 4 and Fig. 3(b) show the performance of different methods on the pathway network. We could see that the performance of our methods peak at k = 30 (which is the same as PPI network), our method surpass the PCF2 by nearly 8 %.

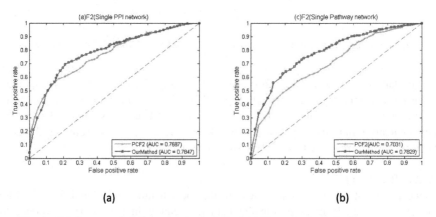

(a) (b)

Fig. 3. ROC curve of different methods (F_2): (a) PPI network (b) Pathway network

3.2.3 Results of F_3

In F_3, a feature vector for a given gene is organized as $R_i = (1, R_{i1}^1, R_{i2}^1, \ldots, R_{i1}^n, R_{i2}^n)$. In this experiment, weapply three different biological networks: PPI network, pathway

Table 4. Results of our method on the pathway network (F_2)

k	weight-feature	number-feature
30	**0.7141**	**0.7829**
50	**0.7099**	**0.7310**
100	**0.7158**	**0.7794**
150	**0.7315**	**0.7637**

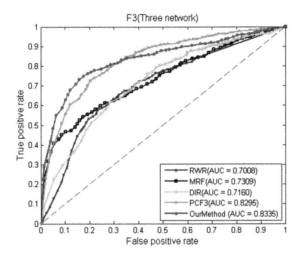

Fig. 4. ROC curve of our method and PCF3 (F_3)

network and gene co-expression network to formulate the feature vector. Chen [31] also organized F_3 with local features from different networks. Figure 4 shows the comparison ROC curves between our method, PCF3 which is mentioned in [31]. We can see clearly that our method has better performance.

4   Conclusion

In this paper, we proposed to apply the random walk algorithm to extract the global feature for a gene. As we know, the random walk algorithm adopts a global perspective to capture interactions, which takes into account the entire network structure. So the global feature will contain more information than the Local feature shown in [31]. We also organize the global feature into three kinds of feature vector F_1, F_2, F_3. In terms of F_1 and F_2, although the AUC score of our method is slightly higher than Chen's method [31] on the PPI network, we could see that on the pathway network, our method outperform much more than Chen. As for F_3, we integrate the global feature from different biological networks and get higher AUC score than other state-of-the-art methods. In conclusion, the global feature is of great significance and efficiency for gene identification.

Acknowledgment. This work is surported by JCYJ20140904154645958 and CXZZ20140904
154910774.

References

1. Wang, K., Li, M., Hakonarson, H.: ANNOVAR: functional annotation of genetic variants
   from high-throughput sequencing data. Nucleic Acids Res. **38**(16), e164 (2010)
2. Pan, Q., Shai, O., Lee, L.J., et al.: Deep surveying of alternative splicing complexity in the
   human transcriptome by high-throughput sequencing. Nat. Genet. **40**(12), 1413–1415 (2008)
3. Stelzl, U., Worm, U., Lalowski, M., et al.: A human protein-protein interaction network: a
   resource for annotating the proteome. Cell **122**(6), 957–968 (2005)
4. Simonis, N., Rual, J., Carvunis, A., et al.: Empirically controlled mapping of the
   Caenorhabditis elegans protein-protein interactome network. Nat. Methods **6**(1), 47–54
   (2009)
5. Consortium A I M: Evidence for network evolution in an Arabidopsis interactome
   map. Science **333**(6042), 601–607 (2011)
6. Gavin, A.C., Aloy, P., Grandi, P., et al.: Proteome survey reveals modularity of the yeast cell
   machinery. Nature **440**(7084), 631–636 (2006)
7. Krogan, N.J., Cagney, G., Yu, H., et al.: Global landscape of protein complexes in the yeast
   Saccharomyces cerevisiae. Nature **440**(7084), 637–643 (2006)
8. Hawkins, R.D., Hon, G.C., Ren, B.: Next-generation genomics: an integrative approach.
   Nat. Rev. Genet. **11**(7), 476–486 (2010)
9. Nielsen, R., Paul, J.S., Albrechtsen, A., et al.: Genotype and SNP calling from
   next-generation sequencing data. Nat. Rev. Genet. **12**(6), 443–451 (2011)
10. Quackenbush, J.: Computational analysis of microarray data. Nat. Rev. Genet. **2**(6), 418–
    427 (2001)
11. Dahlquist, K.D., Salomonis, N., Vranizan, K., et al.: GenMAPP, a new tool for viewing and
    analyzing microarray data on biological pathways. Nat. Genet. **31**(1), 19–20 (2002)
12. Marioni, J.C., Mason, C.E., Mane, S.M., et al.: RNA-seq: an assessment of technical
    reproducibility and comparison with gene expression arrays. Genome Res. **18**(9), 1509–1517
    (2008)
13. Mortazavi, A., Williams, B.A., Mccue, K., et al.: Mapping and quantifying Mammalian
    transcriptomes by RNA-Seq. Nat. Methods **5**(7), 621–628 (2008)
14. Wang, Z., Gerstein, M., Snyder, M.: RNA-Seq: a revolutionary tool for transcriptomics. Nat.
    Rev. Genet. **10**(1), 57–63 (2008)
15. Köhler, S., Bauer, S., Horn, D., et al.: Walking the interactome for prioritization of candidate
    disease genes. AIDS Res. Hum. Retroviruses **21**(4), 314–318 (2005)
16. Wu, X., Jiang, R., Zhang, M.Q., et al.: Network-based global inference of human disease
    genes. Mol. Syst. Biol. **4**(1), 189 (2008)
17. Vanunu, O., Magger, O., Ruppin, E., et al.: Associating genes and protein complexes with
    disease via network propagation. PLoS Comput. Biol. **6**(1), e1000641 (2010)
18. Vidal, M., Cusick, M.E., Barabási, A.L.: Interactome networks and human disease: cell. Cell
    144(6), 986–998 (2011)
19. Aittokallio, T., Schwikowski, B.: Graph-based methods for analysing networks in cell
    biology. Briefings Bioinf. **7**(3), 243–255 (2006)
20. Pržulj, N.: Protein-protein interactions: making sense of networks via graph-theoretic
    modeling. Bioessays News Rev. Mol. Cell. Dev. Biol **33**(2), 115–123 (2011)

21. Hakes, L., Pinney, J.W., Robertson, D.L., et al.: Protein-protein interaction networks and biology–what's the connection? Nat. Biotechnol. **26**(1), 69–72 (2008)
22. Lesage, G., Bader, G.D., Ding, H., et al.: Global mapping of the yeast genetic interaction network: discovering gene and drug function. Science **303**(5659), 808–813 (2004)
23. Dixon, S.J., Costanzo, M., Baryshnikova, A., et al.: Systematic mapping of genetic interaction networks. Annu. Rev. Genet. **43**(43), 601–625 (2009)
24. Costanzo, M., Baryshnikova, A., Bellay, J., et al.: The genetic landscape of a cell. Science **327**(5964), 425–431 (2010)
25. Tanaka, R.: Scale-rich metabolic networks. Phys. Rev. Lett. **94**(16), 168101 (2005)
26. Ravasz, E., Somera, A.L., Mongru, D.A., et al.: Hierarchical organization of modularity in metabolic networks. Science **297**(5586), 1551–1555 (2002)
27. Ma, H., Zeng, A.P.: Reconstruction of metabolic networks from genome data and analysis of their global structure for various organisms. Bioinformatics **19**(2), 270–277 (2003)
28. Prieto, C., Risueño, A., Fontanillo, C., et al.: Human gene coexpression landscape: confident network derived from tissue transcriptomic profiles. PLoS ONE **3**(12), e3911 (2008)
29. Stuart, J.M., Segal, E., Koller, D., et al.: A gene-coexpression network for global discovery of conserved genetic modules. Science **302**(5643), 249–255 (2003)
30. Guo, X., Gao, L., Wei, C., et al.: A computational method based on the integration of heterogeneous networks for predicting disease-gene associations. PLoS ONE **6**(9), e24171 (2011). [SCI:000294686100018] [SCI IF = 4.092, JCR = 2]
31. Chen, B., Li, M., Wang, J., et al.: A logistic regression based algorithm for identifying human disease genes. In: IEEE International Conference on Bioinformatics and Biomedicine. IEEE (2014)
32. Chen, B., Wang, J., Li, M., et al.: Identifying disease genes by integrating multiple data sources. BMC Med. Genomics **7**(Suppl 2), S2 (2014)
33. Chen, Y., Wang, W., Zhou, Y., et al.: In silico gene prioritization by integrating multiple data sources. PLoS ONE **6**(6), e21137 (2011)
34. Burton, P.R., Clayton, D.G., Cardon, L.R., et al.: Genome-wide association study of 14,000 cases of seven common diseases and 3,000 shared controls. Nature **447**(7145), 661–678 (2007)
35. Weinstein, J.N., Collisson, E.A., Mills, G.B., Shaw, K.R., Ozenberger, B.A., Ellrott, K., Shmulevich, I., Sander, C., Stuart, J.M.: The cancer genome atlas pan-cancer analysis project. Nat. Genet. **45**(10), 1113–1120 (2013). Cancer Genome Atlas Research Network
36. Emmertstreib, F., Tripathi, S., Simoes, R.D.M., et al.: The human disease network. Proc. Natl. Acad. Sci. **1**(1), 20–28 (2014)

Multi-stage Feature Extraction in Offline Handwritten Chinese Character Recognition

Xianglian Wu[(⊠)], Chang Shu, and Ning Zhou

University of Electronic Science and Technology of China, Chengdu, China
18215613453@163.com

Abstract. Convolutional neural network (CNN) has achieved tremendous success in handwritten Chinese character recognition (HCCR). However, most CNN-based HCCR research nowadays focus on complicated and deep CNN module, rarely analyzing the whole feature extraction process which has a crucial impact on the final recognition rate. In this paper, the following two questions are answered: (1). Information loss is inevitable on the training stage of complex learning problems, but at which layer does the information loss mainly occur; (2). Different layers have different effects on CNN, what is the best place for multi-stage feature extraction that influences CNN most. We make use of the proposed module in typical CNN and analyze classification results on CASIA-HWDB1.1. It is shown in this paper that, (1). Multi-stage feature extraction achieves better performance on HCCR than single stage feature extraction. (2). Multi-stage feature extraction should be designed at the convolution layer rather than the pooling layer. (3). Multi-stage feature extraction designed at shallow layers outperforms that designed at deeper layers. By analyzing the structure of multi-stage feature extraction, we propose an appropriate CNN approach to HCCR, which achieves a new state-of-the-art recognition accuracy of 91.89 %.

Keywords: Deep learning · Convolutional neural networks · Feature extraction · Handwritten chinese character recognition

1 Introduction

Despite the amount of work and extensive applications on HCCR at the past several decades, it is still a considerable challenge nowadays due to numerous categories, multifarious handwriting styles and semblable characters etc. In traditional offline HCCR, a mass of approaches based on hand-designed feature extractor have been proposed. Some succeeded in obtaining character structural features like strokes, outlines, components, etc. However, there were little improvements with characters which are pitched, twisted or have adhesiveness because of the low anti-jamming capability of these approaches. Others focus on the statistical characteristics of Chinese characters after mathematical transformation. However, these methods show a poor performance with similar characters. Traditional offline HCCR seems to meet a bottleneck in recent years since preferable approach has not been proposed yet.

© Springer Nature Singapore Pte Ltd. 2016
T. Tan et al. (Eds.): CCPR 2016, Part II, CCIS 663, pp. 474–485, 2016.
DOI: 10.1007/978-981-10-3005-5_39

Because of the tedious preprocessing and blindfold feature selection in traditional image recognition, CNN structure is put forward by LeCun, which automatically learns a unique set of features optimized for a given task differ from traditional approaches, and achieves a fairly good performance on handwritten characters. With the blooming of deep learning in recent years, new CNN structures are springing up everywhere. AlexNet [2], MCDNN [3], GoogleNet [4] and VGG [5] etc. have been raised in continuous succession and achieved a fairly good performance on ImageNet classification. Though yielding excellent results, most existing models throw out a huge challenge for training because of their very deep and complicated networks. For instance, GoogleNet proposed by Google at 2014, has up to 22 layers and more than 60 million parameters. It takes more than one week for training. In addition, different kinds of preprocessing techniques are introduced before training to improve the final performance, such as feature extraction techniques like Principle Component Analysis (PCA), Linear Discriminant Analysis (LDA), Scale Invariant Feature Extraction (SIFT), Gabor based features, HOG and so on [6]. These techniques have greatly improved CNNs' performance, at the cost of losing the original meaning of deep learning by introducing too much artificial intervention. This paper mainly devotes to improve the single-column convolution neural network recognition performance by analyzing the feature extraction process.

The remainder of this paper is organized as follows: Sect. 2 briefly introduces CNN theory. Section 3 presents two modules we designed for feature extraction. In Sect. 4, our major work are introduced, a short description on data preparation as well as CNN structure, the experimental results and analysis are given. Feature comparison through visualization on the feature maps are presented in Sect. 5. Conclusions are drawn in Sect. 6.

2   A Brief Introduction of CNN

The Convolution neural network, which is biologically inspired, generally consists of alternating convolution and sub-sampling operations, the last stage of the architecture (closest to the outputs) consists of fully connected layers [7]. CNN is a hierarchical and deep neural network, mining the related spatial image information by strengthening the local connection mode of neural network nodes between adjacent layers. It is directly operated on the two-dimensional original image by local receptive field, weight sharing and pooling, avoiding complex preprocessing and feature extraction of traditional artificial neural networks.

Convolution layer is composed of a group of kernel filters. In this layer, feature maps of the previous layer are convolved with learnable kernels and nonlinear transform is carried out by the activation function to form output feature map [7]. Each output map is likely to associate with multiple input maps. CNN can automatically learn a unique set of features by convolution operation. The convolution operation is given by Eq. (1).

$$x_j^l = f\left(\sum_{i \in M_j} x_i^{l-1} * k_{ij}^l + b_j^l \right) \tag{1}$$

Where x_j^l represents the jth map in the lth the layer, k_{ij}^l is a convolution kernel made by i^{th} feature map in the $(l-1)^{th}$ layer connected to the jth map in the ith layer [7, 15]. An additive bias b_j^l is given in the ith layer, $f(*)$ is a point-wise nonlinear activation function. The output feature maps represent the spatial arrangement of the activation. The CNN obtains convolutional feature maps through a group of kernel filters and convolution operations.

A pooling layer produces down-sampled versions of the input maps. The purpose of pooling layer is to add robustness to small distortions, just like the complex cells in visual perception models. Each output feature map is spatially sub-sampled by a factor horizontally and vertically. Pooling also reduces feature dimension of the convolution output and ignores irrelevant detail components. The process can be expressed in formula (2).

$$x_j^l = f\left(\beta_j^l down(x_j^{l-1}) + b_j^l\right) \tag{2}$$

Where $down(*)$ stands for down-sampling function. The jth map in the $(l-1)^{th}$ layer is sampled by pooling, then the output is given its own multiplicative bias β_j^l and an additive bias b_j^l. $f(*)$ is the nonlinear activation function.

Approximate biological neural nonlinear activation function $f(*)$ is generally rectified linear unit function [2, 8], which promotes a competition among all input data. The models that are strongly inspired by biology have included rectifying nonlinearities, such as positive part, absolute value, or squaring functions [18]. In addition, CNN using stochastic gradient descent algorithm [9] to minimize the mean squared-error loss function, take dropout [10] or maxout [11] measures to improve data sparseness.

In the learning process, the original image is filtered through several layers of convolution and pooling sampling process, the initial edge linear features are constantly integrated into more advanced abstract characteristics, features maps play an important role to the final recognition rates. And feature extraction is the heart of HCCR application.

3   Design of Multi-stage Feature Extractor

The crux of the pattern recognition is features extraction. As is known, good features must have good seperability, reliability and independence. Both low loss of useful information and irrelevance among each feature dim are appreciated. Feature extraction is a process of redundant information removing.

The CNN being used nowadays are mainly with single feature extraction structure. However, single-stage feature extraction suffers loss of useful original information such as local image textures and fine details [12]. Although the concept of multi-level feature extraction has been proposed for several mouths, no detailed analysis of the results was given, nor were there any explanation about the difference between extracted features of multi-stage and single-stage structure. In this paper, we analyze about multi-stage feature extraction on CNN.

Feature extraction error mainly comes from two aspects. One is the estimated variance, due to the fixed window size, the neighborhood size of features maps is limited, resulting in an increasing estimated error of variance value. The other is a mean estimate offset due to the convolution layer parameter deviation.

For a single network, it operates in a sliding-window manner using a fixed kernel size during the convolution process. Loss of original information is inevitable in such situation. If the fixed size is too small, each convolution region has few elements, the essential structure information between feature maps will lose. Otherwise, if the size is too large, details of the image features [13] will be ignored, which violates biological vision of local relevance. Since single-level feature extraction has such drawbacks. We propose a multi-stage feature extraction on convolution layer called Multi-conv module, as is shown in Fig. 1. Differ from the inception module in GoogleNet, the same number of convolution kernels for each convolutional scale is used instead of dimension reduction before convolving by 1×1 convolutions at additional stage, which ensures the fair competition of various scales and prevents features co-adaptation during training. In this paper, multi-stage convolution is followed by multi-scale pooling to produce the same dimensional outputs, twofold multi-scale improves the adaptivity of the Multi-conv module. The previous layer convolved at several scales has both detailed components and approximate structural features. Training with variable-size features increases scale-invariance and reduces over-fitting.

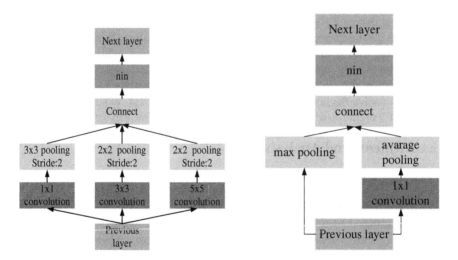

Fig. 1. Multi-conv module   **Fig. 2.** Multi-pool module

The purpose of pooling is to transform joint information into a more usable and representative one that still preserves important information while discarding irrelevant and unimportant details simultaneously. Jarrett et al. has shown that pooling type matters for classification problems. Average-pooling reduces error caused by fixed sliding-window size and keep much background information. Meanwhile, max-pooling reduces estimation error coming from the mean deviation and retain more texture information

(a)   (b)   (c)   (d)

Fig. 3. The character after preprocessing (a). The original sample (b). After local contrast normalization (c). Resized image (d). Distorted image

[13]. Both the two types have drawbacks during training. In average-pooling, all Eigen values in one pooling region are considered coequally even though many elements' magnitude vary widely, which weakens strong activations' effects generated by following nonlinear components since many zero elements are included. Max-pooling only acquires the strongest activation for each pooling region [17], ignoring additional activations in the same pooling region that may be useful. In addition, Matthew D. Zeiler [14] has shown that max pooling easily overfits the training set, which leads to a poor generalization. In this paper, we designed a multi-stage pooling, as shown in Fig. 2, which combines with both pooling types, and is expected to provide more powerful features. Our proposed pooling scheme uses average-pooling and max-pooling to extract features separately, the two results are then integrated, followed by a 1 × 1 convolution layer aiming to solve the high-dimensional problems, which not only reduces the high dimensions of integration output but also increases the non-linearity of CNN.

4   Experiments and Analysis

4.1   Data Preparation

The isolated offline CASIA-HWDB1.1 (DB1.1) [1] database is used for training and testing. It is provided by the National Laboratory of Pattern Recognition (NLPR) for study of character preprocessing and feature extraction methods. HWDB1.1 includes 3, 755 GB2312-80 level-1 Chinese characters and 171 alphanumeric and symbols. Each data set is partitioned into a standard training set (240 writers) and a test set (60 writers). The data sets contain respectively 1, 172, 907 samples. All samples are gray-scaled images with background pixels labeled as 255.

Since the original samples suffer from low image contrast, vague stroke edge and different sizes, during the preprocessing stage of our experiment, all images are rescaled to a size of 48 × 48, local contrast normalization is carried out and the gray value of original image is stretched to [0 255] after normalization. Due to numerous category and various writing styles of handwritten Chinese character, the generalization of recognition model is very important. Sample distortions [19] were used to expand data set and improve invariance. Two matrix templates were created first with random values in the range of [−1, +1], indicating horizontal and vertical random deformation of character points respectively. The templates were then convolved with a standard Gaussian kernel function with a variable deviation σ, with which one could achieve a balance in between fine texture keeping and noise smoothing. The convolved templates were finally applied to the original input image. A padding of 4 pixels with zero value was added to each side of the input image in order to center the following convolutional filters onto image borders. Figure 3 shows results of a character sample after different data preparation process.

4.2 Details of Architecture

The basic SS-ConvNets structure [15, 20] used in our experiments consults the model in the ICDAR offline HCCR competition and takes the similar architecture as Fujitsu team, which consists of ten weighted layers; the first eight layers include four groups of convolutional layers as well as our groups of max-pooling layers; the remaining two layers are fully connected layers. The details of SS-ConvNets are depicted in Fig. 4. The number of convolution layer is [100, 200, 300, 400] respectively, the convolution kernel window size is 3×3 pixels, the pooling layer uses max-pooling method with 2×2 pixels window size. The final classification result is obtained after several turns of alternate convolution and sampling process. Stochastic gradient descent is used as our optimization method, and the data set is shuffled after the training iteration.

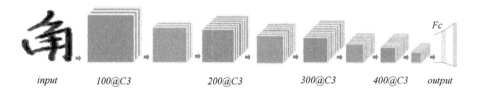

$$input \qquad 100@C3 \qquad\qquad 200@C3 \qquad\qquad 300@C3 \qquad 400@C3 \qquad output$$

Fig. 4. Single-stage CNN structure (SS-ConvNets)

The multi-stage ConvNets make use of the multi-stage modules introduced in Sect. 3. Instead of the native single-stage module, the multi-stage feature extraction structure used on convolution layer is shown in Fig. 1. For the convolution layer, the module is made up of 1×1 and 3×3 convolutions for two-stage or 1×1, 3×3 and 5×5 convolutions for three-stage. Differ from the GoogleNet which obtains a group of feature maps of the same size by padding feature maps after the convolution operation, we put the outputs of different scale convolutions into different scale pooling to reduce their feature dimension, then unit them together by a concat-layer. The corresponding pooling is 2 step-size with no overlapping and 3 step-size with 1 overlapping. The major advantage of our module is that it significantly increases the variation of filtering at each convolution and pooling stage. Thanks to the Multi-stage convolution module, local feature representation can be extracted using complicated convolutional kernel filter size robustly and efficiently.

Comparison is made on CNN structure with single-stage and two-stage feature extraction. A new local multi-stage module shown as Fig. 2 is introduced. Multi-stage feature extraction on convolution layer and the pooling layers separately overcomes the drawbacks of fixed window size. A 2×2 window size is adopted on the pooling layer for both average-pooling and max-pooling, while 1×1 convolutions are applied to the module before average-pooling to reduce the number of parameters as well as to rectify activation since average-pooling generally plays a less important role than max-pooling at character recognition.

4.3   Experiment and Analysis

Our experiments are conducted on an open CNN platform Caffe [16] using a GTX TITAN X GPU card and it costs about 2 days in training such a CNN structure. Fixed learning rate tactics are used in training. More specifically, we first set base_lr = 0.1 when the loss function does not change, then base_lr is set smaller (base_lr = base_lr * 0.1) until there is no significant change in the loss function. For the sake of credibility, ConvNets are trained four times for each model, and the average result is reported. The final experiment result on convolution layer is shown in Fig. 5.

Experiment 1   The conv2 layer of SS-ConvNets is replaced by two and three stage feature modules as mentioned in Sect. 3. The number of each convolution layer's feature maps is set to 200, which is up to 400 and 600 on two and three stage feature modules respectively after concat operation. To better compared with SS-ConvNets, nin layer is followed to reduce the dimension to 200 to keep the same with SS-ConvNets. Figure 5a shows the accuracy comparison on ConvNets with single, two and three feature extraction structure on convolution layer respectively. It is obvious that multi-stage feature extraction structure is better than single-stage structure as discussed in Sect. 3, which proves that convolution with different slide-window size is effective. Besides, we notice that the performance between two-stage and three-stage structure has no obvious difference, while three-stage structure is much more complex and time consuming. Therefore, a proper balance of recognition accuracy and computation cost should be considered. Table 1 shows a comparison of these models in terms of recognition accuracy and the loss on convolution layer using the ICDAR 2013 offline HCCR competition data sets.

Table 1.   Performance of convnets of multi-stage feature extraction

Structure	Loss	Accuracy(%)	Improve(%)
Conv:1-stage(3×3)	0.3734	90.44	n/a
Conv:2-stage($3 \times 3, 5 \times 5$)	0.3417	91.26	0.82
Conv:3-stage($1 \times 1, 3 \times 3, 5 \times 5$)	0.3503	91.60	1.16
Pool:1-stage(2MP)	0.3527	91.11	0.67
Pool:2-stage(2MP + 2AP)	0.3523	91.30	0.86

Experiment 2   Pooling layer performs a down-sampling process on the input maps for dimensionality reduction. No matter average-pooling or max-pooling will inevitably lead to the loss of useful information. So the multi-stage feature extraction on pooling

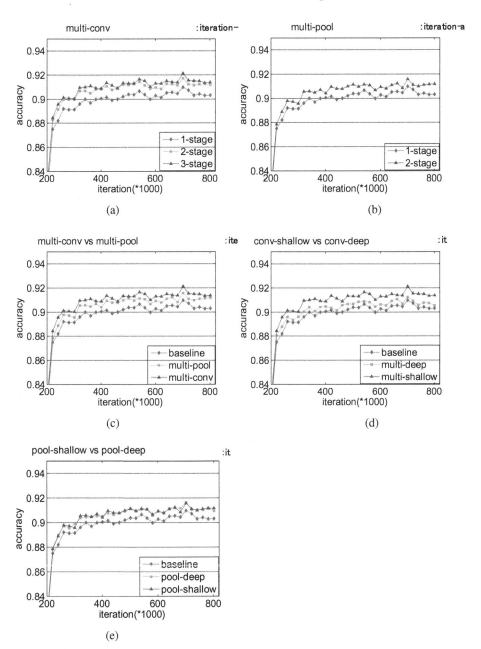

Fig. 5. The contrast in accuracy of different structure on CASIA-OLHWDB1.1. The x-axis indicates the number (*1000) of iteration.

also makes sense. A feature extraction module shown in Fig. 2 has been used to take the place of the first pooling layer in the SS-ConvNets. Because the max- pooling is more advantageous to character recognition, we will emphasis on the max-pooling layer, the num of feature maps in max-pooling is set to 100 while 50 in the average-pooling layer. To compared with SS-ConvNets, a nin dimension reduction operation is following with 100 feature maps. Figure 5b shows the accuracy of different models. Despite the ups and downs, multi-stage pooling structure shows a higher accuracy in every iteration than single-stage structure, indicating that multi-stage structure used on pooling layer could achieve a better recognition performance. Conclusion can be drawn that a multi-stage structure is beneficial to the performance of HCCR models. Details of the experiment are listed in the Table 1.

The comparison result between multi-stage convolutional network and multi-stage pooling network is shown in Fig. 5c. An interesting conclusion can be drawn from the result that the recognition accuracy improvement in the pooling layer is less significant than that of the convolution layer, indicating that the loss of useful information is mainly concentrated on the convolution layer. A reasonable explanation is that the feature extraction operation works mainly at convolution layer.

Experiment 3 In this paper, experiments on shallow CNN layer and deep CNN layer are also carried out. We replace the original single-layer structure on the 3th level pooling layer and the 4th level convolution layer with multi-stage pooling structure and multi-stage convolution layer respectively to construct two new CNN models. In the multi-stage CNN at deep convolution layer, each convolution layers have 400 feature maps, In order to keep the same dimension with the SS-ConvNets, a nin layer is used.,which have 400 feature maps. In the multi-stage CNN at deep pooling layer, max-pooling layers have 300 feature maps and average-pooling layers have 150 feature maps, the nin layer have 300 feature maps. Comparison of the experimental results are shown in Figs. 5d and e.

From Table 2, it is obvious that multi-stage feature extraction in the shallow CNN layer is better than that in the deep layer. A reasonable explanation is that high-level features are combinations of shallow-level features, any small change in shallow-level could cause butterfly effects to deep features. Details are shown in Table 2.

Table 2. Performance of multi-stage feature extraction on shallow and deep layer

	Base-CNN	Muftistage CNN at shallow conv layer	Multi-stage CNN at deep conv layer	Multi-stage CNN at shallow pool layer	Multi-stage CNN at deep pool layer
Loss	0.3734	0.3503	0.3814	0.3523	0.3457
Accuracy	0.9044	0.9160	0.9078	0.9130	0.9110
Improve(%)	n/a	0.82		0.20	

Based on experiments on single stage and multi-stage layer, convolution and pooling layer, as well as shallow and deep layer, a new structure with both multi-stage convolution structure and multi-stage pooling structure is designed. Because of the superiority of shallow multi-stage structure, the two multi-stage structures are placed at the shallow layer in the model. In shallow CNN, a new state-of-the-art of 91.89 % accuracy on the CASIA-HWDB1.1 data set is achieved. The structure is shown in Fig. 6.

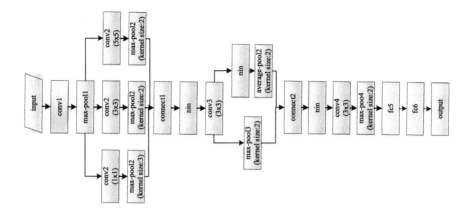

Fig. 6. Multi-stage CNN structure (MS-ConvNets)

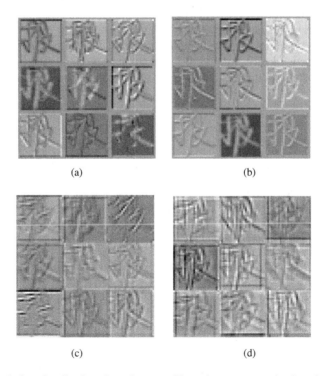

(a)   (b)

(c)   (d)

Fig. 7. Convolution visualizations from the second layer feature map activations for the character image. (a) SS-3 × 3 (b) MS-1 × 1 (c) MS-3 × 3 (d) MS-5 × 5

5   Visualization

Some insight into the mechanism of multi-stage feature extraction can be gained by novel visualization of our trained ConvNets [21]. With the multi-stage structure of the module, the multi-stage ConvNets distributions effectively capture the regularities of the data. To further demonstrate that multi-stage is superior to single-stage structure, the outputs produced by a multi-stage convolutional network are compared with that produced by a single ConvNets distributions, producing the visualizations in Fig. 7, which show the convolutional feature maps produced by 3×3 slide-window on single-stage network. In contrast to Fig. 7a which uses only fixed feature size, Fig. 7(b–d) replace the 3×3 filter size with $(1 \times 1, 3 \times 3, 5 \times 5)$ distributions respectively. It is evident that small-size filter emphasis on texture and details as shown in Fig. 7b, but large-size one laid more particular stress on the outline and structure of characters.

6   Conclusion

In this paper, two new multi-stage feature extraction modules are presented, the SS-ConvNets and MS-ConvNets are analyzed, and the convolution outputs are visualized for the convenience of comparison. Experiments on the CASIA-HWDB1.1 data set show that our best MS-ConvNets is superior to the basic single and ensemble CNN models in terms of both accuracy and storage performance. We discovered that although information loss occurs on both convolution layer and pooling layer, it is severer on the convolution layer. Moreover, our experiments show that the position of multi-stage module in the network structure has a significant impact on the final performance. To place the multi-stage module at shallow layer is important to keep much useful original information.

References

1. Liu, C.L., Yin, F., Wang, D.H., et al.: Online and offline handwritten Chinese character recognition:benchmarking on new databases. Pattern Recogn. **46**(1), 155–162 (2013)
2. Krizhevsky, A., Sutskever, I., Hinton, G.E.: Imagenet classification with deep convolutional neural networks. In: Advances in Neural Information Processing Systems, pp. 1097–1105 (2012)
3. Ciresan, D., Schmidhuber, J.: Multi-column deep neural networks for offline handwritten Chinese character classification. arXiv (2013)
4. Szegedy, C., Liu, W., Jia, Y., et al.: Going deeper with convolutions. arXiv preprint arXiv: 1409.4842 (2014)
5. Simonyan, K., Zisserman, A.: Very deep convolutional networks for large-scale image recognition. arXiv preprint arXiv:1409.1556 (2014)
6. Shah, M., Jethava, G.B.: A literature review on hand written character recognition. Indian Streams Res. J. **3**(2), 1–19 (2013)
7. Bouvrie, J.: Notes on convolutional neural networks (2006)
8. Tripathi, N.: A Survey of Regularization Methods for Deep Neural Network (2014)

9. LeCun, Y., Bottou, L., Bengio, Y., et al.: Gradient-based learning applied to document recognition. Proc. IEEE **86**(11), 2278–2324 (1998)

10. Srivastava, N., Hinton, G., Krizhevsky, A., et al.: Dropout: a simple way to prevent Neural networks from overfitting. J. Mach. Learn. Res. **15**(1), 1929–1958 (2014)

11. Goodfellow, I.J., Warde-Farley, D., Mirza, M., et al.: Maxout networks. arXiv preprint arXiv: 1302.4389 (2013)

12. Sermanet, P., Chintala, S., LeCun, Y.: Convolutional neural networks applied to house numbers digit classification. In: 2012 21st International Conference on Pattern Recognition (ICPR). IEEE, pp. 3288–3291 (2012)

13. He, K., Zhang, X., Ren, S., Sun, J.: Spatial pyramid pooling in deep convolutional networks for visual recognition. In: Fleet, D., Pajdla, T., Schiele, B., Tuytelaars, T. (eds.) ECCV 2014. LNCS, vol. 8691, pp. 346–361. Springer, Heidelberg (2014). doi: 10.1007/978-3-319-10578-9_23

14. Zhong, Z., Jin, L.: High performance offline handwritten Chinese character recognition using GoogLeNet and directional feature maps. arXiv preprint arXiv:1505.04925 (2015)

15. Ciresan, D.C., Meier, U., Schmidhuber, J.: Transfer learning for Latin and Chinese characters with deep neural networks. In: The 2012 International Joint Conference on Neural Networks (IJCNN), pp. 1–6. IEEE (2012)

16. Jia, Y.: Caffe: an open source convolutional architecture for fast feature embedding (2013). http://caffe.berkeleyvision.org/

17. Boureau, Y.L., Ponce, J., LeCun, Y.: A theoretical analysis of feature pooling in visual recognition. In: Proceedings of the 27th International Conference on Machine Learning (ICML-2010), pp. 111–118 (2010)

18. Jarrett, K., Kavukcuoglu, K., Ranzato, M.A., et al.: What is the best multi-stage architecture for object recognition? In: 2009 IEEE 12th International Conference on Computer Vision, pp. 2146–2153. IEEE (2009)

19. Goodfellow, I.J., Bulatov, Y., Ibarz, J., et al.: Multi-digit number recognition from street view imagery using deep convolutional neural networks. arXiv preprint arXiv:1312.6082 (2013)

20. Ciresan, D.C., Meier, U., Gambardella, L.M., et al.: Convolutional neural network committees for handwritten character classification. In: 2011 International Conference on Document Analysis and Recognition (ICDAR). IEEE, pp. 1135–1139 (2011)

21. Yu, W., Yang, K., Bai, Y., et al.: Visualizing and comparing convolutional neural networks. arXiv preprint arXiv:1412.6631 (2014)

22. LeCun, Y.: Generalization and network design strategies, pp. 143–155. Connections in Perspective, North-Holland (1989)

Multi-modal Brain Tumor Segmentation Based on Self-organizing Active Contour Model

Rui Liu, Jian Cheng[✉], Xiaoya Zhu, Hao Liang, and Zezhou Chen

School of Electronic Engineering, University of Electronic Science and Technology of China,
Chengdu, China
chengjian@uestc.edu.cn

Abstract. In this paper, an automatic and practical method based on active contour model (ACM) is proposed for multi-modal brain tumor segmentation. Firstly, we construct a concurrent self-organizing map (CSOM) networks. Then, applying the networks into a local region based ACM framework constructs a SOM based ACM, i.e. self-organizing active contour model (SOAC). Finally, by using SOAC, making tumor segmentation problems to be stated as a process of contour evolution. However, the segmentation task cannot be well performed for single-modal MRI images due to intensity similarities between brain normal tissues and lesions. For highlighting different tissues, between normal and abnormal, using multi-modal MRI information is an effective way to improve segmentation accuracy, obviously. Therefore, we introduce a global difference strategy, which creates a series of difference images from multi-modal MRI images, namely global difference images (GDI). By reorganizing MRI images and GDI, we propose an automatic segmentation method for brain tumor region extraction with multi-modal MRI images based on SOAC. The effectiveness of the method is tested on the real data from BRATS2013 and part of BRATS2015.

Keywords: MRI images · Brain tumor segmentation · SOAC · ACM

1 Introduction

Magnetic resonance imaging (MRI) is non-invasive imaging technique with the ability of revealing various brain tissues at good contrast and high resolution. In brain tumor studies, MRI images have a significant role in diagnosis of brain tumor, which provide detailed information of the brain. For the region of brain, it could be imaging with different contrast by different MR sequence imaging (i.e. multi-modal MRI, such as, T1, T1c, T2, and Flair). With multi-modal MRI images, the existence of abnormal tissues may be detected easily in MR scans, and varieties of methods have proposed [1–5]. However, fast and accurate segmentation of brain tumor is still a challenging task, because lesion regions are defined by variations of intensity that are compared with surrounding normal tissues. Besides, size, extension, and location of tumor vary considerably across patients. However, introducing these priors is important to improve the performance of many segmentation methods [6–8]. Therefore, effectively using information and attributes contained in different modal MRI images plays a crucial role in tumor segmentation.

© Springer Nature Singapore Pte Ltd. 2016
T. Tan et al. (Eds.): CCPR 2016, Part II, CCIS 663, pp. 486–498, 2016.
DOI: 10.1007/978-981-10-3005-5_40

Recently years, a number of semi-automatic [9–11] and fully automatic approaches [12–14] have been proposed. As the name indicates, semi-automatic methods need the intervention of human to initialize them or check the accuracy of the results, even to manually correct the segmentation results. In addition, they are subjected to subjective judgment and prior knowledge of users, which increases the possibility that different users would reach diverse conclusions of the segmentation results on the same occasions. Relative to semi-automatic methods, fully automatic methods model the priors' knowledge into automatic segmentation model by the way of machine learning. More exactly, using supervised or unsupervised learning [15–17] locates pathological tissues rather than manual work. With machine learning development, the fully automated brain tumor segmentation will become more and more popular. Although many segmentation methods have been used into medical image segmentation, such as region growing [18], clustering [19–21] and watershed [22], they are not easily applicable to the image with intensity inhomogeneity. Unfortunately, intensity inhomogeneity is ubiquitous throughout brain MRI images. For solving this problem, support vector machine (SVM) [23], Markov random fields (MRFs) [24, 25], and Artificial Neural Networks (ANNs) [26] were successively introduced into brain tumor segmentation. However, for the image with intensity similarities between lesion and normal tissues, these methods are not able to perform well.

Active Contour Model (ACM) [27], based on level set, is one of the most effective methods for medical image segmentation. For example, CV model [28], DRLSE [29] and localized CV model (LCV) [37]. It has the ability that models arbitrarily complex shape and handles topological changes of the boundary. While the image contains intensity inhomogeneity, traditional ACMs would evolve an object contour away from true object boundary. That is to say, intensity inhomogeneity would cause object leaking in traditional ACM framework. Compared with level set based ACM, SOM-based ACMs have clear superiority in handling the segmentation of images with intensity inhomogeneity. However, how to balance global and local energy terms is still a challenging problem in SOM-based ACMs. For this issue, [31] proposed a robust local-global SOM-based ACM, whereas it uses SOMs only as unsupervised way. To incorporate prior knowledge into SOM-based ACMs, [32] proposed a supervised SOM-based ACM, i.e. SOAC, which learns how to model complex shapes and intensity distributions by training examples. It is capable of handling the images with intensity similarity (e.g. complex normal/abnormal tissue intensity distribution) and has good performance for the images with intensity inhomogeneity.

In this paper, based on existing study achievement, we introduce SOAC model to the segmentation of multi-modal MRI brain tumor images. In the tumor segmentation problem, we firstly use CSOM [33, 34] to model pathological and normal tissues intensity distribution by supervised learning way, and incorporates CSOM into region based ACM framework evolving lesion region contour by Flair, T2 and GDI images. The detailed will be discussed in Chapter 3. The main idea is to use multi-modal MRI images and their difference images to locate and extract tumor region, then taking advantage of the extracted tumor region into next study, which will be displayed in our next paper (i.e. refining edema, active and core of tumor). By using local regional descriptor trained by CSOM, MRI images with intensity inhomogeneity can be efficiently segmented.

Moreover, with difference information of multi-modal MRI images, lesion region can be accurately determined by the method. Particularly, compared with existing ACMs, our proposed mixture model is insensitive to the location of initial contour, and it has good performance on multi-modal brain tumor segmentation.

The reminder of this paper is organized as follows: In Sect. 2, we describe briefly region based ACM, which mainly contains CV and local region-based CV model (LRCV) [30], and CSOM. Section 3 introduces tumor segmentation with SOAC. The experimental results and analysis are displayed in Sect. 4. Finally, Sect. 5 summaries the work of this paper.

2 Region Based ACM and CSOM

In this section, we briefly review some state-of-the-art region based ACMs, and the concurrent self-organizing map.

2.1 Region Based ACM

For object segmentation, its shape and location change across its property. Therefore, using shape and location as prior information becomes impractical. Based on this, most ACMs assume the object have a smooth boundary. Then, using some suitable regularization term to penalize energy functional.

The CV model is one of the most famous and state-of-the-art region based ACM. Starting from initial region, the evolving of contour is created by an unsupervised way with the aim of minimizing energy functional,

$$
\begin{aligned}
E_{cv}(c) = {} & \mu \cdot Length(c) + v \cdot Area(in(c)) \\
& + \lambda^+ \int_{in(c)} \left(I(x) - C^+(c) \right)^2 dx \\
& + \lambda^- \int_{out(c)} (I(x) - C^-(c))^2 dx,
\end{aligned}
\tag{1}
$$

where $I(x)$ is image, c is contour, $in(c)$ and $out(c)$ denote foreground and background. In brain tumor, they are celled tumor region and normal issue region, respectively. μ controls the smoothness of the contour, v penalizes the foreground region. $C^+(c)$ and $C^-(c)$ are the average intensity of foreground and background, respectively. λ^+ and λ^- are control parameters of energy terms of inside and outside of contour.

In CV model, it just assumes the intensity of region in foreground and background uniformly distributed. For MRI images with intensity inhomogeneity, its performance will be inferior to LRCV. In LRCV model, by replacing $C^+(c)/C^-(c)$ of CV model by $c^+(x, c)$ and $c^-(x, c)$, it build a new regional descriptor.

$$
c^+(x, c) = \frac{\int_{in(c)} g_\sigma(x - y) I(y) dy}{\int_{in(c)} g_\sigma(x - y) dy},
\tag{2}
$$

$$c^-(x, c) = \frac{\int_{out(C)} g_\sigma(x - y)I(y)dy}{\int_{out(C)} g_\sigma(x - y)dy}, \tag{3}$$

$$\int_{\mathbb{R}^2} g_\sigma(x)dx = 1, \tag{4}$$

where g_σ is a Gaussian kernel function with width σ. $c^-(x, c)$ and $c^+(x, c)$ will not be constant, i.e. average intensity of regional, but the local weighted average intensity of the image around pixel x.

2.2 Concurrent Self-organizing Map

A self-organizing map is a type of artificial neural networks (ANNs). It is different from other ANNs as they use competitive learning rather than error-correction learning (e.g., back-propagation gradient descent). Like most ANNs, SOMs operate in two modes: training and mapping. Training builds the mapping using input samples, and mapping automatically classifies a new input vector. A basic SOM is composed of an input layer, an output layer, and an intermediate connection layer. There is a neuron for each component of the input vector in the input layer, connecting the neuron of output layer by weights of intermediate connection layer. The learning of the CSOM can be summarized as follows:

(a) Let an input pattern (vector) and the weights of the neurons for each SOM selected and initialized at random, respectively. Then, determine an appropriate topological neighborhood size around "winner" neuron and a suitable initial learning rate.

(b) For each training input vector, compute the Euclidean distance from the input vector to the weights vector, and then determine the winner neuron by competitive learning (i.e. search the minimum distance).

(c) Update the weights vector of winner neuron and the neuron on its topological neighborhood.

(d) Repeat steps (a)–(c), and input another training vectors, until learning converge.

For tumor segmentation, given two sets of training pixels Fg and Bg belonging to a training image, associated with lesion tissues and normal tissues, SOMs can be used as

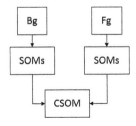

Fig. 1. CSOM model for tumor segmentation.

a concurrent system for pattern classification, shown as Fig. 1, in which each SOM is trained individually by lesion tissues and normal tissues.

3   Tumor Segmentation with SOAC

In this section, we mainly discuss multi-modal brain tumor image segmentation with SOAC. MRI images are different from other images, for solving tumor segmentation task, in Sect. 3.1, we firstly discuss preprocessing, based on the processed images, then introduce how to use these images to extract tumor region (i.e. the whole of tumor) in Sect. 3.2.

3.1   MRI Images Preprocessing

In medical image analysis, brain tumor segmentation is an important technique for extracting pathological tissues from normal tissue in brain MRI images. Under the framework of ACM, pathological tissue extraction can be described as a problem of contour evolution. As discussed above, SOAC is an effective way to address this problem. However, notice that there are inter-scan and intra-scan image intensity variations in MRI, the intensity information can no longer applied directly into SOAC. Therefore, we normalize the intensity of MRI images by a histogram normalization method [35], which defined as,

$$f(x, y) = \frac{GWM - BWM}{h_{max} - h_{min}} \left(h(x, y) - h_{min} \right) + BWM, \tag{5}$$

where $h(x, y)$ is original histogram of initial image, and h_{max} and h_{min} are its maximum and minimum gray scale level separately. $f(x, y)$ is new histogram of the image. GWM and BWM are the normalized upper and lower gray scale level of the new histogram, respectively.

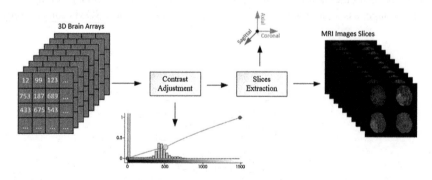

Fig. 2. MRI image contrast adjustment and slices extraction. MRI extracted slices mainly contain axial, coronal, sagittal perspective slice images in Flair, T1, T1c and T2.

After normalizing, MRI images are not very good competent to automatic tumor segmentation. Combing with tumor original character, we use a contrast adjustment by itk-snap [36] to improve lesion significance, displayed in Fig. 2.

3.2 SOAC Model

In brain tumor segmentation, intensity inhomogeneity is one of mainly problems troubled researchers. Local region-based active contour model (LRCV) is one of most popular methods for this problem, which is very robust and effective compared to traditional ACMs. However, there is a limit is that LRCV is not able to exploit supervised examples to model foreground/background intensity distributions. Besides, under without initialization, type of these methods are not capable of accomplishing automatically segmentation task. To eliminate this limit, we adopt new SOM-based LRCVs, named SOAC, which addressed efficiently the limitations of traditional LRCVs, especially for the images that contain intensity inhomogeneity. In the sense that, we proposed a mixture strategy of SOAC for multi-modal tumor segmentation problem.

The SOAC comprises a supervised training session (i.e. training CSOM) and a testing session, which can be performed off-line and on-line, respectively. In the original SOAC, once the training of CSOM has been completed (as describing in Sect. 2.2), the trained networks are applied to LRCV evolving the contour of object. In Fig. 1, after randomly initializing, the weights of neurons in SOMs contained in CSOM are updated by the following learning rule,

$$w_n^+(t+1) = w_n^+(t) + \eta(t)h_{bn}(t)[I^{tr}(x_t^+ - w_n^+(t))], \tag{6}$$

$$w_n^-(t+1) = w_n^-(t) + \eta(t)h_{bn}(t)[I^{tr}(x_t^- - w_n^-(t))], \tag{7}$$

$$\eta(t) = \eta_0 exp(-\frac{t}{\tau_\eta}), \tag{8}$$

$$h_{bn}(t) = exp(-\frac{\| r_b - r_n \|^2}{2r^2(t)}), \tag{9}$$

$$r(t) = r_0 exp\left(-\frac{t}{\tau_r}\right), \tag{10}$$

where $\eta(t)$ is the learning rate, $\eta_0 > 0$ is the initial learning rate. $h_{bn}(t)$ is selected as a Gaussian function centered on the winner neuron, and r_b and r_n are the location vectors in the output neural map of neurons b and n respectively. $r(t) > 0$ is a time-decreasing neighborhood radius, the constant perimeter r_0 is an initial neighborhood radius, and $\tau_r > 0$ is the time constant.

In the sense that, the energy function of SOAC can be defined as [6],

$$E_{SOAC}(c) = \lambda^+ \int_{in(c)} e^+(x, c)dx + \lambda^- \int_{out(c)} e^-(x, c)dx, \tag{11}$$

$$e^+(x,c) = (I(x) - w_b^+(x,c))^2, \tag{12}$$

$$e^-(x,c) = (I(x) - w_b^-(x,c))^2, \tag{13}$$

$$w_b^+(x,c) = argmin \left| w_n^+ - \frac{\int_{in(C)} g_\sigma(x-y)I(y)dy}{\int_{in(C)} g_\sigma(x-y)dy} \right| n \in \{1,2,3\ldots.FN\}, \tag{14}$$

$$w_b^-(x,c) = argmin \left| w_n^- - \frac{\int_{out(C)} g_\sigma(x-y)I(y)dy}{\int_{out(C)} g_\sigma(x-y)dy} \right| n \in \{1,2,3\ldots.BN\}, \tag{15}$$

where formula (14) and (15) can be considered as local regional intensity descriptors of foreground and background, separately.

Using level set function ϕ replaces contour curve c, i.e. c = $\{\emptyset(x) = 0\}$, the energy functional of the SOAC model can be defined as,

$$E_{SOAC}(\phi) = \lambda^+ \int_{\emptyset>0} e^+(x,c)dx + \lambda^- \int_{\emptyset<0} e^-(x,c)dx, \tag{16}$$

where it establishes an explicit dependence relationship between e^+/e^- and ϕ. The parameters λ^+, $\lambda^- \geq 0$, λ are the same with [28]. g_σ in formula (14) and (15) is a Gaussian kernel function, which is used for regulating the current level set function. From the perspective of Heaviside step function $H(\cdot)$, the energy functional can written as follow,

$$E_{SOAC}(\phi) = \lambda^+ \int_\Omega e^+(x,c)H(\Phi(x))dx + \lambda^- \int_\Omega e^-(x,c)(1 - H(\Phi(x)))dx. \tag{17}$$

Then, applying the gradient-descent technique into the energy functional obtains the following PDE,

$$\frac{\partial \phi}{\partial t} = \delta(\phi)\left[-\lambda^+ e^+ + \lambda^- e^-\right], \tag{18}$$

which describes the evolution of the object contour (i.e. tumor contour).

3.3 Tumor Region Segmentation

In brain tumor analysis, the lesion often is divided into early-stage and late-stage, and different tumor types have different forms. For example, gliomas, according to the growth rate of the pathology tissues, can be classified as low-grade and high-grade. Furthermore, there is unequal number of substructures in the different stages of the tumor, which increases the difficulty of localizing the region of the lesion. To perform on extracting the region accurately, we adopt a mixture segmentation strategy of SOAC by use of the information from different modal images (i.e. T1, T1c, T2, Flair and their difference images).

To capture intensity variation and highlight tissues information among the different MR modal images in the determined abnormal tissues region, we introduce global difference images, which are created by difference operators: *DIF* and *WDI*. *DIF* (Difference image feature) represents the simple perceptually intensity differences between multi-modal MRI images, which can be defined as,

$$DIF(i,j) = \mu_1 SubImg(i-j), s.t. \mu_1 \geq 1, \qquad (19)$$

where $SubImg(i-j)$ denotes the result that the image i subtracts the image j. μ_1 regulates the DIF intensity.

For making use of the difference information effectively, we introduce weighted difference image (*WDI*) in mixture of SOAC framework, which denotes a weighted combination between *DIFs* from two different pair of images. The *WDI* is defined as,

$$WDI(i, j, k) = \mu_2 DIF(i,j) + \mu_3 DIF(i, k), s.t. \mu_2 + \mu_3 = 1, \qquad (20)$$

where μ_2 and μ_3 regulate the weight corresponding to *DIF*. Then, *WDI* will create a new image that contains the perceptually features of the image i, j, k.

As shown in Fig. 3, multi-modal MRI images and global difference images are simultaneously used as input of SOAC. Considering that there are ambiguities in the results segmented by SOAC, we adopt a rule of regional fusion scheme to determine the tumor region.

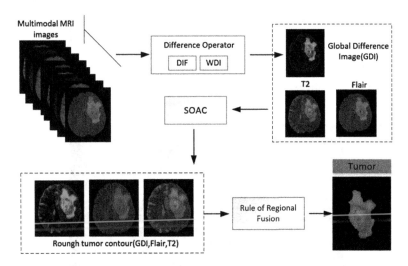

Fig. 3. The framework of multi-modal MRI images tumor segmentation by mixture of SOAC model.

4   Experimental Results and Analysis

In this section, we present numerical experimental results using SOAC, and compare to implementation of some of the state-of-the-art ACMs and level set methods, in handing brain tumor segmentation. For quantitative evolution of these methods performance, we adopt the following measures, which described in BRATS [6],

$$Dice(P,T) = \frac{|P_1 \wedge T_1|}{(|P_1 \wedge T_1|)/2},$$ (21)

$$Sensitivity(P,T) = \frac{|P_1 \wedge T_1|}{|T_1|},$$ (22)

$$Specificity(P,T) = \frac{|P_0 \wedge T_0|}{|T_0|},$$ (23)

where P and T represent the predictions of methods and ground truths, respectively. T_0 is the region of normal area in ground truths, T_1 is the lesion region. Similarly for P_0 and P_1, they are predictions corresponding to tumor and normal region, separately (Table 1).

Table 1. Evaluation summary. Segmentation performance for 25 real high-grade case and 20 real low-grade cases, in term of Dice, specificity and sensitivity about whole tumor region for SOAC, CV, LCV, DRLSE and CSOM.

		LCV	DRLSE	CV	SOAC	CSOM
Dice	Mean	0.8833	0.7836	0.8817	0.9142	0.774
	Med	0.8922	0.8052	0.8905	0.9298	0.8309
Spec	Mean	0.9848	0.9651	0.9653	0.9794	0.9217
Sens	Mean	0.8728	0.8681	0.9303	0.915	0.8678

In Figs. 4, 5 and 6, we organize three groups segmentation results, including axial coronal and sagittal slices. As shown, the mixture of SOAC get the best performance

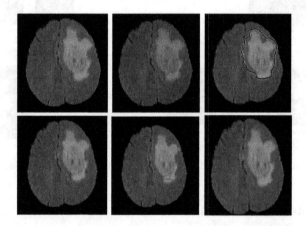

Fig. 4. Segmentation result on one slice of axial perspective about HG type. Arranged in row, from top to bottom and from left to right: the original image with ground truths, and segmentation visual representation of SOAC, CV, LCV, DRLSE and CSOM.

compared to ground truths among CV, LCV, DRLSE, CSOM and SOAC. According to definition of specificity and sensitivity, we know when specificity and sensitivity get bigger value at the same time, the segmentation result will have better performance. By testing on the real dataset of BRATS2013 and part of dataset of BRATS2015, we get the results of Figs. 7 and 8, it shows the effectiveness and robustness of segmentation of SOAC.

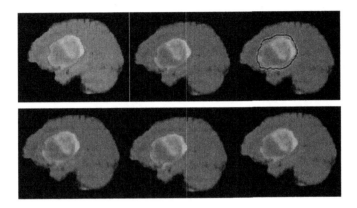

Fig. 5. A comparison on single slice of coronal perspective about LG type. Similarly to Fig. 3, the results correspond to the original image with ground truths, and segmentation visual representation of SOAC, CV, LCV, DRLSE and CSOM.

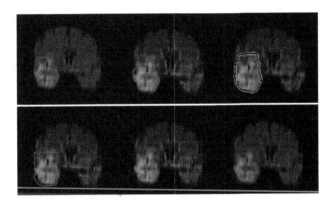

Fig. 6. A segmentation comparison on sagittal perspective slice about HG type in BRATS 2015. The first one in 1st row represents the ground truths, the segmentation results of SOAC, CV, LCV, DRLSE and CSOM display in turn following the ground truth.

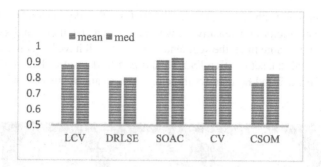

Fig. 7. Average and median dice scores of segmented tumor region for SOAC, CV, LCV, DRLSE and CSOM. By using these quantitative comparisons to evaluate these methods.

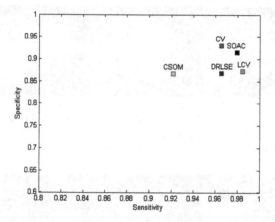

Fig. 8. Sensitivity and specificity scores.

5   Conclusion

Compared to existing ACMs and level set methods, brain tumor segmentation with SOAC has significant advantages for MRI images with intensity inhomogeneity and ill-defined edges (i.e. intensity similarity). More specifically, by using two different SOMs, it increases the discriminative property of SOAC for the images with ill-defined edges, since these SOMs are trained on the edge-map. In addition, relying on SOMs to model intensity distribution of pathological tissues and normal tissues, rather than a particular probability model, it can effectively handle contour evolving of tumor region characterized by different intensities. Therefore, as section four shown, based on these characters, SOAC is very effective for improving the accuracy of tumor segmentation with multi-modal MRI images.

Acknowledgments. This work was supported by the National Science Foundation of China (NO. 61671125 and NO. 61201271), and the State Key Laboratory of Synthetical Automation for Process Industries (NO. PAL-N201401).

References

1. Kharrat, A., Benamrane, N., Ben Messaoud, M., et al.: Detection of brain tumor in medical images. In: IEEE 3rd International Conference on Signals, Circuits and Systems, pp. 1–6 (2009)
2. Lee, M., Street, W.: Dynamic learning of shapes for automatic object recognition. In: 17th Workshop Machine Learning of Spatial Knowledge, pp. 44–49 (2000)
3. Tang, H., Wu, E.X., Ma, Q.Y., et al.: MRI brain image segmentation by multi-resolution edge detection and region selection. Comput. Med. Imaging Graph. **24**(6), 349–357 (2000)
4. Prastawa, M., Bullitt, E., Ho, S., et al.: A brain tumor segmentation framework based on outlier detection. Med. Image Anal. **8**(3), 275–283 (2004)
5. Islam, A., Reza, S., Iftekharuddin, K.M.: Multifractal texture estimation for detection and segmentation of brain tumors. IEEE Trans. Biomed. Eng. **60**(11), 3204–3215 (2013)
6. Menze, B., Reyes, M., Van Leemput, K., et al.: The multimodal brain tumor image segmentation benchmark (BRATS). IEEE Trans. Med. Imaging **34**(10), 1993–2024 (2014)
7. Prastawa, M., Bullitt, E., Moon, N., et al.: Automatic brain tumor segmentation by subject specific modification of atlas priors. Acad. Radiol. **10**(12), 1341–1348 (2003)
8. Zacharaki, E.I., Wang, S., Chawla, S., et al.: Classification of brain tumor type and grade using MRI texture and shape in a machine learning scheme. Magn. Reason. Med. **62**(6), 1609–1618 (2009)
9. White, D.R.R., Houston, A.S., Sampson, W.F.D., et al.: Intra and inter-operator variations in region-of-interest drawing and their effect on the measurement of glomerular filtration rates. Clin. Nucl. Med. **24**(3), 177–181 (1999)
10. Olabarriaga, S.D., Smeulders, A.W.M.: Interaction in the segmentation of medical images: a survey. Med. Image Anal. **5**(2), 127–142 (2001)
11. Foo, J.L.: A survey of user interaction and automation in medical image segmentation methods. Iowa State University, Human Computer Interaction Technical report ISU-HCI-2006-02 (2006)
12. Ho, S., Bullitt, L., Gerig, G.: Level-set evolution with region competition: automatic 3-D segmentation of brain tumors. In: IEEE 16th International Conference on Pattern Recognition, vol. 1, pp. 532–535 (2002)
13. Vijayakumar, C., Gharpure, D.C.: Development of image-processing software for automatic segmentation of brain tumors in MR images. J. Med. Phys Assoc. Med. Physicists India **36**(3), 147 (2011)
14. Wang, Y., Lin, Z.X., Cao, J.G., et al.: Automatic MRI brain tumor segmentation system based on localizing active contour models. Adv. Mater. Res. **219**, 1342–1346 (2011)
15. Lee, C.P., Snyder, W., Wang, C.: Supervised multispectral image segmentation using active contours. In: IEEE International Conference on Robotics and Automation, pp. 4242–4247 (2005)
16. Nabizadeh, N., Kubat, M.: Brain tumors detection and segmentation in MR images. Gabor wavelet vs. statistical features. Comput. Electr. Eng. **45**, 286–301 (2015)
17. Popuri, K., Cobzas, D., Murtha, A., et al.: 3D variational brain tumor segmentation using Dirichlet priors on a clustered feature set. Int. J. Comput. Assist. Radiol. Surg. **7**(4), 493–506 (2012)
18. Deng, W., Xiao, W., Deng, H., et al.: MRI brain tumor segmentation with region growing method based on the gradients and variances along and inside of the boundary curve. In: IEEE 3rd International Conference on Biomedical Engineering and Informatics (BMEI), vol. 1, pp. 393–396 (2010)

19. Phillips, W.E., Velthuizen, R.P., Phuphanich, S., et al.: Application of fuzzy c-means segmentation technique for tissue differentiation in MR images of a hemorrhagic glioblastoma multiforme. Magn. Reason. Imaging **13**(2), 277–290 (1995)
20. Siyal, M.Y., Yu, L.: An intelligent modified fuzzy c-means based algorithm for bias estimation and segmentation of brain MRI. Pattern Recogn. Lett. **26**(13), 2052–2062 (2005)
21. Kannan, S.R.: A new segmentation system for brain MR images based on fuzzy techniques. Appl. Soft Comput. **8**(4), 1599–1606 (2008)
22. Cates, J.E., Whitaker, R.T., Jones, G.M.: Case study: an evaluation of user-assisted hierarchical watershed segmentation. Med. Image Anal. **9**(6), 566–578 (2005)
23. Lee, C.-H., Schmidt, M., Murtha, A., Bistritz, A., Sander, J., Greiner, R.: Segmenting brain tumors with conditional random fields and support vector machines. In: Liu, Y., Jiang, T.-Z., Zhang, C. (eds.) CVBIA 2005. LNCS, vol. 3765, pp. 469–478. Springer, Heidelberg (2005)
24. Zhang, Y., Brady, M., Smith, S.: Segmentation of brain MR images through a hidden Markov random field model and the expectation-maximization algorithm. IEEE Trans. Med. Imaging **20**(1), 45–57 (2001)
25. Held, K., Kops, E.R., Krause, B.J., et al.: Markov random field segmentation of brain MR images. IEEE Trans. Med. Imaging **16**(6), 878–886 (1997)
26. Özkan, M., Dawant, B.M., Maciunas, R.J.: Neural-network-based segmentation of multimodal medical images: a comparative and prospective study. IEEE Trans. Med. Imaging **12**(3), 534–544 (1993)
27. Yan, L.I.: A survey on active contour models for image segmentation. Remote Sens. Inf. **1**, 233–236 (2014)
28. Chan, T.F., Vese, L.A.: Active contours without edges. IEEE Trans. Image Process. **10**(2), 266–277 (2001)
29. Li, C., Xu, C., Gui, C., et al.: Distance regularized level set evolution and its application to image segmentation. IEEE Trans. Image Process. **19**(12), 3243–3254 (2010)
30. Liu, S., Peng, Y.: A local region-based Chan-Vese model for image segmentation. Pattern Recogn. **45**(7), 2769–2779 (2012)
31. Abdelsamea, M.M., Gnecco, G.: Robust local–global SOM-based ACM. Electron. Lett. **51**(2), 142–143 (2015)
32. Abdelsamea, M.M., Gnecco, G., Gaber, M.M.: An efficient self-organizing active contour model for image segmentation. Neuro-Computing **149**, 820–835 (2015)
33. Kohonen, T.: The self-organizing map. Neurocomputing **21**(1), 1–6 (1998)
34. Neagoe, V.E., Ropot, A.D.: Concurrent self-organizing maps for pattern classification. In: IEEE 1st International Conference on Cognitive Informatics, pp. 304–312 (2002)
35. Loizou, C.P., Pantziaris, M., Seimenis, I., et al.: Brain MRI image normalization in texture analysis of multiple sclerosis. In: IEEE 9th International Conference on Information Technology and Applications in Biomedicine, pp. 1–5 (2009)
36. Yushkevich, P.A., Piven, J., et al.: User-guided 3D active contour segmentation of anatomical structures: significantly improved efficiency and reliability. Neuroimage **31**(3), 1116–1128 (2006)
37. Lankton, S., Tannenbaum, A.: Localizing region-based active contours. IEEE Trans. Image Process. **17**(11), 2029–2039 (2008). A Publication of the IEEE Signal Processing Society

A New Subcellular Localization Predictor for Human Proteins Considering the Correlation of Annotation Features and Protein Multi-localization

Hang Zhou[1], Yang Yang[2(✉)], and Hong-Bin Shen[1(✉)]

[1] Key Laboratory of System Control and Information Processing,
Institute of Image Processing and Pattern Recognition, Shanghai Jiao Tong University,
Ministry of Education of China, Shanghai, China
{zhouhang2,hbshen}@sjtu.edu.cn
[2] Department of Computer Science, Shanghai Jiao Tong University, Shanghai 200240, China
yangyang@cs.sjtu.edu.cn

Abstract. Identifying a protein's subcellular localization is meaningful to understand the function of the protein. While experimental method to identify the subcellular localization of proteins will cost a lot of time, it is necessary to utilize computational approaches for dealing with large scale proteins of unknown location. Current predictors mostly consider the annotation-based features but few of them take their correlation into account. Moreover, most of predictors can only deal with single-locational proteins, while a lot of proteins bear multi-locational characteristics, which play important roles in many biological processes. In this paper, we propose a novel prediction method, which extracts features from prior biological knowledge by considering the correlation between annotation terms. The new method can also deal with the multi-localization problem. We compared the performance of the proposed method with other predictors on four datasets. The result shows that our method is outperform than others.

Keywords: Subcellular localization · Multi-label · Correlation · Gene Ontology

1 Introduction

The information of protein subcellular localization is crucial for understanding molecular function and related biological process of proteins. Since it is labor-intensive and time-consuming to identify a protein's cellular compartment by biological experiments, in-silico tools for the prediction of locations are of great necessity in addressing large scale data sets of proteins with unknown locations. According to SWISS-PROT knowledgebase [1] released in January 2012, among the total of 534242 proteins, only 66203 proteins have defined subcellular localization annotations while 247504 proteins have uncertain location annotations. Machine learning-based computational tools, which allow automatic prediction for the proteins with unknown locations by utilizing available subcellular location annotations, have been largely developed for the last decade. Moreover, as protein sequences and various annotation data grow rapidly in public databases,

© Springer Nature Singapore Pte Ltd. 2016
T. Tan et al. (Eds.): CCPR 2016, Part II, CCIS 663, pp. 499–512, 2016.
DOI: 10.1007/978-981-10-3005-5_41

more available information could be used in computational tools to provide more precise predictions, especially for some difficult issues, such as the locations with very few known examples, or the proteins with multiple locations.

The computational prediction methods mainly consist of two types of features. One is annotation-based and the other is sequence-based. Sequence-based features include amino acid composition [2, 3], amino acid pair [4, 5], pseudo-amino-acid composition [6–8], evolutionary information [9, 10] and sorting signals [11, 12]. As for annotation-based features, Chou and Cai pioneered the use of functional domain composition [13] and Gene Ontology (GO) methods [14], and enhanced the prediction accuracy considerably. A lot of following methods incorporate these two kinds of annotation data [15–18]. The annotation data can be regarded as prior domain knowledge, which is essential for precise prediction. Thus the annotation-based methods generally achieve higher accuracy than sequence-based methods when the proteins have sufficient annotations. But because the annotation dataset is usually incomplete in the public databases, many predictors tend to use both the two types of features in order to deal with lack of annotation for some new or unknown proteins.

By using a Bernoulli event model or a multinomial event model, the annotation-based methods often result in an extremely high dimensional feature space. On the contrary, each protein actually contains only a few of the terms or characteristics. According to our statistics, proteins in SWISS-PROT which have at least one GO term are annotated by 6 GO terms on average. The high dimensional feature vectors increase the complexity of machine learning models and also influence the prediction performance.

Such feature vector model is not an ideal model either for a precise prediction or for an efficient computation. Some predictors have noticed this problem. YLoc [12] selects only the GO terms and PROSITE patterns which are typical for particular subcellular locations. Thus, it can reduce unnecessary features and make the result more interpretable, but it may lead to lose of information. WegoLoc [16] assigns a weight for each GO term and it can highlight useful GO terms. It is well known that GO terms are organized in a hierarchical structure in three directed acrylic graphs (DAGs), i.e., biological process (BP), molecular function (MF) and cellular component (CC). The terms are correlated with paths consisting of different types of relationship (e.g. part_of, is_a) in the DAGs. Although nearly all annotation-based methods contain GO features, few of them consider the correlation between GO terms.

In this paper, we construct a predictor for human proteins, and mainly focus on effective extraction of features from annotation data. We select Gene Ontology terms [19] as the source of annotation-based features. In order to consider the hierarchical structure of GO as well as term correlation, we firstly extract all human proteins from SWISS-PROT and retrieve these proteins' GO terms which have experimental support according to GO Database. Then we use these GO terms to build three matrixes corresponds to BP, MF and CC respectively by using Yang's random walks [20]. These matrixes not only represent the relationship between each pair of GO terms, but also show the weights of every GO terms. In addition, current predictors often put the three parts of BP, MF and CC together, which may lose their own information, so we use seven aspects to describe GO terms ({bp, mf, cc}, {bp&mf, bp&cc, mf&cc}, {bp&mf&cc}).

Another issue worthy attention is that many proteins simultaneously exist at multiple subcellular locations. However most of the previous studies simply neglected the multi-locational proteins. Actually, such kind of proteins usually plays important roles in biological processes. What is more, according to SWISS-PROT database released in January 2012, we observed 3129 human proteins with specific subcellular localization and 823 of them have multiple subcellular locations. It means that almost 30 % of human proteins have multiple subcellular locations. Therefore, specific strategies and evaluation measures are needed to deal with the multi-locational cases. Several studies, such as the methods proposed by Cai et al. [21], Chen et al. [22], Yang et al. [23], YLoc [12] and Hum-mPLoc2 [15] can predict proteins with multiple locations. We regard the prediction of multiple locations as a necessary capability of the tools for protein subcellular localization. Thus, in this paper, we include multi-locational proteins in our data set, and develop a classification method based on Binary Relevance [24] to deal with multiple subcellular locations. Our method can predict the following 12 subcellular locations: centrosome, cytoplasm, cytoskeleton, endoplasmic reticulum, endosome, extracell, Golgi apparatus, lysosome, mitochondrion, nucleus, peroxisome and plasma membrane.

We compare our method with other subcellular localization predictors on a new test set named HumT extracted from SWISS-PROT released in May 2015 and other three datasets (animal proteins in BacelLo Database [10], animal proteins in Höglund dataset [25] and DBMLoc Dataset [26]) which were used in YLoc and other subcellular location predictors. The results demonstrate that the prediction performance can be improved significantly benefited by the new feature extraction approach, and the multi-locational proteins can also be predicted with high accuracy.

2   Materials and Methods

2.1   Data Set

Since this study focuses on developing a prediction tool of subcellular locations mainly for human proteins, we construct a benchmark data set named HumB by collected all the qualified human proteins from SWISS-PROT released on January 2012. The following filtering process is conducted to ensure high data quality: (a) Exclude proteins which have no subcellular localization annotation or have uncertain annotation with keywords like "by similarity", "potential" and "probable"; (b) Set a cutoff value of 25 % for sequence identity to eliminate redundant sequences and reduce bias, by using PISCES [27]. Here we consider 12 subcellular localizations, namely centrosome, cyto-plasm, cytoskeleton, endoplasmic reticulum, endosome, extracell, Golgi apparatus, lysosome, mitochondrion, nucleus, peroxisome and plasma membrane. Thus, proteins located with other compartments are filtered out. Finally, we obtain a data set (shown in Table 1) including 3129 human proteins, 2306 of which exist at single subcellular location and others with two or more subcellular locations are also considered in our predictor. A protein with more than one location is regarded as a multi-labeled sample. Therefore, this benchmark data set includes a total of 4229 labels, and each protein has 1.35 labels on average.

Table 1. The HumB and HumT dataset

Subcellular location	HumB	HumT
Centrosome	93	22
Cytoplasm	1039	159
Cytoskeleton	232	41
Endoplasmic reticulum	207	41
Endosome	65	15
Extracellular	336	13
Golgi apparatus	146	20
Lysosome	81	8
Mitochondrion	359	60
Nucleus	1034	105
Peroxisome	36	2
Plasma membrane	601	55
Total protein	3129	379
Total label	4229	541

Besides the benchmark set, we also form an independent dataset named HumT for prediction. The independent dataset is collect from SWISS-PROT release in May 2015 and exclude the proteins which already existing specific subcellular localization annotation from SWISS-PROT release in January 2012. We only collect the human proteins which CC field contain {ECO:0000269}, which represent a type of experimental evidence. In additional, in order to reduce bias, Sequence similarity between HumB and HumT is below 25 %. Finally, this test set consists of 379 human proteins and 541 labels.

In addition, in order to conduct more comprehensive comparisons with the existing cutting-edge prediction tools, we also perform experiments on the animal proteins in BaCelLo database, animal proteins in Höglund data set and DBMLoc Data set. All of these three data sets have been widely used for comparison by other predictors including YLoc.

BaCelLo and Höglund data set include only mono-locational proteins, while DBMLoc data set contains a large proportion of multi-locational proteins.

2.2 Methods

We have developed a machine learning method to solve the protein subcellular localization problem. The classifier consists of both sequence-based features and annotation-based features. The flowchart is showed in Fig. 1.

Sequence-Based Feature. We record the frequencies of 20 amino acids in protein sequences as features, i.e. amino acid composition (AAC). In order to consider evolutional information, we also adopt Position Specific Scoring Matrix (PSSM) in constructing sequence-based features. PSI-Blast [28] is used to get Position Specific Scoring Matrix SPSSM as follows.

MVTPALQMKKPKQFCRRMGQKKQRP
ARAGQPIISSSDAAQAPAEQPIISSSD
AAQAPCPRERCLGPPTTPGPY······

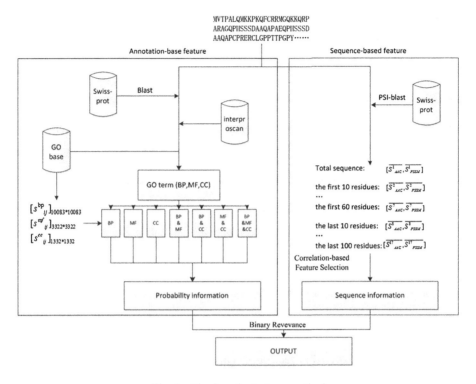

Fig. 1. The flowchart of our method

$$
S_{PSSM} = \begin{pmatrix}
S_{1,1} & S_{1,2} & \cdots & S_{1,20} \\
S_{2,1} & S_{2,2} & \cdots & S_{2,20} \\
\vdots & \vdots & \vdots & \vdots \\
S_{i,1} & S_{i,2} & \cdots & S_{i,20} \\
\vdots & \vdots & \vdots & \vdots \\
S_{L,1} & S_{L,2} & \cdots & S_{L,20}
\end{pmatrix}
\tag{1}
$$

Where $S_{i,j}$ represents the original score that the ith amino acid mutates to jth amino acid during the evolution process and L represents the length of the query sequence. The scores are normalized on each row respectively to reduce bias. Then for each column, an average score is calculated. Finally a 20-Dim vector as normalized PSSM vector for each protein is constructed.

We combine amino acid composition and the normalized PSSM vector to obtain a 40-Dim vector for sequence features.

Moreover, considering that protein subcellular localization information is often implied in the N terminal and C terminal sequence segments, so we select the first 10, 20 … 60 residues as N terminal information and the last 10, 20 … 100 residues as C terminal information. For each part a 40-Dim vector is created using the same method. After concatenating all these 40-Dim vectors, the overall dimensionality is 680. In order to reduce the redundancy in the feature vector, the Correlation-based Feature Selection

[29] (CFS) method is adopted, and it results in a 43-Dim vector, which constitutes the sequence-based part in the final feature vector.

Annotation-Based Feature. Gene Ontology [19] is a structured and controlled gene vocabulary, it consist of three aspects: BP (biological process), MF (molecular function) and CC (cellular component). We download Gene Ontology database (release 146) from ftp://ftp.ebi.ac.uk/pub/databases/GO/goa/human and extract GO features with experimental evidence, it contains Traceable Author Statement (TAS), Inferred from Physical Interaction (IPI), Inferred from Direct Assay (IDA), Inferred from Mutant Phenotype (IMP), Inferred from Expression Pattern (IEP), Inferred from Genetic Interaction (IGI) and Inferred from Experiment (EXP). The pipeline for extracting GO feature is as the following.

Step 1: Build GO terms similarity matrixes. The GO terms are searched for all the human proteins in SWISS-PROT, and a total of 10083 BP terms, 3322 MF terms and 1332 CC terms are obtained. Here we use Yang's random walks [20] to calculate pairwise correlation between terms from BP, MF, and CC, respectively, thus three matrices, namely $[s_{ij}^{bp}]_{10083*10083}$, $[s_{ij}^{mf}]_{3322*3322}$ and $[s_{ij}^{cc}]_{1332*1332}$ are constructed. Yang's random walks method is based on a basic semantic similarity approach, which explores the structure of GO DAG and considers the GO term's uncertainty to improve the measurement of pairwise similarity. In this paper, we choose Information-based Measure [30] as the basic method.

Step 2: Homolog search. For each protein in the data set (including both training and test proteins), its homologous proteins, i.e., the proteins which have more than 50 % sequence identity and 60 % positives with query protein, are searched by BLAST in SWISS-PROT. Then, GO terms annotated for the homologous proteins are retrieved and regarded as the query protein's representative GO terms. Because some newly found sequences may have very few homologous proteins in SWISS-PROT, in order to have more coverage, we also extract query sequence's GO terms by using InterPro 53.0 [31].

Step 3: GO-based feature creation. We calculate the query protein's probability information by making use of these GO terms instead of selecting only the GO terms typical for particular subcellular locations as features. Let $fdst_{i,k}$ denotes the correlation between the ith GO terms in query sequence and the GO set of the kth protein in training set, m denote the number of GO terms in kth protein and $s_{i,j}$ denote the relationship between the ith feature and the jth feature. Here we adopt the integration strategy of max function as shown in Eq. 2.

$$fdst_{i,k} = \max_{1 \le j \le m} s_{i,j}(f_i \in queryseq, f_j \in protein_k) \tag{2}$$

In order to consider BP, MF and CC part's own information and their correlation, we divide these GO terms into seven parts: ({bp, mf, cc}, {bp&mf, bp&cc, mf&cc}, {bp&mf&cc}) and in each part we use Euclidean distance to measure the relationship between the query sequence and proteins in training set. Let sim_k denote the relationship between the query sequence and the kth protein in training set and n denotes the total GO terms in this part (Fig. 2).

$$sim_k = \sqrt{\sum_{i=1}^{n} fdst_{i,k}^2} \tag{3}$$

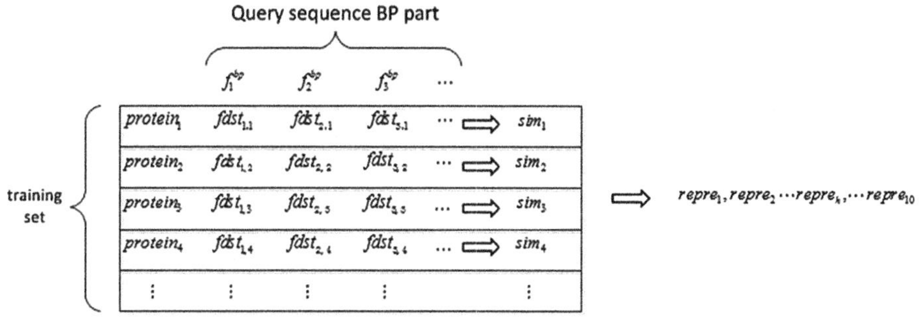

Query sequence BP part

Fig. 2. The procedure of getting BP part's representative proteins

Finally in each part we select ten proteins in the training set ($repre_1, repre_2 \cdots repre_h, \cdots repre_{10}$) as representative proteins which share highest relationship with query sequence. In this method, we set prior probability according to the training data and use Bayesian viewpoint to calculate the probability information of query sequence's subcellular location. The reason is that even though our method use GO terms' correlations and have more coverage, some query proteins still could not find any representative proteins in training set or only have a bit of representative proteins. We add prior probability could reduce the bias if the query sequence lacks sufficient representative proteins in this part. Let pro_a denote the probability of this query sequence locate in ath location, loc_a denote the ath location, num_a denote the number of the training sequence locate in ath subcellular location and num denote the total number of training sequence.

$$pro_a = \frac{\sum_{repre_h \in loc_a} sim_h + num_a / num}{\left(\sum_{repre_h} sim_h + 1 \right)} \tag{4}$$

Multi-label Classification. The sequence-based features and features extracted from GO terms are integrated and $(43 + 12 * 7)$-Dim feature vectors are generated.

These features are fed to SVMs [32]. There are a total of 12 class labels corresponding to 12 subcellular locations. By using a one-versus-rest strategy, 12 binary classifiers are trained. We choose GFO [33] to calculate the optimal G parameter and select the optimal C parameter by 10-fold validation. In the test phase, each test sample obtains a 12-Dim score vector, with each dimension representing the confidence of belonging to a certain subcellular location. If one of the scores is positive, we consider the test sample is in this location and if all the scores are negative, the subcellular localization with maximal score in the vector will be considered as the result.

Evaluation Criteria. In order to evaluate the performance of the proposed method comprehensively, we have used multiple measurements. It should be noted that specific evaluation measures are needed to deal with the multi-locational cases. Thus the measurements used in this paper are tailored for multi-label classification. The first measurement is overall accuracy (ACC) as shown in Eq. 5.

Another major measurement used in this study is F1-score (F1). Overall accuracy index consider every protein has equal weight and F1-score let every class share the same weight. Since F1 is a measurement considering precision and recall comprehensively, in general, F1-score is more suitable to deal with imbalanced dataset.

$$ACC = \frac{1}{Num} \sum_{\{i|i \in A\}} \frac{TP_i}{TP_i + FP_i + FN_i} \tag{5}$$

$$PRE_k = \sum_{\{i|i \in B_k\}} \frac{TP_i}{TP_i + FP_i} \tag{6}$$

$$REC_k = \sum_{\{i|i \in C_k\}} \frac{TP_i}{TP_i + FN_i} \tag{7}$$

$$F1_k = \frac{2 * PRE_k * REC_k}{PRE_k + REC_k} \tag{8}$$

$$F1 = \sum_{k=1}^{M} F1_k / M \tag{9}$$

Where Num denotes the total number of testing proteins, TP_i denotes the number of the ith protein's true positive, FP_i denotes the number of the ith protein's false negative, FN_i denotes the number of the ith protein's false negative, A denotes the testing dataset, B_k denotes the subset of testing set if the proteins are predicted locate in kth subcellular location, C_k denotes the subset of testing set if the proteins locate in kth subcellular location and M denotes the number of subcellular location.

3   Experimental Results

3.1   Comparisons of Different Feature Types

We compare four different feature extraction approaches to demonstrate whether the correlation of annotation-based feature can help improve prediction accuracy, these approaches are namely SEQ+GO, SEQ+GO$_t$, SEQ+GO$_{knn}$ and SEQ. Details about the feature extraction are summarized as follows.

SEQ+GO: the combination of sequence and GO features, i.e. the proposed method in this paper.

SEQ+GO$_t$: in this approach, GO terms from BP, MF and CC are used as a whole set, while SEQ+GO consider 7 different groupings of GO terms.

SEQ+GO$_{knn}$: the combination of sequence and conventional GO features. In this approach GO terms sign a value of one if the query proteins contain this GO terms and in order to conduct a fair comparison, the GO terms are also converted to probabilistic information by the following strategy: the similarity between a query protein and the kth protein in training set is defined as $sim_k = 1 - \text{hit}_k \Big/ \sqrt{num_{query} * num_k}$, where sim$_k$ is the similarity between the query protein and the kth protein in training set, hit$_k$ denotes the number of common GO features of these two proteins, num$_{query}$ denotes the number of the query protein's GO feataure vector, and num$_k$ denotes the number of the kth protein's vector. Then 10-nearest neighbors are used to calculate the probability information.

SEQ: solely make use of sequence-based features.

GO features play an important role in most of these approaches. However, GO database is incomplete, newly found proteins often have rare annotations. In order to reduce bias, we blast their homologs in SWISS-PROT, and only use the GO terms from their homologs (expect the query protein itself) in the following feature extraction procedure. All of these four approaches are tested on the aforementioned four data sets. BaCelLo, Höglund and HumB are evaluated on the paired training and test sets. Because the DBMLoc dataset only contain training set, we make use of nested 5-fold cross-validation to evaluate the prediction performance. The results are showed in Fig. 3.

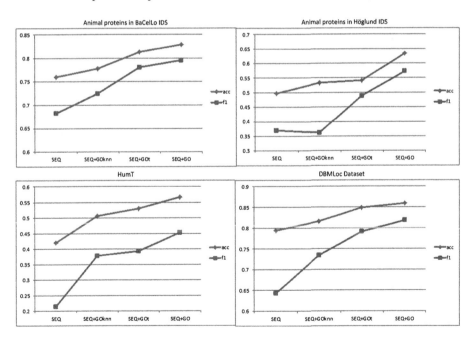

Fig. 3. Result of our method and other three methods in four dataset

Generally, adding more information into the feature vectors improve the prediction performance, which shows the usefulness of the proposed new types of features considering the correlation between GO terms. The SEQ method apparently cannot ensure a good prediction performance. The SEQ+GO$_{knn}$ method which extract GO information in a conventional style, i.e., regardless of the semantic relationship between GO terms, obtains the Acc and F1 around 5 % lower than those of SEQ+GO$_t$. The pair of SEQ +GO$_t$ and   SEQ+GO extracts features in the same way and based on almost the same information source. The only difference is that when extracting GO term similarity, the former one treat GO terms from the three sets (BP, MF and CC) altogether in a single set, while the later one consider all the combinations of these three sets. It seems that the later one contains some redundancy among its features, but it achieves obvious better performance on the four data sets. This may due to two reasons. One reason is that computation of GO set similarity on the whole set does not fully utilize the correlation between different categories of GO terms. The other reason is that reusing GO sets strengthens the impact of GO correlation-based features, making the GO information dominate the feature vector, which is beneficial for the classification in most cases.

3.2 Comparison with Other Predictors

We also compare our method with other state-of-the-art predictors on four data sets. YOC+could only predict nine subcellular localizations while HumT dataset cover 12 subcellular locations, so when we calculate the ACC and F1 scores, we only choose the 9 locations. The results are shown in Table 2. Apparently, both Acc and F1 are greatly improved using the new method. For the two mono-locational data sets, BacelLo's and Höglund's, the Accs of the new method are over 4 % and 6 % higher than the best Accs obtained by other predictors, respectively. For the two multi-locational data sets, DBMLoc and the newly constructed independent test set (HumT), the improvement is even more significant. Both Acc and F1 of the new methods are over 15 % higher than those obtained by YLoc+, which demonstrates the superiority of the new method in dealing with multi-locational proteins.

Table 2. Prediction performance in four dataset (ACC/F1)

	B animal	H animal	HumT	DBMLoc
YLoc-LowRes	0.79/0.75	–	–	–
YLoc-HighRes	0.74/0.69	0.56/0.34	–	–
YLoc+	0.58/0.67	0.53/0.37	0.45/0.34	0.64/0.68
MultiLoc2-LowRes	0.73/0.76	–	–	–
MultiLoc2- HighRes	0.68/0.71	0.57/0.41	–	–
BaCelLo	0.64/0.66	–	–	–
This method	0.83/0.79	0.63/0.57	0.61/0.55	0.86/0.82

YLoc+ and our method could deal with multiple localized proteins, YLoc-HighRes and MultiLoc2-HighRes could predict nine subcellular locations and YLoc-LowRes, MultiLoc2- LowRes and BaCelLo could only predict globular proteins.

There are three possible reasons for our method outperform than YLoc:

- YLoc only considers the annotation-based features which are typical for a location. But in fact there contain a lot of annotation-based feature which may not have high confidence to this location but also could help prediction, our method consider all the features and it will cover more proteins with annotation-based feature.
- YLoc assigned the value of one if the protein contains the feature and our method define the relationship between each feature based on semantic similarity, it could gain more information and find more similar proteins to improve the statistic performance.
- The annotation database increases with time, our method use the newly Gene Ontology database and the database have more coverage.

4   Discussions

In our method we combine sequence-based and annotation-based features for prediction. Annotation features correspond to prior biological knowledge. They often show higher quality than sequence-based features, and seem to dominate the prediction result. However, statistical sequence features also play an indispensable role for the prediction task, especially when the annotation data is incomplete or unreliable. For example, the human protein C9JR72, its homologous proteins have very few annotations, with only one GO term, GO:0005515. Thus sequence features play the leading role.

Most of the existing annotation-based methods adopt binary coding to represent the presence/absence of annotation terms. The major change of the newly proposed feature extraction approach considers correlation between annotation features. Although the semantic similarity between GO terms has been a well-studied subject that various metrics have been proposed [23, 34], it has seldom been used in the construction of features vectors for protein classification problems, since it is not trivial to convert the similarity into feature vectors. In this study, we directly use the similarity between two GO sets as the similarity between their annotated proteins, and find nearest 10 neighbors for each query protein. Then, according to the locations of neighbors and similarities, probabilistic information is computed and used as features. The resulted feature vectors not only contain more information than the conventional method but also have much lower dimensionality.

5   Conclusion

It is very important to identify a protein subcellular localization for understanding the protein functions, and a lot of predictor had been established for this task. But there are still some challenges on improving the performance of the current predictors. Two major challenges are: (a) How to effectively incorporate prior biological knowledge into the prediction tools; (b) How to deal with proteins with multiple locations.

Benefiting by the rapid accumulation of various biological annotation data, the predictors using annotation data for protein subcellular localization have achieved significantly higher accuracy than solely sequence features-based methods. However, the prediction results are not always good, especially when the query protein has rare annotation data. Moreover, exiting annotation-based methods often directly regard each annotation term as a feature without considering the structural characteristics of annotation database or relationship between terms. Therefore, the domain knowledge has not been utilized sufficiently. In this study, we choose Gene Ontology terms as annotation-based features and consider their correlation and hierarchy structure. Using this correlation, we can address the problem that query sequence's annotation is scarce, e.g., GO terms are not found or only occur a few times. Besides the consideration of correlation, a difference between the proposed method and other methods is that we select ten representative proteins and use weighted KNN to measure the probabilistic information for query protein's subcellular locations. Finally, the feature vectors are consist of sequence features and annotation-based probability features, which has a much condensed dimensionality than that obtained by using annotation terms directly.

In additional, the new method is designed to handle multi-locational proteins. The binary relevance method is used for the multi-label classification task. In order to systematically evaluate the proposed predictor, a series of experiments have been conducted on previous published data sets as well as HumT collected by ourselves. We have compared the performance of adding each type of annotation features in our method, and also compared our predictor with other popular predictors. These experiments show apparent superior to the existing predictors on prediction accuracy.

References

1. Boeckmann, B., Bairoch, A., Apweiler, R., et al.: The SWISS-PROT protein knowledgebase and its supplement TrEMBL in 2003. Nucleic Acids Res. **31**(1), 365–370 (2003)
2. Cedano, J., Aloy, P., Perez-Pons, J.A., et al.: Relation between amino acid composition and cellular location of proteins. J. Mol. Biol. **266**(3), 594–600 (1997)
3. Emanuelsson, O., Nielsen, H., Brunak, S., et al.: Predicting subcellular localization of proteins based on their N-terminal amino acid sequence. J. Mol. Biol. **300**(4), 1005–1016 (2000)
4. Park, K.J., Kanehisa, M.: Prediction of protein subcellular locations by support vector machines using compositions of amino acids and amino acid pairs. Bioinformatics **19**(13), 1656–1663 (2003)
5. Nakashima, H., Nishikawa, K.: Discrimination of intracellular and extracellular proteins using amino acid composition and residue-pair frequencies. J. Mol. Biol. **238**(1), 54–61 (1994)
6. Shen, H.B., Chou, K.C.: PseAAC: a flexible web server for generating various kinds of protein pseudo amino acid composition. Anal. Biochem. **373**(2), 386–388 (2008)
7. Chou, K.C., Shen, H.B.: Hum-PLoc: a novel ensemble classifier for predicting human protein subcellular localization. Biochem. Biophys. Res. Commun. **347**(1), 150–157 (2006)
8. Shen, H.B., Chou, K.C.: Hum-mPLoc: an ensemble classifier for large-scale human protein subcellular location prediction by incorporating samples with multiple sites. Biochem. Biophys. Res. Commun. **355**(4), 1006–1011 (2007)
9. Xie, D., Li, A., Wang, M., et al.: LOCSVMPSI: a web server for subcellular localization of eukaryotic proteins using SVM and profile of PSI-BLAST. Nucleic Acids Res. **33**(Suppl. 2), W105–W110 (2005)
10. Pierleoni, A., Martelli, P.L., Fariselli, P., et al.: BaCelLo: a balanced subcellular localization predictor. Bioinformatics **22**(14), e408–e416 (2006)

11. Psort, I.I.: PSORT: a program for detecting sorting signals in proteins and predicting their subcellular localization. J. Mol. Biol. **266**, 594–600 (1997)
12. Briesemeister, S., Rahnenführer, J., Kohlbacher, O.: YLoc—an interpretable web server for predicting subcellular localization. Nucleic Acids Res. **38**(Suppl. 2), W497–W502 (2010)
13. Chou, K.C., Cai, Y.D.: Using functional domain composition and support vector machines for prediction of protein subcellular location. J. Biol. Chem. **277**(48), 45765–45769 (2002)
14. Chou, K.C., Cai, Y.D.: A new hybrid approach to predict subcellular localization of proteins by incorporating gene ontology. Biochem. Biophys. Res. Commun. **311**, 743–747 (2003)
15. Shen, H.B., Chou, K.C.: A top-down approach to enhance the power of predicting human protein subcellular localization: Hum-mPLoc 2.0. Anal. Biochem. **394**(2), 269–274 (2009)
16. Chi, S.M., Nam, D.: WegoLoc: accurate prediction of protein subcellular localization using weighted Gene Ontology terms. Bioinformatics **28**(7), 1028–1030 (2012)
17. Blum, T., Briesemeister, S., Kohlbacher, O.: MultiLoc2: integrating phylogeny and Gene Ontology terms improves subcellular protein localization prediction. BMC Bioinform. **10**(1), 1 (2009)
18. Wan, S., Mak, M.W., Kung, S.Y.: GOASVM: a subcellular location predictor by incorporating term-frequency gene ontology into the general form of Chou's pseudo-amino acid composition. J. Theor. Biol. **323**, 40–48 (2013)
19. Ashburner, M., Ball, C.A., Blake, J.A., et al.: Gene Ontology: tool for the unification of biology. Nat. Genet. **25**(1), 25–29 (2000)
20. Yang, H., Nepusz, T., Paccanaro, A.: Improving GO semantic similarity measures by exploring the ontology beneath the terms and modelling uncertainty. Bioinformatics **28**(10), 1383–1389 (2012)
21. Cai, Y.D., Chou, K.C.: Predicting 22 protein localizations in budding yeast. Biochem. Biophys. Res. Commun. **323**, 425–428 (2004)
22. Chen, K., Lu, B.L., Kwok, J.T.: Efficient classification of multi-label and imbalanced data using min-max modular classifiers. In: Proceedings of the International Joint Conference on Neural Networks (2006)
23. Yang, Yang, Lu, Bao-Liang: Protein subcellular multi-localization prediction using a min-max modular support vector machine. Int. J. Neural Syst. **20**(01), 13–28 (2010)
24. Boutell, M.R., Luo, J., Shen, X., et al.: Learning multi-label scene classification. Pattern Recogn. **37**(9), 1757–1771 (2004)
25. Höglund, A., Dönnes, P., Blum, T., et al.: MultiLoc: prediction of protein subcellular localization using N-terminal targeting sequences, sequence motifs and amino acid composition. Bioinformatics **22**(10), 1158–1165 (2006)
26. Zhang, S., Xia, X., Shen, J., et al.: DBMLoc: a Database of proteins with multiple subcellular localizations. BMC Bioinform. **9**(1), 127 (2008)
27. Wang, G., Dunbrack, R.L.: PISCES: a protein sequence culling server. Bioinformatics **19**(12), 1589–1591 (2003)
28. Altschul, S.F., Madden, T.L., Schäffer, A.A., et al.: Gapped BLAST and PSI-BLAST: a new generation of protein database search programs. Nucleic Acids Res. **25**(17), 3389–3402 (1997)
29. Hall, M.A., Smith, L.A.: Feature selection for machine learning: comparing a correlation-based filter approach to the wrapper. In: FLAIRS Conference, pp. 235–239 (1999)
30. Resnik, P.: Semantic similarity in a taxonomy: an information-based measure and its application to problems of ambiguity in natural language. J. Artif. Intell. Res. (JAIR) **11**, 95–130 (1999)
31. Zdobnov, E.M., Apweiler, R.: InterProScan–an integration platform for the signature-recognition methods in InterPro. Bioinformatics **17**(9), 847–848 (2001)
32. Cortes, C., Vapnik, V.: Support-vector networks. Mach. Learn. **20**(3), 273–297 (1995)

33. Lei, J.B., Yin, J.B., Shen, H.B.: GFO: a data driven approach for optimizing the Gaussian function based similarity metric in computational biology[J]. Neurocomputing **99**, 307–315 (2013)
34. Yu, G., Li, F., Qin, Y., et al.: GOSemSim: an R package for measuring semantic similarity among GO terms and gene products. Bioinformatics **26**(7), 976–978 (2010)

Biomedical Named Entity Recognition Based on Multistage Three-Way Decisions

Hecheng Yu[1,2], Zhihua Wei[1,2(✉)], Lijun Sun[1,2], and Zhifei Zhang[1,3]

[1] Department of Computer Science and Technology, Tongji University,
Shanghai, China
yuhechengcuicui@126.com,
{zhihua_wei,sunlijun,zhifeizhang}@tongji.edu.cn
[2] Key Laboratory of Embedded System and Service Computing,
Ministry of Education, Tongji University, Shanghai 201804, China
[3] Research Center of Big Data and Network Security,
Tongji University, Shanghai 200092, China

Abstract. Biomedical named entity recognition (Bio-NER) is one of the most fundamental tasks in the field of biomedical information extraction. The accuracy of biomedical named entity recognition is crucial to the follow-up research work. This paper presents a method for named entity recognition based on the concept of three-way decisions. The method uses a discriminative approach named conditional random fields (CRFs) to construct models. These models follow the decision-making rule of three-way decision in all stages, the model cannot make decision arbitrarily when the information is incomplete until it gets more information. The experimental results show that our method can improve the performance for biomedical named entity recognition compared with other methods.

Keywords: Biomedical named entity recognition · Conditional random fields · Three-way decisions · Decision-Theoretic Rough Sets

1 Introduction

Biological research literatures have tremendous knowledge and information of interactions and relations among entities. MEDLINE is one of the most important databases for biomedical research and it is an international comprehensive biomedical information bibliographic database and which is the most authoritative international biomedical literature database currently. MEDLINE has over 14 million abstracts, with 300 to 350 thousand new abstracts appearing every year [1]. It pays tremendous cost because all of these resources are largely hand-annotated. Therefore, there is a rising need to develop some automatic techniques to solve problems such as word segmentation, entity recognition, subject classification, word sense disambiguation (WSD) etc. in the biomedical field. However, it is a very challenging task to recognize biomedical entities' names as the diversity and variations of biomedical entity are large, specifications of the naming rules are different, abbreviations and complex forms are used in a large number, the same words in different positions of document may have different

© Springer Nature Singapore Pte Ltd. 2016
T. Tan et al. (Eds.): CCPR 2016, Part II, CCIS 663, pp. 513–524, 2016.
DOI: 10.1007/978-981-10-3005-5_42

meanings. Overall, named entity recognition (NER) in biomedical domain is harder than other domains.

Current approaches for biomedical named entity recognition task can be divided into three categories: dictionary-based approach, rule-based approach and machine learning approach. Compared with dictionary-based and rule-based approaches, machine learning approach can obtain more robust and portable performance. The NER problem is usually considered as a classification problem to identify and classify every word or term in a document into some predefined categories like DNA, protein, RNA etc. Each word or term of the input sequences is represented as some features. The models based on hidden Markov models (HMM) and maximum entropy models are generally used to the NER problem. Conditional random fields (CRFs) are probabilistic model. The biggest difference between CRFs with HMM is that CRFs allow arbitrary non-independent features on the observation sequence. According to the results of the JNLPBA2004 evaluation, the CRFs model is currently the best machine learning approach for biomedical entity recognition.

Though CRFs model performs better than other machine learning methods, it still has weakness: it adopts two-way decisions strategy. Two-way decisions only consider two options of acceptance and rejection (or yes and not). But in practice, because of the inaccuracy or incompleteness of information, it cannot judge to accept or reject. Therefore, Yao proposed the theory of three-way decisions [6], and proved it in some conditions the overall cost of three-way decisions is smaller than that of two-way decisions, and the three-way decisions method is more advantageous in decision making. When the incompleteness of information cannot support to accept or reject, it adopts the third option like deferment. When enough information is obtained, it will judge the deferment to accept or reject. In this paper, a multistage three-way classifier is applied to the biomedical named entity recognition. At different decision-making stages, different information granularities are used respectively. If the information does not meet the current decision-making conditions, a model could not make a decision until it gets enough granularity information.

2   Related Work

Tsujii and Tsuruoka used a combination of a dictionary and a Naive Bias classifier to label the protein, the method gets 66.6 % f-measure [2]. Cohen constructed a dictionary using an online genes resource to recognize genes and proteins and it gets 75.6 % f-measure [3]. Fukuda et al. and Olsson et al. proposed the method based on rules [4, 5]. In 2003, Lee et al. proposed a two-phase biomedical named entity recognition method based on Support Vector Machine and dictionary query preprocessing. The precisions of two phases are 79.9 % and 66.5 %, respectively [6]. Finkel et al. use Maximum Entropy Markov Model (MEMM) [7]. CRF is first proposed by Settles to solve the problem of biomedical named entity recognition. This method is used on GENIA corpus with only a few features and no external resources to achieve 69.9 % F-measure [8]. After that, Tang et al. proposed CRFs based parallel biomedical named entity recognition algorithm. The biggest characteristic of the method is that it employs Map-Reduce framework so it can obtain high performance at the same time it can shorten training time which has important significance for the current big data mining [15]. Sara et al. proposed an integration of condition random fields and maximum entropy classifier to identify biomedical named entity method [9], and this method obtained 81.8 % F-measure, which has considerable improvement comparing to separately use conditional random fields or maximum entropy classifier in the same data set. Ekbal and Saha proposed a stacked ensemble method for biomedical entity recognition problem [10]. The biggest characteristic is that the approach uses a genetic algorithm (GA) based feature selection technique to determine the most relevant feature combination for modeling support vector machine (SVM) and CRF models in the first stage, and in the second stage it integrates the models selected in the first stage stacked. The proposed approach is evaluated on two benchmark datasets of JNLPBA 2004 shared task and GENETAG and yields the overall F-measure values of 75.17 % and 94.70 % for JNLPBA 2004 and GENETAG data sets, respectively.

3 Multistage Three-Way Decisions for Biomedical Named Entity Recognition

3.1 Three-Way Decisions

In traditional two-way decisions, people always have such as acceptance and rejection or yes and no two options. But in solving a lot of practical problems, a compulsive decision to accept or reject may lead to unnecessary costs or consequences. For example, when information or evidence is inadequate, whether acceptance or rejection is not reasonable, that is to say, the cost of making decision is bigger than deferment. As a result, three-way decisions are introduced into the non-committed decision option, thus avoiding the cost of false rejection or false acceptance.

A simple description of three-way decisions: suppose U is a nonempty finite set of objects, and C is a finite set of conditions. Based on the conditional set C, the main task of the three-way decisions is dividing the set U into three pair-wise disjoint regions, called the positive, the negative and the boundary regions, respectively. The three regions can be interpreted as three types of classification rules: rules of acceptance, deferment and rejection for the positive, boundary and negative regions, respectively [11]. In order to realize the three decisions, it needs to introduce an evaluation function of the objects, the value of which is called the decision state value, which reflects the degree of the objects. At the same time, it also needs to introduce the threshold, the values of evaluation function and thresholds can divide the set U into three regions above-mentioned.

In a general named entity recognition task, the model gives the test samples a certain entity classification according to the feature matrix of the training samples. However, for some test samples, the existing models may not be able to assert that they should belong to a certain category, and arbitrary classification will damage the overall performance of the classifier. If these samples can be classified as a separate category when information or evidence is inadequate, it means the machine cannot judge that which category they belong to temporarily, so the decision-making process will be more reasonable. This is in line with the idea of three-way decisions. In the named entity recognition, the classification model is used to return the probability estimate, after that we can use three decisions to deal with the test samples. The corresponding decision rules are shown as follows:

Definition 1. Suppose biomedical entity recognition system is a quaternion S = (U, A, V, f). U is the set of samples, A is the set of samples' features, f is an information function, and V is the values of information function.

Definition 2. Suppose S = (U, A, V, f) is an information table, $\forall x \in U, X \subseteq U$, $Pr(X|[x]) = |[x] \cap X|/|[x]|$. $[x]$ is the equivalence class containing x. $|.|$ denotes cardinal number of set elements, conditional probability $Pr(X|[x])$ representing the probability that an object belongs to C given that the object is described by $[x]$.

Definition 3. Suppose S = (U, A, V, f) is an information table, $X \subseteq U, 0 \leq \beta \leq \alpha \leq 1$. The lower and upper approximations of X are defined by:

$$\begin{cases} \overline{apr}(\alpha, \beta) = \{x \in U|P(X|[x]) > \beta\} \\ \underline{apr}(\alpha, \beta) = \{x \in U|P(X|[x]) \geq \alpha\} \end{cases} \tag{1}$$

According to the pair of approximations, the positive, boundary and negative regions are defined by:

$$\begin{cases} POS_{(\alpha,\beta)}(X) = \{x \in U|P(X|[x]) \geq \alpha\} \\ BND_{(\alpha,\beta)}(X) = \{x \in U|\beta < P(X|[x]) < \alpha\} \\ NEG_{(\alpha,\beta)}(X) = \{x \in U|P(X|[x]) \leq \beta\} \end{cases} \tag{2}$$

It is significant for solving practical decision problems that how to select reasonable alpha and beta value. Yao applied Bayesian decision theory to the derivation of Decision-Theoretic Rough Sets (DTRS). And in DTRS, the value of α and β can be calculated under the condition of the minimum overall risk. The optimal pair of threshold (α, β) with $0 \leq \beta < \alpha \leq 1$ that minimizes the risk or cost is given by:

$$\begin{cases} \alpha = \frac{\lambda_{an} - \lambda_{nn}}{(\lambda_{an} - \lambda_{nn}) + (\lambda_{np} - \lambda_{ap})} \\ \beta = \frac{\lambda_{nn} - \lambda_{rn}}{(\lambda_{nn} - \lambda_{rn}) + (\lambda_{rp} - \lambda_{np})} \end{cases} \tag{3}$$

The meanings of the parameters in Eq. (3) are shown in Table 1. It means the thresholds decided by the risk of different decisions.

Biomedical named entity recognition is a multi-class decision problem, Yao and Zhao [20] provide an approach that converting a multi-class classification problem into a number of two-class classification problems, and to each two-class classification problem to use three-way decision theory. We use the theory to solve our biomedical entity problem. In a finite set of m-class $C = \{c_1, c_2, \cdots, c_m\}$, for each c_j, a two-class classification $\{c_j, c_j^c\}$ can be defined, where $c_j^c = \bigcup_{i \neq j} c_i$. Similar to two-state three-way decisions, we give the loss function for each c_j as follows (Table 2):

Table 1. Loss function for two-state three-way decisions

Decision making	Object state	
	Satisfy the condition	Not satisfy the condition
Accept a	λ_{ap}	λ_{an}
Reject r	λ_{rp}	λ_{rn}
Defer n	λ_{np}	λ_{nn}

Table 2. Loss function for multi-class three-way decisions

	c_1	$\cdots$	c_j	$\cdots$	c_m
a_{P_i}	$\lambda_{P_1} = \lambda\left(a_{P_i}\|c_j^c\right)$	$\cdots$	$\lambda_{P_j} = \lambda\left(a_{P_i}\|c_j\right)$	$\cdots$	$\lambda_{P_m} = \lambda\left(a_{P_i}\|c_j^c\right)$
a_{B_i}	$\lambda_{B_1} = \lambda\left(a_{B_i}\|c_j^c\right)$	$\cdots$	$\lambda_{B_j} = \lambda\left(a_{P_i}\|c_j\right)$	$\cdots$	$\lambda_{B_m} = \lambda\left(a_{B_i}\|c_j^c\right)$
a_{N_i}	$\lambda_{N_1} = \lambda\left(a_{N_i}\|c_j^c\right)$	$\cdots$	$\lambda_{N_j} = \lambda\left(a_{P_i}\|c_j\right)$	$\cdots$	$\lambda_{N_m} = \lambda\left(a_{N_i}\|c_j^c\right)$

In our study, we use the same loss of making any wrong decisions which agrees with Yao's idea.

3.2 Features for Named Entity Recognition

Feature selection plays a significant role in a successful machine learning method. In the first stage, we construct various conditional random field models using different feature combinations, and choose the best model as the final model of the first stage. These features follows are used in the best model. They are not specific to the field of biomedical named entities, and can still be applied in other areas.

Context words: We consider according to experience that words which surround named entities largely carry effective information for the identification of named entities. The current word and its adjacent occur in the context window $w_{i-3}^{i+3} = w_{i-3} \cdots w_{i+3}$, $w_{i-2}^{i+2} = w_{i-2} \cdots w_{i+2}$ and $w_{i-1}^{i+1} = w_{i-1} \cdots w_{i+1}$, which stands for the size 7, 5, 3, respectively, where w_i is the current word. The choice of the size of the window is very significant. The size is too little or too big will lead to inadequate information or the curse of dimensionality.

Word prefix and suffix: Many biomedical named entities have a certain prefix and suffix. Word prefix and word suffix are the character sequences stripped from the leftmost and rightmost positions of the words, separately. We set the feature values to 'undefined' if either the length of w_i is less than or equal to n−1. w_i is a punctuation symbol or if it contains any special symbol or digit.

Infrequent word: According to the experience, the more frequently the words occurring the rarely the named entities are. We make a list, and we use a predetermined threshold to decide a word from the training data whether it can in the list or not. And the value of threshold is determined by the size of the dataset. If a certain word occurs in the list, the feature is defined 1, else 0.

Part-of-Speech information: Part-of-Speech (POS) information is an important feature for entity recognition problem.

Unknown feature: Unknown feature is a feature shows a word appear in the training data or not, and 1 for was seen, 0 for not.

Word normalization: We consider that biomedical named entities in the same category are largely similar in the orthographic. Instead of considering the meaning of each character in a word, wen replace each capitalized character with 'A', small characters with 'a' and all consecutive digits with '0'. For example, 'HIV' is normalized to 'AAA', 'HIV-2' is normalized to 'AAA-0' and 'HIV-1P24' is normalized to 'AAA-0A0'.

N-gram: n continuous words combine to a feature.

Orthographic feature: We define orthographic feature depending upon the contents of the word forms. The feature is multi-values. The feature values are defined corresponding to Greek letters, Roman numbers, single-capital word and etc. It is especially helpful in biomedical domain to use special characters to detect named entities and boundaries of named entities, such as '-'. We also use the value of ATGC sequence and stop words. All the values of orthographic feature are shown in Table 3.

Table 3. Values of orthographic feature

Feature value	Example	Feature value	Example
Greek-Letter	α, β, λ, alpha, beta	alphaDigitAlpha	IL23R, E1A
roman_number	I, II, III	digitAlpha	2×NFkappaB
ATCG-sequence	CCGCCC, AAGAT	alphaDigit	p50, p65
singlecapital	A, B	digitSpecial	12–3
initcapital	Src	hyphen	-
all upper	EBNA, LMP	stopWord	at, in
caps&Digits	32Dc13	digitCommadigit	1,28
capMixAlpha	Src, Ras, Epo	InCap	mAb
lowMixAlpha	mRNA, mAb	digitDotdigit	1.23
all digits	1,123		

During five biomedical named entities, the difference between RNA and other named entities is distinctive. RNA always contains RNA or mRNA. But protein and DNA in biomedical literature sometimes have the same name. It is hard to distinguish them with above features. Cell-type and cell-line are the types of more difficult to distinguish. They have almost the same feature words. Therefore, we cannot distinguish them with feature words. As mentioned before, when information or evidence is inadequate, these samples should be classified as a separate category, so the decision-making process will be more reasonable. In the second stage, we add new features to construct new models to distinguish protein between DNA, cell-type between cell-line separately. The feature words are assigned three classes: high, medium and low classes [13]. When the feature class comes to the same, we mark

Table 4. The feature words and feature classes of DNA and protein

Feature words	Entity types	Feature classes
Prefix, Greek letters	Protein	low
Enhancer, region, level, chromosome, transcription, locus	DNA	medium
Expression, activation, activity, coexpression, factor, phosphorylation	protein	medium
DNA, gene, promoter, element, site	DNA	high
Domain, protein, family, motif	protein	high

Table 5. The feature words and feature classes of cell line and cell type

Feature words	Entity type	Feature classes
Cell, myeloid	cell line	low
Cells, erythroid, granulocytic, T, B	cell type	low
Jurkat, cultures, cultured, colonies	cell line	medium
Progenitor, lineage, thyrocyte, sideroblasts, the possessive of activation, activity and response	cell type	medium
Line	cell line	high
Type	cell type	high

DNA or cell type first. Tables 4 and 5 list the feature words and feature classes for DNA and protein, cell-type and cell-line separately.

3.3 Methodology

The features are divided into A_1 = {context words, word prefix and suffix, Infrequent word, Part-of-Speech (POS) information, unknown token feature, word normalization, n-gram, orthographic feature}, $A_2 = A_1 \cup$ {feature words DNA, protein, cell line and cell type}. In order to ensure the reliability of the experiment, we use 3-fold cross validation to train the training data three times. The implementation process of multistage three-way decision biomedical named entity recognition method is as follows:

Step 1. Construct a lot of conditional random field models with different features combination, use training data to train these classifiers and test data to evaluate these classifiers. We choose the best classifier as classifier A, and the features of best classifier as A1. According to the theory of three-way decisions, classifier A divides samples in each category (e.g. DNA, RNA, protein, cell line, cell type) into three regions: positive, negative and boundary regions. Because the insufficient information, some samples cannot distinguish with each other, and they always in boundary region, and most of them are DNA, protein, cell line and cell type.

Step 2. We construct classifier B to distinguish DNA and protein, cell line and cell type in step2. The feature set of classifier B is A_2.

Step 3. Integrate two base models to form an integrated model. The skeleton of the multistage procedure is shown in Fig. 1.

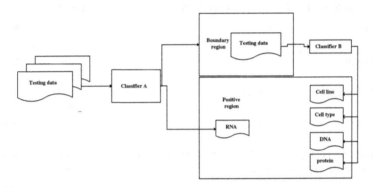

Fig. 1. Integrated model

The pseudo-code of multi-class three-way decisions biomedical named entity recognition method is described as follows:

Step 1 algorithm

Input: training set $X = \{x_1, x_2, \cdots, x_n\}$, attributes A1, m pairs of thresholds $\{\{\alpha_1, \beta_2\}, \{\alpha_2, \beta_2\}, \cdots, \{\alpha_m, \beta_m\}\}$

Output: decision set A $= \{a_1, a_2, \cdots, a_n\}$

Begin:

Compute the probability $p(\vec{y}|\vec{x})$ of a sequence of length k, and compute the product $\alpha_{1,\cdots,k} = \alpha_1 \cdot \alpha_2 \cdots \cdot \alpha_k$ and $\beta_{1,\cdots,k} = \beta_1 \cdot \beta_2 \cdots \cdot \beta_k$.

If: $p(\vec{y}|\vec{x}) \leq \beta_{1,\cdots,k}$, negative region.

Else if: $\alpha_{1,\cdots,k} > p(\vec{y}|\vec{x}) > \beta_{1,\cdots,k}$, boundary region.

Else: $\alpha_{1,\cdots,k} \geq p(\vec{y}|\vec{x})$, positive region.

End.

Algorithm of step 2 is similarity to step 2 except the attributes are A2.

4 Experiments

4.1 Data

A variety of public evaluation of biological named entity recognition has been developed internationally, such as JNLPBA and BioCreative [21]. JNLPBA (the Joint workshop on Natural Language Processing in Biomedicine and its Applications) is an international biological named entity recognition task. In the JNLPBA2004 task, participating systems were asked to identify five named entity classes, namely Protein, DNA, RNA, cell-line and cell-type [12]. Participants can use any approach and knowledge resource to complete their biomedical named entity recognition systems and finally carry out a unified test and evaluation. BioCreative is committed to the retrieval of biological information for Natural Language Processing and the intersection of life sciences. Biocreative2004 evaluation includes a biometric named entity recognition task is to identify a category "NEWGENE", which contains the proteins, DNAs and RNAs. We use the proposed approach on the JNLPBA 2004 shared task datasets.[1] And the training data and testing data were formatted for IOB2 natation, where 'B-xxx' refers to the start of a chunk/single-word named entity of type 'xxx', 'I-xxx' refers to the middle parts of the named entity and 'O' refers to not the named entity.

4.2 Experimental Settings

Like other data mining tasks, we must preprocess the samples before named entity recognition. This study chooses GENIA tagger V3.0.2 to preprocess the data of the original text to get some features.[2] GENIA tagger V3.0.2 is a freely available well-known system and it analyzes English sentences and outputs the base forms, part-of-speech tags, chunk tags, and named entity tags. The accuracy of the GENIA tagger is 98.26 %. The tagger is particularly tuned to biomedical literature such as MEDLINE abstracts. Other features we used are extracted from programming.

In the experiment, the size of context window is 5 (i.e., $w_{i-2}^{i+2} = w_{i-2} \cdots w_{i+2}$), the length of the word prefix and suffix character sequences is 4 (i.e., 8 features), the threshold of Infrequent word is 10, we use Part-of-Speech information of the current and/or the surrounding token(s) as the features, in the training phase, the unknown token feature is set randomly, the value of n in n-gram is 2. This paper use an adaptive learning algorithm for the threshold parameters and loss functions by studying the problem of minimize the decision risk of decision rough set model proposed by [14]. The m pairs of threshold parameters are calculated based on the algorithm can minimize the decision risk loss.

4.3   Performance Evaluation

The classifiers are evaluated in terms of recall, precision and F-measure. Precision is the ratio of the number of correctly recognized named entities to the number of named entities recognized by the classifier, and recall is the ratio of the number of correctly named entities recognized to the number of correct results by the classifier. The calculation formulas of the two are as follows:

$$precision = \frac{TP}{TP+FP} \quad recall = \frac{TP}{TP+FN} \tag{4}$$

where TP (True Positive) represents the number of named entities correctly identified by the system, FP (False Positive) represents the number of names incorrectly recognized, and FN (False Negative) represents the number of named entities which were not identified. From the calculation formulas, it is clear that precision concerns how many of the identified names are correct, but recall concerns how many of the correct results have been identified. These two are contradictory. Therefore, when comparing the performance of two different named entity recognition systems, the comprehensive value of these two indicators, F-measure, is generally used. The calculation formulas of F-measure are as follows:

$$F - measure = \frac{2 \times (recall \times precision)}{precision + recall} \tag{5}$$

4.4   Experimental Results

The overall evaluation results of our proposed approach are presented in Table 6. The approach shows the recall, precision and F-measure values of 80.58 %, 84.58 % and 82.53 %, respectively. We compare the performance of our proposed approach with other biomedical entity recognition approaches on the same dataset in Table 6. We also show the detailed evaluation results of our best individual classifier compared with the best individual classifier of Ekbal and Saha in Table 7. In Table 7 we can see the f-measures of DNA, RNA, protein, cell line and cell type all boost a lot, respectively.

Table 6. Comparison with the stacked ensemble of Ekbal and Saha

Model	Recall (%)	Precision (%)	F-measure (%)
Our proposed system	80.58	84.58	82.53
Ekbal and Saha	75.15	75.20	75.17
Tang et al. [15]	76.01	70.79	73.31
Li et al. [16]	75.71	70.59	73.06
Zhou and Su [17]	75.99	69.42	72.55
Okanohara et al.[18]	72.65	70.35	71.48
Kim and Yoon [19]	69.68	72.77	71.19
Finkel et al. [7]	68.56	71.62	70.06

Table 7. Comparison the best individual classifier with the stacked ensemble of Ekbal and Saha (our/Ekbal's and Saha's)

Class	Recall (%)	Precision (%)	F-measure (%)
Overall	79.80/76.78	84.61/73.10	82.13/74.90
Protein	83.09/82.31	84.21/73.22	83.64/77.50
Cell line	69.45/59.29	85.13/56.62	76.49/57.93
DNA	74.42/74.03	84.15/72.61	78.98/73.31
RNA	78.40/71.83	86.60/72.86	82.29/72.34
Cell type	80.48/69.21	86.78/78.95	83.51/73.76

5   Conclusions

In this paper we have proposed a multistage three-way decisions approach for biomedical named entity recognition. The most important characteristic of our system is that our system combines the theory of three-way decisions and uses the hierarchical feature words. With the application of three-way decisions, when the information is inaccurate or incomplete, the model does not make an arbitrary decision. Some named entities are very similar, such as DNA and protein. We cannot distinguish them when they have the same feature words. So we use divide the feature words into three classes: high, medium and low classes. It helps the model to decide the category the sample belongs to. Our ensemble model has shown the overall F-measure values of 82.53 %. It shows that our proposed approach achieves the greatest performance by comparisons.

In future we would like to add the feature words to distinguish the similar categories to make the classes of feature words more reasonable.

Acknowledgments. The work is partially supported by the National Natural Science Foundation of China (No. 61273304, 61573259), and the program of Further Accelerating the Development of Chinese Medicine Three Year Action of Shanghai (No. ZY3-CCCX-3-6002).

References

1. Finkel, J., Dingare, S., Manning, C., Nissim, M., Alex, B., Grover, C.: Exploring the boundaries: gene and protein identification in biomedical text. BMC Bioinf. **6**, S5 (2005)
2. Tsuruoka, Y., Tsujii, J.: Boosting precision and recall of dictionary-based protein name recognition. In: Proceedings of the ACL 2003 Workshop on Natural Language Processing in Biomedicine, Sapporo, Japan, pp. 41–48 (2003)
3. Cohen, A.M.: Unsupervised gene/protein entity normalization using automatically extracted dictionaries. In: Proceedings of the ACL-ISMB Workshop on Linking Biological Literature, Ontologies and Databases: Mining Biological Semantics, Detroit, MI, pp. 14–24 (2005)
4. Fukuda, K., Tsunoda, T., Tamura, A., et al.: Toward information extraction: identifying protein names from biological of the Pacific Symposium on Biocomputing, Hawai, USA, pp. 705–716 (1998)

5. Olsson, F., Eriksson, G., Franzen, K., et al.: Notions of correctness when evaluating protein name taggers. In: Proceedings of the 19th International Conference on Computational Linguistics, Taipei, Taiwan, pp. 765–771 (2002)
6. Lee, K.J., Hwang, Y.S., Rim, H.C.: Two-phase biomedical NE recognition based on SVMs. In: Proceedings of the ACL Workshop on Natural Language Processing in Biomedicine, Sapporo, Japan, pp. 33–40 (2003)
7. Finkel, J., Dingare, S., Nguyen, H., et al.: Exploiting context for biomedical entity recognition: from syntax to web. In: Proceedings of the Joint Workshop on Natural Language Processing in Biomedicine and Its Applications, Geneva, Switzerland, pp. 89–91 (2004)
8. Settles, B.: Biomedical named entity recognition using conditional random fields and novel feature sets. In: Proceedings of the Joint Workshop on Natural Language Processing in Biomedicine and its Applications, pp. 104–107. Association for Computing Machinery, Geneva (2004)
9. Keretna, S., Lim, C.P., Creighton, D.: Classification ensemble to improve medical named entity recognition. In: 2014 IEEE International Conference on Systems, Man, and Cybernetics, San Diego, CA, USA, pp. 2630–2636 (2014)
10. Ekbal, A., Saha, S.: Stacked ensemble coupled with feature selection for biomedical entity extraction. J. Knowl. Based Syst. **46**, 22–32 (2013)
11. Yao, Y.: An outline of a theory of three-way decisions. In: Yao, J., Yang, Y., Słowiński, R., Greco, S., Li, H., Mitra, S., Polkowski, L. (eds.) RSCTC 2012. LNCS (LNAI), vol. 7413, pp. 1–17. Springer, Heidelberg (2012). doi:10.1007/978-3-642-32115-3_1
12. Jin-Dong, K., Tomoko, O., Yoshimasa T., et al.: Introduction to the bio-entity recognition task at JNLPBA. In: Proceedings of the International Joint Workshop on Natural Language Processing in Biomedicine and Its Applications, pp. 70–75. Association for Computational Linguistics, Geneva (2004)
13. Yang, Z.C.: Research on text mining in biomedical domain. Dalian University of Technology, Dalian (2008). (in Chinese)
14. Jia, X.Y., Li, W.J., Shang, L., et al.: An adaptive algorithm for decision threshold of three-way decisions. J. Electron. **39**, 2520–2525 (2011). (in Chinese)
15. Tang, Z., Jiang, L.G., Yang, L., Li, K.L., Li, K.Q.: CRFs based parallel biomedical named entity recognition algorithm employing MapReduce framework. Cluster Comput. **18**, 493–505 (2015)
16. Li, L., Zhou, R., Huang, D.: Two-phase biomedical named entity recognition using CRFs. Comput. Biol. Chem. **33**(4), 334–338 (2009)
17. Zhou, G.D., Su, J.: Exploring deep knowledge resources in biomedical name recognition. In: Proceedings of the International Joint Workshop on Natural Language Processing in Biomedicine and Its Applications (JNLPBA), pp. 96–99 (2004)
18. Okanohara, D., Miyao, Y., Tsuruoka, Y., Tsujii, J.: Improving the scalability of semi-Markov conditional random fields for named entity recognition. In: Proceedings of the 21st International Conference on Computational Linguistics and 44th Annual Meeting of the ACL, pp. 465–472 (2006)
19. Kim, S., Yoon, J.: Experimental study on a two phase method for biomedical named entity recognition. IEICE Trans. Inf. Syst. **7**(E90–D), 1103–1110 (2007)
20. Yao, Y.Y., Zhao, Y.: Attribute reduction in decision-theoretic rough set models. Inf. Sci. **178**(17), 3356–3373 (2008)
21. Hirschman, L., Yeh, A., Blaschke, C., et al.: Overview of BioCreAtIvE: critical assessment of information extraction for biology. BMC Bioinf. **6**, 1 (2005)

Chinese Image Text Recognition with BLSTM-CTC: A Segmentation-Free Method

Chuanlei Zhai, Zhineng Chen$^{(\boxtimes)}$, Jie Li, and Bo Xu

Interactive Digital Media Technology Research Center, Institute of Automation,
Chinese Academy of Sciences, Beijing 100190, China
{zhaichuanlei2014,zhineng.chen,jie.li,xubo}@ia.ac.cn

Abstract. This paper presents BLSTM-CTC (bidirectional LSTM-Connectionist Temporal Classification), a novel scheme to tackle the Chinese image text recognition problem. Different from traditional methods that perform the recognition on the single character level, the input of BLSTM-CTC is an image text composed of a line of characters and the output is a recognized text sequence, where the recognition is carried out on the whole image text level. To train a neural network for this challenging task, we collect over 2 million news titles from which we generate over 1 million noisy image texts, covering almost the vast majority of common Chinese characters. With these training data, a RNN training procedure is conducted to learn the recognizer. We also carry out some adaptations on the neural network to make it suitable for real scenarios. Experiments on text images from 13 TV channels demonstrate the effectiveness of the proposed pipeline. The results all outperform those of a baseline system.

Keywords: Chinese image text recognition · BLSTM · CTC · Segmentation-free

1 Introduction

With the amount of digital images and videos growing rapidly, it becomes more and more urgent to construct a system that can automatically query and search the multimedia data and then return the interested content. To build such a system, a lot of features and descriptions, i.e. regions, objects and even human face in videos or images, have been applied to index the multimedia data [1]. Apart from this, textual information, such as video caption and characters appearing on the images, provides more valuable indexing information for the reason that it is highly compact and offers a much higher level of semantic information. Therefore, getting the textual information in images and videos becomes vital for such indexing method and retrieval system. Also once we get the textual information, it will supply with more information about the image or video.

© Springer Nature Singapore Pte Ltd. 2016
T. Tan et al. (Eds.): CCPR 2016, Part II, CCIS 663, pp. 525–536, 2016.
DOI: 10.1007/978-981-10-3005-5_43

Benefiting from the high performance of optical character recognition (OCR) systems on the recognition task of document images, some research work focuses on converting the text image to something like document image and then feeds the processed image to the OCR system. However, this method encounters serious performance degradation because of the complex image background and low image resolution. This leads to finding more robust binarization algorithms [2,3] or other skills [4] that can distinguish the text pixels from non-text pixels better. Unfortunately, these attempt could not function well when the text falls in a complex image background or the text image has much lower resolution. What's more, the sophisticated background and low image resolution may do damage to or blur the topological structure which is the main discriminative clues when human beings try to distinguish different Chinese characters.

Considering the circumstances, some researchers abandon such an image text recognition pipeline and they deal with this problem based on some segmentation methods [5–8]. Usually they do the pre-segmentation work using some special designed method. However, because of the complex and varying image background, there are often characters touching each other through the background pixels, thus resulting in incorrectly splitting some chars that are incapable of being recognized. To address this issue, a method called over-segmentation is proposed, and it is used in Chinese handwritten characters recognition [5,6], English and Chinese image text recognition tasks [7,8]. This technique firstly tries to find all the potential cutting positions, then tries to recognize the characters in the segments using some type of method, i.e. deep neural networks [16,23], and at last does some post-processing work by some search method to eliminate the inaccurate cutting positions and transform the output to the recognized sequence. Although such manner achieves considerable performance improvement, some dependency assumptions have been made in the searching or decoding stage and thus, its applicability is constrained [9].

In this paper, we introduce a neural network framework that makes us do not need to explicitly or implicitly do the segmentation work. Thus relieving us of the costly segmentation dependent work (i.e. binarizing the image text, finding all the potential cutting points, cutting the text images etc.). Moreover, we also don't need to make some strong dependency assumptions when we try to get the final recognition result. We call this network structure BLSTM-CTC (bidirectional LSTM-Connectionist Temporal Classification). The similar network framework has been applied in many sequence learning tasks, for example, continuous speech recognition [13] and handwritten text recognition [14,15]. BLSTM is a good sequence learner. Combined with CTC, the framework could model all aspects of a label sequence. Therefore, we no longer need the dependency assumptions which is indispensable in the over-segmentation method. For an given image text, we firstly extract some type of feature sequence and then feed it to the neural network. At last, by some means of decoding we will get the best path corresponding to the final label sequence. The experiment conducted on the text images from 13 TV news videos shows the effectiveness of the proposed pipeline and the results all outperform the method of Bai et al. [8], which we take as the baseline system. In total, our contributions are as follows:

- We firstly introduce BLSTM-CTC to the Chinese image text recognition task. In virtue of the three control gates, long short term memory (LSTM) units remember longer time input information and resolve the vanishing gradient problem in contrast to the traditional RNN unit [10]. In sequence labeling tasks, information both from the past and future is important. Therefore we select the bidirectional LSTM as the hidden layer. We call such network structure BLSTM-CTC.
- We collect a huge amount of news titles and develop a method to generate image texts automatically. To train such a neural network, we need hundreds of thousands text images. Although we can get many news videos, locating the text region and labeling the text manually are both time consuming and spirit consuming. So we conduct a scheme that can automatically generate simulated text images.
- We introduce the sliding window scheme from the domain of continuous speech recognition [13] to generate the feature sequence for a text image. Due to the overlapping of adjacent windows and small window size, we can get more details of the Chinese characters.

Since image text recognition has been researched for a period of time, we will firstly discuss the related work in Sect. 2. In Sect. 3, we will introduce the text image generating method and the recognition system. Experimental configurations and results will be described in Sect. 4. Finally, we conclude our work in the last section.

2 Related Work

Image text recognition problem has not been tackled completely even though it has been pursued for a long time. So far, there has been two kinds of viewpoints thus leading to two kinds of methodologies. Some researchers hold the point that image text is another kind of document image and they propose methods to extract text pixels from non-text pixels, i.e. binarization methods [2,3,11] and stroke width transform method [4] which exploits the inherent characteristic in characters that one character usually has the same stroke width. Such methods dedicate to searching for efficient methods to distinguish text pixels from non-text pixels, then feed the processed image to an OCR engine and finally get the result. On the other hand, some researchers abandon such methodology and they adopt some kind of holistic method, i.e. getting the image text segmented, then constructing some kind of graph and finally getting the result by methods like beam search [7,8,12]. Such methods achieve preferable performance in image text recognition task recently. Zohra et al. [7] build up an image text recognition graph called iTRG which uses a convolution neural network to automatically perform segmentation and recognition and attains the result through a weight calculator and an optimal path search module. Bai et al. [8]

come up with such a pipeline that they firstly binarize the text image, then cut the image at all potential cutting points and combine the cutting pieces dynamically. Finally by combining with a 3-gram language model they utilize a beam search method to get the result and proper cutting points. Yokobayashi et al. [12] propose an adaptive binarization method to segment the character and use an recognition method based on global affine transformation (GAT) correlation for affine-invariant grayscale characters.

Connectionist temporal classification was proposed in 2006 by Alex Graves [9]. Combined with the power of LSTM, it is quickly and successfully applied to several sequence learning tasks, i.e. handwritten text recognition, continuous speech recognition [13]. Graves et al. [14] proposed multidimensional recurrent neural network and successfully applied it in Arabic handwritten recognition. Messina et al. [15] applied multidimensional long short term memory RNN, namely MDLSTM-RNN in Chinese handwritten text recognition and achieved significant performance comparable to the result of the state-of-the-art method. Compared with handwritten text recognition, our text images usually come with complex and varying background. We firstly introduce CTC based method to Chinese image text recognition and find that it functions well in our task.

3   Image Text Generation and the Recognition System

At first, we will introduce the method of generating Chinese image texts which relieves us of labeling a mountain of text images. Then we debate two schemes of extracting feature sequence and at last a bidirectional LSTM neural network and CTC training method will be presented.

3.1   Text Images Generation

In order to train a neural network with large amount of parameters, we need to prepare a certain amount of training images. Whereas we only have several thousand labeled data which is far from enough, so we come up with an scheme to construct text images by simulation. Our objective is to construct as realistic text images as possible. Firstly, we collect a set of news titles from some Chinese news portals. Secondly we get a set of news videos and extract some frames from each video to form an image background library. At last we extract clean born Chinese or English characters from corresponding font files and paste them on the background cropped from the image selected from the image library randomly.

To keep consistent with Bai's work [16], there are 4120 characters. Among these characters are 3755 GB2312-1 standard characters, 245 external characters, 94 printable ASCII symbols and 26 punctuation symbols. However, we find that some full-width punctuation appear with high frequency, i.e.'【' and '】' in the generated training set. Besides, we replace ellipsis with three dots. So we add the two additional symbols and remove ellipsis from the 4120 characters, thus contributing to a 4121-character set. In addition, it is worth stating that we replace full-width punctuation with half-width ones (i.e. the double quotes and

Fig. 1. Examples of the generated text images: the upper ones are with clean background, the below ones are with relatively complex background, which are obtained from our method.

single quotes) in the corresponding target sequence. When selecting the cropped background image, we guarantee that each place of the cropped background where one character appears meets the requirement as [16] says. Unlike [16] we just adopt one font size and for each title we select two fonts from the font library randomly. At last we generate one million image texts for training and ten thousand image texts as validation data which do not participate in the training process. Figure 1 is some examples of clean background and examples with relatively complex background.

3.2   Two Scheme of Extracting Feature Sequence

We explore two kinds of schemes to get the feature sequence of an image. One is gained from [8]. We calculate all potential cutting points, cut the image into segments, resize each segment to a fixed size of 32 * 32 and finally extract a 512-d feature vector. In contrast to [8], we don't group adjacent segments any more in the workflow. The other method is based on a sliding window. We follow the method in the domain of continuous speech recognition. Firstly we scale the image to 32 pixels in vertical direction and scale the image length according to the scaling ratio of the vertical direction. Then we take a lanky window of 32 * 8 sliding from left to right with a step of 4 pixels in horizontal direction and we will discard the last frame if the remaining length is less than 8 pixels. Therefore, this may result in abandoning a very little part of the image which is at most 3 pixels width at the end of the image text and this almost does no damage to the recognition result. We extract a 128-d feature vector in each lanky rectangle region in the end. It should be noted that whatever scheme we adopt we will calculate a gradient direction map of each selected region (i.e. 32 * 32 or

32 * 8) using the Sobel operator. Once we get such a map, we will equally divide the map into 4 * 4 patches and calculate one 8 directional gradient histogram for each patch, thus 512-d vector or 128-d vector come into being. 8 directional gradient histogram feature is proved to be very powerful for Chinese character recognition tasks in [17]. In our first group of experiments we will compare the results of these two feature extracting scheme.

3.3   CTC Training with BLSTM Based RNN

As mentioned above, BLSTM is a good sequence learner and CTC relieves us of the segmentation and alignment work in advance. The framework of our neural network is illustrated in Fig. 2. The network outputs a 4124-d vector and each dimension represents the posterior probability of outputing the corresponding label. We will introduce what these 4124 labels are in the decoding section.

CTC Introduction. CTC can automatically learn the alignment between the input and output sequence [9]. Denoting our training set as S, one training sample input as $X = (x_1, x_2, \cdots, x_T)$ and its corresponding target as $Z = (z_1, z_2, \cdots, z_U)$, the target library as L. $U \leq T$ is a restriction on the training data. Supposing there are K targets, in order to allow the network to emit consecutive same labels and alleviate the strain on the network that it needs to continuously predicting one label before the next label begins, one additional blank labeled as ϕ is added to the target library. So we now have $K + 1$ targets, labeled as L'. Given the training data S, CTC defines an objective function to

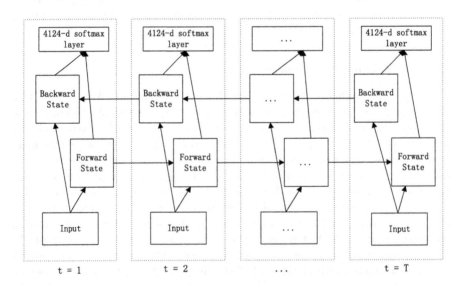

Fig. 2. The BLSTM network used in our system

minimize the negative log likelihood function, namely minimizing the function below:

$$O = -\log \left(\prod_{(X,Z) \in S} p\left(Z \mid X\right) \right) = -\sum_{(X,Z) \in S} \ln p\left(Z \mid X\right) \tag{1}$$

Now we will introduce how to calculate $p\left(Z \mid X\right)$. The output of the BLSTM based RNN is fed to an softmax layer containing $K+1$ nodes, each of which corresponds to one label. The input sequence length is T and at each time t the softmax layer will output $K+1$ dimensional vector y_t representing the posterior probability of each target. Therefore the length of output is T, containing the blank labels. Pick up one label from each output vector, we will get a vector $\pi = (\pi_1, \pi_2, \cdots, \pi_T)$, named as one path. And the probability of such a path is the product of each label probability $y_t^{\pi_t}$:

$$p\left(\pi \mid X\right) = \prod_{t=1}^{T} y_t^{\pi_t} \tag{2}$$

To make a bridge between path π and the target label, CTC defines a map $B : L'^T \to L^{\le T}$ which reflects paths to one target sequence, e.g. $B\left(a\phi\phi ab\phi c\right) = B\left(\phi a\phi abcc\right) = aabc$. In our Chinese image text recognition, this might be $B\left(我我\phi爱中国\right) = B\left(\phi我我爱\phi中国\right) = 我爱中国.$ Therefore, the target sequence probability can be calculated as the sum of corresponding paths' probabilities:

$$p\left(Z \mid X\right) = \sum_{\pi \in B^{-1}(Z)} p\left(\pi \mid X\right) \tag{3}$$

However, there may exist many paths that can be mapped to one same target sequence. Searching the paths is costly and unrealistic. CTC construct a special sequence and two mathematical symbols just like the forward-backward algorithm in HMM to tackle this problem. The details can be found at [9]. Then the error between the output and target can be propagated back through time and the neural network can be trained by some stochastic gradient descent based method.

It should be noted that adaptation work is conducted on the neural network using a few real labeled text images after the training process of the neural network ends. Or saying, we employ an additional adaptation procedure that utilizes some real labeled data which is different from the test data to slightly adjust the parameters of the network. Such method is proved to be effective in many works [18,19] and we adopt such scheme and see the performance improving.

Decoding. In the previous section, we have discussed why CTC objective function can be used to train a recurrent neural network. During the test phase, how to get the recognition result according to the output of the softmax layer of all time from $t = 1$ to $t = T$ is still a question. This can be simply done by best path decoding, namely selecting a label that has the maximum value of the 4124-d output vector at each time t. Once we get the output sequence, the final result can be generated by the previous defined map $B : L'^T \rightarrow L^{\leq T}$. The 4124 labels include the 4121 characters we mentioned above and one additional label, namely 'blank'. Moreover, there are another two symbols required by EESEN [20], our main training and decoding tool. They are 'UNK' and 'space' respectively.

However, this method does not make use of the priors of one language and such a greedy mode could not get the optimal result in most cases. We thus adopt another form of decoding, namely weighted finite-state transducers (WFSTs) and use EESEN [20] to decode the outputs. Our basic unit is the character in the image text recognition task. We need token WFST and lexicon WFST. The token WFST takes sequence of the network output of each time t and outputs a character, while the lexicon WFST maps the characters to words. And finally combined with the 3-order lexicon based language model to construct the final decoding network. The language model is trained by the SRILM [21] toolkit and we adopt the same corpus as [8].

4   Experiments and Results

The experiments are conducted both on the image text generated by machine and a few real labeled text images since the number of labeled text images is far from enough. The basic architecture of the neural network has been showed in Fig. 2. In the hidden layer, a bidirectional LSTM with 400 cells in each direction is applied. Our softmax layer has 4124 output units, including 4121 characters, a blank label and two additional symbols. We conduct two groups of experiments to test the performance of this CTC based BLSTM neural network. One group of experiments determines which of the two schemes of extracting feature sequence is better. The other group is used to test how this framework performs on the labeled image text data in the real world. We use the method of [8] as the baseline system and the same metrics criteria to compare the recognition performance.

4.1   Feature Extraction: Over-Segmentation or Sliding Window

As mentioned in 3.2, we come up with two schemes to extract the feature sequence. We conduct an experiment to verify which scheme is better. In this experiment, we select 400 thousand titles, of which 390 thousand titles as training data, 5 thousand as validation data, 5 thousand as test data respectively. And then we generate the text images of pure black background and finally get the feature sequences in two ways respectively. The experimental condition is the same, i.e., the structure of the network (except for the network input), initial learning rate and so on. The final result shows that the sliding window based method performs better both in the training and testing phases. Figure 3 shows how the recognition accuracy varies as the iterative process goes on. It can be seen that the sliding window based method quickly achieves a high training accuracy of over 90 % and converges gradually, while the training accuracy of the over-segmentation based method increases slowly, and this method finally gets a training accuracy of less than 80 %.

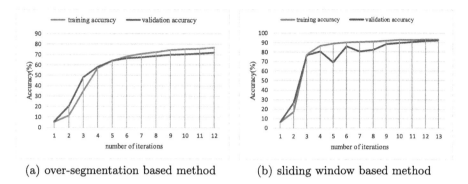

(a) over-segmentation based method   (b) sliding window based method

Fig. 3. Accuracy variation of the two feature extracting schemes during the training process.

4.2 Training with Multifont Image Texts with Complex Background and Evaluation

As illustrated in Fig. 3, the sliding window based feature extracting method performs much better than over-segmentation based method in our task. Therefore in the following experiment we adopt the sliding window based method to extract feature sequences from the text images. As [8], we select news videos from 13 TV channels to construct our test data. In our experiment we adopt all the labeled texts from each video as the test data of the corresponding TV channel, thus including text images of news titles, conversations and information of journalists etc. Among such text images, there are a few ones with very low resolution which are difficult for human beings to recognize. Therefore, the relative degradation of recognition accuracy of both methods is in expectation. Figure 4 are examples of such challenging text images and the source of those images are denoted with their corresponding TV channel name. The news title images can be referred from [8].

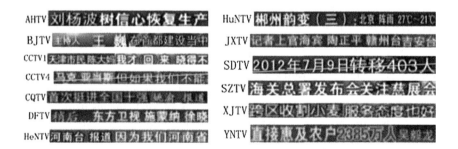

Fig. 4. Challenging examples of different TV channels

To quantitatively analyze how our proposed pipeline performs, we test the text images of each TV channel using our method and [8] respectively. The recognition accuracy is demonstrated in Table 1. From the table, we can see that our method achieves better result in all the TV channels. It is worth noting that in text images with relatively clean background and moderate resolution, both methods perform well. However, when encountering more challenging text images like examples in Fig. 4, our method performs much better than [8], with one times drop in character error rate.

Table 1. The testing data and recognition result.

TV station	Text lines	Bai's [8]	Proposed	TV station	Text lines	Bai's [8]	Proposed
AHTV	109	81.86 %	89.53 %	HuNTV	218	75.47 %	83.74 %
BJTV	158	64.31 %	84.44 %	JXTV	54	65.14 %	81.65 %
CCTV1	134	86.27 %	89.74 %	SDTV	109	81.37 %	87.92 %
CCTV4	132	85.56 %	92.53 %	SZTV	54	75.62 %	84.70 %
CQTV	155	79.87 %	89.93 %	XJTV	128	70.67 %	79.44 %
DFTV	306	85.26 %	87.77 %	YNTV	87	74.30 %	81.91 %
HeNTV	141	89.34 %	94.16 %	ALL	1785	78.01 %	86.73 %

The proposed pipeline outperforms [8] in both relatively clean text images and low resolution text images, where 2.51 % and 20.13 % improvements are observed, respectively. Thus illustrating that our pipeline is more robust than the baseline system. Moreover, with the trained network on our generated text images, we can easily transfer our work to other domain, i.e. scene text probably. Therefore our method is preferable to [8].

5   Conclusion

This paper presents a novel method named BLSTM-CTC to do the Chinese image text recognition task. The peculiarity of the proposed method is that it no longer need any work of binarization or segmentation on the initial text images. Benefiting from the success of CTC on the task of sequence labeling, we applied it on the Chinese image text recognition task which recognizes a sequence of characters as a whole. To train such a network, we employ a machine simulation method that well fullfit the gap between big data requirements and limited number of real labeled data. It generates one million text images using multiple fonts and also 10 thousand images for validation effortlessly. The experiment conducted on text images of 13 TV news videos shows that CTC training method and the method of generating text images are effective and it achieves better result than [8] in more changing scenarios. As mentioned in the experiment, for sequence labeling our proposed method, despite effectiveness, still could not cover all the style that the real image texts exhibit. In the future, we will explore more complete ways of generating representative text images, search more powerful structure of the neural networks and try some other feature extracting method (e.g. taking convolutional layer as feature extractor [22]) to further improve the performance the of the proposed pipeline.

References

1. Liu, G.H., Yang, J.Y.: Content-based image retrieval using color difference histogram. Pattern Recogn. **46**(1), 188–198 (2013)
2. Mishra, A., Alahari, K., Jawahar, C.V.: An MRF model for binarization of natural scene text. In: 2011 International Conference on Document Analysis and Recognition (ICDAR), pp. 11–16. IEEE (2011)
3. Bai, J., Feng, B., Xu, B.: Binarization of natural scene text based on L1-Norm PCA. In: 2013 IEEE International Conference on Multimedia and Expo Workshops (ICMEW), pp. 1–4. IEEE, July 2013
4. Epshtein, B., Ofek, E., Wexler, Y.: Detecting text in natural scenes with stroke width transform. In: 2010 IEEE Conference on Computer Vision and Pattern Recognition (CVPR), pp. 2963–2970. IEEE, June 2010
5. Wang, Q.F., Yin, F., Liu, C.L.: Handwritten Chinese text recognition by integrating multiple contexts. IEEE Trans. Pattern Anal. Mach. Intell. **34**(8), 1469–1481 (2012)
6. Xu, L., Yin, F., Wang, Q.F., Liu, C.L.: An over-segmentation method for single-touching Chinese handwriting with learning-based filtering. Int. J. Doc. Anal. Recogn. (IJDAR) **17**(1), 91–104 (2014)
7. Saidane, Z., Garcia, C., Dugelay, J.L.: The image text recognition graph (iTRG). In: IEEE International Conference on Multimedia and Expo, ICME 2009, pp. 266–269. IEEE, June 2009
8. Bai, J., Chen, Z., Feng, B., Xu, B.: Chinese image text recognition on grayscale pixels. In: 2014 IEEE International Conference on Acoustics, Speech and Signal Processing (ICASSP), pp. 1380–1384. IEEE, May 2014
9. Graves, A., Fernández, S., Gomez, F., Schmidhuber, J.: Connectionist temporal classification: labelling unsegmented sequence data with recurrent neural networks. In: Proceedings of the 23rd International Conference on Machine Learning, pp. 369–376. ACM, June 2006
10. Gers, F.: Long short-term memory in recurrent neural networks (Doctoral dissertation, Universitat Hannover) (2001)
11. Tang, X., Gao, X., Liu, J., Zhang, H.: A spatial-temporal approach for video caption detection and recognition. IEEE Trans. Neural Netw. **13**(4), 961–971 (2002)
12. Yokobayashi, M., Wakahara, T.: Segmentation and recognition of characters in scene images using selective binarization in color space and GAT correlation. In: Proceedings of Eighth International Conference on Document Analysis and Recognition, pp. 167–171. IEEE, August 2005
13. Li, J., Zhang, H., Cai, X., Xu, B.: Towards end-to-end speech recognition for Chinese Mandarin using long short-term memory recurrent neural networks. In: Sixteenth Annual Conference of the International Speech Communication Association (2015)
14. Graves, A., Schmidhuber, J.: Offline handwriting recognition with multidimensional recurrent neural networks. In: Advances in Neural Information Processing Systems, pp. 545–552 (2009)

15. Messina, R., Louradour, J.: Segmentation-free handwritten Chinese text recognition with LSTM-RNN. In: 2015 13th International Conference on Document Analysis and Recognition (ICDAR), pp. 171–175. IEEE, August 2015

16. Bai, J., Chen, Z., Feng, B., Xu, B.: Chinese image character recognition using DNN and machine simulated training samples. In: Wermter, S., Weber, C., Duch, W., Honkela, T., Koprinkova-Hristova, P., Magg, S., Palm, G., Villa, A.E.P. (eds.) ICANN 2014. LNCS, vol. 8681, pp. 209–216. Springer, Heidelberg (2014). doi:10. 1007/978-3-319-11179-7_27

17. Bai, Z.L., Huo, Q.: A study on the use of 8-directional features for online handwritten Chinese character recognition. In: Proceedings of Eighth International Conference on Document Analysis and Recognition, pp. 262–266. IEEE, August 2005

18. Gopalan, R., Li, R., Chellappa, R.: Domain adaptation for object recognition: an unsupervised approach. In: 2011 IEEE International Conference on Computer Vision (ICCV), pp. 999–1006. IEEE, November 2011

19. Patel, V.M., Gopalan, R., Li, R., Chellappa, R.: Visual domain adaptation: a survey of recent advances. Sig. Process. Mag. IEEE 32(3), 53–69 (2015)

20. Miao, Y., Gowayyed, M., Metze, F.: EESEN: End-to-end speech recognition using deep RNN models and WFST-based decoding. arXiv preprint arXiv:1507.08240 (2015)

21. Stolcke, A.: SRILM-an extensible language modeling toolkit. In: Interspeech, September 2002

22. Zhong, Z., Jin, L., Feng, Z.: Multi-font printed Chinese character recognition using multi-pooling convolutional neural network. In: 2015 13th International Conference on Document Analysis and Recognition (ICDAR), pp. 96–100. IEEE, August 2015

23. Bai, J., Chen, Z., Feng, B., Bo, X.: Image character recognition using deep convolutional neural network learned from different languages. In: ICIP 2014, pp. 2560–2564 (2014)

An Unsupervised Change Detection Approach for Remote Sensing Image Using SURF and SVM

Lin Wu[1]($\boxtimes$), Yunhong Wang[2], and Jiangtao Long[1]

[1] Chongqing Communication College, Chongqing 400035, China
wulin@buaa.edu.cn
[2] Beihang University, Beijing 100191, China
yhwang@buaa.edu.cn

Abstract. In this paper, we propose a novel approach for unsupervised change detection by integrating Speeded Up Robust Features (SURF) key points and Support Vector Machine (SVM) classifier. The approach starts by extracting SURF key points from both images and matches them using RANdom SAmple Consensus (RANSAC) algorithm. The matched key points are then viewed as training samples for unchanged class; on the other hand, those for changed class are selected from the remaining SURF key points based on Gaussian mixture model (GMM). Subsequently, training samples are utilized for training a SVM classifier. Finally, the classifier is used to segment the difference image into changed and unchanged classes. To demonstrate the effect of our approach, we compare it with the other four state-of-the-art change detection methods over three datasets, meanwhile extensive quantitative and qualitative analysis of the change detection results confirms the effectiveness of the proposed approach, showing its capability to consistently produce promising results on all the datasets without any priori assumptions.

Keywords: Change detection · Remote sensing image · Speeded Up Robust Features (SURF) · Support Vector Machine (SVM)

1 Introduction

With the development of remote sensing technology, the land changes could now automatically be observed through multi-temporal remote sensing images. Within the past three decades, many change detection methods have been proposed, which could be categorized as either supervised or unsupervised according to the nature of data processing [1]. The former is not widely used because of the absence of ground truth; the latter is thus the focus of change detection study. Unsupervised change detection is a process that conducts a direct comparison of multi-temporal remote sensing images acquired on the same geographical area in order to identify changes that may have occurred [2,3]. Most of the unsupervised methods are developed based on the analysis of the difference image which

© Springer Nature Singapore Pte Ltd. 2016
T. Tan et al. (Eds.): CCPR 2016, Part II, CCIS 663, pp. 537–551, 2016.
DOI: 10.1007/978-981-10-3005-5_44

can be formed either by taking the difference (resp., logarithm difference) of the two images [4,5]. We briefly describe some typical and popular unsupervised methods as well as their limitations in the following.

In general, the widely used unsupervised change detection methods include image differencing [6,7], image rationing [8,9], image regression [10], change vector analysis (CVA) [11,12], etc. Only a single spectral band of the multi-spectral images is taken into account in image differencing, image rationing and image regression methods, while several spectral bands are used at each time in the CVA method. In spite of their relative simplicity and widespread usage, the aforementioned change detection methods exhibit a major limitation: a lack of automatic and nonheuristic techniques for the analysis of the difference image. In fact, in these methods, such an analysis is performed by thresholding the difference image according to empirical strategies [13] or manual trial-and-error procedures, which significantly affect the reliability and accuracy of the final change detection results.

As a result, Bruzzone and Prieto proposed two automatic change detection techniques based on the Bayesian theory in analyzing difference image [14]. The first technique, which is referred to as the expectation maximization (EM-based) approach, allows automatical selection of decision threshold for minimizing the overall change detection error under the assumption that the pixels in difference image are spatially independent. The second technique, which is referred to as the Markov random field (MRF-based) approach, analyzes the difference image by considering the spatial contextual information included in the neighborhood of each pixel. The EM-based approach is free of parameters, whereas the MRF-based approach is parameter-dependent and the spatial contextual information may be affected in the change detection process. Another approach in [15] follows the similar way as in [14]. They analyze difference image in a pixel-by-pixel manner with a complex mathematical model. As the model is too complex, such approaches are not feasible in high resolution remote sensing images.

Afterwards, in [16], a multiscale-based change detection approach was proposed for difference image analysis, which is computed in spatial domain from multi-temporal images and decomposed using undecimated discrete wavelet transform (UDWT). For each pixel in difference image, a multiscale feature vector is extracted using the subbands of the UDWT decomposition and the difference image itself. The final change detection result is obtained by clustering the multiscale feature vectors using the k-means algorithm into two disjoint classes: changed and unchanged. This method, generally speaking, performs well, particularly on detecting adequate changes under strong noise contaminations. However, as it directly use subbands from the UDWT decompositions, this approach has problems in detecting accurate boundaries between changed and unchanged regions. In addition, this method highly depends on the number of scales used in the UDWT decomposition. In [17], an unsupervised change detection approach which is based on fuzzy clustering approach and takes into account spatial correlation between neighboring pixels of the difference image was proposed. Two fuzzy clustering algorithms, namely fuzzy c-means

(FCM) and Gustafson-Kessel clustering (GKC) algorithms have been used for classifying the pixels into changed and unchanged clusters. For clustering purpose, various image features are extracted using the neighborhood information of pixels. Hybridization of FCM and GKC with two other optimization techniques, genetic algorithm (GA) and simulated annealing (SA), is adopted to further enhance the change detection performance. The proposed approach does not require any priori knowledge of distributions of changed and unchanged pixels, but instead it is very sensitive to noise.

More recent studies not only focus on improving existing unsupervised change detection methods but also aim for proposing novel unsupervised change detection methods. In [18], an improved EM-based level set method (EMLS) was proposed to detect changes. Firstly, the distribution of difference image is supposed to satisfy GMM and the EM is then used to estimate the mean values of changed and unchanged pixels in difference image. Secondly, two new energy functions are defined and added into the level set method to detect those changes without initial contours and improve final detection accuracy. Finally, the improved level set approach is implemented to partition pixels into changed and unchanged pixels. In [19], a novel approach for unsupervised change detection based on parallel binary particle swarm optimization (PBPSO) was proposed. This approach operates on a difference image, which is created by using a novel fusion algorithm on multi-temporal remote sensing images, by iteratively minimizing a cost function with PBPSO to produce a final binary change detection mask representing changed and unchanged pixels. Each BPSO of parallel instances is run on a separate processor and initialized with a different starting population representing a set of change detection masks. A communication strategy is applied to transmit data between BPSOs running in parallel. This approach takes the full advantage of parallel processing to improve both the convergence rate and change detection performance. However, the parallel processing using for change detection is expensive and therefore, unattractive.

Almost all the aforementioned unsupervised change detection methods suffers from one limitation listed as follows. Firstly, many methods highly depend on the parameter tuning or priori assumption in modeling the difference image data, which limit the further improvement of change detection accuracy. Secondly, with the increase of the remote sensing data acquisition channels and the scope of remote sensing applications, the now available thresholding methods will be more arduous to establish suitable models or the clustering methods will be more sensitive to noise.

Consequently, within this paper, we aim to propose a general-purpose unsupervised change detection approach which has strong adaptation and robustness with better change detection performance than the conventional unsupervised change detection methods by utilizing SURF key points and SVM classifier. In general, key points which based on SURF are extensively used for image registration [20] and image classification [21]. SURF is a local descriptor extraction method having valuable properties such as invariance to illumination and viewpoint [21]. Moreover, SURF key points is well distributed over the image, and is fast to extract. SVM is a concept in statistics and computer science for a set of related supervised learning methods that analyze data and recognize patterns, used for classification and regression analysis [23]. The SVM learning problem can be expressed as a convex optimization problem, and the global minimum of the objective function can be found by using the known effective algorithm. In comparison, other classification methods (such as classifiers based on rules and artificial neural network) use a strategy based on greedy learning to search the hypothesis space, as a result they generally only get the local optimal solution [24]. Intuitivly, the matched SURF key points from a pair of registered images strongly indicate a location belonging to the unchanged class; conversely, the unmatched SURF key points indicate a candidate changed class at that location, that motivates us to solve the change detection problem based on SURF and SVM. Thus, we are inspired to use the SURF key points as training samples for the SVM classifier which is used to classify pixels in the difference image into changed and unchanged classes. We empirically test the effectiveness using three real-world multi-temporal remote sensing image datasets.

In summary, our key contribution within this paper is that the proposed change detection approach utilizes a novel implementation strategy which is the first time integrating SURF and SVM for remote sensing change detection and the change detection accuracy has been significantly improved.

The remainder of this paper is organized as follows. Section 2 describes the proposed change detection approach. Section 3 provides some experimental results of the proposed approach and compares with some state-of-the-art change detection methods. Finally, this paper is concluded in Sect. 4.

2   Methodology

In this section, we first give an overview of the proposed change detection approach. Afterwards, the key steps of our approach will be described in detail. Finally, to better understand the approach the program code is given.

2.1   An Overview of the Approach

Suppose that there are multi-temporal remote sensing images $I_1 = \{I_1(i,j)|1 \leq i \leq H, 1 \leq j \leq W\}$ and $I_2 = \{I_2(i,j)|1 \leq i \leq H, 1 \leq j \leq W\}$ acquired on the same geographical area at two different times t_1 and t_2 respectively, where H and W are height and width of both images. Then the difference image can be computed by

$$Diff(i,j) = |I_1(i,j) - I_2(i,j)|. \tag{1}$$

Notably, in this work, the absolute value of $Diff$ is used for raw data for change detection. The aim of this work is to classify pixels of the difference image into two categories: changed and unchanged. The procedure of the proposed approach is illustrated in Fig. 1. To sum up, the proposed approach is separated into the following major steps: (1) Extracting SURF key points from both images; (2) Choosing suitable samples from the SURF key points;

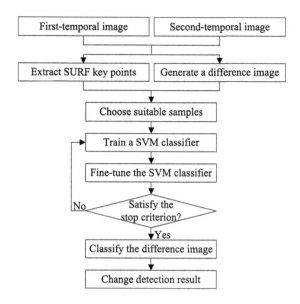

Fig. 1. Flow chart of the proposed change detection approach.

(3) Training and fine-tuning a SVM classifier; (4) Using the trained SVM classifier to classify the difference image. Among them, choosing suitable samples and training a SVM classifier are the core parts of our approach.

2.2   Key Steps of the Approach

(1) Choosing Suitable Samples

Given a pair of targeted image, the SURF algorithm is firstly applied to both images to extract SURF key points, and then matches them using the RANSAC [25] algorithm. The matched SURF key points from a pair of registered images are viewed as training samples for the unchanged class and denoted by S_u. As for the changed class SURF key points, they are selected from the remaining SURF key points based on pixel value in the difference image. In fact, there are a large number of SURF key points remaining and not all of them represent changed class in the difference image, and the thresholding methods can be used for choosing the changed class SURF key points.

An example difference image and the image with matched and unmatched SURF key points are shown in Figs. 2(c) and (d) which illustrate clearly how these points represent the changed and unchanged class well. The matched SURF key points in the difference image are marked with a dot and it is obvious that these fall in unchanged areas. The unmatched SURF key points with higher values in the difference image are marked with a cross and some of these locations indicate that there are changes between both images.

A histogram and a scatter diagram of unmatched SURF key points shown in Fig. 2(d) are demonstrated in Fig. 3. We can find that it is arduous to fairly

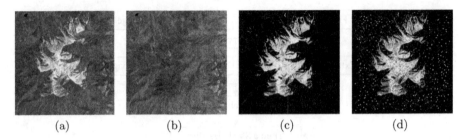

(a)   (b)   (c)   (d)

Fig. 2. (a) and (b) are multi-temporal ASTER images which are sub-regions of Mount Hillers with 400×400 pixels and acquired on 06-05-2005 and 19-07-2005, respectively. (c) is the difference image generated from (a) and (b). (d) is the difference image with SURF key points: green dots for matched points and red crosses for unmatched points. (Color figure online)

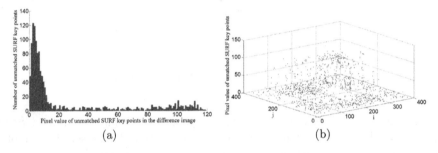

(a)   (b)

Fig. 3. The distribution of unmatched SURF key points shown in Fig. 2(d). (a) is histogram. (b) is scatter diagram.

distinguish the changed and unchanged points using a threshold, some unmatched SURF key points of medium values maybe changed or unchanged pixels. In fact, all the unmatched SURF key points in the difference image are supposed to compose of three types of pixels: changed, unchanged and unlabelled pixels. The points of higher values are easily classified into changed pixels; conversely, the points of lower values are classified into unchanged pixels. The points of medium values are marked as unlabelled pixels. Inspired by this hypothesis, in our approach, the statistical characteristics of the three types of pixels are modeled based on GMM which can then be easily estimated by using the EM algorithm [26], as shown in Eq. 2 and Fig. 4(a).

$$p(x) = \sum_{k=1}^{3} w_k N(x|\mu_k, \sigma_k^2) \tag{2}$$

In Eq. 2, μ_1, μ_2 and μ_3 are the mean of Gaussian distribution of unchanged, unlabelled and changed pixels, respectively; σ_1^2, σ_2^2 and σ_3^2 are the variance of Gaussian distribution of unchanged, unlabelled and changed pixels, respectively; w_1, w_2 and w_3 are the weight of Gaussian distribution of unchanged, unlabelled

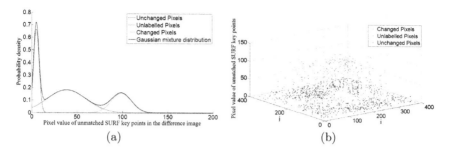

Fig. 4. (a) is the Gaussian mixture distribution of unmatched SURF key points shown in Fig. 2(d). (b) is the classification result of unmatched SURF key points based on GMM.

and changed pixels, respectively. Besides, all the parameters satisfy the following constraints:

$$0 < w_1, w_2, w_3 < 1 \tag{3}$$

$$w_1 + w_2 + w_3 = 1 \tag{4}$$

$$\mu_1 < \mu_2 < \mu_3 \tag{5}$$

Once the GMM are solved we can simply select training samples for the changed class from the unmatched SURF key points by

$$S_c = \{x | x \geq \mu_3 - 3\sigma_3\} \tag{6}$$

where S_c is the training set for changed class.

(2) Training A SVM Classifier

Training a SVM classifier is the other core part of our approach. In our approach, the implementation of SVM relies on the libsvm-mat [27] which is a simple, effective and easy-to-use pattern recognition software package. Libsvm-mat provides a host of default parameters and users only need to provide a handful of parameters to accomplish classification task. Moreover, libsvm-mat can effectively solve the issues of C-support vector classification (C-SVC), nu-support vector classification (nu-SVC), epsilon-support vector regression(epsilon-SVR) and nu-support vector regression (nu-SVR). The frequently used kernel functions of SVM are linear kernel, polynomial kernel, radial basis kernel and perceptron kernel. Kernel function determines the effect of classification and it can be selected freely. The radial basis kernel (RBF) has been adopted by the experimental system in this paper for change detection. RBF is defined as follows:

$$K(x, x_i) = \exp(-\gamma ||x - x_i||^2), \ \gamma > 0 \tag{7}$$

Once the training sets are selected, we train and fine-tune a SVM classifier, subsequently all the remaining pixels in the difference image are classified into changed and unchanged classes.

Input: $I_1 = \{I_1(i,j)|1 \leq i \leq H, 1 \leq j \leq W\}$,
$\quad\quad I_2 = \{I_2(i,j)|1 \leq i \leq H, 1 \leq j \leq W\}$
Output: CM
$Diff = |I_1 - I_2|$;
Extract SURF key points S_1 from I_1;
Extract SURF key points S_2 from I_2;
Extract training samples S_u for the unchanged class from S_1 and S_2 using
RANSAC algorithm;
$S_3 = (S_1 \cup S_2) \backslash S_u$;
Extract training samples S_c for the changed class from S_3 by Equations 2–6;
Train a SVM classifier C_{svm} by S_u and S_c;
while *don't satisfy the stop criterion* **do**
$\quad\quad$ Fine-tune parameters of C_{svm};
$\quad\quad$ Retrain the classifier C_{svm};
end
Classify $Diff$ into CM using C_{svm};
return $CM = \{cm(i,j)|1 \leq i \leq H, 1 \leq j \leq W\}$, where $cm(i,j) \in \{0,1\}$;
$\quad\quad$ **Algorithm 1.** The proposed change detection algorithm.

2.3 Pseudo Code

To better understand our approach the Matlab language pseudo code is shown in Algorithm 1. The input of this algorithm is a pair of remote sensing images which acquired on the same geographical area at two different times t_1 and t_2 respectively. The output of this algorithm is the binary change detection mask CM, in which 0 represents the unchanged pixels and 1 represents the changed pixels.

3 Experimental Results and Analysis

In order to test the effectiveness and adaptability of our approach, experiments are carried out on three publicly available remote sensing image datasets which are provided by Global Land Cover Facility (GLCF) [28]. The GLCF is a center for land cover science focusing on research using remotely sensed satellite data and products to access land cover change from local to global systems.

3.1 Experimental Datasets

The first dataset shown in Fig. 5(a) and (b) is composed of two remote sensing images with 252×250 pixels acquired over the region of Florida, America, by the LandSat sensor in 2000 and 2005, respectively.

The second dataset shown in Fig. 5(c) and (d) is composed of two remote sensing images with 250×250 pixels acquired over the region of Brazil, by the LandSat sensor in 2000 and 2005, respectively.

The third dataset shown in Fig. 5(e) and (f) is composed of two remote sensing images with 255×270 pixels acquired over the region of California, America, by the LandSat sensor in 2000 and 2005, respectively.

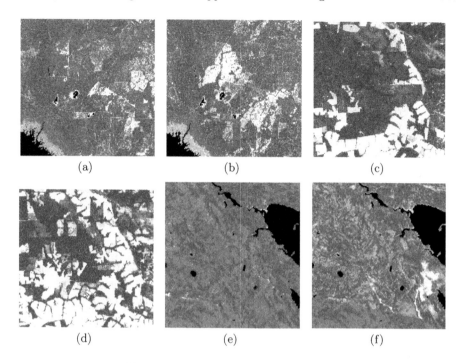

Fig. 5. Datasets used in the experiments.

The main changes among the three datasets are that some forests loss or gain owing to human activities and climate change. Besides, it should be noticed that the registration, atmospheric and radiometric correction of these datasets have been finished by GLCF.

3.2 Evaluation Criteria

To evaluate the effectiveness of our work, following measures are computed and used for comparing the change detection results against the ground truth.

(1) False alarms (*FA*) and false alarm rate (P_{FA}). *FA* is the number of unchanged pixels that were incorrectly detected as changed, and P_{FA} is defined as:

$$P_{FA} = (FA/N_1) \times 100\% \tag{8}$$

(2) Missed alarms (*MA*) and missed alarm rate (P_{MA}). *MA* is the number of changed pixels that were incorrectly detected as unchanged, and P_{MA} is defined as:

$$P_{MA} = (MA/N_0) \times 100\% \tag{9}$$

(3) Total error (*TE*) and total error rate (P_{TE}). *TE* is the sum of *FA* and *MA*, and P_{TE} is defined as:

$$P_{TE} = ((FA + MA)/(N_0 + N_1)) \times 100\% \tag{10}$$

In Eqs. 8–10, N_0 and N_1 are the number of changed and unchanged pixels in the ground truth.

3.3 Change Detection Results and Analysis

In this part, we compare the proposed change detection approach with the following state-of-the-art change detection methods: EM-based method [16], MRF-based method [16], Multiscale-based method [18] and FCM-based method [19].

(1) Results of the First Dataset

In the ground truth of the first dataset, there are 9148 changed and 53852 unchanged pixels, respectively. The quantitative and qualitative change detection results obtained from different methods for the first dataset are shown in Table 1 and demonstrated in Fig. 6, respectively. The proposed approach achieves 7.64 % total error rate (P_{TE}), i.e. 92.36 % correct detection rate. The change detection result from the MRF-based method is very close to that of ours. The correct detection rate of Multiscale-based method is about 3.21 % lower than that of ours. Meanwhile, the change detection results from the EM-based and the FCM-based methods are noisy and are about 8.39 % lower than that of ours.

(2) Results of the Second Dataset

In the ground truth of the second dataset, there are 12053 changed and 50447 unchanged pixels, respectively. The quantitative and qualitative change detection results obtained from different methods for the second dataset are shown in Table 2 and demonstrated in Fig. 7, respectively. Our approach achieves 3.01 % total error rate (P_{TE}), i.e. 96.99 % correct detection rate. The Multiscale-based method obtains suboptimal change detection result with 3.18 % lower than ours. The EM-based and the FCM-based methods obtain very similar change detection results with about 3.85 % lower than ours. Meanwhile, the change detection result from the MRF-based method is rich in false alarms and is about 6.37 % lowers than ours.

(3) Results of the Third Dataset

In the ground truth of the third dataset, there are 1254 changed and 67596 unchanged pixels, respectively. The quantitative and qualitative change detection results obtained from different methods for the third dataset are shown in Table 3

Table 1. Change detection results of the first dataset by different methods. #1-EM, #2-MRF, #3-Multiscale, #4-FCM, #5-Proposed.

Method	TE		P_{FA}	P_{MA}	P_{TE}
	FA	MA			
#1	8678	1511	16.11 %	16.52 %	16.17 %
#2	2611	2625	4.85 %	28.69 %	8.31 %
#3	5159	1677	9.58 %	18.33 %	10.85 %
#4	8613	1396	15.99 %	15.26 %	15.89 %
#5	**2120**	**2695**	**3.94 %**	**29.46 %**	**7.64 %**

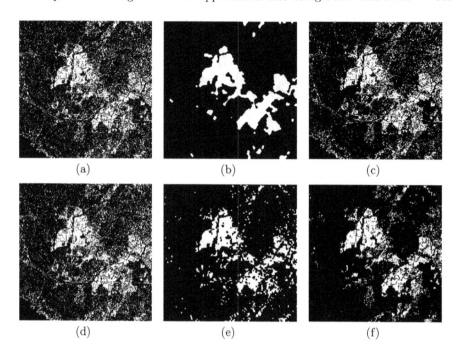

Fig. 6. Change detection results of the first dataset by (a) EM, (b) MRF, (c) Multiscale, (d) FCM, (e) Proposed approach. (f) is the ground truth.

Table 2. Change detection results of the second dataset by different methods. #1-EM, #2-MRF, #3-Multiscale, #4-FCM, #5-Proposed.

Method	TE		P_{FA}	P_{MA}	P_{TE}
	FA	MA			
#1	4207	53	8.34 %	0.44 %	6.82 %
#2	5609	254	11.12 %	2.11 %	9.38 %
#3	3849	19	7.63 %	0.16 %	6.19 %
#4	4274	37	8.47 %	0.31 %	6.90 %
#5	**1072**	**808**	**2.13 %**	**6.70 %**	**3.01 %**

and demonstrated in Fig. 8, respectively. The proposed approach achieves 1.34 % total error rate (P_{TE}), i.e. 98.66 % correct detection rate. The Multiscale-based method obtains suboptimal change detection result with about 6.17 % lower than that of ours. The EM-based and the MRF-based methods obtain very similar change detection results with about 12.25 % lower than that of ours. Meanwhile, the change detection result from the FCM-based method is very noise and is about 21.47 % lower than that of ours.

In summary, the four change detection methods exhibit different performance and none of them can perform well for all the experimental datasets.

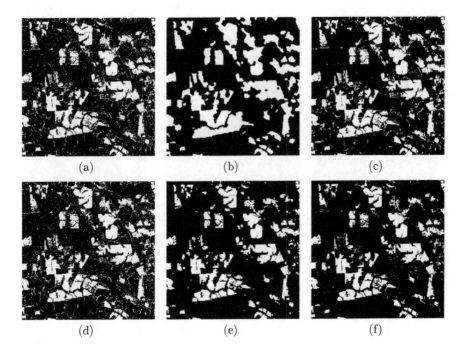

Fig. 7. Change detection results of the second dataset by (a) EM, (b) MRF, (c) Multiscale, (d) FCM, (e) Proposed approach. (f) is the ground truth.

Table 3. Change detection results of the third dataset by different methods. #1-EM, #2-MRF, #3-Multiscale, #4-FCM, #5-Proposed.

Method	TE		P_{FA}	P_{MA}	P_{TE}
	FA	MA			
#1	8830	18	13.06 %	1.44 %	12.85 %
#2	9790	86	14.48 %	6.86 %	14.34 %
#3	5109	59	7.56 %	4.70 %	7.51 %
#4	15692	14	23.21 %	1.12 %	22.81 %
#5	**777**	**145**	**1.15 %**	**11.56 %**	**1.34 %**

For example, the EM-based method performs bad for the first dataset, mainly due to the fact that the unimodal Gaussian model employed in modeling the difference image data fails to provide accurate data model; the FCM-based method performs bad for the third dataset, mainly due to the fact that its performance highly depends on the quality of the initialization data and it easily obtains local optimal solution. In contrast, it is very clear from the aforementioned change detection results on all the datasets that the proposed change detection approach obtains the binary change detection mask more accurately than other methods and does not need any prior knowledge about data distribution.

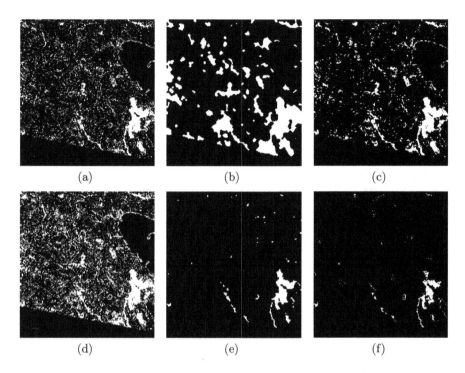

Fig. 8. Change detection results of the third dataset by (a) EM, (b) MRF, (c) Multi-scale, (d) FCM, (e) Proposed approach. (f) is the ground truth.

Our approach essentially belongs to unsupervised change detection, although it has utilized SVM classifier to detect pixel changes. It utilizes a novel implementation strategy which is, to the best of our knowledge, the first time integrating SURF and SVM for remote sensing change detection and the change detection accuracy has been significantly improved.

4   Conclusions

In this paper, we proposed a novel unsupervised change detection approach by integrating SURF and SVM for remote sensing images. Given a pair of images, it firstly generates a difference image and extracts SURF key points from both images. Then the unchanged training samples are selected from the matched SURF key points using RANSAC algorithm; the changed training samples are selected from the unmatched SURF key points based on GMM. We subsequently train and fine-tune a SVM classifier using the changed and unchanged training samples. Finally, the difference image is classified into changed and unchanged classes by the trained SVM classifier. We compared our approach with series of other state-of-the-art methods in the aspect of change detection over three real-world datasets. Empirical results demonstrated that the proposed approach is robust to noise and can consistently produce excellent change detection results on all the datasets without any priori assumptions. Hence, our approach shows a great potential advantages and applications for remote sensing change detection.

As part of our future work, we intend to carry out experiments on different types of remote sensing image datasets and discuss ways to extract the change information directly from the original images rather than the difference image.

References

1. Lu, D., Mausel, P., Brondizio, E., Moran, E.: Change detection techniques. Int. J. Rem. Sens. **25**, 2365–2401 (2004)
2. Melgani, F., Bazi, Y.: Markovian fusion approach to robust unsupervised change detection in remotely sensed imagery. IEEE Trans. Geosci. Rem. Sens. Lett. **3**, 457–461 (2006)
3. Bazi, Y., Melgani, F., Al-Sharari, D.: Unsupervised change detection in multispectral remotely sensed imagery with level set methods. IEEE Trans. Geosci. Rem. Sens. **48**, 3178–3187 (2010)
4. Rosin, D.L.: Thresholding for change detection. Comput. Vision Image Underst. **86**, 79–95 (2002)
5. Prakash, A., Gupta, R.P.: TLand-use mapping and change detection in a coal mining area - a case study in the Jharia coalfield, India. Int. J. Rem. Sens. **19**, 391–410 (1998)
6. Weismiller, R.A., Kristof, S.J., Scholz, D.K., Anuta, P.E., Momin, S.A.: Change detection in coastal zone environments. Photogram. Eng. Rem. Sens. **43**, 1533–1539 (1977)
7. Kapur, J.N., Sahoo, P.K., Wong, A.K.C.: A new method for gray-level picture thresholding using the entropy of the histogram. Comput. Vision Graph. Image Process **29**, 273–285 (1985)
8. Nelson, R.F.: Detecting forest canopy change due to insect activity using Landsat MSS. Photogram. Eng. Rem. Sens. **49**, 1303–1314 (1983)
9. Stow, D.A.: Reducing the effects of misregistration on pixel-level change detection. Int. J. Rem. Sens. **20**, 2477–2483 (1999)
10. Jha, C.S., Unni, N.V.M.: Digital change detection of forest conversion of a dry tropical Indian forest region. Int. J. Rem. Sens. **15**, 2543–2552 (2007)
11. Sohl, T.L.: Change analysis in the United Arab Emirates: an investigation of techniques. Photogram. Eng. Rem. Sens. **65**, 475–484 (1999)
12. Nackaerts, K., Vaesen, K., Muys, B., Coppin, P.: Comparative performance of a modified change vector analysis in forest change detection. Int. J. Rem. Sens. **26**, 839–852 (2005)
13. Fung, T., LeDrew, E.: The determination of optimal threshold levels for change detection using various accuracy indices. Photogram. Eng. Rem. Sens. **54**, 1449–1454 (1988)
14. Bruzzone, L., Prieto, D.: Automatic analysis of the difference image for unsupervised change detection. IEEE Trans. Geosci. Rem. Sens. **38**, 1171–1182 (2000)
15. Kasetkasem, T., Varshney, P.: An image change detection algorithm based on Markov random field models. IEEE Trans. Geosci. Rem. Sens. **40**, 1815–1823 (2002)
16. Celik, T.: Multiscale change detection in multitemporal satellite images. IEEE Trans. Geosci. Rem. Sens. Lett. **6**, 820–824 (2009)

17. Ghosh, A., Mishra, N.S., Ghosh, S.: Fuzzy clustering algorithms for unsupervised change detection in remote sensing images. Inf. Sci. **181**, 699–715 (2011)
18. Hao, M., Shi, W.Z., Zhang, H., Li, C.: Unsupervised change detection with expectation-maximization-based level set. IEEE Trans. Geosci. Rem. Sens. Lett. **11**, 210–214 (2014)
19. Kusetogullari, H., Yavariabdi, A., Celik, T.: Unsupervised change detection in multitemporal multispectral satellite images using parallel particle swarm optimization. IEEE J. Sel. Top. Appl. Earth Obs. Rem. Sens. **8**, 2151–2164 (2015)
20. Teke, M., Temizel, A.: Multi-spectral satellite image registration using scale-restricted SURF. In: 20th International Conference on Pattern Recognition, pp. 2310–2313 (2010)
21. Liang, J., Liu, Q., Ai, Q.S.: Research of image classification based on fusion of SURF and global feature. Comput. Eng. Appl. **49**, 174–177 (2013)
22. Herbert, B., Andreas, E., Tuytelaars, T., Gool, L.V.: SURF: speeded up robust features. Comput. Vis. Image Underst. **110**, 346–359 (2008)
23. Niranjan, M.: Support vector machines: a tutorial overview and critical appraisal. In: IEE Colloquium on Applied Statistical Pattern Recognition, p. 2/1 (1999)
24. Vapnik, V.N.: An overview of statistical learning theory. IEEE Trans. Neural Netw. **10**, 988–999 (1999)
25. Fischler, M.A., Bolles, R.C.: Random sample consensus: a paradigm for model fitting with applications to image analysis and automated cartography. Commun. ACM **24**, 381–395 (1981)
26. Bilmes, J.A.: A gentle tutorial of the EM algorithm and its application to parameter estimation for Gaussian mixture and hidden Markov models. Int. Comput. Sci. Inst. **4**, 1–13 (1998)
27. Chang, C.C., Lin, C.J.: LIBSVM–A Library for Support Vector Machines. http://www.csie.ntu.edu.tw//~cjlin/libsvm/index.html
28. University of Maryland: Global Land Cover Facility. http://www.landcover.org/

Semi-supervised Learning of Bottleneck Feature for Music Genre Classification

Jia Dai[1,2], Wenju Liu[1,2(✉)], Hao Zheng[1,2], Wei Xue[1,2], and Chongjia Ni[3]

[1] NLPR, Institute of Automation, Chinese Academy of Sciences, Beijing, China
{jia.dai,lwj,hzheng,wxue}@nlpr.ia.ac.cn
[2] University of Chinese Academy of Sciences, Beijing, China
[3] School of Mathematic and Quantitative Economics, Shandong University
of Finance and Economics, Shandong, China
cjni_sd@sdufe.edu.cn

Abstract. A good representation of the audio is important for music genre classification. Deep neural networks (DNN) enable a better approach to learn the representation of audio. The representation learned from DNN, which is known as bottleneck feature, is widely used for speech and audio related application. However, in general, it needs a large amount of transcribed data to learn an effective bottleneck feature extractor. While, in reality, the amount of transcribed data is often limited. In this paper, we investigate a semi-supervised learning to train the bottleneck feature for music data. Then, the bottleneck feature is used for music genre classification. Since the target dataset contains few data, which cannot be used train a reliable bottleneck DNN, we train the DNN bottleneck extractor on a large out-of-domain un-transcribed dataset in semi-supervised way. Experimental results show that with the learned bottleneck feature, the proposed system can perform better than the state-of-the-art best methods on GTZAN dataset.

Keywords: Bottleneck · DNN · Semi-supervised · Multilingual · Cross-lingual

1 Introduction

For music genres classification tasks, most conventional methods use a common procedure which consists of two stages: feature extraction and classification. A discriminative feature and a good classifier can often lead to a better classification result [1].

Recently, Deep Neural networks (DNN) have become increasingly popular in many fields [2,3]. Classification systems based on DNN usually have a better performance than others, which also includes audio classification. DNN is widely used especially for the big data cases, in which the amount of available data is large. Music genre classification is a special kind of audio classification. Generally there are two ways to incorporate the DNN into the classification system: using DNN as the classifier [4,5], or training a DNN to learn the feature

© Springer Nature Singapore Pte Ltd. 2016
T. Tan et al. (Eds.): CCPR 2016, Part II, CCIS 663, pp. 552–562, 2016.
DOI: 10.1007/978-981-10-3005-5_45

of audio [6]. The deep architecture enables DNN to learn more invariant and discriminative features. There are many techniques to use a deep architecture to learn the representation of audio data. For example, Sigtia [7] improves feature learning for audio data using DNN. This method provides a significant improvement in training time, and the features learned are better than state-of-art handcrafted features. Andén [8] proposes the deep scattering transform to extract feature from a deep convolution network. This deep scattering feature can perform better when using long frame window and improves the music genre classification performance.

For automatic speech recognition (ASR), the most popular feature learned by using DNN is bottleneck feature. It uses a small hidden layer (bottleneck layer) which is smaller than other hidden layers in DNN to train a feature extractor. Bottleneck feature is used widely and can achieve good results [9,10]. However, there is a limit that we don't always have enough data to train a bottleneck DNN, and little data makes the extractor become over-fitting easily. In order to solve this problem, multilingual bottleneck and cross-lingual bottleneck are proposed [11,12]. The multilingual bottleneck and cross-lingual bottleneck extractors are trained on other languages (rich-resource datasets), and then the trained bottleneck DNN is used to extract bottleneck feature for target dataset with limited amount of data. Experiments show that the bottleneck system trained on other languages outperform the system trained on limited data in ASR [13,14].

Inspired by this, the semi-supervised training for bottleneck feature (ST-BN system) is proposed for music genre classification. Since there is no "other language" for bottleneck DNN training like ASR, we use semi-supervised method to train a bottleneck DNN on a large similar out-of-domain dataset instead. Semi-supervised is an important topic, especially when target dataset is small while large un-transcribed audio data is available [15]. After the low-level feature is extracted, we first use some of the limit target dataset to train an artificial neural network (ANN) to label the large out-of-domain dataset, which is used for semi-supervised training. The out-of-domain dataset is a dataset which is similar to the target dataset. Then the labeled out-of-domain dataset is used to train a bottleneck DNN. At last, the trained bottleneck DNN is used to extract bottleneck feature for target dataset, and the bottleneck feature is used for training and testing the classification system. The experiment results show that our bottleneck feature perform better than the low-level feature.

To the best of our knowledge, in this paper, it is the first work to apply the bottleneck feature to music genre classification. This ameliorates the over-fitting problem of DNN caused by the lacking of training target data. The proposed ST-BN system achieves 94 % average classification result, which is the best classification result on GTZAN dataset [16] compared with other methods.

The rest of this paper is organized as follows. Section 2 describes the prior work related to this paper. Sections 3 and 4 is the feature extraction and proposed ST-BN system respectively. Experiment and results analysis are given in Sect. 5, followed by a conclusion section.

2 Prior Work

The multilingual DNN for music genre classification has been explored in our previous work [17]. It uses an out-of-domain dataset to train a DNN classification model, then transfer it to target dataset. The DNN model trained on out-of-domain dataset can be seen as a pre-training of the target DNN classification model, and the pre-training on out-of-domain dataset improves the performance. In this work, the out-of-domain dataset is used to train a bottleneck extractor in semi-supervised way, and this is different from the pre-training in the previous work. The feature extracted from the trained bottleneck extractor are more discriminative and robust, which can improve the accuracy of "dissimilar" music track. This model is easy to implement as there are a lot of un-transcribed music data available on the Internet. Also, the trained bottleneck extractor can be used to extract music features for other music systems. So, this model can be used more widely than the previous work.

3 Data and Feature

3.1 Data

The GTZAN database is used here as the target dataset to evaluate the proposed system. This dataset has been used by many authors [7,18,19] for music genre classification. GTZAN dataset is divided into ten genres: classical, jazz, blues, metal, pop, rock, country, disco, hip-hop, reggae. Each genre has 100 tracks with 30 s duration for every track. The detail of the database is described as Table 1.

Table 1. Database description

GTZAN		ISMIR	
Genre	Tracks	Genre	Tracks(train/test)
classical	100	Classical	320/320
jazz	100	Electronic	115/114
blues	100	Jazz/Blue	26/26
metal	100	Metal/Punk	45/45
pop	100	Rock/Pop	101/102
rock	100	World	122/122
country	100		
disco	100		
hiphop	100		
reggae	100		
Total	1000	Total	729/729

The ISMIR dataset [20] is used as the large out-of-domain dataset for the semi-supervised training of a bottleneck DNN. It is a large dataset which contains 100 h of music data with 1458 music tracks. The time length of each track is different, ranging from 8 s to 42 min. Those tracks contain six genres: classical, electronic, jazz/blue, metal/punk, rock/pop, and world. In this paper, we do not use these genre labels, and the genre labels of ISMIR dataset is re-labeled by the ANN trained on GTZAN dataset and then used for semi-supervised training. The detail of the database is described as Table 1.

3.2 Scattering Feature

The Mel-Frequency Cepstral Coefficient (MFCC) is widely used in many fields. The scattering feature is an extension of MFCC. It is computed by scattering the signal information along multiple paths, with a cascade of wavelet modulus operators implemented in a deep convolution neural network (CNN). For music, rhythm or beat is an important information to distinguish different music tracks, and short frame features cannot represent this information well. Therefore, long time representation can represent music data better for music genre classification, while MFCC may lose information by averaging spectrogram of long time window. The scattering moments can recover the lost information, which has been proved by [8,21].

Before feature extraction, each audio file in ISMIR and GTZAN has been converted into a 22050 Hz, 16 bit, and single channel WAV file. We extract the scattering feature for GTZAN and ISMIR using ScatNet [22]. In our work, we calculate first-order and second-order time scattering coefficients using a window of 370 ms with half overlap. The parameters for scattering transform are just the same as our previous work in [17]. After that, we stack the 3 consecutive frame features into a long frame feature vector with context information.

4 ST-BN System

4.1 Framework

The proposed ST-BN system framework is illustrated in Fig. 1. The DNN and ANN used in ST-BN are feed-forward neural networks. Karel's DNN in the Kaldi Toolkit [23] is used for DNN or ANN training. The DNN has one input layer, some hidden layers and one output layer. Within each hidden layer, the sigmoid function is used as the active function. For output layer, the soft-max function is used to compute the posterior probability. The learning rate is $8 * 10^{-6}$, and no pre-training is used for DNN training. The cross-entropy function is used as the objective function to optimize the ANN and DNN.

4.2 Semi-supervised Training of the Bottleneck Feature Extractor

A bottleneck DNN extractor is a special DNN which has a narrower hidden layer compared to other hidden layers. The narrow hidden layer is called bottleneck layer, and the bottleneck feature is generated from this narrow bottleneck layer. After we get the scattering feature for target dataset and large out-of-domain dataset, the frame scattering feature of out-of-domain dataset is used for semi-supervised training of the bottleneck DNN. It mainly contains two key stages.

In the first stage, we use scattering feature of target dataset (low-resource data) to label the large out-of-domain dataset. First, we randomly select 70 % (we suggest equal or greater than 70 % since little data cannot train a reliable ANN) of target scattering feature for training an ANN. The structure of ANN is described in Sect. 4.1, and it has one hidden layer with 1024 nodes. We train the ANN for 200 epochs. Then, the scattering feature of out-of-domain dataset is put into the trained ANN. At last, the label of out-of-domain dataset is decided by the posterior probability output of the ANN: the genre index of the max probability is the label.

In the second stage, we train a bottleneck extractor. As in Fig. 1, the bottle-neck DNN has one input layer, two hidden layers followed by a bottleneck layer, a hidden layer and one output layer. First, the frame feature of the labeled out-of-domain dataset is used as the input to train a bottleneck DNN. The training steps is 200. Then, we get the final bottleneck extractor by removing the hidden layer 4 and the output layer (in Fig. 1) of bottleneck DNN.

4.3 BN Feature Extraction and Classification

After we get the bottleneck extractor, we extract the bottleneck features and then do the classification. The scattering feature of target dataset is used as the

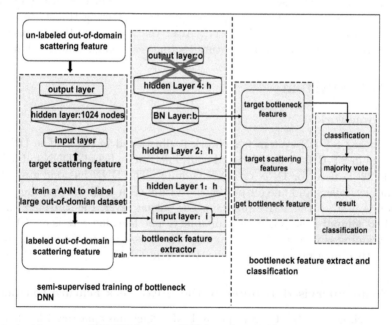

Fig. 1. Architecture of ST-BN System: "BN layer" in the figure is the bottleneck layer.

input of bottleneck extractor, then the output is bottleneck feature. After we get the bottleneck feature of the target dataset, we divide these bottleneck feature into training feature set and testing feature set.

In this paper, support vector machine (SVM) and ANN are used as the classifiers of ST-BN system. The SVM used here is a linear SVM, the training set of bottleneck feature is used as the input to train a liner SVM, and the testing set of bottleneck feature is used to get the frame test predict genre labels. The structure of ANN for classification is the same as the ANN in previous subsection. We use the training set of bottleneck feature to train the ANN, and then use testing set of bottleneck feature to get the soft-max output. After training and testing, we got the posterior probability for each frame feature. Each predict frame label is determined by the genre which has the maximum probability.

After classification, we get the predicted frame genre labels. However, people are always interested in the genre label of the whole music track (a clip of music), rather than the genres of the internal frames. So the majority voting is performed on the frame predict labels of the testing sets to get the label of each music track. The genre label of the whole music clip is decided by the majority of the genre labels on the internal frames. After that we get the genre label of each test music track, and the classification accuracy is computed on the track labels.

5 Experiment and Analysis

5.1 Evaluation Method and Baseline System

In this paper, we use 10-fold cross-validation to evaluate the performance. The features of target database are divided into 10 folds, 9 folds of which are used for training and the remaining one is used to test the classification accuracy. The classification accuracy is evaluated 10 times on the 10 different combinations of training/testing sets. The overall classification accuracy is calculated as the average of 10 independent 10-fold cross-validations.

SVM and DNN are used to build baseline systems: Baseline-SVM and baseline-DNN. The GTZAN dataset is used as target dataset. Frame scattering feature of target dataset is used as the input of baseline systems. Baseline-SVM use radial basis function as kernel function, and the parameters (the "cost" and "gama") for kernel function are obtained by grid search algorithm on the GTZAN training feature sets. This baseline-DNN use the Karel's DNN in Kaldi Toolkit [23], and other parameters are the same as DNN or ANN described in Sect. 4.1. The baseline-DNN are three kinds of DNN with one, two and three hidden layers. Each hidden layer has 1024 nodes, and the training epochs are 1000. After testing, the genre which has the maximum probability is regarded as the frame genre labels. After getting frame labels from the baseline systems, the majority voting is used to get the track labels. The result is reported in Table 2.

5.2 Experiment of ST-BN System

After getting scattering feature of target dataset and large out-of-domain dataset, we start to build the ST-BN system for music genre classification.

Table 2. Classification results of baseline systems

Model name	Classifier(hidden layers)	Average accuracy
baseline-SVM	SVM	88.5 %
baseline-DNN1	DNN(1024)	89.4 %
baseline-DNN2	DNN(1024-1024)	89.7 %
baseline-DNN3	DNN(1024-1024-1024)	90.1 %

In this paper, the ISMIR dataset is used as the large out-of-domain dataset and GTZAN dataset is used as target dataset. We first train 200 epochs to get an ANN to label ISMIR dataset. Then use labeled ISMIR dataset to train 200 epochs to get the bottleneck extractor. At last we use the bottleneck feature as the input of classifier to test the performance. In Table 3, we compare the proposed ST-BN system and other works on GTZAN dataset. The hidden layer structure of bottleneck DNN is "2048-2048-100-2048", and the classifier is linear SVM. From the result we can see that our model can achieve a good result and outperform other approaches.

Table 3. The comparison of different works

Model	Average accuracy
ST-BN system (proposed system)	**94 %**
Dai's work [17]	93.4 %
Pan2009music [24]	92.4 %
And2013deep [8]	91.9 %
Lee2009auto [25]	90.6 %
Eckleafeat [18]	84.3 %
Hena2011uns [19]	83.4 %

5.3   Experiment Analysis

First, we analyze the effectiveness of different classifiers and different structures of bottleneck DNN. The structure of bottleneck DNN can be represented as "i-h-h-b-h-o", just as in Fig. 1. The 'i','h','b','o' respectively represent the number of nodes in input layer, hidden layer, bottleneck layer and output layer. Figure 2 shows the performance of ST-BN system with different 'h', different 'b' and different classifiers. From the figure, we can see that more nodes in bottleneck layer cannot always increase the performance, and more nodes in hidden layer can increase the result in some degree. The number of nodes in bottleneck layer cannot be too large, but also cannot be too small. The best number of node in bottleneck layer for GTZAN feature sets in this paper is 200. SVM perform better than ANN in ST-BN (in Fig. 2), but perform worse than DNN

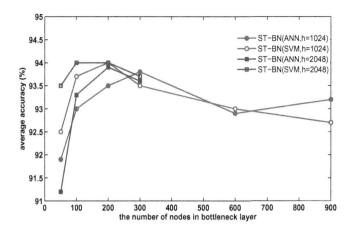

Fig. 2. The classification results of different bottleneck DNN. "SVM" and "ANN" in the legend indicate the classifier used for bottleneck feature classification. "h = 1024" and "h = 2048" represent the nodes in hidden layers.

in the baseline (in Table 2). That's maybe because SVM is more suitable for low dimensional features (The scattering feature has $525 * 3$ dimension, while the bottleneck feature has a much lower dimension). Another reason is that the feature tends to be linear after the processing of bottleneck extractor, therefore a linear SVM is more powerful here.

Then, we analyze how our ST-BN system improves the performance. Figure 3 shows the detailed frame results before majority voting, and they are two fold results chosen from the total ten folds. It indicates the frame rate for each test music track. The total tracks for testing sets is 100, so the track id in the figure is from 1 to 100. In Fig. 3, blue stars represent baseline frame rates in each track, and red triangles represent ST-BN frame rates in each track. A line connect baseline rate and ST-BN rate in a same track. A higher frame rate in a track may led to a correct classification label of the track when perform majority voting. In baseline system, we can see that some tracks have very high frame rate, and some tracks have very low rate. This "un-balanced" situation appears because the limited of training data. It makes the classification model easier to over-fitting. For the tracks which are similar to training tracks, the algorithm will have a good result, but for a "dissimilar" track, it will have a bad result. So in this paper, we use a similar large dataset to training the bottleneck DNN to make the model more stable. It improves the performance by increasing the rate of those low frame rate tracks. From the figure, we can see that many low frame rates are improved. Though some high rates in some tracks are decreased, these rates are still higher enough to produce a correct label for these tracks after majority voting. So the ST-BN system can achieve a better classification performance.

Figure 4 shows the comparison between the baseline systems and ST-BN system. The figure shows the mean accuracy and variance of 10-fold

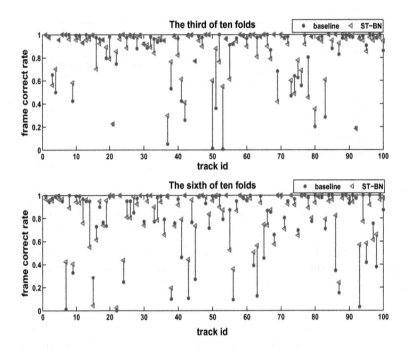

Fig. 3. The comparison of frame correct rate in test music tracks between baseline system and ST-BN system. The baseline system used here is baseline-DNN1 system. The structure of hidden layers in bottleneck DNN used in ST-BN is "1024-1024-100-1024", and the classifier is linear SVM.

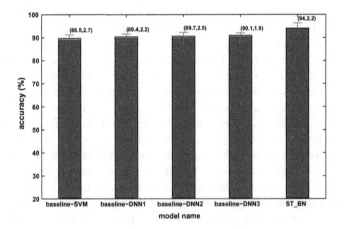

Fig. 4. The comparison of different models. The number in the figure is the mean accuracy and the standard deviation. The hidden layer structure of bottleneck DNN used in ST-BN system is "2048-2048-100-2048".

cross-validation. The accuracy used here is the accuracy after majority voting (track accuracy). From the result, we can see that our model outperforms baseline models.

At last, we analyze the limit of ST-BN system, which uses a large out-of-domain data to improve the performance of target dataset. The limit is that if the target dataset is also a large dataset (e.g. the same dataset as out-of-domain dataset), the result can not be improved so much. That's because that if the target dataset is a large dataset, there will be little "un-balanced" situation or a "dissimilar" tracks.

6 Conclusions

In this paper, we propose a music genre classification approach using bottleneck feature, which is learned at semi-supervised learning framework. This approach can help to learn a more stable feature representation, and the learned bottleneck features using semi-supervised learning from the out of domain data can be used to improve the performance of music genre classification. Experimental results show that the proposed system outperforms the current state-of-the-art best system on GTZAN dataset, and experimental analysis further shows the bottleneck features can improve the classification accuracy of those music tracks which are "dissimilar" to training tracks. Although the proposed approach is similar with "Multilingual bottleneck" and "Cross-lingual bottleneck" in ASR, but to our best knowledge, it is the first work to try to apply the bottleneck feature to music genre classification or audio classification.

Acknowledgements. This research was supported by following two parts: The China National Nature Science Foundation (No. 61573357, No. 61503382No. 61403370, No. 61273267 and No. 91120303, No. 61305027), and technical development project of state grid corporation of China entitled machine learning based Research and application of key technology for multi-media recognition and stream processing.

References

1. Li, T., Ogihara, M., Li, Q.: A comparative study on content-based musicgenre classification. In: Proceedings of the 26th Annual International ACM SIGIR Conference on Research and Development in Informaion Retrieval, SIGIR 2003, pp. 282–289. ACM, New York (2003)

2. Dahl, G.E., Yu, D., Deng, L., Acero, A.: Context-dependent pre-trained deep neural networks for large-vocabulary speech recognition. IEEE Trans. Audio Speech Lang. Process. **20**(1), 30–42 (2012)

3. Mikolov, T., Zweig, G.: Context dependent recurrent neural network language model. In: SLT, pp. 234–239 (2012)

4. Mcloughlin, I., Zhang, H., Xie, Z., Song, Y., Xiao, W.: Robust sound event classification using deep neural networks. IEEE/ACM Trans. Audio Speech Lang. Process. **23**(3), 540–552 (2015)

5. Yang, X., Chen, Q., Zhou, S., Wang, X.: Deep belief networks for automatic music genre classification ntM, vol. 92, no. 11, pp. 2433–2436 (2011)

6. Lu, L., Renals, S.: Probabilistic linear discriminant analysis with bottleneck feature for speech recognition. In: Interspeech (2014)
7. Sigtia, S., Dixon, S.: Improved music feature learning with deep neural networks. In: 2014 IEEE International Conference on Acoustics, Speech and Signal Processing (ICASSP), pp. 6959–6963 (2014)
8. Andén, J., Mallat, S.: Deep scattering spectrum, CoRR abs/1304.6763 (2013)
9. Nguyen, Q.B., Gehring, J., Kilgour, K., Waibel, A.: Optimizing deep bottleneck feature extraction. In: IEEE RIVF International Conference on Computing and Communication Technologies, Research, Innovation, and Vision for the Future (RIVF), pp. 152–156 (2013)
10. Liu, D., Wei, S., Guo, W., Bao, Y., Xiong, S., Dai, L.: Lattice based optimization of bottleneck feature extractor with linear transformation. In: IEEE International Conference on Acoustics, Speech and Signal Processing (ICASSP), pp. 5617–5621 (2014)
11. Vu, N.T., Weiner, J., Schultz, T.: Investigating the learning effect of multilingual bottle-neck feature for ASR. In: Interspeech (2014)
12. Li, J., Zheng, R., Xu, B.: Investigation of cross-lingual bottleneck feature in hybrid ASR system. In: Interspeech (2014)
13. Do, V.H., Xiao, X., Chng, E.S., Li, H.: Kernel density-based acoustic model with cross-lingual bottleneck features for resource limited LVCSR. In: Interspeech (2014)
14. Zhang, Y., Chuangsuwanich, E., Glass, J.: Language id-based training of multilingual stacked bottleneck features. In: Interspeech (2014)
15. Xu, H., Su, H., Chng, E.S., Haizhou, L.: Semi-supervised training for bottleneck feature based DNN-HMM hybrid system. In: Interspeech (2014)
16. Tzanetakis, G., Cook, P.: Musical genre classification of audio signals. IEEE Trans. Speech Audio Process. 10(5), 293–302 (2002)
17. Dai, J., Liu, W., Ni, C., Dong, L., Yang, H.: Multilingual deep neural network for music genre classification. In: Interspeech (2015)
18. Eck, D., Montréal, U.D.: Learning features from music audio with deep belief networks. In: International Society for Music Information Retrieval Conference (ISMIR) (2010)
19. Henaff, M., Jarrett, K., Kavukcuoglu, K., Lecun, Y.: Unsupervised learning of sparse features for scalable audio classification. In: International Society for Music Information Retrieval Conference (ISMIR) (2011)
20. Ismir. http://ismir2004.ismir.net/genre_contest/index.html
21. Anden, J., Mallat, S.: Multiscale scattering for audio classification. In: International Society for Music Information Retrieval Conference (ISMIR), pp. 657–662 (2011)
22. scattering. http://www.di.ens.fr/data/scattering/
23. Kaldi. http://kaldi.sourceforge.net
24. Panagakis, Y., Kotropoulos, C., Arce, G.R.: Music genre classification using locality preserving non-negative tensor factorization and sparse representations. In: International Society for Music Information Retrieval Conference (ISMIR), pp. 249–254 (2009)
25. Lee, C.H., Shih, J.L., Yu, K.M., Lin, H.S.: Automatic music genre classification based on modulation spectral analysis of spectral and cepstral features. IEEE Trans. Multimedia 11(4), 670–682 (2009)

Simultaneous Audio Source Localization and Microphone Placement

Peng Luo$^{(\boxtimes)}$, Jiwen Lu, and Jie Zhou

Department of Automation, Tsinghua University, Beijing, China
Luop13@mails.tsinghua.edu.cn

Abstract. Position information of sound source provides important cues for many audio analysis tasks. In this paper, we present a simple yet effective simultaneous audio source location and microphone placement approach to obtain the position information of stationary sound source. Motivation by the fact that audio source location and microphone placement can help each other, we consider these tasks as a joint inference framework so that more contextual information can be exploited. By fusing geometric properties of the sliding microphone-pair system with an error theory analysis, our location approach can achieve higher accuracy than conventional methods theoretically. Moreover, experimental results on our built real sound source location system are presented to demonstrate the effectiveness of the proposed approach.

Keywords: Sound source localization · Sliding microphone · Microphone placement

1 Introduction

Stationary source location is an important research area in signal processing, such as sound source location and radar monitoring. Generally, we first obtain a number of time differences of arrival (TDOA's) by placing some sensors to capture the emitted signals [1,2], and then deduce the location of the source from TDOA. TDOA-based techniques have been widely studied to locate the source accurately in the literature [3–9]. The observed values of TDOA and an effective TDOA-based location algorithm are both crucial in this procedure. There are roughly two ways to solve this problem: (1) improving sensor devices and sensor array topology to obtain more accurate TDOA values; and (2) introducing more precise algorithms. For the first category, topology structure optimization of sensor array is worthy considering because it has a great influence on sound source location [10]. For example, Gazor and Grenier [11] evaluated and optimized the performance of an sensor array through array geometric configuration. Yang and Scheuing [12] analyzed the Cramer-Rao Bound (CRB) for source location from TDOA and derived properties of the Cramer-Rao Bound to design optimum sensor arrays. However, most conventional methods only considered the topology structure at sensor array design phase, which is not flexible and optimum to all

© Springer Nature Singapore Pte Ltd. 2016
T. Tan et al. (Eds.): CCPR 2016, Part II, CCIS 663, pp. 563–575, 2016.
DOI: 10.1007/978-981-10-3005-5_46

source location situations. For the second category, there are mainly two typical algorithms: maximum likelihood estimator (MLE) [13] and least square estimator (LSE) [14]. The first method has been widely studied in signal processing but still requires prior knowledge of the estimated parameters, which becomes complex considering all kinds of uncertain environments and not practical. The second method has been widely used in source location and achieved good results, which does not need prior knowledge and its solution satisfies closed expression. Owing to LSE properties, many methods have been proposed based on this category. Smith and Abel [15, 16] presented spherical interpolation solution for source location. Chan and Ho [17] proposed a quadratic-correction least squares solution while Huang *et al.* [18] presented a linear-correction least squares method. In 2012, Ho [19] proposed a more precise method which augmented the equation error formulation and imposed a constraint to improve the source location estimation. However, in order to acquire higher location accuracy in noisy environment, four or more sensors are still required, which leads to complex circuit design when taking into account time synchronization for TDOA.

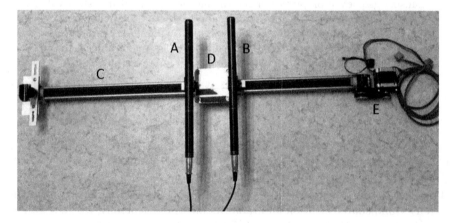

Fig. 1. Sliding microphone-pair system. A, B: microphone; C: track; D: slider; E: stepping motor. Two parallel microphones A, B can slide freely on the track controlled by a stepping motor. After updating sound source position, we obtain a new acquiring position and move the slider to new acquiring position.

In this paper, we propose a new method which is flexible in acquiring data and simple in designing circuit. As shown in Fig. 1, a sliding microphone-pair (our sensor is microphone which is used to acquire audio data) system is devised to overcome the shortcomings of fixed microphone array topology. Besides, we present a location framework (Fig. 2) which utilizes not only the geometric properties of the straight track but also optimal topology structure according to error theory. Unlike fixed microphone array structure, the slider provides variable topology, so that we can search the suitable acquiring position

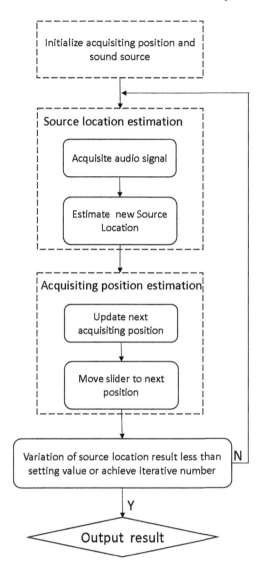

Fig. 2. The flowchart of the proposed source location estimation system which consists of two parts: source location estimation and acquiring position estimation.

by moving the microphone-pair on the track. The contribution of this paper is two-fold:

1. We analyze topology properties of a sliding microphone-pair system and deduce the relationship between sound source position and variance of measurement error, according to the theory, we put forward a method on how to determine the optimal acquiring position on the track.

2. By using the optimal acquiring position, we propose a new framework for sound source location.

The remainder of this paper is organized as follows: Sect. 2 describes source location problem under our location situation and introduces the formulation. Section 3 analyzes the topology structure property of our microphone array, and shows how it affects source location precision and describes our sound source location estimation algorithm. Experiments and evaluations are presented in Sect. 4, and Sect. 5 summarizes the paper finally.

2   Problem Formulation

As described in Sect. 1, the microphone-pair (sensor A and B) is mounted to a straight track, which guarantees the microphone-pair will move straightly and parallelly. When stopping at some positions, the two microphones acquire sound source at the same time. For a stationary source, this setup is equivalent to a linear array with a large number of microphones, but the difference is that in our system the range of microphone-pair is constant and the reference microphone is dynamic.

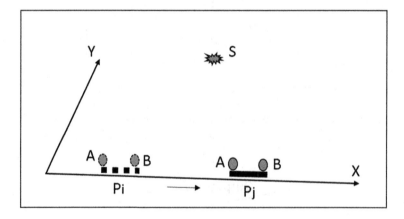

Fig. 3. Geometry relationship between microphone-pairs which are constrained by the track. P_i and P_j denote two different acquiring positions.

We consider the location problem in half R^2, the source position S is $r_s = (x_s, y_s)$. Since all the microphone-pairs are co-linear, we find a coordinate system so that centers of all microphone-pairs are on X-axis (Fig. 3). Thus, after acquiring data at M different positions, the centers of microphone-pairs can be denoted as $p_i = (p_i, 0)$ $i = 1, ..., M$. the difference in the distances of sensor A and B at center position p_i from the source is given by $\delta_i = \|p_{iA} - r_s\| - \|p_{iB} - r_s\| = d_{i,A}(s) - d_{i,B}(s)$. Where $d_{i,*}(s)$ denotes the distance between the source and sensor $*$. Then the noisy TDOA measurement between sensor A and sensor B at center position $\boldsymbol{p}_i$ can be written as:

$$\hat{\tau}_i = \frac{1}{v}\delta_i(\boldsymbol{r}_s) + \epsilon_i. \tag{1}$$

v is the source propagation speed and ϵ_i is the TDOA measurement error.

Thus, for M microphone-pairs the signal model for the TDOA measurements becomes:

$$\hat{\boldsymbol{\tau}} = \frac{1}{v}\boldsymbol{\delta}(\boldsymbol{r}_s) + \boldsymbol{\epsilon}. \tag{2}$$

Where: $\hat{\boldsymbol{\tau}} = \begin{bmatrix} \hat{\tau}_1 \\ \vdots \\ \hat{\tau}_M \end{bmatrix}$, $\boldsymbol{\delta} = \begin{bmatrix} \delta_1 \\ \vdots \\ \delta_M \end{bmatrix}$, $\boldsymbol{\epsilon} = \begin{bmatrix} \epsilon_1 \\ \vdots \\ \epsilon_M \end{bmatrix}$

Then the problem of source location is to estimate the source position S given $\boldsymbol{p}_i$ ($i = 1, ...M$), $\hat{\boldsymbol{\tau}}$ and v.

3 Proposed Approach

In this section, we describe how array topology impacts measurement error and how can we find the optimum acquiring position by moving microphone-pair. First of all, we introduce the Cramer-Rao lower Bound (CRB) property of the sliding microphone-pair system.

3.1 CRB Property

We know that the Cramer-Rao inequality [20] has a lower bound for the variance of any unbiased parameter estimators and it is can be used to evaluate the estimator value. The Cramer-Rao Bound (CRB) can be derived from the inverse of Fisher Information Matrix (FIM), and in our system FIM $\boldsymbol{F}(\boldsymbol{r}_s)$ associated with the estimation of the source location $\boldsymbol{S}$ can be written as:

$$\boldsymbol{F}(\boldsymbol{r}_s) = E[(\frac{\partial}{\partial \boldsymbol{r}_s} lnf(\hat{\tau}|\boldsymbol{r}_s))(\frac{\partial}{\partial \boldsymbol{r}_s} lnf(\hat{\tau}|\boldsymbol{r}_s))^T]. \tag{3}$$

Where E is the expectation on $\hat{\tau}$, $f(\hat{\tau}|\boldsymbol{r}_s)$ is conditional probability density function (PDF) of $\hat{\tau}$ under certain sound source $\boldsymbol{r}_s$. Assuming the measurement error $\boldsymbol{\epsilon}$ is Gaussian distribution with zero mean and independent respectively, the full rank covariance matrix of $\boldsymbol{\epsilon}$ is C_ϵ, then the (2×2) FIM in our system is given by [17]:

$$\boldsymbol{F}(\boldsymbol{r}_s) = \frac{1}{v^2}G^T C_\epsilon^{-1} G. \tag{4}$$

where:

$$G = \frac{\partial}{\partial \boldsymbol{r}_s}\boldsymbol{\delta} = \begin{bmatrix} \frac{\partial \delta_1(\boldsymbol{r}_s)}{\partial x_s} & \frac{\partial \delta_1(\boldsymbol{r}_s)}{\partial y_s} \\ \vdots & \vdots \\ \frac{\partial \delta_M(\boldsymbol{r}_s)}{\partial x_s} & \frac{\partial \delta_M(\boldsymbol{r}_s)}{\partial y_s} \end{bmatrix}. \tag{5}$$

Because the measurement every sliding is independent, then the covariance matrix C_ϵ can be rewritten as:

$$C_\epsilon = \begin{bmatrix} \sigma_{\epsilon 1}^2 & & 0 \\ & \ddots & \\ 0 & & \sigma_{\epsilon M}^2 \end{bmatrix}. \tag{6}$$

input Eqs. (5), (6) into (4), we can deduce that:

$$CRB(\boldsymbol{r_s}) = \begin{bmatrix} \sigma_{x_s}^2 & * \\ * & \sigma_{y_s}^2 \end{bmatrix} = v^2 (G^T C_\tau^{-1} G)^{-1}. \tag{7}$$

$\sigma_{x_s}^2$ and $\sigma_{y_s}^2$ above denote CRB of variance of x_s and y_s, and the $*$ denotes irrelevant terms.

3.2   Optimum Acquiring Position

We note from Eq. (7) that the matrix $\boldsymbol{F(r_s)}^{-1}$ which denotes the lower bound of variance of source coordinate depend not only on variance of measurement but also significantly on the source position and microphone-pair range. It's difficult to calculate the formulate accurately, and now we will deduce how to find the optimum acquiring position on the track.

From Eq. (2), we can obtain the error function is:

$$\epsilon = \hat{\tau} - \frac{1}{v}\delta(\boldsymbol{r_s}). \tag{8}$$

Error function (8) is defined as the difference between the observed TDOA and that generated by the sound source depending upon the unknown parameters x_s and y_s. Then the corresponding LSE criterion can be written as:

$$J_M = \epsilon^T \epsilon = \sum_{i=1}^{M} [\hat{\tau}_i - \frac{1}{v}\delta_i(\boldsymbol{r_s})]^2. \tag{9}$$

To locate the sound source in LSE approach, we attempt to minimize the power of the error function. In Eq. (9), if we could get optimum acquiring position or minimum acquiring error each data acquiring step as far as possible, then the Eq. (9) will be lower and the unknown parameters x_s and y_s will get closer to the real value. According to this idea, we should minimize the error function at each acquiring position. Moreover, we note that the error function consists of random variable, so the optimum acquiring position of $\boldsymbol{p}_i = (p_i, 0)$ $i = 1, ..., M$ can be derived from minimizing the variance of error function, that's:

$$\arg\min[\hat{\tau}_i - \frac{1}{v}\delta_i(\boldsymbol{r_s})]^2 \Leftrightarrow \arg\min D([\hat{\tau}_i - \frac{1}{v}\delta_i(\boldsymbol{r_s})]^2). \tag{10}$$

we have $\epsilon_i \sim N(0, \sigma_i^2)$, therefore:

$$\arg\min[\hat{\tau}_i - \frac{1}{v}\delta_i(\boldsymbol{r_s})]^2 \Leftrightarrow \arg\min D\epsilon_i^2 \Leftrightarrow \arg\min 2\sigma_i^4. \tag{11}$$

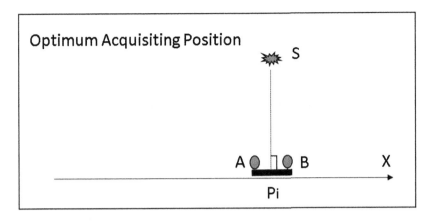

Fig. 4. Optimum acquiring position in 2D half-plane.

Rabinkin gave out the experimental variance of measurement error ϵ [21], and the variance of ϵ is related to microphone-pair inter range, distance and angle from source to center of microphone-pair and reverberation time of room. In our system the inter-sensor range and reverberation time is constant, so the variance of ϵ can be rewritten as:

$$\sigma^2 = C_T(6 \times 10^{-3}|\theta| + 0.73)(9 \times 10^{-2}ln(1 + D/10)). \tag{12}$$

where C_T is constant, D, θ denote the distance and angle from source to center of microphone-pair.

Combining Eqs. (11) and (12), we could deduce that when the connection line of sound source (assuming that the source is a point) and microphone-pair center is perpendicular to the straight track, we may get the optimum acquiring position (see Fig. 4). Besides, if X-coordinate of sound source is out of the X range of track, the optimum acquiring position will be at the endpoint of the track.

3.3   Source Location Estimation

Thanks to the continuity of the slider, the microphone-pair can slide on the track freely which improves the topology of array according to different source. Here we will introduce a method that can guide the microphone-pair to acquire data at optimum acquiring position and output source position after several acquiring steps.

It's a two-step iteration method for finding the optimum acquiring position and source location. We could use the following variational optimization model to get the result from Eq. (9):

$$\min(J_k) = \min[J_{k-1} + \epsilon_k^2]. \tag{13}$$

$$\min(J_{k-1} + \epsilon_k^2) = \min[J_{k-1} + [\hat{\tau}_k - \frac{1}{v}\delta_k(\boldsymbol{r}_s)]^2]. \tag{14}$$

Algorithm 1. Iterative Source Location

Input: iteration number N, parameter inter-sensor range r, reverberation time T and audio propagation velocity v

Output: Source location $\mathbf{S_r}$

 1: Initialize source location S_0, Initialize acquiring position P_0, Initialize location accuracy ϵ
 2: **while** error of adjacent source location result less than ϵ or $n < N$ **do**
 3:     $n = n + 1$;
 4:     Acquire audio source data;
 5:     Update source location S_n;
 6:     Update acquiring position P_n based on current source position;
 7:     Move slider to new acquiring position;
 8: **end while**
 9: **return** $\mathbf{S_r}$

In this optimization, k denotes the kth acquiring position, J_{k-1} is the least square of $k - 1$ times measurement ahead. We can use LSE to minimize J_{k-1} to get an approximate source location, that's:

$$\mathbf{r}_s = \arg\min J_{k-1} = \arg\min \sum_{i=1}^{k-1}[\hat{\tau}_i - \frac{1}{v}\delta_i(\mathbf{r}_s)]^2. \tag{15}$$

At initial phase we assume the sound source is at some known position and the initial acquiring position is at one endpoint of the slider. When getting a coarse $\mathbf{r}_s$ after acquiring, we use the result to guide microphone-pair to next reasonable position to acquire next data according to Eq. (12). Repeating operations above several times when the variation of $\mathbf{r}_s$ is less than the setting value, at last we get the accurate source location, as described in Algorithm 1.

The slider is limited in our system. Therefore, when the real coordinate of sound source is out of range of the track, we could only get the optimum acquiring position at the endpoint of track everytime. To solve this problem, we first judge the approximate position of the source and move the slider to the nearest endpoint, then we move the slider in opposite direction of the source with a constant step size. After getting several TDOAs, we could locate the sound source.

4   Experiments

In this section, we conduct two experiments to verify our algorithm, one is estimating the variance of the TDOA and the other is localization experiment.

4.1   Variance of TDOA

To evaluate the variance of TDOA, we put the microphone-pair at 17 different position of the tracker to acquire audio data which consists of 500 segments of

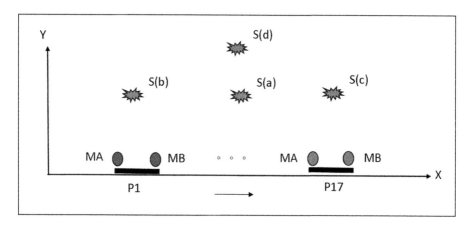

Fig. 5. Four situations of position relationship between the tracker and sound source.

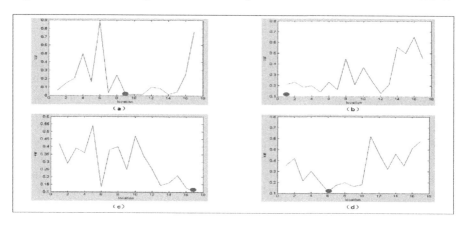

Fig. 6. The results of variance of the four situations.

white noise, then compute all the TDOA of the data and the variance at some position. We list four situations in Fig. 5, the tracker is fixed on the x axis and the sound source is placed at four positions. The results of variance is shown in Fig. 6, the red point denotes the x coordinate of the sound source. From this figure we can conclude that when the connection line of sound source (assuming that the source is a point) and microphone-pair center is perpendicular to the straight track, we may get the optimum acquiring position.

4.2 Localization Experiments

To quantitatively evaluate the accuracy of our algorithm, we compare our sliding method with conventional sensor array method. In our system we acquire the audio data using binaural audio system which consists of left and right channel,

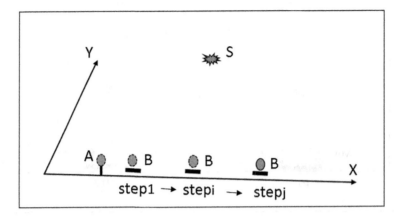

Fig. 7. We use two microphones to simulate the traditional microphone array situation: one microphone is fixed at one endpoint of the track and the other one moves step by step.

so it means that we do not need expensive system to guarantee time synchronization. Besides, in half-plane at least three microphones could locate unique sound source. To simulate conventional microphone array situation using two microphones, we fix one microphone as reference microphone at one endpoint of the track and its coordinate is $(0,0)$, then move the other microphone step by step (our step size is 10 cm, step number is 10) to acquire stationary sound source data (marked as traditional method). Figure 7 describes how we simulate the conventional microphone array situation. We will first carry out experiments using the two methods for different sound source positions, then compare and analyze the results.

We carry out experiments in a $5 \times 3 \times 3$ (meter) room, for our method the inter-range of microphone-pair is 10 cm, length of track is 100 cm, sound source is white noise of one second and fixed in front of the track, the sampling rate is 192 KHz. For traditional method, only one microphone is fixed on the slider and the other one is fixed on one endpoint of the track, the slider will move step by step in 10 cm and stop to acquire audio data at each step.

After acquiring audio data, we use GCC method to calculate TDOA [22]. As described in Sect. 3, our method needs updating source location value when getting a new TDOA. The microphone-pair acquires data at different place, so the reference microphone position is changing each acquiring step. However, conventional location methods such as [17, 18] require a common stationary reference sensor for source location which means these methods are not suitable for sliding microphone array situation. To solve LSE problem, we introduce Levenberg-Marquardt algorithm to get the LSE result. It is proved by experiments that the algorithm is suitable to solve our LSE location problem. We compare various situations and show some results of the experiments.

Table 1. Source location results

Real location (cm)	Our result (cm)	Error (cm)	Method [19] result (cm)	Error (cm)
(0.0, 50.0)	(3.29, 45.93)	5.23	(−4.50, 45.15)	6.62
(50.0, 50.0)	(48.19, 53.80)	4.21	(53.24, 46.56)	4.73
(100.0, 50.0)	(97.84, 45.75)	4.77	(93.82, 45.71)	7.52
(0.0, 100.0)	(4.16, 94.84)	6.63	(−4.36, 103.22)	5.42
(50.0, 100.0)	(53.48, 94.85)	6.21	(53.9, 105.88)	7.06
(100.0, 100.0)	(95.41, 93.68)	7.81	(105.34, 96.29)	6.50
(0.0, 150.0)	(−4.60, 157.16)	8.51	(4.91, 143.56)	8.10
(50.0, 150.0)	(47.85, 143.35)	6.98	(54.52, 158.10)	9.27
(100.0, 150.0)	(96.16, 142.83)	8.13	(106.25, 141.42)	10.61
Mean average		5.85		6.58

4.3 Results and Analysis

Some localization results are shown in Table 1. We also calculate absolute error results in the table. From the table we can draw a conclusion that most results of our method are better than that of traditional array method, If we compare x-coordinate error separately, we can find that the absolute error of x-coordinate is less than that of traditional method. That's because in our system the array move in one dimension space which is more sensitive to x-coordinate. We could image that if our array can move in two dimensional space, there will be a more suitable acquiring position and the source location result will be more precise.

5 Conclusion

In this paper, we propose a new audio location method using a sliding microphone-pair system. As is known to all that topology structure of sensor array has effects on source location. By finding the optimum acquiring position and using iterative method, we could locate the sound source in several acquiring times. Our system only need two microphones and it is simple in circuit design. Moreover, our algorithm achieves good results in practical tests.

In the future, we will work on extending our algorithm to handle continuous sliding scenes or combine with depth estimation camera to realize fusion of audio and video.

Acknowledgment. This work is supported by the National Key Research and Development Program of China under Grant 2016YFB1001001, the National Natural Science Foundation of China under Grants 61225008, 61572271, 61527808, 61373074 and 61373090, the National 1000 Young Talents Plan Program, the National Basic Research Program of China under Grant 2014CB349304, the Ministry of Education of China under Grant 20120002110033, and the Tsinghua University Initiative Scientific Research Program.

References

1. Brandstein, M., Ward, D.: Microphone Arrays: Signal Processing Techniques and Applications. Springer, Heidelberg (2013)
2. Yang, K., Wang, G., Luo, Z.Q.: Efficient convex relaxation methods for robust target localization by a sensor network using time differences of arrivals. IEEE Trans. Signal Process. **57**(7), 2775–2784 (2009)
3. Foy, W.H.: Position-location solutions by Taylor-series estimation. IEEE Trans. Aerosp. Electron. Syst. **12**(2), 187–194 (1976)
4. Torrieri, D.J.: Statistical theory of passive location systems. IEEE Trans. Aerosp. Electron. Syst. **aes–20**(2), 183–198 (1984)
5. Schau, H., Robinson, A.: Passive source location employing spherical surfaces from time-of-arrival differences. IEEE Trans. Acoust. Speech Signal Process. **35**, 1223–1225 (1987)
6. Friedlander, B.: A passive localization algorithm and its accuracy analysis. IEEE J. Oceanic Eng. **12**(1), 234–245 (1987)
7. Fang, B.T.: Simple solutions for hyperbolic and related position fixes. IEEE Trans. Aerosp. Electron. Syst. **26**(5), 748–753 (1990)
8. Abel, J.S.: A divide and conquer approach to least-squares estimation. IEEE Trans. Aerosp. Electron. Syst. **26**(2), 423–427 (1990)
9. Yang, L., Ho, K.C.: An approximately efficient TDOA localization algorithm in closed-form for locating multiple disjoint sources with erroneous sensor positions. IEEE Trans. Signal Process. **57**(12), 4598–4615 (2009)
10. Johnson, D.H., Dudgeon, D.E., Processing, A.S.: Concepts and Techniques. Simon & Schuster, New York (1992)
11. Gazor, S., Grenier, Y.: Criteria for positioning of sensors for a microphone array. IEEE Trans. Speech Audio Process. **3**(4), 294–303 (1995)
12. Yang, B., Scheuing, J.: Cramer-rao bound and optimum sensor array for source localization from time differences of arrival. In: IEEE International Conference on Acoustics, Speech, and Signal Processing, vol. 4, pp. iv/961–iv/964 (2005)
13. Zhang, C., Zhang, Z., Florêncio, D.: Maximum likelihood sound source localization for multiple directional microphones. In: IEEE International Conference on Acoustics, Speech and Signal Processing, ICASSP 2007, vol. 1, p. I-125. IEEE (2007)
14. Ho, K.C.: Bias reduction for an explicit solution of source localization using TDOA. IEEE Trans. Signal Process. **60**(60), 2101–2114 (2012)
15. Smith, J.O., Abel, J.S.: Close-form least-squares source location estimation from range-difference measurements. IEEE Trans. Acoust. Speech Signal Process. **35**(12), 1661–1669 (1988)
16. Abel, J.S., Smith, J.: The spherical interpolation method for closed-form passive source localization using range difference measurements. In: IEEE International Conference on Acoustics, Speech, and Signal Processing, pp. 471–474 (1987)
17. Chan, Y.T., Ho, K.C.: Simple and efficient estimator for hyperbolic location. IEEE Trans. Signal Process. **42**(8), 1905–1915 (1994)
18. Huang, Y., Benesty, J., Elko, G.W., Mersereati, R.M.: Real-time passive source localization: a practical linear-correction least-squares approach. IEEE Trans. Speech Audio Process. **9**(8), 943–956 (2001)
19. Ho, K.C.: Bias reduction for an explicit solution of source localization using TDOA. IEEE Trans. Signal Process. **60**(5), 2101–2114 (2012)

20. Mendel, J.M.: Lessons in Digital Estimation Theory. Prentice-Hall, Englewood Cliffs (1987)
21. Rabinkin, D.V., Renomeron, R.J., French, J.C., Flanagan, J.L.: Optimum microphone placement for array sound capture. J. Acoust. Soc. Am. **101**(5), 227–239 (1997)
22. Knapp, C., Clifford Carter, G.: The generalized correlation method for estimation of time delay. IEEE Trans. Acoust. Speech Signal Process. **24**(4), 320–327 (1976)

A Character-Based Method for License Plate Detection in Complex Scenes

Dingyi Li[1(✉)] and Zengfu Wang[1,2]

[1] Department of Automation, University of Science and Technology of China,
Hefei 230027, China
lidingyi@mail.ustc.edu.cn, zfwang@ustc.edu.cn
[2] Institute of Intelligent Machines, Chinese Academey of Sciences,
Hefei 230031, China

Abstract. License plate detection is a crucial part in license plate recognition systems and is often considered as a solved problem. However, there are still plenty of complex scenes where the current methods are invalidated. In order to increase the performance in these scenes, we propose a novel character-based method to detect multiple license plates in complex images. Firstly, a preprocessing step is performed. Then we use a modified maximally stable extremal region (MSER) based detector called MSER-+ to detect the possible character regions. Some of the regions are removed according to their geographical information. Hierarchical morphology helps to connect candidate MSERs of various sizes. The regions satisfying some geographical limits will be fed into a convolutional neural network (CNN) model for further verification. Extensive experimental results validate that our method works well in a large variety of complex scenes.

Keywords: Detection · MSER-+ · Hierarchical morphology · Deep learning

1 Introduction

License plate recognition systems play an important role in intelligent transportation systems. License plate detection is the most important and difficult step for license plate recognition systems. It is not always easy to detect license plates in complex scenes due to different illumination, weather and background conditions. Sizes and angles changes of the license plates will also affect the detection rate. How to decrease the computational cost for high resolution images is also an important problem.

The detection methods can be classified into several categories according to the features they use, such as shape, edge, color, texture, character and hybrid features [3,8]. However, in complex scenes, some features may be robust while others may be sensitive. Nowadays researchers and companies would like to use higher resolution cameras to increase the license plate recognition rate and/or to recognize multiple license plates. So another important issue for license plate

T. Tan et al. (Eds.): CCPR 2016, Part II, CCIS 663, pp. 576–587, 2016.
DOI: 10.1007/978-981-10-3005-5_47

detection is that the time complexity should be as low as possible to insure the real-time performance in high resolution images.

Li used shape feature to detect license plates [16]. However, edgesplates are not obvious, this method will fail easily. Bai *et al.* [4] and Zheng *et al.* [25] used edge extraction to locate license plate regions. However, these methods are sensitive to unwanted edges thus are not suitable in complex scenes. Dun *et al.* [9] used color feature to locate license plates. Color feature is very sensitive to illumination changes, so they may not be used in complex scenes. Zhou *et al.* proposed an algorithm which is based on principle visual words [26] using character features. Chen *et al.* [6] integrates shape, color and texture features. Their algorithm is more reliable while is computational complex. Yu *et al.* used wavelet transform and empirical mode decomposition (EMD) analysis and achieved a high localization rate in mutilingual datasets [24]. However, it is unable to detect multiple license plates in a single image.

A license plate contains three parts, boundary, plate and characters. Characters are the most important elements of license plates in license plate recognition systems. We also find that character feature is the most robust one to distinguish the license plate region from the background image. There are two major methods for text detection in natural scenes, sliding window based and connected component analysis (CCA) based. Sliding window based method is robust against blur and noise. But its disadvantage is time-consuming. CCA based method is scale-invariant. It is robust to scale, orientation and illumination changes. However, CCA assumes that each character is an individual component, which will suffer from character breakage and character adhesion [19].

The well-known CCA based feature detector Maximally stable extremal regions (MSER) was introduced by Matas *et al.* [17]. Matas *et al.* used MSER for wide baseline-stereo problems. In 2005 they used extremal region (ER) method to detect license plates and other traffic signs [18]. Their ER based method is robust against size changes and multi-orientations. Donoser *et al.* [7] and Nistér *et al.* [21] modified the data structure of MSER by component trees. Their works reduced the time cost of MSER extraction. In recent years, many researchers use MSER based method in text detection in complex scenes [19,20,22,23]. Character feature based algorithms for license plate detection utilize MSER combining with morphology [5], conditional random field model and minimum spanning tree [15] or label-moveable clique [11]. These methods explict high detection rate in simple scenes. However, Bai and Li's method is unable to detect license plates of multiple sizes. In complex scenes, it is not easy to for Gu's strategy to distinguish the correct orders of the characters. [10] combines edge detection, morphology and MSER. This method reduces the computational cost of MSER extraction. However, it is sensitive to unwanted edges and is limited to small plate size ranges.

We propose a novel license plate detection algorithm to extract multiple license plates in a single image. First of all, we use a novel MSER-+ based method to detect the character regions. Then an efficient character elimination method based on geometric information of each MSER-+ is performed. After

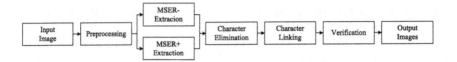

Fig. 1. Overview of the proposed method.

that we use a hierarchical morphology method to connect candidate MSERs. Deep convolutional neural network (CNN) is then utilized to separate the real license plate regions from false regions. Experimental results demonstrate that our method is invariant to illumination, weather, scale and orientation changes. The detection rates retain high in different complex scenes.

The rest of the paper is organized as follows. Section 2 gives the details of the proposed method, containing preprocessing, MSER-+ extracion, character linking and region verification. Section 3 shows some of the implement details and our detection results. Section 4 draws a conclusion (Fig. 1).

2   The Proposed Method

2.1   Preprocessing

Real surveillance systems may suffer from noise and low contrast, therefore a good image preprocessing stage is requisite. First of all, the RGB image are converted to a gray-level image to decrease computational cost. Then a $3*3$ Gaussian filter with standard deviation of 0.8 and a $3*3$ Laplacian sharping filter

$$\begin{pmatrix} 0 & -1 & 0 \\ -1 & 5 & -1 \\ 0 & -1 & 0 \end{pmatrix}$$

(a) Original color image   (b) The preprocessed image

Fig. 2. Image preprocessing.

are used. Gaussian filtering can reduce noise. Laplacian sharping can enhance the edges of all character regions. The results of the preprocessing step are shown in Fig. 2.

2.2 Character Extraction

We use a modified maximally stable extremal regions (MSER) extractor called MSER-+ for license plate detection. We combine the advantages of the state-of-the-art researches on MSER based text extraction algorithms and the prior knowledge of the arrangements of license plates.

MSER is similar to watershed algorithm. According to [17], an extremal region refers to a set of connected pixels. The intensity of all these pixels are larger or smaller than its boundary pixels. MSER is a region of local minimum area variations with threshold changes. We have

$$q(i) = \frac{|Q_{i+\triangle} - Q_{i-\triangle}|}{|Q_i|} \tag{1}$$

where Q_i, $Q_{i+\triangle}$ and $Q_{i-\triangle}$ refers to connected components with the threshold of i, $i + \triangle$ and $i - \triangle$ respectively. $q(i)$ is the gradient of Q_i with the threshold of i. If $q(i)$ is a local minimum value, then the corresponding Q_i is regarded as a MSER. [21] modified the algorithm using

$$q(i) = \frac{|Q_i - Q_{i-\triangle}|}{|Q_{i-\triangle}|} \tag{2}$$

which is more straightforward. [21] also introduced a component tree structure to obtain faster speed.

The procedure of MSER- extraction can distinguish bright regions from dark backgrounds. On the contrary, MSER+ can extract dark regions from bright backgrounds [10]. We extract both MSER- and MSER+ regions. Performing character elimination, we obtain two binary map for MSER- and MSER+ respectively. Different from methods introduced in [5,10,11,15], where MSER- map and MSER+ map are treated separately, we fuse the MSER- and MSER+ on a single binary map. We name this procedure as MSER-+ extraction. On some types of license plates, several white characters are connected with the white boundaries. In these cases, other MSER based approaches will be invalidated since they can't find all separated characters. However, our MSER-+ is still able to find both the separated characters and the separated background regions on these plates so the whole plate region will be extracted.

There are many regions generated by the MSER- and MSER+ extraction procedure. So we need a more efficient strategy to eliminate plenty of disturbances. We set several limits to the candidate regions. Here a denotes the area of the MSER. The minimum enclosing rectangle of each MSER is also calculated. Then a set of geometric parameters of a rectangle

$$\{w, h, \alpha, r_1, r_2\}$$

Fig. 3. Character elimination based on geometric attributes.

are calculated. Where w and h denote the width and height respectively. α is the angle between the shortest edge of the rectangle and the x axis as shown in Fig. 3. r_1 is the aspect ratio of the rectangle as

$$r_1 = \frac{h}{w} \tag{3}$$

r_2 represents the area ratio of the MSER to the corresponding boxes.

$$r_2 = \frac{a}{w * h} \tag{4}$$

We selected thresholds for these parameters according to the formal license plate characters and the number of candidate MSER-+s reduced dramatically. After MSER-+ extraction, a binary map of the contours of the retained MSERs is formulated.

2.3   Character Linking

As shown in Fig. 4, to connect each separated units, we use a hierarchical closed morthology operation. In order to maintain scale invariance, we use four different

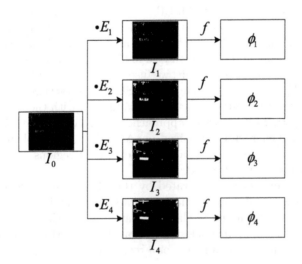

Fig. 4. The flowchart of the character linking procedure.

sizes of the operation element E_i, $i = 1, 2, 3, 4$, which are 5×1, 15×1, 25×1 and 35×1 separately.

Let the input binary map to be I, and the i images obtained after morphology to be I_i, and $\cdot$ represent a morphology procedure, we have

$$I_i = I \cdot E_i, \quad i = 1, 2, 3, 4 \tag{5}$$

Using f to represent a connect component extraction procedure, we have

$$\Phi_i = f(I_i) \tag{6}$$

where Φ_i, $i = 1, 2, 3, 4$ are sets of candidate regions on each map. Note that this step is a coarse detection step, and the results will be modified in the following stages.

2.4   Region Verification

The region verification step consists two parts: coarse verification using geometrical limits and fine verification via deep learning, as demonstrated in Fig. 6. Firstly, based on the structure of the uniform license plate, several simple geometric limitations are added to the detection result.

Fig. 5. Plate region elimination based on several simple geometric limitations.

Area A is the area of the candidate region. The minimum enclosing rectangle of each candidate region is extrated. Then a set of geometric parameters of a rectangle

$$\{W, H, \beta, R_1, R_2\}$$

are calculated where W, H denote the width and height of the rectangle respectively. β is the angle between the shortest edge of the rectangle and the x axis as shown in Fig. 5. R_1 is the aspect ratio of the region, and R_2 represents the area ratio of the candidate region to the corresponding boxes.

$$R_1 = \frac{H}{W} \tag{7}$$

$$R_2 = \frac{A}{W * H} \tag{8}$$

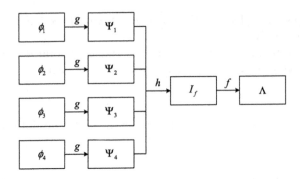

Fig. 6. The flowchart of region verification procedure.

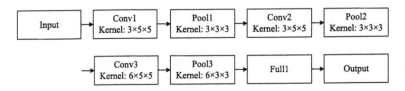

Fig. 7. The structure of our CNN model. Conv, Pool and Full represents a convolution layer, a pooling layer and a fully-connected layer respectively. Pool1, Conv2 and Conv3 are followed by a ReLU layer.

We selected thresholds for these parameters according to the formal license plate characters. After this simple procedure, another fine verification procedure using convolutional neural network is conducted.

Deep learning is very useful in various computer vision problems such as detection, segmentation and recognition [14]. Here, we conducted a convolutional neural network for license plate verification. The structure of our network is shown in Fig. 7.

Our CNN architecture demonstrates promising accuracy in this two-class classification problem. For each Φ_{ij}, if $P(\Phi_{ij}) > 0.9$, then region Φ_{ij} will be restored in the set of Ψ_i. Here $P(\Phi_{ij})$ represents the outputted possibility of our CNN and Φ_{ij} is the jth region in Φ_i which satisfies the previously mentioned geometrical limitations. Here we shift the threshold of the possibility to obtain a higher recall rate. $\Psi_i, i = 1, 2, 3, 4$ are then fused into the same binary map I_f. We also analyse the connected component of I_f, and get several regions

$$\Lambda = f(I_f) \tag{9}$$

where f is the same function mentioned in Sect. 2.3. Λ is the set of license plate regions we acquired ultimately.

3   Experiments

3.1   CNN Training Details

Deep neural network is a data driven model thus the training sets should comprise different situations including illumination changes, blur and distortion. We collected 23500 license plate images and 16000 negative samples. The resolutions of all these images are 136×36. The 16000 negative samples are extracted from the Stanford Cars Dataset [13] using our MSER-+ and hierarchical morphology based extraction method. This car dataset contains various vehicles and street scenes and no Chinese license plates. We implement our model using the Caffe package [12]. Our optimization algorithm in the training phase is the standard stochastic gradient descent (Fig. 8).

Fig. 8. Some of the negative samples extracted from the stanford cars dataset.

3.2   Scene Image Datasets

Since there are few public Chinese license plate images available, we collected four datasets to test our algorithm. Dataset 1 and Dataset 2 are taken in two different parking lots. The resolutions of the images of *Parking lot 1* and *Parking lot 2* are 1280×720 and 1920×1080 respectively. Dataset 3 (*Road* dataset) is taken in a road monitoring system with resolutions of 1920×1080. The fourth dataset (Arbitrary dataset) are of different scenes and of different resolutions ranging from 399×300 to 3264×2488. Images in *Parking lot 1*, *Parking lot 2* and *Road* are taken in three different scenes. *Arbitrary* dataset consists of 131 scenes. All images in *Parking lot 1* and *Parking lot 2* have only a single license plate. In *Road* dataset, 10 images have 2 license plates and the rest 231 images have 1 plate. 149, 18 and 11 images in *Arbitrary* dataset have 1, 2 and 3 license plates respectively. There are 391, 264, 0 and 8 images in *Parking lot 1*, *Parking lot 2*, *Road* and *Arbitrary* separately which are taken at night.

Table 1. Our detection results on different datasets

Dataset	Image number	Plate number	Plate height	Detection rate	Time
Parking lot 1	427	427	25–50 pixels	99.3 %	122 ms
Parking lot 2	752	752	40–100 pixels	99.7 %	220 ms
Road	241	251	10–35 pixels	99.2 %	308 ms
Arbitrary	178	218	10–85 pixels	98.6 %	165 ms

Table 2. Comparisons of the detection rates with other methods

Dataset	RTD [23]	DeepGlint [1]	Vzenith [2]	Plain MSER	Our approach
Parking lot 1	-	-	93.5 %	**99.3 %**	**99.3 %**
Parking lot 2	-	-	99.6 %	97.9 %	**99.7 %**
Road	-	-	94.8 %	97.0 %	**99.2 %**
Arbitrary	86.8 %	67.5 %	97.6 %	94.9 %	**98.6 %**

3.3   Detection Results

We implement our method using C++ with the aid of OpenCV. The experiments are conducted on a PC with a Intel i7-4790 CPU of 3.6 GHz and a 8 GB memory. Some of our successful detection results and failures are illustrated in Figs. 9 and 10 respectively. The statistical analysis is listed in Table 1. In all of the four datasets, our detection rates are above 98.5 %, which indicates that our method is robust to various changes such as illumination, size, orientation and distortion changes. Results in Table 2 demonstrate that our approach is superior to two commercial softwares provided by DeepGlint [1] and Vision-Zenith [2] respectively, the Robust Text Detection (RTD) algorithm [23] and the conventional plain MSER based method. For CNN verification, in *Arbitrary* dataset,

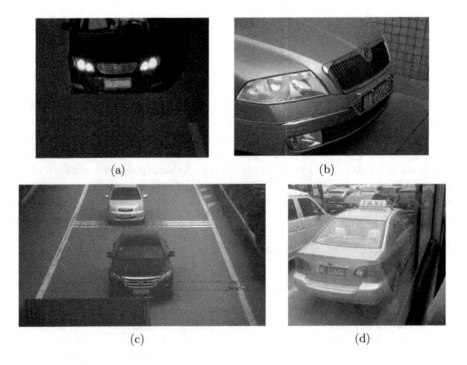

(a)   (b)

(c)   (d)

Fig. 9. Some of our successful results. The images are better viewed by zooming.

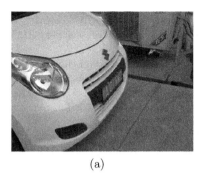

(a)   (b)

Fig. 10. Some of our failures. The plate in (a) is eliminated by our CNN. One of the plate in (b) is too close to another car which leads to a failure. The images are better viewed by zooming.

24.2 % of the false positive regions and only 1 license plate region are eliminated. In Fig. 10(a), although the license plate is extracted by our MSER-+ and the linking procedure, our CNN deletes it since it is of a relatively large tilt angle. In Fig. 10(b), one of the license plate is too close to another vehicle. Since there are plenty of MSER-+s on the neighboring car, the morphology based linking method is greatly influenced. We believe that with stricter geometric limits, the detection rate will be increased at the cost of scale adaption. By optimizing the program and using parallel computing, the processing time will be further decreased.

4   Conclusion

In this paper, we have proposed a novel character-based license plate detection algorithm in complex scenes. Our method combines the invariance of the MSER-+ in character extraction, the efficient hierarchical morphology and the strong classification capacity of deep neural networks. Experimental results show that our approach performs consistently well in plenty of complex scenes. Our future work is expanding license plate detection to generic text detection in natural scenes.

Acknowledgements. This work was supported by the National Natural Science Foundation of China (no. 61472393). The authors would like to thank Zeruo Liu for providing us with plenty of photos taken in complex scenes.

References

1. Deepglint. http://www.deepglint.com/weimu/. Accessed 29 May 2016
2. Vision-zenith. http://www.vzeye.com/. Accessed 29 May 2016

3. Anagnostopoulos, C., Anagnostopoulos, I., Psoroulas, I., Loumos, V., Kayafas, E.: License plate recognition from still images and video sequences: a survey. IEEE Trans. Intell. Transp. Syst. **9**(3), 377–391 (2008)

4. Bai, H., Liu, C.: A hybrid license plate extraction method based on edge statistics and morphology. In: ICPR, pp. 831–834 (2004)

5. Bai, S., Yuan, Y., Zhao, Y.: A license plate detection method based on morphology and maximally stable extremal region. In: Proceedings of the 4th International Conference on Computer and Electrical Engineering (2011)

6. Chen, Z., Liu, C., Chang, F., Wang, G.: Automatic license-plate location and recognition based on feature salience. IEEE Trans. Veh. Technol. **58**(7), 3781–3785 (2009)

7. Donoser, M., Bischof, H.: Efficient maximally stable extremal region (MSER) tracking. In: CVPR, pp. 553–560 (2006)

8. Du, S., Ibrahim, M., Shehata, M., Badawy, W.: Automatic license plate recognition (ALPR): a state-of-the-art review. TCSVT **23**(2), 311–325 (2013)

9. Dun, J., Zhang, S., Ye, X., Zhang, Y.: Chinese license plate localization in multi-lane with complex background based on concomitant colors. IEEE Intell. Transp. Syst. Mag. **7**(3), 51–61 (2015)

10. Gou, C., Wang, K., Yao, Y., Li, Z.: Vehicle license plate recognition based on extremal regions and restricted boltzmann machines. IEEE Trans. Intell. Transp. Syst. **17**(4), 1096–1107 (2016)

11. Gu, Q., Yang, J., Kong, L., Cui, G.: Multi-scaled license plate detection based on the label-moveable maximal mser clique. Opt. Rev. **22**(4), 669–678 (2015)

12. Jia, Y., Shelhamer, E., Donahue, J., Karayev, S., Long, J., Girshick, R., Guadarrama, S., Darrell, T.: Caffe: convolutional architecture for fast feature embedding. In: ACM MM, pp. 675–678 (2014)

13. Krause, J., Stark, M., Deng, J., Fei-Fei, L.: 3D object representations for fine-grained categorization. In: CVPRW, pp. 554–561 (2013)

14. LeCun, Y., Bengio, Y., Hinton, G.: Deep learning. Nature **521**(7553), 436–44 (2015)

15. Li, B., Tian, B., Li, Y., Wen, D.: Component-based license plate detection using conditional random field model. IEEE Trans. Intell. Transp. Syst. **14**(4), 1690–1699 (2013)

16. Li, Q.: A geometric framework for rectangular shape detection. TIP **23**(9), 4139–4149 (2014)

17. Matas, J., Chum, O., Urban, M., Pajdla, T.: Robust wide-baseline stereo from maximally stable extremal regions. Image Vis. Comput. **22**(10), 761–767 (2004)

18. Matas, J., Zimmermann, K.: Unconstrained licence plate and text localization and recognition. In: Proceedings of the 8th IEEE Conference on Intelligent Transportation Systems, pp. 225–230 (2005)

19. Neumann, L., Matas, J.: Real-time scene text localization and recognition. In: CVPR, pp. 3538–3545 (2012)

20. Neumann, L., Matas, J.: Efficient scene text localization and recognition with local character refinement. In: ICDAR, pp. 746–750 (2015)

21. Nistér, D., Stewénius, H.: Linear time maximally stable extremal regions. In: Forsyth, D., Torr, P., Zisserman, A. (eds.) ECCV 2008. LNCS, vol. 5303, pp. 183–196. Springer, Heidelberg (2008). doi:10.1007/978-3-540-88688-4_14

22. Ye, Q., Doermann, D.: Text detection and recognition in imagery: a survey. TPAMI **37**(7), 1480–1500 (2015)

23. Yin, X., Yin, X., Huang, K., Hao, H.: Robust text detection in natural scene images. TPAMI **36**(5), 970–983 (2014)

24. Yu, S., Li, B., Zhang, Q., Liu, C., Meng, M.Q.H.: A novel license plate location method based on wavelet transform and emd analysis. Pattern Recogn. **48**(1), 114–125 (2015)
25. Zheng, D., Zhao, Y., Wang, J.: An efficient method of license plate location. Pattern Recogn. Lett. **26**(15), 2431–2438 (2005)
26. Zhou, W., Li, H., Lu, Y., Tian, Q.: Principal visual word discovery for automatic license plate detection. TIP **21**(9), 4269–79 (2012)

Script Identification Based on HSV Features

Buvajar Mijit[1], Alimjan Aysa[2], Nurbiya Yadikar[1],
Xing-kun Han[1], and Kurban Ubul[1(✉)]

[1] School of Information Science and Engineering, Xinjiang University,
Urumqi, 830046, Xinjiang, China
kurbanu@xju.edu.cn
[2] Network and Information Center, Xinjiang University, Urumqi, 830046, Xinjiang, China

Abstract. Many similar shaped scripts are used all over the world today. Scripts identification with similar shaped characters is one of the difficulties in script identification field and it need to be resolved. However, there are a little report about identification of Central Asian countries and Chinese Minority scripts, which identification of similar scripts. In this paper, a multi-script database was established, which are including 2200 plain document images with different resolution in 11 scripts such as English, Chinese, Arabic, Russian, Uyghur, Mongol, Tibet, Turkish, Kyrgyzstani, Uzbekistani and Tajikistani. Then, HSV features were extracted from each whole page image and they were classified by using BP neural network classifier. After experiment in our system, it is achieved 88.14 % of average identification rate and 99.0 % of highest identification rate in our experiment with the dataset. Experimental results indicated that HSV features were effective feature for identify these scripts.

Keywords: Script identification · HSV features · BP neural network

1 Introduction

Script identification, identify different languages, is text category identification [1–3]. This is because using the same text or ethnic regions may speak different kinds of languages. In recent years, automatic script identification as the front part of the work of the OCR is becoming more popular. Along with the development of computer technology, information processing of minority is gradually becoming necessary work.

In this study, text documents of different scripts were turned into as an image, and then the image is processed by digital image process technology. Since our aim is to identify the text document image classification from different scripts, however, the script identification research can be solved by considered being a typical pattern recognition problem. That being the case, any script identification system has the same structure as pattern recognition system. Script identification technology generally consists of several stages such as document image acquisition, image pre-processing, feature extraction and classification, in which these contents and methods of feature extraction is particularly important.

© Springer Nature Singapore Pte Ltd. 2016
T. Tan et al. (Eds.): CCPR 2016, Part II, CCIS 663, pp. 588–597, 2016.
DOI: 10.1007/978-981-10-3005-5_48

The earliest identification of scripts was in English and Latin [4], and then gradually oriented identification of the East Asian Languages and Latin scripts. Spitz [5] developed an approach for classifying Han-based that it is included Chinese, Japanese, Korean and Latin-based scripts. In this method, Han based script is performed by analysis of the distribution of optical density in the text images and Latin-based languages used a technique based on character shape codes. Ul-Hasan etc. [6] presented a novel methodology for multiple script identification using Long Short-Term Memory (LSTM) networks' sequence-learning capabilities and shape features or bounding boxes of individual characters, developed a database consisting of English based on Greek script. A script identification system which identifies printed Roman, Chinese, Arabic, Devanagari, and Bangia based on text lines have developed by Pal and Chaudhuri [7]. They used different features to identify particular scripts. Shi and Bai et al. [8] have presented DiscCNN, a novel deep learning based method for script identification. This method combines deep features with discriminative mid-level representations, the DiscCNN learns special characteristics of scripts from training data automatically, and DiscCNN achieves state-of-the-art performances on scene, video and document scripts. A successful method of identifies based on bilingual document containing Roman and Tamil scripts method at word level proposed by Dhanya et al. [2]. A method combination of topological, contour and water reservoir concept based features to identify printed Urdu script used by Pal and Sarkar [3]. Obaidullah and Karimet [9] proposed a simple handwritten document script identification technique based on different image transform methods and statistical features with Multilayer Perceptron (MLP) at word level to identify any one of the four popular Indic scripts namely Bangia, Roman, Devanagari and Oriya and from experiment an average accuracy rate of 88.1 % is found. Ferrer et al. [10] proposed a Multiple Training - One Test technique, and has been tested with a script identifier based on the histogram of Local Binary Patterns (LBP) for stroke direction characterization and Least Square Support Vector Machine (LS-SVM) as classifier at text line level. About the techniques available in the area of Offline Script Identification (OSI) for Indic scripts summarized by Singh et al. [11], various feature extraction and classification techniques associated with the OSI of the Indic scripts are discussed. An automatic Handwritten Script Identification (HSI) technique for document images of six popular Indic scripts namely Bangla, Devanagari, Malayalam, Oriya, Roman and Urdu is proposed by Obaidullah et al. [12], and A Block-level approach is followed for the same and initially 34-dimensional feature vector is constructed applying transform based Block level Radon Transform (BRT), Block-level Discrete Cosine Transform (BDCT), Block-level Fast Fourier Transform(BFFT), Block-level Distance Transform(BDT), textural and statistical techniques.

Although, different kinds of approaches have been developed and achieved promising results, however, there is little relatively report for Central Asian countries and Chinese Minority Languages script identification, it is need to solve. In this paper, on the basis of nature on different 11 scripts (such as, English, Chinese, Russian, Arabic, Uyghur, Tajikistan, Mongolian, Tibetan, Turkish, Kyrgyzstani and Uzbekistani) a script Identification method based on HSV features using BP neural network classifier was presented.

This paper organized as follows. HSV Feature extraction is described in Sect. 2; Sect. 3 is given BP neural network based classification; the experimental results and related analysis is presented in Sect. 4; And the last section concludes the paper with future directions.

2   Feature Extraction

In order to be fast and accurate classification, it is need to dig out useful information from the data to be classified. Feature extraction is the key to the identification of scripts, if the extracted features are carefully to be selected. It will benefit to identification rate and identification time of the system. In this paper, HSV features were chosen, the following detailed description of the selected feature, and its extraction algorithm.

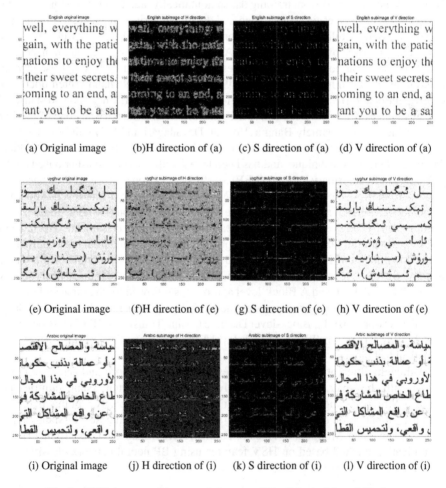

(a) Original image   (b)H direction of (a)   (c) S direction of (a)   (d) V direction of (a)

(e) Original image   (f)H direction of (e)   (g) S direction of (e)   (h) V direction of (e)

(i) Original image   (j) H direction of (i)   (k) S direction of (i)   (l) V direction of (i)

Fig. 1. HSV decomposition sample images of English, Arabic and Uyghur

HSV (Abb. of Hue, Saturation, and Value) color space is based on intuitive nature of color proposed by Stricker and Orengo [13], also known as Hexcone Model. The color information is represented as three kinds of attributes, hue H, saturation S and brightness V.

According to the study on HSV space, generally people able to tell the difference of colors are includes: Red, Yellow, Green, Cyan, Blue, and Magenta. They distributing in the color is not uniform. So the color can be divided into six parts of unequal intervals {Red, Yellow, Green, Cyan, Blue, Magenta} the brightness can be divided into three gray black white unequal interval {Black, Gray, White}. According to the above theory, an image can be divided into three sub-images and extract features from each sub-image. Three direction of some images decomposition image on HSV space as shown in Fig. 1:

As shown above Fig. 1, HSV color space is divided into three different spatial areas. The realization idea of color moment method is an image of any color distributions can be represented by their moments. so with the first moment, second moment, and third moment are enough to express color distribution of the image, color moment has proved to be effective in showing the distribution of colors in an image. Lower order moments feature on HSV space, which means can be calculated as follows:

$$\mu_i = \frac{1}{N} \sum_{j=1}^{N} P_{ij} \tag{1}$$

Variance is:

$$\sigma_i = \left[\frac{1}{N} \sum_{j=1}^{N} (P_{ij} - \mu_i)^2 \right]^{\frac{1}{2}} \tag{2}$$

Third moment was calculated using the following formula:

$$s_i = \left[\frac{1}{N} \sum_{j=1}^{N} (P_{ij} - \mu_i)^3 \right]^{\frac{1}{3}} \tag{3}$$

Where P_{ij} represents i-th color image component appearing probability of on color channels in the j gray pixels. N is the number of pixels in the image (The area of image).

According to the formulas above (1), (2), (3) can calculate mean, variance and third moment on each sub-image. Three image components H, S, V is an image from the first three colors consisting of 9 vectors, lower order moment of HSV space feature vector (mean /variance and the third moment) is represented as follows:

$$HSV = \left[h_\mu, h_\sigma, h_s, s_\mu, s_\sigma, s_s, v_\mu, v_\sigma, v_s \right] \tag{4}$$

As shown above formula (4), we can get the 9 dimensional feature vector finally. And extracted features value from the various sub-image to BP classifier for obtain classification results, following details BP Classifier.

3   Script Identification Using BP Neural Network

Classification is another very important step in script identification. After extracting the image features, it needs to use an algorithm to compare and classify the features which already extracted and stored in the feature database. In this paper, most commonly used classifier which called BP neural network classifier is selected.

Neural network classifier [14], the simplest classification methods where the script identification field in, is a good performance of neural networks. And there are dozens of types BP network learning algorithm, in this paper, gradient descent based method was used, Which this method is a Feed forward network composed of input layer, hidden layer and output layer, as shown in Fig. 2:

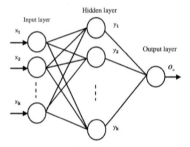

Fig. 2. The basic structure of BP neural network

Assume that, given N samples (x_k, y_k) (where $= 1,2,\dots,N$), output of any one node i is y_{oi}, for one input x_k output of the network is y_k, output of the network is o_{i_o}, When entered the k-th sample, input of node j is:

$$net_{ij}^l = \sum_j w_{ij}^l o_{jk}^{l-m} \tag{5}$$

$$o_{jk}^l = f(net_{jk}^l) \tag{6}$$

Where, o_{jk}^l represents activation function of l-m layer, if, with X representing the input vector and W is the weight vector:

$$X = \left[x_{i1}, x_{i2}, \dots, x_{ik}\right] \tag{7}$$

$$W = \left[w_{i1}, w_{i2}, \dots w_{ik}\right]^T \tag{8}$$

Then, output of the network is:

$$y_k = f(XW) \qquad (9)$$

On the basis of the structure of BP network mentioned above, script identification system based on BP neural network training was established. According to the extracted features herein has been designed a BP network, an input layer, four hidden layer and an output layer. In this paper selected training error is 0.001, training times to 10,000 times. The process of establishing the network structure described as follows:

Step1: Initialize the network and learning parameters.
Step2: Input random samples.
Step3: Calculate the input and output signals of each neuron in the BP network.
Step4: Calculate the difference between the actual output and the expected output.
Step5: Displays results.

In general, the more hidden layer nodes, the more accurate the BP network. Through done several experimentation has been determined the optimum layers and the number of nodes. In this paper, the nodes have been determined to 19. Some experiments have been done, the following experimental results and their analysis.

4   Experimental Results and Analysis

Experimental images are plain-text images scanned from books, magazines and newspapers. Considering the effects of noise and other reasons, the resolution of scanning each image is 400 dpi and the size of each image is 3212 * 4581 was selected. At the same time, it had been stored into script sample database with BMP format. Database contains 11 kinds of scripts mentioned above. Each script contains 200 images, total 2200 experimental images. Experimental sample images of some scripts as shown in Fig. 3 (Due to pages limitations, it lists only eight kinds of image file samples):

Fig. 3. Some experimental sample images

As shown above, based on the database had established, a multi-lingual script identification system was built. The experimental results are given below.

Experimental environment is AMD 3.60 GHZ processor, 4 GB of memory windows 7 environment to Matlab R2014b as an experimental platform for Experiment. Experimental procedures are as following Fig. 4:

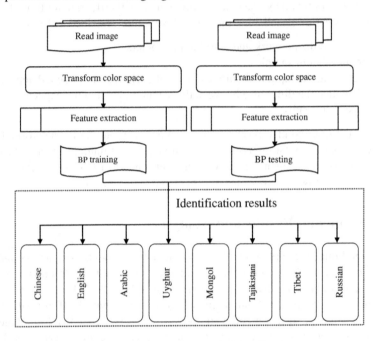

Fig. 4. Flew chart of the script identification

According to the block diagram shown in the above, it reads an image from database firstly, after then it was transformed into color space, and mean, variance and the third moment features are extracted on each H, S, V three direction from HSV space, total 19800 Eigen values, at last using BP classifier classify the extracted features.

Identification results measure by using the recall T_r, which correctly identified numbers of samples ratio total number of samples and error rate T_f is misidentified samples ratio total samples:

$$T_r = \frac{\text{correctly identified samples}}{\text{total samples}} \tag{10}$$

$$T_f = \frac{\text{misidentificated}}{\text{total samples}} \tag{11}$$

To compare the effectiveness of the method adopted herein, based on datasets established, 100 samples were randomly selected as the training, and the remaining 100 samples for testing from each scripts. Experimented 3 times and taken the average of the results,as a comprehensive evaluation of the statistical classification performance of the experimental results. Here is the classification results obtained, shown as following Table 1.

Table 1. Statistics identification results of samples on database

Scripts	En.	Ch.	Ar.	Uy.	Ru.	Mo.	Ti.	Ta.	Tu.	Ky.	Uz.
English	184			1				33	2	26	8
Chinese		198		1			20				
Arabic			190		11		5				
Uyghur		1		188							
Russian	3		5		174	6			5	3	
Mongol		1		3		181					
Tibet			5	4			175				
Tajik	5				2	13		159	8	5	14
Turkish	2								172		1
Kyrgyz	3				5			3		158	17
Uzbek	3			3	8			5	13	7	160
T_r	92 %	99 %	95 %	94 %	87 %	90.5 %	87.5 %	79.5 %	86 %	79 %	80 %
T_f	8.0 %	1.0 %	5.0 %	6.0 %	13 %	9.5 %	12.5 %	20.5 %	14 %	21 %	20 %

It can be seen in Table 1 that, for the Chinese, Tibet, Mongol, Uyghur, HSV feature greatly different scripts, the identification result is quite ideal, up to 99 %. At the same time the error rate is fairly low. The features of Central Asian country scripts similar to each other, for some text images are similar, so it has been lower identification rate, and affect the performance of the system. And to better reflect nature of similar shape scripts, it needs to be found out other effective features. The average results of the experiments are indicated as the following Table 2:

Table 2. The average results of all samples

Database	results
Total samples	2200
Correctly identified	1939
misidentified	167
Recall (%)	88.14 %
Error rate	11.86 %

From above the Table 2, it is concluded that, if it increase the number of scripts to be identified and number of each script containing images, may result in the identification rate is greatly reduced. Because of the features of text is somewhat similar to other scripts, this problem by finding other effective features, and meanwhile, through set multi-level script identification system can resolve this problem.

The advantage of this method is that it can more appropriately describe the scripts nature above mentioned. Limitations HSV characterized in that the input image must be RGB images, or else when the feature extraction cannot extract the corresponding Eigen values. This means that if there is more noise in the image to be classified, it may greatly affect the identification rate.

The advantage of this method is that it is no color space quantization, low dimension of feature vectors. Since the HSV feature is a global feature, so it's not sensitive to changes of orientation and size on image, Therefore, the HSV feature is not well characterized local feature captured in the object of images. To solve this problem, it needs to be extract a few additional features,such as the local features, or other means of effective features. It can be preferably describe all the information which images carried in if it is used feature fusion method.

5   Conclusion and Future Work

In this paper, HSV feature based identification method is proposed to identify different 11 kinds of scripts (such as English, Chinese, Arabic, Uyghur, Russian, Mongol, Tibet, Turkish, Kyrgyzstani, Uzbekistani and Tajikistani) at page level. A multi-script document image database including 11 different kinds of scripts which is contains 2200 images were established firstly. Then, HSV features of each image were extracted, that is, it was transformed into color space, and the mean, variance and the third moment features are extracted on each H, S, V three direction from each image, thus 9 feature from each image and 19800 of total Eigen values extracted from the 2200 images. Finally, BP neural network based classifier was used to classify the extracted features. And it was obtained promising experimental results, that is, 88.14 % of average identification rates and 99 % of highest identification rates were obtained with the dataset. Experimental results show that HSV features is an effective feature for identify these 11 kinds of scripts.

In our future research work, the database will be extended by increasing the type of scripts and the number of each script images. The research work will be concentrated on extracting more efficient features and using other advanced classifying methods to improve the identification rate.

Acknowledgments.   This work was supported by the National Natural Science Foundation of China (No. 61363064, 61563052, 61163028), College Scientific Research Plan Project of Xinjiang Uyghur Autonomous Region (No. XJEDU2013I11), and Special Training Plan Project of Xinjiang Uyghur Autonomous Region's Minority Science and Technological Talents (No. 201323121).

References

1. Gopal, J., Saurabhand, D., Jayanthi, S.: A generalized framework for script identification. Int. J. Doc. Anal. Recogn. (IJDAR) **10**(2), 55–68 (2007)
2. Dhanya, D., Ramakrishnan, A.G.: Script identification in printed bilingual documents. In: Lopresti, D., Hu, J., Kashi, R. (eds.) DAS 2002. LNCS, vol. 2423, pp. 13–24. Springer, Heidelberg (2002)
3. Pal, U., Sarkar, A.: Recognition of printed urdu script. In: Proceedings of the International Conference on Document Analysis and Recognition, Bangalore, pp. 1183–1187 (2003)

4. Spitz, A.L.: Script and language determination from document images. In: 3rd Annual Symposium on Document Analysis and Information Retrieval, Las Vegas, USA, pp. 229–235 (1994)
5. Spitz, A.L.: Determination of the script and language content of document images. IEEE Trans. Pattern Anal. Mach. Intell. **19**(3), 235–245 (1997)
6. Ul-Hasan, A., Afzal, M.Z., Shafait, F., Liwicki, M., Breuel, T.M.: A sequence learning approach for multiple script identification. In: 2015 13th International Conference on Document Analysis and Recognition (ICDAR), Tunis, pp. 1046–1050 (2015)
7. Pal, U., Chaudhuri, B.: Identification of different script lines from multi-script documents. Image Vis. Comput. **20**(13–14), 945–954 (2002)
8. Shi, B., Bai, X., Cong, Y.: Script identification in the wild via discriminative convolutional neural network. Pattern Recogn. **52**, 448–458 (2016)
9. Androutsos, D., Plataniotis, K.N., Venetsanopoulos, A.N.: A novel vector-based approach to color image retrieval using a vector angular-based distance measure. Comput. Vis. Image Underst. **75**(1–2), 46–58 (1999). ISSN 1077-3142
10. Ferrer, M.A., Morales, A., Rodríguez, N., Pal, U.: Multiple training-one test methodology for handwritten word-script identification. In: 14th International Conference on Frontiers in Handwriting Recognition (ICFHR), Heraklion, pp. 754–759 (2014)
11. Pawan, K.S., Ram, S., Mita, N.: Offline script identification from multilingual indic-script documents: a state-of-the-art. Comput. Sci. Rev. **15–16**, 1–28 (2015)
12. Obaidullah, S.M., Das, N., Halder, C., Roy, K.: Indic script identification from handwritten document images-An unconstrained block-level approach. In: 2015 IEEE 2nd International Conference on Recent Trends in Information Systems (ReTIS), Kolkata, pp. 213–218 (2015)
13. Stricker, M., Orengo, M.: Similarity of color images. In: SPIE Storage and Retrieval for Image and Video Databases, vol. 2420, pp. 381–392(1995)
14. Roy, K., Pal, U., Chaudhuri, B.B.: Neural network based word-wise handwritten script identification system for Indian postal automation. In: Proceedings of 2005 International Conference on Intelligent Sensing and Information Processing, pp. 240–245(2005)

Robust Principal Component Analysis Based Speaker Verification Under Additive Noise Conditions

Minghe Wang$^{(\boxtimes)}$, Erhua Zhang, and Zhenmin Tang

School of Computer Science and Engineering,
Nanjing University of Science and Technology, Nanjing 210094, China
{sdwmh, speechstudio}@163.com,
tzm.cs@mail.njust.edu.cn

Abstract. Previous researches show that the approaches based on the total variability space (TVS) followed by Gaussian probabilistic linear discriminant analysis (GPLDA) work effectively for dealing with convolutional noise (such as channel noise) and can bring some degree of gains in term of accuracy under additive noisy environment as well. However they meet difficulty while many types of noises are unseen and non-stationary in real world. To address this issue, we introduce the robust principal component analysis (RPCA) into the TVS modeled speaker verification system, called RPCA-TVS, which regards the noise spectrum as the low-rank component and the speech spectrum as the sparse component in short-time Fourier transform (SFT) domain. The highlighting of this paper is to improve the robustness of speaker verification under additive noisy environment, especially in non-stationary and unseen noise conditions. For evaluating the performance, we designed and generated an additive noisy corpus, based on the TIMIT and NUST603-2014 database, using the NaFT tools with 12 types of noise samples deriving from NOISEX-92 and FREESOUND. Experimental results demonstrate that the proposed RPCA-TVS can achieve better performance than the competing methods at various signal-to-noise ratio (SNR) levels. Especially, RPCA-TVS reduces the equal error rate (EER) by 5.12 % in average than the multi-condition system under additive noise conditions at SNR = 8 dB.

Keywords: Robust speaker verification · Additive noise · Total variability space · Robust principal component analysis

1 Introduction

Over the past decades, text-independent speaker verification technology has been studied actively and the current state-of-the-art speaker verification systems achieve encouraging performance on clean data recorded under quiet environment, but fail to work effectively with noisy speech, especially at low signal-to-noise ratio (SNR).

This work is supported by the National Science Foundation of China (Grand no. 61473154).

T. Tan et al. (Eds.): CCPR 2016, Part II, CCIS 663, pp. 598–606, 2016.
DOI: 10.1007/978-981-10-3005-5_49

However speech signals in real-world scenarios are often noise distorted, the robustness against noise always has been one of the most important challenges so far. A large number of noise robust speaker verification methods have been proposed, and many of them have created significant impact on either research or commercial application. Most of them are generally categorized into four cases: speech enhancement, feature compensation, robust modeling and score compensation [1]. This work is mainly focused on the speech enhancement for denoising of speaker verification.

However, an ideal technique should hold good performance in unseen noise conditions and not be limited to several known noise types [2]. Another difficulty is that many types of noises are non-stationary. In this paper, we aim to design a robust speaker verification system by using total variability space (TVS) model and robust principal component analysis (RPCA) [3]. As a matrix factorization algorithm for recovering underlying low-rank and sparse matrices, RPCA is firstly used in face recognition. Inspired by the encouraging success on singing voice separation using RPCA in [4], we employ RPCA to separate speech and noise for speaker verification. The highlighting of this work is to investigate how to promote the robustness of speaker verification under additive noisy environment, especially in non-stationary and unseen noise conditions.

The rest of this paper is organized as follows: Sect. 2 provides fundamentals and related works of speaker verification. Section 3 discusses how to take advantage of RPCA to recover clean speech signal from noisy speech and construct a new robust speaker verification modeled by TVS. The experimental results are given in Sect. 4. Conclusions are in Sect. 5.

2 Related Work

Recently, most researches in speaker verification have focused on the problem of compensating the mismatch between training and test speech segments caused by the transmission channel. TVS modeled identity vector (i-vector) followed by Gaussian probabilistic linear discriminant analysis (GPLDA) [5–9] has demonstrated high performance and becomes very popular in the text-independent speaker verification system, which can bring some degree of gains in accuracy of speaker verification under additive noisy environment as well as under channel noise condition, even though it is difficult to comprehend intuitively in theory. Another disadvantage is that optimal subspaces for discriminating speakers are noise-level dependent and noise variability leads the i-vectors to shift, causing the noise contaminated i-vectors to form clusters in the i-vector space [7, 10].

Based on digital signal processing technique, a commonly approach to recover speech signals from noisy observations is speech enhancement which estimates and removes the noise spectrum from the input noisy speech spectrum. Although it is certain that speech enhancements can improve the perceptual quality of a noisy speech signal, but it is far from certain that all speech enhancements can always positively affect speaker verification accuracy under various noise conditions, because of that the hidden speaker factor is distorted probably while speech enhancing. Huang [11] introduces RPCA firstly used in face recognition into separating singing voices from music accompaniment, which takes advantage of low-rank, i.e. repetition, of music sound, and sparsity of speech signal in the spectral domain. Actually, many types of noises present the similar repeating structure to music. Furthermore, RPCA has the potential to recover clean speech from distorted speech under various types of noises conditions.

3 RPCA-TVS

Inspired by the encouraging success on singing voice separation using RPCA, we employ RPCA into a speaker verification system. Similar to singing voice separation regards music as low rank component and singing voice as sparse component [11], we take advantage of the correlation of noise spectrum, i.e. low rank, and sparseness of speech spectrum in short-time Fourier transform (SFT) domain to develop a RPCA based denoising scheme for recovering clean speech under various types of noises conditions.

3.1 Denoising Based on RPCA

Most real-life data are sparse or low-rank on a certain basis such as Fourier or Wavelet basis. Therefore, sparse or low-rank representations based methods have achieved high performance and created significant impact on various applications. Based on non-negative matrix factorization (NMF) [12], the spectrum enhancement with sparse representation has been used for speech recognition in noisy environment [13]. The representations describe noisy speech by seeking the sparsest possible linear combination of the exemplar-based audio and noise dictionary. For denoising, the original speech is reconstructed using parts of the dictionary pertaining to only audio. Candès et al. [3] proposed RPCA for recovering low-rank matrices distorted by noise when the noise matrix is sparse enough. RPCA aims at decomposing a data matrix S into $D + E$, where S is a matrix corrupted by errors, D is a low-rank matrix and E is a sparse matrix. When the rank of D is not too large and E is sparse, we can find the solution by optimizing the following problem:

$$
\min_{D,E} rank(D) + \lambda \|E\|_0,
$$
$$
\text{s.t. } S = D + E
\tag{1}
$$

where S, D and $E \in \mathbb{R}^{m \times n}$. $rank$ is the rank function of matrix; $\| \cdot \|_0$ denotes the L_0-norm (number of non-zero values of matrix entries); $\lambda > 0$ is a trade-off parameter between the rank of D and the spasity of E. The approach, Principal Component Pursuit, suggests solving the following convex optimization problem:

$$
\min_{D,E} \|D\|_* + \lambda \|E\|_1,
$$
$$
\text{s.t. } S = D + E
\tag{2}
$$

where $\|\cdot\|_*$ denotes the nuclear norm (sum of singular values) and $\|\cdot\|_1$ denotes the L_1-norm (sum of absolute values of matrix entries).

While achieving simplicity, we set $\lambda = 0.025$ instead of $\lambda = 1/(\max(m, n))^{1/2}$ suggested in [3, 11] and get satisfactory results. In the time domain, for clean speech x distorted by both convolutional noise h and additive noise n generates noisy speech signal y, the distortion formulation [14, 15] is according to,

$$y = x * h + n, \tag{3}$$

where $*$ denotes the convolution operator, h can be discarded for focusing on additive noise issue in this paper. In SFT domain, Eq. (3) will become (4):

$$Y = X + N, \tag{4}$$

where X, Y, N denote the spectrum matrices of clean speech, noisy speech and noise respectively.

In this paper, we would pay more attention to additive noise rather than convolutional noise. From Eqs. (2) and (4), one can find that RPCA model is very suitable to separate clean speech signal X from noisy speech Y under additive noise condition. An important difference between this work and other applications such as face recognition is that in face recognition systems the original face images are regarded as low-rank matrices and errors caused by noise are regarded as sparse component, but in this work, the original voice signal is regarded as sparse portion and additive noise is regarded as low-rank portion. In fact, most noises in real-world have the repeating or quasi-repeating structure that is similar to the rhythm of music. On the other hand, the spectrum of human voice is sparse or quasi-sparse in SFT domain. Therefore, we assume noise as low-rank matrices and think of clean speech as sparse matrices for separating the clean speech signal from noisy speech.

For the given input magnitude of a noisy utterance which is contaminated with the 'keyboardtyping' noise at SNR of 8 dB, we can obtain two output matrices D and E by RPCA. From an example of NUST603-2014 corpus, Fig. 2, we can observe that there are distinct formant structures in the sparse matrix E, which indicates vocal activity. The low-rank matrix D that denotes noise is discarded.

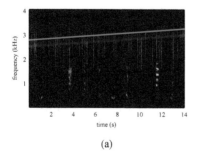
(a)

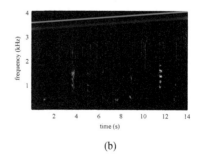
(b)

Fig. 1. RPCA results for two utterances of NUST603-2014 contaminated with noise at SNR = 8 for (a) the original matrix that denotes noisy speech with 'keyboardtyping' noise; (b) the sparse matrix that denotes clean speech recovered

3.2   Feature Extracting with RPCA

At conventional MFCC extracting stage, the log operation makes convolutional noise in the spectral domain to be additive and simple in the log-Mel-filter-bank and cepstral domains, but leads the additive noise in the spectral domain to be very complex in the cepstral domain. Therefore, unlike the majority of traditional noise robust speaker verification approaches dedicate to compensate for the noise impact after MFCC or i-vector extracting [16, 17], not deal with additive noise directly, we embed RPCA based denoising algorithm into MFCC extraction phase, as illustrated in Fig. 2. For given nosy speech, the spectrum of which is firstly generated by preprocessing and SFT. Then RPCA decomposes noisy speech spectrum into two matrices: low-rank matrix and sparse matrix. The former denotes noise component and is discarded. The later denotes recovered speech and is used as the input of Mel-filter group followed by the logarithm. Finally we obtained MFCC by the discrete cosine transformation (DCT). The study of singer identification [14] proves that the sing voice refinement can improve SNR in contrast with the sing voice separation, but for the point view of singer identification, the separated singing voice, i.e. not refined singing voice, is more appropriate than the refined singing voice. In the light of this idea, we directly use the sparse portion obtained by RPCA without refinement.

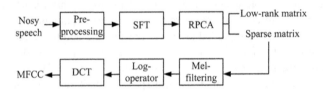

Fig. 2. The framework of MFCC extracting with RPCA

3.3   TVS Model

In recent years, a feature extractor called i-vectors using a simple factor analysis in TVS is proposed in [4–10]. The total factor is a hidden variable, which can be defined by its posterior distribution conditioned to the Baum–Welch statistics for a given utterance. Unlike Joint Factor Analysis (JFA) which models separately inter-speaker and channel variability, the TVS models both speaker and channel information in a single low dimensional subspace directly, which can bring some degree of gains in accuracy of speaker verification under additive noisy environment, even though it is difficult to comprehend intuitively in theory because of the log operation while extracting MFCC.

Generally, a Gaussian mixture model based universal background model (UBM) is estimated using maximum a posterior (MAP) adaptation on universal database in advance. Given an utterance, the TVS-UBM super-vector is defined as follows:

$$M = m + Tx + \varepsilon \tag{5}$$

where M is the means obtained for a given utterance, m represents the means for all speaker's utterances, both of them are estimated by UBM; T is a rectangular matrix of low rank, and x is the total factors, a variable vector, having a standard normal distribution $N(0, I)$; ε is the residual usually caused by other unconcerned variability.

For a given utterance, assuming that the zero order and the first order Baum-Welch statistics are already obtained, the i-vector feature can be extracted according to the way described in [5].

3.4   Score Calculating

GPLDA based i-vector system [7, 9, 10] scoring is calculated using batch likelihood ratio which is computationally very expensive, but more effectively than the traditional cosine similarity scoring. It is proved that GPLDA is effective either in channel noise conditions or in the additive noise conditions. Given two i-vectors Y_1 and Y_2, batch likelihood ratio can be calculated as follows,

$$g = \ln \frac{P(Y_1, Y_2|H_0)}{P(Y_1|H_1)P(Y_2|H_1)}, \tag{6}$$

where g is the score of batch likelihood ratio; H_1: the speakers are same and H_0: the speakers are different.

4   Experimental Results

For evaluating the performance, we design and generate an additive noisy corpus, based on the TIMIT corpus and partial NUST603-2014 data [18], using the NaFT tools with 12 types of noises at SNR of 8, 15 and 25 dB. As described in Table 1, TIMIT contains a total of 6300 English sentences, spoken by 630 speakers, 432 male and 192 female. The Mic part of NUST603-2014 data contains a total of 2961 Chinese utterances, spoken by 423 speakers, 210 male and 213 female respectively. The 12 noise samples are 'babble', 'factory1', 'volvo', 'pink', 'musicbox', 'light-rain', 'airconditioning', 'factory2', 'keyboardtyping', 'happybirds', 'white', and 'wind' noises. Some of them derive from the NOISEX-92 noise database and the others are downloaded from the website[1] freely. For evaluating under unseen noise condition, all noise samples are divided into 2 groups: one group used for training consists of the first 6 and another used for testing contains the rest. The proposed approach is compared against the baseline which uses multi-condition training [19, 20] where clean and noisy utterances are pooled together and other two popular speech enhancements: spectrum subtracting (SS) and Wiener filtering (Wiener) based systems.

Using the training data drawn from the corpuses TIMIT and NUST603-2014, a gender-independent UBM of 512 mixture Gaussians in feature space of 39 dimensional MFCC with first and second derivatives appended are built by adapting UBM with MAP algorithm. The voice activation detection is employed to remove the silence segments in advance. TVS model is estimated simultaneously using the same training

---

[1] available from http://www.freesound.com.

Table 1. Structure of TIMIT corpus and NUST603-2014 data

Corpus	Channel	Number of female	Number of male	Number of utterances
TIMIT (training set)	Mic	326	136	4620
TIMIT (testing set)	Mic	112	56	1680
NUST603-2014 (training set)	Mic	140	142	1974
NUST603-2014 (testing set)	Mic	70	71	987

Table 2. The EER on TIMIT corpus distorted by various types of noises at SNR = 8 dB (%)

Noise type	Multi-condition	SS	Wiener	RPCA-TVS
Airconditioning	30.4	**21.5**	28.5	24.2
Factory2	26.6	**20.1**	31.7	21.6
Keyboardtyping	21.5	19.7	29.7	**17.2**
Happybirds	19.7	20.2	29.8	**17.8**
White	22.7	24.4	27.6	**19.1**
Wind	27.8	24.5	33.1	**23.2**

Table 3. The EER on TIMIT corpus distorted by various types of noises at SNR = 15 dB (%)

Noise type	Multi-condition	SS	Wiener	RPCA-TVS
Airconditioning	20.9	**14.6**	23.2	17.8
Factory2	21.6	16.3	31.7	**16.1**
Keyboardtyping	13.3	14.3	29.7	**12.5**
Happybirds	13.4	15.5	29.8	**13.0**
White	14.7	17.1	27.6	**14.2**
Wind	17.9	16.5	33.0	**15.6**

data set, and then the 400-dimensional i-vectors are extracted. LDA and PLDA are trained by using the noisy utterances noise at various levels in baseline system but using the recovered data via speech enhancement in the other systems, i.e. Wiener, SS and RPCA-TVS. The original utterances (clean speech) and noise contaminated utterances of the testing data set are used for enrolling and testing respectively. The equal error rate (EER) is used for the metrics of evaluation.

The performance of multi-condition training, Wiener filtering, spectrum subtracting and RPCA-TVS based systems are shown in Tables 2 and 3. We can observe that the RPCA-TVS works better than the tree competing methods under several types of noises condition at various SNR. In most cases, the Wiener method shows signs of performance degradation instead of performance improving, thus we arrive at the similar conclusion to singer identification in [14] that refinement is helpless to speaker verification because of the speaker's acoustic character is probably destroyed while refining such as Wiener filtering. Finally, From Tables 2 and 3 and Fig. 1, we can conclude that by taking advantage of the repeating structure of noise and sparsity of clean speech, the proposed RPCA-TVS approach works effectively under additive noise condition even though while the noise is non-stationary and unseen. It is worth mentioning that RPCA-TVS is the best and only one can outperform the baseline in the 'happybirds' noise condition.

5 Conclusion

In this paper, we proposed a novel robust speaker identification approach, termed RPCA-TVS, which employs robust principal component analysis into total variability space modeled speaker verification system. For evaluating the performance, we generated an additive noisy corpus based on the TIMIT and NUST603-2014 data by using the NaFT tools with 12 noise samples deriving from NOISEX-92 database and FREESOUND. We select 6 noise samples for training and the remaining samples for testing independently. Various experiments were conducted on the generated corpus and the results demonstrate that the RPCA-TVS achieves encouraging performance under additive noisy conditions at various SNR levels, even though under non-stationary and unseen noise conditions. Especially, the proposed RPCA-TVS reduces the equal error rate by 5.12 % in average than the multi-condition system under additive noise conditions at SNR of 8 dB.

References

1. Lei, Y., Burget, L., Ferrer, L., et al.: Towards noise-robust speaker recognition using probabilistic linear discriminant analysis. In: IEEE International Conference on Acoustics, Speech, and Signal Processing (ICASSP), Kyoto, pp. 4253–4256 (2012)
2. Sun, M., Zhang, X., Van Hamme, H., et al.: Unseen noise estimation using separable deep auto encoder for speech enhancement. IEEE/ACM Trans. Audio Speech Lang. Process. **24**(1), 93–104 (2016)
3. Candès, E.J., Li, X., Ma, Y., et al.: Robust principal component analysis. J. ACM **58**(3), 1–73 (2011)
4. Dat, T.T., Jin, Y.K., Kim, H.G., et al.: Robust speaker verification using low-rank recovery under total variability space. In: International Conference on IT Convergence and Security, Kuala Lumpur, pp. 1–4 (2015)
5. Dehak, N., Kenny, P., Dehak, R., et al.: Front-end factor analysis for speaker verification. IEEE/ACM Trans. Audio Speech Lang. Process. **19**(4), 788–798 (2011)
6. Li, W., Fu, T., Zhu, J.: An improved i-vector extraction algorithm for speaker verification. EURASIP J. Audio Speech Music Process. **2015**(1), 1–9 (2015)
7. Li, N., Mak, M.W.: SNR-invariant PLDA modeling in nonparametric subspace for robust speaker verification. IEEE/ACM Trans. Audio Speech Lang. Process. **23**(10), 1648–1659 (2015)
8. Kanagasundaram, A., Dean, D., Sridharan, S., et al.: I-vector based speaker recognition using advanced channel compensation techniques. Comput. Speech Lang. **28**(1), 121–140 (2014)

9. Jiang, Y., Lee, K.A., Wang, L.B.: PLDA in the I-SUPERVECTOR space for text-independent speaker verification. EURASIP J. Audio Speech Music Process. **1–13**, 2014 (2014)

10. Mak, M.W., Pang, X., Chien, J.T.: Mixture of PLDA for noise robust i-vector speaker verification. IEEE/ACM Trans. Audio Speech Lang. Process. **24**(1), 130–142 (2016)

11. Huang, P.S., Chen, S.D., Smaragdis, P., et al.: Singing-voice separation from monaural recordings using robust principal component analysis. In: IEEE International Conference on Acoustics, Speech, and Signal Processing (ICASSP), Kyoto, pp. 57–60 (2012)

12. Lee, D.D., Seung, H.S.: Algorithms for non-negative matrix factorization. Adv. Neural Inf. Process. Syst. **13**, 556–562 (2001)

13. Gemmeke, J.F., Virtanen, T., Hurmalainen, A.: Exemplar-based sparse representations for noise robust automatic speech recognition. IEEE/ACM Trans. Audio Speech Lang. Process. **19**(7), 2067–2080 (2011)

14. Hu, Y., Liu, G.: Separation of singing voice using nonnegative matrix partial co-factorization for singer identification. IEEE/ACM Trans. Audio Speech Lang. Process. **23**(4), 643–653 (2015)

15. Li, J., Deng, L., Gong, Y., et al.: An overview of noise-robust automatic speech recognition. IEEE/ACM Trans. Audio Speech Lang. Process. **22**(4), 745–777 (2014)

16. Kheder, W.B., Matrouf, D., Bonastre, J.F., et al.: Additive noise compensation in the i-vector space for speaker recognition. In: IEEE International Conference on Acoustics, Speech and Signal Processing (ICASSP), Brisbane, 35–39 (2015)

17. Gonzalez-Rodriguez, J.: Evaluating automatic speaker recognition systems: An overview of the NIST speaker recognition evaluations (1996-2014). Loquens **1**(1), 1–15 (2014)

18. Wang, M.H., Chen, Y., Tang, Z.M., et al.: I-vector based speaker gender recognition. In: IEEE Advanced Information Technology, Electronic and Automation Control Conference (IAEAC), Chongqing, pp. 729–732 (2015)

19. Avila, A.R., Sarria-Paja, M., Fraga, F.J., et al.: Improving the performance of far-field speaker verification using multi-condition training: The case of GMM-UBM and i-vector systems. In: Fifteenth Conference of the International Speech Communication Association (ISCA), Singapore, pp. 1096–1100 (2014)

20. Mekonnen, B.W., Dufera, B.D.: Noise robust speaker verification using GMM-UBM multi-condition training. In: AFRICON, Addis Ababa, pp. 1–5 (2015)

Recurrent Neural Network Based Language Model Adaptation for Accent Mandarin Speech

Hao Ni[1(✉)], Jiangyan Yi[1], Zhengqi Wen[1], and Jianhua Tao[1,2]

[1] National Laboratory of Pattern Recognition, Institute of Automation,
Chinese Academy of Sciences, Beijing, 100190, China
{hao.ni,jiangyan.yi,zqwen,jhtao}@nlpr.ia.ac.cn
[2] CAS Center for Excellence in Brain Science and Intelligence Technology,
Chinese Academy of Sciences, Beijing, 100190, China

Abstract. In this paper, we propose to adapt the recurrent neural network (RNN) based language model to improve the performance of multi-accent Mandarin speech recognition. N-gram based language model can be easily applied to speech recognition system, but it is hard to describe the long span information in a sentence and arises a serious phenomenon of data sparsity. Instead, RNN based language model can overcome these two shortcomings, but it will take a long time to decode directly. Taking these into consideration, this paper proposes a method which combines these two types of language model (LM) together and adapts the RNN based language model to rescore lattices for different accents of Mandarin speech. The architecture of the adapted RNN LM is accent-specific top layers and shared hidden layer. The accent-specific top layers are used to adapt different accents and the shared hidden layer stores history information, which can be seen as a memory layer. Experiments on the RASC863 corpus show that the proposed method can improve the performance of accented Mandarin speech recognition over the baseline system.

Keywords: Multi-accent · Speech recognition · RNN language model · Adaptation

1 Introduction

Statistical language model (LM) is a crucial component for automatic speech recognition (ASR) system, which models the distribution of word sequences. The traditional approach for language model is N-gram model, which estimates the word's distribution directly from the relative word frequencies with some smoothing techniques. The N-gram model often suffers from the data sparsity problem and hard to describe the long span information in a sentence. However, neural network language model (NNLM) [1, 2] can represent the non-linear relationship between input words and word probabilities through projecting input words into a continuous space and estimating word probabilities in a softmax layer. It is efficient to overcome the problem of data sparsity. In addition, recurrent neural network (RNN) LM [3–5] is proposed as a variant of general

© Springer Nature Singapore Pte Ltd. 2016
T. Tan et al. (Eds.): CCPR 2016, Part II, CCIS 663, pp. 607–617, 2016.
DOI: 10.1007/978-981-10-3005-5_50

NNLM, which can break through the limit of fixed context length information and then capture long span information. Therefore, it is superior to the NNLM in ASR system.

It is very crucial to use the large-scale text corpora to train the universal model. But it is still difficult to cover all the domains. So, it often leads to a deterioration of the recognition performance in mismatched conditions. In the multi-accent Mandarin speech recognition task, the speakers often come from different regions with different speaking styles. It is difficult to cover all situations since the original corpora and the general language model cannot always meet the requirement. There is a way to solve the problem by adapting LM to improve the performance and make up the deficiency.

Previously, there are several works on multi-domain adaptation. In [6] an NNLM adaptation scheme is proposed by cascading an additional layer between the projection an hidden layer. This scheme provides a direct adaptation of NNLMs via a non-linear, discriminative transformation to a new domain. In [7], a domain dependent element-wise multiplication layer is set between projection and hidden layers of NNLM. The model includes the adaptation layer from beginning, many parameters of the adaptation layer are tied across domains. In [8], the architecture is RNN LM, the adaptation layer is added between recurrent layer and output layer, a domain dependent parameter act as one of the inputs of adaptation layer. They think the architecture only need a very small number of domain-specific parameters which enables the model to adapt to domains with little data without the danger of overfitting. On the other hand, in [9], the domain information is fed as an additional input feature to the RNN LM.

The method presented in this paper differs from previous works as follows. An adapted layer or additional feature is needed in previous work. However, it is not necessary in our model. In terms of model architecture, inspired by multi-task learning [10, 11], the architecture in our method is accent-specific top layers and shared hidden layer. The accent-specific top layers are used to adapt different accents and the shared hidden layer stores the history information, which can be seen as the memory layer. We adapt our model on top layers only, and improve the performance of accented Mandarin speech recognition over the baseline system.

The reminder of this paper is organized as follows. In Sect. 2, class based RNN LM is introduced. Section 3 presents the RNN LM adaptation method. The details about our experiments and results are shown in Sect. 4. We conclude the paper in Sect. 5.

2 Class Based RNN Language Model

RNN is a class of artificial neural network where connections between units form a directed cycle. This creates an internal state of the network which allows it to exhibit dynamic temporal behavior. Unlike feed-forward neural networks, RNN can use their internal memory to process arbitrary sequences of inputs. This makes it applicable in speech recognition.

The class-based recurrent neural network is proposed to compute LM probabilities $P_{RNN}(w_i|c_i, s_{i-1})$ in [4]. Its architecture is described in Fig. 1. The input layer concatenates the 1-of-N representation of the previous word w_{i-1} with the previous state of the hidden layer s_{i-2}, which represents the full history vector for word w_i. The hidden layer compresses the information of these two inputs and computes a new representation s_{i-1} using a sigmoid activation to achieve non-linearity. The output layer w_i using a softmax activation to represent the probability distribution of the next word and the state of the hidden layer in the previous time step has the same dimensionality as w_{i-1}.

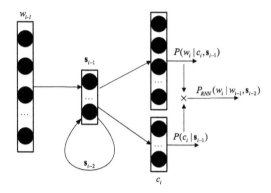

Fig. 1. Architecture *of class based recurrent neural network.*

To reduce the computational cost, a classification layer is added together with the output layer. Each word in the output layer is classified to a unique class based on the frequency counts. In this work, we choose 500 classes, which means the words that correspond to the first 0.2 % of the unigram probability distribution would be mapped to class 1 and the next 0.2 % would be mapped to class 2. The LM probability assigned to a word is factorized onto two individual terms, as:

$$P_{RNN}\left(w_i|w_{i-1}, s_{i-2}\right) = P\left(w_i|c_i, s_{i-1}\right)P(c_i|s_{i-1}) \tag{1}$$

Class-based RNN LM can be trained by back-propagation though time (BPTT) algorithm. [12] describes the BPTT algorithm in detail. Training a RNN is more difficult than the feed-forward neural network since gradient explosion maybe occurred [13]. A simple solution to the exploding gradient problem is to truncate values of the gradients. In our experiments, the maximum size of gradients of errors accumulated in the hidden neurons are limited to $[-20, 20]$, which make the training more stable.

3 Proposed RNN Language Model Adaptation Method

In this paper, we use model adaptation method to adapt RNN LM with the accent-specific top layers. This method is motivated by training the RNN with the shared hidden layer in the multi-task learning structure. Multi-accent speech recognition can adopt the similar method in training the RNN LM. In this paper, four accent-specific tasks learn together with the shared hidden layer. This often leads to a better model for all the four tasks, because it allows the learner to use the commonality among the tasks [11].

Figure 2 shows the architecture of the proposed multi-accent Mandarin speech adaptation model. The model is built on the base of trained Fig. 1 and then copy the top layer N times. In the model, the input is the 1-of-N representation of the previous word. Hidden layer is shared across all the accented speeches, as the global linguistic feature transformation crosses and serves all training accented speeches. Conversely, each accented speech has its own output layer. Each output layer is divided into two parts (words and classes), the classes part is designed to reduce computational cost.

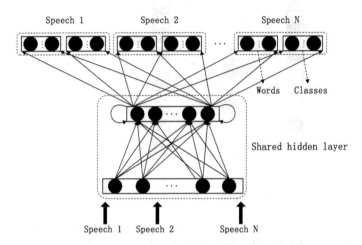

Fig. 2. Architecture of multi-accent RNN LM with region-specific layer and shared hidden layer.

The shared hidden layer, in the multi-accent speech RNN LM, can be treated as a global linguistic feature transformation universal to all training speeches. The shared hidden layer can also be used to transform the linguistic feature for a new accented Mandarin speech. By fixing the hidden layer and only updating the accent-specified layer, supervised adaptation can be achieved with only limited adaptation data.

In our work, we use two pass decoding method. In the first pass, we use a modified Kneser-Ney smoothed 3-gram [14] and in the second pass, RNN LM linearly interpolated with N-grams is used as follows:

$$P\left(w_i|h_1^{i-1}\right) = \alpha P_{RNN}\left(w_i|h_1^{i-1}\right) + (1-\alpha)P_{NG}\left(w_i|h_1^{i-1}\right) \qquad (2)$$

where α is used to control the weight, $P_{RNN}\left(w_i|h_1^{i-1}\right)$ is for RNN LM distribution and $P_{NG}\left(w_i|h_1^{i-1}\right)$ is for N-gram.

4   Experiments and Result

4.1   Data Description

We use the open source Eesen [15] speech recognition toolkit for features extraction, acoustic model training and decoding. N-gram model is trained by SRILM [16]. Our experiments are conducted on RASC863 [17], from which the spontaneous part is selected, about 53.5 h. The corpus consists of four different kinds of accented Mandarin speech. The speakers of different accents are native residents from Chongqing, Guangzhou, Shanghai, and Xiamen respectively. Moreover, balance in terms of the age, sex, and educational background has been considered. The statistics of the training data in the corpus are listed in Table 1.

Table 1. Statistics of acoustic training data.

Accent	#speakers	#utterances	#hours
Chongqing	200	7117	13.0
Guangzhou	200	7659	15.5
Shanghai	200	7401	13.5
Xiamen	200	6115	11.5
Total	800	28292	53.5

For different accent, the corpus partitioned into three parts (training set, validation set, test set) according to 8:1:1 respectively.

We train our N-gram and RNN LM on the same dataset crawled from the Internet, which contains one billion sentences and takes up space of 0.6 GB. The dataset is collected almost by written language, such as People's Daily.

4.2   Training

Acoustic Model Training. We adopt deep RNNs as the acoustic model, and the Long Short-Term Memory (LSTM) [18] units as the RNN building blocks. Unlike in the hybrid approach, the acoustic model is trained by using frame-level labels with respect to the cross-entropy (CE) criterion. Instead, we adopt the connectionist temporal classification (CTC) [19–21] objective function to automatically learn the alignments between speech frames and their label (phonemes) sequences. In our experiments, we adopt 3 layers bi-RNNs and each layer contains 320 LSTM units. The input feature is 120 dimensions. The output layer is a softmax layer, which contains 65 units covering with 61 phones and $<$ lau $>$, $<$ nsn $>$, $<$ spn $>$, $<$ blk $>$.

RNN Language Model Training. Training RNN LM includes two phases: general model training and model adaptation. In the general training phase, RNN LM is trained with BPTT. Networks are trained in several epochs, in which all data from training corpus are sequentially presented. Weights are initialized to small values (random Gaussian noise with zero mean and unit variance). The input layer contains 240078 units, involved with 239078 words and 1000 hidden units from previous step. The output layer is 239578 units, which contain words and 500 classes. BPTT order is set to 3 which mean the error is propagated through recurrent connections back in time for three steps.

The learning rate starts from 0.1 and remains unchanged until the drop of log-perplexity on the validation set between two consecutive epochs falls below 0.3 %. Then the learning rate is decayed by a factor of 0.5 at each of the subsequent epochs. The whole learning process terminates when the perplexity fails to decrease by 0.3 % between two successive epochs. The training finished after 11 epochs.

The adaptation process for the RNN LM is a one-iteration retraining on the accent-specific training set. The learning rate is fixed to 0.025, which is equal to the last epoch for general model training. In the following, rescoring is performed a second time using the adapted RNN LM, and an improved recognition output is obtained.

Decoding. Normally, the speech recognition system output the most likely word sequence directly for given acoustic signal, but it is often advantageous to preserve more information for subsequent processing steps. What's more, it is time consuming to decode directly using RNN based language model. In the experiment we use two pass decoding method to get results. The decoding steps are as follows:

- Decoding utterances form WFST, produce lattices.
- Extract n-best lists from lattices.
- Compute sentence-level scores using N-gram.
- Perform weighted linear interpolation of log-scores given by N-grams and RNN LM.
- Re-rank the n-best lists using the new LM scores.

4.3　Baseline Model

Different Modeling Techniques for LM. To evaluate the performance of single N-gram, single RNN and fusion model (N-gram + RNN), we do experiment on different model. When $\alpha = 0.0$, means N-gram LM is used for lattice rescoring only and $\alpha = 1.0$ means RNN LM is used. Otherwise, fusion model is used. We donate n = 10 in n-best lists and choose the highest-score one from it as result.

- The last row of Table 2 shows, when only RNN LM is used, we achieve 1.43 %, 1.50, 0.55 %, 1.26 % absolute accuracy improves respectively than only N-gram ($\alpha = 0.0$) is used, the result shows that the performance of the RNN LM does better than N-gram in speech recognition.

Table 2. Performance comparison of single N-gram, single RNN and fusion model. (Accuracy %)

α	Chongqing	Guangzhou	Shanghai	Xiamen
0.0	64.51	66.01	74.43	66.33
0.3	66.01	67.82	75.45	67.52
0.4	66.07	67.75	75.43	67.67
0.5	66.12	67.83	75.17	67.63
0.6	66.10	67.76	75.23	67.62
1.0	65.94	67.61	74.98	67.59

- Comparing RNN LM with fusion model, we report results for N-gram linear-interpolation with RNN LM with weight 0.5, We achieved 1.61 %, 1.82 %, 0.74 % and 1.30 % absolute accuracy improvement respectively for four accent speech.

In summary, fusion model significantly outperforms the other two models while N-gram gets worst results.

Different Scale of n-best Lists. To compare the performance of different scale n-best lists, we select n = 0, 10, 100 and 1000 to rescore from lattice. Table 3 report that with the increase of n, we can get higher accuracy, however, the computational complexity will increase rapidly. With the increase of n, search space increase rapidly, we can find better decoding path, but the search time increased dramatically. When the n increases to a certain amount, the accuracy does not improve obviously, this is because the search space is enough, most of high probability paths are included in the search space. The best performance is obtained when we set n to 1000. Therefore, we choose n = 1000 in the baseline model.

Table 3. Performance of different scale n-best lists. (Accuracy %)

n	Chongqing	Guangzhou	Shanghai	Xiamen
0	64.51	66.01	74.43	66.33
10	66.12	67.83	75.45	67.67
100	66.68	68.71	76.04	68.58
1000	67.20	69.11	76.25	69.05

4.4   Adapted Model

Different Scale of n-best Lists for Adaptation. For the different scale of n-best lists in adapted model, Table 4 shows 1-iteration adaptation for different scale n-best lists and Table 5 for 2-iteration. From the two experiments, we can get:

Table 4. Performance of different scale n-best lists for 1-iteration. (Accuracy %)

n	Chongqing	Guangzhou	Shanghai	Xiamen
0	64.51	66.01	74.43	66.33
10	66.65	68.57	76.21	68.18
100	67.49	69.55	76.80	69.68
1000	68.26	70.12	77.23	70.15

Table 5. Performance of different scale n-best lists for 2-iteration. (Accuracy %)

n	Chongqing	Guangzhou	Shanghai	Xiamen
0	64.51	66.01	74.43	66.33
10	66.72	68.75	76.27	68.75
100	67.56	69.74	76.94	69.84
1000	68.24	70.36	77.31	70.33

- With the increasing of n, the performance gets better.
- The best performance achieved when n is set to 1000.

- When n is set to 10 and 100, all four accents perform better in 2-iteration adaptation than 1-interation. When $n = 1000$, except for Chongqing accent, other accents perform better.

In summary, bigger n should be chosen to achieve better performance.

Time Consumption with Different Scale of n. Table 6 shows the time consumption of rescore with different n. The second column is the size of test set. The next three columns are time consumption for n = 10, 100 and 1000 respectively. When n increased from 10 to 100, rescore time increased by less than 2 times, but when n increased to 1000, rescore time increased more than 14 times.

Table 6. Time consumption with different n for rescore.

Accent(test)	Size(h)	n = 10(h)	n = 100(h)	n = 1000(h)
Chongqing	1.29	0.41	0.97	6.61
Guangzhou	1.56	0.44	1.16	7.19
Shanghai	1.40	0.40	0.98	6.80
Xiamen	1.18	0.41	0.88	5.67
Total	5.43	1.66	3.99	26.27

To balance the time consumption and accuracy, 100 is a good choice in practical system.

Different Iterations for Adaptation. We use fusion model and adapts the RNN based language model to rescore lattices for different accents of Mandarin speech. For model adaptation, we first compare the influence of different adapt-iterations.

Figure 3 shows adapted 1-iteration can get obvious improvement than baseline (without adaptation). However, 2-iterations adaptation gets slight improvement than 1-iteration. What is more, for Chongqing accented Mandarin, the accuracy becomes worse using 2-iterations adaptation. We can conclude that 1-interation adaptation is enough to achieve good performance and also can avoid overfitting.

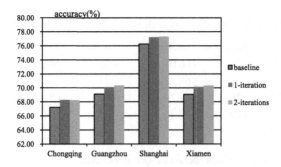

Fig. 3. Performance of different iterations.

After RNN adapted for 1-iteration, we use the trained model to compute accuracy for all four kinds of accent speeches in the test set. Figure 4 shows the accuracy for four different accent speech with RNN LM weight from 0.0 to 1.0. In addition, the experiment is made under the condition that n is set to 1000 in n-best lists.

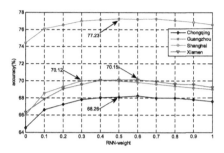

Fig. 4. Performance of RNN LM with one-iteration adaptation.

We can observe that with the RNN LM, the accuracy is higher than without RNN LM for all four accent-speeches. We can conclude that using adapted RNN LM can obviously improve the performance of accented Mandarin speech recognition task, more detail it achieves biggest improvement for 12.09 % on Guangzhou accented speech and the least on Chongqing for 10.57 % relative improvement on word error rate than N-gram based system. However, for RNN LM based system, we achieved at most 4.12 % relative improvement, which is on Shanghai speech. The lowest improvement has only 3.23 % on Chongqing speech. One possible reason is the utterances used to adaptation not enough. We have only about 5600 utterances available for each accented speech. In addition, Chongqing accented speech is similar to standard Mandarin. The improvement of Chongqing accented speech is not obviously than others.

5   Conclusions

We have presented an effective way to improve the performance of multi-accent Mandarin speech recognition system. In our proposed method, we combine N-gram with RNN LM together and adapt the RNN based language model for different accents of Mandarin speech. The architecture of the RNN LM is accent-specific top layers with the shared hidden layer. The accent-specific top layers are used to adapt to different accents and the shared hidden layers stores history information. Experiment results show that proposed method can improve the performance of accented Mandarin speech recognition over the baseline systems.

For future works, we plan to use regularized adaptation methods to avoid overfitting, when the adaptation set is small. Moreover, we want to adopt LSTM-RNN based language model adaptation for other domain text resources (such as spontaneous and written speech) to improve the performance of ASR task.

Acknowledgements. This work is supported by the National Natural Science Foundation of China (NSFC) (No. 61425017, No. 61403386, No. 61305003, No. 61233009, No. 61273288).

References

1. Bengio, Y., Schwenk, H., Senécal, J.S., Morin, F., Gauvain, J.L.: Neural probabilistic language models. In: Jain, L.C., Holmes, D.E. (eds.) Innovations in Machine Learning, pp. 137–186. Springer, Heidelberg (2006)
2. Schwenk, H.: Continuous space language models. Comput. Speech Lang. **21**(3), 492–518 (2007)
3. Mikolov, T., Karafiát, M., Burget, L., Cernocký, J., Khudanpur, S.: Recurrent neural network based language model. In: The Proceedings of Interspeech (2010)
4. Mikolov, T., Kombrink, S., Burget, L., Černocký, J.H., Khudanpur, S.: Extensions of recurrent neural network language model. In: 2011 IEEE International Conference on Acoustics, Speech and Signal Processing (ICASSP), pp. 5528–5531. IEEE, May 2011
5. Stefan, K., Tomas, M., Martin, K., Lukas, B.: Recurrent neural network based language modeling in meeting recognition. In: Proceedings of the Annual Conference of the International Speech Communication Association (2011)
6. Park, J., Liu, X., Gales, M.J., Woodland, P.C.: Improved neural network based language modelling and adaptation. In: Interspeech, pp. 1041–1044, September 2010
7. Alumäe, T.: Multi-domain neural network language model. In: Interspeech, pp. 2182–2186
8. Tilk, O., Alumäe, T.: Multi-domain recurrent neural network language model for medical speech recognition. In: Baltic HLT, pp. 149–152, September 2014
9. Chen, X., Tan, T., Liu, X., Lanchantin, P., Wan, M., Gales, M.J., Woodland, P.C.: Recurrent neural network language model adaptation for multi-genre broadcast speech recognition. In: Proceedings of ISCA Interspeech, Dresden, Germany, pp. 3511–3515 (2015)
10. Caruana, R.: Multitask Learn. Mach. Learn. **28**(1), 41–75 (1997)
11. Huang, J.T., Li, J., Yu, D., Deng, L., Gong, Y.: Cross-language knowledge transfer using multilingual deep neural network with shared hidden layers. In: 2013 IEEE International Conference on Acoustics, Speech and Signal Processing (ICASSP), pp. 7304–7308. IEEE, May 2013
12. Boden, M.: A guide to recurrent neural networks and back-propagation. The Dallas project, SICS Technical report (2002)
13. Bengio, Y., Simard, P., Frasconi, P.: Learning long-term dependencies with gradient descent is difficult. IEEE Trans. Neural Netw. **5**(2), 157–166 (1994)
14. Chen, S.F., Goodman, J.: An empirical study of smoothing techniques for language modeling. In: Proceedings of the 34th annual meeting on Association for Computational Linguistics, pp. 310–318. Association for Computational Linguistics, June 1996
15. Miao, Y., Gowayyed, M., Metze, F.: EESEN: end-to-end speech recognition using deep RNN models and WFST-based decoding. arXiv preprint arXiv:1507.08240 (2015)
16. Stolcke, A.: SRILM-an extensible language modeling toolkit. In: Interspeech, vol. 2002, p. 2002, September 2002
17. RASC863: 863 annotated 4 regional accent speech corpus, Chinese Academy of Social Sciences (2003). http://www.chineseldc.org/doc/CLDC-SPC-2004-005/intro.htm
18. Graves, A., Jaitly, N., Mohamed, A.R.: Hybrid speech recognition with deep bidirectional LSTM. In: 2013 IEEE Workshop on Automatic Speech Recognition and Understanding (ASRU), pp. 273–278. IEEE, December 2013

19. Graves, A., Jaitly, N.: Towards end-to-end speech recognition with recurrent neural networks. In: Proceedings of the 31st International Conference on Machine Learning (ICML 2014), pp. 1764–1772 (2014)
20. Graves, A., Fernández, S., Gomez, F., Schmidhuber, J.: Connectionist temporal classification: labelling unsegmented sequence data with recurrent neural networks. In: Proceedings of the 23rd International Conference on Machine Learning, pp. 369–376. ACM, June 2006
21. Li, J., Zhang, H., Cai, X., Xu, B.: Towards end-to-end speech recognition for chinese mandarin using long short-term memory recurrent neural networks. In: Sixteenth Annual Conference of the International Speech Communication Association (2015)

Emotion Recognition

Audio-Video Based Multimodal Emotion Recognition Using SVMs and Deep Learning

Bo Sun[1], Qihua Xu[1,2], Jun He[1(✉)], Lejun Yu[1], Liandong Li[1], and Qinglan Wei[1]

[1] College of Information Science and Technology, Beijing Normal University,
Beijing, People's Republic of China
{tosunbo,hejun,yulejun}@bnu.edu.cn, bnulee@hotmail.com,
qlwei@mail.bnu.edu.cn
[2] School of Business, Northwest Normal University, Lanzhou, People's Republic of China

Abstract. In this paper, we explored a multi-feature based classification framework for the Multimodal Emotion Recognition Challenge, which is part of the Chinese Conference on Pattern Recognition (CCPR 2016). The task of the challenge is to recognize one of eight facial emotions in short video segments extracted from Chinese films, TV plays and talk shows. In our framework, both traditional methods and Deep Convolutional Neural Network (DCNN) methods are used to extract various features. With different features, different classifiers are trained to predict video emotion labels. Moreover, a decision-level fusion method is explored to aggregate these different prediction results. According to the results on the competition database, our method shows better effectiveness on Chinese facial emotion.

Keywords: Emotion recognition · Spatio-temporal information · Deep learning · Decision-level fusion · Deep convolutional neural network

1 Introduction

With the development of multimedia technology and the wide application of video surveillance, video-based emotion recognition has been an active research topic in some fields such as human computer interaction, machine learning, pattern recognition, etc. At the early stage, researchers focused mostly on emotion analysis from single static facial images under constrained circumstances [1], which may not satisfy the demands of real life. As human emotional expression is a continuous process, the research should turn to identify the human emotion from video. Given a video with a human subject, facial expressions, movements and activities [2] are all beneficial for a computer understanding his/her emotion. In this paper we present a synthetic classification framework, with which different modal features are utilized and their prediction label are fused.

The Multimodal Emotion Recognition Challenge in the 2016 Chinese Conference on Pattern Recognition (CCPR 2016) [3] is an audio-video based emotion classification competition, whose goal is to compare multimedia processing and machine learning methods for multimodal emotion recognition. The official data used in this challenge is the Chinese Natural Audio-Visual Emotion Database (CHEAVD) [4], which contains

© Springer Nature Singapore Pte Ltd. 2016
T. Tan et al. (Eds.): CCPR 2016, Part II, CCIS 663, pp. 621–631, 2016.
DOI: 10.1007/978-981-10-3005-5_51

141 min spontaneous emotional segments extracted from 238 speakers from films, TV plays and talk shows. The video segments are lasting approximately 1–19 s and also feature an audio track, which may contain speech and background music. The discrete emotion labels include eight categories, namely Happy, Angry, Surprise, Disgust, Neutral, Worried, Anxious and Sad. The whole emotion database contains 2852 video segments, among which, 1981 for training, 243 for validation, and 628 for testing. All video labels for the training and validation set are provided.

Observing the database, we found it is much more complex than acted emotional database. In addition, according to the labels, the eight emotion categories are not well balance distributed on the training set and validation set. For the video emotion classification, not only the spatial information of features, but also the temporal information of features needs to be considered. Therefore, several strategies are used to extract various features, including: (1) an open-source toolkit openSMILE [5] was used to extract audio features, (2) a traditional approach was utilized to extract Local Binary Patterns from Three Orthogonal Planes (LBP-TOP) [6] features, (3) a deep convolutional neural network (DCNN) model was trained to extract visual features in single frames, and (4) a Long-term Recurrent Convolutional Network (LRCN) [7] model was trained to learns spatio-temporal features.

The remainder of the paper is organized as follows: Sect. 2 introduces the related literature. Section 3 gives our emotion recognition approach, including the various feature extraction methods, classifiers and details of model structures. In Sect. 4, we introduced the entire experiments we have done, in which the parameters settings and the recognition evaluation accuracy of the sub challenges are available. Then the final conclusion is given in Sect. 5.

2   Related Work

Emotion recognition has attracted growing attention. This research includes several important aspects, namely face detection, facial feature extraction and classification results fusion. Since Viola et al. [8] proposed the boosted cascade of simple features, the Viola-Jones face detector using Haar feature has become the most popular and effective tool for face detection. However, this detector is relatively weak and it cannot meet the more complex environment. In recent years, a number of CNN methods [9, 10] have been proposed to automatically learn advanced feature. In uncontrolled environments, these CNN-based face detectors have higher detection performance and the ability to discriminate.

CNN-based methods not only can be used to detect human face, but also can be used to learn the facial features. Kahou et al. [2] presented an application of Recurrent Neural Networks (RNNs) for modelling spatio-temporal information via aggregation of facial features to perform emotion recognition in the case of video. This method was the winning submission in the 2014 EmotiW challenge. Kim et al. [11] proposed a framework based on committee machines of deep CNNs with Exponentially-Weighted Decision Fusion for Static Facial Expression Recognition in the wild (SFEW). This approach was demonstrated on the SFEW competition data and achieved a high recognition accuracy outperforming the baseline.

Our previous work [12, 13] has achieved good performance in the emotion recognition challenge. In [12], we extracted various features including SIFT, LBP-TOP, PHOG, LPQ-TOP and audio feature, and proposed a novel hierarchical classifier fusion method to combine different classification results based on these features. In [13], We explored another decision-level fusion network for multimodal features. The network is evaluated on EmotiW 2015 AFEW and SFEW datasets and gain very promising achievement on the testing sets.

In this paper, for the specified challenge data, we explored a multi-feature classification framework for video emotion classification in the complex environment. In this framework, both traditional methods and deep convolutional neural network methods are used to extract various features, such as LBPTOP feature, DCNN feature and LRCN feature. Different features contain specific information and have different abilities to distinguish. Corresponding to these features, different classifiers are trained to predict video labels. In order to improve the recognition performance, we combine these different prediction results using a decision-level fusion network proposed in [13].

3 Our Approach

Considering the uneven distribution of the data set, we propose a multi-feature classification framework, which have the ability to take the advantage of both traditional hand-crafted feature extraction methods and unsupervised feature learning methods. The whole framework is shown in Fig. 1.

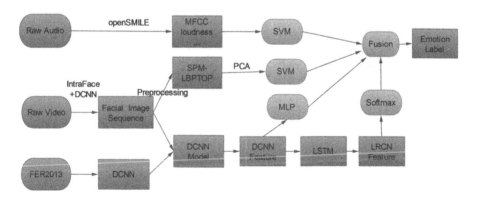

Fig. 1. Pipeline of our framework for emotion recognition

3.1 Datasets and Preprocessing

The Multimodal Emotion Recognition Challenge of CCPR 2016 is evaluated on the dataset CHEAVD, a Chinese natural emotional audio-visual database published by the National Laboratory of Pattern Recognition Institute of Automation at Chinese

Academy of Sciences, which have been annotated for different emotions. In our frame-work, the facial image were extracted by IntraFace toolkit [14], where the OpevCV's Viola & Jones face detector is applied for face detection and initialization of the Intraface tracking library. The 49 facial landmarks generated by Intraface are used for face aligning and the size of each facial image is set to 100×100 pixels. When the facial image extraction failed by the Intraface, another Deep Convolutional Neural Networks (DCNN) method [15] used to replace it, so that we can always get a number of aligned facial images from a given video segments.

In addition to official video dataset, we also used an additional dataset to train a deep convolutional neural network model for frame-based facial expression feature extrac-tion. The additional dataset is a large static facial image dataset named Facial Expression Recognition (FER) [16], which contains 35,887 images with seven facial expression categories: Happy, Angry, Surprise, Disgust, Neutral, Fear and Sad. In order to maintain consistent emotional categories with the challenge data, we weeded out the images of Fear class from the FER database, and added some images of Worried class and Anxiety class which harvested from Google image search then cleaned and labeled by hand. Finally, we obtained an improved additional database.

Considering the difference between the two datasets, we performed pre-processing including min-max normalization and histogram equalization on each facial image to make the datasets compatible. See Fig. 2 for an example.

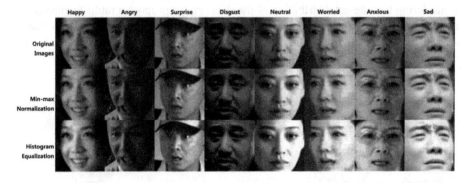

Fig. 2. Different types of facial expression images and the corresponding preprocessed images for the video and multimodal sub-challenges

3.2   Multimodal Features

Audio Feature. The audio features are extracted by the openSMILE toolkit with the extended Geneva Minimalistic Acoustic Parameter Set (eGeMAPS) [17]. The open-SMILE, an open-source Speech & Music Interpretation by Large Space Extraction, is a fast and flexible audio feature extractor. It can extract different types of low-level descriptors (LLD), and apply various functionals to these descriptors via a text-based feature set. The feature set eGeMAPS consists of Minimalistic Parameter Set and Extended Minimalistic Parameter Set. The former contains 18 LLDs, a total of 62 parameters, and the latter contains 7 LLDs, a total of 26 parameters. In total, the eGeMAPS contains 88 parameters.

Thus, for each video, we get a 88-dimensional audio feature, followed by a linear SVM [18] for classification. The hyper-parameters for the SVM are set via iteration search.

SPM LBP-TOP Feature. The Local Binary Patterns from Three Orthogonal Planes (LBP-TOP) [6] is a dynamic texture descriptors which extracts co-occurrence features by concatenating LBP histograms computed from three orthogonal planes: XY, XT, and YT. The XY plane provides spatial related texture information, and others provide temporal related texture information. Experiments show that it has a high recognition performance for acted facial expression. In our framework, we combine Spatial Pyramid Matching [19] with LBP-TOP descriptors to extract dynamic texture feature, in which the number of pyramids was set to four. Then, the final SMP LBP-TOP feature is a 15045-dimensional vector. Principal Component Analysis (PCA) was applied on the extracted feature for dimensionality reduction. Similarly, linear SVM was used for SPM LBP-TOP feature classification.

DCNN Feature. In addition to the traditional hand-crafted feature, we also automatically extract features by deep learning. We designed a DCNN architecture to train models on the improved FER dataset. In this architecture, a data layer takes batches of pre-processed facial images as input and performs a random cropping into smaller 42×42 sub-images, then five stages with different layers were used to extract DCNN feature. The odd stages include a convolutional layer, a batch normalization layer, a rectified linear unit (ReLU) activation function layer and a max pooling layer. Compared with the odd stages, there is no pooling layer in the even stages. All convolutional layers use the Gaussian model with the same parameters for the weight initialization. The outputs of the last stage are fed to three fully-connected layers. The first two fully-connected layers have 1024 and 512 hidden units, respectively, and both followed by a ReLU layer and a dropout layer with 0.5 dropout rate. The last fully-connected layer has eight hidden units which is equal to the number of emotion classes. The final layer of this architecture is a softmax layer which output eight probabilities. The whole DCNN architecture is presented in Fig. 3.

Training was implemented using Convolutional Architecture for Fast Feature Embedding (CAFFE) [20], a deep learning framework developed by the Berkeley Vision and Learning Center (BVLC). During total 100 epochs (44900 iterations), we selected a model yielding the max validation performance as the best DCNN model.

Since video segments may not have the same number of frames, we count the number of each video frame, and then calculate the average. If the number of video frame is less than the average, we simply repeated frames. Conversely, if the number of video frame is greater than the average, we remove some redundant frames according to the correlation between the frames. By processing, we get a fixed number of frames for each video. Then, facial images are detected and cropped from the frame sequence. Here, the pre-softmax output was extracted as per-frame DCNN feature. After concatenating all frame-based features, we yield a fixed length vector as video representation. Finally, the

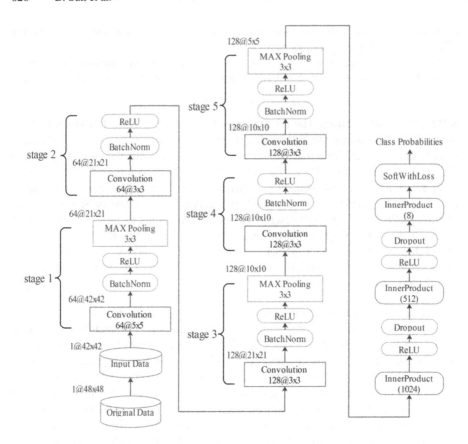

Fig. 3. The DCNN architecture for training model

vector representations of videos together with corresponding emotion labels are used to train a multilayer perceptron (MLP), which acts as a classifier. The MLP is composed of two hidden layers and one softmax layer. The hidden layers with sigmoid activation function have 128 and 8 hidden units, respectively. The softmax layer has an 8-class probability output.

LRCN Feature. This DCNN feature does not take into account the temporal information of image sequence. Therefore, we adopted a Long-term Recurrent Convolutional Network (LRCN) to extract another spatio-temporal feature. LRCN consisting of DCNN and Long Short-Term Memory (LSTM) [21] is the state-of-art model for sequence analysis since it uses memory cells to store information so that it can exploit long range dependencies in the data [22]. We designed an LRCN architecture consisting of two stages: feature extraction stage and emotion classification stage. The first stage includes an input layer and a LSTM layer. The input layer feeds the DCNN feature orderly to the LSTM layer, where each time-step yields an output. After that, three types of results can be used as a video feature: concatenating the output sequence, the output of the last time-step and the average of output sequence. Through experimental contrast, the last time-step encoding has the best classification performance. The second stage includes a fully-connected layer and a softmax layer. The fully-connected layer has the effect of dimension reduction and the softmax yields an 8-dimensional probability vector. The architecture is depicted in Fig. 4.

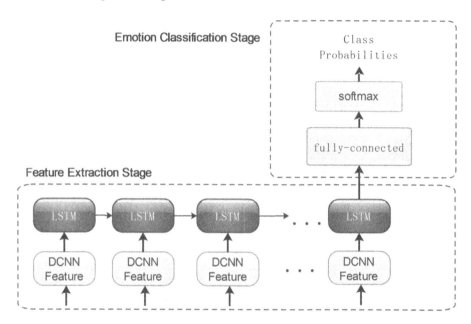

Fig. 4. The LRCN architecture for emotion classification of video and multimodal sub-challenges

We used Keras [23], a minimalist and highly modular neural networks library to achieve this process. During total 60 epochs, we adopted Adam optimizer [24] with default parameters and a batchsize of 64 sequences. Early-stopping method was used based on the challenge validation set to further avoid overfitting.

3.3 Classification Results Fusion

For each video, different features extracted by different methods corresponding to different classification results. The each classification result of each video segment is an 8-class probability vector, so we can easily aggregate all of the classification results. Utilizing a decision-level fusion method to aggregate these different results, we can get higher classification accuracy. In our classification framework, we explored two kinds of fusion methods: a majority vote algorithm and a weight shared fusion network proposed in [13]. Experiments show that the classification performance of the latter is better than the former.

4   Experiments

We extracted different types of features by various methods introduced in Sect. 4. The audio feature and SMP LBP-TOP feature can be extracted directly, while the DCNN feature and the LRCN feature require a pre-training DCNN model. We trained the model on the improved FER dataset with Stochastic Gradient Descent (SGD) and a learning rate of 0.001. During the entire training, the training loss curve and validation accuracy curve are shown in Fig. 5. Validation accuracy reached the maximum value at the 88th epoch. So we select this epoch model as the final DCNN model to extract feature.

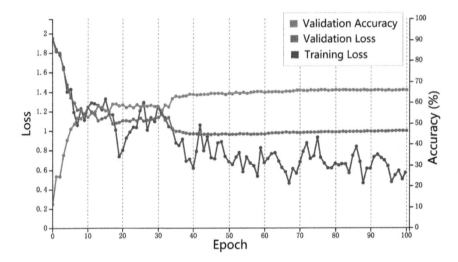

Fig. 5. The training loss curve and validation accuracy curve (Color figure online)

As classes are unbalanced, this challenge select macro average precision (MAP) as the primary measure to rank the results, and secondly the accuracy (ACC) [3]. Calculating formulas of macro average precision and accuracy are given in Eqs. (1)–(3), respectively.

$$macro\ average\ precision(MAP) = \frac{1}{s} \times \sum_{i=1}^{s} P_i, \tag{1}$$

$$P_i = \frac{TP_i}{TP_i + FP_i}, \tag{2}$$

$$accuracy = \frac{\sum_{i=1}^{s} TP_i}{\sum_{i=1}^{s} (TP_i + FP_i)}, \tag{3}$$

where s denotes the number of emotion categories. P_i is the precision of the ith emotion class. TP_i and FP_i denote the number of correct classification and the number of wrong classification in the ith emotion class, respectively.

A comparison of prediction accuracy and macro average precision (MAP) on validation set with different classifiers is shown in Table 1.

Table 1. Validation accuracy and macro average precision for all features

Features	Classifiers	Train ACC (%)	Val ACC (%)	MAP (%)
Audio feature	Linear SVM	48.86	38.27	16.84
SPM LBP-TOP	Linear SVM	84.51	42.39	21.32
DCNN feature	MLP	72.36	42.80	20.87
LRCN feature	Softmax	53.23	48.97	24.50
Majority vote	–	–	49.34	24.62
Fusion network	–	–	51.85	25.55

From Table 1 we clearly see the deep learning features are superior to traditional hand-crafted features. Audio feature achieved 38.27 % classification accuracy and 16.84 % macro average precision, which has the worst recognition rate. LRCN feature achieved 48.97 % classification accuracy and 24.50 % macro average precision, significantly higher than the SPM LBP-TOP feature. Applying two different fusion methods to aggregate these features, we get the best multimodal classification accuracy of 51.85 % and macro average precision of 25.55 %. The confusion matrix of the best result on the validation set is showed in Table 2. The result demonstrated that uneven distribution of training data will have a strong impact on the final recognition result. On the training set, most of samples belong to neutral category. Therefore, most the emotion prediction results also tend to neutral category. Other categories lack of training samples, so it's difficult to identify.

Table 2. Confusion matrix of the best prediction result on validation set

		Actual Classes							
		Hap.	Ang.	Sur.	Dis.	Neu.	Wor.	Anx.	Sad
Predict Classes	Hap.	23	1	1	0	4	1	2	5
	Ang.	1	18	1	1	4	0	3	3
	Sur.	0	0	0	0	0	0	0	0
	Dis.	0	0	0	0	0	0	0	0
	Neu.	14	25	5	4	79	9	5	17
	Wor.	0	0	0	0	0	0	0	0
	Anx.	0	0	0	0	0	0	0	0
	Sad	0	5	1	0	3	1	1	6

In our five submissions, we picked the best macro average precision of each sub-challenge and listed them in Table 3. The confusion matrix of the multimodal sub-challenge prediction result is shown in Table 4.

Table 3. Test accuracy (ACC) and macro average precision (MAP) for three sub-challenges

Sub-challenges	ACC (%)	MAP (%)
Audio	27.87	26.33
Video	24.68	28.35
Multimodal	19.90	25.35

Table 4. Confusion matrix of the multimodal sub-challenge prediction result on test set

		Actual Classes							
		Hap.	Ang.	Sur.	Dis.	Neu.	Wor.	Anx.	Sad
Predict Classes	Hap.	32	9	12	5	34	10	13	8
	Ang.	3	11	5	10	20	16	22	4
	Sur.	0	1	4	0	1	0	1	0
	Dis.	0	0	0	1	1	0	0	0
	Neu.	19	22	18	15	63	42	40	45
	Wor.	1	13	9	5	11	7	13	11
	Anx.	0	4	0	0	0	1	1	1
	Sad	1	13	3	6	12	10	13	6

5   Conclusions

In this paper, we explored a multi-feature based classification framework for the Multi-modal Emotion Recognition Challenge of CCPR 2016. In this framework, both traditional methods and deep convolutional neural network methods are used to extract various features, such as SPM LBP-TOP feature, DCNN feature and LRCN feature. For each feature, we trained a classifier which has different discriminative ability for emotions classification. In order to improve the recognition performance, we aggregate different classification results using a decision-level fusion method. The framework is evaluated on the challenge dataset and gain very promising achievement. In future work, we will further seek methods to resolve the difficulties of emotion recognition due to the uneven distribution of samples.

Acknowledgements. This work is supported by the National Education Science Twelfth Five-Year Plan Key Issues of the Ministry of Education (DCA140229).

References

1. Pantic, M., Rothkrantz, L.J.M.: Automatic analysis of facial expressions: the state of the art. IEEE Trans. Pattern Anal. Mach. Intell. **22**(12), 1424–1445 (2000)

2. Ebrahimi Kahou, S., et al.: Recurrent neural networks for emotion recognition in video. In: ACM International Conference on Multimodal Interaction (2015)
3. Li, Y., et al.: MEC 2016: the multimodal emotion recognition challenge of CCPR 2016. In: Chinese Conference on Pattern Recognition (CCPR), Chengdu, China (2016)
4. Bao, W., et al.: Building a Chinese natural emotional audio-visual database. In: International Conference on Signal Processing (2015)
5. Eyben, F., et al.: Recent developments in openSMILE, the munich open-source multimedia feature extractor. In: ACM International Conference on Multimedia (2013)
6. Zhao, G., Pietikinen, M.: Dynamic texture recognition using local binary patterns with an application to facial expressions. IEEE Trans. Pattern Anal. Mach. Intell. **29**(6), 915–928 (2007)
7. Donahue, J., et al.: Long-term recurrent convolutional networks for visual recognition and description. In: Proceedings of the IEEE Conference on Computer Vision and Pattern Recognition (2015)
8. Viola, P., Jones, M.: Rapid object detection using a boosted cascade of simple features. Proc. CVPR **1**, 511 (2001)
9. Li, H., et al.: A convolutional neural network cascade for face detection. In: Computer Vision and Pattern Recognition (2015)
10. Szarvas, M., et al.: Multi-view face detection using deep convolutional neural networks. In: Proceedings of Intelligent Vehicles Symposium. IEEE (2015)
11. Kim, B.K., et al.: Hierarchical committee of deep CNNs with exponentially-weighted decision fusion for static facial expression recognition. In: ACM on International Conference on Multimodal Interaction (2015)
12. Sun, B., et al.: Combining multimodal features with hierarchical classifier fusion for emotion recognition in the wild. In: The International Conference (2014)
13. Sun, B., et al.: Combining multimodal features within a fusion network for emotion recognition in the wild. In: ACM on International Conference on Multimodal Interaction (2015)
14. Xiong, X., Torre, F.D.L.: Supervised descent method and its applications to face alignment. In: IEEE Conference on Computer Vision & Pattern Recognition (2013)
15. Farfade, S.S., Saberian, M.J., Li, L.J.: Multi-view face detection using deep convolutional neural networks. In: Proceedings of Intelligent Vehicles Symposium. IEEE (2015)
16. Carrier, P., et al.: FER-2013 face database. Technical report, 1365, Université de Montréal (2013)
17. Eyben, F., et al.: The geneva minimalistic acoustic parameter set (GeMAPS) for voice research and affective computing. IEEE Trans. Affect. Comput. **12**(2), 190–202 (2016)
18. Fan, R.E., et al.: LIBLINEAR: a library for large linear classification. J. Mach. Learn. Res. **9**(12), 1871–1874 (2010)
19. Lazebnik, S., Schmid, C., Ponce, J.: Beyond bags of features: spatial pyramid matching for recognizing natural scene categories. In: IEEE Computer Society Conference on Computer Vision and Pattern Recognition (2006)
20. Jia, Y., et al.: Caffe: convolutional architecture for fast feature embedding. Eprint Arxiv, pp. 675–678 (2014)
21. Hochreiter, S., Schmidhuber, J.: Long short-term memory. Neural Comput. **9**(8), 1735–1780 (1997)
22. Chen, S., Jin, Q.: Multi-modal dimensional emotion recognition using recurrent neural networks. In: International Workshop on Audio/visual Emotion Challenge (2015)
23. Chollet (2015). GitHub. https://github.com/fchollet/keras. Accessed 10 June 2016
24. Kingma, D., Ba, J.: Adam: a method for stochastic optimization. Computer Science (2015)

Emotion Recognition in Videos via Fusing Multimodal Features

Shizhe Chen[1], Yujie Dian[1], Xinrui Li[1], Xiaozhu Lin[1],
Qin Jin[1(✉)], Haibo Liu[2], and Li Lu[2]

[1] Multimedia Computing Laboratory, School of Information,
Renmin University of China, Beijing, People's Republic of China
{cszhe1,dianyujie-blair,lialialia,infinity0a,qjin}@ruc.edu.cn
[2] Tencent Inc., Beijing, People's Republic of China
{geneliu,adolphlu}@tencent.com

Abstract. Emotion recognition is a challenging task with a wide range of applications. In this paper, we present our system in the CCPR 2016 multimodal emotion recognition challenge. Multimodal features from acoustic signals, facial expressions and speech contents are extracted to recognize the emotion of the character in the video. Among them the facial CNN feature is the most discriminative feature for emotion recognition. We train SVM and random forest classifiers based on each type of features and utilize early and late fusion to combine the different modality features. To deal with the data unbalance issue, we propose to adapt the probability thresholds for each emotion class. The macro precision of our best multimodal fusion system achieves 50.34 % on the testing set, which significantly outperforms the baseline of 30.63 %.

Keywords: Emotion recognition · Multimodal features fusion · CNN Features

1 Introduction

Automatic video emotion recognition is a challenging task, which has a wide range of applications such as human-computer interaction, e-learning and mental health care.

Numerous emotional datasets have been collected to promote the emotion recognition research. However, most of the past datasets such as Berlin speech dataset [1] and CK+ facial expression dataset [2] fall short in three aspects. Firstly, the data is collected in a constrained experimental environment and can not well reflect natural emotions expressed in real life. Secondly most of the datasets only contain single modality like speeches or faces. Also, the size of the dataset is quite small. Recent studies have addressed on these issues and collected several enlarged natural multimodal emotional datasets. These datasets are usually used in competitions to improve the standards in the field. The AVEC 2015 emotional challenge [3] provides audio, video and physiologic data collected in

© Springer Nature Singapore Pte Ltd. 2016
T. Tan et al. (Eds.): CCPR 2016, Part II, CCIS 663, pp. 632–644, 2016.
DOI: 10.1007/978-981-10-3005-5_52

human remote interactions to predict continuous dimensional affects. EmotiW challenges [4] focus on discrete emotion recognition in the wild using emotional movie clips. In this paper, we present our methods on the Chinese Natural Emotional Audio-Visual Database (CHEAVD) [5] in the CCPR Multimodal Emotion Challenge (MEC 2016).

The task of MEC 2016 is to predict the emotion of each video clip in the dataset. The video clips are extracted from Chinese movies, soap operas and TV shows. The difficulties of the task result from the wide variety of changes in the videos such as context scene, subjects, poses, illuminations and so on. So the fusion of multimodal features can comprehensively capture the emotional information from different aspects and thus would be more robust to predict the emotion labels.

In this task, we extract various multimodal features from aural, visual and speech content. Acoustic features include statistical acoustic features and MFCC Bag-of-Audio-Words. The facial expression features contain hand-crafted dense SIFT features and deep convolution neural network (CNN) features. The pretrained word vectors and character vectors are used as the text features to represent the semantic meaning of the speech content. SVM and random forest are used as classifiers for each unimodal feature. The facial CNN feature is the most discriminative feature in all extracted features. Early and late fusions are applied over different combinations of these features. To deal with the unbalance problem in the dataset, we also post-process the predicted labels by adjusting probability thresholds of the emotions. The framework of the proposed method is shown in Fig. 1. The best performances (macro precision of 50.34 %) on the testing dataset is achieved by late fusion of features from the three modalities, which shows the importance of the fusion of multi-modalities.

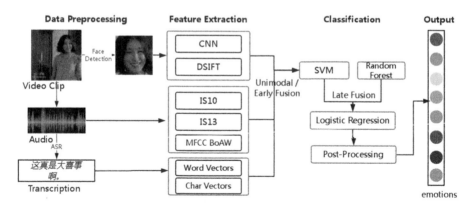

Fig. 1. The framework of multimodal feature fusion

The paper is organized as follows. Section 2 introduces some related work. Section 3 describes all the multi-modal features we extract in the challenge.

Section 4 presents our experimental results. Section 5 concludes the paper and presents our future work.

2   Related Work

In the literature of emotion recognition, previous works have explored various features from aural, facial and text modalities.

For acoustic features, Chen et al. [6] applied Bag-of-Audio-Words and GMM supervectors over low-level acoustic features to generate sentence-level features. Xia et al. [7] proposed a modified Denoising Autoencoder (DAE) framework which considered gender information to better capture emotion specific features. Deng et al. [8] addressed the cross domain emotion recognition and trained autoencoder with different decoder layers for different domains to extract robust features across speech domain. Huang et al. [9] learned emotion-salient features using CNN with novel loss functions.

For facial features, LBP-TOP, Dense SIFT, HOG and various hand-crafted features are wildly used for facial emotion recognition [10,11]. Yao et al. [12] extracted AU-aware features for each pair of emotions and encoded the facial feature relations with graph structure to represent discriminative facial features. Kim et al. [13] proposed an unsupervised segmentation method to reduce speech-related variability for temporal facial emotion recognition. Jung et al. [14] trained two networks from facial features. The first deep network extracts temporal appearance features from image sequences, while the other deep network extracts temporal geometry features from temporal facial landmark points. They jointly finetuned the two networks and achieved great improvement.

For text features, the Bag-of-Words representation is one of the most widely used features. It processes texts without considering the word order, the semantic structure or the grammar. Another commonly used feature is term frequency-inverse document frequency (TF-IDF), which can evaluate the importance of words. Word vectors are the hottest representations in recent studies [15]. It projects words into a high dimensional vector space, which can capture the semantic meanings of each word.

Fusion strategies for different modalities in previous works can be divided into 3 categories, namely feature-level (early) fusion, decision-level (late) fusion and model-level fusion [16]. Early fusion uses the concatenated features from different modalities as input features for classifiers. It has been widely used in the literature to successfully improve performance [17]. However, it suffers from the curse of dimensionality. Late fusion eliminates some disadvantages of early fusion. It combines the predictions of different modalities and trains a second level model such as RVM [18]. But it ignores interactions and correlations between different modality features. Model-level fusion is a compromise between the two extremes. The implementation of model-level fusion depends on the specific classifiers. For example, for neural networks, model-level fusion could be concatenation of different hidden layers from different modalities [19]. For kernel classifiers, model-level fusion could be kernel fusion [20].

3 Multimodal Features

Emotions are conveyed through multi-modalities. In this section, we present all the multimodal features we extract from the videos including aural, visual and speech to represent the emotion from different aspects.

3.1 Aural Modality Features

Statistical Acoustic Features: Statistical acoustic features are proved to be very effective in audio emotion recognition. We use the open-source toolkit OpenSMILE [21] to extract two kinds of statistical acoustic features IS10 and IS13, which use the configuration in INTERSPEECH 2010 [22] and 2013 Paralinguistic challenge [23] respectively. Low-level acoustic features such as energy, pitch, jitter and shimmer are first extracted over a short-time window. And then statistical functions like mean, max are applied over the set of low-level features to get the fixed dimensional sentence-level features. The dimensionality of the IS10 feature is 1582 and that of IS13 is 6373.

MFCCBoAW: The Mel-frequency Cepstral Coefficients (MFCCs) [24] are used as the low-level feature and encoded into the sentence-level features using the Bag-of-Audio-Words (BoAW) [25] strategy. Firstly an acoustic codebook is generated by K-means clustering algorithm with $K = 1024$ on the training set. We then assign the MFCCs in one sentence to the discrete set of codewords in the codebook, thus providing a histogram of codewords counts. L1-norm is used to get the probability distributions on the codebook for each sentence.

3.2 Visual Modality Features

Face Pre-processing: We use Deformable Parts Model proposed in [26] to detect faces every 5 frames in the videos and then align faces with facial landmarks extracted by methods in [27,28]. Faces are scaled into the size of 128×128.

Dense SIFT: Scale Invariant Feature Transform (SIFT) [29] is wildly used as the image feature, which is invariant to image scale and rotation. In this task, we first extract dense SIFT descriptors in every 7×7 patch with one shift pixel. We encode the local features using the Bag-of-Feature (BoF) [30] framework. A codebook with 1024 codewords is generated by K-means clustering and then each local SIFT feature can be represented by the distributions on its top-10 nearest codewords. Then we use the Spatial Pyramid Matching (SPM) [31] method to encode the set of SIFT features in a image. To be specific, a image is partitioned into blocks with different levels (1×1, 2×2, 4×4). In each image block, the transformed SIFT BoF local features are applied with mean pooling to generate the fixed dimensional block features. The procedure is illustrated in Fig. 2. All features in each block are concatenated as the image feature. The video features are obtained using max pooling over all the facial images features, the dimensionality of which is 21504. We call this feature as DSIFT.

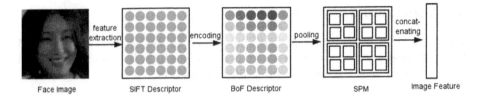

Fig. 2. The pipeline of DSIFT feature

CNN Feature: The convolution neural networks (CNN) [32] have shown the state-of-the-art performance in many visual tasks such as object detection, scene recognition etc. It benefits from its hierarchical network architecture and parameters sharing in different local patches. In this work, we build a CNN model on the Facial Expression Recognition dataset (FER2013) [33] which contains over 35,887 images with seven basic expressions: angry, disgust, fear, happy, sad, surprise and neutral. Although the label set is slightly different from the CCPR emotion set, we use the CNN mainly for feature extraction since CNN features are proved to have great generalization ability to similar tasks. The CNN architecture is shown in Fig. 3. We use 25,887 images from FER2013 as training set and the remained 1000 images are used as validation set to select the best model. Data augmentation and dropout are used to prevent overfitting. We extract the outputs of fc6 layer and apply mean pooling over all detected faces in a video as our CNN features.

3.3   Text Features

We use Baidu Speech Recognition API to get speech content transcriptions for each video. For clips without transcripts, we simply pad zeroes as the text feature. We first train a word2vec model [34] using Wikipedia Chinese corpus with genism [35]. Therefore, each character or word is represented by a dense semantic vector with dimensionality of 400. Then we average all the word vectors in one

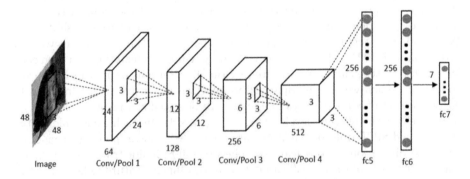

Fig. 3. The structure of the convolution neural network

Table 1. The distribution of emotion categories of training and validation sets (number of video clips)

	Ang	Anx	Dis	Wor	Sur	Neu	Hap	Sad	Total
Train	399	89	45	93	67	725	307	256	1981
Validation	49	11	5	11	8	90	38	31	243

transcript to represent the sentence-level text feature. Since the transcripts are quite short, we also apply mean pooling over character vectors in the sentence as the char vector feature.

4 Experiments

4.1 Dataset

The CCPR multimodal emotion dataset contains 2852 video clips from movies and TV shows, which are divided into 3 parts: 1981 for training, 243 for validation and 628 for testing. The task is to classify each video clip into one emotion label from eight possible emotion categories (including angry, anxious, disgust, surprise, worried, sad, happy and neutral). The distribution of emotion categories of training and validation sets are presented in Table 1. Since the data is quite unbalanced, macro precision is used as the evaluation metrics in the competition.

4.2 Experimental Setup

Following standard data split, we randomly select 15% of the training data as local validation to select best parameters for the classifiers. And we report experiment results on the original validation set (243 video clips) in the following results. Although macro precision is the evaluation metrics in the competition, we find that selecting models using macro F1 is more robust due to the relative small size of the local validation set. For example, if one model accidentally classified one and only one video clip to a certain correct category but is not as good as other models in overall performance, the macro precision of this model may still have a significant boost which makes the selection risky. We therefore use macro F1 to select the best model.

4.3 Unimodal Results

We use SVM [36] and random forest [37] as our unimodal classifiers. Hyper parameters of the models are selected according to macro F1 on our local validation set using grid search. For SVM, RBF kernel is applied and the cost is searched from 2^{-2} to 2^{10}. For Random Forest, the number of trees is set to be 500 and the depth of the tree is searched from 2 to 20.

Table 2. Unimodal performance on the validation set

Modality	Feature	SVM		Random forest	
		MAP	MaF1	MAP	MaF1
Visual	CNN	**41.14**	**29.86**	**27.71**	**23.68**
	DSIFT	17.35	17.67	15.16	14.30
Acoustic	IS10	23.72	22.45	24.51	22.05
	IS13	19.64	19.80	22.33	20.12
	mfccBoAW	20.91	21.10	20.51	18.07
Text	wordvec	19.41	21.02	17.02	17.66
	charvec	21.80	18.79	20.51	16.06

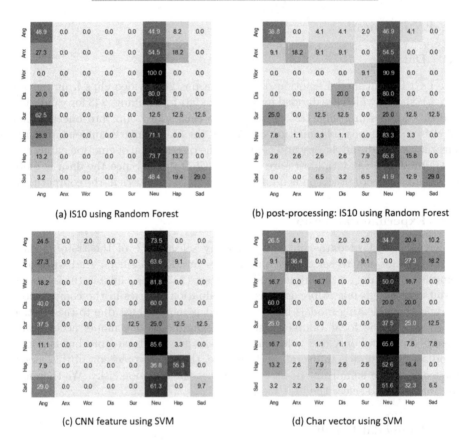

(a) IS10 using Random Forest

(b) post-processing: IS10 using Random Forest

(c) CNN feature using SVM

(d) Char vector using SVM

Fig. 4. Confusion matrix of unimodal features on the validation set

The performance of unimodal features on the validation set are listed in Table 2. As we can see from the table, the CNN feature significantly outperforms other unimodal features with the highest macro precision of 41.14 %, which

Table 3. Early fusion performance on the validation set

Modality	Feature	SVM		Random forest	
		MAP	MaF1	MAP	MaF1
In modality	IS10-mfccBoAW	21.13	20.63	24.02	21.57
	wordvec-charvec	20.55	19.28	25.02	17.08
Across modalities	IS10-CNN	30.43	28.27	30.54	27.04
	mfccBoAW-CNN	32.43	30.48	30.05	26.80
	IS13-DSIFT	27.24	25.58	26.06	22.67
	mfccBoAW-CNN-wordvec	**34.97**	**31.82**	31.68	27.66
	mfccBoAW-CNN-charvec	31.51	29.97	**40.22**	**33.41**

demonstrates the effectiveness and great generalization ability of the CNN model since the model is trained on a foreigners faces dataset with slight different emotion labels. DSIFT features achieves comparably lowest performance. This might relate to the high dimensionality of DSIFT which makes it easy to overfit. We also tried to apply PCA on DSIFT but did not observe performance improvements.

IS10 performs best among the acoustic features followed by mfccBoAW. The inferior performance of IS13 might also relate to its high dimensionality. For text features, the word vector features are slightly stable than char vectors with higher macro F1 but achieves lower macro precision. In general, visual modality performs the best and acoustic modality the second best, which matches with the observations in the literature. Text features are not comparable because of the inaccurate transcriptions and insufficient training data. Confusion matrices on the validation set of IS10 using random forest, CNN feature and char vector using SVM are presented in Figs. 4(a), (c) and (d) respectively. The acoustic IS10 feature can distinguish angry emotion best and the CNN feature can recognize happy emotion best. The text features can recognize some minority emotions well such as anxious and worried. Neutral emotion class achieves the highest recall for all modalities because of its dominant numbers in the dataset.

4.4 Multimodal Fusion Results

Since different features are complementary for emotion recognition, we explore early and late fusion methods to fully make use of all the multimodal features. We use '−' to denote the early fusion (feature concatenation) and '+' to denote the late fusion (probability concatenation).

For early fusion, we only fuse features with compatible dimensions. In Table 3, we can see that the early fusion of features from the same modality does not bring additional gain, which shows that the features from the same modality have less complementary characteristics. However, combining features from the three different modalities can boost the performance a lot and mfccBoAW-CNN-charvec

Table 4. Late fusion performance on the validation set

Feature	MAP	MaF1
mfccBoAW+CNN	27.31	27.10
mfccBoAW+CNN+(IS13-Dsift)	29.44	28.00
mfccBoAW+CNN+(IS13-Dsift)+wordvec	31.01	30.46
mfccBoAW+CNN+(IS13-Dsift)+charvec	29.86	29.06
IS10+CNN+(IS13-DSIFT)	41.55	33.05
IS10+CNN+(IS13-DSIFT)+wordvec	28.95	27.83
IS10+CNN+(IS13-DSIFT)+charvec	35.59	30.51
IS13+CNN	30.71	27.82
IS13+CNN+DSIFT	35.00	30.71
IS13+CNN+DSIFT+wordvec	**42.14**	**34.17**
IS13+CNN+DSIFT+charvec	30.29	29.12

Table 5. Best multimodal results on the validation and testing sets

	Val		Test	
	ACC	MAP	ACC	MAP
IS13+CNN+DSIFT+wordvec	53.90	42.14	31.85	50.34
Multi-modal baseline	16.05	13.40	21.18	30.63

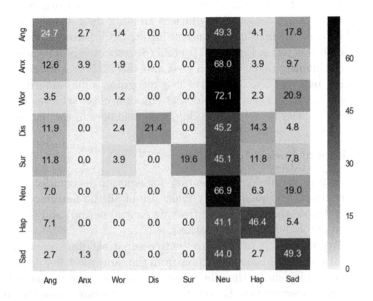

Fig. 5. Confusion matrix of the best submission on testing set

achieves macro F1 of 33.41 % which is 9.73 % higher than using CNN features (23.68 %) alone with random forest.

For late fusion, we use the logistic regression as the second level classifier in order to avoid overfitting on the local validation set. The probabilities of the SVM and random forest are both used as the late fusion features for each modality feature. The late fusion of features from the same modality also has no improvements which is similar to the early fusion. The performance of late fusion of different modalities are presented in Table 4. Late fusion can achieve better performance than early fusion in most combinations because early fusion usually suffers from the curse of dimensionality.

4.5 Prediction Post-processing

Since the dataset is quite unbalanced where only a few samples are annotated with anxious, disgust, worried and surprise emotions, the classifiers have a preference to predict the video sample with the dominant emotion labels. However, if the prediction probabilities of the model is accurate, the sample with top ranking probability of the minority emotion in the dataset is more likely to be classified correctly even if this probability is less than the probability of its majority emotion label. So we decide not to classify video samples into the label with maximum probability but to adjust the probability thresholds for each emotion classes. The probability thresholds are selected for each type of features according to the performance on the validation set. Figure 4(b) shows the confusion matrix of IS10 features using random forest, which greatly improve the performance by classify the minority emotions correctly with macro precision changing from 24.5 % to 37.6 % and macro F1 changing from 22.0 % to 30.5 %.

4.6 Challenge Submission

The best submission of our system in MEC 2016 is the late fusion of feature IS13, CNN, DSIFT and word vector with post-processing, which also achieves the best performance on the validation set. Table 5 shows the comparison of our system and the baseline multimodal system. Our emotion recognition system outperforms the baseline over 19.71 % on macro precision. The confusion matrix is provided in Fig. 5. Although most of the video samples are misclassified into the neutral emotion, happy and sad still achieve good performance and some minority classes like disgust and surprise even achieve 100 % precision.

5 Conclusions and Future Work

In this paper, we investigate various multimodal features from aural, face image and text modalities and two kinds of fusion strategies for the MEC 2016 multimodal emotion recognition challenge. Among all the extracted unimodal features, the facial CNN feature is the most discriminative features for emotion recognition. The late fusion of different modality features significantly boost the recognition performance. The promising result on testing set further shows the effectiveness of our proposed methods. In the future, we will continue to extract more discriminative features especially from the facial expression modality and utilize some personalized fusion methods. Also, we will explore modelling the temporal information for video emotion recognition.

Acknowledgments. This work is supported by National Key Research and Development Plan under Grant No. 2016YFB1001202. This work is also partially supported by Tencent Inc.

References

1. Burkhardt, F., Paeschke, A., Rolfes, M., Sendlmeier, W.F., Weiss, B.: A database of German emotional speech. Interspeech **5**, 1517–1520 (2005)
2. Lucey, P., Cohn, J.F., Kanade, T., Saragih, J., Ambadar, Z., Matthews, I.: The extended cohn-kanade dataset (ck+): a complete dataset for action unit and emotion-specified expression. In: 2010 IEEE Computer Society Conference on Computer Vision and Pattern Recognition-Workshops, pp. 94–101. IEEE (2010)
3. Ringeval, F., Schuller, B., Valstar, M., Jaiswal, S., Marchi, E., Lalanne, D., Cowie, R., Pantic, M.: AV+ EC 2015: the first affect recognition challenge bridging across audio, video, and physiological data. In: Proceedings of the 5th International Workshop on Audio/Visual Emotion Challenge, pp. 3–8. ACM (2015)
4. Abhinav Dhall, O.V., Murthy, R., Goecke, R., Joshi, J., Gedeon, T.: Video and image based emotion recognition challenges in the wild: Emotiw 2015. In: Proceedings of the 2015 ACM on International Conference on Multimodal Interaction, pp. 423–426. ACM (2015)
5. Li, Y., Tao, J., Schuller, B., Shan, S., Jiang, D., Jia, J.: MEC 2016: the multimodal emotion recognition challenge of CCPR 2016. In: Chinese Conference on Pattern Recognition (CCPR), Chengdu, China (2016)
6. Chen, S., Jin, Q., Li, X., Yang, G., Jieping, X.: Speech emotion classification using acoustic features. In: 2014 9th International Symposium on Chinese Spoken Language Processing (ISCSLP), pp. 579–583. IEEE (2014)
7. Xia, R., Deng, J., Schuller, B., Liu, Y.: Modeling gender information for emotion recognition using denoising autoencoder. In: 2014 IEEE International Conference on Acoustics, Speech and Signal Processing (ICASSP), pp. 990–994. IEEE (2014)
8. Deng, J., Xia, R., Zhang, Z., Liu, Y., Schuller, B.: Introducing shared-hidden-layer autoencoders for transfer learning and their application in acoustic emotion recognition. In: 2014 IEEE International Conference on Acoustics, Speech and Signal Processing (ICASSP), pp. 4818–4822. IEEE (2014)
9. Huang, Z., Dong, M., Mao, Q., Zhan, Y.: Speech emotion recognition using CNN. In: Proceedings of the 22nd ACM International Conference on Multimedia, pp. 801–804. ACM (2014)
10. Jianlong, W., Lin, Z., Zha, H.: Multiple models fusion for emotion recognition in the wild. In: Proceedings of the 2015 ACM on International Conference on Multimodal Interaction, Seattle, WA, USA, 09–13 November 2015, pp. 475–481 (2015)
11. Sun, B., Li, L., Zhou, G., Xuewen, W., He, J., Lejun, Y., Li, D., Wei, Q.: Combining multimodal features within a fusion network for emotion recognition in the wild. In: Proceedings of the 2015 ACM on International Conference on Multimodal Interaction, Seattle, WA, USA, 09–13 November 2015, pp. 497–502 (2015)

12. Yao, A., Shao, J., Ma, N., Chen, Y.: Capturing AU-aware facial features and their latent relations for emotion recognition in the wild. In: Proceedings of the 2015 ACM on International Conference on Multimodal Interaction, pp. 451–458. ACM (2015)

13. Kim, Y., Provost, E.M.: Say cheese vs. smile: reducing speech-related variability for facial emotion recognition. In: Proceedings of the 22nd ACM International Conference on Multimedia, pp. 27–36. ACM (2014)

14. Jung, H., Lee, S., Yim, J., Park, S., Kim, J.: Joint fine-tuning in deep neural networks for facial expression recognition. In: Proceedings of the IEEE International Conference on Computer Vision, pp. 2983–2991 (2015)

15. Xue, B., Chen, F., Shaobin, Z.: A study on sentiment computing and classification of sina weibo with word2vec. In: 2014 IEEE International Congress on Big Data, pp. 358–363. IEEE (2014)

16. Chung-Hsien, W., Lin, J.-C., Wei, W.-L.: Survey on audiovisual emotion recognition: databases, features, and data fusion strategies. APSIPA Trans. Signal Inf. Process. **3**, e12 (2014)

17. Rozgic, V., Ananthakrishnan, S., Saleem, S., Kumar, R., Vembu, A.N., Prasad, R.: Emotion recognition using acoustic and lexical features. In: 13th Annual Conference of the International Speech Communication Association, INTERSPEECH 2012, Portland, Oregon, USA, 9–13 September 2012, pp. 366–369 (2012)

18. Huang, Z., Dang, T., Cummins, N., Stasak, B., Le, P., Sethu, V., Epps, J.: An investigation of annotation delay compensation and output-associative fusion for multimodal continuous emotion prediction. In: The International Workshop on Audio/Visual Emotion Challenge (2015)

19. Zuxuan, W., Jiang, Y.-G., Wang, J., Jian, P., Xue, X.: Exploring inter-feature and inter-class relationships with deep neural networks for video classification. In: Proceedings of the ACM International Conference on Multimedia, MM 2014, Orlando, FL, USA, 03–07 November 2014, pp. 167–176 (2014)

20. Chen, J., Chen, Z., Chi, Z., Hong, F.: Emotion recognition in the wild with feature fusion and multiple kernel learning. In: Proceedings of the 16th International Conference on Multimodal Interaction, ICMI 2014, Istanbul, Turkey, 12–16 November 2014, pp. 508–513 (2014)

21. Eyben, F., Llmer, M., Schuller, B.: Opensmile: the Munich versatile and fast opensource audio feature extractor. In: ACM International Conference on Multimedia, MM, pp. 1459–1462 (2010)

22. Schuller, B., Batliner, A., Steidl, S., Seppi, D.: Recognising realistic emotions and affect in speech: state of the art and lessons learnt from the first challenge. Speech Commun. **53**(9–10), 1062–1087 (2011)

23. Schuller, B., Steidl, S., Batliner, A., Vinciarelli, A., Scherer, K., Ringeval, F., Chetouani, M., Weninger, F., Eyben, F., Marchi, E.: The interspeech 2013 computational paralinguistics challenge: social signals, conflict, emotion, autism. In: INTERSPEECH 2013, Conference of the International Speech Communication Association, pp. 148–152 (2013)

24. Davis, S.B.: Comparison of parametric representations for monosyllabic word recognition in continuously spoken sentences. Read. Speech Recogn. **28**(4), 65–74 (1990)

25. Pancoast, S., Akbacak, M.: Softening quantization in bag-of-audio-words. In: 2014 IEEE International Conference on Acoustics, Speech and Signal Processing, ICASS 2014, pp. 1370–1374 (2014)

26. Mathias, M., Benenson, R., Pedersoli, M., Gool, L.: Face detection without bells and whistles. In: Fleet, D., Pajdla, T., Schiele, B., Tuytelaars, T. (eds.) ECCV 2014. LNCS, vol. 8692, pp. 720–735. Springer, Heidelberg (2014). doi:10.1007/978-3-319-10593-2_47

27. Zhang, Z., Luo, P., Loy, C.C., Tang, X.: Facial landmark detection by deep multi-task learning. In: Fleet, D., Pajdla, T., Schiele, B., Tuytelaars, T. (eds.) ECCV 2014. LNCS, vol. 8694, pp. 94–108. Springer, Heidelberg (2014). doi:10.1007/978-3-319-10599-4_7

28. Zhang, Z., Luo, P., Loy, C.C., Tang, X.: Learning deep representation for face alignment with auxiliary attributes. IEEE Trans. Pattern Anal. Mach. Intell. 38(5), 918–930 (2016)

29. Lowe, D.G.: Distinctive image features from scale-invariant keypoints. Int. J. Comput. Vision 60(2), 91–110 (2004)

30. Csurka, G., Dance, C., Fan, L., Willamowski, J., Bray, C.: Visual categorization with bags of keypoints. In: Workshop on Statistical Learning in Computer Vision, ECCV, Prague, vol. 1, pp. 1–2 (2004)

31. Jianlong, W., Lin, Z., Zha, H.: Multiple models fusion for emotion recognition in the wild. In: Proceedings of the 2015 ACM on International Conference on Multimodal Interaction, pp. 475–481. ACM (2015)

32. LeCun, Y., Bengio, Y.: Convolutional networks for images, speech, and time series. In: The Handbook of Brain Theory and Neural Networks, vol. 3361, no. 10 (1995)

33. Tang, Y.: Deep learning using support vector machines. CoRR, abs/1306.0239, 2 (2013)

34. Mikolov, T., Chen, K., Corrado, G., Dean, J.: Efficient estimation of word representations in vector space. arXiv preprint arXiv:1301.3781 (2013)

35. Rehurek, R., Sojka, P.: Software framework for topic modelling with large corpora. In: Proceedings of the LREC 2010 Workshop on New Challenges for NLP Frameworks. Citeseer (2010)

36. Hsu, C.-W., Lin, C.-J.: A comparison of methods for multiclass support vector machines. IEEE Trans. Neural Netw. 13(2), 415–425 (2002)

37. Liaw, A., Wiener, M.: Classification and regression by randomforest. R News 2(3), 18–22 (2002)

Feature Learning via Deep Belief Network for Chinese Speech Emotion Recognition

Shiqing Zhang[✉], Xiaoming Zhao, Yuelong Chuang, Wenping Guo, and Ying Chen

Institute of Intelligent Information Processing, Taizhou University, Taizhou, China
tzczsq@163.com

Abstract. Speech emotion recognition is an interesting and challenging subject due to the emotion gap between speech signals and high-level speech emotion. To bridge this gap, this paper present a method of Chinese speech emotion recognition using Deep belief networks (DBN). DBN is used to perform unsupervised feature learning on the extracted low-level acoustic features. Then, Multi-layer Perceptron (MLP) is initialized in terms of the learning results of hidden layer of DBN, and employed for Chinese speech emotion classification. Experimental results on the Chinese Natural Audio-Visual Emotion Database (CHEAVD), show that the presented method obtains a classification accuracy of 32.80 % and macro average precision of 41.54 % on the testing data from the CHEAVD dataset on speech emotion recognition tasks, significantly outperforming the baseline results provided by the organizers in the speech emotion recognition sub-challenges.

Keywords: Deep learning · Deep belief networks · Speech emotion recognition · Feature learning

1 Introduction

During the past two decades, massive efforts have been made to recognize human emotions from emotional speech signals, i.e., called speech emotion recognition. At present, speech emotion recognition has attracted much interest in various fields such as signal processing, pattern recognition, artificial intelligence, etc., since it can be applied to human-machine interactions [1, 2].

Feature extraction is a critical step to bridge the emotion gap between speech signals and high-level speech emotion. Up to now, a variety of features have been employed for speech emotion recognition [3, 4]. These features can be roughly divided into four categories: (1) acoustic features, such as prosody features, voice quality features as well as spectral features, (2) language features, such as lexical information, (3) context information such as subject, gender, culture influences, (4) hybrid features such as the integration of two or three features abovementioned. However, these hand-designed features, there is no agreement that which is the best one sufficiently and efficiently characterizing emotion in speech signals. In addition, these hand-designed features were low-level, hence may not be reliable enough to efficiently characterize the subjective

© Springer Nature Singapore Pte Ltd. 2016
T. Tan et al. (Eds.): CCPR 2016, Part II, CCIS 663, pp. 645–651, 2016.
DOI: 10.1007/978-981-10-3005-5_53

emotion in complicated scenarios. It is thus important to develop automatic feature learning algorithms for speech emotion recognition.

In recent years, deep learning [5], which is multi-layered with a deep architecture, has attracted extensive attentions in machine learning, signal processing, artificial intelligence and pattern recognition. Deep belief networks (DBN) [6], as a representative method of deep learning, exhibits a strong ability of unsupervised feature learning. In recent years, DBN has been successfully applied for acoustic modeling [7], natural language understanding [8], speech recognition [9], speech emotion recognition [10] and so on.

To address the emotion gap, in this work we present a method of speech emotion recognition by combining DBN with a multi-layer perceptron (MLP) model. Different from [10], in which a complex Generalized Discriminant Analysis (GerDA) based on DBN is developed to learn discriminative features, this paper presents a simple and promising speech emotion classification method which directly integrates DBN with MLP. To evaluate the performance of the presented method, speech emotion recognition experiments are conducted on the Chinese Natural Audio-Visual Emotion Database (CHEAVD) [11], built from Chinese movies and TV programs. Experiment results show the effectiveness of the presented method.

2   Deep Belief Networks

Deep belief networks (DBN) [6] is built by stacking a number of restricted Boltzmann machine (RBM).

2.1   RBM

Restricted Boltzmann machine (RBM) is comprised of one layer of stochastic hidden units and one layer of stochastic visible units. RBM can be regarded to be a bipartite graph, where all visible units are connected to all hidden units. Note that visible-visible or hidden-hidden connections in an RBM do not exist.

Suppose the model parameters θ and the energy function $E(v, h; \theta)$, the joint distribution $P(v, h; \theta)$ over the visible units v and the hidden units h in a RBM is denoted by

$$P(v, h; \theta) = \frac{\exp(-E(v, h; \theta))}{Z}, \tag{1}$$

where $Z = \sum_v \sum_h \exp(-E(v, h; \theta))$ is a normalization factor. The energy function $E(v, h; \theta)$ in a Bernoulli (visible)-Bernoulli (hidden) RBM is represented as

$$E(v, h; \theta) = -\sum_{i=1}^{I} \sum_{j=1}^{J} \omega_{ij} v_i h_j - \sum_{i=1}^{I} b_i v_i - \sum_{j=1}^{J} a_j h_j, \tag{2}$$

where ω_{ij} represents the symmetric interaction term between the visible unit v_i and the hidden unit h_j, b_i and a_j the bias terms, and I and J are individually the amounts of the visible units and the hidden units.

In a RBM, the RBM weights can be updated via this rule:

$$\Delta \omega_{ij} = E_{data}(v_i h_j) - E_{model}(v_i h_j), \tag{3}$$

where $E_{data}(v_i h_j)$ denotes the expectation value over in the training data and $E_{model}(v_i h_j)$ denotes that same expectation value over the distribution represented by the given model.

2.2 DBN

DBN is built by stacking a variety of RBMs which are learned layer-by-layer from bottom-up. Figure 1 gives an illustration of DBN structure. From Fig. 1, we can see that the output of the lowest layer of DBN is used as the input of the next layer, and then the output of the next layer is subsequently employed as the input of the higher level's layer. For DBN's training, two steps are needed: pre-training and fine-tuning. Hinton et al. [5] has shown that a greedy learning method, i.e., so-called contrastive divergence (CD), is an effective way to train DBN. This CD training method has a good property that it can effectively improve the training data's lower bound of likelihood probability. Moreover, it is unsupervised to implement this CD training method.

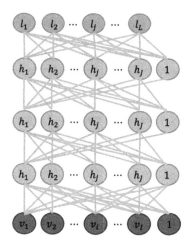

Fig. 1. An illustration of DBN

DBN is widely used for unsupervised feature learning tasks due to its discriminative feature learning ability. Note that when DBN is used for classification, its pre-training results can be adopted to initialize the weights of a traditional neural network, like multi-layer perceptron (MLP). To this end, the obtained weights of pre-trained DBN are used as the weights of MLP. Then the initialized MLP can be employed to perform discriminatively fine-tuning by using the back propagating (BP) algorithm.

3   Our System Structure

Figure 2 depicts our system structure of speech emotion recognition by combining DBN and MLP (abbreviated as DBN + MLP). Our method consists of four critical steps: acoustic feature extraction, DBN feature learning, MLP initialization and emotion classification, as described below detailedly.

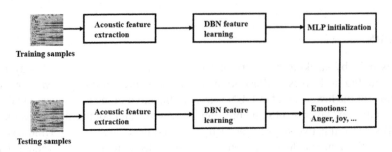

Fig. 2. Speech emotion recognition based on DBN and MLP

First, acoustic feature extraction is to extract the relevant acoustic features from affective speech signals, characterizing human emotion expression. Second, DBN feature learning, such as DBN's pre-training and fine-tuning, aims at implementing feature learning on the extracted low-level acoustic features. After performing DBN feature learning, high-level feature representation is produced. The produced high-level feature representation can be reflected in each hidden layer of DBN. Third, the MLP initialization aims to employ the learned high-level feature representation in each hidden layer of DBN to initialize the MLP classification model. The initialized MLP model has the same parameters as DBN, such as the number of hidden layers, the nodes of each hidden layer, and the weights of each hidden layer. Finally, we take the initialized MLP model as a classifier to distinguish emotions, such as anger, joy, sadness, etc.

4   Experiments

To testify the performance of the presented DBN + MLP method on speech emotion recognition tasks, speech emotion recognition experiments are conducted on the Chinese Natural Audio-Visual Emotion Database (CHEAVD) [11].

4.1   Dataset

The used Chinese Natural Audio-Visual Emotion Database (CHEAVD) [11] includes 2852 spontaneous emotional segments (140 min) extracted from films, TV plays and talk shows. It contains 238 speakers, aging from child to elderly, in which 52.5 % are male speakers, 47.5 % are female speakers. In the emotion recognition challenge, these 2852 segments are split into three parts: training set, validation set and testing set. Namely, 'Train', 'Val' and 'Test' sets have 1981, 243 and 628 clips, respectively. The duration of these segments is from 1 s to 19 s and the average duration is 3.3 s. Eight emotion classes, i.e., angry, happy, sad, worried, anxious, surprise, disgust and neutral are included in the challenge. The sampling rate of audio data extracted from the movie is 41 kHz.

4.2 Experiment Setup

For acoustic feature extraction, we employed a well-evolved set for automatic recognition of paralinguistic phenomena – the one of the INTERSPEECH 2013 Computational Paralinguistics Evaluation, i.e., the 2013 ComParE acoustic feature set [12]. This feature set consists of 6373 low-level descriptors (LLDs) including energy, spectral, and voicing related LLDs. We adopt TUM's open-source openSMILE feature extractor [13] and the corresponding configuration file – the IS13 ComParE.conf. More details about these 6373 features can be found in [14]. All the extracted acoustic features were normalized with one variance and zero mean.

We employ the DeeBNet toolbox, available at http://ceit.aut.ac.ir/~keyvanrad/DeeBNet%20Toolbox.html, to implement DBN experiments. We use a 6373-2000-1000-8 DBN structure for speech emotion classification due to its good performance. The number of epochs is 200, while the other parameters of DBN is set to be default in DeeBNet toolbox. We perform experiments with MATLAB2014 on a NVIDIA GTX TITAN X GPU.

4.3 Results and Analysis

Table 1 presents the recognition results of our DBN + MLP on the Val and Test data from the CHEAVD database. Since emotion classes on the CHEAVD database are unbalanced, the classification accuracy (ACC) and macro average precision (MAP) are presented to evaluate the performance of the used method, as described below.

$$\text{micro average precision (MAP)} = \frac{1}{s} \times \sum_{i=1}^{s} P_i, \tag{4}$$

$$P_i = \frac{TP_i}{TP_i + FP_i}, \tag{5}$$

$$\text{accuracy (ACC)} = \frac{\sum_{i=1}^{s} TP_i}{\sum_{i=1}^{s} (TP_i + FP_i)}, \tag{6}$$

where s denotes the number of the emotion classes, P_i represents the precision of the ith emotion class. TP_i and FP_i are the number of true positive prediction and the false positive prediction of the ith emotion class, respectively. Note that the baseline method [11] is that Random Forest with 100 trees is used for emotion classification, based on the extracted 88 acoustic features.

Table 1. Classification accuracy (in %) and average precision (in %) for Val and Test sets for speech emotion recognition

	Val		Test	
	ACC (%)	MAP (%)	ACC (%)	MAP (%)
Baseline [11]	41.98	30.02	24.36	24.02
Ours	**47.33**	**32.39**	**32.80**	**41.54**

The results in Table 1 indicate that our presented DBN + MLP method gives an accuracy of 47.33 %, and macro average precision of 32.39 % on the Val data, and an accuracy of 32.80 %, and macro average precision of 41.54 % on the Test data. More importantly, on the Test data our method makes 8.44 % improvement of classification accuracy, and 17.52 improvement of macro average precision, in comparison with the baseline method. This demonstrates the advantages of our presented DBN + MLP method on speech emotion recognition tasks. This is because that DBN is able to learn the high-level discriminative features from the low-level acoustic features for emotion classification due to its strong unsupervised feature learning ability.

To further investigate the recognition accuracy per emotion, the confusion matrix of eight emotion recognition results achieved by our DBN + MLP on the Test data is given in Table 2, in which the bold numbers denote the recognition samples for each emotion. The confusion matrix in Table 2 shows that surprise and disgust, were classified well with an accuracy of 80 % and 85.71 %, respectively, whereas the other six emotions were recognized badly. This is because these emotions may be very confused for each other.

Table 2. Confusion matrix of recognition results on the Test data (Hap-happy, Ang-angry, Wor-worried, Anx-anxious, Sur-surprise, Dis-disgust, Neu-neutral)

	Hap	Sad	Ang	Wor	Anx	Sur	Dis	Neu
Hap	**22**	21	11	16	19	11	3	25
Sad	3	**26**	8	19	14	2	0	17
Ang	5	2	**29**	4	14	5	11	12
Wor	0	0	0	**1**	1	0	1	0
Anx	0	0	1	0	**1**	1	1	1
Sur	0	0	0	1	4	**24**	0	1
Dis	0	0	0	0	2	0	**18**	1
Neu	26	26	24	45	48	8	8	**85**

5   Conclusions

This paper presents a simple and effective method of speech emotion recognition based on DBN and MLP. Experimental results on the CHEAVD database show the promising performance of the presented DBN + MLP method on speech emotion recognition tasks. This is because that DBN owns a strong ability of unsupervised feature learning to produce high-level discriminative feature representation from low-level acoustic features. In our future work, it is interesting to investigate the performance of some more advanced deep learning methods for speech emotion recognition.

Acknowledgments. This work is supported by Zhejiang Provincial Natural Science Foundation of China under Grant No. LY16F020011, and No. LY14F020036, National Natural Science Foundation of China under Grant No. 61203257 and No. 61272261.

References

1. Zeng, Z., Pantic, M., Roisman, G., Huang, T.: A survey of affect recognition methods: audio, visual, and spontaneous expressions. IEEE Trans. Pattern Anal. Mach. Intell. **31**(1), 39–58 (2009)
2. Zixing, Z., Coutinho, E., Jun, D., Schuller, B.: Cooperative learning and its application to emotion recognition from speech. IEEE/ACM Trans. Audio Speech Lang. Process. **23**(1), 115–126 (2015)
3. El Ayadi, M., Kamel, M.S., Karray, F.: Survey on speech emotion recognition: features, classification schemes, and databases. Pattern Recogn. **44**(3), 572–587 (2011)
4. Anagnostopoulos, C.-N., Iliou, T., Giannoukos, I.: Features and classifiers for emotion recognition from speech: a survey from 2000 to 2011. Artif. Intell. Rev. **43**(2), 155–177 (2015)
5. Hinton, G.E., Salakhutdinov, R.R.: Reducing the dimensionality of data with neural networks. Science **313**(5786), 504–507 (2006)
6. Hinton, G.E., Osindero, S., Teh, Y.-W.: A fast learning algorithm for deep belief nets. Neural Comput. **18**(7), 1527–1554 (2006)
7. Li, X., Yang, Y., Pang, Z., Wu, X.: A comparative study on selecting acoustic modeling units in deep neural networks based large vocabulary Chinese speech recognition. Neurocomputing **170**, 251–256 (2015)
8. Sarikaya, R., Hinton, G.E., Deoras, A.: Application of deep belief networks for natural language understanding. IEEE Trans. Audio Speech Lang. Process. **22**(4), 778–784 (2014)
9. Lu, Z., Quo, D., Garakani, A.B., Liu, K., May, A.: A comparison between deep neural nets and Kernel acoustic models for speech recognition. In: 2016 IEEE International Conference on Acoustics, Speech and Signal Processing (ICASSP), Shanghai, China, pp. 5070–5074 (2016)
10. Stuhlsatz, A., Meyer, C., Eyben, F., Zielke, T., Meier, G., Schuller, B.: Deep neural networks for acoustic emotion recognition: raising the benchmarks. In: 2011 IEEE International Conference on Acoustics, Speech and Signal Processing (ICASSP), Prague, pp. 5688–5691 (2011)
11. Li, Y., Tao, J., Schuller, B., Shan, S., Jiang, D., Jia, J.: MEC 2016: the multimodal emotion recognition challenge of CCPR 2016. In: 2016 Chinese Conference on Pattern Recognition (CCPR), Chengdu, China (2016)
12. Schuller, B., Steidl, S., Batliner, A., Vinciarelli, A., Scherer, K., Ringeval, F., Chetouani, M., Weninger, F., Eyben, F., Marchi, E., Mortillaro, M., Salamin, H., Polychroniou, A., Valente, F., Kim, S.: The INTERSPEECH 2013 computational paralinguistics challenge: social signals, conflict, emotion, autism. In: INTERSPEECH 2013, Lyon, France (2013)
13. Eyben, F., Weninger, F., Gross, F., Schuller, B.: Recent developments in openSMILE, the Munich open-source multimedia feature extractor. In: Proceedings of the 21st ACM International Conference on Multimedia, New York, USA, pp. 835–838 (2013)
14. Weninger, F., Eyben, F., Schuller, B.W., Mortillaro, M., Scherer, K.R.: On the acoustics of emotion in audio: what speech, music, and sound have in common. Front. Emot. Sci. **4**(292), 1–12 (2013)

The University of Passau Open Emotion Recognition System for the Multimodal Emotion Challenge

Jun Deng[1]([✉]), Nicholas Cummins[1], Jing Han[1], Xinzhou Xu[1,2], Zhao Ren[3],
Vedhas Pandit[1], Zixing Zhang[1], and Björn Schuller[1]

[1] Chair of Complex and Intelligent Systems, University of Passau, Passau, Germany
jun.deng@uni-passau.de
[2] Technische Universität München, Munich, Germany
[3] Northwestern Polytechnical University, Xi'an, People's Republic of China

Abstract. This paper presents the University of Passau's approaches for the Multimodal Emotion Recognition Challenge 2016. For audio signals, we exploit Bag-of-Audio-Words techniques combining Extreme Learning Machines and Hierarchical Extreme Learning Machines. For video signals, we use not only the information from the cropped face of a video frame, but also the broader contextual information from the entire frame. This information is extracted via two Convolutional Neural Networks pre-trained for face detection and object classification. Moreover, we extract facial action units, which reflect facial muscle movements and are known to be important for emotion recognition. Long Short-Term Memory Recurrent Neural Networks are deployed to exploit temporal information in the video representation. Average late fusion of audio and video systems is applied to make prediction for multimodal emotion recognition. Experimental results on the challenge database demonstrate the effectiveness of our proposed systems when compared to the baseline.

Keywords: Multimodal emotion recognition · Bag-of-audio-words · Transfer learning · Long short-term memory · Convolutional neural networks

1 Introduction

Emotion recognition 'in the wild' is attracting growing interest due to its practical importance in many real-world applications, such as human-computer interaction (HCI), e-learning, and health care. Despite a large number of existing research efforts to collect and analyse spontaneous or in the wild emotion databases in English, French or German [5], there have only been a small number of similar investigations undertaken on Chinese databases [2]. To advance spontaneous emotion recognition in the Chinese context, the *Multimodal Emotion Recognition Challenge* 2016 (MEC) provides a common benchmark database consisting of audiovisual clips taken from Chinese movies and TV programs [16].

© Springer Nature Singapore Pte Ltd. 2016
T. Tan et al. (Eds.): CCPR 2016, Part II, CCIS 663, pp. 652–666, 2016.
DOI: 10.1007/978-981-10-3005-5_54

In this paper, we present our approaches to the audio, video, and multimodal emotion recognition tasks introduced in this challenge [16].

Given the variety of acoustic events that can potentially occur within the selected audiovisual clips present in the challenge data, *Bag-of-Audio-Words* (BoAW) represents a potentially robust audio representation of the signal. This technique has been successfully used in similar emotion recognition tasks [20,22,23]. Inspired by this success, we create BoAW features based on five different *Low-Level-Descriptors* (LLDs) sets available from our open source toolkit openSMILE [7] and investigate their suitability for in-the-wild emotion detection.

Neural Networks based classification is widely used in audio-based emotion detection systems [3,8,17]. *Extreme Learning Machines* (ELMs) are a feedforward neural networks with a single hidden layer, currently gaining considerable interest in the machine learning community. This is due in part to their fast training and ease of implementation [11]. ELMs have shown competitive performance compared to *Support Vector Machines* (SVMs) and *Deep Neural Networks* (DNNs) in similar tasks [8,14] and are a key component in our audio-based system.

Many of the latest video recognition approaches are based on the features extracted from deep *Convolutional Neural Networks* (CNNs). However, the major challenge of applying this technique in the emotion recognition field is the lack of adequate training data. We overcome this challenge by using CNN models pre-trained on a large scale of data. The basic idea is to leverage the pre-trained model as a feature extractor for the new dataset at hand. We first make use of a pre-trained CNN model, referred to as *VGGFace* [19], to extract features relevant to the facial expressions present in a video frame. In addition to the features from the face in a video frame, we leverage another pre-trained CNN model, referred to as *VGG* [28], to extract features relevant to the broader contextual information. Further, recent work [33] has verified the significance of Facial *Action Units* (AUs) as features for emotion recognition. Thus, we also incorporate AU features into our video-based system.

The remainder of this paper is organised as follows. First, the Sects. 2 to 4 present each of the models used for different modalities. Next, Sect. 5 briefly introduces the MEC 2016 data and presents the results on the data. Finally, in Sect. 6 we conclude this paper and highlight future work directions.

2   Audio Systems

2.1   Feature Representations

BoAW audio feature representations are gaining popularity in many paralinguistic classification tasks [20,22,23]. BoAW involves generating a fixed length audio representation of each clip by first identifying a set of audio words, and then quantising (bagging) the original feature space, with respect to the generated codebook, to form a histogram representation of each data. The final BoAW representation represents the frequency of each previously identified audio word in a given instance [23].

Our basic framework to extract BoAW representations is depicted in Fig. 1. During training, acoustic LLDs are extracted from the audio files and are standardised to zero mean and unit standard deviation. The next step is to generate the codebook. Work presented in [22] indicates that the codebooks generated via random sampling of the extracted LLDs offer similar emotion recognition performance to codebooks generated using k-means clustering. Given this result, we also employ a random sampling of the extracted LLDs to generate our codebooks.

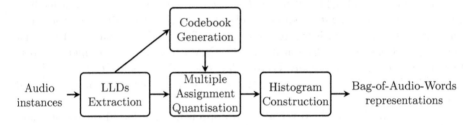

Fig. 1. Overview of the *Bag-of-Audio-Words* (BoAW) generation framework used in our audio systems. Figure adapted from [18].

After the codebook is built, a multi-assignment quantisation technique is applied to map LLDs from each frame to the first d closest audio-words, as measured by Euclidean distance. Preliminary experiments in [22] show that, for speech emotion recognition, the multi-assignment $(d > 1)$ outperforms the uni-assignment $(d = 1)$; therefore this paradigm is employed for all our extracted BoAW representations. A histogram is generated by calculating the counts of occurrence of each audio-word in all frames of one audio file. Finally, to generate a BoAW representation of a file, its corresponding relative counts are normalised to sum to one. This final step is undertaken to help minimise effects relating to disparities caused by various lengths of the files present in the MEC dataset. Note that, for the test data the LLDs of each frame are mapped to the audio words from the codebook pre-generated during the training phase.

2.2   Classifiers

Classification is performed in our audio systems either using a SVM or ELM back-end. SVMs are used due to their proven ability to handle a small dataset, relative lack of computational expense and established software implementations. Further, SVMs can also be regarded as a de-facto classifier for audio-based emotion detection systems [25].

As previously mentioned, ELMs have demonstrated competitive performance in similar audio based classification tasks [14]. The ELM is a single-hidden-layer

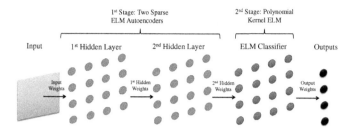

Fig. 2. The *Hierarchical Extreme Learning Machine* (HELM) framework consists of two phases: firstly two ELM based auto-encoders are used to form a sparse representation of the input feature space followed by a *Extreme Learning Machine* (ELM) classifier. Figure adapted from [30].

feedforward neural network which is exceptionally fast to train as the weights and biases corresponding to the hidden layer are randomly assigned [11] and never tuned. The basic theory behind ELMs, as introduced in [12], is that the first layer (the inputs weights and biases) can be regarded as carrying an unsupervised feature mapping and the only learning being performed is between the output of the second layer (the output weights) and the label matrix. This is achieved efficiently using a least-squares solution. The generalised output function $f(\mathbf{x})$ of an ELM is given by:

$$f(\mathbf{x}) = \sum_{i=1}^{L} \beta_i h_i(\mathbf{x}) = \mathbf{H}\beta, \tag{1}$$

where $\mathbf{x}$ is the input feature space, $\mathbf{H}$ a non-linear feature mapping (hidden layer) and $\beta = [\beta_1, \ldots, \beta_L]$ the (learnt) output weight vector. $h_i(\mathbf{x})$, the output of the ith hidden node can be obtained using a range of different activation functions, i.e., Sigmoid, Hyperbolic tangent, Gaussian, etc. As mentioned, β is found by minimising the least squared error:

$$\min_{\beta} \|\mathbf{H}\beta - \mathbf{T}\|^2, \tag{2}$$

where $\mathbf{T}$ is the training target matrix. For further details on the ELM, the reader is referred to [10,11].

A wide variety of variants to the basic ELM structure have been proposed which include: *Regularised ELMs* (RELM), *Kernel ELMs* (KELM), and *Hierarchical*-ELM (HELM) which can be regarded as the ELM analogue to deep learning [11,30]. Given the effectiveness of both DNNs and ELMs for performing speech-based emotion classification [8], an aim of this paper is to explore the suitability of the HELM framework for in-the-wild audio based emotion classification.

The HELM framework is an extension of the original ELM that allows ELMs to operate in a similar manner to a multilayer perceptron [30]. As seen in Fig. 2, the HELM framework consists of three layers; two ELM based auto-encoders and an ELM classifier. The role of the two ELM-based autoencoder layers is to form a sparse hierarchical representation of the original input feature space; the aim of this step is to exploit hidden information present in the training data. Sparsity is achieved by enforcing an ℓ_1 penalty during ELM training:

$$\min_{\beta} \left\{ \|\mathbf{H}\beta - \mathbf{X}\|^2 + \|\beta\|_{\ell_1} \right\}. \tag{3}$$

Each hidden layer in the HELM is an independent module and $\mathbf{H}$ is a randomly initialized output which does not require fine tuning.

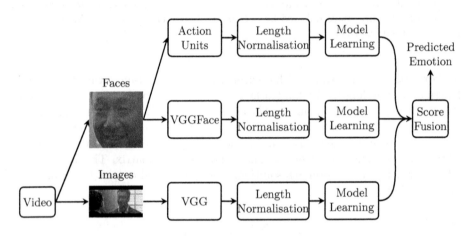

Fig. 3. Video emotion recognition system architecture.

In the original HELM framework, standard ELM classifiers was proposed to perform the classification step; however, as per the hidden layers, this module is independent and can be replaced with any ELM variant. Initial experiments (results not given) indicated that the use of a Polynomial Kernel ELM improved overall system performance. The closed-form solution for β is given by:

$$\beta = \left(\mathbf{H}^T \mathbf{H} + \frac{\mathbf{I}}{C} \right)^{-1} \mathbf{H}^T \mathbf{T}, \tag{4}$$

where $\mathbf{I}$ is the identity matrix, C is the regularization coefficient, and $\mathbf{H}$ is the given polynomial transformation of the input feature space. For further detail on the KELM and HELM, the reader is referred to [10] and [30] respectively.

3 Video Systems

Given a video sequence, Fig. 3 illustrates the architecture of our proposed video emotion recognition system. Like many other recent approaches for video analysis, CNNs can be directly applied to learn salient features from the input image. Our video system makes use of two CNN models (i.e. *VGG* [28] and *VGGFace* [19]) pre-trained for object classification and face recognition on a large scale of data. Here, the idea is to leverage the pre-trained models as feature extractors to provide features from each frame image and cropped face. In addition, we exploit facial action units as complementary features to enhance the video system. *Length normalisation* techniques such as *max* or *temporal k-max pooling* on frames are employed in order to give a fixed length input to a following classifier (i.e., SVMs or *Long Short-Term Memory Recurrent Neural Networks* (LSTM RNNs)), making it easy to perform model learning. Finally, an average decision rule is used to aggregate the scores predicated by the different models.

3.1 CNN Features

Deep CNNs are currently the most dominant approach in both video action recognition [32] and video emotion recognition [33]; this is due to their overwhelming accuracy. Deep CNNs trained on natural images exhibit an interesting phenomenon: the features learned from bottom to top layers are from *general* to *specific*. On the one hand, the first layer learns the features that are similar to Gabor filters and colour blobs. On the other hand, the higher-level layers are usually well trained for specific datasets and tasks. Consequently, the outputs of higher-level layers are widely chosen for recognition tasks because they combine all the general features into a rich image representation [4]. Thus, a deep CNN pre-trained on a large scale of image data can be used as a feature extractor for a task of interest.

A number of deep CNN architectures, which were originally proposed for image classification tasks, are popularly applied to directly extract deep CNN features from input pixels in a variety of computer vision tasks. These architectures include VGG-16, VGG-19 [28], AlexNet [15], and GooLeNet-22 [29]. In this challenge, we select the representative VGG-16 network (*VGG*), which consists of 13 convolutional layers, and 3 fully connected layers. Specifically, we use the 'FC7' features (i.e., the last feature layer, 4096 dimensions); FC7 is the most widely used deep feature extraction method for other computer vision tasks.

Facial-based features are known to be well suited for emotion recognition. Therefore, in addition to the pre-trained VGG model, we use VGGFace [19] to extract visual face descriptors of each frame. The VGGFace network has the same network architecture as VGG, but was trained on a very large-scale face data (2.6M images, 2.6k people). Hence, VGGFace tends to yield visual features with a more specific focus on faces than VGG.

Further, for CNN features in the video system, we want not only use the broad contextual information from the entire frame, but also the more specific information from the face. To this end, as illustrated in Fig. 3, the feature extraction of our proposed system achieves both an image-based video representation and a face-based video representation by using VGG and VGGFace, respectively.

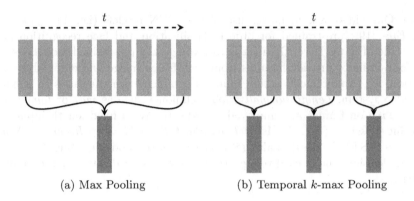

(a) Max Pooling   (b) Temporal k-max Pooling

Fig. 4. Illustration of max pooling and temporal k-max pooling, which are used for length normalisation in order to provide fixed length feature vectors for a later emotion classifier.

3.2 Action Units

Facial action units (AUs), which reflect facial muscle movements, are of importance in nonverbal behaviour and emotion recognition systems [21]. For example, [33] recently presented the use of AUs, resulting in a superior multimodal emotion recognition system in the wild; this result encouraged us to look into AUs in this challenge. The aim of AUs is to provide features that are complementary to the CNN features. We estimate action units for each video frame by using the OpenFace toolkit [1], resulting in 14 AU intensity factors and 6 AU occurrence factors extracted for each frame.

3.3 Length Normalisation

Length normalisation methods encode video sequence data into a fixed-length vector video representation by pooling all the descriptors from all the frames. Max pooling over video frames, as shown in Fig. 4, is typically used in video emotion recognition and is considered as the default length normalisation method tested in this paper. Such a pooling method results in one identical video representation, which simplifies model learning.

Max pooling, however, ignores all temporal information within the video. This information has been found important for distinguishing between different emotions. Therefore, we also test temporal k-max pooling in order to preserve valuable temporal information. The temporal k-max pooling, shown in Fig. 4, is applied to the frame-level features (e. g., CNN features) where the whole frame sequence is divided into k sub-sequences in a temporal manner and a max pooling step is used over frames in each sub-sequence. Note that, temporal k-max pooling corresponds to max pooling when k is equal to 1.

3.4   Temporal Modelling with LSTM RNNs

Since temporal k-max pooling captures valuable time varying information within the video sequences, it enables us to perform temporal modelling. A large number of previous works suggest that LSTM RNNs are good at exploiting temporal information [31,34,35]. Therefore, in addition to SVMs used in model learning, LSTM RNNs are also used to leverage temporal information.

The LSTM RNN model uses one or multiple LSTM blocks [9,35]. Every memory block consists of self-connected linear memory cells $\mathbf{c}$ and three multiplicative gate units: an input gate $\mathbf{i}$, a forget gate $\mathbf{f}$, and an output gate $\mathbf{o}$. Given an input $\mathbf{x}_t$ at the time step t, the activations of the input gate $\mathbf{i}_t$, the forget gate $\mathbf{f}_t$, the output gate $\mathbf{o}_t$, the candidate state value $\mathbf{g}_t$, the memory cell state $\mathbf{c}_t$ are separately computed by the following equations:

$$\mathbf{i}_t = \text{sigm}(\mathbf{W}_{ix}\mathbf{x}_t + \mathbf{W}_{ih}\mathbf{h}_{t-1} + \mathbf{b}_i), \tag{5}$$

$$\mathbf{f}_t = \text{sigm}(\mathbf{W}_{fx}\mathbf{x}_t + \mathbf{W}_{fh}\mathbf{h}_{t-1} + \mathbf{b}_f), \tag{6}$$

$$\mathbf{o}_t = \text{sigm}(\mathbf{W}_{ox}\mathbf{x}_t + \mathbf{W}_{oh}\mathbf{h}_{t-1} + \mathbf{b}_o), \tag{7}$$

$$\mathbf{g}_t = \tanh(\mathbf{W}_{gx}\mathbf{x}_t + \mathbf{W}_{gh}\mathbf{h}_{t-1} + \mathbf{b}_g), \tag{8}$$

$$\mathbf{c}_t = \mathbf{f}_t \odot \mathbf{c}_{t-1} + \mathbf{i}_t \odot \mathbf{g}_t, \tag{9}$$

$$\mathbf{h}_t = \mathbf{o}_t \odot \tanh(\mathbf{c}_t), \tag{10}$$

where $\mathbf{W}$ is a weight matrix of the mutual connections; $\mathbf{h}_t$ represents the output of the hidden block; $\mathbf{b}$ indicates the block bias, $\odot$ indicates the convolution operation.

4   Multimodal Systems

Our multimodal systems are based on the aforementioned audio and video systems (see Sects. 2 and 3). Late fusion or decision level fusion are adopted simply because it has been constantly proven efficient in multimodal emotion recognition tasks [33]. Specifically, we select a simple yet powerful, and widely used *average rule* to fusion scores from different models.

5   Results

5.1   The MEC 2016 Data and Evaluation Metrics

The MEC 2016 data are a subset of the Chinese Natural Emotional Audio-Visual Database (CHEAVD) that consists of video clips from a variety of Chinese movies and TV programs. This subset was chosen with the aim to provide natural emotion data close to real-world environments [2,16]. The challenge data contain samples labelled in eight emotional states: *angry, anxious, disgust, happy, neutral, sad, surprise* and *worried*. In total, there are 2 852 examples, which are partitioned into a *Training* (Tr.) set (1 981), a *Validation* (Val.) set (243), and a *Test* set (628). The full details of the challenge data can be found in [16]. As the dataset is unbalanced, *Macro Average Precision* (MAP) is used as the primary metric in this challenge.

Table 1. Audio results (in %) on the MEC 2016 test data. The last 3 runs combined the validation (Val.) set and the training (Tr.) set to form a larger training set.

Methods	MAP (Accuracy)
Audio baseline	24.02 (24.36)
Run 1: BoAW-SVM (Tr.)	19.04 (25.16)
Run 2: BoAW-HELM (Tr.)	30.61 (25.16)
Run 3: BoAW-SVM (Tr. + Val.)	23.95 (25.80)
Run 4: BoAW-SVM (Tr. + Val., excluded talk-show data)	23.54 (25.32)
Run 5: BoAW-SVM + Up-sampling (Tr. + Val.)	**36.11 (32.17)**

5.2   Audio Results

A wide range of preliminary experiments were performed to establish a suitable BoAW and SVM combination; the aim of this testing was to establish the suitability of BoAW for in-the-wild emotion classification. In preliminary experiments, we tested five distinct LLD sets: the 18 LLDs of the GeMAPS feature set [6], the 23 LLDs of eGeMAPS [6], the 16 LLDs and their corresponding delta regression coefficients of the INTERSPEECH 2009 Emotion Challenge feature set (IS09-Emotion) [26], the 38 LLDs and their deltas of the INTERSPEECH 2010 Paralinguistics Challenge set (IS10-Paraling) [24], and the 65 LLDs and their deltas of the ComParE feature set [27]. To optimise the codebook dimension, we chose three different sizes, i. e., 500, 1 000, 2 000. To optimise the number of assignments, LLDs of each frame were mapped to the $d = [25, 50, 100]$ closest words.

Table 2. Confusion matrix of the best audio system on the test set in the 8-way (i. e., Angry (Ang), Anxious (Anx), Disgust (Di), Happy (Ha), Neutral (Ne), Sad (Sa), Surprise (Su), Worried (Wo)) emotion recognition.

		Predicted Labels							
		Ang	**Anx**	**Di**	**Ha**	**Ne**	**Sa**	**Su**	**Wo**
	Ang	33	1	5	11	8	15	0	2
	Anx	0	26	1	4	3	3	3	2
True Labels	**Di**	24	3	3	18	1	34	3	0
	Ha	9	2	3	13	4	18	4	3
	Ne	14	5	2	21	12	40	8	1
	Sa	32	4	3	26	5	67	2	3
	Su	13	0	4	10	14	15	16	1
	Wo	3	1	1	8	2	4	0	32

These experiments (results not given) revealed that the best combination was found to be BoAW formed from the IS10-Paraling feature set (Codebook size 500, $d = 25$), in combination with a polynomial SVM (Degree: 1, Cost: 1). This set-up gave a validation MAP of 32.62 %. Despite this relatively strong performance on the validation set, this audio system did not perform well on the test set, achieving a MAP of 19.04 %.

As with the BoAW-SVM system, a wide range set of preliminary experiments was performed to establish a robust set-up for the BoAW-HELM system. Again, the IS10-Paraling feature set was identified as the most suitable for forming the BoAW representation (Codebook size 500, $d = 100$). Further initial testing also revealed the benefits of applying *Canonical Correlation Analysis* (CCA) feature selection [13], before HELM training. It was observed during these initial tests that it was easy for the ELM systems to overfit to their training set. Therefore to establish a robust set-up, system configurations were chosen which performed well under both validation (train on training set, tested on validation set) and pseudo-test conditions (trained on training and validation sets, tested the extra set of labelled data released by the challenge organisers). This testing set-up revealed a single layer polynomial kernel ELM (Degree: 5, C: 0.2) set-up was able to achieve a MAP of 33.78 %, and our BoAW-HELM system (Degree: 9, C: 500) was able to achieve a MAP of 37.88 % representing a 12 % relative improvement over the single layer ELM system. However, as with the BoAW-SVM system, the strong performance of the HELM system in validation did not generalise onto the test set where it achieved a MAP of 30.61 %; which is a relative increase of 27 % over the challenge test set baseline.

To help minimise the effects related to potential overfitting in the BoAW-SVM and the BoAW-HELM audio systems, we re-trialled the BoAW-SVM system combining the training set and the validation set to form a larger training set with 2 473 instances. We then performed a series of ten-fold cross validation preliminary experiments on the combined training set and found that the BoAW representation (Codebook size 2000, $d = 25$), of IS09-Emotion in conjunction with polynomial SVM (Degree: 1, Cost: 0.04) produced the best ten-fold cross validation performance; this set-up was used for the rest audio systems. This system achieved a MAP of 45.99 % under the cross validation condition; however, disappointingly it achieved a MAP of 23.95 % on the test set.

Because the MEC dataset includes both movie clips and talk-show data, the training set potentially exhibits an unwanted source of noise. Therefore, to investigate this potential effect, our fourth audio system excludes the talk-show data from the whole training data and uses the cleaned training data for modelling. This system achieved a MAP of 74.66 % under the cross validation condition. However, this system obtained a MAP of only 23.54 % on the test set.

Inspired by the baseline systems using sampling techniques to balance the training data with talk-show data, we up-sampled the full training set before the model learning for the fifth audio system. The final submission system yields a MAP of 54.66 % under the cross validation condition and a test set MAP of 36.11 %. This is a relative increase of 50 % over the challenge baseline. Table 1 summarises the performance of our five audio systems on the test set and the confusion matrix of the fifth audio system is presented in Table 2.

5.3 Video Results

As mentioned, our videos systems make full use of the VGG video representation, the VGGFace video representation, as well as the AU video representation in conjunction with SVMs and LSTM. A large number of preliminary experiments were performed to identify for the best set-up on the validation set. Consequently, we found that the VGG-based SVM system, the VGGFace-based SVM system, and the AU-based SVM system obtain a validation MAP of 17.17 %, 33.31 %, and 30.16 %, respectively, which are all higher than the video baseline. It is worth noting that the VGG-based SVM system is surprisingly competitive with the baseline video system; this indicates that CNN video representations derived from whole frames are as informative as hand-crafted video representation derived from faces. Average pooling was also tested as a length normalisation method. However, it resulted in worse performance than max pooling, and hence, was not used in our emotion recognition systems.

Encouraged by the preliminary experiments above, we next investigated the combination of the VGGFace video representation, the temporal k-max pooling, and LSTM on the given training and validation sets. We trained the LSTM

Table 3. Video results (in %) on the MEC 2016 test data.

Methods	MAP (Accuracy)
Video baseline	34.28 (19.59)
Run 1: VGG (max), VGGFace ($k=1,3$), AU (Average late fusion)	42.04 (35.89)
Run 2: VGG (max), VGGFace ($k=1,3,5,7$), AU (Majority voting)	45.21 (35.03)
Run 3: VGGFace ($k = 3$) (LSTM)	**53.43** (32.17)
Run 4: VGG (max), VGGFace ($k=1,3,5$), AU (Average late fusion)	43.34 (35.67)
Run 5: VGG (max), VGGFace ($k=1,3,5,7$), AU (Average late rule)	44.36 (34.13)

Table 4. Confusion matrix of the best video system on the test set.

True Labels \ *Predicted Labels*	**Ang**	**Anx**	**Di**	**Ha**	**Ne**	**Sa**	**Su**	**Wo**
Ang	16	0	0	3	0	44	12	0
Anx	3	22	0	1	0	8	7	1
Di	9	0	1	3	0	38	35	0
Ha	3	0	0	26	0	18	9	0
Ne	13	0	0	5	1	56	28	0
Sa	14	0	2	7	0	93	25	1
Su	9	0	0	3	0	40	20	1
Wo	2	0	2	2	0	15	7	23

network by feeding the k-max pooling video representations from the VGGFace descriptors to the network and using the last sequence to produce the class prediction. We found that the VGGFace-based LSTM system obtains the best validation macro precision of 47.17 % when $k = 3$ for k-max pooling.

Furthermore, we realised that the strong performance of a video system in validation may not generalise to the test set. Hence, we decided to use five-fold cross validation on the training set plus the validation set to select a robust video model. As with the audio, we also excluded the talk-show data in an attempt to clean the training data for model learning. Using this set-up, we obtain a cross validation macro precision of 37.47 %, 52.05 %, and 23.55 % for the VGG-based, VGGFace-based, and AU-based systems, respectively.

Using the set-up established in our preliminary experiments, all our five submission systems achieve very notable performance on the test data. Table 3 shows their results on the MEC 2016 test data. Four of our submissions used a Radial Basis Function SVM (Gamma: 2^{-12}, Cost: 10) as this set-up obtained consistently good cross-validation performance across different video features. Our other system (Run 3, Table 3) used LSTM and temporal k-max pooling, where the network has one hidden layer with 256 hidden units; $k = 3$ for k-max pooling achieved our best video-system MAP of 53.43 %, which is a relative increase of

Table 5. Multimodal results (in %) on the MEC 2016 test data. Each multimodal system was found by performing average late fusion of the equivalent audio and video systems.

Methods	MAP (Accuracy)
Multimodal baseline	30.63 (21.18)
Average late fusion of 'Run 1 from Table 1' and 'Run 1 of Table 3'	**49.43** (34.87)
Average late fusion of 'Run 2 from Table 1' and 'Run 2 of Table 3'	44.78 (34.39)
Average late fusion of 'Run 3 from Table 1' and 'Run 3 of Table 3'	36.70(28.34)
Average late fusion of 'Run 4 from Table 1' and 'Run 4 of Table 3'	43.55 (35.35)
Average late fusion of 'Run 5 from Table 1' and 'Run 5 of Table 3'	43.82 (34.24)

Table 6. Confusion matrix of the best multimodal system on the test set.

	Predicted Labels							
	Ang	**Anx**	**Di**	**Ha**	**Ne**	**Sa**	**Su**	**Wo**
Ang	8	1	0	4	0	54	8	0
Anx	1	22	0	1	0	11	7	0
Di	5	0	1	5	0	54	20	1
Ha	1	0	0	31	0	21	3	0
Ne	3	0	0	8	0	70	21	1
Sa	9	0	0	12	0	106	10	5
Su	2	1	0	4	1	45	20	0
Wo	0	0	0	5	0	13	2	31

True Labels

56 % over the challenge baseline. Table 4 presents the confusion matrix for this system.

5.4 Multimodal Results

Table 5 shows the performance of each of our submitted runs for the multimodal task. Each result has been found by performing average late fusion of the equivalent audio and video systems, e. g., Run 1 in Table 5 is the fusion of Audio System Run 1 (from Table 1) and Video System Run 1 (from Table 3). It can been seen from Table 5 that all of our systems outperformed the baseline multimodal system. Table 6 depicts the confusion matrix of the best multimodal system which yielded a MAP of 49.43 %; a relative increase of 61 % over the challenge baseline.

6 Conclusions

This paper presented the University of Passau's audio, video and multimodal systems for submission to the Multimodal Emotion Recognition Challenge 2016. For our audio systems, we investigated the effectiveness of a BoAW representation in a combination with ELMs, Hierarchical ELMs, and SVMs. Disappointingly, the strong performance of the audio systems during system development did not generalise to testing. For our video-based systems we leveraged LSTM RNNs based on visual features extracted from deep CNNs and facial action units. All of our videos systems outperformed the challenge baseline, indicating the benefits of using video features extracted from CNNs pre-trained for object recognition or face recognition, for emotion classification. Our multimodal results highlight the benefit of using late fusion. As with our video system, all of our multimodal approaches outperformed the challenge baseline.

Future audio work will explore the benefits of using different techniques to form BoAW codebook and the advantages offered by different Neural Network based classifiers. To further improve the performance of our video systems, we will investigate the use of Bidirectional Long Short-Term Memory Recurrent Neural Networks. We will also consider sampling methods for audiovisual emotion recognition as a way to balance the training data.

Acknowledgments. This work has been partially supported by the BMBF IKT2020-Grant under grant agreement No. 16SV7213 (EmotAsS), the European Communitys Seventh Framework Programme through the ERC Starting Grant No. 338164 (iHEARu), and the EU's Horizon 2020 Programme agreement No. 688835 (DE-ENIGMA), and the European Union's Horizon 2020 Programme through the Innovative Action No. 645094 (SEWA). It was further partially supported by research grants from the China Scholarship Council (CSC) awarded to Xinzhou Xu.

References

1. Baltrušaitis, T., Robinson, P., Morency, L.P.: OpenFace: an open source facial behavior analysis toolkit. In: Proceedings of WACV, Lace Placid, USA, pp. 1–10 (2016)

2. Bao, W., Li, Y., Gu, M., Yang, M., Li, H., Chao, L., Tao, J.: Building a Chinese natural emotional audio-visual database. In: Proceedings of ICSP, Hangzhou, China, pp. 583–587. IEEE (2014)
3. Deng, J., Zhang, Z., Marchi, E., Schuller, B.: Sparse autoencoder-based feature transfer learning for speech emotion recognition. In: Proceedings of ACII, pp. 511–516. IEEE (2013)
4. Donahue, J., Jia, Y., Vinyals, O., Hoffman, J., Zhang, N., Tzeng, E., Darrell, T.: DeCAF: a deep convolutional activation feature for generic visual recognition. In: Proceedings of ICML, Beijing, China, pp. 647–655 (2014)
5. El Ayadi, M., Kamel, M.S., Karray, F.: Survey on speech emotion recognition: features, classification schemes, and databases. Pattern Recognit. **44**(3), 572–587 (2011)
6. Eyben, F., Scherer, K., Schuller, B., Sundberg, J., André, E., Busso, C., Devillers, L., Epps, J., Laukka, P., Narayanan, S., Truong, K.: The Geneva minimalistic acoustic parameter set (GeMAPS) for voice research and affective computing. IEEE Trans. Affect. Comput. **7**(2), 190–202 (2016)
7. Eyben, F., Weninger, F., Groß, F., Schuller, B.: Recent developments in openS-MILE, the Munich open-source multimedia feature extractor. In: Proceedings of the 21st ACM International Conference on Multimedia, MM 2013, pp. 835–838. ACM (2013)
8. Han, K., Yu, D., Tashev, I.: Speech emotion recognition using deep neural network and extreme learning machine. In: Proceedings of INTERSPEECH, Singapore, pp. 223–227 (2014)
9. Hochreiter, S., Schmidhuber, J.: Long short-term memory. Neural Comput. **9**(8), 1735–1780 (1997)
10. Huang, G., Huang, G.B., Song, S., You, K.: Trends in extreme learning machines: a review. Neural Netw. **61**, 32–48 (2015)
11. Huang, G.B.: What are extreme learning machines? Filling the gap between Frank Rosenblatt's Dream and John von Neumann's Puzzle. Cognitive Comput. **7**(3), 263–278 (2015)
12. Huang, G.B., Zhu, Q.Y., Siew, C.K.: Extreme learning machine: theory and applications. Neurocomputing **70**(1–3), 489–501 (2006)
13. Kaya, H., Eyben, F., Salah, A.A.: CCA based feature selection with application to continuous depression recognition from acoustic speech features. In: 2014 IEEE International Conference on Acoustics, Speech and Signal Processing (ICASSP), Florence, Italy, pp. 3757–3761. IEEE (2014)
14. Kaya, H., Salah, A.A.: Combining modality-specific extreme learning machines for emotion recognition in the wild. In: Proceedings of the 16th International Conference on Multimodal Interaction, ICMI 2014, pp. 487–493. ACM (2014)
15. Krizhevsky, A., Sutskever, I., Hinton, G.E.: ImageNet classification with deep convolutional neural networks. In: Proceedings of NIPS, Lake Tahoe, USA, pp. 1106–1114 (2012)
16. Li, Y., Tao, J., Schuller, B., Shan, S., Jiang, D., Jia, J.: MEC 2016: the multimodal emotion recognition challenge of CCPR 2016. In: Proceedings of CCPR, Chengdu, China (2016). 11 pages
17. Mao, Q., Dong, M., Huang, Z., Zhan, Y.: Learning salient features for speech emotion recognition using convolutional neural networks. IEEE Trans. Multimedia **16**(8), 2203–2213 (2014)
18. Pancoast, S., Akbacak, M.: Bag-of-audio-words approach for multimedia event classification. In: Proceedings of INTERSPEECH, Portland, USA, pp. 2105–2108 (2012)

19. Parkhi, O.M., Vedaldi, A., Zisserman, A.: Deep face recognition. In: Proceedings of BMVC, Swansea, UK, pp. 41.1–41.12 (2015)
20. Pokorny, F., Graf, F., Pernkopf, F., Schuller, B.: Detection of negative emotions in speech signals using bags-of-audio-words. In: Proceedings of ACII, Xi'an, China, pp. 879–884 (2015)
21. Sariyanidi, E., Gunes, H., Cavallaro, A.: Automatic analysis of facial affect: a survey of registration, representation, and recognition. IEEE Trans. Pattern Anal. Mach. Intell. **37**(6), 1113–1133 (2015)
22. Schmitt, M., Ringeval, F., Schuller, B.: At the border of acoustics and linguistics: bag-of-audio-words for the recognition of emotions in speech. In: Proceedings of INTERSPEECH, San Francsico, USA (2016). 5 pages
23. Schmitt, M., Schuller, B.W.: Openxbow-introducing the passau open-source cross-modal bag-of-words toolkit. CoRR abs/1605.06778 (2016)
24. Schuller, B., Steidl, S., Batliner, A., Burkhardt, F., Devillers, L., Müller, C., Narayanan, S.S.: The INTERSPEECH 2010 paralinguistic challenge. In: Proceedings of INTERSPEECH, pp. 2794–2797 (2010)
25. Schuller, B., Batliner, A.: Computational Paralinguistics: Emotion, Affect and Personality in Speech and Language Processing. Wiley Publishing, Chichester (2013)
26. Schuller, B., Steidl, S., Batliner, A.: The INTERSPEECH 2009 emotion challenge. In: Proceedings of INTERSPEECH, Brighton, UK, pp. 312–315 (2009)
27. Schuller, B., Steidl, S., Batliner, A., Vinciarelli, A., Scherer, K., Ringeval, F., Chetouani, M., Weninger, F., Eyben, F., Marchi, E., Mortillaro, M., Salamin, H., Polychroniou, A., Valente, F., Kim, S.: The INTERSPEECH 2013 computational paralinguistics challenge: social signals, conflict, emotion, autism. In: Proceedings of INTERSPEECH, Lyon, France, pp. 148–152 (2013)
28. Simonyan, K., Zisserman, A.: Very deep convolutional networks for large-scale image recognition. CoRR (2014). http://arxiv.org/abs/1409.1556
29. Szegedy, C., Liu, W., Jia, Y., Sermanet, P., Reed, S.E., Anguelov, D., Erhan, D., Vanhoucke, V., Rabinovich, A.: Going deeper with convolutions. In: Proceedings of CVPR, Boston, USA, pp. 1–9 (2015)
30. Tang, J., Deng, C., Guang, G.B.: Extreme learning machine for multilayer perceptron. IEEE Trans. Neural Netw. Learn. Syst. **27**(4), 809–821 (2015)
31. Wöllmer, M., Kaiser, M., Eyben, F., Schuller, B., Rigoll, G.: LSTM-modeling of continuous emotions in an audiovisual affect recognition framework. Image Vision Comput. **31**(2), 153–163 (2013). Special Issue on Affect Analysis in Continuous Input
32. Xu, Z., Yang, Y., Hauptmann, A.G.: A discriminative CNN video representation for event detection. In: Proceedings of CVPR, Boston, USA, pp. 1798–1807 (2015)
33. Yao, A., Shao, J., Ma, N., Chen, Y.: Capturing au-aware facial features and their latent relations for emotion recognition in the wild. In: Proceedings of ICMI, Seattle, USA, pp. 451–458. ACM (2015)
34. Zhang, Z., Pinto, J., Plahl, C., Schuller, B., Willett, D.: Channel mapping using bidirectional long short-term memory for dereverberation in hands-free voice controlled devices. IEEE Trans. Consum. Electron. **60**(3), 525–533 (2014)
35. Zhang, Z., Ringeval, F., Han, J., Deng, J., Marchi, E., Schuller, B.: Facing realism in spontaneous emotion recognition from speech: feature enhancement by autoencoder with LSTM neural networks. In: Proceedings of INTERSPEECH, San Francsico, CA (2016). 5 pages

MEC 2016: The Multimodal Emotion Recognition Challenge of CCPR 2016

Ya Li[1(✉)], Jianhua Tao[1,2], Björn Schuller[3,4,5], Shiguang Shan[6],
Dongmei Jiang[7], and Jia Jia[8]

[1] Institute of Automation, Chinese Academy of Sciences, Beijing, People's Republic of China
yli@nlpr.ia.ac.cn
[2] University of Chinese Academy of Sciences, Beijing, People's Republic of China
[3] Chair of Complex and Intelligent Systems, University of Passau, Passau, Germany
[4] Department of Computing, Imperial College London, London, UK
[5] Harbin Institute of Technology, Harbin, People's Republic of China
[6] Institute of Computing Technology, Chinese Academy of Sciences,
Beijing, People's Republic of China
[7] Northwestern Polytechnical University, Xi'an, People's Republic of China
[8] Tsinghua University, Beijing, People's Republic of China

Abstract. Emotion recognition is a significant research filed of pattern recognition and artificial intelligence. The Multimodal Emotion Recognition Challenge (MEC) is a part of the 2016 Chinese Conference on Pattern Recognition (CCPR). The goal of this competition is to compare multimedia processing and machine learning methods for multimodal emotion recognition. The challenge also aims to provide a common benchmark data set, to bring together the audio and video emotion recognition communities, and to promote the research in multimodal emotion recognition. The data used in this challenge is the Chinese Natural Audio-Visual Emotion Database (CHEAVD), which is selected from Chinese movies and TV programs. The discrete emotion labels are annotated by four experienced assistants. Three sub-challenges are defined: audio, video and multimodal emotion recognition. This paper introduces the baseline audio, visual features, and the recognition results by Random Forests.

Keywords: Audio-visual corpus · Features · Multimodal fusion · Challenge · Emotion · Affective computing

1 Introduction

Automatic emotion recognition is a technology to identify the human's emotional states by analyzing and processing human speech, facial expression, body gesture and physiological signals, amongst others. As an important branch of artificial intelligence, emotion recognition can be widely utilized in human-computer interaction, diagnosis and monitoring of disease, public security and other fields. In recent years, with the development of psychology, physiology, neurology and computer science, both speech-based emotion recognition and vision-based emotion recognition have made remarkable progress. Therefore, this challenge focuses on audio-visual emotion recognition.

© Springer Nature Singapore Pte Ltd. 2016
T. Tan et al. (Eds.): CCPR 2016, Part II, CCIS 663, pp. 667–678, 2016.
DOI: 10.1007/978-981-10-3005-5_55

Emotion recognition methods can be classified on the basis of the emotion labeling method. The early methods and databases used the "universal" (big) six emotions (angry, disgust, fear, happy, sad and surprised) often plus neutral. Traditionally, emotion recognition has been performed on laboratory controlled data. While undoubtedly worthwhile at the time, such laboratory controlled data poorly represents the environment and conditions faced in real-world situations. However, with the growing demands from applications, more and more work has been carried out on natural and spontaneous multimodal emotion corpora. Databases and methods which are closer to real-world environments (such as indoor, outdoor, different color backgrounds, occlusion and background clutter) have been recently introduced. Such databases (e.g., Semaine [1]) use continuous labeling in the valence and arousal and further dimensions. Acted Facial Expressions in the Wild (AFEW) [2] and Static Facial Expressions in the Wild (SFEW) [3] are recent emotion databases representing (closer to) real-world scenarios.

To provide a common platform for emotion recognition researchers, challenges such as the Audio/Visual Emotion Challenges (AVEC) [4] and the INTERSPEECH Emotion Challenge [5] and its predecessors at Interspeech, the Facial Expression Recognition & Analysis (FERA) [6], Emotion Challenge in the Wild Challenge (EmotiW) [7] or further related ones such as tasks in the MediaEval series have been organized. These are mostly based on spontaneous databases [1]. In these challenges, both EmotiW and FERA use categorical emotion labeling, where FERA targets facial expression recognition on "universal" five emotion classes on acted facial expression data, and EmotiW aims to compare the performance of their methods on audio-visual data. The INTERSPEECH emotion challenge focuses on emotion recognition from speech, and the FAU Aibo Emotion Corpus is used as the basis database. In contrast, AVEC is organized to evaluate *multimodal* emotion recognition. The features used in AVEC are not only speech, facial data, but recently also physiological signals [8]. AVEC uses dimensionally labeled emotion data, and the raters annotated dimensions including *Activity/Arousal*, *Expectation*, *Power*, and *Valence* usually in continuous time and continuous value. Similar to the AVEC effort, the Multimodal Emotion Recognition Challenge (MEC 2016) aims to provide a common benchmark data set, to bring together the audio and video emotion recognition communities, and to promote the research in multimodal emotion recognition, however, especially for Chinese data, since emotional expression varies across different languages and cultures.

In MEC 2016, we use discrete labels to classify the emotion and define three sub-challenges, which are audio (only), video (only) and multimodal emotion recognition in order to attract research groups with different foci in the affective computing society. All sub-challenges seek participation from researchers and industrials whoever work on emotion recognition to create, extend and validate their methods on data in real-world conditions. The baseline audio and video features provided by the organizers are free to use either all or part of them. Participants are encouraged to extract their own features and apply their own classification algorithms. For each sub-challenge, participants will be allowed at most five trials to upload their results for evaluation on the test set.

Each registered team should submit a paper introducing the results and the methods that they use, which will be peer-reviewed. Only entries into MEC 2016 accompanied by a qualified paper will be eligible for challenge participation. In their paper, the authors may report on the results obtained on the training and validation sets, but only the results on the test set will be taken into account in the final competition.

Since emotion recognition has received increasing attention, we also want to take this opportunity to promote the development of affective computing by encouraging the participants to share their work. The participants should submit executable files used in this challenge and documentation to OpenPR (http://www.openpr.org.cn/), and share with other researchers. The Open Pattern Recognition (OpenPR) project is an open source platform for sharing algorithms of image processing, computer vision, natural language processing, pattern recognition, machine learning and related fields. OpenPR was initiated in 2009 under the BSD license, and is currently supported by the National Laboratory of Pattern Recognition, Institution of Automation, Chinese Academy of Sciences. From 2009 to now, OpenPR has become a valuable resource in pattern recognition, and the codes contained have been downloaded more than 50,000 times.

The remainder of this paper is organized as follows. Section 2 describes the database used in this challenge. Sections 3 and 4 present the features and the baseline experiments. Finally, Sect. 5 concludes this paper.

2   Multimodal Emotional Database

One of the major needs of the affective computing society is the constant requirement of emotional data. The existing emotional corpora could be divided into three types: simulated/acted, elicited and natural(-istic) corpora [9, 10]. Recently, the demand for real application forces emotion researchers to put more effort on natural and spontaneous emotion data. Ideally, a corpus should be collected from our daily life which includes natural and spontaneous emotion. But because of copyright and privacy issues, several of the existing natural(-istic) emotion corpora were collected from films and TV programs (Yu et al. 2001). Although movies and TV programs are often shots in controlled environments, they are significantly closer to real-world environments than the lab-recorded datasets due to highly varying and often adverse conditions. Some of the successful examples are the Belfast Natural Database [11], the Vera am Mittag German Audio-visual Emotional Speech Database (VAM) [12], the EmoTV Database [13] and the SAFE (Situation Analysis in a Fictional and Emotional) Corpus [14]. But, none of these corpora is in Chinese. Since emotion expression has some specific characteristics for different languages in different cultures, in the challenge, we choose the CASIA Chinese Natural Emotional Audio-Visual Database (CHEAVD) [15], which aims to provide a basic Chinese resource for the research on multimodal multimedia interaction.

CHEAVD contains 140 min spontaneous emotional segments extracted from films, TV plays and talk shows. 238 speakers, aging from child to elderly, are included in this database. The partition of the recordings with respect to gender is as follows: 52.5 % are male subjects, 47.5 % are female subjects. A discrete emotion annotation strategy is adopted, and 26 non-prototypical emotional states, including the basic six, were labeled by four native speakers. Pairwise kappa coefficients were calculated to evaluate the annotation consistency, which are shown in Table 1. In contrast to most other available

Fig. 1. Selected screenshots of videos in the database.

emotional databases, multiple-emotion labels and fake/suppressed emotion labels are provided too. However, to simplify this challenge, the average of the four annotations is adopted as the unique final label of each segment by majority vote, and we only select the top eight major emotion classes, namely, *angry*, *happy*, *sad*, *worried*, *anxious*, *surprise*, *disgust* and *neutral* in the challenge (Fig. 1).

Table 1. The pairwise kappa coefficients of the four annotations.

Annotators	A1	A2	A3	A4
A1		0.58	0.55	0.43
A2	0.58		0.52	0.41
A3	0.55	0.52		0.42
A4	0.43	0.41	0.42	

The data used for this challenge contains the original video data, the extracted audio from the video, and the corresponding emotion label. The sampling rate of video data is 25 frames per second, and the sampling rate of audio data extracted from the movie is 44.1 kHz. The background of the video data and audio data is complex and mostly close to a real life environment. The dataset in particular addresses the issue of emotion recognition in real world conditions.

Table 2. Number of instances for eight emotion-classes.

	'Train'	'Val'	'Test'	Total
Neutral	725	90	142	957
Angry	399	49	73	521
Happy	307	38	56	401
Sad	256	31	75	362
Worried	93	11	86	190
Anxious	89	11	103	203
Surprise	67	8	51	126
Disgust	45	5	42	92
Sum	1981	243	628	2852

In order to assess the emotion recognition performance, the 2852 segments used in this challenge are divided into three sets: training set, validation set and testing set. These 'Train', 'Val' and 'Test' sets contain 1981, 243 and 628 clips, respectively. The duration of these segments is from 1 s to 19 s and the average duration is 3.3 s. The number of the instances for eight classes is shown in Table 2.

In the challenge, the labeled 'Train' and 'Val' sets are shared along with the unlabeled 'Test' set. Participants can train their modals on the training set, and validate their modals on the validation set. At the end of the competition, participants should upload their emotion recognition results on the test set no more than five time as outlined above. The best recognition result will be considered as their final score.

3 Features

3.1 Acoustic Features

Many features for emotion recognition from speech have been explored. Prosodic features, i.e., pitch, durations and intensity were widely used in emotion recognition from speech in the past [16]. In addition, the variation of voice quality can reveal a speaker's emotional state [17], and the timbre of speech is determined by laryngeal characteristics. The common investigated voice quality features include (logarithmic) harmonic-to-noise ratio (HNR), jitter, or shimmer. Spectral and cepstral features are also used in acoustic emotion recognition and can be extracted in a number of ways including formants, linear prediction (spectral or cepstral) coefficients (LPC) and MFCC, etc. Usually, the statistics of the low level feature vectors, e.g., mean, standard deviation, maximum, minimum, coefficients are used to provide features on a suprasegmental basis. Increasingly, one constructs a high dimensional acoustic feature space by systematic brute-forcing – usually thousands of features.

In this challenge, we chose the most common and at the same time promising feature types and functionals covering prosodic, spectral and voice quality features. For the highest transparency we utilize the open source openSMILE toolkit [18] to extract the audio features. These are the extended GeMAPS (*eGeMAPSv01a.conf*) features as also used in AV+EC 2015 [8]. The low level descriptors (LLDs) of the audio features are shown in Table 3.

The dimension of the audio features vector after functional application is 88.

Table 3. Acoustic features of the multimodal emotion recognition challenge of CCPR 2016

Energy & spectral low-level descriptors (26)
Sum of auditory spectrum (loudness), α ratio (50–1000 Hz/1–5 kHz)[a], Energy slope (0–500 Hz, 0.5–1.5 kHz)[a], Hammarberg index[a], MFCC 1–4[b], Spectral flux[b]

Voicing related low-level descriptors (16)
F_0 (linear & semi-tone), Formants 1, 2, 3 (freq., bandwith, ampl.), Harmonic difference H1–H2, H1–H3, Log.HNR, jitter (local), shimmer (local)

[a] Computed on voiced and unvoiced frames respectively;

[b] Computed on voiced, unvoiced and all frames respectively.

3.2 Visual Features

Facial expression plays an important role in human communication. Facial features can be divided into two groups: appearance features and shape features. Appearance features reflect the texture of the face, like wrinkles, facial bumps and so on. The shape features reflect the position information of the face. In order to extract the facial features accurately, data pre-processing is performed, which can include grey processing, face detection, face correction and face normalization.

We utilize the tracking algorithm and toolkit [19] to detect the human face. At first, the face detection method put forward by Viola and Jones [20] is employed to initialize the face tracking algorithm. Then, we use the tracking algorithm to detect human face. If the tracking algorithm fails, we utilize a face detection method to initialize the tracking algorithm again. In conclusion, the tracking algorithm makes use of the correlation between each frame, which can accelerate the face detection, and also increase the accuracy of the position of the facial points detection.

We utilize an affine transformation to correct the (processed) human face. An affine transformation can realize the rotation, scaling and translation of the original face image. Through the observation of the position of the facial points that the face detection returns, we can find that the 20th and 29th of facial points respectively represent the corner of the left eye and the corner of the right eye. Utilizing the position of two points, we can get the amount of rotation, scaling and translation to correct the original face image.

In the end, we scale the detected face into 100×100, which can make the feature extraction more convenient.

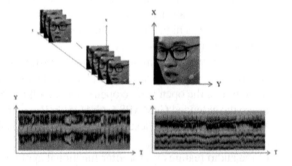

Fig. 2. Three dimensions of LBP-TOP.

Local Binary Pattern - Three Orthogonal Planes (LBP-TOP) is a popular descriptor of dynamic texture (DT) in computer vision (Fig. 1). For highest transparence, we utilize the open Matlab code created by Huang, which is based on the algorithm of Zhao and Pietikainen [21]. LBP-TOP considers patterns in three orthogonal planes: XY, XT and YT, and concatenates the pattern co-occurrences in these three directions. The local binary pattern (LBP) descriptor assigns binary labels to pixels by thresholding the neighborhood pixels with the central value. In order to reflect the location information of the LBP-TOP features, we divide the image into 2×2 blocks based on the code provided by Huang. The three orthogonal patterns are histogrammed separately, and concatenated into a single feature histogram. At the same time, we utilize uniform LBP to reduce the dimension of feature vector, and the uniform LBP has 59 possibilities. Therefore, the dimension of LBP-TOP features is 708 ($59 \times 3 \times 2 \times 2$). The mean pooling technique is applied to the feature sequence, and thus a 708-dimension feature vector is obtained for each segment.

4 Baseline Experiments

For computing the baseline results, we decided to entirely rely on a further "standard" publicly available tool which is widely used in the community: the WEKA 3.6 Data Mining Toolkit. This ensures easy reproducibility of the results. Frequently, SVMs are utilized in emotion recognition; however, we find that the performance of the modal trained by Random Forests is relatively better than SVM in the baseline experiments. Therefore, we chose Random Forests classification for the emotion recognition baseline in this challenge. The number of the tree is 100. The depth of the tree is varied in order to find the best emotion recognition model.

In the baseline experiments, Principal Components Analysis (PCA) with the Principal Components computed on the training data is used to reduce the dimension of visual features to 100. The implementation of PCA follows the suggestion of the tutorials in Unsupervised Feature Learning and Deep Learning (UFLDL)[1]. After PCA, We also normalize the features into the range of [0, 1] in the experiments.

In the sub-challenge of the multimodal emotion recognition, we utilize feature level fusion to get the audio-visual emotion recognition results. In the feature level fusion, the different features are directly concatenated to create a new feature vector. Then, a new prediction model on the basis of the feature vector is trained by a Random Forest to predict the final emotion label.

As classes are unbalanced, this challenge chooses macro average precision (MAP) as the primary measure to rank the results, and secondly the accuracy (ACC). Calculating macro average precision and accuracy as employed here are given in Eqs. (1)–(3), respectively. The baseline results are listed in Table 4.

$$\text{macro average precision (MAP)} = \frac{1}{s} \times \sum_{i=1}^{s} P_i, \tag{1}$$

$$P_i = \frac{TP_i}{TP_i + FP_i}, \tag{2}$$

$$\text{accuracy} = \frac{\sum_{i=1}^{s} TP_i}{\sum_{i=1}^{s} (TP_i + FP_i)}, \tag{3}$$

---

[1] http://ufldl.stanford.edu/wiki/index.php/Implementing_PCA/Whitening.

where s represents the number of the emotion classes. TP_i and FP_i represent the number of true positive prediction and the false positive prediction of the ith emotion class, respectively. P_i is the precision of the ith emotion class.

Table 4. Classification accuracy (ACC, in %) and macro average precision (MAP, in %) for the Val and Test sets for audio, video and multimodal emotion recognition

	Val			Test		
	Depth of the tree	ACC (%)	MAP (%)	Depth of the tree	ACC (%)	MAP (%)
Audio	10	41.98	**30.02**	10	24.36	24.02
Video	19	28.81	11.46	19	19.59	**34.28**
Multimodal	19	37.03	15.93	19	24.52	24.53

On the Val set, the baseline system of the audio emotion challenge gives 41.98 % classification accuracy and 30.02 % macro average precision, which resembles the best recognition rate. The baseline system of the video emotion challenge achieves 28.81 % classification accuracy and 11.46 % macro average precision, which has the worst recognition rate in the three sub challenges. As for the multimodal emotion recognition, the system achieves 37.03 % classification accuracy and 15.93 % macro average precision on the Val set.

On the Test set which contains 628 video clips, the audio channel gives 24.36 % classification accuracy and 24.02 % macro average precision, which resembles the worst recognition rate. The video channel gives 19.59 % classification accuracy and 34.28 % macro average precision, which has the best recognition rate. Multimodal emotion recognition by feature level fusion gives 24.52 % classification accuracy and 24.53 % macro average precision.

Table 5. Classification accuracy (ACC, in %) and macro average precision (MAP, in %) for the Val and Test sets for audio, video and multimodal emotion recognition after random sampling.

	Val			Test		
	Depth of the tree	ACC (%)	MAP (%)	Depth of the tree	ACC (%)	MAP (%)
Audio	5	18.93	19.51	5	21.97	20.48
Video	19	12.35	10.71	19	21.02	30.41
Multimodal	20	16.05	13.40	20	21.18	30.63

However, by checking the confusion matrix of each challenge, we found that the recognition rate is zero for some low-frequent occurred emotion states, namely, *worried*, *anxious*, *surprise*, and *disgust*. The classifiers tend to favor the majority emotion classes, which will not be considered as a "good" classifier. Therefore, we use sampling techniques to reconstruct a relatively balanced training corpus, considering that the training set is not balanced while the test set is relatively balanced. The random sampling is performed by Weka 3.6 filter (*SpreadSubsample*). The maximum count of each emotion class in the training set is set as 100. All other parameters in this filter are the default ones. Three new models built by Random Forests are rebuilt for each sub-challenge, and the results are shown in Table 5.

On the Val set, the audio-only baseline system gives 18.93 % classification accuracy and 19.51 % macro average precision, which is the best recognition rate. The video emotion recognition system achieves 12.35 % classification accuracy and 10.71 % macro average precision, which is the worst recognition rate. The multimodal emotion recognition system achieves 16.05 % classification accuracy and 13.40 % macro average precision on the Val set. On the Test set which contains 628 video clips, audio only gives 21.97 % classification accuracy and 20.48 % macro average precision, which is the worst recognition rate. Video only gives 21.02 % classification accuracy and 30.41 % macro average precision. Feature level fusion gives 21.18 % classification accuracy and 30.63 % macro average precision, which is the best recognition rate.

The confusion matrices of the sub-challenges are shown in Tables 6, 7 and 8 respectively.

Table 6. Confusion matrix: audio baseline system with sample balancing method

	Ha	Ang	Su	Di	Ne	Wo	Anx	Sa
Ha	15	4	0	0	32	0	0	5
Ang	20	26	0	0	22	0	0	5
Su	20	13	5	0	16	0	0	2
Di	6	12	0	0	21	0	0	3
Ne	30	4	0	0	103	0	0	5
Wo	21	4	0	0	55	0	0	6
Anx	24	16	0	0	60	0	1	2
Sa	22	5	0	0	40	0	0	8

Table 7. Confusion matrix: video baseline system with sample balancing method

	Ha	Ang	Su	Di	Ne	Wo	Anx	Sa
Ha	5	17	0	0	29	0	0	5
Ang	4	18	0	0	43	0	0	8
Su	0	11	1	0	33	0	0	6
Di	1	14	0	2	23	0	0	2
Ne	4	36	0	0	88	0	0	14
Wo	2	18	0	0	54	0	0	12
Anx	3	27	0	0	69	0	0	4
Sa	1	9	0	0	56	0	0	9

Table 8. Confusion matrix: multimodal baseline system with sample balancing method

	Ha	Ang	Su	Di	Ne	Wo	Anx	Sa
Ha	18	7	2	0	9	3	5	12
Ang	20	16	4	0	6	4	9	14
Su	12	4	10	0	8	2	7	8
Di	7	7	1	2	7	3	12	3
Ne	27	10	12	0	38	9	9	37
Wo	25	1	5	0	21	5	6	23
Anx	28	15	4	0	29	6	10	11
Sa	17	3	4	0	12	3	2	34

5   Conclusion

Providing a common benchmark data set will highly likely promote the research in affective computing. Although there are some emotion recognition challenges, we believe this challenge will make up for a valuable addition to the landscape of the existing challenges, since this is the first one to use a *Chinese* multimodal emotional corpus, namely CHEAVD. The emotion expression depends on the language and culture, and it is interesting to find out the state-of-the-art of emotion recognition for Chinese data. CHEAVD is a multimodal emotional corpus selected from Chinese movies and TV programs, which is mostly close to our daily life. Three sub-challenges are organized, i.e., audio, video and multimodal emotion (sub-) challenges. The acoustic features and visual features are extracted by open source toolkits. The baseline experiments were carried out by Random Forests with another open toolkit, and the macro average baseline precisions are about 30 %. It is exciting that Long Short-Term Memory Recurrent Neural Networks (LSTM-RNN), which are a dominant state-of-the-art machine learning method in today's emotion recognition, is adopted by many of the participants, and a considerable improvement was achieved. This paper focuses on the introduction of the baselines, data and protocols for the sub-challenges. Nevertheless, the results from all the participants show that emotion in the daily life is still an open-problem and it still has various difficulties, which needs to be researched upon. In addition, we will extend the emotion database with more audio video data and more emotion classes in the future to promote the research on affective computing.

Acknowledgement. This work is supported by the National High-Tech Research and Development Program of China (863 Program) (No. 2015AA016305), the National Natural Science Foundation of China (NSFC) (No. 61305003, No. 61425017), the Strategic Priority Research Program of the CAS (Grant XDB02080006), and partly supported by the Major Program for the National Social Science Fund of China (13 & ZD189).

We thank the data providers for their kind permission to make their data for non-commercial, scientific use. Due to space limitations, providers' information is available in http://www.speakit.cn/. The corpus can be freely achieved at ChineseLDC, http://www.chineseldc.org.

References

1. McKeown, G., Valstar, M., Cowie, R., Pantic, M., Schröder, M.: The semaine database: annotated multimodal records of emotionally colored conversations between a person and a limited agent. IEEE Trans. Affect. Comput. **3**, 5–17 (2012)
2. Dhall, A., Goecke, R., Lucey, S., Gedeon, T.: Collecting large, richly annotated facial-expression databases from movies. IEEE Multimedia **19**, 34–41 (2012)
3. Dhall, A., Goecke, R., Lucey, S., Gedeon, T.: Static facial expression analysis in tough conditions: data, evaluation protocol and benchmark. In: IEEE International Conference on Computer Vision Workshops, ICCV 2011 Workshops, Barcelona, Spain, pp. 2106–2112, November 2011
4. Schuller, B., Valstar, M., Eyben, F., McKeown, G., Cowie, R., Pantic, M.: AVEC 2011-the first international audio/visual emotion challenge. In: Affective Computing and Intelligent Interaction, pp. 415–424 (2011)
5. Schuller, B., Steidl, S., Batliner, A.: The interspeech 2009 emotion challenge. In: Interspeech, pp. 312–315 (2009)
6. Valstar, M.F., Jiang, B., Mehu, M., Pantic, M., Scherer, K.: The first facial expression recognition and analysis challenge. In: 2011 IEEE International Conference on Automatic Face and Gesture Recognition and Workshops (FG 2011), pp. 921–926 (2011)
7. Dhall, A., Ramana Murthy, O., Goecke, R., Joshi, J., Gedeon, T.: Video and image based emotion recognition challenges in the wild: Emotiw 2015. In: Proceedings of the 2015 ACM on International Conference on Multimodal Interaction, pp. 423–426 (2015)
8. Ringeval, F., Schuller, B., Valstar, M., Jaiswal, S., Marchi, E., Lalanne, D., et al.: AV+EC 2015: the first affect recognition challenge bridging across audio, video, and physiological data. In: Proceedings of the 5th International Workshop on Audio/Visual Emotion Challenge, pp. 3–8 (2015)
9. Ververidis, D., Kotropoulos, C.: Emotional speech recognition: resources, features, and methods. Speech Commun. **48**, 1162–1181 (2006)
10. Wu, C.-H., Lin, J.-C., Wei, W.-L.: Survey on audiovisual emotion recognition: databases, features, and data fusion strategies. APSIPA Trans. Signal Inf. Process. **3**, 12 (2014)
11. Douglas-Cowie, E., Campbell, N., Cowie, R., Roach, P.: Emotional speech: towards a new generation of databases. Speech Commun. **40**, 33–60 (2003)
12. Grimm, M., Kroschel, K., Narayanan, S.: The Vera am Mittag German audio-visual emotional speech database. In: International Conference on Multimedia Computing and Systems/ International Conference on Multimedia and Expo, pp. 865–868 (2008)
13. Devillers, L., Cowie, R., Martin, J.C., Douglas-Cowie, E., Abrilian, S., Mcrorie, M.: Real life emotions in French and English TV video clips: an integrated annotation protocol combining continuous and discrete approaches. In: International Conference on Language Resources and Evaluation, pp. 1105–1110 (2006)
14. Clavel, C., Vasilescu, I., Devillers, L., Richard, G., Ehrette, T., Sedogbo, C.: The SAFE corpus: illustrating extreme emotions in dynamic situations. In: First International Workshop on Emotion: Corpora for Research on Emotion and Affect, pp. 76–79 (2006)
15. Bao, W., Li, Y., Gu, M., Yang, M., Li, H., Chao, L., et al.: Building a Chinese natural emotional audio-visual database. In: 2014 International Conference on Signal Processing, pp. 583–587 (2014)
16. Cowie, R., Douglas-Cowie, E., Tsapatsoulis, N., Votsis, G., Kollias, S., Fellenz, W., et al.: Emotion recognition in human-computer interaction. IEEE Signal Process. Mag. **18**, 32–80 (2001)

17. Gobl, C., Chasaide, A.N.: The role of voice quality in communicating emotion, mood and attitude. Speech Commun. **40**, 189–212 (2003)
18. Eyben, F., Weninger, F., Gross, F., Schuller, B.: Recent developments in openSMILE, the Munich open-source multimedia feature extractor. In: Proceedings of the 21st ACM International Conference on Multimedia, pp. 835–838 (2013)
19. Xiong, X., Torre, F.D.L.: Supervised descent method and its applications to face alignment. In: IEEE Conference on Computer Vision and Pattern Recognition, pp. 532–539 (2013)
20. Viola, P., Jones, M.J.: Robust real-time object detection. Int. J. Comput. Vision **57**, 87 (2001)
21. Zhao, G., Pietikinen, M.: Dynamic texture recognition using local binary patterns with an application to facial expressions. IEEE Trans. Pattern Anal. Mach. Intell. **29**, 915–928 (2007)

Video Based Emotion Recognition Using CNN and BRNN

Youyi Cai, Wenming Zheng$^{(\boxtimes)}$, Tong Zhang, Qiang Li, Zhen Cui, and Jiayin Ye

Key Laboratory of Child Development and Learning Science, Ministry of Education, Research Center for Learning Science, Southeast University, Nanjing 210096, China
{yy_cai,wenming_zheng,tongzhang,qiangli, zhen.cui,yeying}@seu.edu.cn

Abstract. Video-based Emotion recognition is a rather challenging computer vision task. It not only needs to model spatial information of each image frame, but also requires considering temporal contextual correlations among sequential frames. For this purpose, we propose a hierarchical deep network architecture to extract high-level spatial-temporal features. In this architecture, two classic deep neural networks, convolutional neutral networks (CNN) and bi-directional recurrent neutral networks (BRNN), are employed to respectively capture facial textural characteristics in spatial domain and dynamic emotion changes in temporal domain. We endeavor to coordinate the two networks by optimizing each of them, so as to boost the performance of the emotion recognition. In the challenging competition, our method achieves a promising performance compared with the baselines.

Keywords: Convolutional neutral networks (CNN) · Bi-directional recurrent neutral networks (BRNN) · Emotion recognition

1 Introduction

In recent years, emotion recognition has become a significant topic in the field of artificial intelligence along with the rapid progress of the technologies in human-computer interaction and affective computing. Specifically, recognizing human facial expressions in image sequences has attracted much attention due to its wide applications, such as human-computer interaction, disease surveillance and diagnosis, public security and other fields.

Facial expression recognition can be divided into two categories, namely, image sequence based method and static image based method, where the facial expressions in video is a spatial-temporal process. To well model the facial expressions, the spatial dependencies within each frame and contextual dependencies between different frames should be taken into consideration. With the consideration of temporal information, the image sequence based method always outperforms the static image based method [1–4].

Recently, many methods have been proposed for facial expression recognition based on image sequences, including hand-crafted feature-based method and deep

© Springer Nature Singapore Pte Ltd. 2016
T. Tan et al. (Eds.): CCPR 2016, Part II, CCIS 663, pp. 679–691, 2016.
DOI: 10.1007/978-981-10-3005-5_56

learning based method. For hand-crafted feature-based methods, one kind of the methods is based on local spatial-temporal descriptors, e.g. LBP-TOP [5] and HOG3D [6]. However, in these methods, only low-level information can be obtained while complex variations over those mid-level facial action parts are hardly to be characterized. Jain et al. [7] investigated a method named Latent-Dynamic Conditional Random Fields (LDCRFs) for video based facial expression recognition. The results show that temporal variations within shapes are significant for recognizing expressions especially for subtle facial motions like sad and anger. Wang et al. [8] proposed a new approach using the Interval Temporal Bayesian Network (ITBN) to capture the complex temporal relations among facial muscles for facial expression recognition from video sequences. Deep learning has shown excellent performance in various visual applications and some researchers have also attempted to use deep learning methods for video based facial expression recognition. In [9], the deformable facial action part model for dynamic expression analysis is learned based on the 3D Convolutional Neural Networks (3D CNN) which is named 3D CNN-DAP. Wöllmer et al. [10] explored a technique based on context modeling using Long Short-Term Memory (LSTM) neutral network to improve the audiovisual emotion recognition accuracy. Another work is DTAGN (deep temporal appearance-geometry network) in [11], the proposed framework is based on two different models with the first deep network extracting temporal appearance features from image sequences and the other extracting temporal geometry features from temporal facial landmark points. A boost integration method is applied to combine the two networks so as to improve the recognition rates.

As deep learning methods have achieved great success, our method is also based on deep learning techniques in order to recognize facial expressions. Currently, well-known deep learning methods such as Convolutional Neural Networks (CNN) show remarkable performance in image processing due to their strong ability of automatically extracting discriminative representations from single image in multiple tasks such as image classification [12, 13], object detection [14–17], emotion recognition [18–21] and face recognition [22, 23], where the spatial dependencies within each image are well modeled. For the temporal dependencies, RNN [24, 25] provides a very elegant way of dealing with sequential data that embodies correlations between data points that are close in the sequences. The parameters of RNN are less than LSTM [26] which makes it easier to train. However, a regular RNN has to be outlined in the previous section. To overcome this limitation, a bi-directional recurrent neural network (BRNN) is proposed [27] that can be trained using all available input information in the past and future of a specific frame.

Inspired by the deep learning methods aforementioned, in this paper we propose a method of combining the CNN and BRNN hierarchically to boost the performance of sequence emotion recognition. The proposed architecture is shown in Fig. 1. Consider that the images in the sequence are with facial expressions, thus we fine-tuned the robust VGG_Face16 model [13, 22] used in face recognition to extract discriminative features of the facial expressions within each frame. Then the extracted features are reformed to a sequence as the input of BRNN which makes the classification. BRNN captures long-temporal dependencies by scanning the temporal sequences forward and backward. In each BRNN layer, the previous states are connected to the current one so that the network is inherently deep and able to retain all the past inputs and discover

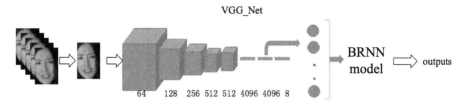

VGG_Net

64   128   256 512 512 4096 4096 8

BRNN model ⇨ outputs

Fig. 1. Framework of our method. Given a sequence, using CNN feature from the VGG_Net to form a sequence feature as the input of BRNN model and make the emotion recognition.

correlations among the sequences. Thus the proposed network is able to well model the spatial and temporal dependencies.

Our algorithm is tested on the Chinese Natural Audio-Visual Emotion Database (CHEAVD) [28] in 2016 Chinese Conference on Pattern Recognition (CCPR), which covers various situations in realistic scenes selected from Chinese movies and TV programs for different emotions. The experimental results show that our algorithm is superior to the baseline method.

This paper is organized as follows. In the next section, we give a detailed introduction of our method, including the process of feature extraction with VGG_Face16 model and classification with the BRNN model. The Sect. 3 describes our experimental framework. Finally, we analyze the results compared with baseline experiments and draw a conclusion of work in our paper.

2   Our Method

In this section, we detail our proposed method of combining CNN and BRNN for facial expression recognition. Moreover, the processes of feature extraction and classification of the two networks are also described.

2.1   Fine-Tuning VGG_FACE16 Model

We pick the VGG_Face16 model which did exceptionally well in deep face recognition [22] to extract image features. The model is fine-tuned with images in CHEAVD. The network architecture is illustrated in Fig. 2.

The VGG_Net [13, 22] is a deep network comprising five stacks of ConvNet, three fully-connected layers and one softmax layer. In other words, it consists of thirteen convolutional layers and three fully-connected layers. All convolution layers are followed by a rectification layer (ReLU) and a max pooling layer. The resulting vector from the last fully connected (FC) layer is regarded as an input of the softmax layer to compute the class probabilities. The model is regularized using weight decay and dropout, which is applied after the first two FC layers to avoid overfitting.

2D convolution is performed at the convolutional layers to extract features from local neighborhood on feature maps in the previous layer [29]. Then an additive bias is

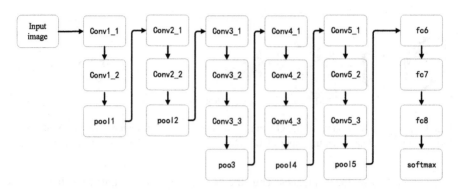

Fig. 2. Network architecture of CNN. The network is derived from the VGG_Face16 model

applied and the result is passed through a sigmoid function. Formally, the value of unit at position (x, y) in the jth feature map in the ith layer, denoted as v_{ij}^{xy}, is given by

$$v_{ij}^{xy} = ReLU\left(b_{ij} + \sum_m \sum_{p=0}^{P_i-1} \sum_{q=0}^{Q_i-1} w_{ijm}^{pq} v_{(i-1)m}^{(x+p)(y+p)}\right) \tag{1}$$

Where $ReLU(.)$ is the Rectified Linear Units function, b_{ij} is the bias for this feature map, m indexes over the set of feature maps in the $(i-1)^{th}$ layer connected to the current feature map, w_{ijm}^{pq} is the value at the position (p, q) of the kernel connected to the kth feature map, P_i and Q_i are the height and width of the kernel, respectively.

In our implementation, the challenge dataset CHEAVD includes sequences of eight classes. We fine-tuned the basic VGG_Face16 model by replacing the original N-class (N denotes the class number) softmax layer with a new M-class softmax layer, where N is 4096, M is set to 8. For fine-tuning, we randomly initialize the last two FC layers with certain number of nodes according to the dataset in the pre-trained model. We keep the first 15 convolution layers, while the output dimensions of the last fully connected layer is reset to 8 depending on the number of emotion classes. Given a face image, we apply the convolution layers of a pre-trained model to compute the convolutional feature map of the entire image. The VGG_Net firstly processes the image with detected face from selected samples with several convolutional and max pooling layers to produce a convolution feature map with a fixed-length feature vector. Then, we get features of fixed dimension from the FC7 as an image feature.

The training images of fine-tuning the VGG_Net come from the videos in CHEAVD. The image samples are considered to have the same emotion types with the video where they are cropped from. However, as the intensities of emotion in the initial few frames and the last few frames of a video are usually not so high, choosing all frames as training samples may reduce the performance. For this reason, we choose the frames from the middle part of every video with fixed proportion as the training images because we found that the frames in the middle part of a video always contain high intensities of emotion.

Generally, the number of frames of each video is variable, so combining the feature of every frame is not reasonable considering that the input dimension is usually fixed in a deep network. For a video, we compute the difference values of features between each two frames in neighbor according to the time order. By this strategy, we then rank the difference values. In the CHEAVD dataset, there are many sequences in which more than 100 frames contain little expression variation, these frames do little contribution to emotion recognition. So we choose certain number of frames with most high values of difference to form a feature sequence of fixed length. With the certain frames selected, we rearrange the selected frames in time order to keep the successive relationship of the frames. The selected frames might not be continuous, but they contain rich temporal emotion variations which are very representative of a sequence and prove to be helpful for emotion recognition. If the number of total frames in one video is less than the fixed length of the feature sequence, the number of frames will be increased with the copies of the first frame and then we rearrange the frames according to time. In this way, the copies will not break the temporal relationship of the sequence features. Consequently, for the videos in CHEAVD, we are able to extract robust feature sequences of the same length which will be the input of the BRNN network described later on.

2.2 Bi-Directional RNN

In this section, we introduce the architecture of Bi-directional RNN (BRNN) [27]. BRNN is employed to learn temporal dependencies between the past and future frames. A fully connected layer is used to gather the outputs and learn a sequence representation followed by an 8-class softmax layer for classification.

Original RNN is a network with memory and deep in time which is developed for modeling the dependencies in time sequences. Therefore, RNN model is greatly suitable for our sequence classification task with the advantage of encoding contextual information for sequences [30, 31]. Here we employ a bi-directional recurrent neutral network to simultaneously capture forward and backward dynamic transforms of sequence, i.e., two RNN are respectively used to traverse the temporal sequence in a forward or backward behavior. A BRNN structure and its expansion of full neutral network are shown in Fig. 3. Suppose that sequential representations are denoted as x and the length is T, then the temporal BRNN layer can be written as

$$h_t^f = f_h\left(W_{hh}^f h_{t-1}^f + W_{ih}^f x_t^f + b^f\right) \qquad (2)$$

$$h_t^b = f_h\left(W_{hh}^b h_{t-1}^b + W_{ih}^b x_t^b + b^b\right) \qquad (3)$$

$$o_t = V^f h_t^f + V^b h_t^b \qquad (4)$$

$$x_t^b = x_{T-t+1}^f \qquad (5)$$

where $\left\{W_{hh}^f,\ W_{ih}^f,\ b^f\right\}$ and $\left\{W_{hh}^b,\ W_{ih}^b,\ b^b\right\}$ are the learned weight matrices parameters for recurrently traversing the sequences scanned forward and backward respectively. x_t, h_t^f, h_t^b, o_t are respectively denoted as the input nodes, hidden nodes for the forward scanned network, hidden nodes for the backward scanned network, and the

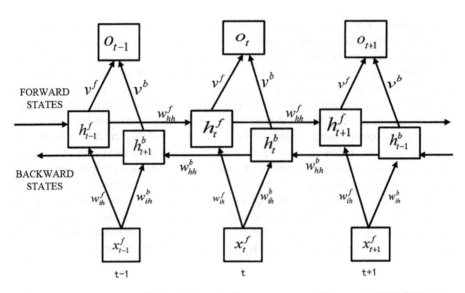

Fig. 3. General structure of the bi-directional recurrent neutral network (BRNN) shown unfolded in time for three times steps.

corresponding output nodes at the state $t \in [1, \ldots, T]$. b^b and b^f are the bias terms and f_h are defined as the non-linear activation function.

Fully Connected Layers. Finally, the output nodes of BRNN layer, denoted as $O = [o_1, \ldots, o_t, \ldots o_T]$, are fed into softmax layer for emotion classification. We ultilize the fully connected layers to collect the all the hidden units outputs in the BRNN layers, and then connect them with the final sequence labels. One fully connected layer and a softmax layer is applied in our network.

$$y = f_o(W_{ho}O + b_o) \tag{6}$$

$$O = [o_1, \ldots, o_t, \ldots o_T] \tag{7}$$

where W_{ho} is the fully connected layer weight matrices. O is the concatenation of all its sequential states o_t, $t \in [1, \ldots, T]$, b_o is the bias terms, f_o is softmax, the predicted class label is y.

As the last output node memorizes this necessary emotion information while adaptively forgetting those external noises in the evolutional of sequence network. Thus, we only choose the last output of state T, o_T as the final sequence characteristic, i.e.,

$$P(i|X) = \exp(o_{Ti}) / \sum_{k=1}^{C} \exp(o_{Tk}) \tag{8}$$

where $P(i|X)$ represents the probability for the input X being predicted as the ith class. In the objective function, the cross entropy loss is defined, which can be represented as:

$$L = -\sum_{i=1}^{N} \sum_{c=1}^{C} y_i^c log P(c|I^i) \tag{9}$$

where L denotes the cross entropy loss, N is the number of training samples, I^i denotes the ith training samples of the training set which is denoted as I, and y_i denotes the label of the ith training sample.

BRNN Optimization. The BRNN can principally be optimized by back propagation through time (BPTT) [24, 25, 31–33] as used in RNN, but the process of forward and backward pass are more complicated. The BPTT is a transformation of back propagation for time sequences. During BPTT, the forward and backward pass are done over the unfolded the bi-directional recurrent nets. There are three procedures in BPTT. The first one is to pass forward running all the input data for one time slice through the BRNN and determine all predicted outputs. The next step is to calculate the part of the objective function derivative for the time slice used in the forward pass. The last step is to update weights.

3 Experiments and Results

In the following section, we first describe how we extend the samples to test performance and present some parameters setting of VGG and BRNN. Then, we validate our network and compare the performance of the best setup with the method of baseline on test set.

3.1 Dataset and Data Augmentation

The challenge dataset CHEAVD contains 140 min spontaneous emotional segments extracted from films, TV plays and talk shows. 238 speakers, aging from child to elderly, are included in this database. It consists of eight emotions sequences which are angry, anxious, disgust, happy, neutral, sad, surprise, and worry. The dataset mostly contains with sequences with a whole people in the images. Furthermore, the dataset was built by movies clips and TV shows which are complicated in a variety of real-world scenes. Thus, we employ two steps to effectively preprocess the complex dataset. The first one is to detect faces in every frame from the sequences. We use the state-of-the-art Faster-RCNN [15, 17] detector for face detection which is of good performance in detection. The next procedure is to crop the face detected by Faster-RCNN as our samples. When meeting some faces hard to detect, we pick them out and crop them manually.

In the period of fine-tuning VGG for extracting image features, since the initial and last few frames in every emotion sequence are mostly neutral, which is useless for recognition. So we rear-range the samples by the strategy of choosing images from the middle part of every video about 50 % of the total frames. However, the number of images in various class is not equal, for the purpose of balancing the samples of each class by single image, we utilize the dataset of Multi-PIE [34] for augmentation. The number of subjects in the Multi-PIE is 337 with 4 recording sessions. Totally, over 750,000 images of high resolution are captured under 15 view points and 19 illumination conditions. Subjects were recorded in a certain circumstance with a well-equipped setup. The Multi-PIE dataset contains six emotion classes, respectively, neutral, smile, surprise, squint, disgust and scream, which includes expressions of disgust, neutral, sad, happy, and surprise that we need. Finally, there are about 23000 training images in each class after the images of Multi-PIE are added to the challenge dataset.

When training using BRNN, the number of videos belonging to each class is also not balanced enough. Furthermore, in order to better classify unseen data, a large amount of training data covering various situations are required. To balance the data and reduce overfitting on the training data, part of the samples in the validation set are used as the training samples while the other parts are used to evaluate the performance of different parameters. Moreover, two forms of data augmentation are employed to enlarge the dataset. The first form of data augmentation is to generate sequence rotations, the angles of rotating the image include $\{\pm 7, \pm 15\}$. The second form of data augmentation is mirroring. These two forms are used to make the training data more complicated and diversified.

After augmentation, finally the numbers of training sequences of angry, anxious, disgust, happy, neutral, sad, surprise and worry are 798, 801, 585, 614, 723, 768, 737 and 744 respectively.

3.2   Parameters Setting

When fine-tuning the public released VGG_Face16 model that is pre-trained on deep face recognition, a new 8-class softmax layer is used to take place the original one based on the emotion classes. And then we get a fixed dimension of 4096 from the FC7 as an image feature. After extracting image feature, we make an image sequence into a fixed length. Consequently, from every video, we obtain a sequence feature of 50 frames (50×4096 dimensional) with the strongest emotion which will be the input of the BRNN network. The fixed-length of frames is beneficial for data storage and batch training. In the experiments, we have tried to choose 40, 50, 60, 70, 80 frames as the length of a sequence. Finally, we choose the length of 50 with the consideration of balancing the final accuracy result and the training speed. For a training image, in order to keep the optimization algorithm stable, the input to the network is a fixed-size 224×224 RGB face image with the average face image subtracted. On the training set, we compute the average face image from each pixel. We use a min-batch size of 32. We start with a learning rate of 0.005 and double the learning rate for biases. We use momentum of 0.9 and weight decay of 0.0015. All FC layers except the last 8-dimentional FC layer are with a dropout ratio of 50 %.

When applying BRNN for training, in our experiment, the input of BRNN is the sequence feature of 50×4096 dimensions derived from the VGG_Net. In our BRNN layers, the number of the input, hidden and output nodes are set to be 50, 400 and 8 respectively. The non-linear transformation f_h and f_o are all set to ReLU functions. We fine-tuned the basic BRNN model by transforming the original softmax layer with a new 8-class softmax layer. We set the batch-size to 256 and fix learning rates to 0.005. We follow the default momentum term weight 0.9 and the weight decay factor 0.0005.

3.3   Evaluation Protocols

The primary measure used in this challenge is macro average precision (MAP) [28], and secondly the accuracy (ACC). Calculating formulas of macro average precision and accuracy here are given in Eqs. (10)–(12), respectively.

$$\text{macro average precision (MAP)} = \frac{1}{s} \sum_{i=1}^{s} P_i \tag{10}$$

$$P_i = \frac{TP_i}{TP_i + FN_i} \tag{11}$$

$$\text{accuracy(ACC)} = \frac{\sum_{i=1}^{s} TP_i}{\sum_{i=1}^{s} (TP_i + FN_i)} \tag{12}$$

where s is the number of the emotion labels, and P_i is the precision of each emotion class i. FN_i and TP_i represent the number of wrong prediction and the number of correct prediction in the ith emotion class.

3.4   Results

By exploiting the pre-trained very deep VGG_Face16 model and BRNN, we achieve a promising result in the preprocessed CHEAVD. In this section, we compare our method with the method of baseline [28].

On the test set which contains 628 video clips, Table 1 shows the accuracy and macro average precision for classification of eight emotions based on sequences. When comparing our method with the baseline, we can see that our framework is capable of getting better performance both in accuracy and average precision than baseline. Our method achieves 31.85 % in classification accuracy and 37.63 % in macro average

Table 1. Sequence classification accuracy(in %) and macro average precision(in %) for video comparison with baseline

Methods	Test	
	ACC (%)	MAP (%)
Our method (CNN + BRNN)	31.85	37.63
Baseline	19.59	34.28
Baseline (after sampling)	21.02	30.41

precision. It outperforms the baseline with nearly 12 % in accuracy and 4 % in average precision respectively.

The baseline method also uses sampling techniques to reconstruct a relatively balanced training corpus which is consistent with the test set. By comparing with the result after sampling, our method leads the baseline about 10 % in accuracy and 7 % in macro average precision.

From this comparison, it can be observed that our method achieves comparable results to the baseline both in the accuracy and macro average precision. One main reason for the good performance of our method is that it uses the deep networks to extract high-level spatial-temporal features instead of the traditional hand-crafted visual features like HOG3D and LBP-TOP, and we believe that the deep networks we utilized work well for emotion recognition.

The confusion matrices of the video based emotion recognition of our work and the baseline method are shown in Figs. 4 and 5 respectively. By comparing the two confusion matrices, we can see that recognition rates in our method distribute more evenly for eight different emotion classes while the recognition rates of worried, anxious, surprise, and disgust in baseline method are nearly zero. In other words, the baseline classifier is not a good classifier which can only partly recognize three emotions, neutral (61.97 %), angry (24.66 %) and sad (12 %). But in our method, the recognition rates are more average ranging from 15.57 % (anxious) to 74.07 % (disgust). Furthermore, our method behaves well in recognizing emotions of angry (74.07 %), happy (46.27 %), surprise (54.48 %), worried (33.33 %), angry (22.37 %) and sad (26.88 %), which means our method is more suitable for multiclass emotion recognition. The highest recognition rate of our method is 74.07 % for disgust while the baseline is only 4.76 % proving that our method appears to be highly effective for recognizing expression of disgust. However, the recognition rate of our method for

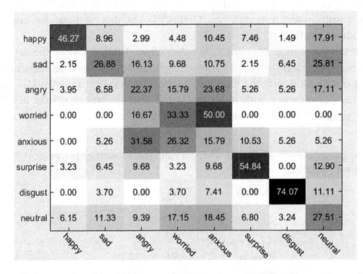

Fig. 4. Confusion matrix: video emotion recognition of our method with sample equalization on test set

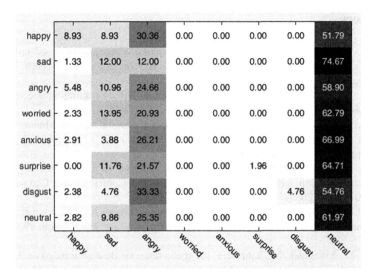

Fig. 5. Confusion matrix: video baseline system with sample balancing method on test set

neutral is 27.51 % which is less than the baseline (61.97 %) and the lowest recognition rate of our method is 15.79 % for anxious.

To sum up, our method on the challenge dataset CHEAVD yields superior performance to baseline in most classes of emotion. And it is also more balanced in recognizing eight emotion classes by checking the confusion matrices. But our method is not good enough in recognizing emotion like anxious and angry which is remained to be further improved. It may due to the fact that anxious and angry themselves are hard to recognize even for human and they can be easily confused with other emotions. In this case, most of the anxious and angry samples are casually mistaken as worried and disgust.

4   Conclusion

In this work, we present a facial expression recognition framework consisting of CNN and BRNN that collaborate with each other for emotion recognition based on sequences in CHEAVD. CNN performs image feature extraction of facial expressions and BRNN models contextual dependencies for sequence representations of different emotions. By combing the CNN and BRNN, we learn more powerful emotion feature representations in spatial-temporal domain which have been confirmed to be effective for improving the accuracy of emotion classification. Experimental results over the challenge dataset demonstrate better performance of our framework compared with the baseline.

Acknowledgement. This work was supported by the National Basic Research Program of China under Grant 2015CB351704, the National Natural Science Foundation of China (NSFC) under Grants 61231002 and 61572009, the Natural Science Foundation of Jiangsu Province under Grant BK20130020.

References

1. Zheng, W., Zhou, X., Xin, M.: Color facial expression recognition based on color local features. In: 2015 IEEE International Conference on Acoustics, Speech and Signal Processing (ICASSP), pp. 1528–1532 (2015)
2. Zheng, W.: Multi-view facial expression recognition based on group sparse reduced-rank regression. IEEE Trans. Affect. Comput. **5**(1), 71–85 (2014)
3. Zheng, W., Tang, H., Lin, Z., Huang, T.S.: Emotion recognition from arbitrary view facial images. In: Maragos, P., Paragios, N., Daniilidis, K. (eds.) ECCV 2010, Part VI. LNCS, vol. 6316, pp. 490–503. Springer, Heidelberg (2010)
4. Zheng, W., Tang, H., Lin, Z., et al.: A novel approach to expression recognition from non-frontal face images. In: IEEE 12th International Conference on Computer Vision, pp. 1901–1908 (2009)
5. Zhao, G., Pietikainen, M.: Dynamic texture recognition using local binary patterns with an appli-cation to facial expressions. IEEE Trans. Pattern Analy. Mach. Intell. **29**(6), 915–928 (2007)
6. Klaser, M., Marszałek, M., Schmid, C.: A spatio-temporal descriptor based on 3d-gradients. In: BMVC 2008-19th British Machine Vision Conference. British Machine Vision Association, vol. 275, pp. 1–10 (2008)
7. Jain, S., Hu, C., Aggarwal, J.: Facial expression recognition with temporal modeling of shapes. In: ICCV Workshops, pp. 1642–1649 (2011)
8. Wang, Z., Wang, S., Ji, Q.: Capturing complex spatio-temporal relations among facial muscles for facial expression recognition. In: Proceedings of the IEEE Conference on Computer Vision and Pattern Recognition, pp. 3422–3429 (2013)
9. Liu, M., Li, S., Shan, S., Wang, R., Chen, X.: Deeply learning deformable facial action parts model for dynamic expression analysis. In: Cremers, D., Reid, I., Saito, H., Yang, M.-H. (eds.) ACCV 2014. LNCS, vol. 9006, pp. 143–157. Springer, Heidelberg (2015)
10. Wöllmer, M., Kaiser, M., Eyben, F., et al.: LSTM-modeling of continuous emotions in an audiovisual affect recognition framework. Image Vis. Comput. **31**(2), 153–163 (2013)
11. Jung, H., Lee, S., Yim, J., et al.: Joint fine-tuning in deep neural networks for facial expression recognition. In: Proceedings of the IEEE International Conference on Computer Vision, pp. 2983–2991 (2015)
12. Krizhevsky, A., Sutskever, I., Hinton, G.: Imagenet classification with deep convolutional neu-ral networks. In: Advances in Neural Information Processing Systems, pp. 1097–1105 (2012)
13. Simonyan, K., Zisserman, A.: Very deep convolutional networks for large-scale image recognition. arXiv preprint arXiv:1409.1556 (2014)
14. Szegedy, C., Liu, W., Jia, Y., et al.: Going deeper with convolutions. In: Proceedings of the IEEE Conference on Computer Vision and Pattern Recognition, pp. 1–9 (2015)
15. Girshick, R.: Fast r-cnn. In: Proceedings of the IEEE International Conference on Computer Vision, pp. 1440–1448 (2015)
16. Ouyang, W., Luo, P., Zeng, X., et al.: Deepid-net: multi-stage and deformable deep convolutional neural networks for object detection. arXiv preprint arXiv:1409.3505 (2014)
17. Ren, S., He, K., Girshick, R., et al.: Faster R-CNN: towards real-time object detection with region proposal networks. In: Advances in Neural Information Processing Systems, pp. 91–99 (2015)
18. Zhang, T., Zheng, W., Cui, Z., et al.: A deep neural network driven feature learning method for multi-view facial expression recognition. IEEE Trans. Multimed. **99**, 1 (2016)

19. Kahou, S., Pal, C., Bouthillier, X., et al.: Combining modality specific deep neural networks for emotion recognition in video. In: Proceedings of the 2013 ACM on International Conference on Multimodal Interaction, pp. 543–550 (2013)

20. Dhall, A., Ramana, M., Goecke, R., et al.: Video and image based emotion recognition challenges in the wild: Emotiw 2015. In: Proceedings of the 2015 ACM on International Con-ference on Multimodal Interaction, pp. 423–426 (2015)

21. Kahou, S., Bouthillier, X., Lamblin, P., et al.: EmoNets: multimodal deep learning approaches for emotion recognition in video. J. Multimodal User Inter. **10**, 1–13 (2015)

22. Parkhi, O., Vedaldi, A., Zisserman, A.: Deep face recognition. Br. Mach. Vis. Conf. **1**(3), 6 (2015)

23. Sun, Y., Wang, X., Tang, X.: Deep learning face representation from predicting 10,000 classes. In: Proceedings of the IEEE Conference on Computer Vision and Pattern Recognition, pp. 1891–1898 (2014)

24. Zuo, Z., Shuai, B., Wang, G., et al.: Convolutional recurrent neural networks: learning spatial dependencies for image representation. In: Proceedings of the IEEE Conference on Computer Vision and Pattern Recognition Workshops, pp. 18–26 (2015)

25. Jaeger, H.: Tutorial on training recurrent neural networks, covering BPPT, RTRL, EKF and the "echo state network" approach. GMD-Forschungszentrum Informationstechnik (2002)

26. Hochreiter, S., Schmidhuber, J.: Long short-term memory. Neural Comput. **9**(8), 1735–1780 (1997)

27. Schuster, M., Paliwal, K.: Bidirectional recurrent neural networks. IEEE Trans. Signal Process. **45**(11), 2673–2681 (1997)

28. Li, Y., Tao, J., Schuller, B., Shan, S., Jiang, D., Jia, J.: MEC 2016: the multimodal emotion recognition challenge of CCPR 2016. In: Chinese Conference on Pattern Recognition (CCPR), Chengdu, China (2016)

29. Ji, S., Xu, W., Yang, M., et al.: 3D convolutional neural networks for human action recognition. IEEE Trans. Pattern Analy. Mach. Intell. **35**(1), 221–231 (2013)

30. Graves, A., Jaitly, N.: Towards end-to-end speech recognition with recurrent neural networks. In: Proceedings of the International Conference on Machine Learning, pp. 1764–1772 (2014)

31. Veeriah, V., Zhuang, N., Qi, G.: Differential recurrent neural networks for action recognition. In: Proceedings of the IEEE International Conference on Computer Vision, pp. 4041–4049 (2015)

32. Sutskever, I.: Training recurrent neural networks. University of Toronto (2013)

33. Cuéllar, M.P., Delgado, M., Pegalajar, M.C.: An application of non-linear programming to train recurrent neural networks in time series prediction problems. In: Chen, C.S., Cordeiro, J., Filipe, J., Seruca, I. (eds.) Enterprise Information Systems VII, pp. 95–102. Springer, Heidelberg (2007)

34. Gross, R., Matthews, I., Cohn, J., et al.: Multi-pie. Image Vis. Comput. **28**(5), 807–813 (2010)

Audio Visual Recognition of Spontaneous Emotions In-the-Wild

Xiaohan Xia[1,2], Liyong Guo[1,2], Dongmei Jiang[1,2(✉)], Ercheng Pei[1,2],
Le Yang[1,2], and Hichem Sahli[3,4]

[1] NPU-VUB Joint AVSP Lab, School of Computer Science,
Northwestern Polytechnical University (NPU), Xi'an, China
jiangdm@nwpu.edu.cn
[2] Shaanxi Key Laboratory on Speech and Image Information Processing,
Xi'an 710072, China
[3] NPU-VUB Joint AVSP Lab, Department ETRO, Vrije Universiteit Brussel (VUB),
Pleinlaan 2, 1050 Brussels, Belgium
[4] Interuniversity Microelectronics Centre, Kepeldreef 75, 3001 Heverlee, Belgium
hsahli@vub.ac.be

Abstract. In this paper, we target the CCPR 2016 Multimodal Emotion Recognition Challenge (MEC 2016) which is based on the Chinese Natural Audio-Visual Emotion Database (CHEAVD) of movies and TV programs showing (nearly) spontaneous human emotions. Low level descriptors (LLDs) are proposed as audio features. As visual features, we propose using histogram of oriented gradients (HOG), local phase quantisation (LPQ), shape features and behavior-related features such as head pose and eye gaze. The visual features are post processed to delete or smooth the all-zero feature vector segments. Single modal emotion recognition is performed using fully connected hidden Markov models (HMMs). For multimodal emotion recognition, two schemes are proposed: in the first scheme the normalized probability vectors from the HMMs are input to a support vector machine (SVM) for final recognition. For the second scheme, the final emotion is estimated using audio or video features depending if the face has been detected on the full video. Moreover, to make full use of the labeled data and to overcome the problem of unbalanced data, we use the training set and validation set together to train the HMMs and SVMs with parameters optimized via cross-validation experiments. Experimental results on the test set show that the macro average precisions (MAPs) of audio, visual, and multimodal emotion recognition reach 42.85 %, 54.24 %, and 53.90 %, respectively, which are much higher than the corresponding baseline results of 24.02 %, 34.28 %, and 30.63 %.

Keywords: LLD · HOG · LPQ · Shape · Fully connected HMM

1 Introduction

Automatic recognition of human emotions has been a very active research field in recent years. For the categorical emotion recognition, the INTERSPEECH

© Springer Nature Singapore Pte Ltd. 2016
T. Tan et al. (Eds.): CCPR 2016, Part II, CCIS 663, pp. 692–706, 2016.
DOI: 10.1007/978-981-10-3005-5_57

2009 Emotion Challenge [1] and INTERSPEECH 2013 Computational Paralinguistics Challenge (COMPARE) [2] organized emotion recognition from speech, and the Facial Expression Recognition and Analysis challenge (FERA) [4] targeted facial expression recognition. By providing the audio visual multi-modal emotion databases along with the continuous labels, the Audio-Visual Emotion Challenges (AVEC 2012–2015 [5,6,19,20]) helped speeding up the development of new frameworks for continuous affect recognition. Traditional emotion recognition has been performed on data captured in constrained lab-controlled environment. Recently, the Emotion Recognition In The Wild Challenge (EmotiW) [7,8] focused on emotion recognition in real-world conditions with varied scenarios. Through these challenges, various single modal and multimodal emotion recognition models have been proposed.

For the audio features, the AVEC 2011–2015 and INTERSPEECH 2009–2015 challenges utilized the open source software openSMILE [21] to extract Low-Level Descriptors (LLDs) features as baseline feature set. Such features included the Energy and spectral related LLDs (such as loudness, Spectral Flux, Spectral Centroid) and voicing related LLDs (such as logarithm Harmonics-to-Noise Ratio, Jitter, Shimmer). Functionals, consisting of mean, standard deviation, kurtosis, skewness, minimum and maximum value etc., have been applied on these LLDs to obtain global audio features. These feature sets have been proven to be very effective in the above challenges.

Facial expression recognition has been studied extensively in the past decade. Various hand-crafted visual features, such as the Scale Invariant Feature Transform (SIFT) [10], Dense SIFT [9], Local Binary Patterns (LBP) and its extensions like Local Gabor Binary Patterns (LGBP), LGBP-TOP (Three Orthogonal Planes) [4,19] and Volume LBP [7], Local Phase Quantization (LPQ) and LPQ-TOP [11], Histogram of Oriented Gradients (HOG) and Pyramid HOG (PHOG) [12,13], have been proposed and proven to be effective in emotion recognition. In very recent years, convolutional neural networks (CNN) and Deep CNN have been proposed to learn facial features [9,14], providing very promising results.

The early categorical emotion recognition methods mainly adopted statistical models and neural network models, such as support vector machine (SVM), hidden Markov model (HMM), Extreme Learning Machines (ELM), logistic regression, multiple kernel learning, etc. Along with the rapid development of deep learning, in recent years, new emotion recognition methods have been proposed. For example, [17] utilized deep neural networks (DNNs) to extract high level features from raw audio data, then fed them into an ELM to identify utterance-level emotions in speech. [15] proposed a CNN architecture together with a data perturbation and voting method to further increase the recognition performance of CNN. Dealing with the small size of the EmotiW challenge dataset, Ng et al. [16] followed a transfer learning approach for deep CNN architectures to overcome the over-fitting problem, by utilizing a two-stage supervised fine-tuning process. This approach was motivated by the fact that such models (deep CNNs) have the tendency to overfit the training data when the training dataset is small [16], as such the DNN and CNN are known to achieve significant gains when the training dataset is very large.

The MEC 2016 [25] is based on the Chinese Natural Audio-Visual Emotion Database (CHEAVD) [18], in which the video clips are collected from movies, talk shows and TV plays showing (nearly) spontaneous human emotions. The videos have been recorded in-the-wild with variability in scenes, backgrounds, illumination conditions, head poses, occlusion etc. MEC 2016 provides very limited training data with only 1981 video clips of 1 s to 19 s for 8 emotions. Moreover, the emotion classes are not well balanced. In the training set, neutral has 725 samples, while disgust has only 45 samples, and in the validation set, neutral has 90 samples, but surprise and disgust has only 8 and 5 samples, respectively, which will influence the validation experiments significantly. In addition, the data categories are also not balanced. There are no talk show clips in the training set, and no TV play clips in the validation set.

Due to the very limited training data, deep learning models might be very prune to be over-fitting. Therefore in this paper, we adopt the traditional HMM and SVM as recognition models, but try to seek more discriminative audio visual features, and overcome the data unbalance problem to improve the emotion recognition performance.

The remainder of this paper is organized as follows. The audio features are addressed in Sect. 2. The visual features as well as the post-processing are described in Sect. 3. Section 4 addresses the single modal and multimodal fusion emotion recognition schemes. Section 5 analyzes the experimental results and finally conclusions are drawn in Sect. 6.

2   Audio Features

MEC 2016 provides the extended GeMAPS based baseline audio features as those used in AV+EC 2015 [20], which correspond to a 88-dimensional functional feature vector on 42 LLDs extracted by the open source openSMILE toolkit [21]. In this paper, we extend the 65 LLDs of the INTERSPEECH 2014 Computational Paralinguistics Challenge [3], by adding the first dimension of the Mel Filterbank Cepstral Coefficients (MFCC 0), the first order derivatives of all the LLDs, as well as the second order derivatives of MFCC 0–14. The resulting 147 LLDs, extracted by openSMILE, are listed in Table 1, where $\triangle$ denotes the first order derivatives.

3   Visual Features

MEC 2016 adopts the LBP-TOP as baseline visual features, which have been proved to be promising in emotion recognition [19, 20]. However, for each video clip, the provided LBP-TOP feature vector is the mean over the whole image sequence which ignores the dynamic emotion information embedded in the face images. In fact, the temporal changes of face shape and appearance are very important information for discriminating different emotions. Therefore in this

Table 1. Low-Level Descriptors (LLDs) features of audio

8 energy related LLD
Sum of auditory spectrum (loudness) $+\triangle$
Sum of RASTA-filtered auditory spectrum $+\triangle$
RMS Energy $+\triangle$, Zero-Crossing Rate $+\triangle$
127 spectral LLD
RASTA-filtered auditory spectrum, bands 1–26 (0–8 kHz) $+\triangle$
MFCC 0–14 $+\triangle + \triangle\triangle$, Spectral Centriod $+\triangle$
Spectral energy 250–650 Hz, 1 k–4 kHz $+\triangle$
Spectral Roll Off Point 0.25, 0.50, 0.75, 0.90 $+\triangle$
Spectral Flux, Entropy, Variance, Skewness, Kurtosis,
Slope, Psychoacoustic Sharpness, Harmonicity $+\triangle$
12 voicing related LLD
F0 (SHS + Viterbi smoothing), Probability of voicing $+\triangle$
logarithmic HNR, Jitter (local, delta), Shimmer (local) $+\triangle$

paper, besides the baseline LBP-TOP features, we extract the HOG features, face shape features, as well as head pose and eye gaze features as the dynamic visual features.

3.1 The Baseline LBP-TOP Features

In MEC 2016, the baseline LBP-TOP features have been extracted as follows. The Supervised Descent Method (SDM) [23] is firstly adopted to track the human faces in the video clips, initialized using the Viola and Jones face detector [22], then an affine transformation is adopted to rotate and normalize the detected face images. Finally the processed face images are scaled to the size of 100 * 100, on which the LBP-TOP features are extracted using the open source Matlab code of Huang based on the algorithm of Zhao and Pietikainen [24]. For a segment of face images, a LBP-TOP feature vector of 708 dimension is obtained, and finally the mean pooling technique is applied on the feature sequence of a video clip, resulting in a suprasegmental LBP-TOP feature vector of 708 dimension.

3.2 Features Extracted by OpenFace

OpenFace [26] is the first open source tool capable of facial landmark detection, head pose estimation, facial action unit recognition, and eye-gaze estimation. It uses the Conditional Local Neural Fields (CLNF) [27] for facial landmark detection and tracking, in which Point Distribution Model (PDM) captures landmark shape variations, while patch experts capture local appearance variations of each landmark. Instead of detecting all the 68 facial landmarks together, OpenFace extends the CLNF model by training separate sets of point distribution and patch expert models for eyes, lips and eyebrows. In addition, it employs a face validation step by using a three layer convolutional neural network (CNN) to predict the expected landmark detection error, and uses multiple initialization hypotheses at different orientations and picks the model with the best converged likelihood. These measures guarantee accurate tracking of the landmarks in difficult in-the-wild images, and in case that the video clips are very long which may lead to drift or the person being out of the scene.

- **Histogram of Oriented Gradients (HOG) features.** Before extracting the HOG features, OpenFace uses a similarity transform from the currently detected landmarks to a representation of frontal landmarks from a neutral expression, resulting in a 112 * 112 pixel face image. 4464 HOG features are extracted on the aligned face image, consisting of 12 * 12 blocks of 31 dimensional histograms. For feature dimension reduction, we use a principle component analysis (PCA) model trained on the training set and validation set of MEC 2016. By applying the PCA model to the HOG features and keeping 80 % variability, the dimension of the final HOG features is reduced to 425.
- **Shape Features.** 40 shape features, consisting of 34 non-rigid and 6 rigid shape parameters, are obtained by the PDM trained on LFPW [28] and Helen [29] training sets.
- **Head Pose Features.** As both the 2D and 3D coordinates of 68 landmarks on the face have been obtained from the CLNF model, 6 head pose features can be estimated from the 3D landmarks by projecting them to the image using orthographic camera projection. When the camera calibration parameters are unknown, OpenFace uses a rough estimate based on the image size.
- **Eye Gaze Features.** A CLNF model is firstly trained to detect the eye-region landmarks including eyelids, iris and pupil, then the eye gaze vector from the 3D eyeball center to the pupil location is computed individually for each eye. For each image, the eye gaze is output as 4 vectors, the first two vectors are in world coordinate space describing the gaze direction of both eyes, the second two vectors describe the gaze in head coordinate space. As the first two vectors in world coordinate space are strongly related to the latter two vectors in head coordinate space, we take only the first two vectors and make average over them, resulting in a 3 dimension eye gaze feature vector.

Finally, as the 6 head pose features and 3 eye gaze features are also landmarks related shape features, we concatenate them with the 40 extracted shape features, and denote the resulting 49 features as "SHAPE" features.

3.3   LPQ Features

Local Phase Quantisation (LPQ) [11] features are extracted on the aligned 112 * 112 pixel face images, which are the outputs of OpenFace. The LPQ descriptor extracts local phase information using the 2-D short term Fourier transform (STFT) computed over a rectangular M-by-M neighbourhood at each pixel position of the image. The phase information in the Fourier coefficients is recorded by examining the signs of the real and imaginary parts of each component in the Fourier transform. This process is done by using a simple scalar quantiser. The resulting eight bit binary coefficients are represented as integers using binary coding, and a histogram of these values from all positions is composed to form a 256-dimensional feature vector. In our implementation, we adopt the code from [30] to extract the LPQ features.

3.4 Post-processing of the Aligned Face Images and Visual Features

The CHEAVD real-world video clips have been collected from movies, talk shows and TV plays, captured under challenging conditions. Even though OpenFace adopts a face validation step in case that the person may leave the scene in a long video, as well as multiple initialization hypotheses at different orientations in case of difficult in-the-wild images, it fails in detecting faces in some images of the CHEAVD database. For these images, OpenFace will output all-zero matrices for the 112 * 112 pixel aligned face images, and hence all-zero vectors for the HOG features and SHAPE features. The LPQ features will also be all-zero vectors.

Segments of all-zero feature vectors, with different lengths, appear discontinuously in the feature sequences, which will influence significantly the emotion recognition performance. Therefore, for each video clip, we select from the feature sequence a representative sub-sequence with detected faces and apply a linear interpolation to estimate the features where face detection failed.

4 Single Modal and Multimodal Emotion Recognition

For the single modal emotion recognition, as the baseline LBP-TOP features are suprasegmental features, we adopt SVM models with Radial basis function (RBF) kernels to train the emotion models, using the WEKA 3.6 Data Mining Toolkit [31]. For the dynamic features, such as the LLD features of audio, HOG features, SHAPE features, and LPQ features of video, emotion models are trained using HMMs. However, instead of the left-to-right topology which is commonly used in speech recognition applications, a fully connected topology (Ergodic HMM) [32] is adopted here for emotion recognition. We train one HMM for each emotion class and recognize the most probable class during the test stage. All HMMs are trained using the HTK Toolbox [33].

For the multimodal emotion recognition, we propose two schemes.

- **Scheme 1 - SVM fusion.** The LLD, HOG, SHAPE, and LPQ features are input into their corresponding eight HMMs (each for one emotion class), respectively. For each feature sequence, the output logarithm probabilities from the HMMs, which are normally very negative numbers, compose a 8-dimensional vector. A positive constant is added to the logarithm probabilities to make them more close to zero, then the exponential function is performed and the resulting vector is then divided by its sum for normalization. The four normalized probability vectors, from LLD, HOG, SHAPE and LPQ, respectively, are input into a trained SVM for the final recognition of emotion.

- **Scheme 2 - Complementarity.** For one audio video clip, emotion recognition is done using the LLD audio features in case that OpenFace fails to detect/track any faces, otherwise, the visual feature, which provides the best recognition results via the cross validation experiments on the combined training and validation sets, is used for recognition.

5   Experiments and Analysis

Experiments have been carried out on the Chinese Natural Audio Visual Emotion Database (CHEAVD) provided by the MEC 2016. Eight emotion classes, namely angry, happy, sad, worried, anxious, surprise, disgust and neutral, have been considered in the challenge. The 2852 audio video clips are divided into training set (Train), validation set (Val) and test set (Test), with 1981, 243 and 628 clips, respectively. The emotion classes in this challenge are not well balanced. In the training set, neutral has 725 samples, angry has 399 samples, while disgust has only 45 samples. In the validation set, neutral has 90 samples, but surprise has only 8 samples and disgust has 5 samples. This could significantly influence the validation experiments. In addition, the data categories in the training set and validation set are also not balanced. There are no talk show clips in the training set, and no TV play clips in the validation set. The test set is relatively more balanced.

MEC 2016 adopts macro average precision (MAP) as the primary measure to rank the results, and secondly the accuracy (ACC). To analyze the results in more details, we also calculate the macro average recall (MAR) and macro average F1 (MAF) from the results. These measures are calculated using the following equations.

$$P_i = \frac{TP_i}{TP_i + FP_i}, MAP = \frac{1}{s}\sum_{i=1}^{s} P_i \tag{1}$$

$$R_i = \frac{TP_i}{TP_i + FN_i}, MAR = \frac{1}{s}\sum_{i=1}^{s} R_i, ACC = \frac{\sum_{i=1}^{s} TP_i}{\sum_{i=1}^{s}(TP_i + FP_i)} \tag{2}$$

$$F1_i = \frac{2P_i R_i}{P_i + R_i}, MAF = \frac{1}{s}\sum_{i=1}^{s} F1_i \tag{3}$$

where TP_i (FN_i) is the number of correctly (not correctly) recognized samples of class i, FP_i is the number of samples from other classes that are recognized as class i, s is the number of classes.

5.1   Audio Emotion Recognition Results

We firstly perform emotion recognition experiments from audio, using the training set to train the HMMs and test their performance on the validation set. The

HMMs with 7 states and 5 Gaussian mixtures obtain the highest MAP, corresponding results of the above measures are listed in Table 2. One can see that the performance is barely satisfactory with MAP as 25.03 % and MAF as 19.46 %. This may be due to the fact that the data categories are not well balanced in the training and validation sets.

To reduce the influence of data unbalance, we combine the training set and validation set in one pool, and perform 3-fold cross-validation emotion recognition experiments. For each trial, 1981 LLD feature sequences have been randomly chosen as the training data (Train_cross), and the rest 243 sequences as the test data (Val_cross). The test data of the three trials are not overlapping. The final results are calculated by averaging the measures of the 3 trials. Experiments have been carried out with different HMM structures, the one with the highest mean MAP is regarded as the best structure. The obtained results are listed in Table 2. One can see that after mixing the training set and validation set, the MAP is improved from 25.03 % to 30.88 % and accuracy from 20.58 % to 31.36 %. This may own to two reasons: (1) the categories of video clips, such as talk show and TV play, are more balanced in the training data and test data. (2) for some emotions with very few samples in the original validation set, the number of test samples is increased.

Table 2. Audio emotion recognition results (%)

Training data	Test data	(State, Gauss)	MAP	ACC	MAR	MAF
Train	Val	(7, 5)	25.03	20.58	21.76	19.46
Train_cross	Val_cross	(6, 17)	30.88	**31.36**	**26.95**	**28.91**
Train+Val	Test	(6, 17)	**42.85**	27.55	23.46	21.71

In the emotion recognition experiment on the test set, we use all the data in the original training set and validation set to train the HMMs. The numbers of hidden states and Gaussian mixtures are decided referring to the best HMM structure which obtains the highest MAP in the cross-validation experiments above. The results are shown in Table 2. One can see that the MAP reaches 42.85 %, with accuracy as 27.55 %. The confusion matrix on the test set is shown in Tabel 3, together with the precision and recall of each emotion class, whose name is abbreviated as its first three characters. It is worth noticing that: (1) since the training samples of neutral are much more than those of the other emotions, the HMM of neutral is fully trained compared with other HMMs, therefore most of the other emotions are wrongly recognized as neutral, which reduces the recalls of the other emotions. (2) disgust and surprise obtain quite high precisions but low recalls. On the contrary, sad obtains a relatively high recall but low precision, showing that other emotions are prone to be recognized as sad, but not disgust or surprise.

Table 3. Confusion matrix on the test set - audio modality

	Predicted classes									Recall(%)
Ground truth classes		Ang	Sad	Hap	Dis	Sur	Wor	Anx	Neu	
	Ang	**15**	15	12	0	0	5	1	25	20.55
	Sad	1	**41**	3	0	0	4	0	26	54.67
	Hap	1	19	**6**	0	0	10	0	20	10.71
	Dis	8	6	4	**1**	1	8	0	14	2.38
	Sur	4	9	6	0	**13**	4	2	13	25.49
	Wor	1	26	8	0	0	**11**	0	40	12.79
	Anx	5	22	12	0	1	17	**2**	44	1.94
	Neu	1	39	9	0	0	8	1	**84**	59.15
Precision(%)		41.67	23.16	10.00	100.00	86.67	16.42	33.33	31.58	
F1(%)		27.53	32.54	10.34	4.65	39.39	14.38	3.67	41.18	

5.2 Visual Emotion Recognition Results

In the visual emotion recognition experiments, the video clips in the training set
and validation set, whose visual feature sequences are all-zero vector sequences,
have been removed. As such we removed 282 samples from the 1981 training
samples and 26 samples from the 243 validation data.

We firstly train the HMMs on the training set using different visual features, and compare their emotion recognition performances on the validation
set. Table 4 lists for each feature type, its best results with the highest MAP.
We also report between brackets the corresponding numbers of hidden states
and Gaussian mixtures of the HMMs. The results of the suprasegmental LBP-
TOP features based on SVM are also listed for comparison. One can notice that
compared with the LPQ features, SHAPE features, and LBP-TOP features, the
HOG features indeed obtain the best performance on all measures.

Table 4. Visual emotion recognition results on the validation set (%)

Feature	(State, Gauss)	MAP	ACC	MAR	MAF
HOG	(6, 21)	**34.18**	**47.00**	**19.63**	**19.62**
LPQ	(8, 7)	26.22	17.51	11.87	13.21
SHAPE	(8, 43)	27.57	43.78	17.84	17.62
LBP-TOP	SVM	10.02	25.10	10.27	8.94

3-fold cross validation experiments have also been carried out using the visual
features. The same procedure as used for the audio features is applied. For each
feature type, the mean measures of the 3 trials are listed in Table 5, together with
the numbers of hidden states and Gaussian mixtures of the HMM, which obtains
the highest mean MAP. One can notice that: (1) compared with the results in
Table 4, the overall performance is improved with more balanced training data

and test data. (2) compared with the LPQ features, SHAPE features and LBP-TOP features, the HOG features again obtain the best performance.

Table 5. Cross validation emotion recognition results from video(%)

Feature	(State, Gauss)	MAP	ACC	MAR	MAF
HOG	(8, 17)	**43.40**	**42.77**	**27.74**	**28.96**
LPQ	(8, 45)	34.91	37.25	25.75	26.22
SHAPE	(8, 37)	35.45	35.55	22.64	22.84
LBP-TOP	SVM	33.76	39.92	22.85	22.61

Finally, for each feature type, using the HMM structure which obtains the highest mean MAP in the cross-validation experiment, we train the HMMs (one emotion class each) on the training set and validation set after taking out the samples of zero vector sequences. The trained HMMs are then adopted to recognize the emotions of the video clips on the test set. However, in the test set, there are 42 video clips on which OpenFace fails to detect the faces. For these clips, we input their suprasegmental LBP-TOP features into the SVM models, trained on the training set and validation set, for recognition. The final results on the test set are listed in Table 6, one can see that the HOG features, together with the LBP-TOP features for the 42 video clips whose faces can't be tracked, obtain the best performance, with MAP reaching 54.24 %, and accuracy reaching 31.05 %.

Table 6. Visual emotion recognition results on the test set(%)

Feature	(State, Gauss)	MAP	ACC	MAR	MAF
HOG (LBP-TOP)	(8, 17)	**54.24**	**31.05**	**30.92**	**31.42**
LPQ (LBP-TOP)	(8, 45)	33.51	25.96	27.26	27.60
SHAPE (LBP-TOP)	(8, 37)	47.70	25.96	27.49	27.82

The confusion matrix on the test set corresponding to the HOG (LBP-TOP) features is listed in Table 7, together with the precision, recall, and F1 of each emotion class. One can notice that disgust and surprise, having only 50 and 75 training samples, respectively, obtain high recall, precision and F1 values. On the other hand, for the other emotions (except for neutral), as they are likely to be recognized as neutral, their recalls are relatively low. For example, the recall of anxious is only 0.97 % even though its precision is 100 %. As F1 is a more objective measure, the F1 values in Table 7 are compared with those in Table 3, we can see that for some emotions, e.g., angry and sad, the audio features are more discriminative, while for some emotions like disgust, surprise,

Table 7. Confusion matrix on the test set - visual modality

Ground truth classes		Predicted classes								Recall(%)
		Ang	Sad	Hap	Dis	Sur	Wor	Anx	Neu	
	Ang	15	3	3	0	0	0	0	52	20.55
	Sad	14	10	3	0	0	0	0	48	13.33
	Hap	6	4	19	0	0	0	0	27	33.93
	Dis	3	0	1	23	0	0	0	15	54.76
	Sur	4	0	6	0	27	0	0	14	52.94
	Wor	21	10	4	0	0	1	0	50	1.16
	Anx	16	7	7	0	0	1	1	71	0.97
	Neu	24	8	10	0	0	1	0	99	69.72
Precision(%)			14.56	23.81	35.85	100.00	100.00	33.33	100.00	26.33	
F1(%)			17.05	17.09	34.86	70.77	69.23	2.25	1.92	38.22	

and happy, the visual features are more discriminative. For worried and anxious, they are very difficult to be recognized correctly either from audio or from video, because they are quite complicated emotions. On the other hand, for neutral, the recognition performances from audio and video are very close to each other.

5.3   Multi-modal Emotion Recognition

For the multimodal emotion recognition, we fuse the audio visual features using the proposed scheme 1 and scheme 2, respectively, in which the HMMs and SVMs are trained on the combined training and validation sets. The recognition results on the test set are listed in Table 8. As it can be noticed, the performance of scheme 1, obtained by fusing the output probabilities of HMMs using SVM, is worse than the single modal emotion recognition performance. This may be due to the fact that the output probability vectors from HMMs are not very discriminative for emotion recognition.

Table 8. Multimodal emotion recognition results on the test set(%)

Fusion scheme	Features	MAP	ACC	MAR	MAF
Scheme 1 (SVM)	HOG+LPQ+SHAPE+LLD	33.56	24.68	27.75	26.93
Scheme 2 (complementarity)	HOG+LLD	**53.90**	31.05	31.34	31.88

For scheme 2, the difference between it and the HOG (LBP-TOP) features in Table 6 is on the 42 video clips of the test set whose faces can't be detected. For these clips, scheme 2 uses the recognition results from the audio (LLD) features based on the HMMs, while in the video modality recognition, the recognition results are from the suprasegmental LBP-TOP features based on the SVMs. Compared with the HOG (LBP-TOP) features in Table 6, the MAP of scheme 2 reduces from 54.24 % to 53.90 %, but the MAF increases a little bit from 31.42 % to 31.88 %. The confusion matrix of scheme 2 on the test set is shown in Table 9.

Table 9. Confusion matrix on the test set - multimodal

Ground truth classes		Predicted classes								Recall(%)
		Ang	Sad	Hap	Dis	Sur	Wor	Anx	Neu	
	Ang	15	4	4	0	0	0	0	50	20.55
	Sad	14	12	3	0	0	0	0	46	16.00
	Hap	5	5	19	0	0	0	0	27	33.93
	Dis	3	0	1	22	0	0	0	16	52.38
	Sur	4	1	6	0	30	0	0	10	58.82
	Wor	20	13	3	0	0	1	0	49	1.16
	Anx	16	10	7	0	0	1	1	68	0.97
	Neu	23	13	10	0	0	1	0	95	66.90
Precision(%)			15.00	20.69	35.85	100.00	100.00	33.33	100.00	26.32	
F1(%)			17.34	18.05	34.86	68.75	74.07	2.25	1.92	37.77	

Comparing with that in Table 7, one can notice that 2 more samples of sad, and 3 more samples of surprise, are correctly recognized, while the numbers of correctly recognized samples of disgust and neutral are reduced by 1 and 4, respectively. On the other hand, the recall of sad is increased from 13.33 % to 16.00 %, but its precision is reduced from 23.81 % to 20.69 %, showing that with the audio features, the 42 clips are prone to be recognized as sad.

6   Conclusions

In this paper, owning to the very limited and unbalanced training set and validation set, HMMs and SVMs have been adopted for audio visual emotion recognition. Meanwhile, we try to seek discriminative audio visual features and take into account the data unbalance. An extended set of LLD audio features have been proposed, and OpenFace is adopted to detect the face images and extract HOG features as well as shape features including head pose and eye gaze. LPQ features have also been extracted from the aligned face images. Single modal and multimodal emotion recognition experiments have been carried out on the benchmark database of MEC 2016. To make full use of the labeled data and overcome the unbalance problem of the training set and validation set, we combine the two data sets together to train the HMMs and SVMs, whose structure parameters have been decided by cross validation experiments. Experimental results on the test set show that the proposed audio/visual emotion recognition scheme, and the multimodal recognition scheme making use of the audio and visual features complementarily, obtain much higher MAPs and accuracies than the baseline results. In the future work, we will investigate emotion recognition models which can efficiently handle unbalanced data.

Acknowledgments. This work is supported by the National Natural Science Foundation of China (grant 61273265), the Research and Development Program of China (863 Program) (No. 2015AA016402), and the VUB Interdisciplinary Research Program through the EMO-App project. We would like to express our thanks to the team members Xunqin Yin, Meng Zhang and Qian Lei who helped processing the data.

References

1. Schuller, B., Steidl, S., Batliner, A.: The INTERSPEECH 2009 emotion challenge. In: Proceedings of Interspeech, pp. 312–315, Brighton (2009)
2. Schuller, B., et al.: The INTERSPEECH 2013 computational paralinguistics challenge: social signals, conflict, emotion, autism. In: Proceedings of Interspeech, pp. 148–152, Lyon (2013)
3. Schuller, B., Steidl, S., Batliner, A., Epps, J., Eyben, F., Ringeval, F., Marchi, E., Zhang, Y.: The INTERSPEECH 2014 computational paralinguistics challenge: cognitive and physical load. In: Proceedings of Interspeech 2014, Singapore (2014)
4. Valstar, M., Jiang, B., Mehu, M., Pantic, M., Scherer, K.: The first facial expression recognition and analysis challenge. In: Proceedings of IEEE International Conference Automatic Face and Gesture Recognition, pp. 921–926, Ljubljana (2011)
5. Schuller, B., Valster, M., Eyben, F., Cowie, R., Pantic, M.: AVEC 2012: the continuous audio/visual emotion challenge. In: Proceedings of the 14th ACM International Conference on Multimodal Interaction, pp. 449–456. ACM, USA (2012)
6. Valstar, M., Schuller, B., Smith, K., Eyben, F., Jiang, B., Bilakhia, S., Schnieder, S., Cowie, R., Pantic, M.: AVEC 2013: the continuous audio/visual emotion and depression recognition challenge. In: Proceedings of the 3rd ACM International Workshop on Audio/Visual Emotion Challenge, pp. 3–10. ACM, Spain (2013)
7. Dhall, A., Goecke, R., Joshi, J., Sikka, K., Gedeon, T.: Emotion recognition in the wild challenge 2014: baseline, data and protocol. In: Proceedings of the 2014 ACM on International Conference on Multimodal Interaction, pp. 461–466, Istanbul, Turkey (2014)
8. Dhall, A., Ramana Murthy, O., Goecke, R., Joshi, J., Gedeon, T.: Video and image based emotion recognition challenges in the wild: Emotiw 2015. In: Proceedings of the 2015 ACM on International Conference on Multimodal Interaction, pp. 423–426, Seattle (2015)
9. Liu, M., Wang, R., Li, S., Shan, S., Huang, Z., Chen, X.: Combining multiple kernel methods on Riemannian manifold for emotion recognition in the wild. In: Proceedings of the 16th International Conference on Multimodal Interaction, pp. 494–501, Istanbul (2014)
10. Kaya, H., Gürpinar, F., Afshar, S., Salah, A.A.: Contrasting and combining least squares based learners for emotion recognition in the wild. In: Proceedings of the 17th International Conference on Multimodal Interaction, pp. 459–466, Seattle (2015)
11. Jiang, B., Valstar, M., Martinez, B., Pantic, M.: A dynamic appearance descriptor approach to facial actions temporal modeling. IEEE Trans. Cybern. **44**(2), 161–174 (2014)
12. Dhall, A., Asthana, A., Goecke, R., Gedeon, T.: Emotion recognition using PHOG and LPQ features. In: Ninth IEEE International Conference on Automatic Face and Gesture Recognition (FG 2011), pp. 21–25, Santa Barbara (2011)
13. Sikka, K., Dykstra, K., Sathyanarayana, S., Littlewort, G., Bartlett, M.: Multiple Kernel learning for emotion recognition in the wild. In: Proceedings of the 15th ACM on International Conference on Multimodal Interaction, pp. 517–524, Sydney (2013)
14. Yao, A., Shao, J., Ma, N., Chen, Y.: Capturing AU-aware facial features and their latent relations for emotion recognition in the wild. In: Proceedings of the 2015 ACM on International Conference on Multimodal Interaction, pp. 451–458, Seattle (2015)

15. Zhiding, Y., Zhang, C.: Image based static facial expression recognition with multiple deep network learning. In: Proceedings of the 2015 ACM on International Conference on Multimodal Interaction, Seattle (2015)
16. Ng, H.-W., Nguyen, V.D., Vonikakis, V., Winkler, S.: Deep learning for emotion recognition on small datasets using transfer learning. In: Proceedings of the 2015 ACM on International Conference on Multimodal Interaction, pp. 443–449, Seattle (2015)
17. Han, K., Dong, Y., Tashev, I.: Speech emotion recognition using deep neural network and extreme learning machine. In: Proceedings of Interspeech, Singapore (2014)
18. Bao, W., et al.: Building a Chinese natural emotional audio-visual database. In: 2014 International Conference on Signal Processing. IEEE Press, Hangzhou (2014)
19. Valstar, M.F., Schuller, B.W., Smith, K., Almaev, T.R., Eyben, F., Krajewski, J., Cowie, R., Pantic, M.: AVEC 2014: 3D dimensional affect and depression recognition challenge. In: Proceedings of the 4th International Workshop on Audio/Visual Emotion Challenge (AVEC). ACM MM, Orlando, USA (2014)
20. Ringeval, F., Schuller, B., Valstar, M., Jaiswal, S., Marchi, E., Lalanne, D., Cowie, R., Pantic, M.: AV+EC 2015 - the first affect recognition challenge bridging across audio, video, and physiological data. In: Proceedings of the 5th International Workshop on Audio/Visual Emotion Challenge (AVEC). ACM MM, Brisbane, Australia (2015)
21. Eyben, F., Weninger, F., Gross, F., Schuller, B.: Recent developments in openSMILE, the Munich open-source multimedia feature extractor. In: Proceedings of the 21st ACM International Conference on Multimedia, pp. 835–838, New York (2013)
22. Viola, P., Jones, M.J.: Robust real-time object detection. Int. J. Comput. Vis. **57**(2), 137–154 (2004)
23. Xiong, X., Torre, F.D.L.: Supervised descent method and its applications to face alignment. In: Proceedings of IEEE Conference on Computer Vision and Pattern Recognition, pp. 532–539, Portland, USA (2013)
24. Zhao, G., Pietikinen, M.: Dynamic texture recognition using local binary patterns with an application to facial expressions. IEEE Trans. Pattern Anal. Mach. Intell. **29**, 915–928 (2007)
25. Li, Y., Tao, J., Schuller, B., Shan, S., Jiang, D., Jia, J.: MEC 2016: the multimodal emotion recognition challenge of CCPR 2016. In: Chinese Conference on Pattern Recognition (CCPR), Chengdu, China (2016)
26. Baltrušaitis, T., Robinson, P., Morency, L.-P.: OpenFace: an open source facial behavior analysis toolkit. In: Proceedings of IEEE Winter Conference on Applications of Computer Vision, New York, USA (2016)
27. Baltrušaitis, T., Morency, L.-P., Robinson, P.: Constrained local neural fields for robust facial landmark detection in the wild. In: Proceedings of 2013 IEEE International Conference on Computer Vision Workshops, pp. 354–361, Sydney, Australia (2013)
28. Belhumeur, P.N., Jacobs, D.W., Kriegman, D.J., Kumar, N.: Localizing parts of faces using a consensus of exemplars. IEEE Trans. Pattern Anal. Mach. Intell. **35**(12), 2930–2940 (2013)
29. Le, V., Brandt, J., Lin, Z., Bourdev, L., Huang, T.S.: Interactive facial feature localization. In: Proceedings of 12th European Conference on Computer Vision, pp. 679–692, Florence, Italy (2012)
30. http://www.cse.oulu.fi/wsgi/CMV/Downloads. Accessed 28 July 2016

31. http://prdownloads.sourceforge.net/weka/weka-3-6-14.zip. Accessed 28 July 2016
32. Wöllmer, M., Metallinou, A., Eyben, F., Narayanan, S.S.: Context-sensitive multimodal emotion recognition from speech and facial expression using bidirectional LSTM modeling. In: Proceedings of INTERSPEECH 2010, Makuhari, Chiba (2010)
33. Young, S., Evermann, G., Gales, M., Hain, T., Kershaw, D., Liu, X., Moore, G., Odell, J., Ollason, D., Povey, D., Valtchev, V., Woodland, P.: The HTK Book. Entropic Cambridge Research Laboratory, Cambridge (2006)

The SYSU System for CCPR 2016 Multimodal Emotion Recognition Challenge

Gaoyuan He[1], Jinkun Chen[1], Xuebo Liu[1], and Ming Li[1,2(✉)]

[1] SYSU-CMU Shunde International Joint Research Institute, Foshan, China
[2] School of Electronics and Information Technology,
SYSU-CMU Joint Institute of Engineering, Sun Yat-sen University,
Guangzhou, China
`liming46@mail.sysu.edu.cn`

Abstract. In this paper, we propose a multimodal emotion recognition system that combines the information from the facial, text and speech data. First, we propose a residual network architecture within the convolutional neural networks (CNN) framework to improve the facial expression recognition performance. We also perform video frames selection to fine tune our pre-trained model. Second, while the text emotion recognition conventionally deal with the clean perfect texts, here we adopt an automatic speech recognition (ASR) engine to transcribe the speech into text and then apply Support Vector Machine (SVM) on top of bag-of-words (BoW) features to predict the emotion labels. Third, we extract the openSMILE based utterance level feature and MFCC GMM based zero-order statistics feature for the subsequent SVM modeling in the speech based subsystem. Finally, score level fusion was used to combine the multimodal information. Experimental results were carried on the CCPR 2016 Multimodal Emotion Recognition Challenge database, our proposed multimodal system achieved 36 % macro average precision on the test set which outperforms the baseline by 6 % absolutely.

Keywords: Multimodal emotion recognition · Residual network · Speech recognition · Text emotion recognition

1 Introduction

A computer with powerful emotion recognition intelligence has a wide range of applications, such as human-computer interaction, psychological research, video recommendation services, etc. Although there are many existing works on this topic, understanding human emotion precisely is still a challenging task for researchers. First, signals from multiple modalities provide complementary information about emotion states. How to make full use of this information to make an accurate decision has been a difficult point. Second, there are many variances in the emotion datasets, such as age, gender, identity, background, etc. Third, human emotion can be characterized in the continue Valence-Arousal

© Springer Nature Singapore Pte Ltd. 2016
T. Tan et al. (Eds.): CCPR 2016, Part II, CCIS 663, pp. 707–720, 2016.
DOI: 10.1007/978-981-10-3005-5_58

space [1]. Labelling the data with categorical classes can result in ambiguous and inconsistent labels among multiple evaluators.

Although there remain a number of difficulties in solving the emotion recognition problem, many existing works have been proposed to improve the performance. Many of those early works focus on single modality or "lab-controlled" environments [2,3]. Recent works on emotion recognition mainly focus on recognizing the emotional states in more real and spontaneous environments with multimodal signals [4]. In this paper, we first utilize three emotion recognition subsystems that analyzing video, text, and speech signals, respectively. Then we fuse these three subsystems on the score level to further enhance the performance.

For the video based subsystem, traditional approaches are based on hand-engineered features, such as Local Binary Pattern (LBP) [5] and Local Binary Patterns on Three Orthogonal Planes (LBP-TOP) [6] features. They are proved to work well on certain datasets, but lack the generality on other datasets [7]. Recently, deep learning methods have achieved state-of-the-art results in many computer vision tasks [8,9]. Therefore, researchers also start to apply the deep learning methods on the emotion recognition task [10]. In this paper, we propose a residual network [11] architecture within the convolutional neural networks (CNN) framework to improve the facial expression recognition performance. In addition, we also propose a method to select certain video frames to fine tune our pre-trained model.

For the text based subsystem, some machine learning based methods have been proposed [12]. Conventionally, in multimodal emotion recognition, the text emotion recognition mainly deals with perfect texts or manually transcribed subtitles while the speech processing focuses on the acoustic features rather than the semantic meanings of speech utterances. And in recent years, the automatic speech recognition (ASR) has boomed and been potentially practical with higher accuracy, which means that the ASR error rate is not the principal detrimental factor any more. All these factors motivated us to integrate the emotion recognition on ASR generated text into our system, which contributes to better practicality and robustness. We use millions of short film subtitles and short conversation sentences to build a N-gram language model for rescoring the ASR lattices. And more importantly, we proposed a hierarchical classifier constructed with two Support Vector Machines to avoid the over fitting problem on small datasets.

The remainder of the paper is organized as follows. The related works are explained in Sect. 2. Our proposed methods are described in Sect. 3. Experimental results are presented in Sect. 4 while conclusions and future works are provided in Sect. 5.

2   Related Works

2.1   The Video Based Subsystem

Convolutional neural networks (CNN) has already generated state-of-the-art performance in a wide range of computer vision fields. Novel modified CNN models have also been proposed in recent years, for example, the bottleneck structure in Residual-Net [11] has been proposed to reduce the number of parameters and reduce the computational complexity.

Recently, deep learning method is also heavy used in emotion recognition task. In particular, Yu [10] proposed an innovative voting framework to combine multiple CNN models to get the final decision. In addition, Yu used the FER [13] dataset to train a model firstly, then fine tune the model on SFEW [14]. Kahou [15] proposed a hybrid CNN-RNN architecture for facial expression recognition in video. A CNN model was pre-trained on additional face expression datasets to extract facial features as the input of a simple recurrent neural network (RNN) model. He also extracted audio features and activity features from videos as multiple modalities. Two different fusion strategies were proposed to take advantage of these different modalities, operating on the feature and decision level, respectively.

2.2 The Text Based Subsystem

Some machine learning based methods for text emotion recognition have been proposed, such as naive Bayes (NB), maximum entropy (ME) classification and support vector machine (SVM) [12], which are generally based on bag-of-words (BoW) model [16] in word-level studies. The emotions or sentiments associated with topics is based on topic model [17], which introduces an intermediate topical layer into latent Dirichlet allocation (LDA) [18,19]. In the task of text emotion recognition, the methods based on BoW model work well on short texts or small datasets while the methods based on topic model have better performances on long documents. In this challenge, since the speech utterances are short, we employ SVM based on BoW feature to handle the multi-class text based emotion recognition task. Moreover, we adopt a Mandarin ASR system to automatically decode the speech utterances into text as the input of our text based emotion recognition system. Some works related to emotion recognition on ASR generated text have been reported [20]. However, for mandarin, there are very few prior works on the idea of using automatically transcribed text for emotion recognition and participating the visual-audio multimodal fusion.

2.3 The Speech Based Subsystem

It has been shown in [21,22] that speech emotion recognition can be modeled at various levels, such as acoustic, prosodic, phonetic and linguistic levels. Due to the different aspects of modeling, combining different classification methods with different features can significantly improve the overall performance. The openSMILE toolkit [23] has widely been used to perform extraction of utterance level acoustic and prosodic features followed by the Support Vector Machine (SVM) for the subsequent classification on the spontaneous short utterances [24].

Despite the openSMILE features, several Gaussian Mixture Model (GMM) based supervectors have also been proposed as features for paralinguistic speaker states recognition [25,26]. These supervectors originally were proposed for speaker verification and language identification tasks but also performed well in the paralinguistic challenges. However, when the duration of speech utterance is very short (e.g. less than 2 s), the performance of those supervectors replyiproceeding13ng on the first-order Baum-Welch statistics (mean supervector, i-vector, Maximum Likelihood Linear Regression (MLLR) supervector, etc. [25]) drops as there are not enough feature frames to calculate the sufficient statistics. The zero-order statistics based posterior probability feature achieves better performance in these short duration scenarios with limited training data [25].

It is worth noting that deep learning has been applied on the speech emotion recognition recently [27]. However, due to the lack of large scale mandarin emotional speech database for model training, in this work we only extract the conventional openSMILE based utterance level feature and MFCC GMM based zero-order statistics feature, then use SVM to perform the classification.

3   Methods

The Multimodal Emotion Recognition Challenge (MEC) is part of the 2016 Chinese Conference on Pattern Recognition (CCPR). The task of this competition is to classify the given audio and video into 8 emotion categories, i.e., "happy", "sad", "angry", "surprise", "disgust", "worried", "anxious", and "neutral". The dataset is the Chinese Natural Audio-Visual Emotion Database (CHEAVD) [28], which contains 140 min spontaneous emotional segments extracted from Chinese movies and TV programs. The 2852 samples are divided into three sets: training set, validation set and testing set, containing 1981, 243 and 628 clips, respectively. More details about the database and the challenge can be found in [29]. The emotion recognition we built consisted of three subsystems, which are video, text, and speech based emotion recognition. In this section, we will respectively explain the methods used in each subsystem in details.

3.1   The Video Based Subsystem

In addition to the video data, the organizers also provide face images extracted by IntraFace toolkit [30], where the OpenCV's Viola and Jones face detector is applied for face detection and initialization of the Intraface tracking library. We refer to this dataset as CHEAVD-faces. Each video clip contains multiple frames, but a particular emotion state may only occur in some certain frames. Thus labeling all frames with the same emotion label will introduce noise. Therefore, two additional emotion datasets of static images, Facial Expression Recognition dataset (FER2013) [13] and Static Facial Expression in the wild dataset (SFEW) [14], were used to pre-train our CNN model. In addition, we do not use all frames of each competition video clip to fine-tune our model, but select a portion of frames.

3.1.1   Pre-processing

Since CHEAVD-faces, FER2013 and SFEW use different face detection and alignment techniques. Thus we re-aligned all datasets to FER2013 using the method proposed by Kahou [15]. First, we detected five facial keypoints(left eye center, right eye center, nose tip, left mouse corner and right mouse corner) for all images on the FER2013, SFEW, and CHEAVD-faces training set using the method in [9]. Second, for each dataset we computed the mean shape by averaging the coordinates of keypoints. Third, we use a similarity transformation between mean shapes to map all datasets to FER2013. Finally, all three dataset images are preprocessed with standard histogram equalization, followed by normalization with mean and standard deviation. CHEAVD-faces validation and test sets were mapped using the same transformation learned from the training set.

3.1.2   The Residual Network Architecture

We pre-train our residual network on the FER+SFEW datasets. Figure 1 shows an overview of the networks architecture. In our CNN model, the stride of every convolutional layers is 1. The size of first convolutional filter is 5×5, and the number of first convolutional layer's channels are 64. After the first layer, the size of each convolutional filer in main branch is 3×3, the size of each convolutional filter in another branch is 1×1. The stride length of each pooling layer is 2, and the window size for each pooling layer is 3×3. We use average pooling layer in our CNN model, because after standard histogram equalization, the global features of images are more distinct than local features. There are three fully connected layers at the end. To avoid over-fitting, we add dropout after each fully connected layer.

3.1.3   Networks Pre-training on FER+SFEW

We use a deep learning library named Keras [31] to pre-train our CNN model on the FER+SFEW datasets. The stochastic gradient descent (SGD) method is taken to optimize our loss function. The batch size is set to 128. The initial learning rate is set to 0.01 while the minimum learning rate is set to 0.0005.

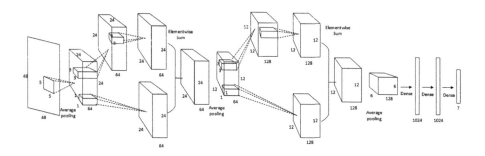

Fig. 1. The residual network for emotion recognition

The loss and trained network parameters in each epoch are recorded. If the training loss keeps increasing in more than five consecutive epochs, the learning rate is reduced by half. We then select the model with the best accuracy on held-out development data as our pre-train model.

3.1.4   Networks Fine-Tuning on CHEAVD-Faces

The pre-trained model achieved 65.65 % accuracy on the FER test dataset. As previously stated, as labeling all frames in one video clip with the same emotion label will introduce noise, we do not use all image frames to fine-tune our model. Figures 2 and 3 show the method that we use to select training CHEAVD-faces dataset for model fine-tuning. For each video in CHEAVD training dataset, we use our pre-trained model to predict all frames of the video. Our pre-trained model will produce an estimated label of seven emotions categories for every face image frame that is fed to the model, we select the most common one as the final label of the video. If the label is different from the ground truth, we use all the image frames of this video to fine-tune our model. Otherwise, we selected the frames with estimated labels matching with the ground truth, discard the frames with estimated labels are different from the ground truth. What needs illustrating is that since FER and SFEW both have seven categories, but CHEAVD-faces has eight categories. In this step, we merge the worried category and anxious category into fear category.

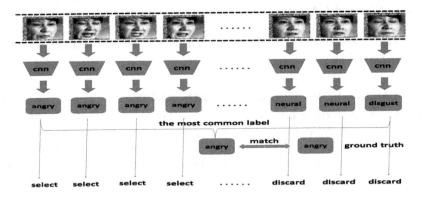

Fig. 2. The estimated label is as same as ground truth

Next we fine-tune our CNN model on the selected training CHEAVD-faces dataset. As previously stated, these three datasets have different categories. Therefore, we change the size of last fully connected layer from seven to eight and only retrain the weights of fully connected layers in the pre-trained model. Because the macro average precision (MAP) is the final criterion of this competition, the validation accuracy is based on MAP. We not only record MAP but also confusion matrix at each epoch. We finally choose the fine-tuning model with relatively high MAP and relatively balanced confusion matrix. On the test

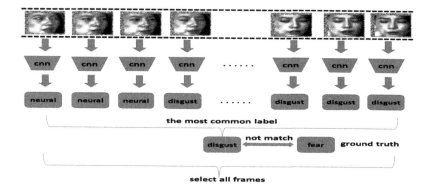

Fig. 3. The estimated label is different with ground truth

dataset, our fine-tuned model will produce an estimated label of eight emotions categories for each face image frame that is fed to the model, we select the most common one as the final label of the test video clip.

3.2   The Text Based Subsystem

Although some films are dubbed and have subtitles, we recognize the utterances from speech signals instead of using the existing subtitles. This is for better practicality and applicability in emotion recognition on any speech signals without word describing data provided. The following paragraphs explain the ASR procedure on speech signals and emotion recognition on film subtitles.

3.2.1   Speech Recognition on CHEAVD-Audio

We build our Mandarin automatic speech recognition system based on the KALDI toolbox [34]. To decrease the word error rate in ASR, we crawl about fifteen million sentences of short film subtitles from the Internet, then a special N-Gram language model is built with these subtitles and is applied in the ASR rescoring. To clarify, the subtitles of the movies or TVs in CHEAVD dataset are not contained in the training set for ASR language model. In order to measure the performance of our ASR engine, first, we manually transcribe 200 audio waves provided by CHEAVD dataset into text to form the speech evaluation data and ground truth. Second, we call the speech recognition service of iFLY-TEk[1] and Microsoft[2], as well as our ASR engine to generate three versions of ASR recognized text. Third, we calculate the word error rate[3] as the metric of speech recognition.

As the experimental results of the validation dataset shown in Table 1, Sect. 4, the iFLYTEK speech recognition service achieves the best performance among

---

the three ASR engines and our engine ranks the second. However, the differences of emotion recognition results on ASR decoded texts, corresponding to ASR engine of iFLYTEK and our laboratory respectively, are nearly negligible. Thus, we choose our in-house speech recognition engine in this challenge.

3.2.2   Emotion Recognition on Short Text

Considering the film subtitles are very colloquial and may carry emotional meanings, we extract 267043 short conversations with emotional labels from thousands of Chinese novels to fit and train the classifier. First, we manually create a dictionary containing a set of adjectives which have emotional tendencies and classify these adjectives into 36 emotion categories. The set of 8 emotional categories in the challenge, i.e., "happy", "sad", "angry", "worried", "anxious", "surprise", "disgust" and "neutral", is a subset of the 36 categories. Second, we find out all the conversions which have the structures fulfilling the pattern of "someone [says] {ADVERB} something", where ADVERB can be any adverb, to qualify the verbs such as "say", "talk", "speak", etc. Third, we select the conversation if the adverb has a strong emotional tendency, and choose the corresponding adjective as the emotional label of the conversation. Finally, we obtain 267043 conversation sentences with 360 different labels approximately and classify all conversation sentences into 36 emotional categories according to the dictionary we built.

We propose a hierarchical classifier constructed with two SVMs to recognize emotions on the ASR generated text. With the aforementioned conversations dataset, we calculate the term frequencies as the text feature and train the first stage SVM classifier, which acts as a tandem feature extractor. For each utterance u_i from the CHEAVD-audio data, either from train dataset, validation dataset or test dataset, is sent to the first stage SVM classifier to generate a 36-dimension feature vector $P_i = (p_0, p_2, ..., p_{35})$, where p_j ($0 \leq j \leq 35$) is the probability of the c_j class in the 36 categories. Furthermore, another backend SVM classifier is trained with the 36 dimensional tandem features to perform 8-class classification. The flow of emotion recognition on ASR generated text is shown as Algorithm 1.

In the beginning, we chose the classifiers of NB and SVM respectively and trained classifiers with BoW feature of the training dataset directly. And then we predicted the labels of the samples in test dataset. Unfortunately, this yielded over fitted results. Most of the samples were predicted as "neutral" for the reason that forty percent of samples in training set are "neutral". The small sample size of dataset can also result in over fitting. With the two stage hierarchical SVM method on ASR generated text, we avoid the over fitting problem and improve both the accuracy and average macro precision of text emotion recognition.

3.3   The Speech Based Subsystem

Due to the lack of large scale mandarin emotional speech database for model training, in this work, we only extract the conventional openSMILE based

---

Algorithm 1. Flow of emotion recognition on film subtitles

Require:

The datasets of short conversations and its labels, Cs, Lb_cs;

The utterances of ASR generated text of train dataset and its labels, U_trn, Lb_trn;

The utterances of ASR generated text of test dataset, U_tst;

Ensure:

The labels of the ASR generated text of test set, Lb_tst;

1: Train the text SVM classifier $text\_CLF$ with Cs, Lb_cs;

2: Input U_trn and U_tst to $text\_CLF$, get feature matrix $mat\_trn_{|U\_trn|\times 36}$ and $mat\_tst_{|U\_tst|\times 36}$ respectively;

3: Train another SVM classifier vec_CLF with feature matrix $mat\_trn_{|U\_trn|\times 36}$ and labels Lb_trn;

4: Find the optimal parameters (C, γ) for RBF kernel function with 5-fold cross validation and grid search;

5: Classify the vectors in test feature matrix $mat\_tst_{|U\_tst|\times 36}$ via vec_CLF, get the labels of test set Lb;

6: **return** Lb;

---

utterance level feature and MFCC GMM based zero-order statistics feature, then adopt SVM to perform the classification.

The utterance level openSMILE features provided by the challenge organizers are extracted by openSMILE with the extended Geneva Minimalistic Acoustic Parameter Set (eGeMAPS) [29].

For each utterance in the training and validation sets, zero-order statistics feature extraction is performed using the Universal Background Model (UBM) trained by the training dataset. Given a frame-based MFCC feature $\boldsymbol{x_t}$ and the GMM-UBM λ with M Gaussian components (each component is defined as λ_i),

$$\lambda_i = \{w_i, \boldsymbol{\mu_i}, \boldsymbol{\Sigma_i}\}, i = 1, \cdots, M, \tag{1}$$

the occupancy posterior probability is calculated as follows,

$$P(\lambda_i | \boldsymbol{x_t}) = \frac{w_i p_i(\boldsymbol{x_t} | \boldsymbol{\mu_i}, \boldsymbol{\Sigma_i})}{\Sigma_{j=1}^{M} w_j p_j(\boldsymbol{x_t} | \boldsymbol{\mu_j}, \boldsymbol{\Sigma_j})}. \tag{2}$$

This posterior probability can also be considered as the fraction of this feature $\boldsymbol{x_t}$ coming from the ith Gaussian component which is also denoted as partial counts. The larger the posterior probability, the better this Gaussian component can be used to represent this feature vector. The zero-order statistics supervector is defined as follows,

$$\boldsymbol{b} = [b_1, b_2, \cdots, b_M], b_i = \frac{y_i}{T} = \frac{1}{T} \Sigma_{t=1}^{T} P(\lambda_i | \boldsymbol{x_t}), \tag{3}$$

$$MGPP_{feature} = \sqrt{\boldsymbol{b}}. \tag{4}$$

Equation 3 is for calculating the zero-order Baum-Welch statistics and is exactly the same as the weight updating equation in the expectation-maximization (EM) algorithm in GMM training. In order to apply Bhattacharyya probability product (BPP) kernel [36], we adopt $\sqrt{b}$ as our zero-order statistics features [25].

Based on the above mentioned features, we employed the LibLinear toolbox [35] for the backend linear kernel SVM modeling. To reduce the data imbalance problem, we set the SVM weight parameter of the ith emotion class as the inverse ratio of its samples vs the neutral class $(\#neutral)/(\#i^{th}class)$. Finally, the two SVM prediction scores with OpenSMILE features and zero-order statistics features are linear fused as the outputs of our speech based subsystem.

3.4   Score Level Fusion

Due to the limited amount of training data, we simply employed the weighted summation fusion approach with parameters tuned by cross validation. When the evaluation was performed on the testing set of the CHEAVD database, both the training and validation sets were used for modeling and the weight vector was exactly the same as the one tuned on the validation set. It is worth noting that other advanced score fusion approaches, like the logistic regression method in the popular FoCal toolkit [37], can also be adopted here to increase the performance which is a topic for our future work.

4   Experiment Results

As classes are unbalanced, this challenge choose macro average precision (MAP) as the primary measure to rank the results, and secondly the accuracy. The formula of MAP and accuracy are given in Eqs. 5 and 8, respectively.

$$macro\ average\ precision = \frac{1}{s} \times \sum_{i=1}^{s} P_i \tag{5}$$

$$P_i = \frac{TP_i}{TP_i + FN_i} \tag{6}$$

$$R_i = \frac{TP_i}{TP_i + FP_i} \tag{7}$$

$$accuracy = \frac{\sum_{i=1}^{s} TP_i}{\sum_{i=1}^{s}(TP_i + FN_i)} \tag{8}$$

In these four formulas, s represent the number of the emotion labels. FP_i, FN_i TP_i represent the number of false positive , the number of false negative and the number of true positive in the ith emotion class, respectively. P_i is the precision of the ith emotion class and R_i is the recall of the ith emotion class [29].

The Table 1 presents the word error rate and emotion recognition result on validation dataset. The ASR engine of iFLYTEK has the lowest word error rate and our engine ranks the second. As for the accuracy and MAP, we get the similar emotion recognition result on the text generated by ASR of iFLYTEK and ours respectively.

Table 1. Word error rate and emotion recognition results of validation set

	Our tools	iFLYTEK API	Microsoft API
Word error rate	32.26 %	24.78 %	47.94 %
Emotion accuracy	40.1 %	40.2 %	38.8 %
Average macro precision	46.4 %	45.8 %	42.9 %

There are 1981, 243 and 628 samples in the train, validation and test dataset respectively in this challenge. We use 10-folder cross-validation method to select our model. Since the validation dataset is very unbalanced, we find that the performance on validation dataset does not have representativeness. But the test dataset is relatively balanced, we only compare our system with the baseline system on the test dataset.

The Table 2 lists the accuracy and MAP of our system and the baseline system. In each subsystem, our methods are better than the baseline. And the fusion system gives 29.93 % classification accuracy and 36.42 % MAP, which has the best recognition rate. It shows that our fusion method can truly improve the performance.

Table 2. Comparison of baselines and our result

	Accuracy(%)			MAP(%)		
	Baseline-1	Baseline-2	Our result	Baseline-1	Baseline-2	Our result
Video	19.59	21.02	**27.38**	34.28	30.41	**36.56**
Text			**27.10**			**33.84**
Audio	24.36	21.91	**26.11**	24.02	20.48	**25.98**
Fusion	24.52	21.18	**29.93**	24.53	30.63	**36.42**

The Figs. 4 and 5 show the confusion matrices of our system and the baseline system respectively. In the baseline system, the video and audio subsystem confusion matrices are very unbalanced. Such as, for audio confusion matrix of the baseline system, the precision and recall of disgust and worried categories are both 0 %. The precision of angry and anxious categories are both 100 %, but the recall of angry and anxious are 9.80 % and 0.97 % respectively. But for our audio confusion matrix, none of the precision or recall of these categories is 0 % or 100 %. Therefore, our system is much more robust than the baseline system.

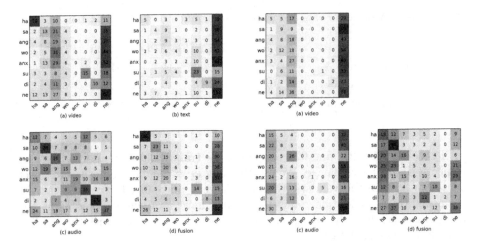

Fig. 4. Our system   **Fig. 5.** Baseline system

5   Conclusion

In this paper, we propose a multimodal emotion recognition system that combines the information from the facial, text and speech data. The residual network architecture within the CNN framework and video frames selection for fine tuning can significantly improve the facial expression recognition performance. We extract the openSMILE based utterance level feature and MFCC GMM based zero-order statistics feature for the subsequent SVM modeling in the speech based subsystem and these two features are complimentary. For the ASR decoded text subsystem, the hierarchical classifier constructed with SVMs performs well and provide a new idea to avoid over fitting on short texts and small dataset. In this challenge, our multimodal system achieves better macro average precision in comparison to baseline, which proves its effectiveness.

Acknowledgement. This research was funded in part by the National Natural Science Foundation of China (61401524), Natural Science Foundation of Guangdong Province (2014A030313123), the Fundamental Research Funds for the Central Universities (15lgjc10) and National Key Research and Development Program (2016YFC0103905).

References

1. Scherer, K.R.: What are emotions? And how can they be measured? Soc. Sci. Inf. **44**(4), 695–729 (2005)
2. Lyons, M., Akamatsu, S., Kamachi, M., Gyoba, J.: Coding facial expressions with gabor wavelets. In: Third IEEE International Conference on Automatic Face and Gesture Recognition, Proceedings, pp. 200–205 (1998)

3. Gross, R., Matthews, I., Cohn, J., Kanade, T., Baker, S.: Multi-pie. Image Vision Comput. **28**(5), 807–813 (2010)
4. Dhall, A., Goecke, R., Joshi, J., Sikka, K., Gedeon, T.: Emotion recognition in the wild challenge 2014: baseline, data and protocol. In: Proceedings of the 16th International Conference on Multimodal Interaction, pp. 461–466 (2014)
5. Shan, C., Gong, S., McOwan, P.W.: Facial expression recognition based on local binary patterns: a comprehensive study. Image Vision Comput. **27**(6), 803–816 (2009)
6. Zhao, G., Pietikainen, M.: Dynamic texture recognition using local binary patterns with an application to facial expressions. IEEE Trans. Pattern Anal. Mach. Intell. **29**(6), 915–928 (2007)
7. Dhall, A., Goecke, R., Lucey, S., Gedeon, T.: Collecting large, richly annotated facial-expression databases from movies. IEEE Multimedia **19**(3), 34–41 (2012)
8. Krizhevsky, A., Sutskever, I., Hinton, G.: Imagenet classification with deep convolutional neural networks. In: Advances in Neural Information Processing Systems (NIPS), pp. 1106–1114 (2012)
9. Sun, Y., Wang, X., Tang, X.: Deep convolutional network cascade for facial point detection. In: IEEE Conference on Computer Vision and Pattern Recognition (CVPR), pp. 3476–3483 (2013)
10. Yu, Z., Zhang, C.: Image based static facial expression recognition with multiple deep network learning. In: Proceedings of the 2015 ACM on International Conference on Multimodal Interaction, pp. 435–442 (2015)
11. He, K., Zhang, X., Ren, S., Sun, J.: Deep residual learning for image recognition. arXiv preprint arXiv:1512.03385 (2015)
12. Pang, B., Lee, L., Vaithyanathan, S.: Thumbs up? Sentiment classification using machine learning techniques. In: Proceedings of the ACL-02 Conference on Empirical Methods in Natural Language Processing, vol. 10, pp. 79–86 (2002)
13. Goodfellow, I.J., Erhan, D., Carrier, P.L., Courville, A., Mirza, M., et al.: Challenges in representation learning: a report on three machine learning contests. In: International Conference on Neural Information Processing, pp. 117–124 (2013)
14. Dhall, A., Goecke, R., Lucey, S., Gedeon, T.: Static facial expression analysis in tough conditions: data, evaluation protocol and benchmark. In: 2011 IEEE International Conference on Computer Vision Workshops (ICCV Workshops), pp. 2106–2112 (2011)
15. Kahou, S.E., Michalski, V., Konda, K., Memisevic, R., Pal, C.: Recurrent neural networks for emotion recognition in video. In: Proceedings of the 2015 ACM on International Conference on Multimodal Interaction, pp. 467–474 (2015)
16. Zhang, Y., Jin, R., Zhou, Z.H.: Understanding bag-of-words model: a statistical framework. Int. J. Mach. Learn. Cybern. **1**(1–4), 43–52 (2010)
17. Wallach, H.M.: Topic modeling: beyond bag-of-words. In: Proceedings of the 23rd International Conference on Machine learning, pp. 977–984. ACM (2006)
18. Blei, D.M., Ng, A.Y., Jordan, M.I.: Latent Dirichlet allocation. J. Mach. Learn. Res. **3**(Jan), 993–1022 (2003)
19. Ramage, D., Hall, D., Nallapati, R., et al.: Labeled LDA: a supervised topic model for credit attribution in multi-labeled corpora. In: Proceedings of the 2009 Conference on Empirical Methods in Natural Language Processing, Association for Computational Linguistics, vol. 1, pp. 248–256 (2009)
20. Metze, F., Batliner, A., Eyben, F., Polzehl, T., Schuller, B., Steidl, S.: Emotion recognition using imperfect speech recognition. ISCA (2010)

21. Anagnostopoulos, C.N., Iliou, T., Giannoukos, I.: Features and classifiers for emotion recognition from speech: a survey from 2000 to 2011. Artif. Intell. Rev. **43**(2), 155–177 (2015)
22. El Ayadi, M., Kamel, M.S., Karray, F.: Survey on speech emotion recognition: features, classification schemes, and databases. Pattern Recogn. **44**(3), 572–587 (2011)
23. Eyben, F., Wöllmer, M., Schuller, B.: OpenSMILE: the Munich versatile and fast open-source audio feature extractor. In: Proceedings of the 18th ACM International Conference on Multimedia, pp. 1459–1462 (2010)
24. Schuller, B., Steidl, S., Batliner, A.: The INTERSPEECH 2009 emotion challenge. In: INTERSPEECH, pp. 312–315 (2009)
25. Li, M., Han, K.J., Narayanan, S.: Automatic speaker age and gender recognition using acoustic and prosodic level information fusion. Comput. Speech Lang. **27**(1), 151–167 (2013)
26. Li, M., Metallinou, A., Bone, D., Narayanan, S.: Speaker states recognition using latent factor analysis based eigenchannel factor vector modeling. In: 2012 IEEE International Conference on Acoustics, Speech and Signal Processing (ICASSP), pp. 1937–1940 (2012)
27. Trigeorgis, G., Ringeval, F., Brueckner, R., Marchi, E., Nicolaou, M.A., Zafeiriou, S.: Adieu features? End-to-end speech emotion recognition using a deep convolutional recurrent network. In: 2016 IEEE International Conference on Acoustics, Speech and Signal Processing (ICASSP), pp. 5200–5204 (2016)
28. Bao, W., Li, Y., Gu, M., Yang, M., Li, H., Chao, L., Tao, J.: Building a Chinese natural emotional audio-visual database. In: 2014 12th International Conference on Signal Processing (ICSP), pp. 583–587 (2014)
29. Li, Y., Tao, J., Schuller, B., Shan, S., Jiang, D., Jia, J.: MEC 2016: the multimodal emotion recognition challenge of CCPR 2016, submitted to CCPR 2016
30. Xiong, X., De la Torre, F.: Supervised descent method and its applications to face alignment. In: Proceedings of the IEEE Conference on Computer Vision and Pattern Recognition (CVPR), pp. 532–539 (2013)
31. Chollet, F.: Keras. Github (2015). https://github.com/fchollet/keras
32. Mohri, M., Pereira, F., Riley, M.: Weighted finite-state transducers in speech recognition. Comput. Speech Lang. **16**(1), 69–88 (2002)
33. Hsu, C.W., Chang, C.C., Lin, C.J.: A practical guide to support vector classification (2003)
34. Povey, D., Ghoshal, A., Boulianne, G., Burget, L., Glembek, O., Goel, N., et al.: The Kaldi speech recognition toolkit. In: IEEE 2011 Workshop on Automatic Speech Recognition and Understanding (2011)
35. Fan, R.E., Chang, K.W., Hsieh, C.J., Wang, X.R., Lin, C.J.: LIBLINEAR: a library for large linear classification. J. Mach. Learn. Res. **9**(Aug), 1871–1874 (2008)
36. Jebara, T., Kondor, R., Howard, A.: Probability product kernels. J. Mach. Learn. Res. **5**, 819–844 (2004)
37. Brümmer, N.: FoCal multi-class: toolkit for evaluation, fusion and calibration of multi-class recognition scores–tutorial and user manual– (2007). http://sites.google.com/site/nikobrummer/focalmulticlass

Transfer Learning of Deep Neural Network for Speech Emotion Recognition

Ying Huang$^{(\boxtimes)}$, Mingqing Hu$^{(\boxtimes)}$, Xianguo Yu, Tao Wang,
and Chen Yang

Media Intelligence Group, IQIYI, Beijing, China
{yinger,humingqing,yuxianguo,wtao,yangchen}@qiyi.com

Abstract. Deep learning has made great impact in several areas, where large size dataset usually plays a great role in its effect. As emotion recognition task usually lacks in labelled data, transfer learning is proposed to initiate from models of relevant classification task with amount of data and fine tune some part of network with target emotion domain data. An ensemble method based on posterior probability difference is performed to take advantage of different networks. Experiments show that method in this paper outperforms others like random forest and ranks top two in audio subtask of the CCPR multimedia emotion recognition challenge.

Keywords: Speech emotion recognition · Deep neural network · Transfer learning · Feed forward network

1 Introduction

Speech emotion recognition has gathered more and more attention. Traditional methods more focuses on how to perform feature extraction, and selection with domain knowledge [1, 2] and then adopt standard classifiers, such as SVM, HMM, random forest for classification [1, 3].

In recent years, deep learning, emerging as a promising method, has achieved the state-of-the-art performance in many tasks, such as speech recognition and image recognition [5, 6] and outperformed traditional methods which use hand-crafted features and standard classifiers. It was also introduced in the tasks of emotion recognition. For instance, RBM is pre-trained and its parameters are used to initialize a deep neural network [8–10].

Deep neural network, as an example of deep learning, usually requires a large amount of labelled data. With more labelled data collected, better performance achieved. But in many tasks, labelled data is hard or costly to collect. For example, in the task of emotion recognition, labels are hard to get, as its categories are difficult to define, and results from different annotators could be in conflict, making a voting strategy usually used. Transfer learning, as a method to cope with the labelling problem, can borrow information from a related source domain, where a large amount of labelled data is available, to a target domain, where few or no labelled data is available. For example, in cross language speech recognition, experiments show that speech from one language can be used to train model in another language by sharing

© Springer Nature Singapore Pte Ltd. 2016
T. Tan et al. (Eds.): CCPR 2016, Part II, CCIS 663, pp. 721–729, 2016.
DOI: 10.1007/978-981-10-3005-5_59

the parameters of hidden layers [14]. In image recognition, by evaluating relation between a source domain (with a large number of data) and a target domain (with a small number of data) and partially reuse the output layer of source network for fine tuning, classification performance can be improved for the target task [13, 16]. In sound event classification, as speech and sound show similar acoustic patterns, speech data is used to pre-train models for deep neural network and achieves better performance for sound event task [15]. In [17], feature transfer learning based on a sparse autoencoder is proposed for speech emotion recognition, which uses a single-layer autoencoder to find a common structure in small target data and then apply such structure to reconstruct source data in order to transfer useful knowledge from source data into a target task. Inspired by these results and considering lacking of labeled data to train a deep neural network for emotion recognition, we proposed a transferring learning method in which knowledge transfers from a model of a speaker classification task (large amount of data available) to a model of emotion task (small size of labelled data).

The remainder of the paper is organized as follows: in Sects. 2.1 and 2.2 a transfer learning framework based on forward network is proposed; In order to fully use forward networks with various structures, an ensemble method based on posterior probabilities outputted by each structure is presented in Sect. 2.3; Sect. 3 introduces the dataset and feature set, and also present the experiments and results; finally, in Sect. 4 conclusions are drawn and future work is discussed.

2   Proposed Method

2.1   Deep Feed Forward Network (DFFN) for Classification

A feed forward network defines a mapping from an input feature vector to an output label [7]:

$$Y = f(X, \theta) \tag{1}$$

For example a feed forward network has N layers $f^1(X), f^2(X), \cdots, f^N(X)$, which are connected in a chain form $f^N(\cdots f^3(f^2(f^1(X))))$. In this case, $f^1(X)$ is called the first layer of the network, $f^N(X)$ is called the final layer, and so on. These chain structures are the most frequently used structures of neural networks.

A sigmoid function over a linear model, with parameter θ consisting of weight w and bias b, is chosen as the activation function for each hidden layer:

$$f(X; w, b) = \text{sigmod}(X^T w + b) = \frac{1}{1 + \exp(-(X^T w + b))} \tag{2}$$

In order to derive a probability representation, a softmax function over a linear model is used for the output layer, where the probability belonging to class i is denoted as:

$$\hat{y}_i = P(y = i/X) = \text{softmax}\left(X^T w_i + b_i\right) = \frac{exp\left(X^T w_i + b_i\right)}{\sum_j exp\left(X^T w_j + b_j\right)} \tag{3}$$

In the training process, cross entropy on softmax layer is used to measure and optimize the network's performance.

$$J = -\sum_j y_j \log\left(\hat{y}_j\right) \tag{4}$$

2.2   Transfer Learning of DFFN

It is observed that features like energy/amplitude, frequency parameters, and spectral parameters are highly related to emotion state [3]. These features also exhibits discriminative power between different speakers, and between genders of speakers [2]. It means that these features can be reused in these tasks, and by fine tuning network parameters, we can borrow information from speaker classification and gender classification task into the task of emotion recognition. Performance of this transferring is dependent on the relevance between source and target domain. Higher relevance will result in more reused information from source task and achieve better performance. In the following, two kinds of models, respectively trained from the above two source tasks, i.e., speaker classification and genre classification, are used as the initialization of emotion recognition task.

As the number of categories can be different between source task and target task. Following method is used to transfer model, as depicted in Fig. 1:

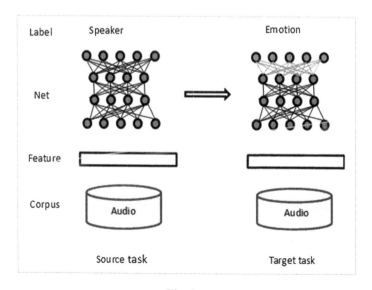

Fig. 1.

First, as the size of top layer, i.e., output layer, is determined by the number of categories of a specific task, the layer and its parameters connecting to the hidden layer, is removed when transferring from the source task to the target task and the rest of networks is retained and kept unchanged.

In the target task, a new output layer is added, whose size is determined by the number of categories of the target task and parameters connecting to the new-added layer are random initialized.

Then, the new model from the above step is fine-tuned on the target task.

2.3   Ensemble of Different Network

For each utterance, a posterior probability vector, as shown in Eq. (3), is outputted from a deep neural network, with each entry representing the probability of the utterance belonging to a category. Usually, the category with top entry value is chosen as the target label, seldom considering the entry second to top. Intuitively, the difference between these top two entries could be an indicator of confidence of the category with top entry value. If the entry second to top is closer to the top one, it means that category with top entry value has less confidence and there could be a mistake to choose it as the target label. The larger the probability difference, the more reliable the category with top entry value.

Based on this observation, an ensemble strategy, based on probability difference metric, is proposed, in which multiple deep neural networks with various hidden-layer structure are trained, and the category from structure outputting a maximal difference for an utterance is chosen as the target label.

For example, suppose that two kinds of networks have been trained, with hidden layers respectively being 1024:1024 and 2048:2048. Let $y_{1,i}^{1024}, y_{2,j}^{1024}$ denote the top two posterior of network 1024:1024, and $y_{1,m}^{2048}, y_{2,n}^{2048}$ denote the top two posterior of network 2048:2048, where i, j, m, n are labels. The output of ensemble can be obtained with the following strategy:

$$l = \begin{cases} i(y_{1,i}^{1024} - y_{2,j}^{1024}) \geq (y_{1,m}^{2048} - y_{2,n}^{2048}) \\ m(y_{1,i}^{1024} - y_{2,j}^{1024}) < (y_{1,m}^{2048} - y_{2,n}^{2048}) \end{cases} \tag{5}$$

3   Experiment and Result

3.1   Dataset and Feature

In order to evaluate the performance of the proposed method for emotion classification, the dataset CHEAVD [12] is used. CHEAVD contains 140 min spontaneous emotional segments extracted from films, and includes 8 kinds of emotion classes, namely, angry, happy, sad, worried, anxious, surprise, disgust and neutral.

For two source tasks, we use the same in-house dataset, which is annotated with labels respectively corresponding to speakers and genre. The dataset contains with a balanced gender distribution among 1105 speakers, where each one has about 400 utterances.

For both source and target tasks, we uses feature set of Interspeech 2012 challenge [2]. Open-source openSMILE feature extractor is used for feature extraction [4]. Computational network toolkit of Microsoft is used to train deep neural network [11].

For performance evaluation, macro average precision (MAP) is chosen as the primary measure, and secondly the accuracy (ACC) [12]. The whole dataset is randomly split into a training set and a validation set with ratio 9:1. As the size of validation set is so small, the splitting is repeated for 100 times, and the results are averaged over the 100 times in order to obtain a reliable estimation of performance. Moreover, the performance on an unpublished test set is also reported.

3.2   Source Task

We train FFN on two kinds of tasks, i.e., gender classification (GC), and speaker classification (SC). For each task, two kinds of network structures with different hidden layers are used.

Table 1 shows that for both of source tasks, as there is amount of labelled data, the classification results from deep feed forward network are quite good with error rate less than 2 %. Gender task is obviously less difficult, which achieves better performance with error rate less than 0.5 %.

Table 1.

Network structure (hidden layer)	Speaker task Error rate (%)	Gender task Error rate (%)
1024:1024	1.84	0.27
2048:2048	1.23	0.29

3.3   Target Task

After the deep neural networks are trained for the source task, its final layer is replaced with a new added output layer for the target task. The parameters of the new-added output layer is random initialized, and the rest of network parameters are reused as initialization values for the target task. Then the corpus of the target task is used to fine tune the network. Performance can be evaluated on the validation set and the optimal network structures can be searched in a grid manner. Experiments show that promising results can be obtained from following network structures. (1) For network with hidden layers of 2048:2048, only the output layer has been fine-tuned; and (2) For network with hidden layers of 1024:1024, with all parameters updated. In order to illustrate the effectiveness of the proposed method, outputs of the last hidden layer of the network structure 1024:1024 are visualized in Fig. 2 for both the network trained in source domain and fine-tuned network on the target task.

From the above figure (a) (b), we can see that after fine-tuning the source network, the hidden layer parameters becomes more separable on target emotion task.

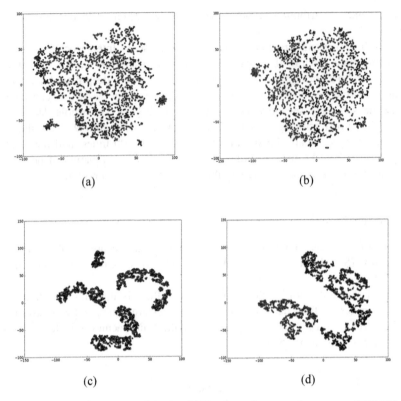

(a)   (b)

(c)   (d)

Fig. 2. Two-dimension plot of the last hidden layer for network structure 1024:1024

Figure (b) (d) show that network fine-tuned from speaker classification task are more separable on the whole than that from gender classification. It can be concluded that more classification labels of source task help the network to converge on the target task. If the source task has too few labels, some class may be well be classified, but others are difficult to separate, thus resulting in worse performance on the whole.

Tables 2 and 3 lists detailed experiments results of DFFN transferred from different source tasks. It can be seen than model transferred from speaker classification performs better than that from gender classification. The best result is achieved from ensemble of two kinds of DFFN transferred from speaker classification. The best average precision is 35.72 %, which is 5.7 % higher than the baseline, the method of

Table 2. Recognition results of network from SC on eval set

Method	ACC (%)	MAP (%)
Baseline	41.98	**30.02**
DFFN (1024:1024)	46.93	**34.16**
DFFN (2048:2048)	50.01	**35.00**
DFFN (Ensemble)	49.46	**35.72**

Table 3. Recognition results of network from GC on eval set

Method	ACC (%)	MAP (%)
Baseline	41.98	30.02
DFFN (1024:1024)	43.17	19.03
DFFN (2048:2048)	44.81	20.90
DFFN (Ensemble)	45.24	20.81

which is random forest method [12]. The best overall accuracy is 50.01 %, which is about 8 % higher than the baseline. The confusion matrix of eight classes are shown in Table 4.

Table 4. Confusion matrix of ensemble DFFN on the test data set

	Ha	Sa	Ang	Wo	Anx	Su	Di	Ne
Ha	20	7	12	17	19	5	3	23
Sa	11	48	10	23	14	4	2	35
Ang	2	3	30	4	30	1	11	10
Wo	1	0	0	1	0	0	0	0
Anx	1	0	2	0	2	2	0	1
Su	1	0	0	1	0	31	1	0
Di	2	0	2	0	3	1	20	1
Ne	18	17	17	40	35	7	5	72

Results on the unpublished test data set is listed in Table 5. Consistent improvement is achieved by ensemble of DFFN transferred from speaker classification. The average precision outperforms the baseline result by over 20 %. The accuracy outperforms the baseline by over 10 %.

Table 5. Results of ensemble DFFN on test data set

Method	ACC (%)	MAP (%)
baseline	24.36	**24.02**
DFFN (Ensemble)	35.67	**44.22**

4 Conclusion and Future Work

A transfer learning method is proposed for speech emotion recognition, where model from speaker recognition task is used as a source domain. Experiments show that, transfer learning achieves better performance than traditional methods such as random forest. In the future, other methods can be tried to achieve better performance. For example, some nodes in the output layer of the original model can be reserved and chosen to represent output classification on target task, while the rest are discarded, like that in [13]. Also, we can append a new layer without removing the original output layer. New ensemble strategy can be investigated in further.

References

1. Schuller, B., Steidl, S., Batliner, A.: The INTERSPEECH 2009 emotion challenge. In: Interspeech, Conference of the International Speech Communication Association, pp. 312–315 (2009)
2. Schuller, B., Steidl, S., Batliner, A.: The INTERSPEECH 2012 speaker trait challenge. In: Interspeech, Conference of the International Speech Communication Association (2012)
3. Eyben, F., Scherer, K., Schuller, B.: The Geneva minimalistic acoustic parameter set (GeMAPS) for voice research and affective computing. IEEE Trans, Affect. Comput. **12**(2), 190–202 (2016)
4. Eyben, F., Wollmer, M., Schuller, B.: openSMILE - the Munich versatile and fast open-source audio feature extractor. In: International Conference on Multimedia, pp. 1459–1462 (2010)
5. Hinton, G., Deng, L., Dong, Yu., Dahl, G., Mohamed, A.-r., Jaitly, N., Senior, A., Vanhoucke, V., Nguyen, P., Sainath, T., Kingsbury, B.: Deep neural networks for acoustic modeling in speech recognition. IEEE Signal Process. Magaz. **29**(6), 82–97 (2012)
6. Krizhevsky, A., Sutskever, I., Hinton, G.E.: ImageNet classification with deep convolutional neural networks. In: Advances in Neural Information Processing Systems vol. 25, no. 2 (2012)
7. Goodfellow, I., Bengio, Y., Courville, A.: Deep Learning. MIT Press, Cambridge (2016)
8. Albornoz, E.M., Sánchez-Gutiérrez, M., Martinez-Licona, F., Rufiner, H.L., Goddard, J.: Spoken emotion recognition using deep learning. In: Bayro-Corrochano, E., Hancock, E. (eds.) CIARP 2014. LNCS, vol. 8827, pp. 104–111. Springer, Heidelberg (2014)
9. Stuhlsatz, A., Meyer, C., Eyben, F., Zielke, T., Meier, G., Schuller, B.: deep neural networks for acoustic emotion recognition: raising the benchmarks. In: IEEE International Conference on Acoustics, vol. 125, no. 3, pp. 5688–5691 (2011)
10. Niu, J., Qian, Y., Yu, K.: Acoustic emotion recognition using deep neural network. In: 2014 9th International Symposium on Chinese Spoken Language Processing (ISCSLP), pp. 128–132. IEEE (2014)
11. Agarwal, A., Akchurin, E., Basoglu, C., Chen, G., Cyphers, S., Droppo, J., Eversole, A., Guenter, B., Hillebrand, M., Hoens, R., Huang, X., Huang, Z., Ivanov, V., Kamenev, A., Kranen, P., Kuchaiev, O., Manousek, W., May, A., Mitra, B., Nano, O., Navarro, G., Orlov, A., Padmilac, M., Parthasarathi, H., Peng, B., Reznichenko, A., Seide, F., Seltzer, M.L., Slaney, M., Stolcke, A., Wang, H., Wang, Y., Yao, K., Yu, D., Zhang, Y., Zweig, G.: An introduction to computational networks and the computational network toolkit. Microsoft Technical Report MSR-TR-2014-112 (2014)

12. Li, Y., Tao, J., Schuller, B., Shan, S., Jiang, D., Jia, J.: MEC 2016: the multimodal emotion recognition challenge of CCPR 2016. In Chinese Conference on Pattern Recognition (CCPR), Chengdu, China (2016)
13. Sawada, Y., Kozuka, K.: Transfer learning method using multi-prediction deep Boltzmann machines for a small scale dataset. In: IAPR International Conference on Machine Vision Applications (2015)
14. Huang, J.-T., Li, J., Yu, D, Deng, L., Gong, Y.: Cross-language knowledge transfer using multilingual deep neural network with shared hidden layers. In: IEEE International Conference on Acoustics, Speech, and Signal Processing (ICASSP), pp. 7304–7308 (2013)
15. Lim, H., Kim, M.J., Kim, H.: Cross-acoustic transfer learning for sound event classification. In: IEEE International Conference on Acoustics, Speech, and Signal Processing (ICASSP) (2016)
16. Ding, Z., Nasrabadi, N.M., Fu, Y.: Task-driven deep transfer learning for image classification. In: IEEE International Conference on Acoustics, Speech, and Signal Processing (ICASSP) (2016)
17. Deng, J., Zhang, Z., Marchi, E., Schuller, B.: Sparse autoencoder-based feature transfer learning for speech emotion recognition. In: Humaine Association Conference on Affective Computing and Intelligent Interaction (ACII), pp. 511–516 (2013)

Author Index

Printed in the United States
By Bookmasters